Interdisciplinary Applied Mathematics

Volume 21

Problems in engineering, computational science, and the physical and biological sciences are using increasingly sophisticated mathematical techniques. Thus, the bridge between the mathematical sciences and other disciplines is heavily traveled. The correspondingly increased dialog between the disciplines has led to the establishment of the series: *Interdisciplinary Applied Mathematics*.

The purpose of this series is to meet the current and future needs for the interaction between various science and technology areas on the one hand and mathematics on the other. This is done, firstly, by encouraging the ways that mathematics may be applied in traditional areas, as well as point towards new and innovative areas of applications; and, secondly, by encouraging other scientific disciplines to engage in a dialog with mathematicians outlining their problems to both access new methods and suggest innovative developments within mathematics itself.

The series will consist of monographs and high-level texts from researchers working on the interplay between mathematics and other fields of science and technology.

Interdisciplinary Applied Mathematics

Volumes published are listed at the end of this book.

Springer

New York
Berlin
Heidelberg
Hong Kong
London
Milan
Paris
Tokyo

Tamar Schlick

Molecular Modeling and Simulation

An Interdisciplinary Guide

With 147 Full-Color Illustrations

Tamar Schlick
Department of Mathematics and Chemistry
Courant Institute of Mathematical Sciences
New York University
New York, NY 10012
USA
schlick@nyu.edu

Cover illustration: © Wayne Thiebaud/Licensed by VAGA, New York, NY. Courtesy of the Allan Stone Gallery, NYC.

Mathematics Subject Classification (2000): 92–01, 92Exx, 92C40

Library of Congress Cataloging-in-Publication Data
Schlick, Tamar.
Molecular modeling and simulation: an interdisciplinary guide / Tamar Schlick.
p. cm. — (Interdisciplinary applied mathematics ; 21)
Includes bibliographical references and index.
ISBN 0-387-95404-X (alk. paper)
1. Biomolecules—Models. 2. Biomolecules—Models—Computer simulation. I. Title.
II. Interdisciplinary applied mathematics ; v. 21.
QD480 .S37 2002
572′.33′015118—dc21 2002016003

ISBN 0-387-95404-X Printed on acid-free paper.

Printed in the United States of America

9 8 7 6 5 4 3 2 1 SPIN 10860583

Typesetting: Pages created by author using a Springer TEX macro package.

www.springer-ny.com

Springer-Verlag New York Berlin Heidelberg
A member of BertelsmannSpringer Science+Business Media GmbH

About the Cover

Molecular modelers are artists in some respects. Their subjects are complex, irregular, multiscaled, highly dynamic, and sometimes multifarious, with diverse states and functions. To study these complex phenomena, modelers must apply computer programs based on precise algorithms that stem from solid laws and theories from mathematics, physics, and chemistry.

Many of Wayne Thiebaud's landscape paintings, like *Reservoir Study* shown on the cover, embody this productive blend of nonuniformity with orderliness. Thiebaud's body of water — organic, curvy, and multilayered (reminiscent of a cell) — is surrounded by apparently ordered fields and land sections. Upon close inspection, the multiplicity in perspectives and interpretations emerges. This artwork thus mirrors the challenging crossdisciplinary interplay, as well as blend of science and art, central to biomolecular modeling.

To my grandparents — Fanny and Iancu Iosupovici, Lucy and Charles Schlick — whose love and courage I carry forever.

Book URLs

For Text:
monod.biomath.nyu.edu/index/book.html

For Course:
monod.biomath.nyu.edu/index/course/IndexMM.html

Preface

> Science is a way of looking, reverencing. And the purpose of all science, like living, which amounts to the same thing, is not the accumulation of gnostic power, the fixing of formulas for the name of God, the stockpiling of brutal efficiency, accomplishing the sadistic myth of progress. The purpose of science is to revive and cultivate a perpetual state of wonder. For nothing deserves wonder so much as our capacity to experience it.

Roald Hoffman and Shira Leibowitz Schmidt, in *Old Wine, New Flasks: Reflections on Science and Jewish Tradition* (W.H. Freeman, 1997).

Challenges in Teaching Molecular Modeling

This textbook evolved from a graduate course termed *Molecular Modeling* introduced in the fall of 1996 at New York University. The primary goal of the course is to stimulate excitement for molecular modeling research — much in the spirit of Hoffman and Leibowitz Schmidt above — while providing grounding in the discipline. Such knowledge is valuable for research dealing with many practical problems in both the academic and industrial sectors, from developing treatments for AIDS (via inhibitors to the protease enzyme of the human immunodeficiency virus, HIV-1) to designing potatoes that yield spot-free potato chips (via transgenic potatoes with altered carbohydrate metabolism). In the course of writing this text, the notes have expanded to function also as an introduction to the field for scientists in other disciplines by providing a global perspective into problems and approaches, rather than a comprehensive survey.

As a textbook, my intention is to provide a framework for teachers rather than a rigid guide, with material to be supplemented or substituted as appropriate for the audience. As a reference book, scientists who are interested in learning about biomolecular modeling may view the book as a broad introduction to an exciting new field with a host of challenging, interdisciplinary problems.

The intended audience for the course is beginning graduate students in medical schools and in all scientific departments: biology, chemistry, physics, mathematics, computer science, and others. This interdisciplinary audience presents a special challenge: it requires a broad presentation of the field but also good coverage of specialized topics to keep experts interested. Ideally, a good grounding in basic biochemistry, chemical physics, statistical and quantum mechanics, scientific computing (i.e., numerical methods), and programming techniques is desired. The rarity of such a background required me to offer tutorials in both biological and mathematical areas.

The introductory chapters on biomolecular structure are included in this book (after much thought) and are likely to be of interest to physical and mathematical scientists. Chapters 3 and 4 on proteins, together with Chapters 5 and 6 on nucleic acids, are thus highly abbreviated versions of what can be found in numerous texts specializing in these subjects. The selections in these tutorials also reflect some of my group's areas of interest. Because many introductory and up-to-date texts exist for protein structure, only the basics in protein structure are provided, while a somewhat more expanded treatment is devoted to nucleic acids.

Similarly, the introductory material on mathematical subjects such as basic optimization theory (Chapter 10) and random number generators (Chapter 11) is likely to be of use more to readers in the biological / chemical disciplines. General readers, as well as course instructors, can skip around this book as appropriate and fill in necessary gaps through other texts (e.g., in protein structure or programming techniques).

Text Limitations

By construction, this book is very broad in scope and thus no subjects are covered in great depth. References to the literature are only representative. The material presented is necessarily selective, unbalanced in parts, and reflects some of my areas of interest and expertise. This text should thus be viewed as an attempt to introduce the discipline of molecular modeling to students and to scientists from disparate fields, and should be taken together with other related texts, such as those listed in Appendix C, and the representative references cited.

The book format is somewhat unusual for a textbook in that it is nonlinear in parts. For example, protein folding is introduced early (before protein basics are discussed) to illustrate challenging problems in the field and to interest more advanced readers; the introduction to molecular dynamics incorporates illustrations that require more advanced techniques for analysis; some specialized topics are also included throughout. For this reason, I recommend that students reread certain parts of the book (e.g., first two chapters) after covering others (e.g., the biomolecular tutorial chapters). Still, I hope most of all to grab the reader's attention with exciting and current topics.

Given the many caveats of introducing and teaching such a broad and interdisciplinary subject as molecular modeling, the book aims to introduce selected biomolecular modeling and simulation techniques, as well as the wide range of biomolecular problems being tackled with these methods. Throughout these presentations, the central goal is to develop in students a good understanding of the inherent approximations and errors in the field so that they can adequately assess modeling results. Diligent students should emerge with basic knowledge in modeling and simulation techniques, an appreciation of the fundamental problems — such as force field approximations, nonbonded evaluation protocols, size and timestep limitations in simulations — and a healthy critical eye for research. A historical perspective and a discussion of future challenges are also offered.

Dazzling Modeling Advances Demand Perspective

The topics I chose for this course are based on my own unorthodox introduction to the field of modeling. As an applied mathematician, I became interested in the field during my graduate work, hearing from Professor Suse Broyde — whose path I crossed thanks to Courant Professor Michael Overton — about the fascinating problem of modeling carcinogen/DNA adducts.

The goal was to understand some structural effects induced by certain compounds on the DNA (deduced by energy minimization); such alterations can render DNA more sensitive to replication errors, which in turn can eventually lead to mutagenesis and carcinogenesis. I had to roam through many references to obtain a grasp of some of the underlying concepts involving force fields and simulation protocols, so many of which seemed so approximate and not fully physically grounded. By now, however, I have learned to appreciate the practical procedures and compromises that computational chemists have formulated out of sheer necessity to obtain answers and insights into important biological processes that cannot be tackled by instrumentation. In fact, approximations and simplifications are not only tolerated when dealing with biomolecules; they often lead to insights that cannot easily be obtained from more detailed representations. Furthermore, it is often the neglect of certain factors that teaches us their importance, sometimes in subtle ways.

For example, when Suse Broyde and I viewed in the mid 1980s her intriguing carcinogen/modified DNA models, we used a large Evans and Sutherland computer while wearing special stereoviewers; the hard-copy drawings were ball and stick models, though the dimensionality projected out nicely in black and white. (Today, we still use stereo glasses, but current hardware stereo capabilities are much better, and marvelous molecular renderings are available). At that time, only small pieces of DNA could be modeled, and the surrounding salt and solvent environment was approximated. Still, structural and functional insights arose from those earlier works, many of which were validated later by more comprehensive computation, as well as laboratory experiments.

Book Overview

The book provides an overview of three broad topics: (a) biomolecular structure and modeling: current problems and state of computations (Chapters 1–6); (b) molecular mechanics: force field origin, composition, and evaluation techniques (Chapters 7–9); and (c) simulation techniques: conformational sampling by geometry optimization, Monte Carlo, and molecular dynamics approaches (Chapters 10–13). Chapter 14 on the similarity and diversity problems in chemical design introduces some of the challenges in the growing field related to combinatorial chemistry (Chapter 14).

Specifically, Chapters 1 and 2 give a historical perspective of biomolecular modeling, outlining progress in experimental techniques, the current computational challenges, and the practical applications of this enterprise — to convey the immense interest in, and support of, the discipline. *Since these chapters discuss rapidly changing subjects (e.g., genome projects, disease treatments), they will be updated as possible on the text website.* General readers may find these chapters useful as an introduction to biomolecular modeling and its applications.

Chapters 3 and 4 review the basic elements in protein structure, and Chapter 5 similarly presents the basic building blocks and conformational flexibility in nucleic acids. Chapter 6 presents additional topics in nucleic acids, such as DNA sequence effects, DNA/protein interactions, departures from the canonical DNA helix forms, RNA structure, and DNA supercoiling.

The second part of the book begins in Chapter 7 with a view of the discipline of molecular mechanics as an offspring of quantum mechanics and discusses the basic premises of molecular mechanics formulations. A detailed presentation of the force field terms — origin, variation, and parameterization — is given in Chapter 8. Chapter 9 is then devoted to the computation of the nonbonded energy terms, including cutoff techniques, Ewald and multipole schemes, and continuum solvation alternatives.

The third part of the book, simulation algorithms,[1] begins with a description of optimization methods for multivariate functions in Chapter 10, emphasizing the

---

[1]The word *algorithm* is named after the ninth-century Arab mathematician al-Khwarizmi (nicknamed after his home town of Khwarizm, now Khiva in the Uzbek Republic), who stressed the importance of methodical procedures for solving problems in his algebra textbook. The term has evolved to mean the systematic process of solving problems by machine execution.

tradeoff between algorithm complexity and performance. Basic issues of Monte Carlo techniques, appropriate to a motivated novice, are detailed in Chapter 11, such as pseudorandom number generators, Gaussian random variates, Monte Carlo sampling, and the Metropolis algorithm. Chapters 12 and 13 describe the algorithmic challenges in biomolecular dynamics simulations and present various categories of integration techniques, from the popular Verlet algorithm to multiple-timestep techniques and Brownian dynamics protocols. Chapter 14 outlines the challenges in similarity and diversity sampling in the field of chemical design, related to the new field of combinatorial chemistry.

The book appendices complement the material in the main text through homework assignments, reading lists, and other information useful for teaching molecular modeling.

Instructors may find the sample course syllabus in Appendix A helpful. Important also to teaching is an introduction to the original literature; a representative reading list of articles used for the course is collected in Appendix B. An annotated general reference list is given in Appendix C.

Selected biophysics applications are highlighted through the homework assignments (Appendix D). Humor in the assignments stimulates creativity in many students. These homeworks are a central component of learning molecular modeling, as they provide hands-on experience, extend upon subjects covered in the chapters, and expose the students to a wide range of current topics in biomolecular structure. Advanced students may use these homework assignments to learn about molecular modeling through independent research.

Many homework assignments involve a molecular modeling software package. I selected the Insight program in conjunction with our Silicon Graphics computer laboratory, but other suitable modeling programs can be used. Students also learn other basic research tools (such as programming and literature searches) through the homeworks.

Our memorable "force field debate" (see homework 7 in Appendix D) even brought the AMBER team to class in white lab coats, each accented with a name tag corresponding to one of AMBER's original authors. The late Peter Kollman would have been pleased. Harold Scheraga would have been no less impressed by the long list of ECEPP successes prepared by his loyal troopers. Martin Karplus would not have been disappointed by the strong proponents of the CHARMM approach. I only hope to have as much spunk and talent in my future molecular modeling classes.

Extensive use of web resources is encouraged, while keeping in mind the caveat of lack of general quality control. I was amazed to find some of my students' discoveries regarding interesting molecular modeling topics mentioned in the classroom, especially in the context of the term project, which requires them to find outstanding examples of the successes and/or failures of molecular modeling.

Interested readers might also want to glance at additional course information as part of my group's home page, monod.biomath.nyu.edu/. Supplementary text information (such as program codes and figure files) can also be obtained.

To future teachers of molecular modeling who plan to design similar assignments and material, I share with you my following experience regarding student reactions to this discipline: what excited students the most about the subject matter and led to enthusiasm and excellent feedback in the classroom were the rapid pace at which the field is developing, its exciting discoveries, and the medical and technological breakthroughs made possible by important findings in the field.

In more practical terms, a mathematics graduate student, Brynja Kohler, expressed this enthusiasm succinctly in the introduction to her term project:

> As I was doing research for this assignment, I found that one interesting article led to another. Communication via e-mail with some researchers around the world about their current investigations made me eagerly anticipate new results. The more I learned the more easy it became to put off writing a final draft because my curiosity would lead me on yet another line of inquiry. However, alas, there comes a time when even the greatest procrastinator must face the music, and evaluate what it is that we know and not linger upon what we hope to find out.

Future teachers are thus likely to have an enjoyable experience with any good group of students.

Acknowledgments

I am indebted to Jing Huang for her devoted assistance with the manuscript preparation, file backups, data collection, and figure design. I also thank Wei Xu and Mulin Ding for important technical assistance. I am grateful to my other devoted current and former group members who helped read book segments, collect data, prepare the figures found throughout this book, and run to libraries throughout New York City often: Karunesh Arora, Danny Barash, Paul Batcho, Dan Beard, Mulin Ding, Hin Hark Gan, Jennifer Isbell, Joyce Noah, Xiaoliang Qian, Sonia Rivera, Adrian Sandu, Dan Strahs, Dexuan Xie, Linjing Yang, and Qing Zhang. Credits for each book figure and table are listed on the text's website.

I thank my colleagues Ruben Abagyan, Helen Berman, Dave Case, Jonathan Goodman, Andrej Sali, and Harold Scheraga, who gave excellent guest lectures in the course; and my course assistants Karunesh Arora, Margaret Mandziuk, Qing Zhang, and Zhongwei Zhu for their patient, dedicated assistance to the students with their homework and queries.

I am also very appreciative of the following colleagues for sharing reprints, information, and unpublished data and/or for their willingness to comment on segments of the book: Lou Allinger, Nathan Baker, Mike Beer, Helen Berman, Suse Broyde, John Board, Dave Beveridge, Ken Breslauer, Steve Burley, Dave Case, Philippe Derreumaux, Ron Elber, Eugene Fluder, Leslie Greengard, Steve Harvey, Jan Hermans, the late Peter Kollman, Robert Krasny, Michael Levitt, Xiang-Jun Lu, Pierre L'Ecuyer, Neocles Leontis, the late Shneior Lifson, Kenny

Lipkowitz, Jerry Manning, Andy McCammon, Mihaly Mezei, Jorge Nocedal, Wilma Olson, Michael Overton, Vijay Pande, Dinshaw Patel, Harold Scheraga, Shulamith Schlick, Klaus Schulten, Suresh Singh, Bob Skeel, A. R. Srinivasan, Emad Tajkhorshid, Yuri Ushkaryov, Wilfred van Gunsteren, Arieh Warshel, Eric Westhof, Weitao Yang, and Darren York. Of special note are the extremely thorough critiques which I received from Lou Allinger, Steve Harvey, Jerry Manning, Robert Krasny, Wilma Olson, and Bob Skeel; their extensive comments and suggestions led to enlightening discussions and helped me see the field from many perspectives. I thank my colleague and friend Suse Broyde for introducing me to the field and for reading nearly every page of this book's draft.

To my family — parents Haim and Shula, sisters Yael and Daphne, aunt Cecilia, and especially Rick and Duboni — I am grateful for tolerating my long months on this project.

Finally, I thank my excellent students for making the course enjoyable and inspiring.

Tamar Schlick

New York, New York
June 10, 2002

Prelude

> Every sentence I utter must be understood not as an affirmation but as a question.
>
> Niels Bohr (1885–1962).

Only rarely does science undergo a dramatic transformation that can be likened to a tectonic rumble, as its character is transfigured under the weights of changing forces. *We are now in such an exciting time.* The discovery of the DNA double helix in the early 1950s prefigured the rise of molecular biology and its many offspring in the next half century, just as the rise of Internet technology in the 1980s has molded, and is still reshaping, nearly every aspect of contemporary life. With completion of the first draft of the human genome sequence trumpeting the beginning of the twenty-first century, triumphs in the biological sciences are competing with geopolitics and the economy for prominent-newspaper headlines. The genomic sciences now occupy the center stage, linking basic to applied (medical) research, applied research to commercial success and economic growth, and the biological sciences to the chemical, physical, mathematical and computer sciences.

The subject of this text, molecular modeling, represents a subfield of this successful marriage. In this text, I attempt to draw to the field newcomers from other disciplines and to share basic knowledge in a modern context and interdisciplinary perspective. Though many details on current investigations and projects will undoubtedly become obsolete as soon as this book goes to press, the basic foundations of modeling will remain similar. Over the next decades, we will surely witness a rapid growth in the field of molecular modeling, as well as many success stories in its application.

Contents

List of Figures

List of Tables

Acronyms, Abbreviations, and Units

A

A	adenine (purine nitrogenous base)
Å	angstrom (10^{-10} m)
AdMLP	adenovirus major late promoter (protein)
AIDS	acquired immune deficiency syndrome
Ala (A)	alanine
Arg (R)	arginine
Asp (D)	asparagine
Asn (N)	aspartic acid
AS	Altona/Sundaralingam (sugar description)
ATP	adenosine triphosphate (energy source)
AZT	zidovudine (AIDS drug)

B

bp	base pair
bps	base pairs
BAC	bacterial artificial chromosome
BOES	Born-Oppenheimer energy surfaces
BPTI	bovine pancreatic trypsin inhibitor
BSE	bovine spongiform encephalopathy ('mad cow disease')

C

cm	centimeter (10^{-2} m)
C	cytosine (pyrimidine nitrogenous base)

CAP	catabolite gene activator protein
CASP	Critical Assessment of Techniques for Protein Structure Prediction
CG	Conjugate gradient method (for minimization)
CJD	Creutzfeld-Jakob disease (brain disorder, human version of BSE)
CN	Crigler-Najjar (debilitating disease, gene therapy applications)
CP	Cremer/Pople (sugar description)
CPU	central processing units
Cys (C)	cysteine

D

DFT	density functional theory (quantum mechanics approach)
DH	Debye-Hückel
DNA	deoxyribonucleic acid (also A-, B-, C-, D-, P-, S-, T-, and Z-DNA)
DOE	Department of Energy

E

erg	energy unit (10^{-7} J)
EM	electron microscopy

F

fs	femtosecond (10^{-15} s)
FFT	Fast Fourier Transforms

G

G	guanine (purine nitrogenous base)
Gln (Q)	glutamine
Glu (E)	glutamic acid
Gly (G)	glycine
GSS	Gerstmann-Straussler-Scheinker disease (brain disorder similar to CJD)

H

HDV	hepatitis delta helper virus
His (H)	histidine
HIV	human immunodeficiency virus
HMC	hybrid Monte Carlo
HTH	helix/turn/helix (motif)
Hz	hertz (inverse second)

I

Ile (I)	isoleucine
IHF	integration host factor (protein)

K

kbp	kilobase pairs
kcal/mol	kilocalories per mole (energy unit)
kDa	kilodaltons (mass unit used for proteins)
KR	Kirkwood-Riseman

L

Leu (L)	leucine
Lys (K)	lysine
LCG	linear congruential generator

M

m	meter
mgr	minor groove
ms	millisecond (10^{-3} s)
μs	microsecond (10^{-6} s)
mm	millimeter (10^{-3} m)
MAD	multiple isomorphous replacement (crystallography technique)
MC	Monte Carlo
MD	molecular dynamics
Met (M)	methionine
Mgr	major groove
MIR	multiwavelength anomalous diffraction (crystallography technique)
MLCG	multiplicative linear congruential generator
MTS	multiple-timestep methods (for MD)

N

nm	nanometer (10^{-9} m)
ns	nanosecond (10^{-9} s)
NCBI	National Center for Biotechnology Information
NASA	National Aeronautics and Space Administration
NDB	nucleic acid database (ndbserver.rutgers.edu/)
NIH	National Institutes of Health
NMR	nuclear magnetic resonance
NSF	National Science Foundation

O

OTC	ornithine transcarbamylase (chronic ailment, gene therapy applications)

P

pn	picoNewton (force unit)
ps	picosecond (10^{-12} s)
PB	Poisson-Boltzmann

PBE	Poisson-Boltzmann equation
PC	principal component
PCA	principal component analysis
PCR	polymerase chain reaction
PDB	protein databank (www.rcsb.org/pdb)
Phe (F)	phenylalanine
PIR	Protein Information Resource (pir.georgetown.edu)
PME	particle-mesh Ewald
PNA	peptide nucleic acid (DNA mimic)
Pro (P)	proline
PrP^{C}	prion protein cellular (harmless)
PrP^{Sc}	harmful isoform of PrP^{C}, causes scrapie in sheep
Pur	purine (base)
Pyr	pyrimidine (base)

Q

QM	quantum mechanics
QN	quasi Newton method (for minimization)
QSAR	quantitative structure/activity relationships

R

RCSB	Research Collaboratory for Structural Bioinformatics (www.rcsb.org)
RMS (rms)	root-mean-square
RMSD	root-mean-square deviations
RNA	ribonucleic acid (also cRNA, gRNA, mRNA, rRNA, snRNA, tRNA)
RT	reverse transcriptase (AIDS protein)

S

s	second
Ser (S)	serine
SAR	structure/activity relationships
SCF	self-consistent field (quantum mechanical approach)
SCOP	structural classification of proteins (scop.mrc-lmb.cam.ac.uk/scop/)
SD	steepest descent method (for minimization)
SGI	Silicon Graphics Inc.
SNPs	single-nucleotide polymorphisms (“snips”)
SRY	sex determining region Y (protein)
STS	single-timestep methods (for MD)
SVD	singular value decomposition

T

T	thymine (pyrimidine nitrogenous base)
Thr (T)	threonine

Trp (W)	tryptophan
Tyr (Y)	tyrosine
TBP	TATA-box DNA binding protein (transcription regulator)
TE	transcription efficiency
TMD	targeted molecular dynamics
TN	truncated Newton method (for minimization)
2D	two-dimensional
3D	three-dimensional

U

U	uracil (pyrimidine nitrogenous base)
URL	uniform resource locator
UV	ultraviolet spectroscopy

V

Val (V)	valine

W

WC	Watson/Crick base pairing

1

Biomolecular Structure and Modeling: Historical Perspective

Chapter 1 Notation

Symbol	Definition
Vectors	
$\mathbf{h}$	unit cell identifier (crystallography)
$\mathbf{r}$	position
$F_{\mathbf{h}}$	structure factor (crystallography)
$\phi_{\mathbf{h}}$	phase angle (crystallography)
Scalars	
d	distance between parallel planes in the crystal
$I_{\mathbf{h}}$	intensity, magnitude of structure factor (crystallography)
V	cell volume (crystallography)
θ	reflection angle (crystallography)
λ	wavelength of the X-ray beam (crystallography)

. . . physics, chemistry, and biology have been connected by a web of causal explanation organized by induction-based theories that telescope into one another. . . . Thus, quantum theory underlies atomic physics, which is the foundation of reagent chemistry and its specialized offshoot biochemistry, which interlock with molecular biology — essentially, the chemistry of organic macromolecules — and hence, through successively higher levels of organization, cellular, organismic, and evolutionary biology. . . . Such is the unifying and highly productive understanding of the world that has evolved in the natural sciences.

Edward O. Wilson: "Resuming the Enlightenment Quest", in *The Wilson Quarterly*, Winter 1998.

1.1 A Multidisciplinary Enterprise

1.1.1 Consilience

The exciting field of modeling molecular systems by computer has been steadily drawing increasing attention from scientists in varied disciplines. In particular, modeling large biological polymers — proteins, nucleic acids, and lipids — is a truly multidisciplinary enterprise. Biologists describe the cellular picture; chemists fill in the atomic and molecular details; physicists extend these views to the electronic level and the underlying forces; mathematicians analyze and formulate appropriate numerical models and algorithms; and computer scientists and engineers provide the crucial implementational support for running large computer programs on high-speed and extended-communication platforms. The many names for the field (and related disciplines) underscore its cross-disciplinary nature: computational biology, computational chemistry, in silico biology, computational structural biology, computational biophysics, theoretical biophysics, theoretical chemistry, and the list goes on.

As the pioneer of sociobiology Edward O. Wilson reflects in the opening quote, some scholars believe in a unifying knowledge for understanding our universe and ourselves, or *consilience*[1] that merges all disciplines in a biologically-grounded framework [1]. Though this link is most striking between genetics and human behavior — through the neurobiological underpinnings of states of mind and mental activity, with shaping by the environment and lifestyle factors — such a unification that Wilson advocates might only be achieved by a close interaction among the varied scientists at many stages of study. The genomic era has such immense ramifications on every aspect of our lives — from health to technology to law — that it is not difficult to appreciate the effects of the biomolecular revolution on our 21st-century society.

In biomolecular modeling, a multidisciplinary approach is important not only because of the many aspects involved — from problem formulation to solution — but also since the best computational approach is often closely tailored to the

[1] *Consilience* was coined in 1840 by the theologian and polymath William Whewhell in his synthesis *The Philosophy of the Inductive Sciences*. It literally means the *alignment*, or *jumping together*, of knowledge from different disciplines. The sociobiologist Edward O. Wilson took this notion further recently by advocating in his 1998 book *Consilience* [1] that the world is orderly and can be explained by a set of natural laws that are fundamentally rooted in biology.

biological problem. In the same spirit, close connections between theory and experiment are essential: computational models evolve as experimental data become available, and biological theories and new experiments are performed as a result of computational insights. (See [2, 3, 4], for example, in connection to the characterization of protein folding mechanisms).

Although few theoreticians in the field have expertise in experimental work as well, the classic example of Werner Heisenberg's genius in theoretical physics but naiveté in experimental physics is a case in point: Heisenberg required the resolving power of the microscope to derive the uncertainty relations. In fact, an error in the experimental interpretations was pointed out by Niels Bohr, and this eventually led to the 'Copenhagen interpretation of quantum mechanics'.

If Wilson's vision is correct, the interlocking web of scientific fields rooted in the biological sciences will succeed ultimately in explaining not only the functioning of a biomolecule or the workings of the brain, but also many aspects of modern society, through the connections between our biological makeup and human behavior.

1.1.2 What is Molecular Modeling?

Molecular modeling is the science and art of studying molecular structure and function through model building and computation. The model building can be as simple as plastic templates or metal rods, or as sophisticated as interactive, animated color stereographics and laser-made wooden sculptures. The computations encompass *ab initio* and semi-empirical quantum mechanics, empirical (molecular) mechanics, molecular dynamics, Monte Carlo, free energy and solvation methods, structure/activity relationships (SAR), chemical/biochemical information and databases, and many other established procedures. The refinement of experimental data, such as from nuclear magnetic resonance (NMR) or X-ray crystallography, is also a component of biomolecular modeling.

I often remind my students of Pablo Picasso's statement on art: *"Art is the lie that helps tell the truth."*. This view applies aptly to biomolecular modeling. Though our models represent a highly-simplified version of the complex cellular environment, systematic studies based on tractable quantitative tools can help discern patterns and add insights that are otherwise difficult to observe. *The key in modeling is to develop and apply models that are appropriate for the questions being examined with them.*

The questions being addressed by computational approaches today are as intriguing and as complex as the biological systems themselves. They range from understanding the equilibrium structure of a small biopolymer subunit, to the energetics of hydrogen-bond formation in proteins and nucleic acids, to the kinetics of protein folding, to the complex functioning of a supramolecular aggregate. As experimental triumphs are being reported in structure determination — from ion channels to single-molecule biochemistry[2] — modeling approaches are needed to fill in many gaps and to build better models and theories that will ultimately make (testable) predictions.

1.1.3 Need For Critical Assessment

The field of biomolecular modeling is relatively young, having started in the 1960s, and only gained momentum since the mid 1980s with the advent of supercomputers. Yet the field is developing with dazzling speed. Advances are driven by improvements in instrumentational resolution and genomic and structural databases, as well as in force fields, algorithms for conformational sampling and molecular dynamics, computer graphics, and the increased computer power and memory capabilities. These impressive technological and modeling advances are steadily establishing the field of theoretical modeling as a partner to experiment and a widely used tool for research and development.

Yet, as we witness the tantalizing progress, a cautionary usage of molecular modeling tools is warranted, as well as a critical perspective of the field's strengths and limitations. This is because the current generation of users and application scientists in the industrial and academic sectors may not be familiar with some of the caveats and inherent approximations in biomolecular modeling and simulation approaches. Indeed, the tools and programs developed by a handful of researchers thirty years ago have now resulted in extensive profit-making software for genomic information, drug design, and every aspect of modeling. More than ever, a comprehensive background in the methodology framework is necessary for sound studies in the exciting era of computational biophysics that lies on the horizon.

[2]Examples of recent triumphs in biomolecular structure determinations include elucidation of the **nucleosome** — essential building block of the DNA/protein spools that make up the chromosomal material [5]; **ion channel proteins** — regulators of membrane electrical potentials in cells, thereby generating nerve impulses and controlling muscle contraction, hormone production, and cardiac rhythm [6, 7, 8, 9, 10]; and the **ribosome** — the cell's protein-synthesis factory, the machine bundle of 54 proteins and three RNA strands that moves along messenger RNA and synthesizes polypeptides. The complete 70*S* ribosome system was first solved at low [11] and moderate [12] resolution; its larger [13, 14, 15]) and smaller subunits [16, 17, 18, 19] were then solved at moderate resolution (see perspective in [20]). Other important examples of experimental breakthroughs involve **overstretched DNA** — as seen in single-molecule force versus extension measurements [21], and competing folding and unfolding **pathways for proteins** — as obtained by kinetic studies using spectroscopic probes (e.g., [22, 23]).

Table 1.1 Structural Biology Chronology

1865	Genes discovered by Mendel
1910	Genes in chromosomes shown by Morgan's fruitfly mutations
1920s	Quantum mechanics theory develops
1926	Early reports of crystallized proteins
1930s	Reports of crystallized proteins continue and stimulate Pauling & Corey to compile bond lengths and angles of amino acids
1944	Avery proves genetic transformation via DNA (not protein)
1946	Molecular mechanics calculations reported (Westheimer, others)
1949	Sickle cell anemia identified as 'molecular disease' (Pauling)
1950	Chargaff determines near-unity A:T and G:C ratios in many species
1951	Pauling & Corey predict protein α-helices and β-sheets
1952	Hershey & Chase reinforce genetic role of DNA (phage experiments)
1952	Wilkins & Franklin deduce that DNA is a helix (X-ray fiber diffraction)
1953	**Watson & Crick report the structure of the DNA double helix**
1959	Myoglobin & hemoglobin deciphered by X-ray (Kendrew & Perutz)
1960s	Systematic force-fields develop (Allinger, Lifson, Scheraga, others)
1960s	Genetic code deduced (Crick, Brenner, Nirenberg, Khorana, Holley, coworkers)
1969	Levinthal paradox on protein folding posed
1970s	Biomolecular dynamics simulations develop (Stillinger, Karplus, others)
1970s	Site-directed mutagenesis techniques developed by M. Smith; restriction enzymes discovered by Arber, Nathans, and H. Smith
1971	Protein Data Bank established
1974	t-RNA structure reported
1975	Fifty solved biomolecular structures available in the PDB
1977	DNA genome of the virus ϕX174 (5.4 kb) sequenced; soon followed by human mitochondrial DNA (16.6 kb) and λ phage (48.5 kb)
1980s	Dazzling progress realized in automated sequencing, protein X-ray crystallography, NMR, recombinant DNA, and macromolecular synthesis
1985	PCR devised by Mullis; numerous applications follow
1985	NSF establishes five national supercomputer centers
1990	International Human Genome Project starts; spurs others
1994	RNA hammerhead ribozyme structure reported; other RNAs follow
1995	First non-viral genome completed (bacterium *H. influenzae*), 1.8 Mb
1996	Yeast genome (*Saccharomyces cerevisiae*) completed, 13 Mb
1997	Chromatin core particle structure reported; confirms earlier structure
1998	Roundworm genome (*C. elegans*) completed, 100 Mb
1998	Crystal structure of ion channel protein reported
1998	Private Human Genome initiative competes with international effort
1999	Fruitfly genome (*Drosophila melanogaster*) completed (Celera), 137 Mb
1999	Human chromosome 22 sequenced (public consortium)
1999	IBM announces petaflop computer to fold proteins by 2005
2000	**First draft of human genome sequence announced**, 3300 Mb
2000	Moderate-resolution structures of ribosomes reported
2001	First annotation of the human genome (February)
2002	First draft of rice genome sequence, 430 Mb (April)

1.1.4 Text Overview

This text aims to provide this critical perspective for field assessment while introducing the relevant techniques. Specifically, the elementary background for biomolecular modeling will be introduced: protein and nucleic-acid structure tutorials (Chapters 3–6), overview of theoretical approaches (Chapter 7), details of force field construction and evaluation (Chapters 8 and 9), energy minimization techniques (Chapter 10), Monte Carlo simulations (Chapter 11), molecular dynamics and related methods (Chapters 12 and 13), and similarity/diversity problems in chemical design (Chapter 14).

As emphasized in this book's Preface, given the enormously broad range of these topics, depth is often sacrificed at the expense of breadth. Thus, many specialized texts (e.g., in Monte Carlo, molecular dynamics, or statistical mechanics) are complementary, such as those listed in Appendix C; the representative articles used for the course (Appendix B) are important components. For introductory texts to biomolecular structure, biochemistry, and biophysical chemistry, see those listed in Appendix C, such as [24, 25, 26, 27, 28]. For molecular simulations, a solid grounding in classical statistical mechanics, thermodynamic ensembles, time-correlation functions, and basic simulation protocols is important. Good introductory texts for these subjects, including biomolecular applications are [29, 30, 31, 32, 33, 34, 35, 36, 37, 38].

The remainder of this chapter and the next chapter provide a historical context for the field's development. Overall, this chapter focuses on a historical account of the field and the experimental progress that made biomolecular modeling possible; chapter 2 introduces some of the field's challenges as well as practical applications of their solution.

Specifically, to appreciate the evolution of biomolecular modeling and simulation, we begin in the next section with an account of the milieu of growing experimental and technical developments. Following an introduction to the birth of molecular mechanics (Section 1.2), experimental progress in protein and nucleic-acid structure is described (Section 1.3). A selective reference chronology to structural biology is shown in Table 1.1.

The experimental section of this chapter discusses separately the early days of biomolecular instrumentation — as structures were emerging from X-ray crystallography — and the modern era of technological developments — stimulating the many sequencing projects and the rapid advances in biomolecular NMR and crystallography. Within this presentation, separate subsections are devoted to the techniques of X-ray crystallography and NMR and to the genome projects.

Chapter 2 continues this perspective by describing the computational challenges that naturally emerge from the dazzling progress in genome projects and experimental techniques, namely deducing structure and function from sequence.

Problems are exemplified by protein folding and misfolding. (Students unfamiliar with basic protein structure are urged to re-read Chapter 2 after the protein minitutorial chapters). The sections that follow mention some of the exciting and important biomedical, industrial, and technological applications that lend enormous practical utility to the field. These applications represent a tangible outcome of the confluential experimental, theoretical, and technological advances.

Since the material presented in these introductory chapters is changing rapidly (e.g., the status of the genome projects, theoretical and instrumentational progress), the author anticipates periodic updating and placement on the text web page.

1.2 The Roots of Molecular Modeling in Molecular Mechanics

The roots of molecular modeling began with the notion that molecular geometry, energy, and various molecular properties can be calculated from mechanical-like models subject to basic physical forces. A molecule is represented as a mechanical system in which the *particles* — atoms — are connected by *springs* — the bonds. The molecule then rotates, vibrates, and translates to assume favored conformations in space as a collective response to the inter- and intramolecular forces acting upon it.

The forces are expressed as a sum of harmonic-like (from Hooke's law) terms for **bond-length** and **bond-angle** deviations from reference equilibrium values; trigonometric **torsional terms** to account for *internal rotation* (rotation of molecular subgroups about the bond connecting them); and **nonbonded van der Waals and electrostatic potentials**. See Chapter 8 for a detailed discussion of these terms, as well as of more intricate cross terms.

1.2.1 The Theoretical Pioneers

Molecular mechanics arose naturally from the concepts of molecular bonding and van der Waals forces. The Born-Oppenheimer approximation assuming fixed nuclei (see Chapter 7) followed in the footsteps of quantum theory developed in the 1920s. While the basic idea can be traced to 1930, the first attempts of molecular mechanics calculations were recorded in 1946. Frank Westheimer's calculation of the relative racemization rates of biphenyl derivatives illustrated the success of such an approach. However, computers were not available at that time, so it took several more years for the field to gather momentum.

In the early 1960s, pioneering work on development of systematic force fields — based on spectroscopic information, heats of formation, structures of small compounds sharing the basic chemical groups, other experimental data, and quantum-mechanical information — began independently in the laboratories of the late Shneior Lifson at the Weizmann Institute of Science (Rehovot, Israel)

Table 1.2. The evolution of molecular mechanics and dynamics.

Period	System and Size[a]	Trajectory Length[b] [ns]	CPU Time/Computer[c]
1973	Dinucleoside (GpC) in vacuum (8 flexible dihedral angles)	—	—
1977	BPTI, vacuum (58 residues, 885 atoms)	0.01	
1983	DNA, vacuum, 12 & 24 bp (754/1530 atoms)	0.09	several weeks each Vax 780
1984	GnRH, vacuum (decapeptide, 161 atoms)	0.15	
1985	Myoglobin, vacuum (1423 atoms)	0.30	50 days VAX 11/780
1985	DNA, 5 bp (2800 atoms)	0.50	20 hrs Cray X-MP
1989	Phospholipid Micelle ($\approx$ 7,000 atoms)	0.10	
1992	HIV protease (25,000 atoms)	0.10	100 hrs. Cray Y-MP
1997	Estrogen/DNA (36,000 atoms, multipoles)	0.10	22 days HP-735 (8)
1998	DNA, 24 bp (21,000 atoms, PME)	0.50	1 year, SGI Challenge
1998	β-heptapeptide in methanol ($\approx$ 5000/9000 atoms)	200	8 months, SGI-Challenge (3)
1998	Villin headpiece (36 residues, 12,000 atoms, cutoffs)	1000	4 months, 256-proc. Cray T3D/E
1999	bc_1 complex in phospholipid bilayer (91,061 atoms, cutoffs)	1	75 days, 64 450-MHz-proc. Cray T3E
2001	C-terminal β-hairpin of protein-G (177 atoms, implicit solvent)	38000	$\sim$ 8 days, 5000 proc. Folding@home megacluster
2002	channel protein in lipid membrane (106,189 atoms, PME)	5	30 hrs, 500 proc. LeMieux terascale system; 50 days, 32 proc. Linux (Athlon)

[a]The examples for each period are representative. The first five systems are modeled in vacuum and the others in solvent. Except for the dinucleoside, simulations refer to molecular dynamics (MD). The two system sizes for the β-heptapeptide [39] reflect two (temperature-dependent) simulations. See text for definitions of abbreviations and further entry information.

[b]The 38 μs β-hairpin simulation in 2001 represents an ensemble (or aggregate) dynamics simulation, as accumulated over several short runs, rather than one long simulation [40].

[c]The computational time is given where possible; estimates for the vacuum DNA, heptapeptide, β-hairpin, and channel protein simulations [41, 39, 40, 42] were kindly provided by M. Levitt, W. van Gunsteren, V. Pande, and K. Schulten, respectively.

[43], Harold Scheraga at Cornell University (Ithaca, New York), and Norman Allinger at Wayne State University (Detroit, Michigan) and then the University of Georgia (Athens). These researchers began to develop force field parame-

ters for families of chemical compounds by testing calculation results against experimental observations regarding structure and energetics.

In the early 1970s, Rahman and Stillinger reported the first molecular dynamics work of a polar molecule, liquid water [44, 45]; results offered insights into the structural and dynamic properties of this life sustaining molecule. Rahman and Stillinger built upon the simulation technique described much earlier (1959) by Alder and Wainwright but applied to hard spheres [46].

In the late 1970s, the idea of using molecular mechanics force fields with energy minimization as a tool for refinement of crystal structures was presented [47] and developed [48]. This led to the modern versions employing simulated annealing and related methods [49, 50].

It took a few more years, however, for the field to gain some 'legitimacy'.[3] In fact, these pioneers did not receive much general support at first, partly because their work could not easily be classified as a traditional discipline of chemistry (e.g., physical chemistry, organic chemistry). In particular, spectroscopists criticized the notion of transferability of the force constants, though at the same time they were quite curious about the predictions that molecular mechanics could make. In time, it indeed became evident that force constants are not generally transferable; still, the molecular mechanics approach was sound since nonbonded interactions are included, terms that spectroscopists omitted.

Ten to fifteen more years followed until the first generation of biomolecular force fields was established. The revitalized idea of molecular dynamics in the late 1970s propagated by Martin Karplus and colleagues at Harvard University sparked a flame of excitement that continues with full force today with the fuel of supercomputers. Most programs and force fields today, for both small and large molecules, are based on the works of the pioneers cited above (Allinger, Lifson, and Scheraga) and their coworkers. The water force fields developed in the late 1970s/early 1980s by Berendsen and coworkers (e.g., [51]) and by Jorgensen and coworkers [52] (SPC and TIP3P/TIP4P, respectively) laid the groundwork for biomolecular simulations in solution.

Peter Kollman's legacy is the development and application of force field methodology and computer simulation to important biomolecular, as well as medicinal, problems such as enzyme catalysis and protein/ligand design [53]; his group's free energy methods and combined quantum/molecular mechanics approaches have opened many new doors of applications. With Kollman's untimely death in May 2001, the community mourns the loss of a great leader and innovator.

[3]Personal experiences shared by Norman L. Allinger on those early days of the field form the basis for the comments in this paragraph. I am grateful for his sharing these experiences with me.

1.2.2 Biomolecular Simulation Perspective

Table 1.2 and Figures 1.1 and 1.2 provide a perspective of biomolecular simulations. Specifically, the selected examples illustrate the growth in time of system complexity (size and model resolution) and simulation length. The three-dimensional (3D) rendering in Figure 1.1 shows 'buildings' with heights proportional to system size. Figure 1.2 offers molecular views of the simulation subjects and extrapolations for long-time simulations of proteins and cells based on [54].

Representative Progress

Starting from the first entry in the table, **dinucleoside GpC** (guanosine-3′, 5′-cytidine monophosphate) posed a challenge in the early 1970s for finding all minima by potential energy calculations and model building [55]. Still, clever search strategies and constraints found a correct conformation (dihedral angles in the range of helical RNA and sugar in C3′-endo form) as the lowest energy minimum. *Global optimization remains a difficult problem!* (See Chapter 10).

The small protein **BPTI** (Bovine Pancreatic Trypsin Inhibitor) was the subject of a **1977** pioneering dynamic simulation applied to a protein [56]. It showed substantial atomic fluctuations on the picosecond timescale.

The 12 and 24-base-pair (bp) **DNA** simulations in **1983** [41] were performed in vacuum without electrostatics, and that of the DNA pentamer system in 1985, with 830 water molecules and 8 sodium ions and full electrostatics [57]. Stability problems for nucleic-acids emerged in the early days — unfortunately, in some cases the strands untwisted and separated [41]. Stability became possible with the introduction of scaled phosphate charges in other pioneering nucleic-acid simulations [58, 59, 60] and the introduction a decade later of more advanced treatments for solvation and electrostatics; see [61], for example, for a discussion.

The linear **decapeptide** GnRH (gonadotropin-releasing hormone) was studied in **1984** for its pharmaceutical potential, as it triggers LH and FSH hormones [62].

The 300 ps dynamics simulation of the protein **myoglobin** in **1985** [63] was considered three times longer than the longest previous MD simulation of a protein. The results indicated a slow convergence of many thermodynamic properties.

The large-scale **phospholipid** aggregate simulations in **1989** [64] was an ambitious undertaking: it incorporated a hydrated micelle (i.e., a spherical aggregate of phospholipid molecules) containing 85 LPE molecules (lysophosphatiadyl-ethanolamine) and 1591 water molecules.

The **HIV protease** system simulated in solution in **1992** [65] captured an interesting flap motion at the active site. See also Figure 2.5 and a discussion of this motion in the context of protease inhibitor design.

The **1997 estrogen/DNA simulation** [66] sought to understand the mechanism underlying DNA sequence recognition by the protein. It used the multipole electrostatic treatment, crucial for simulation stability, and also parallel processing for speedup [67].

The **1998 DNA** simulation [68] used the alternative, Particle Mesh Ewald (PME) treatment for consideration of long-range electrostatics (see Chapter 9) and uncovered interesting properties of A-tract sequences.

The **1998 peptide** simulation in methanol used periodic boundary conditions (defined in Chapter 9) and captured reversible, temperature-dependent folding [39]; the 200 ns time reflects four 50 ns simulations at various temperatures.

The **1998** 1 μs **villin-headpiece** simulation (using periodic boundary conditions) [69] was considered longer by three orders of magnitude than prior simulations. A folded structure close to the native state was approached; see also [70].

The solvated protein bc_1 **embedded in a phospholipid bilayer** [71] was simulated in **1999** for over 1 ns by a 'steered molecular dynamics' algorithm (45,131 flexible atoms) to suggest a pathway for proton conduction through a water channel. As in villin, the Coulomb forces were truncated.

By **2002**, an aquaporin membrane channel protein in the glycerol conducting subclass (*E. coli* **glycerol channel, GlpF**) in a lipid membrane (106,189 total atoms) was simulated for 5 ns (as well as a mutant) with all nonbonded interactions considered, using the PME approach [42]. The simulations suggested details of a selective mechanism by which water transport is controlled; see also [72] for simulations examining the glycerol transport mechanism.

By early 2002, the longest simulation published was 38 μs, but for aggregate (or ensemble) dynamics — usage of many short trajectories to simulate the microsecond timescale — set for the C-terminal β-**hairpin from protein G** (16 residues) in **2001** [40]. Whereas the continuous 1 μs villin simulation required months of dedicated supercomputing, the β-hairpin simulation (177 atoms, using implicit solvation and Langevin dynamics) was performed to analyze folding kinetics on a new distributed computing paradigm which employs personal computers from around the world (see Folding@home, folding.stanford.edu and [73]). About 5000 processors were employed and, with the effective production rate of 1 day per nanosecond per processor, about 8 days were required to simulate the 38 μs aggregate time.

Trends

Note from the table and figure the transition from simulations in vacuum (first five entries) to simulations in solvent (remaining items). Observe also the steady increase in simulated system size, with a leap increase in simulation lengths made only recently.

Large system sizes or long simulation times can only be achieved by sacrificing other simulation aspects. For example, truncating long-range electrostatic interactions makes possible the study of large systems over short times [71], or small systems over long times [69]. Using implicit solvent and cutoffs for electrostatic interactions also allows the simulation of relatively small systems over long times [40]. In fact, with the increased awareness of the sampling problem in dynamic simulation (see Chapter 12), we now see the latter trend more often, namely studying smaller solvated molecular systems for longer times; one long simulation is often replaced by several trajectories, leading to overall better sampling statistics.

For recent reviews and perspectives on dynamics simulations, see [74, 75], for example. The former discusses progress to date and future challenges in long-timescale simulations of peptides and proteins in solution, and the latter summarizes progress in various macromolecular systems (including membranes and channels) and simulation methodologies, including drug design applications.

Duan *et al.* make an interesting 'fanciful' projection on the computational capabilities of modeling in the coming decades [54]: they suggest the feasibility, in 20 years, of simulating a second in the life-time of medium-sized proteins and, in 50–60 years, of following the entire life cycle of an *E. Coli* cell (1000 seconds or 20 minutes, for 30 billion atoms). This estimate was extrapolated on the basis of two data points — the 1977 BPTI simulation [56] and the 1998 villin simulation [69, 70] discussed above — and relied on the assumption that computational power increases by a factor of 10 every 3–4 years. These projections are displayed by entries for the years 2020 and 2055 in Figure 1.2.

1.3 Emergence of Biomodeling from Experimental Progress in Proteins and Nucleic Acids

At the same time that molecular mechanics developed, tremendous progress on the experimental front also began to trigger further interest in the theoretical approach to structure determination.

1.3.1 Protein Crystallography

The first records of crystallized polypeptides or proteins date back to the late 1920s / early 1930s (1926: urease, James Sumner; 1934: pepsin, J. D. Bernal and Dorothy Crowfoot-Hodgkin; 1935: insulin, Crowfoot-Hodgkin). However, only in the late 1950s did John Kendrew (Perutz' first doctoral student) and Max Perutz succeed in deciphering the X-ray diffraction pattern from the crystal structure of the protein (1958: myoglobin, Kendrew; 1959: hemoglobin, Perutz). This was possible by Perutz' crucial demonstration (around 1954) that structures of proteins can be solved by comparing the X-ray diffraction patterns of a crystal of a native protein to those associated with the protein bound to heavy atoms like mercury (i.e., by 'isomorphous replacement'). The era of modern structural biology began with this landmark development.

As glimpses of the first X-ray crystal structures of proteins came into view, Linus Pauling and Robert Corey began in the mid-1930s to catalogue bond lengths and angles in amino acids. By the early 1950s, they had predicted the two basic structures of amino acid polymers on the basis of hydrogen bonding patterns: α helices and β sheets [76, 77]. As of 1960, about 75 proteins had been crystallized,

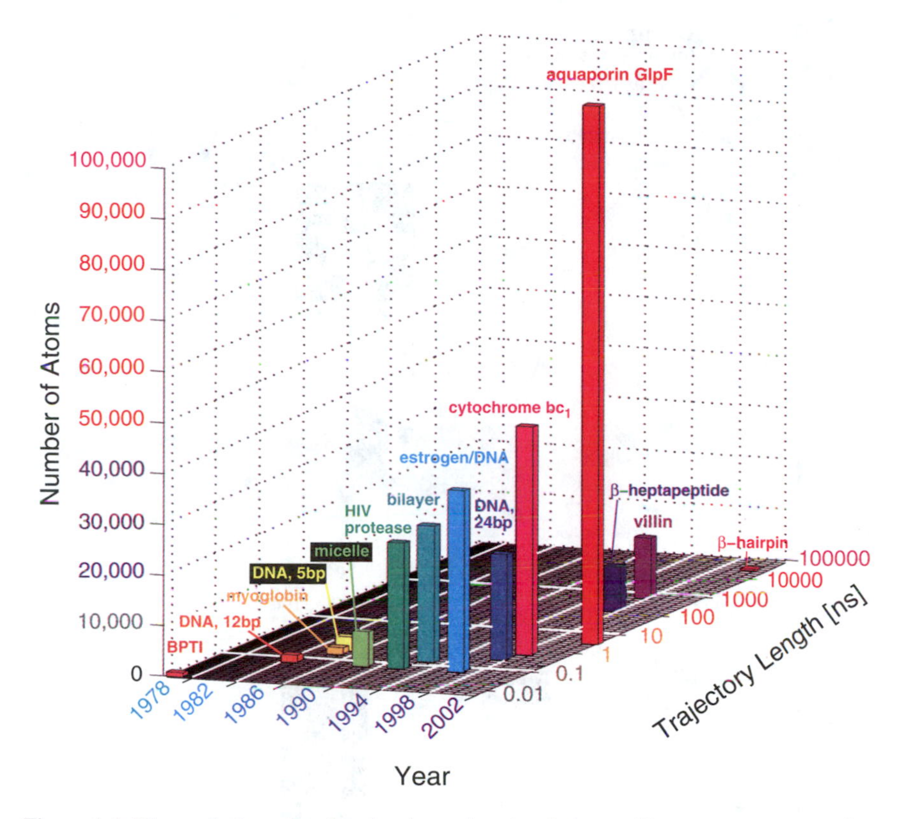

Figure 1.1. The evolution of molecular dynamics simulations with respect to system sizes and simulation lengths (see also Table 1.2).

and immense interest began on relating the sequence content to catalytic activity of these enzymes.

By then, the exciting new field of molecular biology was well underway. Perutz, who founded the Medical Research Council Unit for Molecular Biology at the Cavendish Laboratory in Cambridge in 1947, created in 1962 the Laboratory of Molecular Biology there. Perutz and Kendrew received the Nobel Prize for Chemistry for their accomplishments in 1962.[4]

[4]See the formidable electronic museum of science and technology, with related lectures and books that emerged from Nobel-awarded research, on the website of the Nobel Foundation (www.nobel.se). This virtual museum was recently constructed to mark the 100th anniversary in 2001 of Alfred B. Nobel's legacy.

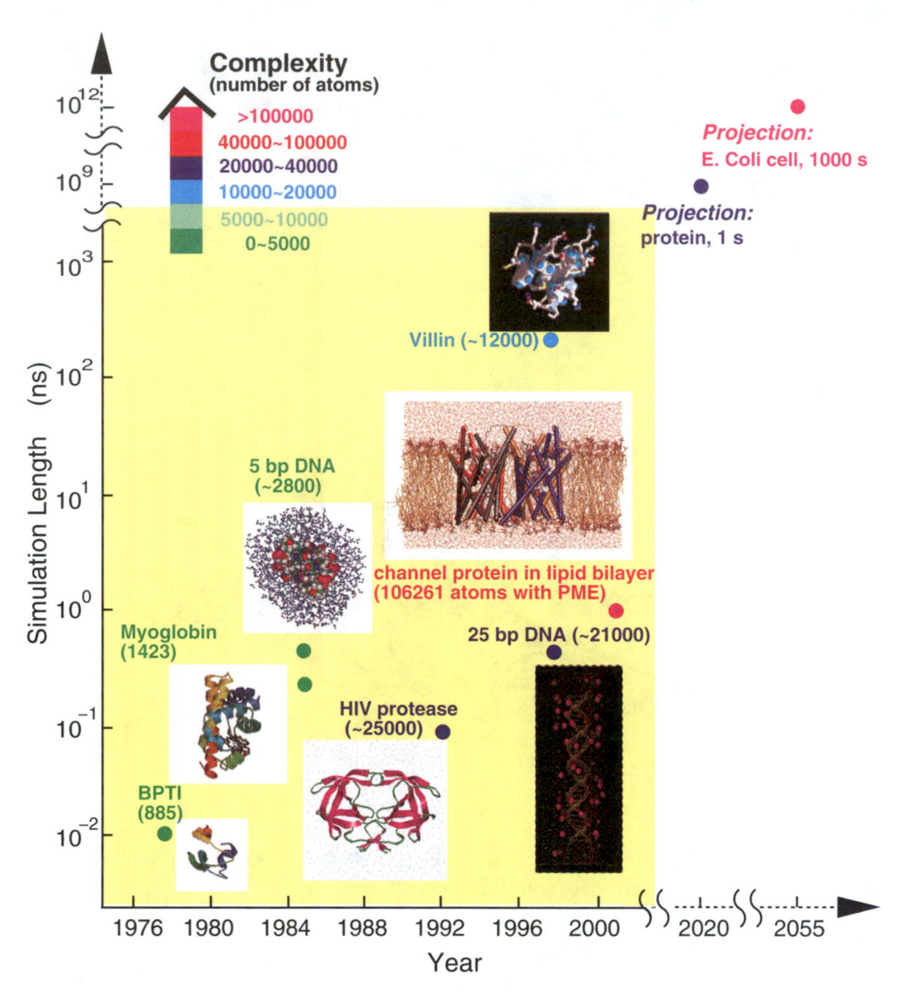

Figure 1.2. The evolution of molecular dynamics simulations with respect to simulation lengths (see also Table 1.2 and Figure 1.1). The data points for 2020 and 2055 represent extrapolations from the 1977 BPTI [56] and 1998 villin [70, 69] simulations, assuming a computational power increase by a factor of 10 every 3–4 years, as reported in [54].

1.3.2 DNA Structure

Momentum at that time came in large part from parallel experimental work that began in 1944 in the nucleic acid world and presaged the discovery of the DNA double helix.

Inspired by the 1928 work of the British medical officer Fred Griffith, Oswald Avery and coworkers Colin MacLeod and Maclyn McCarty studied pneumonia infections. Griffith's intriguing experiments showed that mice became fatally ill upon infection from a live but harmless (coatless) strain of pneumonia-causing bacteria mixed with the DNA from heat-killed pathogenic bacteria; thus, the

DNA from heat-killed pathogenic bacteria transformed live harmless into live pathogenic bacteria. Avery and coworkers mixed DNA from virulent strains of pneumococci with harmless strains and used enzymes that digest DNA but not proteins. Their results led to the cautious announcement that the 'transforming agent' of traits is made exclusively of DNA.[5]

Their finding was held with skepticism until the breakthrough, Nobel prize-winning phage experiments of Alfred Hershey and Martha Chase eight years later, which demonstrated that only the nucleic acid of the phage entered the bacterium upon infection, whereas the phage protein remained outside.[6]

Much credit for the transforming agent evidence is due to the German theoretical physicist and Nobel laureate Max Delbrück, who brilliantly suggested to use bacterial viruses as the model system for the genome demonstration principle. Delbrück shared the Nobel Prize in Physiology or Medicine in 1969 with Hershey and Salvador Luria for their pioneering work that established bacteriophage as the premier model system for molecular genetics.

In 1950, Erwin Chargaff demonstrated that the ratios of adenine-to-thymine and guanine-to-cytosine bases are close to unity, with the relative amount of each kind of pair depending on the DNA source. These crucial data, together with the X-ray fiber diffraction photographs of hydrated DNA taken by Rosalind Franklin and Raymond Gosling (both affiliated with Maurice Wilkins who was engaged in related research [78]), led directly to Watson and Crick's ingenious proposal of the structure of DNA in 1953. The photographs were crucial as they suggested a helical arrangement.

Although connecting these puzzle pieces may seem straightforward to us now that the DNA double helix is a household word, these two ambitious young Cambridge scientists deduced from the fiber diffraction data and other evidence that the observed base-pairing specificity, together with steric restrictions, can be reconciled in an *anti-parallel* double-helical form with a sugar-phosphate backbone and nitrogenous-bases interior. Their model also required a key piece of information from the organic chemist Jerry Donahue regarding the *tautomeric* states of the bases.[7] Though many other DNA forms besides the classic Crick and Watson (B-DNA) form are now recognized, including triplexes and quadruplexes, the B-form is still the most prevalent under physiological conditions.

---

[5]Interested readers can visit the virtual gallery of Profiles in Science at www.profiles.nlm.nih.gov/ for a profile on Avery.

[6]A wonderful introduction to the rather recluse Hershey, who died at the age of 88 in 1997, can be enjoyed in a volume edited by Franklin W. Stahl titled *We can sleep later: Alfred D. Hershey and the origins of molecular biology* (Cold Spring Harbor Press, New York, 2000). The title quotes Hershey from his letter to contributors of a volume on *bacteriophage* λ which he edited in 1971, urging them to complete and submit their manuscripts!

[7]Proton migrations within the bases can produce a *tautomer*. These alternative forms depend on the dielectric constant of the solvent and the pH of the environment. In the bases, the common *amino* group (–N–H_2) can tautomerize to an *imino* form (=N–H), and the common *keto* group (–C=O) can adopt the *enol* state (=C–O–H); the fraction of bases in the rare imino and enol tautomers is only about 0.01% under regular conditions.

RNA crystallography is at a more early stage, but has recently made quantum leaps with the solution of several significant RNA molecules (see Chapter 6) [79, 80]. These developments followed the exciting discoveries in the 1980s that established that RNA, like protein, can act as an enzyme in living cells. Sidney Altman and Thomas Cech received the 1989 Nobel Prize in Chemistry for their discovery of RNA biocatalysts, *ribozymes*.

The next two subsections elaborate upon the key techniques for solving biomolecular structures: X-ray crystallography and NMR. We end this section on experimental progress with a description of modern technological advances and the genome sequencing projects they inspired.

1.3.3 The Technique of X-ray Crystallography

Much of the early crystallographic work was accomplished without computers and was inherently very slow. *Imagine calculating the Fourier series by hand!* Only in the 1950s were direct methods for the phase problem developed, with a dramatic increase in the speed of structure determination occurring about a decade later.

Structure determination by X-ray crystallography involves analysis of the X-ray diffraction pattern produced when a beam of X-rays is directed onto a well-ordered crystal. Crystals form by vapor diffusion from purified protein solutions under optimal conditions. See [28, 81] for overviews.

The diffraction pattern can be interpreted as a reflection of the primary beam source from sets of parallel planes in the crystal. The diffracted spots are recorded on a detector (electronic device or X-ray film), scanned by a computer, and analyzed on the basis of Bragg's law[8] to determine the unit cell parameters.

Each such recorded diffraction spot has an associated amplitude, wavelength, and phase; all three properties must be known to deduce atomic positions. Since the phase is lost in the X-ray experiments, it must be computed from the other data. This central obstacle in crystal structure analysis is called the *phase problem* (see Box 1.1). Together, the amplitudes and phases of the diffraction data are used to calculate the electron density map; the greater the resolution of the diffraction data, the higher the resolution of this map and hence the atomic detail derived from it.

Both the laborious crystallization process [82] and the necessary mathematical analysis of the diffraction data limit the amount of accurate biomolecular

---

[8]The Braggs (father William-Henry and son Sir William-Lawrence) observed that if two waves of electromagnetic radiation arrive at the same point in phase and produce a maximal intensity, the difference between the distances they traveled is an integral multiple of their wavelengths. From this they derived what is now known as *Bragg's law*, specifying the conditions for diffraction and the relation among three key quantities: d (distance between parallel planes in the crystal), λ (the wavelength of the X-ray beam), and θ (the reflection angle). Bragg's condition requires that the difference in distance traveled by the X-rays reflected from adjacent planes is equal to the wavelength λ. The associated relationship is $\lambda = 2d \sin\theta$.

data available. Well-ordered crystals of biological macromolecules are difficult to grow, in part because of the disorder and mobility in certain regions. Crystallization experiments must therefore screen and optimize various parameters that influence crystal formation, such as temperature, pH, solvent type, and added ions or ligands.

The phase problem was solved by direct methods for small molecules (roughly ≤ 100 atoms) by Jerome Karle and Herbert Hauptman in the late 1940s and early 1950s; they were recognized for this feat with the 1985 Nobel Prize in Chemistry. For larger molecules, biomolecular crystallographers have relied on the method pioneered by Perutz, Kendrew and their coworkers termed *multiple isomorphous replacement* (MIR).

MIR introduces new X-ray scatters from complexes of the biomolecule with heavy elements such as selenium or heavy metals like osmium, mercury, or uranium. The combination of diffraction patterns for the biomolecule, heavy elements or elements or metals, and biomolecule/heavy-metal complex offers more information for estimating the desired phases. The differences in diffracted intensities between the native and derivative crystals are used to pinpoint the heavy atoms, whose waves serve as references in the phase determination for the native system.

To date, advances in the experimental, technological, and theoretical fronts have dramatically improved the ease of crystal preparation and the quality of the obtained three-dimensional (3D) biomolecular models [28, last chapter]. Techniques besides MIR to facilitate the phase determination process — by analyzing patterns of heavy-metal derivatives using *multi-wavelength anomalous diffraction* (MAD) or by molecular replacement (deriving the phase of the target crystal on the basis of a solved related molecular system) [83, 84] have been developed. Very strong X-ray sources from synchrotron radiation (e.g., with light intensity that can be 10,000 times greater than conventional beams generated in a laboratory) have become available. New techniques have made it possible to visualize short-lived intermediates in enzyme-catalyzed reactions at atomic resolution by time-resolved crystallography [85, 86, 87]. And improved methods for model refinement and phase determination are continuously being reported [88]. Such advances are leading to highly refined biomolecular structures[9] (resolution ≤ 2 Å) at much greater numbers [89], even for nucleic acids [90].

[9] The resolution value is similar to the quantity associated with a microscope: objects (atoms) can be distinguished if they are separated by more than the resolution value. Hence, the lower the resolution value the more molecular architectural detail that can be discerned.

Box 1.1: The Phase Problem

The mathematical phase problem in crystallography [91, 92] involves resolving the phase angles $\phi_{\mathbf{h}}$ associated with the structure factors $F_{\mathbf{h}}$ when only the intensities (squares of the amplitudes) of the scattered X-ray pattern, $I_{\mathbf{h}} = |F_{\mathbf{h}}|$, are known. The structure factors $F_{\mathbf{h}}$, defined as

$$F_{\mathbf{h}} = |F_{\mathbf{h}}| \exp(i\phi_{\mathbf{h}}) , \tag{1.1}$$

describe the scattering pattern of the crystal in the Fourier series of the electron density distribution:

$$\rho(\mathbf{r}) = \frac{1}{V} \sum_{\mathbf{h}} F_{\mathbf{h}} \exp(-2\pi i \mathbf{h} \cdot \mathbf{r}). \tag{1.2}$$

Here $\mathbf{r}$ denotes position, $\mathbf{h}$ identifies the three defining planes of the unit cell (e.g., h, k, l), V is the cell volume, and $\cdot$ denotes a vector product. See [93], for example, for details.

1.3.4 The Technique of NMR Spectroscopy

The introduction of NMR as a technique for protein structure determination came much later (early 1960s), but since 1984 both X-ray diffraction and NMR have been valuable tools for determining protein structure at atomic resolution.

Nuclear magnetic resonance is a versatile technique for obtaining structural and dynamic information on molecules in solution. The resulting 3D views from NMR are not as detailed as those that can result from X-ray crystallography, but the NMR information is not static and incorporates effects due to thermal motions in solution.

In NMR, powerful magnetic fields and high-frequency radiation waves are applied to probe the magnetic environment of the nuclei. The local environment of the nucleus determines the frequency of the resonance absorption. The resulting NMR spectrum contains information on the interactions and localized motion of the molecules containing those resonant nuclei.

The absorption frequency of particular groups can be distinguished from one another when high-frequency NMR devices are used (*"high resolution NMR"*). This requirement for nonoverlapping signals to produce a clear picture limits the protein sizes that can be studied by NMR to proteins with masses less than $\sim$ 35 kDa at this time. The biomolecular NMR future is bright, however, with novel strategies for isotopic labeling of proteins [94] and solid-state NMR techniques, the latter of which may be particularly valuable for structure analysis of membrane proteins.

As in X-ray crystallography, advanced computers are required to interpret the data systematically. NMR spectroscopy yields a wealth of information: a network of distances involving pairs of spatially-proximate hydrogen atoms. The distances are derived from Nuclear Overhauser Effects (NOEs) between neighboring hydrogen atoms in the biomolecule, that is, for atom pairs separated by less than 5–6 Å.

To calculate the 3D structure of the macromolecule, these NMR distances are used as conformational restraints in combination with various supplementary information: primary sequence, reference geometries for bond lengths and bond angles, chirality, steric constraints, spectra, and so on. A suitable energy function must be formulated and then minimized, or surveyed by various techniques, to find the coordinates that are most compatible with the experimental data. See [49] for an overview. Such deduced models are used to back calculate the spectra inferred from the distances, from which iterative improvements of the model are pursued to improve the matching of the spectra. Indeed, the difficulty of using NMR data for structure refinement in the early days can be attributed to this formidable refinement task, formally, an over-determined or under-determined global optimization problem.[10]

The pioneering efforts of deducing peptide and protein structures in solution by NMR techniques were reported between 1981 and 1986; they reflected year-long struggles in the laboratory. Only a decade later, with advances on the experimental, theoretical, and technological fronts, 3D structures of proteins in solution could be determined routinely for monomeric proteins with less than 200 amino acid residues. See [95, 96] for texts by modern NMR pioneers, [97] for a historical perspective of biomolecular NMR, and [49, 98, 99] for recent advances.

Today's clever methods have been designed to facilitate such refinements, from formulation of the target energy to conformational searching, the latter using tools from distance geometry, molecular dynamics, simulated annealing, and many hybrid search techniques [100, 49, 97, 101]. The ensemble of structures obtained is not guaranteed to contain the "best" (global) one, but the solutions are generally satisfactory in terms of consistency with the data. The recent technique of residual dipolar coupling also has great potential for structure determination by NMR spectroscopy without the use of NOE data [102, 103].

1.4 Modern Era of Technological Advances

1.4.1 From Biochemistry to Biotechnology

The discovery of the elegant yet simple DNA double helix not only led to the birth of molecular biology; it led to the crucial link between biology and chemistry. Namely, the genetic code relating triplets of RNA (the template for protein synthesis) to the amino acid sequence was decoded ten years later, and biochemists began to isolate enzymes that control DNA metabolism.

[10]Solved NMR structures are usually presented as sets of structures since certain molecular segments can be over-determined while others under-determined. The better the agreement for particular atomic positions among the structures in the set, the more likely it is that a particular atom or component is well determined.

One class of those enzymes, restriction endonucleases, became especially important for recombinant DNA technology. These molecules can be used to break huge DNA into small fragments for sequence analysis. Restriction enzymes can also cut and paste DNA (the latter with the aid of an enzyme, ligase) and thereby create spliced DNA of desired transferred properties, such as antibiotic-resistant bacteria that serve as informants for human insulin makers. The discovery of these enzymes was recognized by the 1978 Nobel Prize in Physiology or Medicine to Werner Arber, Daniel Nathans, and Hamilton O. Smith.

Very quickly, X-ray, NMR, recombinant DNA technology, and the synthesis of biological macromolecules improved. The 1970s and 1980s saw steady advances in our ability to produce, crystallize, image, and manipulate macromolecules. Site-directed mutagenesis developed in 1970s by Canadian biochemist Michael Smith (1993 Nobel laureate in Chemistry) has become a fundamental tool for protein synthesis and protein function analysis.

1.4.2 PCR and Beyond

The polymerase chain reaction (PCR) devised in 1985 by Kary Mullis (winner of the 1993 Chemistry Nobel Prize, with Michael Smith) and coworkers [104] revolutionized biochemistry: small parent DNA sequences could be amplified exponentially in a very short time and used for many important investigations. DNA analysis has become a standard tool for a variety of practical applications. Noteworthy classic and current examples of PCR applications are collected in Box 1.2. See also [105] for stories on how genetics teaches us about history, justice, diseases, and more.

Beyond amplification, PCR technology made possible isolation of gene fragments and their usage to clone whole genes; these genes could then be inserted into viruses or bacterial cells to direct the synthesis of biologically active products. With dazzling speed, the field of bioengineering was born. Automated sequencing efforts continued during the 1980s, leading to the start of the International Human Genome Project in 1990, which spearheaded many other sequencing projects (see next section).

Macromolecular X-ray crystallography and NMR techniques are also improving rapidly in this modern era of instrumentation, both in terms of obtained structure resolution and system sizes [106]. Stronger X-ray sources, higher-frequency NMR spectrometers, and refinement tools for both data models are leading to these steady advances. The combination of instrumentational advances in NMR spectroscopy and protein labeling schemes is suggesting that the size limit of protein NMR may soon reach 100 kDa [107, 94].

In addition to crystallography and NMR, cryogenic electron microscopy (cryo-EM) contributes important macroscopic views at lower resolution for proteins that are not amenable to NMR or crystallography (see Box 1.3 and Figure 1.3) [108, 109].

Together with recombinant DNA technology, automated software for structure determination, and supercomputer and graphics resources, structure determination at a rate of one biomolecule per day (or more) is on the horizon.

Box 1.2: PCR Application Examples

- **Medical diagnoses of diseases and traits**. DNA analysis can be used to identify gene markers for many maladies, like cancer (e.g., BRCA1/2, *p53* mutations), late Alzheimer's or Parkinson's disease. A classic story of cancer markers involves Vice President Hubert Humphrey, who was tested for bladder cancer in 1967 but died of the disease in 1978. In 1994, after the invention of PCR, his cancerous tissue from 1976 was compared to a urine sample from 1967, only to reveal the same mutations in the *p53* gene, a cancer suppressing gene, that escaped the earlier recognition. Sadly, if PCR technology had been available in 1967, Humphrey may have been saved.

- **Historical analysis**. DNA is now being used for genetic surveys in combination with archaeological data to identify markers in human populations.[11] Such analyses can discern ancestors of human origins, migration patterns, and other historical events [110]. These analyses are not limited to humans; the evolutionary metamorphosis of whales has recently been unraveled by the study of fossil material combined with DNA analysis from living whales [111].

  Historical analysis by French and American viticulturists also recently showed that the entire gene pool of 16 classic wines can be conserved by growing only two grape varieties: *Pinot noir* and *Gouais blanc*. Depending on your occupation, you may either be comforted or disturbed by this news . . . .

  PCR was also used to confirm that the fungus that caused the Irish famine (since potato crops were devastated) in 1845–1846 was caused by the fungus *P. infestans*, a water mold (infected leaves were collected during the famine) [112]. Studies showed that the Irish famine was not caused by a single strain called US-1 which causes modern plant infections, as had been thought. Significantly, the studies taught researchers that further genetic analysis is needed to trace recent evolutionary history of the fungus spread.

- **Forensics and crime conviction**. DNA profiling — comparing distinctive DNA sequences, aberrations, or numbers of sequence repeats among individuals — is a powerful tool for proving with extremely high probability the presence of a person (or related object) at a crime, accident, or another type of scene. In fact, from 1989 through early 2002, 106 prisoners have been exonerated in the U.S., 12 from death row, and many casualties from disasters (like plane crashes and the 11 September 2001 New York World Trade Center terrorist attacks) were identified from DNA analysis of assembled body parts. In this connection, personal objects analyzed for DNA — like a black glove or blue dress — made headlines as crucial

[11] Time can be correlated with genetic markers through analysis of mitochondrial DNA or segments of the Y-chromosome. Both are genetic elements that escape the usual reshuffling of sexual reproduction; their changes thus reflect random mutations that can be correlated with time.

'imaginary witnesses'[12] in the O.J. Simpson and Lewinsky/Clinton affairs. In fact, a new breed of high-tech detectives is emerging with modern scientific tools; see www.uio.no/~mostarke/forens_ent/forensic_entomology.html, for example, for the use of bugs in crime research, and the 11 August 2000 issue of *Science* (volume 289) for related news articles.

- **Family lineage / paternity identification**. DNA fingerprinting can also be used to match parents to offspring. In 1998, DNA from the grave confirmed that President Thomas Jefferson fathered at least one child by his slave mistress, Sally Hemmings, 200 years ago. The remains of Tsar Nicholas' family, executed in 1918, were recently identified by DNA. In April 2000, French historians with European scientists solved a 205-year-old mystery by analyzing the heart of Louis XVII, preserved in a crystal urn, confirming that the 10-year old boy died in prison after his parents Marie Antoinette and Louis XVI were executed, rather than spirited out of prison by supporters (Antoinette's hair sample is available). Similar post-mortem DNA analysis proved false a paternity claim against Yves Montand.

See also the book by Reilly [105] for many other examples.

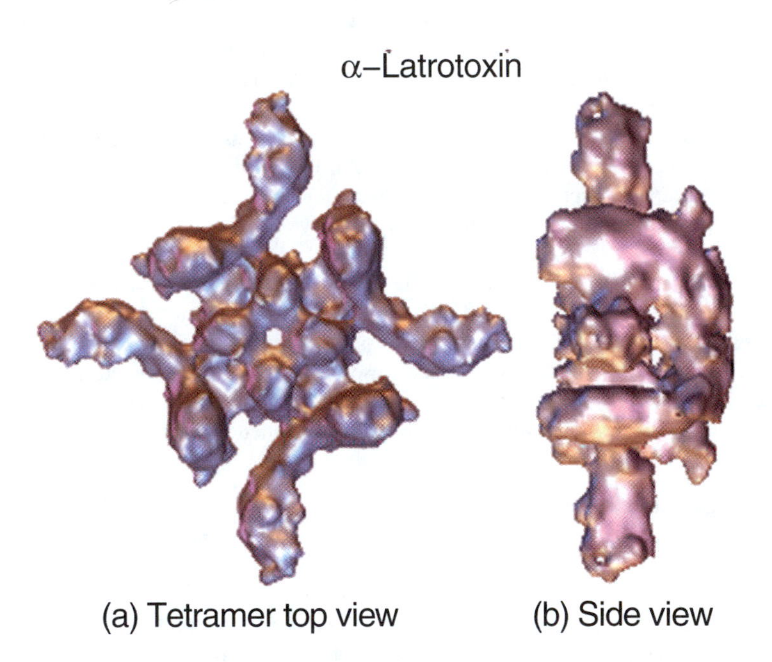

Figure 1.3. Top and side cryo-EM views of the 520-kDa tetramer of α-latrotoxin solved at 14 Å resolution [113]. Images were kindly provided by Yuri Ushkaryov.

[12] George Johnson, "OJ's Blood and The Big Bang, Together at Last", *The New York Times*, Sunday, May 21, 1995.

Box 1.3: Cryogenic Electron Microscopy (Cryo-EM)

Proteins that are difficult to crystallize or study by NMR may be studied by cryogenic electron microscopy (cryo-EM) [108, 114, 109]. This technique involves imaging rapidly-frozen samples of randomly-oriented molecular complexes at low temperatures and reconstructing 3D views from the numerous planar EM projections of the structures. Adequate particle detection imposes a lower limit on the subject of several hundred kDa, but cryo-EM is especially good for large molecules with symmetry, as size and symmetry facilitate the puzzle gathering (3D image reconstruction) process.

Though the resolution is low compared to crystallography and NMR, new biological insights may be gained, as demonstrated for the ribosome [115, 20] and the recent cryo-EM solution of the 520-kDa tetramer of α-latrotoxin at 14 Å resolution [113], as shown in Figure 1.3. (This solution represents an experimental triumph, as the system is relatively small for cryo imaging).

This toxic protein in the venom of black widow spiders (so called because the cruel females eat their mates!) forms a tetramer only in the presence of divalent cations. This organization enables the toxin to adhere to the lipid bilayer membrane and form channels through which neurotransmitters are discharged. This intriguing system has long been used to study mechanisms of neurotransmitter discharge (synaptic vesicle exocytosis), by which the release of particles too large to diffuse through membranes triggers responses that lead to catastrophic neuro and cardiovascular events.

With faster computers and improvements in 3D reconstruction algorithms, cryo-EM should emerge as a greater contributor to biomolecular structure and function in the near future.

1.5 Genome Sequencing

1.5.1 Projects Overview: From Bugs to Baboons

Spurred by this dazzling technology, thousands of researchers worldwide have been, or are now, involved in hundreds of sequencing projects for species like the cellular slime mold, roundworm, zebrafish, cat, rat, pig, cow, and baboon. Limited resources focus efforts into the seven main categories of genomes besides *Homo sapiens*: viruses, bacteria, fungi, *Arabidopsis thaliana* ('the weed'), *Drosophila melanogaster* (fruitfly), *Caenorhabditis elegans* (roundworm), and *M. musculus* (mouse). For an overview of sequence data, see www.ncbi.nlm.nih.gov/entrez/query.fcgi?db=Genome and for genome landmarks, readers can search the online collection available on www.sciencemag.org/feature/plus/sfg/special/index.shtml. The Human Genome Project is described in the next section.

The first completed genome reported was of the bacterium *Haemophilus influenzae*) in 1995 (see also Box 1.4). Soon after came the yeast genome (*Saccharomyces cerevisiae*) (1996, see genome-www.stanford.edu/Saccharomyces/), the bacterium *Bacillus subtilis* (1997), and the tuberculosis bacterium (*Mycobacterium tuberculosis*) in 1998. Reports of the worm, fruitfly, mustard plant, and rice genomes (described below) represent important landmarks, in addition to the human genome (next section).

Roundworm, *C. elegans* (1998)

The completion of the genome deciphering of the first multicellular animal, the one-millimeter-long roundworm *C. elegans*, made many headlines in 1998 (see the 11 December 1998 issue of *Science*, volume 282, and www.wormbase.org/). It reflects a triumphant collaboration of more than eight years between Cambridge and Washington University laboratories.

The nematode genome paves the way to obtaining many insights into genetic relationships among different genomes, their functional characterization, and associated evolutionary pathways. A comparison of the worm and yeast genomes, in particular, offers insights into the genetic changes required to support a multicellular organism. A comparative analysis between the worm and human genome is also important. Since it was found that roughly one third of the worm's proteins (> 6000) are similar to those of mammals, automated screening tests are already in progress to search for new drugs that affect worm proteins that are related to proteins involved in human diseases. For example, diabetes drugs can be tested on worms with a mutation in the gene for the insulin receptor.

Fruitfly, *Drosophila* (1999)

The deciphering of most of the fruitfly genome in 2000 by Celera Genomics, in collaboration with academic teams in the Berkeley and European *Drosophila* Genome projects, made headlines in March 2000 (see the 24 March 2000 issue of *Science*, volume 287, and www.fruitfly.org), in large part due to the groundbreaking "annotation jamboree" employed to assign functional guesses to the identified genes.

Interestingly, the million-celled fruitfly genome has fewer genes than the tiny, 1000-celled worm *C. elegans* (though initial reports of the number of worm's genes may have been over-estimated) and only twice the number of genes as the unicellular yeast. This is surprising given the complexity of the fruitfly — with wings, blood, kidney, and a powerful brain that can compute elaborate behavior patterns. Like some other eukaryotes, this insect has developed a nested set of genes with alternate splicing patterns that can produce more than one meaning from a given DNA sequence (i.e., different mRNAs from the same gene). Indeed, the number of core proteins in both fruitflies and worms is roughly similar (8000 vs. 9500, respectively).

Fly genes with human counterparts may help to develop drugs that inhibit encoded proteins. Already, one such fly gene is *p53*, a tumor-suppressor gene that, when mutated, allows cells to become cancerous. The humble baker's yeast proteins are also being exploited to assess activity of cancer drugs.

Mustard Plant, *Arabidopsis* (2000)

Arabidopsis thaliana is a small plant in the mustard family, with the smallest genome and the highest gene density so far identified in a flowering plant (125 million base pairs and roughly 25,000 genes). Two out of the five chromosomes of *Arabidopsis* were completed by the end of 1999, and the full genome (representing 92% of the sequence) published one year later, a major milestone for genetics. See the 14 December 1999 issue of *Nature*, volume 408, and www.arabidopsis.org/, for example.

This achievement is important because gene-dense plants (25,000 genes versus 19,000 and 13,000 for brain and nervous-system containing roundworm and fruitfly, respectively) have developed an enormous and complex repertoire of genes for the needed chemical reactions involving sunlight, air, and water. Understanding these gene functions and comparing them to human genes will provide insights into other flowering plants, like corn and rice, and will aid in our understanding of human life. Plant sequencing analysis should lead to improved crop production (in terms of nutrition and disease resistance) by genetic engineering and to new plant-based ingredients in our medicine cabinets. For example, engineered crops that are larger, more resistant to cold, and that grow faster have already been produced.

Arabidopsis's genome is also directly relevant to human biological function, as many fundamental processes of life are shared by all higher organisms. Some common genes are related to cancer and premature aging. The much more facile manipulation and study of those disease-related genes in plants, compared to human or animal models, is a boon for medical researchers.

Interestingly, scientists found that nearly two-thirds of the *Arabidopsis* genes are duplicates, but it is possible that different roles for these apparently-duplicate genes within the organism might be eventually found. Others suggest that duplication may serve to protect the plants against DNA damage from solar radiation; a 'spare' could become crucial if a gene becomes mutated. Intriguingly, the plant also has little "junk" (i.e., not gene coding) DNA, unlike humans.

The next big challenge for *Arabidopsis* aficionados is to determine the function of every gene by precise experimental manipulations that deactivate or overactivate one gene at a time. For this purpose, the community launched a 10-year gene-determination project (a "2010 Project") in December 2000. Though guesses based on homology sequences with genes from other organisms have been made (for roughly one half of the genes) by the time the complete genome sequence was reported, much work lies ahead to nail down each function precisely. This large number of "mystery genes" promises a vast world of plant biochemistry awaiting exploration.

Rice (2002)

The second largest genome sequencing project — for the rice plant (see a description of the human genome project below) — has been underway since the late 1990s in many groups around the world. The relatively small size of the rice genome makes it an ideal model system for investigating the genomic sequences of other grass crops like corn, barley, wheat, rye, and sugarcane. Knowledge of the genome will help create higher quality and more nutritious rice and other cereal crops. Significant impact on agriculture and economics is thus expected.

By May 2000, a rough draft (around 85%) of the rice genome (430 million bases) was announced, another exemplary cooperation between the St. Louis-based agrobiotechnology company Monsanto (now part of Pharmacia) and a University of Washington genomics team.

In April 2002, two groups (a Chinese team from the Bejing Genomics Institute led by Yang Huanming and the Swiss agrobiotech giant Syngenta led by Stephen Goff) published two versions of the rice genome (see the 5 April 2002 issue of *Science*, volume 296), for the rice subspecies *indica* and *japonica*, respectively. Both sequences contain many gaps and errors, as they were solved by the whole-genome 'shotgun' approach (see Box 1.4), but represent the first detailed glimpse into a world food staple. A more accurate and complete version of the rice genome is expected by the end of 2002 by the International Rice Genome Sequencing Project, a public international consortium led by Japan, who has been sequencing *japonica* since 1997.

Other Organisms

Complete sequences and drafts of many genomes are now known. (check websites such as www.ncbi.nlm.nih.gov/entrez/query.fcgi?db=Genome for status reports). Included are bacterial genomes of a microbe that can survive environments lethal for most organisms and might turn useful as a metabolizer of toxic waste (*D. radiodurans* R1); a nasty little bacterium that causes diseases in oranges, grapes, and other plants (*Xylella fastidiosa*, decoded by a Brazilian team); and the bugs for human foes like cholera, syphilis, tuberculosis, malaria, the plague, and typhus. Proteins unique to these pathogens are being studied, and disease treatments will likely follow (e.g., cholera vaccine).

Implications

The genomic revolution and the comparative genomics enterprises now underway will not only provide fundamental knowledge about the organization and evolution of biological systems in the decades to come [116] but will also lead to medical breakthroughs.

Already, the practical benefits of genomic deciphering have emerged (e.g., [117, 118]). A dramatic demonstration in 2000 was the design of the first vaccine to prevent a deadly form of bacterial meningitis using a two-year gene-hunting process at Chiron Corporation. Researchers searched through the computer database of all the bacterium's genes and found several key proteins that in laboratory experiments stimulated powerful immune responses against all known strains of the *Neisseria meningitidis* Serogroup B Strain MC58 bug [119].

The mammalian mouse genome sequence is considered crucial for annotating the human genome and also for testing new drugs for humans. The Mouse Genome Sequencing Consortium (MGSC) formed in late fall of 2000 follows in the footsteps of the human genome project [120]. Though coming in the backdrop of the human genome, draft versions of the mouse genome were announced by private and public consortia in 2001 and 2002, respectively.

1.5.2 The Human Genome

The International Human Genome Project was launched in 1990 to sequence all three billion bases in human DNA [120]. The public consortium has contributions from many parts of the world (such as the United States, United Kingdom, Japan, France, Germany, China, and more) and is coordinated by academic centers funded by NIH and the Wellcome Trust of London, headed by Francis Collins and Michael Morgan (with groups at the Saenger Center near Cambridge, UK, and four centers in the United States); see www.nhgri.nih.gov/HGP/.

In 1998, a competing private enterprise led by Craig Venter's biotechnology firm Celera Genomics and colleagues at The Institute for Genomic Research (TIGR), both at Rockville, Maryland (owned by the PE Corporation; see www.celera.com), entered the race. Eventually, this race to decode the human genome turned into a collaboration in part, not only due to international pressure but also because the different approaches for sequencing taken by the public and private consortia are complementary (see Box 1.4 and related articles [121, 122, 123] comparing the approaches for the human genome assembly).

Milestones

A first milestone was reached in December 1999 when 97% of the second smallest chromosome, number 22, was sequenced by the public consortium (the missing 3% of the sequence is due to 11 gaps in contiguity); see the 2 December 1999 issue of *Nature*, volume 402. Though small (43 million bases, $< 2\%$ of genomic DNA), chromosome 22 is gene rich and accounts for many genetic diseases (e.g., schizophrenia).

Chromosome 21, the smallest, was mapped soon after (11 May 2000 issue of *Nature*, volume 405) and found to contain far fewer genes than the 545 in chromosome 22. This opened the possibility that the total number of genes in human DNA is less than the 100,000 previously estimated. Chromosome 21 is best known for its association with Down's syndrome; affected children are born with three rather than two copies of the chromosome. Learning about the genes associated with chromosome 21 may help identify genes involved in the disease and, eventually, develop treatments. See a full account of the chromosome 21 story in www.sciam.com/explorations/2000/051500chrom21/.

Completion of the first draft of the human genome sequence project broke worldwide headlines on 26 June 2000 (see, for example, the July 2000 issue of *Scientific American*, volume 283). This draft reflects 97% of the genome cloned and 85% of it sequenced accurately, that is, with 5 to 7-fold redundancy.

Actually, the declaration of the 'draft' status was arbitrary and even fell short of the 90% figure set as target. Still, there is no doubt that the human genome represents a landmark contribution to humankind, joined to the ranks of other 'Big Science' projects like the Manhattan project and the Apollo space program. The June 2000 announcement also represented a 'truce' between the principal players of the public and private human genome efforts and a commitment to continue to work together for the general cause.

A New York Times editorial by David Baltimore on the Sunday before the Monday announcement was expected underscored this achievement, but also emphasized the work that lies ahead:

> The very celebration of the completion of the human genome is a rare day in the history of science: an event of historic significance is recognized not in retrospect, but as it is happening .... While it is a moment worthy of the attention of every human, we should not mistake progress for a solution. There is yet much work to be done. It will take many decades to fully comprehend the magnificence of the DNA edifice built over four billion years of evolution and held in the nucleus of each cell of the body of each organism on earth.
>
> David Baltimore, *New York Times*, 25 June 2000.

Baltimore further explains that the number of proteins, not genes, determines the complexity of an organism. The gene number should ultimately explain the complexity of humans. Already, in June 2000, the estimated number of total genes (50,000) is not too far away from the number in a fly (14,000) or a worm (18,000). Several months after the June announcement, this estimate was reduced to 30,000–40,000 (see the 15 February 2001 issue of *Nature*, volume 409, and the 16 February 2001 issue of *Science*, volume 291). This implies an 'equivalence' of sorts between each human and roughly two flies .... However, the estimated number has been climbing since then; see [124] for a discussion of the shortcomings of the gene identification process on the basis of the available genomic sequence. Intriguing recent findings that human cells make far more RNA than can be accounted for by the estimated 30,000 to 40,000 human genes may also suggest that the number of human genes is larger [125, 126]. Another explanation to this observation is that hidden levels of complexity exist in the human genome; for example, there may be more genes than previously thought that produce RNA as an end product in itself (rather than as messenger for protein production as for other known roles of RNA). Clearly, the final word on the number of human genes and the conserved genes that humans share with flies, mice or other organisms awaits further studies.

Some chromosome segments of the human genome will likely be impossible to characterize (at least with current technology) as they are too repetitive; fortunately, these segments may be relatively insignificant for the genome's overall function. With determination of the human genome sequence for several individuals, variations (polymorphisms) will be better understood (important for research aimed at designer drugs). Undoubtedly, the determination of sequences for 1000 major species in the next decade will shed further insights on the human genome.

Box 1.4: Different Sequencing Approaches

Two synergistic approaches have been used for sequencing. The public consortium's approach relies on a 'clone-by-clone' approach: breaking DNA into large fragments, cloning each fragment by inserting it into the genome of a bacterial artificial chromosome (BAC), sequencing the BACs once the entire genome is spanned, and then creating a physical map from the individual BAC clones. The last part — rearranging the fragments in the order they occur on the chromosome — is the most difficult. It involves resolving the overlapped fragments sharing short sequences of DNA ('sequence-tagged sites').

The alternative approach pioneered by Venter's Celera involves reconstructing the entire genome from small pieces of DNA without a prior map of their chromosomal positions. The reconstruction is accomplished through sophisticated data-processing equipment. Essentially, this gargantuan jigsaw puzzle is assembled by matching sequence pieces as the larger picture evolves.

The first successful demonstration of this piecemeal approach was reported by Celera for decoding the genome of the bacterium *Haemophilus influenzae* in 1995. This bacterium has a mere 1.8 million base pairs with estimated 1700 genes, versus three billion base pairs for human DNA with at least 30,000 genes. The sequence of *Drosophila* followed in 1998 (140 million base pairs, 13,000 estimated genes) and released to the public in early 2000 (see the 24 March 2000 issue of *Science*, volume 287).

This 'shotgun' approach has been applied to the human genome, more challenging than the above organisms for two reasons. The human genome is larger — requiring the puzzle to be formed from $\sim$ 70 million pieces — and has many more repeat sequences, complicating accurate genome assembly. For this reason, the public data were incorporated into the whole genome assembly [122]. The whole-genome shotgun approach has also been applied to obtain a draft of the mouse genome (2001), as well as the rice genome sequence as reported by two teams in April 2002.

The two approaches are complementary, since the rapid deciphering of small pieces by the latter approach relies upon the larger picture generated by the clone-by-clone approach for overall reconstruction. See a series of articles [121, 122, 123] scrutinizing those approaches as applied to the human genome assembly.

A Gold Mine of Biodata

The most up-to-date information on sequencing projects can be obtained from the U.S. National Center for Biotechnology Information (NCBI) at the U.S. National Library of Medicine, which is developing a sophisticated analysis network for the human genome data.

For information, see the Human Genome Resources Guide www.ncbi.nlm.nih.gov/genome/guide/human (click on Map Viewer), the U.S. National Human Genome Research Institute's site www.nhgri.nih.gov/, that of Department of Energy (DOE) at www.ornl.gov/hgmis/, the site of the University of California at Santa Cruz at genome.ucsc.edu/, and others.[13]

Since 1992, NCBI has maintained the GenBank database of publicly available nucleotide sequences (www.ncbi.nlm.nih.gov). A typical GenBank entry includes information on the gene locus and its definition, organism information, literature citations, and biological features like coding regions and their protein translations. Many search and analysis tools are also available to serve researchers.

As the sequencing of each new human chromosome is being completed, the biological revolution is beginning to affect many aspects of our lives [127], perhaps not too far away from Wilson's vision of consilience. A 'gold mine' of biological data is now amassing, likened to "orchards . . . just waiting to be picked".[14] This rich resource for medicine and technology also provides new foundations, as never before, for computational applications.

Consequently, in fifty years' time, we anticipate breakthroughs in protein folding, medicine, cellular mechanisms (regulation, gene interactions), development and differentiation, history (population genetics, origin of life), and perhaps new life forms, through analysis of conserved and vital genes as well as new gene products. See the 5 October 2001 issue of *Science* (volume 294) for a discussion of new ideas, projects, and scientific advances that followed since the sequencing of the human genome.

Among the promising medical leaps are personalized and molecular medicine, perhaps in large part due to the revolutionary DNA microarray technology (see [128] and Box 1.5) and gene therapy. Of course, *information is not knowledge*, but rather a road that can lead to perception. Therefore, these aforementioned achievements will require concerted efforts to extract information from all the sequence data concerning gene products.

---

[13] Some useful web sites for genomic data include www.arabidopsis.org, www.ncbi.nlm.nih.gov/Sitemap/index.html, the Agricultural Genome Information System, *Caenorhabditis elegans* Genetics and Genomics, Crop Genome Databases at Cornell University, FlyBase, The Genome Database, Genome Sequencing Center (Washington University), GenomeNet, U.S. National Agricultural Library, Online Mendelian Inheritance in Man, *Pseudomonas aeruginosa* Community Annotation Project, *Saccharomyces* Genome Database, The Saenger Centre, Taxonomy Browser, and UniGene.

[14] B. Sinclair, in *The Scientist*, 19 March 2001.

Many societal, ethical, economic, legal, and political issues will also have to be addressed with these developments. Still, like the relatively minor Y2K (Year 2000) anxiety, these problems could be resolved in stride through multidisciplinary networks of expertise.

For more on the ethical, legal, and social implications of human genome research, visit www.nhgri.nih.gov/ELSI.

In a way, sequencing projects make the giant leap directly from *sequence to function* (possible only when a homologous sequence is available whose function is known). However, the crucial middle aspect — *structure* — must be relied upon to make systematic functional links. This systematic interpolation and extrapolation between sequence and structure relies and depends upon advances in biomolecular modeling, in addition to high-throughput structure technology ('the human proteomics project').

The next chapter introduces some current challenges in modeling macromolecules and mentions important applications in medicine and technology.

Box 1.5: Genomics & Microarrays

DNA microarrays — also known as gene chips, DNA chips, and biochips — are becoming marvelous tools for linking gene sequence to gene products. They can provide, in a single experiment, an expression profile of many genes [128]. As a result, they have important applications to basic and clinical biomedicine. Particularly exciting is the application of such genomic data to *personalized medicine* or *pharmacogenomics* — prescribing medication based on genotyping results of both patient and any associated bacterial or viral pathogen [129].

Essentially, each microarray is a grid of DNA oligonucleotides (called *probes*) prepared with sequences that represent various genes. These probes are directed to a specific gene or mRNA samples (called *targets*) from tissues of interest (e.g., cancer cells). Binding between probe and target occurs if the RNA is complementary to the target nucleic acid. Thus, probes can be designed to bind a target mRNA if the probe contains certain mutations. Single nucleotide polymorphisms or SNPs, which account for 0.1% of the genetic difference among individuals, can be detected this way [130].

The hybridization event — amount of RNA that binds to each cell grid — reflects the extent of gene expression (gene activity in a particular cell). Such measurements can be detected by fluorescence tagging of oligonucleotides. The color and intensity of the resulting base-pair matches reveal gene expression patterns.

Different types of microarray technologies are now used (e.g., using different types of DNA probes), each with strengths and weaknesses. The technique of principal component analysis (PCA, see Chapter 14) has shown to be useful in analyzing microarray data (e.g., [131]). Technical challenges remain concerning verification of the DNA sequences and ensuring their purity, amplifying the DNA samples, and quantitating the results accurately. For example, false positives or false negatives can result from irregular target/probe binding (e.g., mismatches) or from self-folding of the targets, respectively. The problem of accuracy of the oligonucleotides has stimulated various companies to develop appropriate design techniques. Affymetrix Corporation, for example, has developed technology for designing silicon chips with oligonucleotide probes synthesized directly onto them, with thousands of human genes on one chip. All types of DNA microarrays rely on substantial computational analysis of the experimental data to determine absolute or relative patterns of gene expression.

Such patterns of gene expression (induction and repression) can prove valuable in drug design. An understanding of the affected enzymatic pathway by proven drugs, for example, may help screen and design novel compounds with similar effects. This potential was demonstrated for the bacterium *M. tuberculosis*, based on experimental profiles obtained before and after exposure to the tuberculosis drug isoniazid [132].

For further information on microarray technology and available databases, see www.gene-chips.com, www.ncbi.nlm.nih.gov/geo, ihome.cuhk.edu.hk/~b400559/array.html, and industry.ebi.ac.uk/~alan/MicroArray, for example.

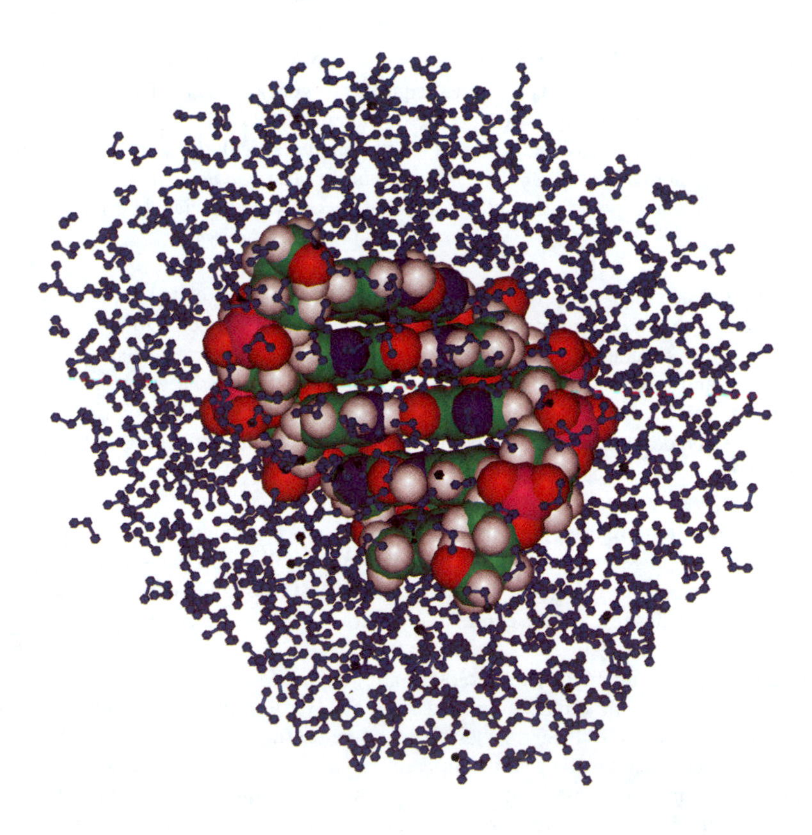

2

Biomolecular Structure and Modeling: Problem and Application Perspective

All things come out of the one, and the one out of all things. Change, that is the only thing in the world which is unchanging.

Heraclitus of Ephesus (550–475 BC).

2.1 Computational Challenges in Structure and Function

2.1.1 Analysis of the Amassing Biological Databases

The experimental progress described in the previous chapter has been accompanied by an increasing desire to relate the complex three-dimensional (3D) shapes of biomolecules to their biological functions and interactions with other molecular systems. Structural biology, computational biology, genomics, proteomics, bioinformatics, chemoinformatics, and others are natural partner disciplines in such endeavors.

Structural biology focuses on obtaining a detailed structural resolution of macromolecules and relating structure to biological function.

Computational biology was first associated with the discipline of finding similarities among nucleotide strings in known genetic sequences, and relating these relationships to evolutionary commonalities; the term has grown, however, to encompass virtually all computational enterprises to molecular biology problems [43].

Comparative genomics — the search and comparison of sequences among species — is a natural outgrowth of the sequencing projects [116]. So are *structural and functional genomics*, the characterization of the 3D structure and biological function of these gene products [133, 134, 135, 136, 137].

In the fall of 2000, the U.S. National Institute of General Medical Sciences (NIGMS) launched a structural genomics initiative by funding seven research groups aiming to solve collectively the 3D structures of 10,000 proteins, each representing a protein family, over the next decade. This important step toward the assembly of a full protein library requires improvements in both structural biology's technology and methodology. (See the November 2000 supplement issue of *Nature Structural Biology*, volume 7, devoted to structural genomics and the progress report [137]). Ultimately, the functional identification of unknown proteins is likely to dramatically increase our understanding of human disease processes and the development of structure-based drug therapies.

Proteomics is another current buzzword defining the related discipline of protein structure and function (see [138] for an introduction), and even *cellomics* has been introduced.[1] Cellomics reflects the expanded interest of gene sequencers in integrated cellular structure and function. The *Human Proteomics Project* — a collaborative venture to churn out atomic structures using high-throughput and robotics-aided methods based on NMR spectroscopy and X-ray crystallography, rather than sequences — may well be on its way.

New instruments that have revolutionized genomics known as DNA microarrays, biochips, or gene expression chips (introduced in Chapter 1 and Box 1.5) allow researchers to determine which genes in the cell are active and to identify gene networks. A good introduction to these areas is volume 10 of the year 2000, pages 341–384, of *Current Opinion in Structural Biology* entitled "Sequences and Topology. Genome and Proteome Informatics", edited by P. Bork and D. Eisenberg.

The range of genomic sciences also continues [139] to the *metabolome*, the endeavor to define the complete set of metabolites (low-molecular cellular intermediates) in cells, tissues, and organs. Experimental techniques for performing these integrated studies are continuously being developed. For example, yeast geneticists have developed a clever technique for determining whether two proteins interact and, thereby by inference, participate in related cellular functions [140]. Such approaches to proteomics provide a powerful way to discover functions of newly identified proteins. DNA chip technology is also thought to hold the future of individualized health care now coined personalized medicine or *pharmacogenomics*; see Chapter 14 and Box 1.5.

[1] A glossary of biology disciplines coined with "ome" or "omic" terms can be found at http://www.genomicglossaries.com/content/omes.asp.

It has been said that current developments in these fields are *revolutionary rather than evolutionary*. This view reflects the clever exploitation of biomolecular databases with computing technology and the many disciplines contributing to the biomolecular sciences (biology, chemistry, physics, mathematics, statistics, and computer science). *Bioinformatics* is an all-embracing term for many of these exciting enterprises [141, 142] (structural bioinformatics is an important branch); *chemoinformatics* has also followed (see Chapter 14) [143]. Some genome-technology company names are indicative of the flurry of activity and grand expectations from our genomic era. Consider titles like Genetics Computer Group, Genetix Ltd., Genset, Protana, Protein Pathways, Inc., Pyrosequencing AB, Sigma-Genosys, or Transgenomic Incorporated.

Although the number of sequence databases has grown very rapidly and exceeds the amount of structural information,[2] the 1990s saw an exponential rise of structural databases as well. From only 50 solved 3D structures in the Protein Data Bank (PDB) in 1975, to 500 in 1988, another order of magnitude was reached in 1996 (around 5000 entries). In fact, the rate of growth of structural information is approaching the rate of increase of amino acid sequence information (see Figure 2.1 and Table 2.1). It is no longer a rare event to see a new crystal structure on the cover of *Nature* or *Science*. Now, on a weekly basis, groups compete to have their newly-solved structure on the cover of prominent journals, including the newer, colorful publications like *Nature Structural Biology* and *Structure*. The fraction of NMR-deduced structures deposited in the PDB is also rising steadily, reflecting roughly 15% by the end of 2001 (for updated information, see www.rcsb.org/pdb/holdings.html).

This trend, coupled with tremendous advances in genome sequencing projects [145], argues strongly for increased usage of computational tools for analysis of sequence/structure/function relationships and for structure prediction. Thus, besides genomics-based analyses and comparisons, important are accurate, reliable, and rapid theoretical tools for describing structural and functional aspects of gene products. (See [136], for example, for computational challenges in genomics).

2.1.2 Computing Structure From Sequence

One of the most successful approaches to date on structure prediction comes from *homology modeling* (also called comparative modeling) [146, 147].

In general, a large degree of sequence similarity often suggests similarity in 3D structure. It has been reported, for example, that a sequence identity of greater than 40% usually implies more than 90% 3D-structure overlap (defined as percentage of C^{α} atoms of the proteins that are within 3.5 Å of each other in a rigid-body alignment; see definitions in Chapter 3) [148]. Thus, sequence sim-

[2] In 1991, it was pointed out [144] that the amount of 3D protein information is lagging by orders of magnitude behind the accessible sequence data [144].

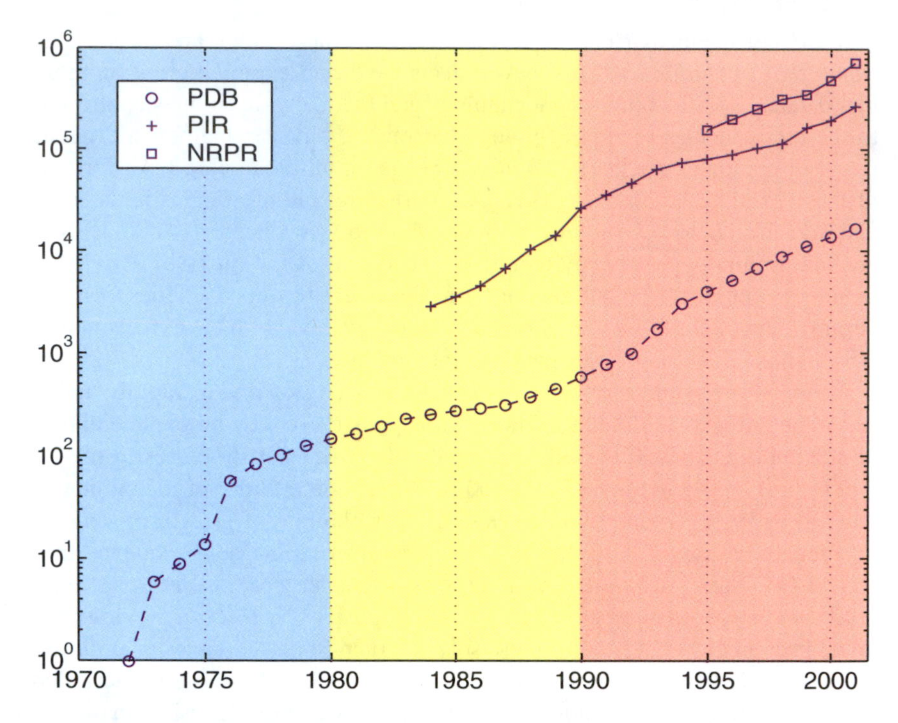

Figure 2.1. The growth of protein sequence databases (named PIR and NRPR) versus structural database of macromolecules (PDB). See Table 2.1 and www.dna.affrc.go.jp/htdocs/growth/P-history.html. The International Protein Information Resource (PIR) of protein sequences (pir.georgetown.edu) is a comprehensive, annotated public-domain database reflecting a collaboration among the United States (Georgetown University Medical Center in Washington, D.C.), Germany (Munich Information Center for Protein Sequences in Martinsried), and Japan (International Protein Sequence Database in Tsukuba). PIR is extensively cross referenced and linked to many molecular databases of genes, genomes, protein structures, structural classification, literature, and more. The NRPR database represents merged, non redundant protein database entries from several databases: PIR, SWISS-PROT, Genpept, and PDB.

ilarity of at least 50% suggests that the associated structures are similar overall. Conversely, small sequence similarity generally implies structural diversity.

There are many exceptions, however, as demonstrated humorously in a contest presented to the protein folding community (see Box 2.1). The *myoglobin* and *hemoglobin* pair is a classic example where large structural, as well as evolutionary, similarity occurs despite little sequence similarity (20%). Other exceptional examples and various sequence/structure relationships are discussed separately in Chapter 3, as well as Homework 6 and [149] for example.

More general than prediction by sequence similarity is structure prediction *de novo* [147], a *Grand Challenge* of the field, as described next.

Table 2.1. Growth of protein sequence databases.

Year	PDB[a]	PIR[b]	NRPR[c]
1972	1		
1973	6		
1974	9		
1975	14		
1976	58		
1977	85		
1978	104		
1979	128		
1980	150		
1981	168		
1982	197		
1983	234		
1984	258	2898	
1985	280	3615	
1986	296	4612	
1987	318	6796	
1988	382	10527	
1989	459	14372	
1990	597	26798	
1991	790	36150	
1992	1006	47234	
1993	1727	64760	
1994	3091	75511	
1995	4056	82066	159808
1996	5240	91006	204123
1997	6833	105741	258272
1998	8942	117482	324237
1999	11364	168808	360674
2000	14063	198801	497787
2001	16973	274514	744991

[a]From the Protein Data Bank (PDB), www.rcsb.org/pdb/holdings.html.

[b]From the Protein International Resource (PIR), www-nbrf.georgetown.edu/cgi-bin/dbinfo.

[c]From the 'Non Redundant Proteins database merged Regular release', www.dna.affrc.go.jp/htdocs/growth/P-history.html.

2.2 Protein Folding – An Enigma

2.2.1 'Old' and 'New' Views

There has been much progress on the protein folding challenge since Cyrus Levinthal first posed the well-known "paradox" named after him; see [152]

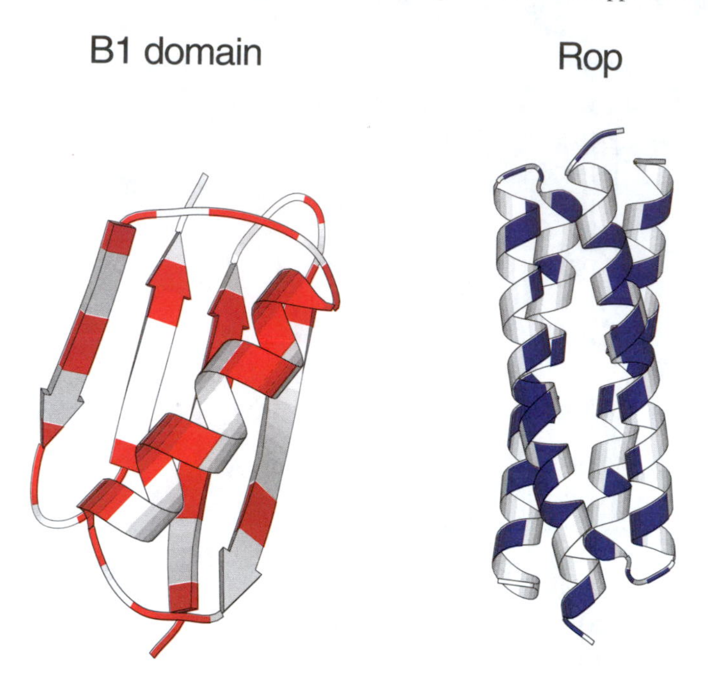


```
B1     MTYKLILNGKTLKGETTTEAVDAATAEKVFKQYANDNGVDGEWTYDDATKTFTVTE
Janus  MTKKAILALNTAKFLRTQAAVLAAKLEKLGAQEANDNAVDLEDTADDLYKTLLVLA
Rop    GTKQEKTALNMARFIRSQTLTLLEKLNELDADEQADICESLHDHADELYRSCLARF
```

Figure 2.2. Ribbon representations of the B1 domain of IgG-binding protein G and the Rop monomer (first 56 residues), which Janus resembles [150], with corresponding sequences. Half of the protein G β domain (B1) residues were changed to produce Janus in response to the Paracelsus challenge (see Box 2.1 and [151]). The origin of the residues is indicated by the following color schemes: residues from B1: red; residues from Rop: blue; residues in both: green; residues in neither: **black**. While experimental coordinates of protein G and Rop are known, the structure of Janus was deduced by modeling. The single-letter amino acid acronyms are detailed in Table 3.1, Chapter 3.

for a historical perspective. Levinthal suggested that well defined folding pathways might exist [153] since real proteins cannot fold by exhaustively traversing through their entire possible conformational space [154]. (See [155, 156, 157] for a recent related discussion of whether the number of protein conformers depends exponentially or nonexponentially on chain length). Levinthal's paradox led to the development of two views of folding — the 'old' and the 'new' — which have been merging of late [158, 159, 23].

The old view accents the existence of a specific folding pathway characterized by well-defined intermediates. The new version emphasizes the rugged, heterogeneous multidimensional energy landscape governing protein folding, with many competing folding pathways [160]. Yet, the boundary between the two views is pliant and the intersection substantial [152]. This integration has resulted from a variety of information sources: theories on funnel-shaped energy landscape (e.g., [161, 159, 162]); folding and unfolding simulations of simplified models (e.g., [163, 164, 165, 166, 167, 4]), at high temperatures or low pH concentrations (e.g., [23, 22]); NMR spectroscopic experiments that monitor protein folding intermediates (e.g., [168, 2]); predictions of secondary and/or tertiary structure on the basis of evolutionary information [169]; and statistical mechanical theories.

Such studies suggest that while wide variations in folding pathways may occur, 'fast folders' possess a unifying pattern for the evolution of native-structure contacts. In particular, the pathway and other kinetic aspects governing the folding of a particular protein depend on the free energy landscape, which is temperature-dependent. Namely, the folding ensemble is sensitive to shallow energy traps at lower temperatures. Thus, changes in temperature can substantially change the folding kinetics.

2.2.2 *Folding Challenges*

The great progress in the field can also be seen by evaluations of biannual prediction exercises held in Asilomar, California (termed CASP for Critical Assessment of Techniques for Protein Structure Prediction) [170, 171] (see PredictionCenter.llnl.gov, and special issues of the journal *Proteins*: vol. 23, 1995; Suppl. 1, 1997; Suppl. 3, 1999, and Suppl. 5, 2001).

The CASP organizers assign specific proteins for theoretical prediction that protein crystallographers and NMR spectroscopists expect to complete by the next CASP meeting. Prediction assessors then consider several categories of structural prediction tools, for example: comparative (homology) modeling, fold recognition (i.e., based on a library of known protein folds), and *ab initio* prediction (i.e., using first principles). The quality of the prediction has been characterized in terms of C^α root-mean-square (RMS) deviations, the best of which is in the range of 2–6 Å; the lowest values have been obtained from the best comparative modeling approaches.

To the fourth CASP meeting in December 2000, two additional competitive experiments were added, dealing with protein structure prediction and rational drug design: CAFASP2 (assessing automatic methods for predicting protein structure), and CATFEE (evaluating methods for protein ligand docking); see predictioncenter.llnl.gov/casp4/Casp4.html). These additions reflect the rapid developments of many genome efforts and the rise of computational biology as an important discipline of biology, medicine, and biotechnology.

Results of the CASP4 meeting indicated considerable progress in fully-automated approaches for structure prediction and hence promise that computer modeling might become an alternative to experimental structure determination. See [172] for an overview with articles collected on CASP4 in that special issue of *Proteins* (Suppl. 5, 2001), as well as a related discussion of field progress and prospects [173] and of the limitations and challenges in the comparative modeling section of CASP4 [174, 175, 176].

Modeling work in the field is invaluable because it teaches us to ask, and seek answers to, systematic questions about sequence/structure/function relationships and about the underlying forces that stabilize biomolecular structures. Still, given the rapid improvements in the experimental arena, the pace at which modeling predictions improve must be expeditious to make a significant contribution to protein structure prediction from sequence. To this end, computational biologists are soliciting candidates for the "Ten Most Wanted" proteins that they will attempt to solve (see www.doe-mbi.ucla.edu/TMW/).

The annual Johns Hopkins Coolfront Protein Folding meetings also reflect the great progress made in this exciting field. As gleaned from reports of the fourth [177] and fifth [178] meetings, progress is rapid on many theoretical and experimental fronts. Particularly exciting are emerging areas of study dealing with protein folding in membranes and protein folding diseases (e.g., from misfolding or chaperone interference; see below). Furthermore, progress on both experimental ultrafast methods for studying protein folding kinetics and theoretical predictions based on sequence and structural genomics is impressive.

Box 2.1: Paracelsus Challenge

In 1994, George Rose and Trevor Creamer posed a challenge, named after a 16th-century alchemist: change the sequence of a protein by 50% or less to create an entirely different 3D global folding pattern [151]. Though this challenge might sound not particularly difficult, imagine altering at most 50% of the ingredients for a chocolate cake recipe so as to produce bouillabaisse instead! Rose and Creamer offered a reward of $1000 to entice entrants.

The transmutation was accomplished four years later by Lynne Regan and coworkers [150], who converted the four-stranded β-sheet B1 domain of protein G — which has the β sheets packed against a single helix — into a four-helix bundle of two associating helices called Janus (see Figure 2.2). These contestants achieved this wizardry by replacing residues in a β-sheet-encoding domain (i.e., those with high β-sheet-forming propensities) with those corresponding to the four-helix-bundle protein Rop (repressor of primer). Other modifications were guided by features necessary for Rop stability (i.e., internal salt bridge), and the combined design was guided by energy minimization and secondary-structure prediction algorithms.

The challenge proposers, though delighted at the achievement they stimulated, concluded that in the future only tee-shirt prizes should be offered rather than cash!

2.2.3 Folding by Dynamics Simulations?

While molecular dynamics simulations are beginning to approach the timescales necessary to fold small peptides [39] or small proteins [69], we are far from find-

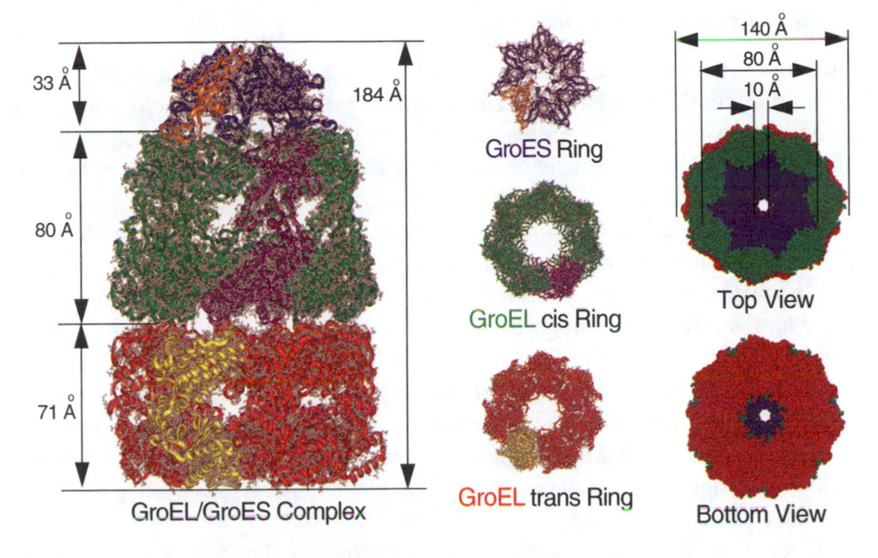

Figure 2.3. The bullet-shaped architecture of the GroEL/GroES chaperonin/co-chaperonin complex sequence. Overall assembly and dimensions are shown from a side view (left). The top ring is the GroES 'cap', and the other layers are GroEL rings. Sidechains are shown in grey. As seen from the top and bottom views (right), a central channel forms in the interior, conducive to protein folding. The protein is organized as three rings that share a 7-fold rotational axis of symmetry (middle), where GroEL contains 14 identical proteins subunits assembled in two heptameric rings, and GroES contains 7 smaller identical subunits in its heptamer ring.

ing the Holy Grail, if there is one [179]. Indeed, IBM's ambitious announcement in December 1999 of building a 'Blue Gene' *petaflop* computer (i.e., capable of 10^{15} floating-point operations per second) for folding proteins by 2005 depends on the computational models guiding this ubiquitous cellular process.[3] For example, some proteins require active escorts to assist in their folding *in vivo*. These *chaperone* molecules assist in the folding and rescue misbehaved polymers. Though many details are not known about the mechanisms of chaperone assistance (see below), we recognize that chaperones help by guiding structure assembly and preventing aggregation of misfolded proteins. For an overview of chaperones, see [180, 181], for example, and the 1 August 2001 issue of the *Journal of Structural Biology*, volume 135, devoted to chaperones.

[3]This $100 million program to build the world's most powerful supercomputer depends on a million processors running in parallel with the PIM (processor in memory) design. This architecture puts the memory and processor on the same chip: a total of 40,000 chips with 25 or more processors per chip. Besides the crucial role of software in the success leading to protein folding 'in silico', it is a challenge to eliminate bottlenecks in this enormous parallel system that might result between memory and processors when sequential (rather than parallel) computations must be performed.

2.2.4 *Folding Assistants*

Current studies on chaperone-assisted folding, especially of the archetypal chaperone duo, the *E. Coli* bacterial chaperonin GroEL and its cofactor GroES, are providing insights into the process of protein folding [182, 183] (see Figure 2.3 and Box 2.2). The rescue acts of chaperones depend on the subclass of these escorts and the nature of the protein being aided. Some chaperones can assist a large family of protein substrates, while others are more restrictive (see Box 2.2); detailed structural explanations remain unclear. Many families of chaperones are also known, varying in size from small monomers (e.g., 40 or 70 kDa for DnaJ and DnaK of Hsp70) to large protein assemblies (e.g., 810 kDa for GroEL or 880 kDa for the GroEL/GroES complex).

The small assistants bind to short runs of hydrophobic residues[4] to delay premature folding and prevent aggregation. Larger chaperones are likely needed to prevent aggregation of folded compact intermediates in the cell termed 'molten globules', requiring a complex trap-like mechanism involving co-chaperones (see also Box 2.2).

Recent work also suggests that small chaperones in the PapD-like superfamily — which directs the folding of various surface organelles — facilitate folding by providing direct *steric* information to their substrates, in addition to capping their interactive surfaces [184, 185]. In certain assemblies (pilus subunits called pilins), the chaperone donates a β-strand that completes the fold of the pilin to promote correct folding and binding to neighboring subunits.

Since macromolecular crowding affects aggregation and diffusion properties of proteins, uncertainties remain regarding the interpretation of *in vitro* experiments on chaperone-assisted folding. Different models for chaperone activity have also been proposed: a 'pathway' route for direct folding, and a 'network' model, involving iterative folding. They differ by the extent to which unfolded proteins are released to the cellular medium while they remain vulnerable to aggregation [186, 187].

2.2.5 *Unstructured Proteins*

Though our discussion has focused on the concept of native folds, not all proteins are intrinsically structured [188]. The intrinsic lack of structure can be advantageous, for example in binding versatility to different targets or in ability to adapt different conformations. Unfolded or non-globular structures are recognized in connection with regulatory functions, such as binding of protein domains to specific cellular targets [188, 189]. Examples include DNA and RNA-binding regions of certain protein complexes (e.g., basic region of leucine zipper protein GCN4, DNA-binding domain of NFATC1, RNA recognition regions of the HIV-1 Rev protein). Here, the unstructured regions become organized only upon binding to

[4]The terms *hydrophobic* ('water-hating') and *hydrophilic* ('water-loving') characterize water-soluble and water-insoluble molecular groups, respectively.

the DNA or RNA target. This folding flexibility offers an evolutionary advantage, which might be more fully appreciated in the future, as more gene sequences that code for unstructured proteins are discovered and analyzed.

Box 2.2: Studies on Protein Escorts

The archetypal chaperone GroEL is a member of a chaperone class termed *chaperonins*; hsp60 of mitochondria and chloroplasts is another member of this class. These chaperones bind to partially-folded peptide chains and assist in the folding with the consumption of ATP. The solved crystal structure of GroEL/GroES [190] suggests beautifully, in broad terms, how the large central channel inside a barrel-shaped chaperone might guide protein compaction in its container and monitor incorrect folding by diminishing aggregation (see Figure 2.3). The two-ringed GroEL chaperonin (middle and bottom levels in Figure 2.3) attaches to its partner chaperonin GroES (top ring) upon ATP binding, causing a major conformational change; the size of GroEL nearly doubles, and it assumes a cage shape, with GroES capping over it. This capping prevents the diffusion of partially folded compact intermediates termed 'molten globules' and offers them another chance at folding correctly.

Experiments that track hydrogen exchange in unfolded rubisco protein by radioactive tritium (a hydrogen isotope) labeling suggest how misfolded proteins fall into this cavity and are released: a mechanical stretching force triggered by ATP binding partially or totally unfolds the misfolded proteins, eventually releasing the captive protein [191].

These results also support an *iterative annealing* (or *network* model) for chaperone-guided folding, in which the process of forceful unfolding of misfolded molecules, their trapping in the cavity, and their subsequent release is iterated upon until proper folding.

The identification of preferential substrates for GroEL *in vivo* [192], namely multidomain *E. Coli* proteins with complex $\alpha\beta$ folds, further explains the high-affinity interactions formed between the misfolded or partially folded proteins and binding domains of GroEL. These proteins require the assistance of a chaperone because assembly of β-sheet domains requires long-range coordination of specific contacts (not the case for formation of α-helices). Natural substrates for other chaperones, like Eukaryotic type II chaperonin CCT, also appear selective, favoring assistance to proteins like actin [193].

However, such insights into folding kinetics are only the tip of the iceberg. Chaperone types and mechanisms vary greatly, and the effects of macromolecular crowding (not modeled by *in vitro* experiments) complicate interpretations of folding mechanisms *in vivo*. Unlike chaperonins, members of another class of chaperones that includes the heat-shock protein Hsp70 bind to exposed hydrophobic regions of newly-synthesized proteins and short linear peptides, reducing the likelihood of aggregation or denaturation. These are classified as 'stress proteins' since their amount increases as environmental stresses increase (e.g., elevated temperatures). Other chaperones are known to assist in protein translocation across membranes.

2.3 Protein Misfolding – A Conundrum

2.3.1 *Prions and Mad Cows*

Further clues into the protein folding enigma are also emerging from another puzzling discovery involving certain proteins termed *prions*. These misfolded proteins — triggered by a conformational change rather than a sequence mutation — appear to be the source of infectious agent in fatal neurodegenerative diseases like bovine spongiform encephalopathy (BSE) or 'mad cow disease' (identified in the mid 1980s in Britain), and the human equivalent Creutzfeld-Jacob disease (CJD).[5] The precise mechanism of protein-misfolding induced diseases is not known, but connections to neurodegenerative diseases, which include Alzheimer's, are growing and stimulating much interest in protein misfolding [194].

Stanley Prusiner, a neurology professor at the University of California at San Francisco, coined the term prion to emphasize the infectious source as the protein ('proteinaceous'), apparently in contradiction to the general notion that nucleic acids must be transferred to reproduce infectious agents. Prusiner won the 1998 Nobel Prize in Physiology or Medicine for this *"pioneering discovery of an entirely new genre of disease-causing agents and the elucidation of the underlying principles of their mode of action"*.

Prions add a new symmetry to the traditional roles long delegated to nucleic acids and proteins! Since the finding in the 1980s that nucleic acids (catalytic RNAs) can *catalyze* reactions — a function traditionally attributed to proteins only — the possibility that certain proteins, prions, *carry genetic instructions* — a role traditionally attributed to nucleic acids — completes the duality of functions to both classes of macromolecules.

2.3.2 *Infectious Protein?*

Is it possible for an ailment to be transmitted by 'infectious proteins' rather than viruses or other traditional infectious agents? The prion interpretation for the infection mechanism remains controversial for lack of clear molecular explanation. In fact, one editorial article stated that "*whenever prions are involved, more open questions than answers are available*" [195]. Yet the theory is winning more converts with laboratory evidence that an infectious protein that causes mad cow disease also causes a CJD variant in mice [196]. These results are somewhat frightening because they suggest that the spread of this illness from one species to another is easier than has been observed for other diseases.

The proteinaceous theory suggests that the prion protein (see Figure 2.4) in the most studied neurodegenerative prion affliction, *scrapie* (long known in sheep and goats), becomes a pathologic agent upon conversion of one or more of its α-helical regions into β-regions (e.g., parallel β-helix [197]); once this confor-

[5]See information from the UK Department of Health on www.doh.gov.uk, the UK CJD Surveillance Unit at www.cjd.ed.ac.uk, and the CJD Disease Foundation at cjdfoundation.org.

mational change occurs, the conversion of other cellular neighbors proceeds by a domino-like mechanism, resulting in many abnormally-folded molecules which eventually reap havoc in the mammal. This protein-only hypothesis was first formulated by J.S. Griffith in 1967, but Prusiner first purified the hypothetical abnormal protein thought to cause BSE. New clues are rapidly being added to this intriguing phenomenon (see Box 2.3).

Both the BSE and CJD anomalies implicated with prions have been linked to unusual deposits of protein aggregates in the brain. (Recent studies on mice also open the possibility that aberrant proteins might also accumulate in muscle tissue). It is believed that a variant of CJD has caused the death of dozens of people in Britain (and a handful in other parts of the world) since 1995 who ate meat infected with BSE, some only teenagers. Recent studies also suggest that deaths from the human form of mad cow disease could be rising significantly and spreading within Europe as well as to other continents.

Since the incubation period of the infection is not known — one victim became a vegetarian 20 years before dying of the disease — scientists worry about the extent of the epidemic in the years to come. The consequences of these deaths have been disastrous to the British beef industry and have led indirectly to other problems (e.g., the 2001 outbreak of foot-and-mouth disease, a highly infectious disease of most farm animals except horses). The panic has not subsided, as uncertainties appear to remain regarding the safety of various beef parts, as well as sheep meat, and the possible spread of the disease to other parts of the world.

2.3.3 Other Possibilities

Many details of this intriguing prion hypothesis and its associated diseases are yet to be discovered and related to normal protein folding. Some scientists believe that a lurking virus or virino (small nonprotein-encoding virus) may be involved in the process, perhaps stimulating the conformational change of the prion protein, but no such evidence has yet been found. Only creation of an infection *de novo* in the test tube is likely to convince the skeptics, but the highly unusual molecular transformation implicated with prion infection is very difficult to reproduce in the test tube.

Box 2.3: Prion: Structural Evidence

The detailed structural picture associated with the prion conformational change is only beginning to emerge as new data appear [198]. In 1997, Kurt Wütrich and colleagues at the Swiss Federal Institute of Technology in Zurich reported the first NMR solution structure of the 208-amino acid glycoprotein "prion protein cellular" PrP^C anchored to nerve cell membranes. The structure reveals a flexibly disordered assembly of helices and sheets (see Fig. 2.4). This organization of the harmless protein might help explain the conversion process to its evil isoform PrP^{Sc}. It has been suggested that chaperone molecules may bind to PrP^C and drive its conversion to PrP^{Sc} and that certain membrane proteins may also be involved in the transformation.

In early 1998, a team from the University of California at San Francisco discovered a type of prion, different from that associated with mad cow disease, that attaches to a major structure in neuron cells and causes cells to die by transmitting an abnormal signal. This behavior was observed in laboratory rats who quickly died when a mutated type of prion was placed into the brains of newborn animals; their brains revealed the abnormal prions stuck within an internal membrane of neuron cells. The researchers believe that this mechanism is the heart of some prion diseases. They have also found such abnormal prions in the brain tissue of patients who died from a rare brain disorder called Gerstmann-Straussler-Scheinker disease (GSS) — similar to Creutzfeld-Jacob disease (CJD) — that destroys the brain.

Important clues to the structural conversion process associated with prion diseases were further offered in 1999, when a related team at UCSF, reported the NMR structure of the core segment of a prion protein rPrP that is associated with the scrapie prion protein PrP^{Sc} [199, 200]. The researchers found that part of the prion protein exhibits multiple conformations. Specifically, an intramolecular hydrogen bond linking crucial parts of the protein can be disrupted by a single amino acid mutation, leading to different conformations. This compelling evidence on how the molecule is changed to become infectious might suggest how to produce scrapie-resistant or BSE-resistant species by animal cloning.

Prion views from several organisms (including human and cow) have been obtained [201], allowing analyses of species variations, folding, and misfolding relationships; see [197], for example. This high degree of similarity across species is shown in Figure 2.4.

Still, until prions are demonstrated to be infectious *in vivo*, the proteinaceous hypothesis warrants reservation. Clues into how prions work may emerge from parallel work on yeast prions, which unlike their mammalian counterparts do not kill the organism but produce transmitted heritable changes in phenotype; many biochemical and engineering studies are underway to explore the underlying mechanism of prion inheritance.

2.3.4 Other Misfolding Processes

There are other examples of protein misfolding diseases (e.g., references cited in [202]). The family of amyloid diseases is a notable example, of which Alzheimer's is a member. For example, familial amyloid polyneuropathy is a heritable condition caused by the misfolding of the protein transthyretin. The amyloid deposits that result interfere with normal nerve and muscle function.

As in mad cow disease, a molecular understanding of the misfolding process may lead to treatments of the disorders. In the case of familial amyloid polyneuropathy, research has shown that incorporating certain mutant monomers in the tetramer protein transthyretin reduces considerably the formation of amyloid deposits (amyloid fibrils); moreover, incorporating additional mutant monomers can prevent misfolding entirely [202]. These findings suggest potential therapeutic strategies for amyloid and related misfolding disorders. See also [203] for a new pharmacological approach for treating human amyloid diseases by using a small-molecule drug that targets a protein present in amyloid deposits; the drug links two pentamers of that protein and leads to its rapid clearance by the liver.

Recent studies also suggest that misfolded proteins generated in the pathway of protein folding can be dangerous to the cell and cause harm (whether or not they convert normal chains into misfolded structures, as in prion diseases) [204, 205]. The cellular mechanisms associated with such misfolded forms and aggregates are actively being pursued.

2.3.5 *Deducing Function From Structure*

Having the sequence and also the 3D structure at atomic resolution, while extremely valuable, is only the beginning of understanding biological function. How does a complex biomolecule accommodate its varied functions and interactions with other molecular systems? How sensitive is the 3D architecture of a biopolymer to its constituents?

Despite the fact that in many situations protein *structures* are remarkably stable to tinkering (mutations), their *functional* properties can be quite fragile. In other words, while a protein often finds ways to accommodate substitutions of a few amino-acids so as not to form an entirely different overall folding motif [206], even the most minute sequence changes can alter biological activity significantly. Mutations can also influence the *kinetics* of the folding pathway.

An example of functional sensitivity to sequence is the altered transcriptional activity of various protein/DNA complexes that involve single base changes in the TATA-box recognition element and/or single protein mutations in TBP (TATA-Box binding protein) [207]. For example, changing just a single residue in the common nucleotide sequence of TATA-box element, TA**T**AAAAG, to TA**A**AAAAG impairs binding to TBP and hence disables transcriptional activity.

In principle, theoretical approaches should be able to explain these relations between sequence and structure from elementary physical laws and knowledge of basic chemical interactions. In practice, we are encountering immense difficulty pinpointing what Nature does so well. After all, the notorious "*protein folding*" problem is a challenge to us, not to Nature.

Much work continues on this active front.

2.4 From Basic to Applied Research

An introductory chapter on biomolecular structure and modeling is aptly concluded with a description of the many important practical applications of the field, from food chemistry to material science to drug design. A historical perspective on drug design is given in Chapter 14. Here, we focus on the current status of drug development as well as other applied research areas that depend strongly on progress in molecular modeling. Namely, as biological structures and functions

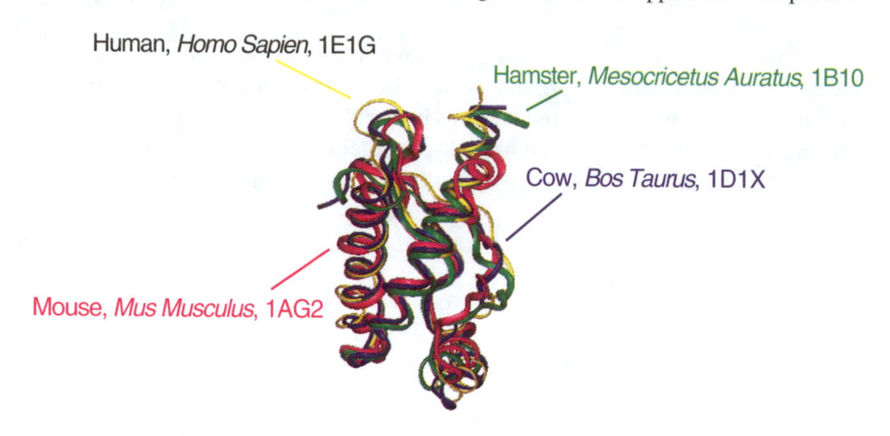

Figure 2.4. Structure of the prion protein.

are being resolved, natural disease targets that affect the course of disease can be proposed. Other biological and polymer targets, such as the ripening genes of vegetables and fruit or strong materials, can also be manipulated to yield benefits to health, technology, and industry.

2.4.1 Rational Drug Design: Overview

The concept of systematic drug design, rather than synthesis of compounds that mimic certain desired properties, is only about 50 years old (see Chapter 14). Gertrude Elion and George Hitchings of Burroughs Wellcome, who won the 1988 Nobel Prize in Physiology or Medicine, pioneered the field by creating analogues of the natural DNA bases in an attempt to disrupt normal DNA synthesis. Their strategies eventually led to a series of drugs based on modified nucleic-acid bases targeted to cancer cells. Today, huge compound libraries are available for systematic screening by various combinatorial techniques, robotics, other automated technologies, and various modeling and simulation protocols (see Chapter 14).

Rational pharmaceutical design has now become a lucrative enterprise. The sales volume for the world's best seller prescription drug in 1999, *prilosec* (for ulcer and heartburn), exceeded six billion dollars. A vivid description of the climate in the pharmaceutical industry and on Wall Street can be found in *The Billion-Dollar Molecule: One Company's Quest for the Perfect Drug* [208]. This thriller describes the racy story of a new biotech firm for drugs to suppress the immune system, specifically the discovery of an alternative treatment to *cyclosporin*, medication given to transplant patients. Since many patients cannot tolerate cyclosporin, an alternative drug is often needed.

Tremendous successes in 1998, like Pfizer's anti-impotence drug *viagra* and Entre-Med's drugs that reportedly eradicated tumors in mice, have generated much excitement and driven sales and earnings growth for drug producers. A glance at the names of biotechnology firms is an amusing indicator of the hope and prospects of drug research: Biogen, Cor Therapeutics, Genetech, Genzyme, Immunex, Interneuron Pharmaceuticals, Liposome Co., Millennium Pharmaceuticals, Myriad Genetics, NeXstar Pharmaceuticals, Regeneron Pharmaceuticals, to name a few. Yet, both the monetary cost and development time required for each successful drug remains very high [209].

2.4.2 A Classic Success Story: AIDS Therapy

HIV Enzymes

A spectacular example of drugs made famous through molecular modeling successes are inhibitors of the two viral enzymes *HIV protease* (HIV: human immunodeficiency virus) and *reverse transcriptase* for treating AIDS, acquired immune deficiency syndrome.

This world's most deadly infectious disease is caused by an insidious retrovirus. Such a virus can convert its RNA genome into DNA, incorporate this DNA into the host cell genome, and then spread from cell to cell. To invade the host, the viral membrane of HIV must attach and fuse with the victim's cell membrane; once entered, the viral enzymes reverse transcriptase and integrase transform HIV's RNA into DNA and integrate the DNA into that of the host [118].

Current drugs inhibit enzymes that are key to the life cycle of the AIDS virus (see Figure 2.5). **Protease inhibitors** like indinavir block the activity of proteases, protein-cutting enzymes that help a virus mature, reproduce, and become infectious [210]. **Reverse transcriptase** (RT) inhibitors block the action of an enzyme required by HIV to make DNA from its RNA [211].

AIDS Drug Development

A typical drug cocktail is the triplet drug combination of the protease inhibitor indinavir with the two nucleoside analogue RT inhibitors AZT (zidovudine, or 3′-azido-3′-deoxythymidine) and 3TC. More than one drug is needed because mutations in the HIV enzymes can confer drug resistance; thus, acting on different sites as well as on different HIV proteins increases effectiveness of the therapy.

Two types of RT blockers are *nucleoside analogues* and *non-nucleoside* inhibitors. Members of the former group like AZT interfere with the HIV activity by replacing a building block used to make DNA from the HIV RNA virus with an inactive analog and thereby prevent accurate decoding of the viral RNA. Non-nucleoside RT inhibitors (e.g., *nevirapine*, *calanolide* molecules, and *sustiva* under development) are designed to bind with high affinity to the active site of reverse transcriptase and therefore physically interfere with the enzyme's action.

Design of such drugs was made possible in part by molecular modeling due to the structure determination of the HIV protease by X-ray crystallography in 1989 and RT a few years later [212]. Figure 2.5 shows molecular views of these HIV enzymes complexed with drugs.

Besides the HIV protease and reverse transcriptase, a third target is the HIV integrase, which catalyzes the integration of a DNA copy of the viral genome into

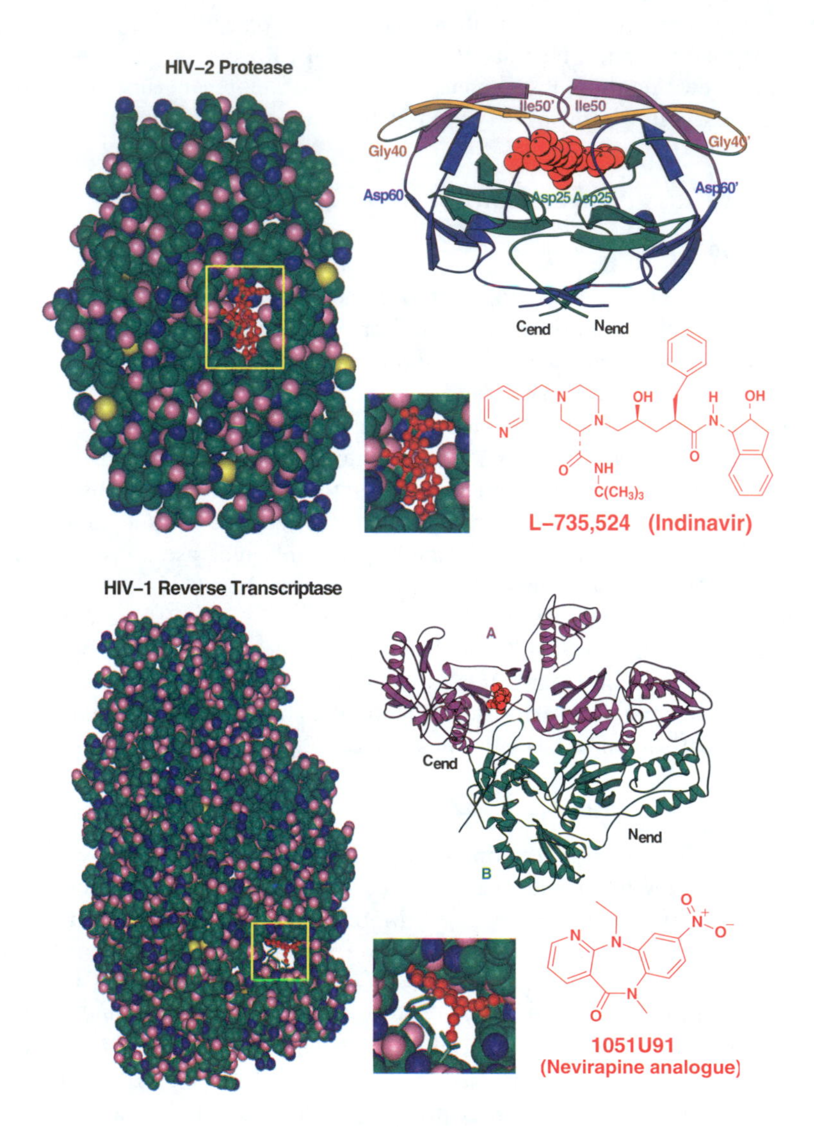

Figure 2.5. Examples of AIDS drug targets — the HIV protease inhibitor and reverse transcriptase (RT) — with corresponding designed drugs. The protease inhibitor indinavir (crixivan) binds tightly to a critical area of the dimer protease enzyme (HIV-2, 198 residues total shown here [210]), near the flaps (residues 40 to 60 of each monomer), inducing a conformational change (flap closing) that hinders enzyme replication; intimate interactions between the ligand and enzyme are observed in residues 25 and 50 in each protease monomer. The non-nucleoside RT inhibitor 1051U91 (a nevirapine analogue), approved for use in combination with nucleoside analogue anti-HIV drugs like AZT, binds to a location near the active site of RT that does not directly compete with the oligonucleotide substrate. The large RT protein of 1000 residues contains two subdomains (A and B).

the host cell chromosomes. Scientists at Merck have identified 1,3-diketo acid integrase inhibitors that block strand transfer, one of the two specific catalytic functions of HIV-1 integrase [213]; this function has not been affected by previous inhibitors. This finding paves the way for developing effective integrase inhibitors.

AIDS Drug Limitations

Much progress has been made in this area since the first report of the rational design of such inhibitors in 1990 [214] (see [209] for a review). In fact, the dramatic decline of AIDS-related deaths by such drug cocktails can be attributed in large part to these new generation of designer drugs (see Box 2.4) since the first introduction of protease inhibitors in 1996.

Indeed, the available drug cocktails of protease inhibitors and nucleoside analogues RT inhibitors have been shown to virtually suppress HIV, making AIDS a manageable disease. However, the cocktails are not a cure. The virus returns once patients stop the treatment. The mechanisms of drug resistant mutations and the interactions among them are not clearly understood [211].

In addition, the many sufferers worldwide cannot afford the virus suppressing drugs; the drug-cocktail regimen is complex, requiring many daily pills taken at multiple times and separated from eating, most likely for life; serious side effects also occur.

Lurking Virus

As mentioned, even available treatments cannot restore the damage to the patient's immune system; the number of T-cell (white blood cells), which that HIV attaches itself to, is still lower than normal (which lowers the body's defenses against infections), and there remain infected immune cells that the drugs cannot reach because of integration. Thus, new drugs are being sought to interrupt the first step in the viral life cycle — binding to a co-receptor on the cell surface to rid the body of the cell's latent reservoirs of the HIV virus, to chase the virus out of cells where it hides for subsequent treatment, or to drastically reduce the HIV reservoir so that the natural immune defenses can be effective. New structural and mechanistic targets are currently being explored (see Box 2.4).

A better understanding of the immune-system mechanism associated with AIDS, for example, may help explain how to prime the immune system to recognize an invading AIDS virus. Fusion (or entry) inhibitors define one class of drugs that seek to prevent HIV from entering the cell membrane. The AIDS virus has apparently developed a complex, tricky, and multicomponent-protection infection machinery.

Besides integrase and fusion inhibitors, among the newer drugs to fight AIDS being developed are immune stimulators and antisense drugs. The former stimulate the body's natural immune response, and the latter mimic the HIV genetic code and prevent the virus from functioning.

Vaccine?

Still, many believe that only an AIDS vaccine offers true hope against this deadly disease. More than one vaccine may be needed, since the HIV virus mutates and replicates quickly.[6] Still, it is hoped that therapeutic vaccination in combination with anti-HIV-1 drug treatment, even if it fails to eradicate infection, will suppress AIDS infection and the rate of transmission, and ultimately decrease the number of AIDS deaths substantially.

For a comprehensive overview of the biology of AIDS, the HIV life cycle, current status of the AIDS pandemic, and efforts for treating AIDS, see *Nature Insight* in the 19 April 2001 issue of *Nature* (Volume 410). This review was written at the 20 years after the first hints of the disease were reported in the summer of 1981, in clusters of gay men in large American cities; these groups exhibited severe symptoms of infection by certain pneumonias combined with those from Kaposi's sarcoma (KS) cancer.

Box 2.4: Fighting AIDS

AIDS drugs attributed to the success of molecular modeling include *AZT* (zidovudine) sold by Bristol-Myers Squibb, and the newer drugs *viracept* (nelfinavir) made by Agouron Pharmaceuticals, *crixivan* (indinavir) by Merck & Company, and *amprenavir* discovered at Vertex Pharmaceuticals Inc. and manufactured by Glaxo Wellcome. Amprenavir, in particular, approved by the U.S. Government in April 1999, is thought to cross the 'blood-brain barrier' so that it can attack viruses that lurk in the brain, where the virus can hide. This general class of inhibitors has advanced so rapidly that drug-resistant AIDS viruses have been observed.

Current structural investigations are probing the structural basis for the resistance mechanisms, which remain mysterious, particularly in the case of nucleoside analogue RT inhibitors like AZT [216]. The solved complex of HIV-1 reverse transcriptase [217] offers intriguing insights into the conformational changes associated with the altered viruses that influence the binding or reactivity of inhibitors like AZT and also suggests how to construct drug analogues that might impede viral resistance.

Basic research on the virus's process of invading host cells — by latching onto receptors (e.g., the CD4 glycoprotein, which interacts with the viral envelope glycoprotein, gp120, and the transmembrane component glycoprotein, gp41), and co-receptors (e.g., CCR5 and CXCR4) — may also offer treatments, since developments of disease intervention and vaccination are strongly aided by an understanding of the complex entry of HIV into cells; see [218] for example.

---

[6] For example, there is an enormous variation in the HIV-1 envelope protein. It has also been found that nearly all of non-nucleoside reverse transcriptase inhibitors can be defeated by site-directed mutation of tyrosine 181 to cysteine in reverse transcriptase. For this reason, the derivatives of *calanolide A* under current development are attractive drug targets because they appear more robust against mutation [215].

The HIV virus uses a spear-like agent on the virus' protein coat to puncture the membrane of the cells which it invades; vaccines might be designed to shut the chemical mechanism or stimuli that activate this invading harpoon of the surface protein. The solved structure of a subunit of gp41, for example, has been exploited to design peptide inhibitors that disrupt the ability of gp41 to contact the cell membrane [219]. A correlation has been noted, for example, between co-receptor adaptation and disease progression.

Novel techniques for gene therapy for HIV infections are also under development, such as internal antibodies (*intrabodies*) against the Tat protein, a vehicle for HIV infection of the immune cells; it is hoped that altered T-cells that produce their own anti-Tat intrabody would lengthen the survival time of infected cells or serve as an HIV 'dead-end'.

Other clues to AIDS treatments may come from the finding that HIV-1 originally came from a subspecies of chimpanzees [220]. Since chimps have likely carried the virus for hundreds of thousands of years but not become ill from it, understanding this observation might help fight HIV-pathogeny in humans. Help may also come from the interesting finding that a subset of humans have a genetic mutation (32 bases deleted from the 393 of gene CCR5) that creates deficient T-cell receptor; this mutation intriguingly slows the onset of AIDS. Additionally, a small subset of people is endowed with a huge number of helper (CD4) T-cells which can coordinate an attack on HIV and thus keep the AIDS virus under exquisite control for many years; such people may not even be aware of the infection for years.

2.4.3 Other Drugs

Another example of drug successes based on molecular modeling is the design of potent *thrombin inhibitors*. Thrombin is a key enzyme player in blood coagulation, and its repressors are being used to treat a variety of blood coagulation and clotting-related diseases. Merck scientists reported [221] how they built upon crystallographic views of a known thrombin inhibitor to develop a variety of inhibitor analogues. In these analogues, a certain region of the known thrombin inhibitor was substituted by hydrophobic ligands so as to bind better to a certain enzyme pocket that emerged crucial for the fit. Further modeling helped select a subset of these ligands that showed extremely compact thrombin/enzyme structures; this compactness helps oral absorption of the drug. The most potent inhibitor that emerged from these modeling studies has demonstrated good efficacy on animal models [221].

Other examples of drugs developed in large part by computational techniques include the *antibacterial agent* norfloxacin of Kyorin Pharmaceuticals (noroxin is one of its brand names), *glaucoma treatment* dorzolamide ("trusopt"/Merck), *Alzheimer's disease treatment* donepezil ("aricept"/ Eisai), and *migraine medicine* zolmitriatan ("zomig") discovered by Wellcome and marketed by Zeneca [209]. The headline-generating drug that combats impotence (*viagra*) was also found by a rational drug approach. It was interestingly an accidental finding: the compound had been originally developed as a drug for hypertension and then angina.

There are also notable examples of *herbicides and fungicides* that were successfully developed by statistical techniques based on linear and nonlinear regression and classical multivariate analysis (or QSAR, see Chapter 14): the herbicide metamitron — a bestseller in 1990 in Europe for protecting sugar beet crops — was discovered by Bayer AG in Germany.

2.4.4 A Long Way To Go

With an annual yield of about 50 new approved pharmaceutical agents that has become accepted in the last couple years, we are enjoying improved treatments for cancer, AIDS, heart disease, Alzheimer and Parkinson's disease, migraine, arthritis, and many more ailments. Yet the average cost of $500 million and time of 12–15 years required to develop a single drug remains extremely high. It can now be hoped that through the new fields of knowledge-based biological information, like *bioinformatics* [141, 142] and *chemoinformatics* [143], computers will reduce drastically these costs. Perhaps such revolutionary advances in drug development, expected in the next decade, will also alleviate the industry's political problems, associated with inadequate availability of drugs to the world's poor population.

Improved modeling and library-based techniques, coupled with robotics and high-speed screening, are also likely to increase the demand for faster and larger-memory computers.

"In a marriage of biotech and high tech," writes the New York Times reporter Andrew Pollack on 10 November 1998, *"computers are beginning to transform the way drugs are developed, from the earliest stage of drug discovery to the late stage of testing the drugs in people"*. Chapter 14 in this text points to some of these computational challenges.

2.4.5 Better Genes

Looking beyond drugs, gene therapy is another approach that is benefiting from key advances in biomolecular structure/function studies. Gene therapy attempts to compensate for defective or missing genes that give rise to various ailments — like hemophilia, the severe combined immune deficiency SCID, sickle-cell anemia, cystic fibrosis, and Crigler-Najjar (CN) syndrome — by trying to coerce the body to make new, normal genes. This regeneration is attempted by inserting replacement genes into viruses or other vectors and delivering those agents to the DNA of a patient (e.g., intravenously). However, delivery control, biological reliability, as well as possible unwelcome responses by the body against the foreign invader remain technical hurdles. See Box 2.5 for examples of gene therapy.

The first death in the fall of 1999 of a gene-therapy patient treated with the common fast-acting weakened cold virus adenovirus led to a barrage of negative publicity for gene therapy.[7] However, the first true success of gene therapy was reported four months later: the lives of infants who would have died of the severe immune disorder SCID (and until then lived in airtight bubbles to avoid the risk of infection) were not only saved, but thus far are able to live normal lives following gene therapy treatments that restore the ability of a gene essential to make T cells [222] (see Box 2.5).

Though this medical advance appears just short of a miracle, it remains to be seen how effective gene therapy will be on a wide variety of diseases. Still, given that gene therapy is a young science in a state of continuous flux, results to date indicate a promising future for the field [223].

A related technique for designing better genes is another relatively new technique known as *directed molecular evolution*. Unlike protein engineering, in which natural proteins are improved by making specific changes to them, directed evolution involves mutating genes in a test tube and screening the resulting ('fittest') proteins for enhanced properties. Companies specializing in this new Darwinian mimicking (e.g., Maxigen, Diversa, and Applied Molecular Evolution) are applying such strategies in an attempt to improve the potency or reduce the cost of existing drugs, or improve the stain-removing ability of bacterial enzymes in laundry detergents. Beyond proteins, such ideas might also be extended to evolve better viruses to carry genes into the body for gene therapy or evolve metabolic pathways to use less energy and produce desired nutrients (e.g., carotenoid-producing bacteria).

Box 2.5: Gene Therapy Examples

A prototype disease model for gene therapy is hemophilia, whose sufferers lack key blood-clotting protein factors. Specifically, Factor VIII is missing in hemophilia A patients (the common form of the disease); the much-smaller Factor IX is missing in hemophilia B patients (roughly 20% of hemophiliacs in the United States).

Early signs of success in treatment of hemophilia B using adeno-associated virus (a vector not related to adenovirus, which is slower acting and more suitable for maintenance and prevention) were reported in December 1999. However, introducing the much larger gene needed for Factor VIII, as required by the majority of hemophiliacs, is more challenging. Here, the most successful treatments to date only increase marginally this protein's level. Yet even those minute amounts are reducing the need for standard hemophilia treatment

[7]The patient of the University of Pennsylvania study was an 18-year old boy who suffered from ornithine transcarbamylase (OTC) deficiency, a chronic disorder stemming from a missing enzyme that breaks down dietary protein, leading to accumulation of toxic ammonia in the liver and eventually brain and kidney failure. The teenager suffered a fatal reaction to the adenovirus vector used to deliver healthy DNA rapidly. Autopsy suggests that the boy might have been infected with a second cold virus, parvovirus, which could have triggered serious disorders and organ malfunction that ultimately led to brain death.

(injections of Factor IX) in these patients.

Larger vectors to stimulate the patient's own cells to repair the defective gene are thus sought, such as retroviruses (e.g., lentiviruses, the HIV-containing subclass), or non-virus particles, like chimeraplasts (oligonucleotides containing a DNA/RNA blend), which can in theory correct point mutations by initiating the cell's DNA mismatch repair machinery.

An interesting current project involving chimeraplasts is being tested in children of Amish and Mennonite communities to treat the debilitating Crigler-Najjar (CN) syndrome. Sufferers of this disease lack a key enzyme which break down the toxic waste product bilirubin, which in the enzyme's absence accumulates in the body and causes jaundice and overall toxicity. Children with CN must spend up to 18 hours a day under a blue light to clear bilirubin and seldom reach adulthood, unless they are fortunate to receive and respond to a liver transplant. Chimeraplasty offers these children hope, and might reveal to be safer than the adenovirus approach, but the research is preliminary and the immune response is complex and mysterious.

Recent success was reported for treating children suffering from the severe immune disorder SCID type XI [222]. Gene therapy involves removing the bone marrow from infants, isolating their stem-cells, inserting the normal genes to replace the defective genes, and then re-infusing the stem cells into the blood stream. As hoped, the inserted stem cells were able to generate the cells needed for proper immune functioning in the patients, allowing the babies to live normal lives. Though successful till now, the real test is a long-term monitoring of the children to see whether the new immune systems will deteriorate over time or continue to function properly.

Success in any such gene therapy endeavors would lead to enormous progress in treating inherited diseases caused by point mutations.

2.4.6 Designer Foods

From our farms to medicine cabinets to supermarket aisles, designer foods are big business.

As examples of these practical applications, consider the transgenic organisms designed to manufacture medically-important compounds: bacteria that produce *human insulin*, and the food product *chymosin to make cheese*, a substitute for the natural rennet enzyme traditionally extracted from cows' stomachs. Genetically modified bacteria, more generally, hold promise for administering drugs and vaccines more directly to the body (e.g., the gut) without the severe side effects of conventional therapies. For example, a strain of the harmless bacteria *Lactococcus lactis* modified to secrete the powerful anti-inflammatory protein interleukin-10 (IL-10) has shown to reduce bowel inflammation in mice afflicted with inflammatory bowel disease (IBD), a group of debilitating ailments that includes Crohn's disease and ulcerative colitis.

The production of drugs in genetically-altered plants — "biopharming" or "molecular pharming" — represents a growing trend in agricultural biotechnology. The goal is to alter gene structure of plants so that medicines can be grown on the farm, such as to yield an edible vaccine from a potato plant against hepatitis B, or a useful antibody to be extracted from a tobacco plant. As in bioengineered foods, many obstacles must be overcome to make such technologies effective as medicines, environmentally safe, and economically profitable. Proponents of molecular pharming hope eventually for far cheaper and higher yielding drugs.

Genetically-engineered crops are also helping farmers and consumers by improving the taste and nutritional value of food, protecting crops from pests, and enhancing yields. Examples include the roughly one-half of the soybean and one-third of the corn grown in the United States, sturdier salad tomatoes,[8] corn pollen that might damage monarch butterflies, papaya plants designed to withstand the papaya ringspot virus, and caffeine-free plants (missing the caffeine gene) that produce decaffeinated cups of java.

Closer to the supermarket, one of the fastest growing category of foods today is *nutraceuticals* (a.k.a. functional foods or pharmaceuticals), no longer relegated only to health-food stores. These foods are designed to improve our overall nutrition as well as to help ward off disease, from cancer prevention to improved brain function. See Box 2.6 for examples.

The general public (first in Europe and now in the United States) has resisted genetically-modified or biotech crops, and this was followed by several blockades of such foods by leading companies, as well as global biosafety accords to protect the environment. Protesters have painted these products as unnatural, hazardous, evil, and environmentally dangerous ('Frankenfoods').[9]

With the exception of transferred allergic sensitivities — as in Brazil nut allergies realized in soybeans that contained a gene from Brazil nuts — most negative reactions concerning *food safety* are not scientifically well-grounded in this writer's opinion. In fact, not only do we abundantly use various sprays and chemicals to kill flies, bacteria, and other organisms in our surroundings and on the farm; each person consumes around 500,000 kilometers of DNA on an average day! Furthermore, there are many potential benefits from genetically-

[8]The *Flavr Savr* tomato that made headlines when introduced in 1993 contained a gene that reduces the level of the ripening enzyme polygalacturonase. However, consumers were largely disappointed: though beautiful, the genetically engineered fruit lacked taste. This is because our understanding of fruit ripening is still limited; a complex, coordinated series of biochemical steps is involved — modifying cell wall structure, improving texture, inducing softening, and producing compounds in the fruit that transform flavor, aroma, and pigmentation. Strawberries and other fruit are known to suffer similarly from the limitations of our understanding of genetic regulation of ripening and, perhaps, also from the complexity of human senses! See [224], for example, for a recent finding that a tomato plant whose fruit cannot ripen carries a mutation in a gene encoding a transcription factor.

[9]Amusing Opinion/Art ads that appeared in The New York Times on 8 May 2000 include provocative illustrations with text lines like "GRANDMA'S MINI-MUFFINS are made with 100% NATURAL irradiated grain and other ingredients"; "TOTALLY ORGANIC Biomatter made with Nucleotide Resequencing"; "The Shady Glen Farms Promise: Our Food is fresh from the research labs buried deep under an abandoned farm". [Note: the font size and form differences here are intentional, mimicking the actual ads].

engineered foods, like higher nutrients and less dependency on pesticides, and these considerations might win in the long run. Still, environmental effects must be carefully monitored so that genetically-altered food will succeed in the long run (see Box 2.6 for possible problems).

Box 2.6: Nutraceuticals Examples

The concept of fortified food is not new. Vitamin-D supplemented milk has eradicated rickets, and fortified breakfast cereals have saved many poor diets. In fact, classic bioengineering has been used for a long time to manipulate genes through conventional plant and animal inter-breeding. But the new claims — relying on our increased understanding of our body's enzymes and many associated vital processes — have been making headlines. ("Stressed Out? Bad Knee? Try a Sip of These Juices.", J.E. Barnes and G. Winter, *New York Times*, Business, 27 May 2001). Tea brews containing sedative roots like kava promise to tame tension and ease stress. Fruit-flavored tonics with added glucosamine (building block of cartilage) are claimed to soothe stiff knees of aging bodies. Fiber-rich grains are now touted as heart-disease reducers. Herb-coated snacks, like potato corn munchies coated with ginkgo biloba, are advertised as memory and alertness boosters.

With this growing trend of designer foods, the effect of these manipulations on our environment demands vigilant watch. This is because it is possible to create 'super-resistant weeds' or genetically-improved fish that win others in food or mate competitions. This potential danger emerges since, unlike conventional cross-breeding (e.g., producing a tangelo from a tangerine and grapefruit), genetic engineering can overcome the species barrier — by inserting nut genes in soybeans or fish genes in tomatoes, for example. This newer type of tinkering can have unexpected results in terms of toxins or allergens which, once released to the environment, cannot be stopped easily. For example, the first genetically-modified animal to reach American dinner plates is likely to be a genetically-altered salmon endowed with fortified genes that produce growth hormones, making the fish grow twice as fast as normal salmon. The effect of these endowed fish on the environment is yet unknown.

Popular examples of fortified food products with added vitamins and minerals (e.g., calcium and vitamin E) that also help protect against osteoporosis are orange juice, specialty eggs, and some vegetarian burritos. Other designer disease-fighting foods include drinks enriched with echinacea to combat colds; juices filled with amino acids and herbs claimed to boost muscle and brain function; margarines containing plant stanol esters (from soybean or pine trees) to fight heart disease and cancer, as well as green teas enriched with ginseng and other herbs; superyogurts to enhance the immune system; and tofu and yams to combat hot flashes. Such functional foods are also touted to lower cholesterol, provide energy, fight off depression, or to protect against salmonella and E. coli poisoning (e.g., yogurt fortified with certain bacteria).

Will Ginkgo Biloba chips, Tension Tamer cocktails, or Quantum Punch juice become part of our daily diet (and medicine cabinet) in this millennium?

2.4.7 *Designer Materials*

New specialty materials are also being developed in industry with the needed thermochemistry and kinetic properties. Examples are enzymes for detergents, adhesives and coatings, or photography film. Fullerene nanotubes (giant linear fullerene chains that can sustain enormous elastic deformations [225]), formed from condensed carbon vapor, have many potential applications. These range from architectural components of bridges and buildings, cars, and airplanes to heavy-duty shock absorbers, to components of computer processors, scanning microscopes, and semiconductors.

Long buckyball nanotubes have even been proposed as elements of 'elevators' to space in the new millennium [225]. These applications arise from their small size (their thickness is five orders of magnitude smaller than human hair), amazing electronic properties, and enormous mechanical strength of these polymers. In particular, these miniscule carbon molecules conduct heat much faster than silicon, and could therefore replace the silicon-based devices used in microelectronics, possibly overcoming current limitations of computer memory and speed.

2.4.8 *Cosmeceuticals*

Cosmeceutical companies are also rising — companies that specialize in design of cosmetics with bioactive ingredients (such as designer proteins and enzymes), including cosmetics that are individually customized (by *pharmacogenomics*) based on genetic markers, such as single nucleotide polymorphisms (SNPs). Most popular are products for sun or age-damaged skin containing alpha hydroxy acids (mainly glycolic and lactic acid), beta hydroxy acids (e.g., salicylic acid), and various derivatives of vitamin A or retinol (e.g., the tretinoin-containing *Retin-A* and *Renova* topical prescriptions). Besides reducing solar scars and wrinkling, products can also aid combat various skin diseases. Many of these compounds work by changing the metabolism of the epidermis, for example by increasing the rate of cell turnover, thereby enhancing exfoliation and the growth of new cells. New cosmeceuticals contain other antioxidants, analogues of various vitamins (A, D, and E), and antifungal agents.

The recent information gleaned from the Human Genome Project can help recognize changes that age and wrinkle skin tissue, or make hair or teeth gray. This in turn can lead to the application of functional genomics technology to develop agents that might help rejuvenate the skin, or color only target gray hair or tooth enamel. Computational methods have an important role in such developments by screening and optimizing designer peptides or proteins. Such biotechnology research to produce products for personal care will likely rise sharply in the coming years.

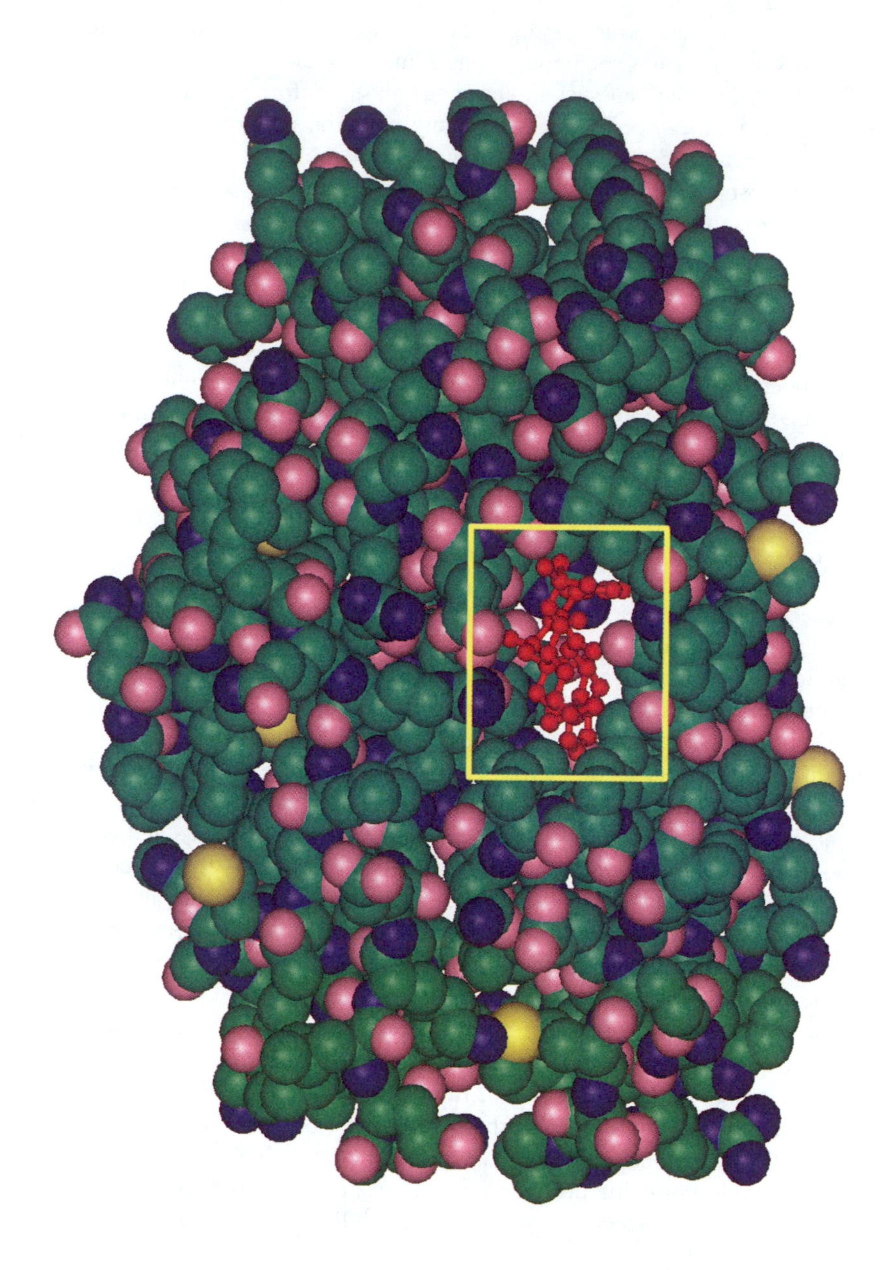

3

Protein Structure Introduction

Chapter 3 Notation

SYMBOL	DEFINITION
C^{α}	α-Carbon
τ	dihedral angle
ϕ	{N–C^{α}} rotation about peptide bond
χ_1–χ_4	rotamer dihedral angles in amino acid sidechains
ψ	{C^{α}–C}=O rotation about peptide bond
ω	C_1^{α}–{C–N}–C_2^{α} rotation

Life is the mode of existence of proteins, and this mode of existence essentially consists in the constant self-renewal of the chemical constituents of these substances.

Friedrich Engels, 1878 (1820–1895).

3.1 The Machinery of Life

3.1.1 From Tissues to Hormones

The term "protein" originates from the Greek word *proteios*, meaning "primary" or "of first rank". The name was adapted by Jöns Berzelius in 1838 to emphasize the importance of this class of molecules. Indeed, proteins play crucial, life-sustaining biological roles, both as constituent molecules and as triggers of physiological processes for all living things. For example, proteins provide the architectural support in muscle tissues, ligaments, tendons, bones, skin, hair, organs, and glands. Their environment-tailored structures make possible the coordinated function (motion, regulation, etc.) in some of these assemblies.

Proteins also provide the fundamental services of transport and storage, such as of oxygen and iron in muscle and blood cells. The first pair of solved protein structures **hemoglobin** and **myoglobin**, serve as the crucial oxygen carriers in vertebrates. Hemoglobin is found in red blood cells and is the chief oxygen carrier in the blood (it also transports carbon dioxide and hydrogen ions). Myoglobin is found in muscle cells, where it stores oxygen and facilitates oxygen movement in muscle tissue. The sperm whale depends on myoglobin in its muscle cells for large amounts of oxygen supplies during long underwater journeys.

Proteins further play crucial regulatory roles in many basic processes fundamental to life, such as reaction catalysis (e.g., digestion), immunological and hormonal functions, and the coordination of neuronal activity, cell and bone growth, and cell differentiation.

Given this enormous repertoire, Berzelius could not have coined a better name!

3.1.2 Size and Function Variability

Protein molecules come in a wide range of sizes and have evolved many functions. The major classes of proteins include *globular*, *fibrous*, and *membrane* proteins. Globular proteins are among the most commonly studied group. Newly found ribosomal proteins form a characteristic class of proteins that can be ordered as globular proteins, with disordered extensions.

To suit their environment and function, fibrous proteins (e.g., the collagen molecule in skin and bones), which are generally insoluble in aqueous environments, are extended in shape, whereas globular proteins tend to be compact. **Collagen** is a left-handed helix with a quaternary structure made of collagen fibrils aggregated in a parallel superhelical arrangement. See [226] for the crystal structure of a collagen-like peptide with a biologically relevant sequence (also shown in Figure 3.9) and summary of collagen structures elucidated to date. The globular protein **myoglobin** (see Figure 3.12) is highly compact, organized as 75% α-helices. Similarly, **hemoglobin** is a tetramer composed of four polypeptide chains held by noncovalent interactions; each subunit of hemoglobin in humans is very similar to myoglobin. Both proteins bind oxygen molecules through a central heme group.

There certainly are some very large proteins such as the muscle protein **titin** of about 27000 amino acids (and mass of 3000 kDa), but the average protein contains several hundred residues. The size of polypeptides can be determined from gel electrophoresis experiments: the rate of migration of the molecule is inversely proportional to the logarithm of its length. The mass of a polypeptide or protein can be estimated from mobility-to-mass relationships established for reference

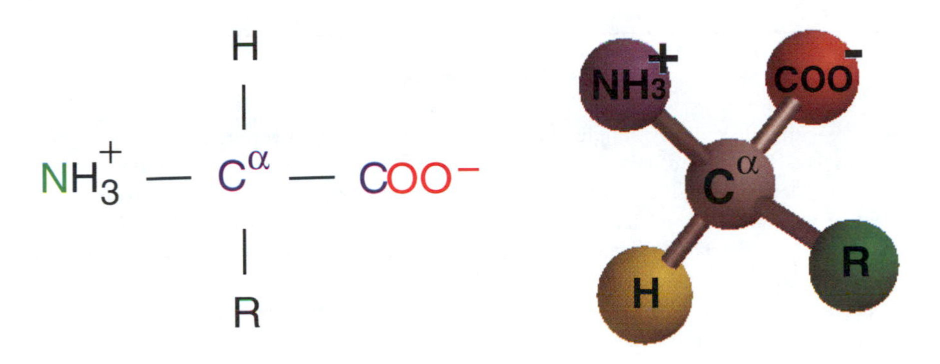

Figure 3.1. (left) The general formula for an amino acid, and (right) the spatial tetrahedral arrangement of an L-amino acid. The mirror image, a D-amino acid, is rare in proteins in Nature, if it exists.

proteins and by mass spectrometry measurements. Equilibrium ultracentrifugation [24] is another favored technique for determining various macromolecular features, including molecular weight, on the basis of transport properties.

3.1.3 Chapter Overview

This chapter introduces the bare basics in protein structure: the amino acid building blocks, primary sequence variations, and the framework for describing conformational flexibility in polypeptides. Included also is an introduction to the more advanced topic of sequence similarity and relation to structure, in the section on variations in protein sequences. *Students are encouraged to return to this subsection after reading Chapters 3 and 4.* Chapter 4 continues to describe secondary, supersecondary, and tertiary structural elements in proteins, as well as protein classification.

The protein treatment in these chapters is brief in comparison to the minitutorial on nucleic acids. Readers should consult the many excellent texts on protein structure (see books listed in Appendix C), like that by Branden and Tooze, and review introductory chapters in biochemistry texts like Stryer's. The 1999 text by Fersht [26], in particular, is a comprehensive description of the state-of-the-art in protein structure, and also reviews recent advances and insights from theoretical approaches.

Box 3.1: Water Structure

Water — that deceptively simple molecule composed of two hydrogens and one oxygen — displays highly unusual and complex properties that are far from fully understood. Perhaps because of those properties — like the contraction of ice when it melts, large heat of vaporization, and large specific heat — water is the best of all solvents and a fundamental substance to sustain life. Solvent organization and reorganization are crucial to the stability of proteins, nucleic acids, saccharides, and other molecular systems. The energetic and kinetic aspects of water structure are difficult to pinpoint by experiment and simulation because of the range of timescales associated with water motions, from the fast perturbations of order 0.1 ps to the slow proton exchanges of millisecond order.

Important to the understanding of solvation structure and dynamics in the vicinity of macromolecules is the tendency of water to form *hydrogen bonds* [227] (see also Box 3.2 for a definition of a hydrogen bond). In ice, the ordered crystal structure of water molecules, each oxygen is surrounded by a tetrahedron of four other oxygen atoms, with one hydrogen between each oxygen pair. In liquid water, many water molecules are engaged in such a hydrogen-bonded network, but the network is highly dynamic, with hydrogen-bonded partners changing rapidly.

This local organization of liquid water can easily be observed from experimental and computed radial distribution functions (e.g., O–O and O–H distances), which reflect the degree of occupancy of neighbors from a central oxygen or hydrogen molecule. Thus, for example, the highest peak in the O–O radial distribution function at a distance of about 2.9 Å corresponds to the first solvation shell, in which the four near-neighbor oxygens of the central oxygen molecule can be found at room temperature.

Figure 3.2 illustrates the structure of water clusters as computed by minimizing the potential energy composed of bond length, bond angle, and intermolecular Coulomb and van der Waals terms (see Chapter 8 for energy terms discussion). Such hydrogen bonds form ubiquitously in the environment of biomolecules. Water molecules penetrate into the grooves of nucleic acid helices, aggregate around hydrophilic, or water-soluble, segments of proteins (which cluster at the protein surface) and stabilize solute conformations through various hydrogen bonds and bridging arrangements. The dynamic nature of both water structure and biomolecules gives rise to the concept of hydration shells; see Chapter 6 in the context of DNA. That is, the solvent structure around the solute is multilayered, with the first hydration shell associated with water molecules in direct contact with the solute and the outermost layer as the bulk solvent.

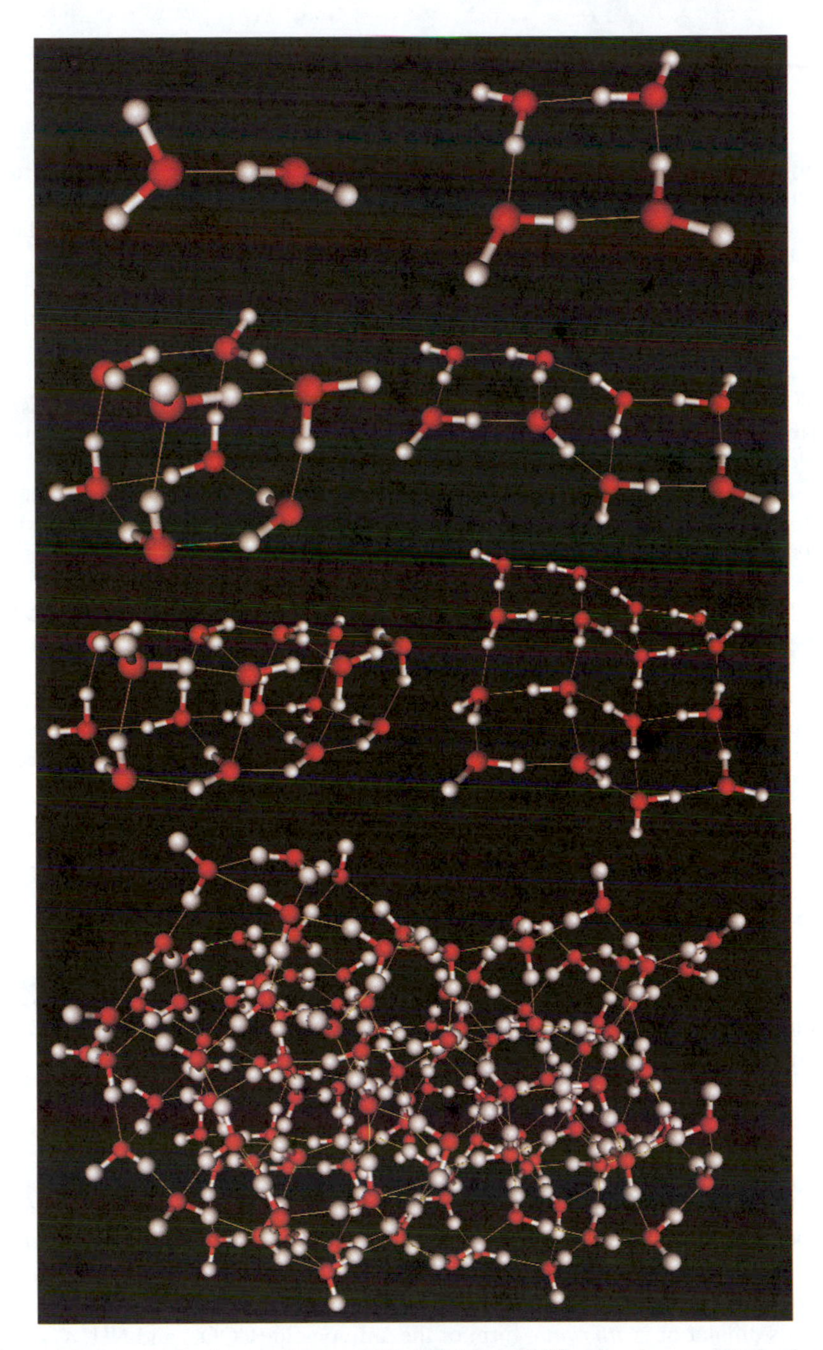

Figure 3.2. Structures of water clusters of 2, 4, 8, 16, and 125 molecules as minimized with the CHARMM force field from initial coordinates computed in [228]. Two geometries are shown for both the 8 and 16-molecule clusters; the lower energy structures (by roughly 4 and 6%, respectively) are associated with the more compact, cube-like shapes (left side in both cases). The tetrahedral structure of water is apparent in the inner molecules of the larger systems, where each molecule is hydrogen bonded to four others.

Box 3.2: Hydrogen Bonds

A hydrogen bond is an attractive, weak electrostatic (noncovalent) bond [227]. It forms when a hydrogen atom covalently binds to an electronegative atom and is electrostatically attracted to another (typically electronegative) atom. The atom to which the hydrogen atom (H) is covalently bound is considered the hydrogen *donor* (D), and the other atom is the hydrogen *acceptor* (A). Thus, the hydrogen bond is stabilized by the Coulombic attraction between the partial negative charge of the A atom and the partial positive charge of H. See Figure 3.2 for examples in water clusters.

In biological polymers, the donor and acceptor atoms are either nitrogens or oxygens. For example, in protein helices and sheets, the D–H $\cdots$ A sequence is N–H $\cdots$ O=C. In the nucleic-acid base pair of adenine–thymine, the two D–H $\cdots$ A sequences are N–H $\cdots$ O and N–H $\cdots$ (see Nucleic Acid chapters for details). Non-classical, weaker hydrogen bonds have been noted in biological systems (e.g., protein/DNA complexes), involving a carbon instead of one electronegative atom: C–H $\cdots$ O [229].

The strength of a hydrogen bond can be characterized by two geometric quantities which govern the hydrogen bond energy: colinearity of the D–H $\cdots$ A atoms, and optimal H $\cdots$ A (or D $\cdots$ A) distance. The ideal, strongest hydrogen bond often has its three atoms colinear. The strength of a hydrogen bond is several kilocalories per mole, compared to about 0.6 kcal/mol for thermal energy at room temperature, but the exact value remains uncertain (e.g., [230]). However, the formation of a network of hydrogen bonds in macromolecular systems leads to a cooperative effect that enhances stability considerably [227].

3.2 The Amino Acid Building Blocks

Proteins and polypeptides are composed of linked amino acids. That amino acid composition of the polymer is known as the *primary structure* or sequence for short.

3.2.1 Basic C^α Unit

Each amino acid consists of a central tetrahedral carbon known as the alpha (α) carbon (C^α) attached to four units: a hydrogen atom, a protonated amino group (NH_3^+), a dissociated carboxyl group (COO^-), and a distinguishing sidechain, or R group (see Figure 3.1).

This dipolar or *zwitterionic* form of the amino acid (COO^- and NH_3^+) is typical for neutral pH (pH of 7). The un-ionized form of an amino acid corresponds to COOH and NH_2 end groups. Different combinations involving ionized/un-ionized forms for each of the side groups can occur for the amino acid depending on the pH of the solution.

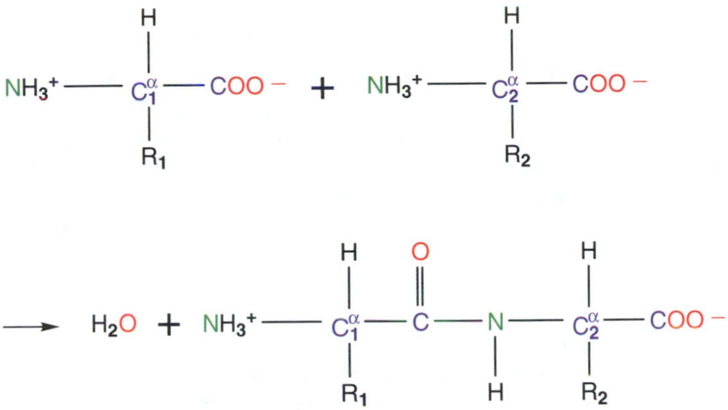

$$\mathrm{NH_3^+}-\underset{\displaystyle \mathrm{R_1}}{\overset{\displaystyle \mathrm{H}}{\mathrm{C_1^{\alpha}}}}-\mathrm{COO^-} + \mathrm{NH_3^+}-\underset{\displaystyle \mathrm{R_2}}{\overset{\displaystyle \mathrm{H}}{\mathrm{C_2^{\alpha}}}}-\mathrm{COO^-}$$

$$\longrightarrow \mathrm{H_2O} + \mathrm{NH_3^+}-\underset{\displaystyle \mathrm{R_1}}{\overset{\displaystyle \mathrm{H}}{\mathrm{C_1^{\alpha}}}}-\overset{\displaystyle \mathrm{O}}{\overset{\|}{\mathrm{C}}}-\underset{\displaystyle \mathrm{H}}{\mathrm{N}}-\underset{\displaystyle \mathrm{R_2}}{\overset{\displaystyle \mathrm{H}}{\mathrm{C_2^{\alpha}}}}-\mathrm{COO^-}$$

Figure 3.3. Formation of a dipeptide by joining two amino acids.

The tetrahedral arrangement about C^{α} makes possible two mirror images for the molecule. Only the L-isomer ("left-handed" from the Latin word *levo*) is a constituent of proteins on earth (see Figure 3.1). This asymmetry is not presently understood, but one explanation is that this imbalance is related to an asymmetry in elementary particles.

3.2.2 *Essential and Nonessential Amino Acids*

There are 20 naturally-occurring amino acids. Among them, humans can synthesize about a dozen. The remaining 9 amino acids must be ingested through our diet; these are termed *essential amino acids*. They are: histidine, isoleucine, leucine, lysine, methionine, phenylalanine, threonine, tryptophan, and valine.

Meat eaters need not be too concerned about a balanced diet of those nutrients, since animal flesh is a complete source of the essential amino acids. In contrast, vegetarians, particularly vegans — who omit all animal products like eggs and dairy in addition to meat, poultry, and fish — must perform a delicate balancing act to ensure that their bodies can synthesize all the basic proteins essential to good health. See Box 3.3 for a discussion of essential amino acids and balanced diets.

Box 3.3: Protein Chemistry and Vegetarian Diets

Vegetarians must take care to combine foods from three basic groups which are complementary with respect to their supply of these amino acids: (a) rice and grains like oats, wheat, corn, cereals, and breads; (b) legumes and soy-products; and (c) nut products (cashews, almonds, and various nut butters). Peanuts are technically members of the legume family rather than nuts.

Notable vegetarian combinations are: rice or grains (low in or lacking isoleucine, lysine, and threonine) with beans (rich in isoleucine and leucine and, in the case of lima-beans, also lysine); cereals with leafy vegetables; corn, wheat, or rye (low in or lacking tryptophan and lysine) with soy protein/soybeans (rich in isoleucine, tryptophan, lysine, methionine, and valine); corn with nuts or seeds (rich in methionine, isoleucine, and leucine); bread/wheat with peanut butter (rich in valine and tryptophan); and potatoes (limited methionine and leucine) with onions, garlic, lentils and egg or fish (if permitted), all of which are rich in methionine.

A classic Native-American dish of acorn squash stuffed with wild rice, quinoa, and black beans is a superior mixture of nutrients. The plant quinoa, prepared like a grain, is actually a fruit and moreover a complete protein (rich in lysine and other amino acids), as well as rich in vitamins E and B, fiber, and the minerals calcium, phosphorus, and iron. Nuts also contain several vitamins and minerals that protect against heart disease, like folate, and calcium, magnesium, and potassium, which also protect against high blood pressure.

Contrary to an existing myth, such food group combinations need not be eaten at the same meal to guarantee a complete source of the essential amino acids; a daily approach suffices. Given all these food sources, vegetarians — even vegans who carefully comply — will not be deficient in protein. However, nutrients that present a challenge to vegans and lacto-ovo vegetarians are vitamin B_{12} (deficiencies of which can cause nerve damage), found in fortified cereals, and zinc (needed for protein synthesis, healing wounds, and immunity), available in fortified cereals, soy-based foods, and dairy products.

Given the intimate relationship between protein chemistry and good nutrition, readers of this text should be healthy as well as smart!

Interestingly, the requirements for vegetarian diets are now of crucial interest to NASA researchers: future astronauts who will spend extended periods of time in space stations (on Mars, Jupiter, or the Moon) will have to depend on hydroponically-grown plant crops for nearly all their protein requirements, as well as vitamins, minerals, and fiber. Research is now in progress on how best to select a limited set of plants that can adapt to growing in nutrient-enriched water (rather than soil) and in small spaces. At the same time, this selection must meet the basic dietary requirements of space-station scientists, as well as satisfy their culinary taste and demand for variety [231].

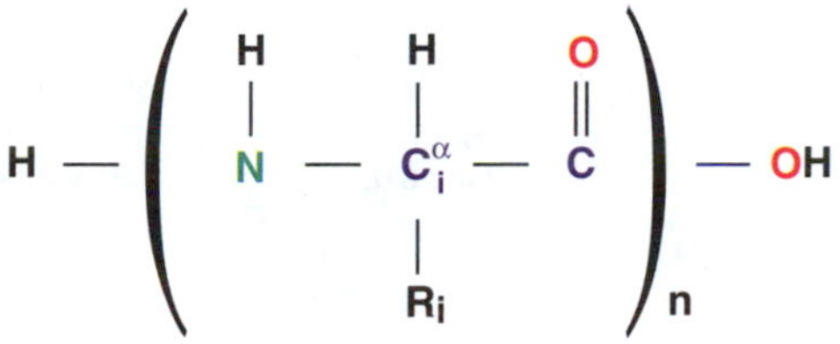

$$H - \left(\begin{array}{c} H \\ | \\ N \end{array} - \begin{array}{c} H \\ | \\ C_i^{\alpha} \\ | \\ R_i \end{array} - \begin{array}{c} O \\ \| \\ C \end{array} \right)_n - OH$$

Figure 3.4. The repeating formula for a polypeptide.

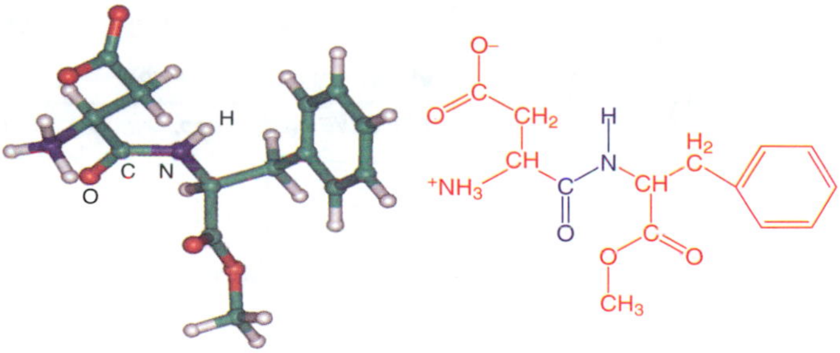

Aspartame

Figure 3.5. The dipeptide aspartame.

3.2.3 *Linking Amino Acids*

A polypeptide is formed when amino acids join together. Namely, the carboxyl carbon of one amino acid joins the amino nitrogen of another amino acid to form the peptide (C–N) bond with the release of one water molecule (Figure 3.3). The general repeating formula for a polypeptide is shown in Figure 3.4. When the amino acid residue is proline, its C^{α} is linked to the nitrogen of the peptide backbone through the proline ring.

A model of **aspartame**, a dipeptide of aspartic acid and phenylalanine, is shown in Figure 3.5. It was discovered accidentally in 1965 by a 'careless' chemist who licked his fingers during his laboratory work. To his surprise, a substance 100–200 times sweeter than sucrose was discovered. Because it is a kind of protein, aspartame is metabolized in the body like proteins and is a source of amino acids. (*This should not, however, be taken as an endorsement for diet soft drinks as a source of nutrients!*)

The synthesis of polypeptides and proteins occurs in a cellular structure, the ribosome, *in vivo*. Synthesis *in vitro* is facile for 100–150 residues but much more involved for longer chains.

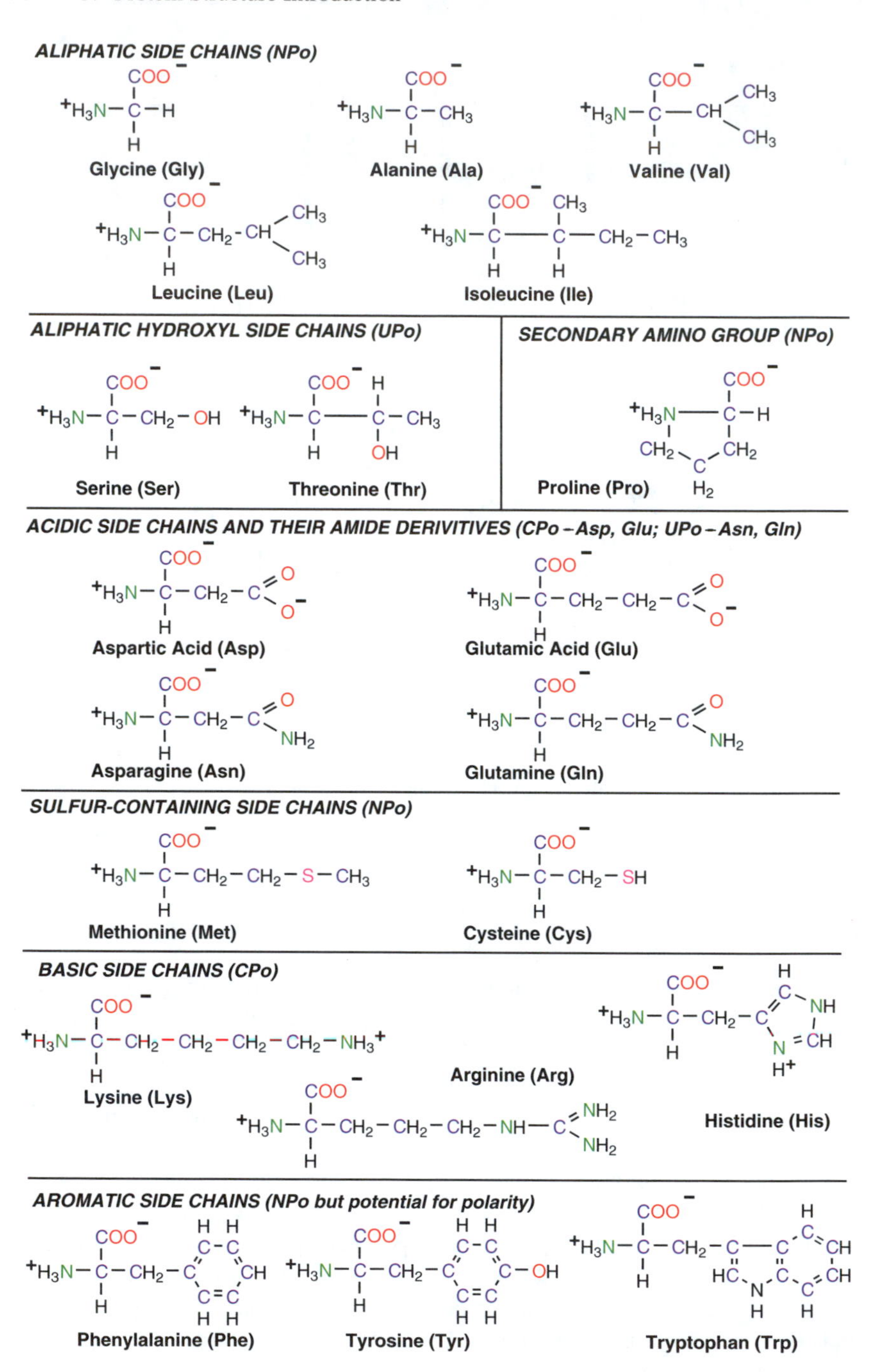

Figure 3.6. The chemical formulas of the 20 natural amino acids as found in neutral pH (pH of 7). The acronyms **NPo**, **UPo**, **CPo** denote, respectively, nonpolar, uncharged polar, and charged polar amino acids.

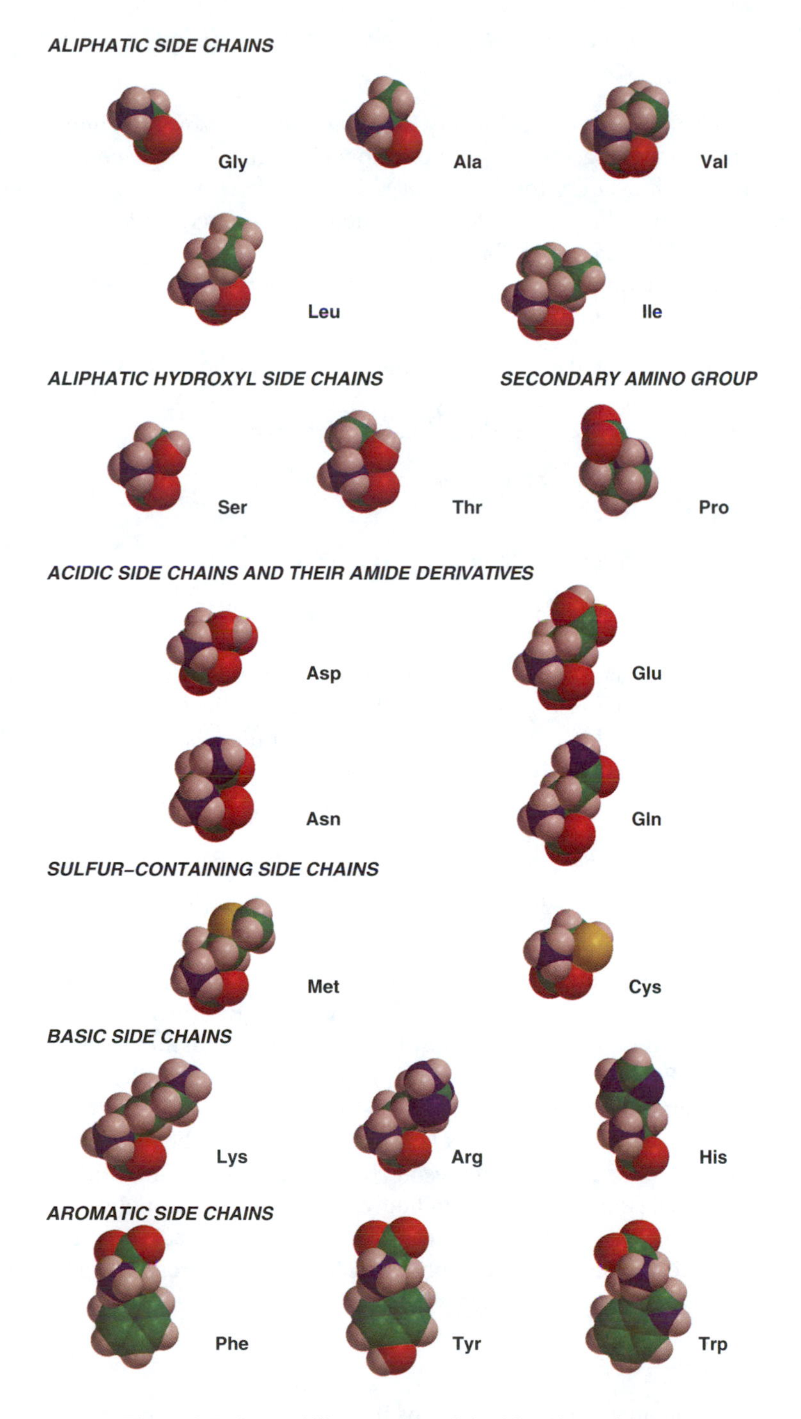

Figure 3.7. Space-filling models of the 20 amino acids.

3.2.4 The Amino Acid Repertoire: From Flexible Glycine to Rigid Proline

The chemical formulas of the twenty L-amino acids are shown in Figure 3.6, with the corresponding space-filling models shown in Figure 3.7. The commonly used three-letter abbreviation for each acid is illustrated, as well as a grouping into amino acid subfamilies. A one-letter mnemonic is also used to identify sequences of amino acids, as shown in Table 3.1.

A broader classification than indicated in the figures consists of the following three groups:

- **NPo**: amino acids with strictly *nonpolar* (hydrophobic, or water insoluble) side chains:
  Ala, Val, Leu, Ile, Phe, Pro, Met, Gly, Trp, Tyr;
- **CPo**: amino acids with *charged polar* residues:
  Asp, Glu, His, Lys, Arg; and
- **UPo**: amino acids with *uncharged polar* side chains:
  Ser, Thr, Cys, Asn, Gln.

Each amino acid has a unique combination of properties — size, polarity, cyclic constituents, sulfur constituents, etc. — that critically affects the noncovalent and covalent (i.e., disulfide bonds) interactions that form and stabilize protein three-dimensional (3D) architecture. These interactions originate from electrostatic, van der Waals, hydrophobic, and hydrogen-bonding forces. These properties are described in turn for these amino-acid classes.

Aliphatic R: Gly, Ala, Val, Leu, Ile

Glycine, alanine, valine, leucine, and isoleucine can be classified as nonpolar. Glycine, the simplest of the amino acids, is first in the aliphatic-sidechain subgroup. Each member in this family has a sidechain (R) which increases in bulk and branching design. Glycine is most flexible and hence an important constituent of proteins. For example, glycine is a major component of the α-helix of the protein α-keratin, which makes up hair and wool, as well as the β-sheet of the polypeptide β-keratin, which is silk (see Figure 3.9). Since the increasing aliphatic substitutions in this family increases the bulkiness of the amino acid, the overall conformational flexibility correspondingly decreases within a polypeptide. However, the conformational variability of each of these amino acids increases due to the *rotameric* variations of the amino acid (roughly, different 3D arrangements about central bonds within the sidechain — see Section 3.4).

Rigid Proline

Proline is a nonpolar amino acid as well. In contrast to glycine residues, which allow a great deal of conformational flexibility about the backbone (i.e., wide range of sterically-permissible rotations ϕ and ψ about the peptide bond — see Section 3.4), flexibility in proline residues is largely limited, due to the cyclic nature of its sidechain.

Aliphatic Hydroxyl R: Ser, Thr

Serine and threonine contain aliphatic hydroxyl groups and are considered uncharged polar, capable of forming hydrogen bonds (see Box 3.2).

Acidic R and Amide Derivatives: Asn, Gln, Asp, Glu

Similarly, asparagine and glutamine possess amide groups and are also considered uncharged polar with potential for hydrogen-bond formation. Their acidic analogs, aspartic acid and glutamic acid, are negatively charged (intrinsic pH of around 4) and thus considered charged polar, but the polar end of their sidechains is separated from C^{α} by hydrophobic CH_2 groups.

Basic R: Lys, Arg, His

Lysine, arginine, and histidine have basic sidechains and are thus in the charged polar category of amino acids. Lysine and arginine — the longest amino acids — are positively charged at physiological concentrations (that is, sidechain pH of 10–12), whereas histidine charge can be both positive or negative depending on its environment. This duality in histidine stems from its imidazole ring, which is in the physiological range of pH. For this reason, histidine residues serve as good metal binders and are often found in the active sites of proteins.

Aromatic R: Phe, Tyr, Trp

The amino acids with aromatic sidechains — phenylalanine, tyrosine, and tryptophan — have significant potential for electrostatic interactions due to an electron deficit in the ring hydrogen atoms. Phenylalanine is highly hydrophobic while the other two can be considered mildly hydrophobic, since their aromaticity is juxtaposed with polar properties (hydroxyl group of tyrosine and indole-ring nitrogen of tryptophan). The aromatic rings of this amino acid family also have potential for electron transfer. They can all be classified as nonpolar, though the mild hydrophobicity of tyrosine often warrants its classification as an uncharged polar amino acid.

Sulfur-Containing R: Met, Cys

Finally, nonpolar cysteine and methionine contain sulfur in their sidechains and are thus hydrophobic. Cysteine, in particular, is very reactive and binds to heavy metals. It has an important role in protein conformations through its unique ability to form *disulfide bonds* between two cysteine residues. Disulfide bonds are covalent but reversible and are thought to be important in many cases by directing a protein to its native structure and maintaining this functionally-important state.

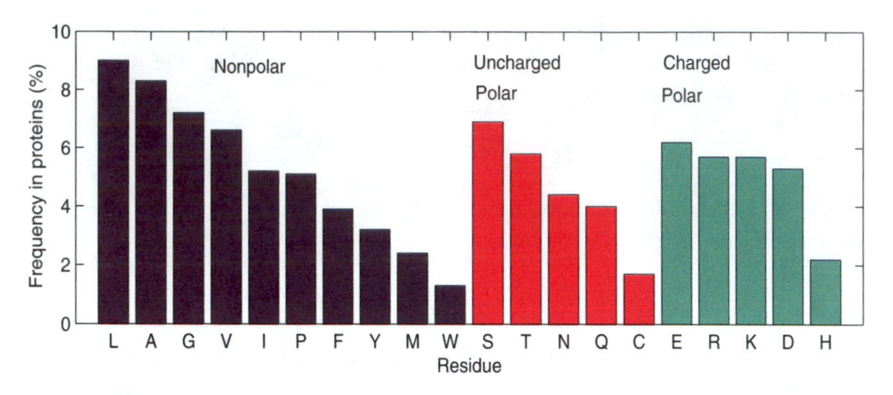

Figure 3.8. Amino acid frequencies as computed from 971 unrelated proteins [232]. See Table 3.1 for the one-letter key to amino acid abbreviations.

3.3 Sequence Variations in Proteins

From the constituent library of twenty natural amino acids, 20^N sequence combinations for an N-residue peptide are possible, an enormous number when N is several hundred. However, natural evolution has favored certain sequences more than others. Sequence similarity is an important factor in indicating common evolutionary ancestry of proteins, as discussed below. It is therefore widely used as a tool for classifying proteins into families, as well as for relating sequence to structure and predicting structure from sequence (*homology modeling* [146, 147], as introduced in Chapter 2).

3.3.1 Globular Proteins

In most proteins, the twenty amino acids occur at roughly similar frequencies. Notable exceptions occur for certain amino acids like methionine, which is frequently found at the N-terminus of the peptide since it serves as the amino acid initiator of synthesis, or special groups of proteins, such as membrane or fibrous proteins.

Table 3.1 shows the frequency of occurrence of amino acid residues in the PDB40 dataset of 971 domains of unrelated proteins with a sequence identity of 40% or less [232], and Figure 3.8 displays the data as histograms. We see that nonpolar Ala and Leu (boldfaced entries in the table) have the highest percentages (above 8%) within the representative protein database. The lowest frequencies (4% and below) occur for Trp (aromatic sidechain), Cys (sulfur-containing sidechain), His and Met (the other sulfur-containing sidechain), Tyr and Phe (aromatic sidechains also), and Gln.

Table 3.1. Amino acid frequencies in proteins based on the data of [232]. Bold and italics types are used, respectively, for the highest (>8%) and lowest (< 2.5%) frequencies.

Amino Acid	Freq. [%]
Alanine (Ala, A)	**8.44**
Arginine (Arg, R)	4.72
Asparagine (Asp, D)	5.98
Aspartic acid (Asn, N)	4.68
Cysteine (Cys, C)	*1.69*
Glutamine (Gln, Q)	3.70
Glutamic acid (Glu, E)	6.16
Glycine (Gly, G)	7.95
Histidine (His, H)	*2.19*
Isoleucine (Ile, I)	5.49
Leucine (Leu, L)	**8.28**
Lysine (Lys, K)	5.80
Methionine (Met, M)	*2.18*
Phenylalanine (Phe, F)	4.03
Proline (Pro, P)	4.61
Serine (Ser, S)	6.08
Threonine (Thr, T)	5.87
Tryptophan (Trp, W)	*1.47*
Tyrosine (Tyr, Y)	3.61
Valine (Val, V)	6.94

3.3.2 Membrane and Fibrous Proteins

Membrane proteins are embedded in a dynamic lipid bilayer environment, where mobility is more restricted. They therefore have more hydrophobic residues than globular proteins, which favor polar groups on the exterior surface. Since membrane proteins are particularly difficult to crystallize, simulation work is especially important in this area to understand their detailed function.

Fibrous, or structural, proteins tend to have repetitive sequences. The triple-stranded collagen helix, for example, is composed of repeating triplets which include glycine as the first residue and often proline as one or both of the other residues of the triplet. A model of collagen is shown in Figure 3.9.

Since collagen is needed to rebuild joint cartilage, there are important practical applications to skin and bone ailments. For example, a gelatin-containing powdered drink mix called *Knox NutraJoint* is being touted as a dietary supplement that helps maintain healthy joints and bones. (Gelatin is rich in two amino acids, glycine and proline, that make up collagen.) Even though our bodies make these two amino acids, manufacturers claim that this gelatin-containing supplement may be helpful in decreasing the progression of osteoarthritis, a condition caused by cartilage deterioration.

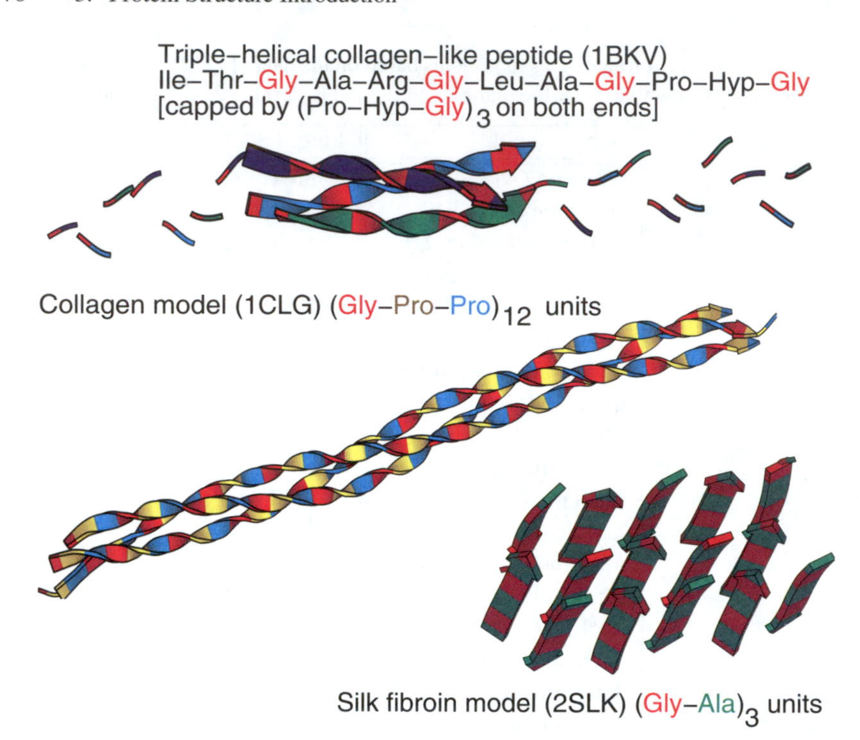

Figure 3.9. Models of the fibrous proteins collagen (triple helix) and silk, along with a crystallographically-determined collagen-like peptide (Hyp denotes hydroxyproline).

Another use of collagen is in a skin product used to heal wounds such as from venous skin ulcers, burns, and skin surgery. In May 1998 the FDA approved Apilgraf, a product for treating venous skin ulcers made of human skin cells mixed with collagen cells from cattle.

Silk is another example of a fibrous protein with a repetitive sequence. The product of many insects and spiders, silk is the polypeptide β-keratin composed largely of glycine, alanine, and serine residues, with smaller amounts of other amino acids such as glutamine, tyrosine, leucine, valine, and proline. The softness, flexibility, and high tensile strength of silk stems from its unique arrangement of loose hydrogen-bonding networks in the form of β-sheets connected by β-turns, a mixture of both highly-ordered and less densely-packed regions. Figure 3.9 shows a model of the repetitive β-sheet network of silk (without connecting regions).

3.3.3 Emerging Patterns from Genome Databases

As genome sequencing projects are completed, interesting findings on enzyme sequences also emerge. For example, the genome of the tuberculosis bacterium (completed in 1998 by the Wellcome Trust Genome Campus of the Saenger Centre in collaboration with the Institut Pasteur in Paris) revealed surprisingly that, unlike other bacteria, repetitive gene families of glycine-rich proteins exist in *M. tuberculosis*; these approximately 10% of the enzyme-coding sequences are associated with gene families involved in anaerobic respiratory functions.

3.3.4 Sequence Similarity

Sequence Similarity Generally Implies Structure Similarity

As mentioned above, sequence similarity generally implies structural, functional, and evolutionary commonality. Thus, for example, if we were to scan the Protein Databank (PDB) randomly and find two proteins with low sequence identity (say less than 20%), we could reasonably propose that they also have little structural similarity. Such an example is shown in Figure 3.12 for the **cytochrome/barstar pair**. Similarly, large sequence similarity generally implies structural similarity (see introduction in 2.1.2 of Chapter 2).

In general, small mutations (e.g., single amino acid substitutions) are well tolerated by the native structure, even when they occur at critical regions of secondary structure. The small protein **Rop** (Repressor of primer), which controls the mechanism of plasmid replication, provides an interesting subject to both this *sequence-implies-structure* paradigm, and to exceptions to this rule (discussed below).

Rop is a dimer, with each monomer consisting of two antiparallel α-helices connected by a short turn; it dimerizes to form a four-helix bundle as active form, as shown in Figure 3.10. (Fold details and motifs are discussed in the next chapter). Recall that Rop was used as the basis for solving Paracelsus challenge (Chapter 2) because the α-helix motif was thought to be quite stable.

The high stability of Rop emerged surprisingly from experiments of Castagnoli *et al.* [206]. When these researchers deleted just a few residues in a key turn region that produces the overall bundle fold in the native Rop structure, they expected one long contiguous helix to form. Instead, their tinkering produced a small variation of the original bundle motif. Apparently, the four-helix bundle motif is so stable that a *new turn* was formed from residues that used to be part of the α-helix backbone! Thus, the original bundle motif, though slightly smaller, was maintained in the mutants. This is seen in Figure 3.10, which displays the wildtype enzyme structure (a) and those of two mutants in the above cited study (b and c).

Though this experiment supports the general notion that protein *structures* are remarkably stable to tinkering (mutations), we emphasize that *functional* properties of proteins are fragile and quite sensitive to sequence changes.

Exceptions Exist

There are many exceptions, however, to this simple sequence/structure/ function relationship.

Namely, examples exist where despite *large sequence similarity* there is *small structural and functional similarity*. A classic example of this relationship is the disease sickle-cell anemia, where a minute substitution in sequence leads to al-

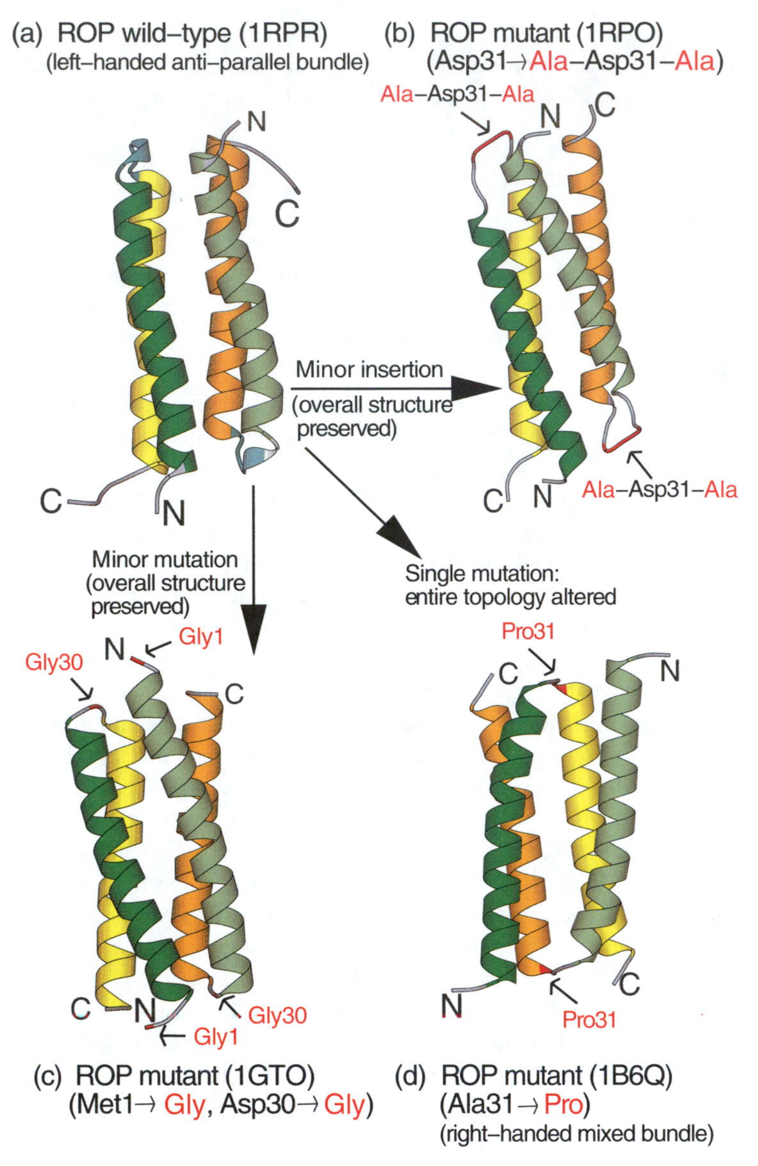

Figure 3.10. The protein **Repressor of Primer** (63 residues per monomer) provides interesting examples of the paradigm of structure inference by sequence similarity: the four-helix bundle motif of the wildtype (a) can be both *structurally stable*, i.e., resistant to mutations — as shown by the two variants in (b) and (c) — or *structurally fragile* and highly sensitive to mutations, caused by proline substitution at the turn region — as shown in (d), a mutant with an entirely different topology [233]. In (b), two Ala residues were inserted at both sides of the amino acid Asp in the loop region. In (c), the Asp residue connecting the two α-helices of each monomer was mutated to Gly, and Met1 was changed to Gly [206].

tered function with devastating consequences. This abnormality results from the replacement of the highly-polar glutamate residue in **hemoglobin** by the nonpolar amino acid valine. This key substitution at the surface of the protein leads to an entirely different quaternary structure for this multidomain red-blood pigment protein. This is because the markedly altered structure affects the solubility of oxygenated hemoglobin and leads to a clumping of the deoxygenated form of the molecule (HbS instead of HbA).

Conversely, examples exist where despite *small sequence similarity* there is *large structural*, and even functional and evolutionary, similarity. A classic example for this relationship is the **myoglobin/hemoglobin** pair of proteins (see Figure 3.12). These proteins only share 20% of the sequence. However, as oxygen-carrying molecules, they share structural, functional, and evolutionary similarity. Proteins in the **calmodulin** family are also known to display a great deal of structural variability for similar sequences [171] (see Figure 3.11).

More generally, changes in 3D architecture (despite a nontrivial degree of sequence similarity) can result from a variety of factors, as follows.

- Mutations in *critical regions* of the proteins, such as active sites and ligand binding sites, can change 3D structures dramatically. Such an example is shown for the pair of **immunoglobulins** in Figure 3.12.

- Mutations in regions that *connect two secondary-structural elements* can also be responsible for structural divergence, as in the helix-loop-helix motif of the **EF-hand family**, and the connecting loops in helix bundles.

  Figure 3.11 illustrates this principle for the two EF-hand calcium-binding proteins **calmodulin** and **sarcoplasmic calcium-binding** protein: one is overall extended in shape while the other is more compact [234].

  Helix bundles are sensitive to mutations in loop or turn regions that connect different helices, to the extent that a single amino acid substitution (alanine to proline) can change the topology of a homodimeric 4–helical bundle protein from the canonical left-handed all-antiparallel form to a right-handed mixed parallel and antiparallel bundle [233]. Figure 3.10(d) shows this different resulting topology of the **Rop four-helix bundle** subject to the single mutation Ala31→Pro at the turn region.

- Structural variations can be observed in the same system determined at different *environmental conditions* such as solvent or crystal packing. The same **T4-lysozyme** mutants in Figure 3.12 (100% sequence similar) display intriguing mobility, adopting 5 different crystal conformations [235] due to a hinge bending motion.

- *Multidomain proteins* can adapt quaternary structures that depend sensitively on the number of subunits and/or on the sequence.

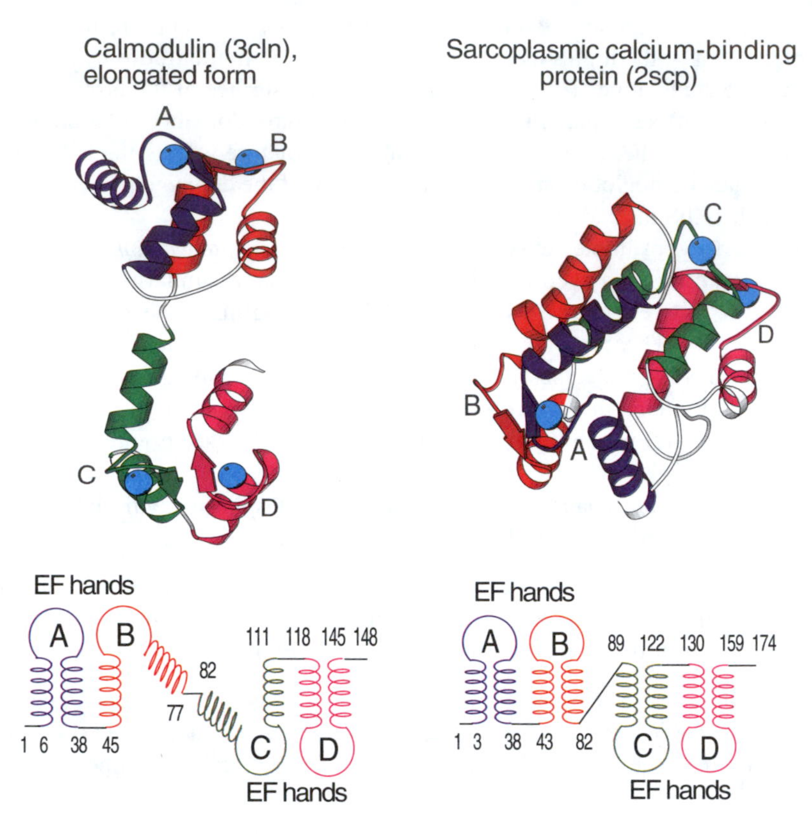

Figure 3.11. Structural variability despite large similarity in the protein secondary-structural elements is illustrated for two calcium-binding proteins — **calmodulin** (148 residues) and **sarcoplasmic calcium binding protein** (174 residues) [234] — due to different overall 3D arrangement of the shared motifs. Though sharing only 30% of the sequence, both proteins are made of two repeating units, each consisting of two EF hand motifs. Each hand motif contains two helical regions surrounding a calcium binding loop (crystal-bound calciums are rendered as large spheres; only three are bound to 2scp).

3.4 Protein Conformation Framework

3.4.1 The Flexible ϕ and ψ and Rigid ω Dihedral Angles

Polypeptides can have a wide variety of *conformations*, i.e., 3D structures differing only in rotational orientations about covalent bonds.[1] This type of rotational flexibility is characterized by a *dihedral angle*, which measures the relative orientation of four linked atoms in a molecule, $i - j - k - l$. A dihedral angle for a 4-atom sequence that is not necessarily covalently bonded can also be used for

[1] see Chapter 7, Subsection 7.4.1, for the related definition of *configuration*.

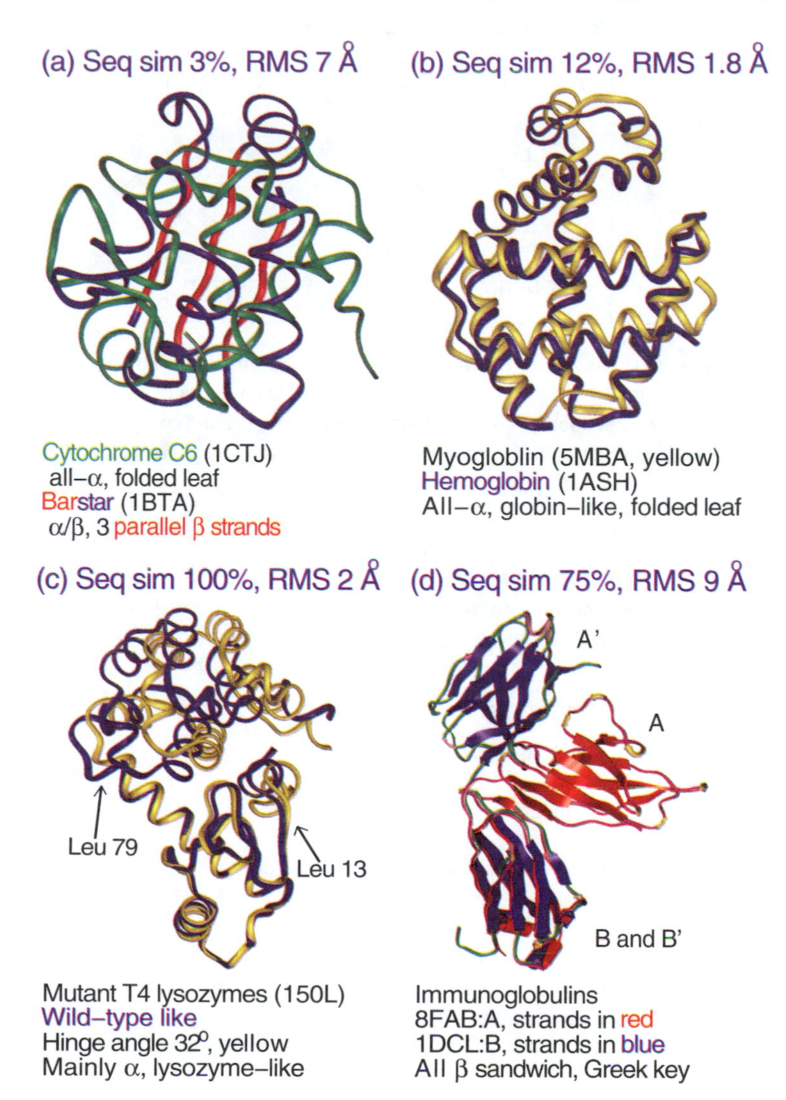

Figure 3.12. Various examples of sequence/structure relationships in proteins: (a) Low sequence similarity (3% for alignment of 72% of the residues) generally implies low structure similarity (**cytochrome C6** versus **barstar**). Still, exceptions are found. For example, in (b), despite low (12%) sequence similarity, there is large structure and function similarity (**hemoglobin** and **myoglobin**); conversely, despite high sequence similarity, there can be structural diversity, due to (c) hinge bending in two **lysozyme mutants** (Met → Ile in residue 6) or (d) different orientation of one of the two subunits in two **immunoglobulins**. The lysozyme mutant displays 5 different crystal conformations, one similar to the wild-type (shown in blue) and others overall very similar except for a different hinge-bending angle (see defining arrows); the form with largest bend (32°) is shown in yellow. The two immunoglobulins differ markedly in tertiary organization due mainly to differences in the linker domain between the A and B subunits of each protein.

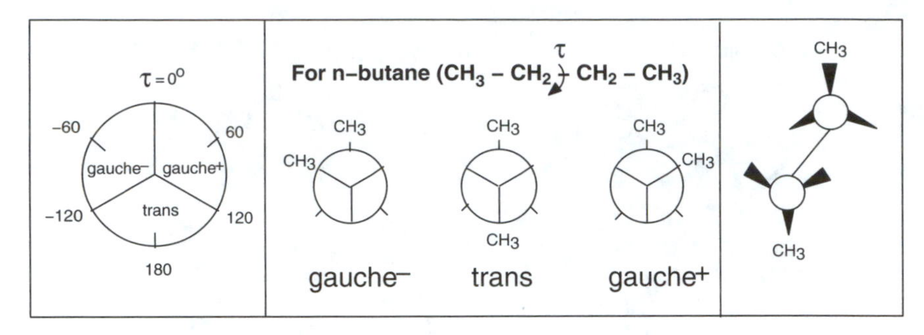

Figure 3.13. *Gauche* (g) and *trans* (t) dihedral-angle orientations for n-butane: (left) classification wheel; (middle) simple Newman projections that illustrate the three favored orientations of the two end methyl groups about the central C–C bond (perpendicular to the plane of the paper); and (right) the *trans* conformation, which has the least steric clashes.

special terms in the potential energy function; see Chapter 8 for examples. See Box 3.4, Figure 3.14, and the equations in Appendix D, under the Addendum to Assignment 8 for a definition of a dihedral angle.

Box 3.4: Dihedral Angle

The dihedral angle τ_{ijkl} defined for a sequence of linked atoms i–j–k–l (Figure 3.14) is the angle between the normal to the plane of atoms i–j–k and the normal to the plane of atoms j–k–l. The sign of τ_{ijkl} is determined by the triple product $(a \times b) \cdot c$, where a, b and c are the interatomic distance vectors for atoms $i \to j$, $j \to k$ and $k \to l$, respectively.

Strictly defined, the related *torsion angle*, $\widehat{\tau}$ is the angle between the two planes defined by i–j–k and j–k–l. Thus, $\tau + \widehat{\tau} = \pi$ (180°). However, the terms torsion and dihedral angle are often used interchangeably. We will often use *dihedral angle* to refer to the numerical value of the angle, and *torsion angle* or *torsional potential* when we discuss general properties of these rotations.

When the dihedral angle is 0°, the four atoms i–j–k–l are coplanar and atoms i and l coincide in their projections onto the plane normal to the j–k bond; this orientation is defined as *cis* or *syn*. When the dihedral angle is 180°, the atoms are coplanar but atoms i and l lie opposite one another in the projection onto the plane normal to the j–k bond; such an orientation is defined as *trans* or *anti*. More generally, angular regions convenient to describe protein and nucleic acid conformations are the following: *cis* ($\approx$ 0°), *trans* ($\approx$ 180°) and $\pm$ *gauche* ($\approx \pm$60°). Another common terminology is: *syn* ($\approx$ 0°), *anti* ($\approx$ 180°), $\pm$ *synclinal* ($\approx \pm$60°), and $\pm$ *anticlinal* ($\approx \pm$120°). See Figure 3.13 for a simple illustration for n-butane.

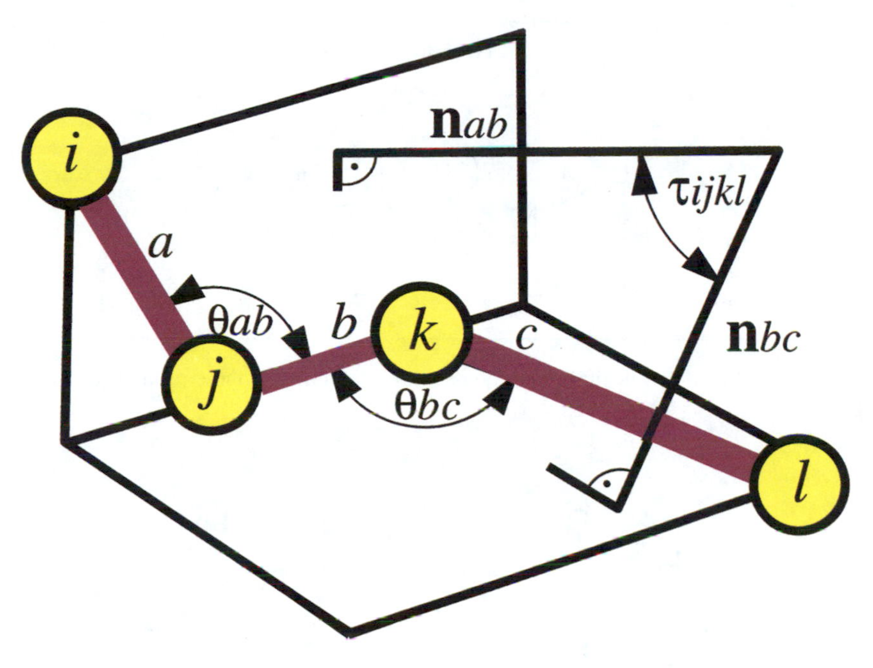

Figure 3.14. Definition of a dihedral angle $\tau_{ijkl} = \cos^{-1}(\mathbf{n}_{ab} \cdot \mathbf{n}_{bc})$, the angle between the two normals spanned by atoms i, j, k and j, k, l.

While the peptide group (Figure 3.3) is relatively rigid — it has 40% double-bond character — there is a great deal of flexibility about each of the single bonds along the backbone, $\{N–C^\alpha\}$ and $\{C^\alpha–C\}$=O. The two angles ϕ and ψ dihedral angles are used to define rotations about the bond between the nitrogen and C^α of the mainchain and between C^α and the carbonyl carbon, respectively (Figure 3.15).

The dihedral angle angle ω defines the rotation about the peptide bond, namely for the atomic sequence C_1^α–{C–N}–C_2^α, where C_1 and C_2 are the α-carbons of two adjacent amino acids. Because of the partial double-bond character of the peptide bond and the steric interactions between adjacent sidechains, ω is typically in the *trans* configuration: $\omega = 180^\circ$.[2] In this orientation, all four atoms lie in the same plane, with the distance between C_1^α and C_2^α as large as possible (see Figure 3.13 for a definition of various dihedral angle orientations).

[2]Non-trivial deviations from planar peptide bonds can be shown by theory and experiment (e.g., as reviewed in [236]). A statistical survey of peptide and protein databases verified that the distribution of rotation angles (or energies associated with peptide bond rotations) follows Boltzmann statistics [237].

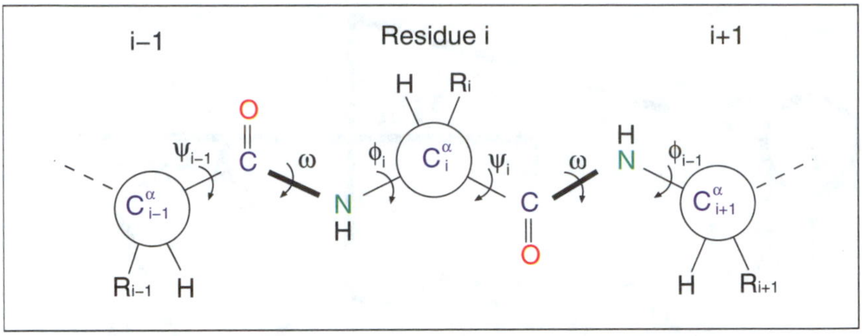

Figure 3.15. Rotational flexibility in polypeptides: definition of the ϕ, ψ and ω dihedral angles.

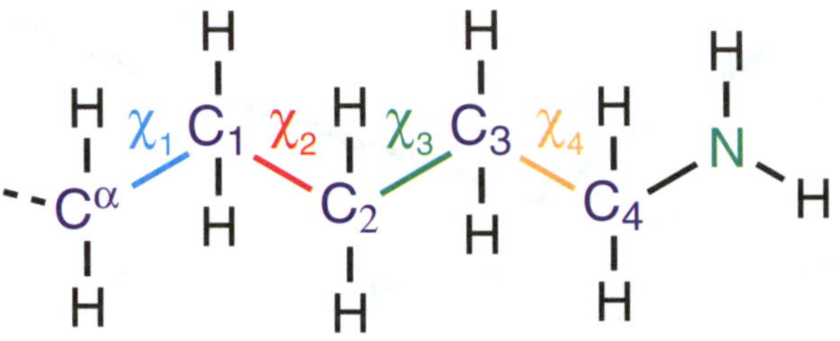

Figure 3.16. Lysine's four rotamers defined by torsional variables χ_1 through χ_4.

3.4.2 Rotameric Structures

Besides the $\{\phi, \psi\}$ flexibility associated with the two backbone bonds involving C^α, multiple conformations are possible for 18 of the 20 amino acids when the sidechain geometries differ (excluded are glycine and alanine). *Rotameric* structures of amino acids (and hence proteins) are those that have the same $\{\phi, \psi\}$ angles but differ in the sidechain conformations. The dihedral angles used to define sidechain rotations are denoted by χ, with subscripts used as needed (see Figure 3.17).

For example, in lysine, whose sidechain has four carbons (see Figure 3.16), dihedral angles χ_1 through χ_4 denote the rotations about bonds C^α–C_1, C_1–C_2, C_2–C_3, and C_3–C_4, respectively (see also Figure 3.17 for other amino acids). Rotameric structures for polypeptides and proteins depend on the environment of the polymer and on the secondary and tertiary structures.

3.4.3 Ramachandran Plots

The feasible combinations of the ϕ and ψ angles are limited due to steric hindrance. That is, only certain combinations are typically observed, with some dependence on residue size and shape. Glycine is unique in its flexibility — it is

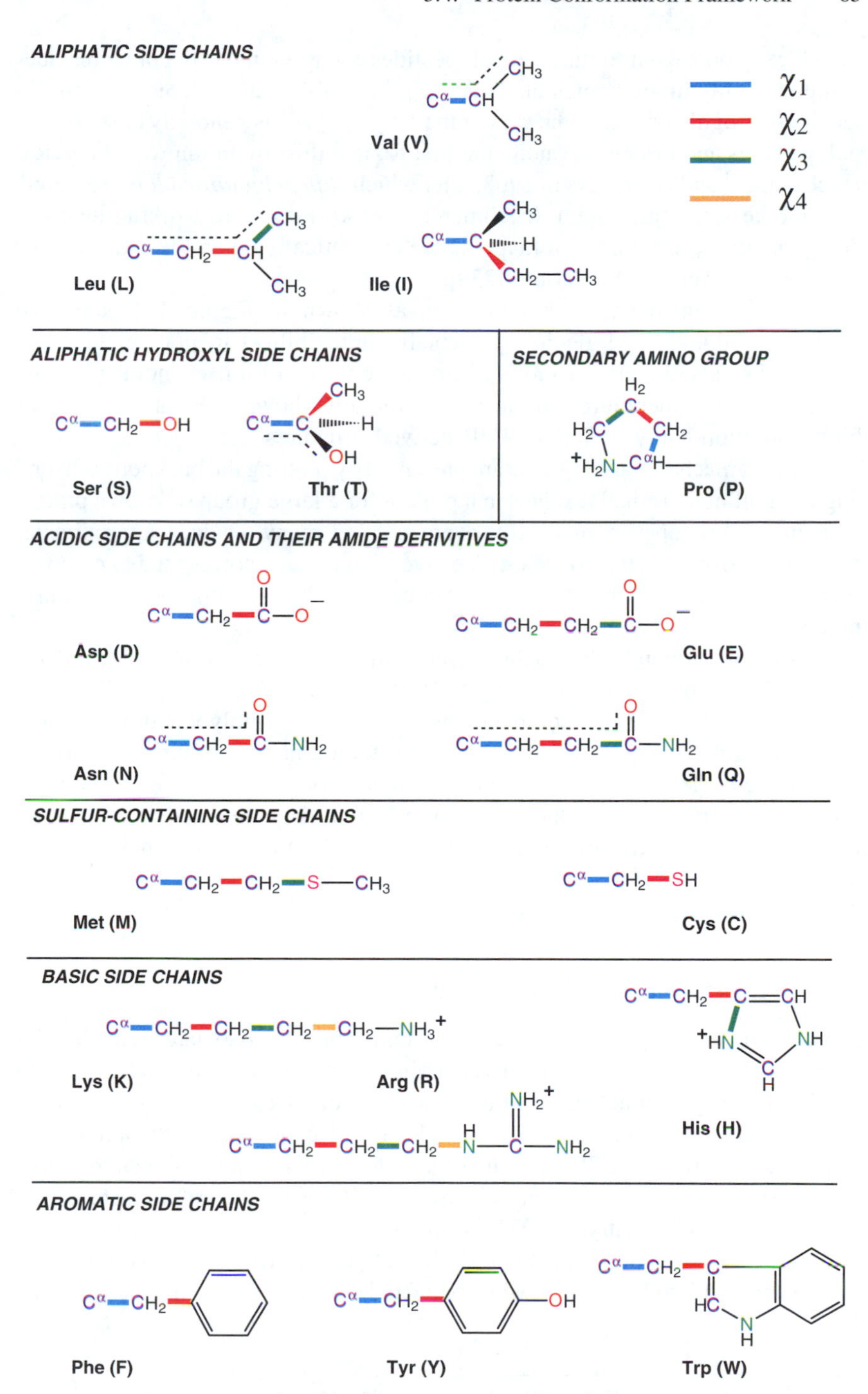

Figure 3.17. Rotameric notation used for 18 of the 20 amino acids is illustrated using different colors for χ_1–χ_4, as shown in the top right key.

therefore a good agent for turns in polypeptides and proteins — but other residues exhibit a highly limited range of sterically-permissible ϕ and ψ combinations. In fact, only roughly one tenth the area of the $\{\phi, \psi\}$ space is generally observed for polypeptides and proteins. Among the first to note this limitation were G.N. Ramachandran[3] and coworkers in 1963, after which *Ramachandran plots* are called. Around the same time, John Schellman and coworkers were working independently along the same lines of mapping the energetically favorable and excluded regions for protein conformations [239].

These diagrams in the $\{\phi, \psi\}$ space, as shown in Figure 3.18, are used to describe this $\{\phi, \psi\}$ flexibility (actually inflexibility) in polypeptides and proteins. See also Figure 3.19 for a comparative view of Ramachandran plots derived from the moderate-resolution X-ray structures shown in Figure 3.18 versus high-resolution X-ray as well as NMR-derived structures.

Often, Ramachandran diagrams are presented by plotting the backbone dihedral angles of all nonterminal residues in a protein for a large group of known protein structures. This superimposed view, averaged over many residues, approximates protein conformational tendencies. The favorable regions correspond to common secondary-structure elements, such as helices and sheets, with finer motifs also noted.

A program written by E. Abola for calculating ϕ-ψ plots is dihdrl.f, available on RCSB: ftp://bnlarchive.rcsb.org/pub/pdb_software/dihdrl/.

In addition to favorable combinations of ϕ and ψ in polypeptides, the side-chain dihedral angle χ_1 has been found to cluster around one of three conformers known as *gauche*$^+$ (or g$^+$, $\chi_1 = +60^\circ$), *gauche*$^-$ (or g$^-$, $\chi_1 = -60^\circ$), and *trans* (or t, $\chi_1 = 180^\circ$). These are the favored orientations about tetrahedral atoms (Figure 3.13). Some dependence of χ_1 on the residue's ϕ and ψ values has also been noted.

3.4.4 *Conformational Hierarchy*

Most natural proteins adopt specific 3D structures that are associated with their biological activity. Of course, proteins are dynamic, but typical thermal fluctuations and local configurational arrangements revolve around a specific globally-folded structure. One of the hallmarks of biomolecular structure is that the amino-acid sequence determines the 3D structure of a protein. This was first shown by Christian B. Anfinsen and his colleagues in the early 1960s [240]. Anfinsen shared the Nobel Prize in Chemistry in 1972 for his work on ribonuclease — connecting the amino acid sequence to the biologically active conformation — with Stanford Moore and William H. Stein — who connected ribonuclease's chemical structure

[3]For lovers of scientific history, the 1998 biography of this renowned Indian molecular biophysicist (1922–2001) is recommended [238]. His peppered poetry is an added bonus. (My favorite poem is number 9, on Superhelical Twisting and Replication of D N A [238, p. 159].)

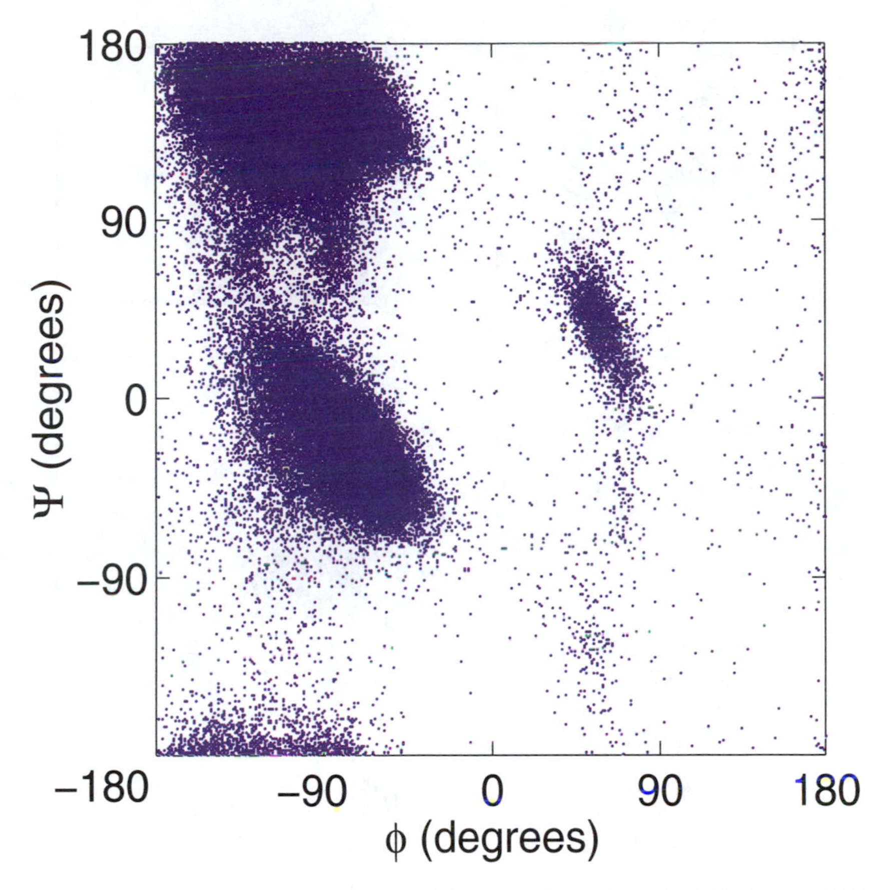

Figure 3.18. Ramachandran plots, obtained from a subset of the PDB40 dataset [232], corresponding to X-ray protein structures with resolution of 2.5 Å or better (470 proteins, 95778 total residues plotted, with proline and glycine excluded).

to its catalytic activity.[4] In Anfinsen's work, the protein ribonuclease was denatured by destroying its hydrogen-bonding network as well as intrinsic disulfide bonds. The researchers observed that the protein spontaneously refolded into its native state in a short time, regaining all its enzymatic activity. Of course, we recognize now that accessory chaperone molecules may be necessary to assist in the folding of many large proteins *in vivo*, as discussed in Chapter 2.

Four basic levels are used to describe protein structure:

---

[4]Readers are invited to browse the electronic museum of Profiles in Science at www.profiles.nlm.nih.gov/ for a glimpse not only of Anfinsen's scientific activities but also of his other hobbies and interests.

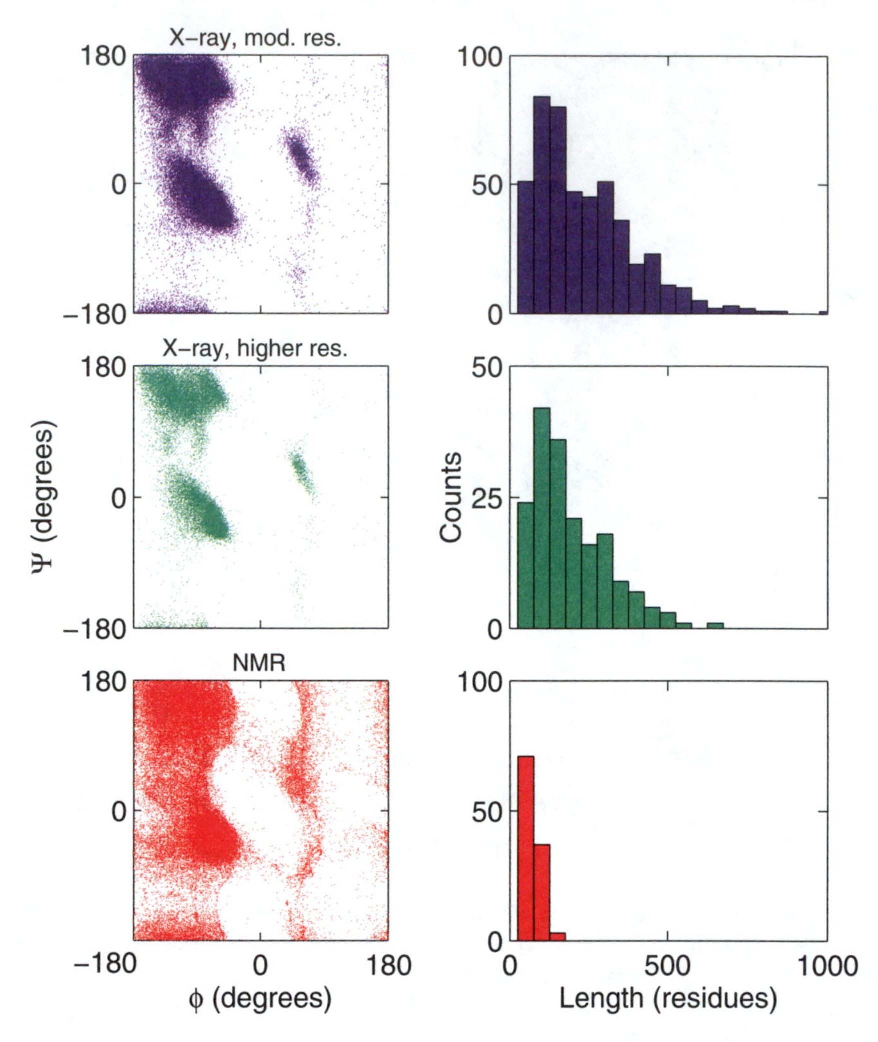

Figure 3.19. Three sets of Ramachandran plots based on the PDB40 dataset [232], corresponding to: (top) X-ray protein structures with resolution of 2.5 Å or better (470 proteins, 95,778 total residues plotted, with proline and glycine excluded); (middle) X-ray protein structures with resolution of 1.8 Å or better (183 proteins, 29,758 total residues plotted, proline and glycine excluded); and (bottom) NMR-derived structures (113 proteins, 84,719 total residues plotted, proline and glycine excluded). For each subset of structures, the length distribution is also shown.

- *primary structure* — the sequence of amino acids;
- *secondary structure* — regular local structural patterns such as α-helices and β-sheets, or combination motifs thereof (*supersecondary structure*);

- *tertiary structure* — the 3D arrangement of all atoms in the polypeptide chain in space; and
- *quaternary structure* (used for large proteins with independent subunits) — the complete 3D interaction network among the different subunits.

The next chapter describes in turn the secondary and supersecondary, tertiary, and quaternary structure of proteins.

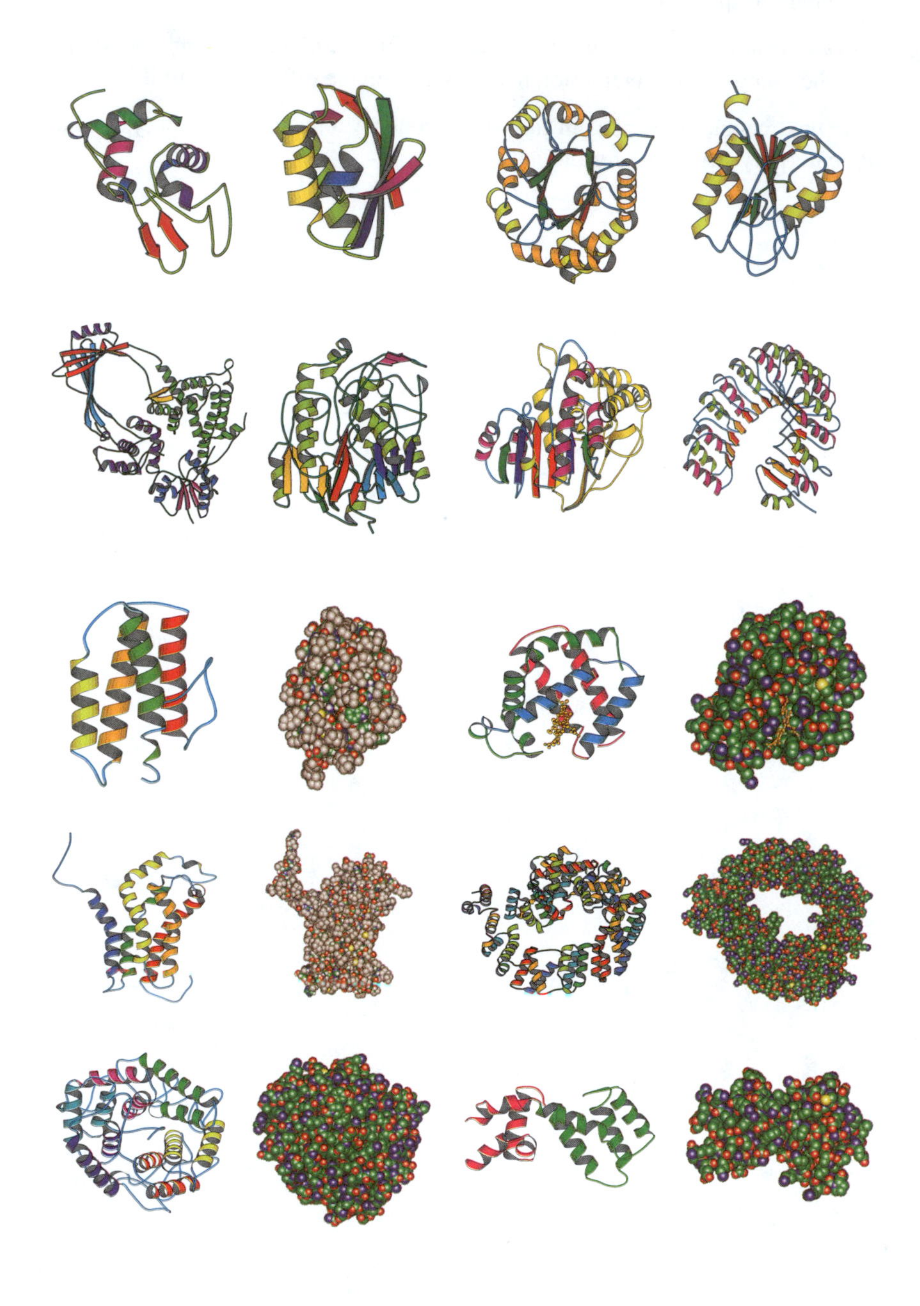

4
Protein Structure Hierarchy

Chapter 4 Notation

SYMBOL	DEFINITION
α_R	classic right-handed α-helix
α/β	protein class
$\alpha+\beta$	protein class
β-sheet	aggregating amino-acid strands
β_2	hairpin motif
β_4	Greek key motif
C^α	α-Carbon
π	helix form looser than α_R
ϕ	$\{N\text{–}C^\alpha\}$ rotation about peptide bond
ψ	$\{C^\alpha\text{–}C\}$=O rotation about peptide bond
3_{10}	helix form tighter than α_R

Try to learn something about everything and everything about something.

Thomas Henry Huxley (1825–1895).

4.1 Structure Hierarchy

The complexity of protein structures requires a description of their structural components. This chapter describes the elements of protein secondary structure — regular local structural patterns — such as helices, sheets, turns, and loops. Helices and sheets tend to fall into specific regions in the $\{\phi, \psi\}$ space of the Ramachandran plot (see Figures 3.18 and 3.19). The corresponding width and shape of each region reflects the spread of that motif as found in proteins.

Following this description of each secondary structural element, we discuss the basic four *classes* of protein supersecondary or tertiary structure (the 3D spatial architecture of a protein), namely α-proteins, β-proteins, α/β-proteins, and $\alpha+\beta$-proteins. This is followed by a presentation of the *fold* motifs for each such class. Classes and folds are at the top of protein structure classification, as introduced in the last section. Describing these folds and structural motifs is far from an exact science, so variations in some of these aspects are common.

4.2 Helices: A Common Secondary Structural Element

4.2.1 *Classic α-Helix*

In the classic, right-handed α-helix (α_R), a *hydrogen-bonding* network connects each backbone carbonyl (C=O) oxygen of residue i to the backbone hydrogen of the NH group of residue $i + 4$ (see Figure 4.1). This hydrogen-bonding provides substantial stabilization energy.

The regular spiral network of the α-helix is ubiquitous in proteins. It is associated with a $\{\phi, \psi\}$ pair of about $\{-60^\circ, -50^\circ\}$. The resulting helix has 3.6 residues per turn, and each residue occupies approximately 1.5 Å in length. The helix may be curved or kinked depending on the amino acid sequence, as well as on solvation and overall packing effects. Such distortions are reflected by the $\{\phi, \psi\}$ distribution around the α_R region in typical Ramachandran plots. **Hemoglobin**, **myoglobin**, **bacteriorhodopsin**, **human lysozyme**, **T4 lysozyme**, **Trp repressor**, and **repressor-of-primer (Rop)** are all examples of proteins that are virtually entirely α-helical. See Figures 4.2 and 4.3 for illustrations of such α-proteins (see below) and Figure 3.10 for Rop.

An α-helix is associated with a dipole moment: the amino terminus of the helix has a positive charge and the carboxyl end has a negative charge clustered about it. Thus, residues that are negatively charged on the amino end and positively-charged on the carboxyl end stabilize the helix; residues with the opposite charge allocation destabilize the helix.

Experimental and theoretical work has shown that both intrinsic and extrinsic (inter-residue interactions) factors are important for helix formation in proteins. Residues with restricted sidechain conformations, due to long or bulky groups, are poorer α-helix participants than other residues. Glutamine, methionine, and

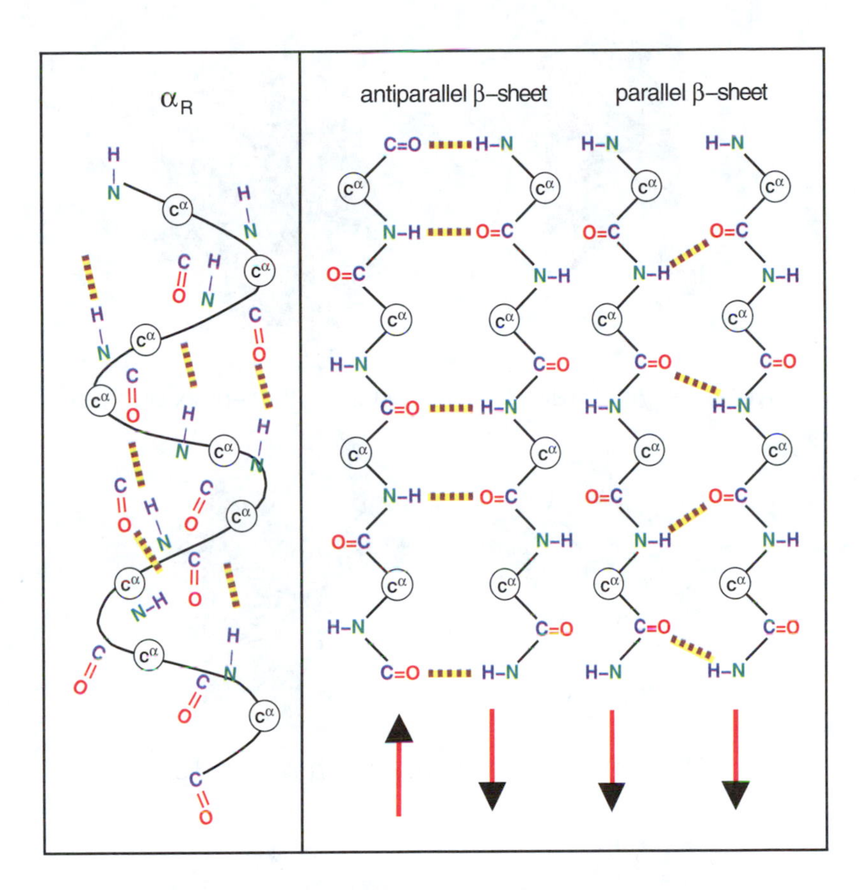

Figure 4.1. Hydrogen bonding patterns in the classic α-helix (α_R), with the ribbon tracing the α-carbons (left), anti-parallel β-sheet (middle), and parallel β-sheet (right).

leucine favor α-helix formation, while valine, serine, aspartic acid, and asparagine tend to destabilize α-helices (e.g., due to steric and electrostatic considerations).

4.2.2 3_{10} and π Helices

There are more common variants of the α-helix motif that are typically not stable in solution but can play a part in overall protein structure. These include the tighter 3_{10} and looser π helices, with $\{\phi, \psi\}$ angles around $\{-50^\circ, -25^\circ\}$ and $\{-60^\circ, -70^\circ\}$, respectively.

The tighter 3_{10} helix of three residues per turn (instead of 3.6 in the classic α-helix) involves hydrogen bonds between residues i and $i+3$ instead of i and $i+4$ as in α_R. There are 10 atoms within the hydrogen bond; hence the nomenclature

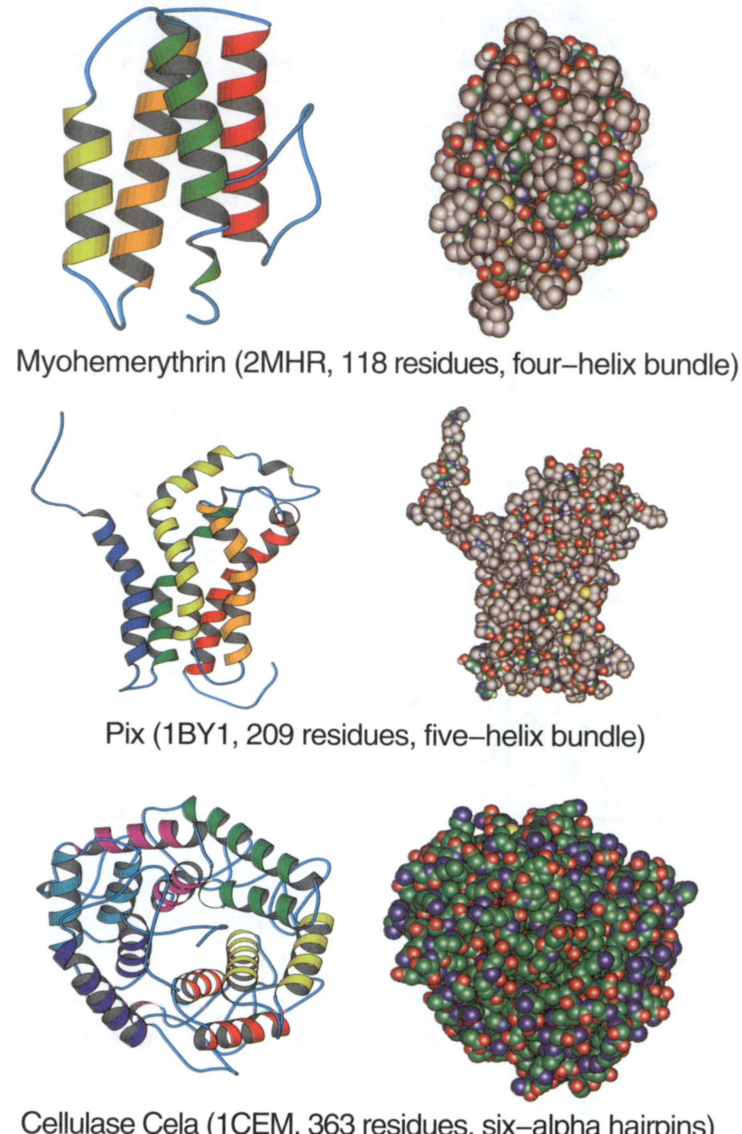

Figure 4.2. Examples of α proteins: **myohemerythrin**, **pix**, and **cellulase cela**.

3_{10}. The more loosely coiled π helix has hydrogen bonds between residues i and $i+5$ of the polypeptide.

Because of their close packing, 3_{10} helices generally form for a few residues only, often at the C-terminus end of classic α-helices where the helix tends to tighten. Similarly, the π helix occurs rarely since the backbone atoms are so loosely packed that they leave a hole.

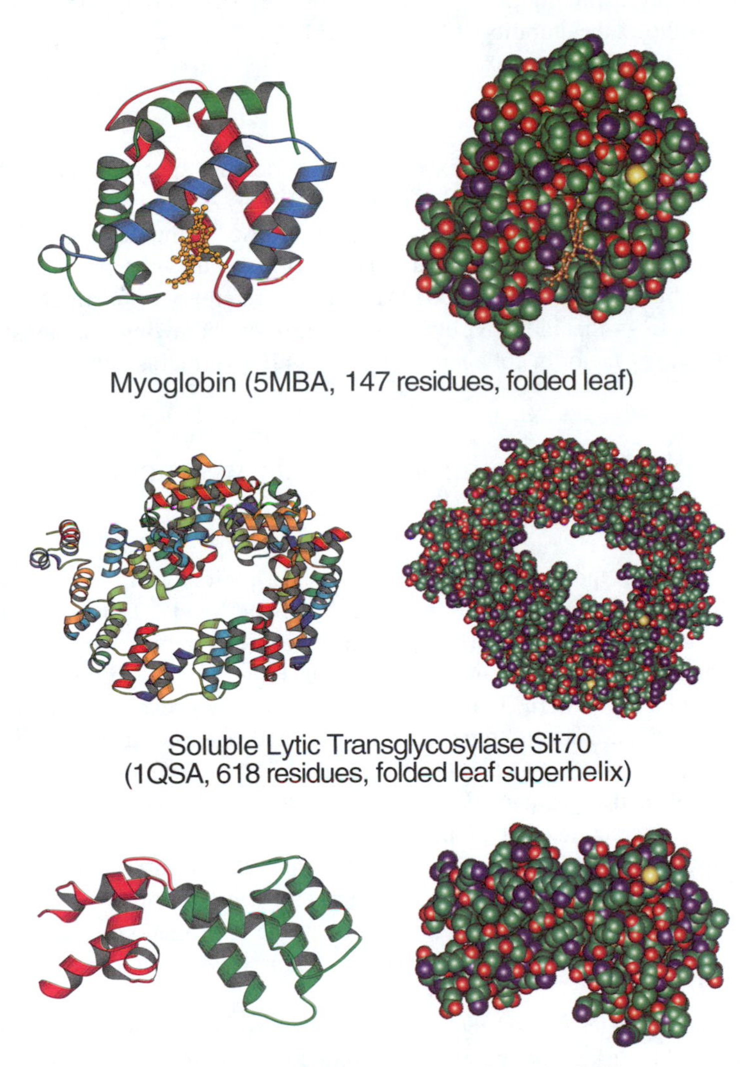

Figure 4.3. Examples of α proteins: **myoglobin**; **soluble lytic transglycosylate** protein of bacterial muramidase, in the N-terminal region of the enzyme muramidase in bacterial cell walls; and **guanine nucleotide-binding protein**, an irregular α-helical protein with a fold containing a 4-helix bundle with left-handed twist.

4.2.3 *Left-Handed α-Helix*

A left-handed α-helix is theoretically possible, with $\{\phi, \psi\} = \{+60^\circ, +60^\circ\}$. However, this motif is generally unstable. The chirality preference for α-helices follows the chirality of L-amino acids.

4.2.4 *Collagen Helix*

The triple-stranded **collagen** helix is often considered a specific secondary element. It is associated with $\{\phi, \psi\} = \{-60^\circ, +125^\circ\}$. A large body of structural data has suggested that extensive hydration networks in the collagen triple helix (among the protein residues and with water molecules) are responsible for collagen stability and assembly (see [241, 226] and references cited therein). A recent hypothesis — that inductive effects by electron-withdrawing residue moieties might play a key factor in collagen's stability [242] — remains to be proven.

4.3 β-Sheets: A Common Secondary Structural Element

Another common motif is a β-sheet. These sheet regions form by aggregating amino-acid strands, termed β-strands, via hydrogen bonds. Typical lengths of β-strands are 5–10 residues. The aggregation can occur in a parallel or anti-parallel orientation of the strands, as shown in Figure 4.1, each with a distinct hydrogen bonding pattern. Each such β-strand has two residues per turn and can be considered a special type of helix. The hydrogen-bond crosslinking between strands — alternating C=O $\cdots$ H–N and N–H $\cdots$ O=C — is such that the sheet has a pleated appearance. Thus, in comparison to α-helices, β-sheets require connectivity interactions that are much longer in range.

For parallel β-sheets, $\phi \approx -120^\circ$ and $\psi \approx +115^\circ$. For anti-parallel β-sheets, $\phi \approx -140^\circ$ and $\psi \approx +135^\circ$. As for α-helices, the ring of proline does not adapt well into β-sheets since it cannot participate in the hydrogen-bond network between strands. Valine, isoleucine, and phenylalanine have been found to enhance β-sheet formation.

Often, at the edges of β-sheets, an additional residue that cannot be included in the normal hydrogen-bonding pattern produces a *β-bulge* of the extra residue. Figures 4.4 and 4.5 show the structures of proteins that are mostly β-sheets.

4.4 Turns and Loops

Other common structural motifs in proteins are turns and loops.

Turns (also called β-turns or reverse turns) occur in regions of sharp reversal of orientation, such as the junction of two anti-parallel β-strands. Such motifs are

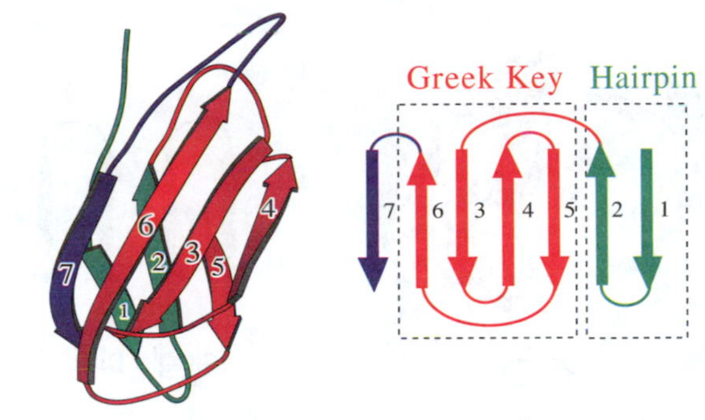

Fibronectin (1TTG, 94 residues, β sandwich)

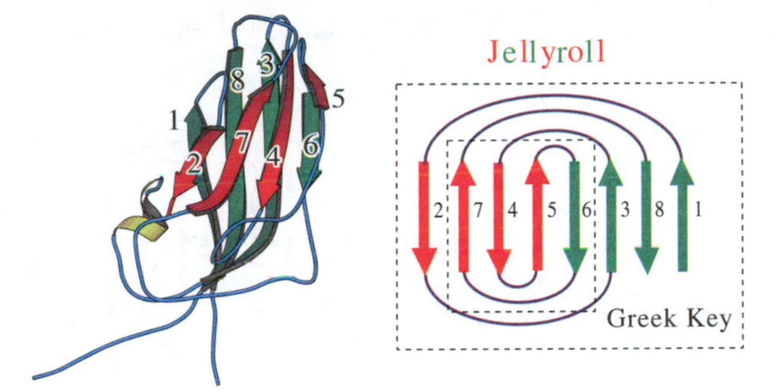

Satellite Panicum Mosaic Virus (1STM, 157 residues)

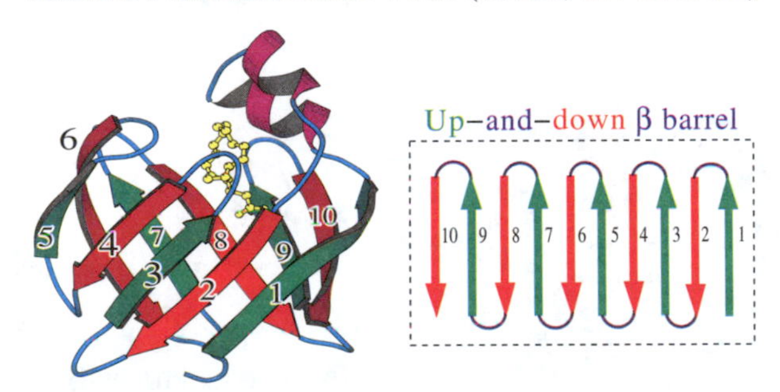

Fatty Acid Binding Protein (1HMS, 132 residues, β barrel)

Figure 4.4. Examples of β proteins and common motifs: **fibronectin**, β-sandwich illustrating hairpin and Greek key motifs; coat protein of **satellite panicum mosaic virus**; and **fatty acid binding protein**, up-and-down β-barrel.

classified as turns based on distance criteria (e.g., the C^α atoms of residues i and $i+3$ are less than 7 Å distant).

Loops occur often in short (five residues or less) regions connecting various motifs. Loop regions that connect two adjacent anti-parallel β-strands are known as hairpin loops. Short hairpin loops are found at protein surfaces.

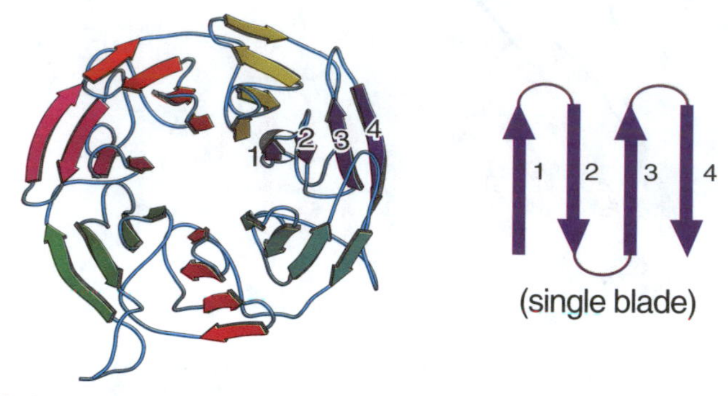

Galactose Oxidase (1GOF, 639 residues, 7–bladed β–propeller)

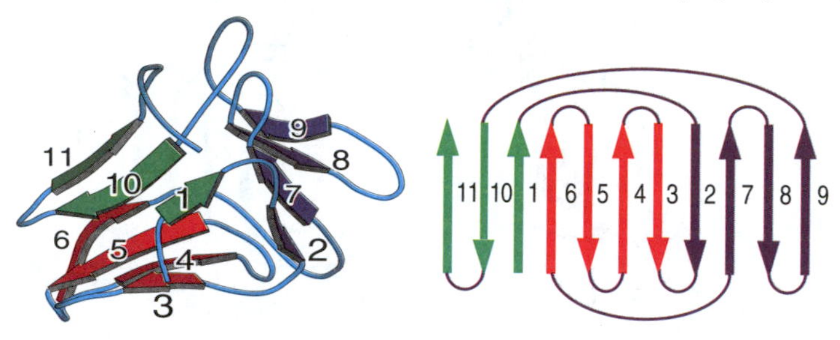

Agglutinin (1BWU, 106 residues, β–prism)

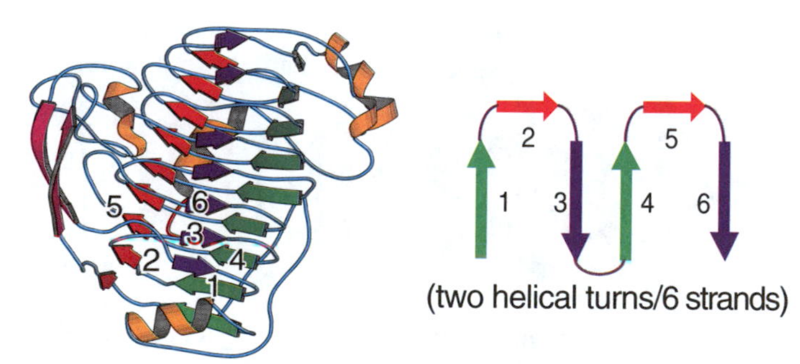

Pectin Lyase A (1IDK, 359 residues, right–handed β–helix)

Figure 4.5. Examples of β proteins and common motifs: **galactose oxidase**, **agglutinin**, and **pectin lyase A**.

The majority of turns and loops lies on the protein surface because of solvation considerations. They are important elements that allow, and possibly drive, protein compaction. Most loops interact with solvent and are highly hydrophilic (water soluble). Since protein core regions are more stable than short connective elements of helices and strands, evolutionary differences among homologous sequences are often localized to loop and turn regions. Non-coding regions (introns) are similarly found in genes that correspond to loops and turns in protein structures.

4.5 Formation of Supersecondary and Tertiary Structure

4.5.1 Complex 3D Networks

The secondary structural elements described above often combine into simple motifs that occur frequently in protein structures. Such motifs (or folds) are also called *supersecondary structure*. Examples are β hairpin (β-loop-β units), Greek key, and β-α-β units (see below).

Supersecondary and tertiary structures of proteins can be described by the specific topological arrangement of the secondary or supersecondary structural motifs. Although the 3D architecture of a protein can be a complex composite of various secondary and supersecondary structural motifs, the majority of the residues — roughly 90% — are found to be involved in secondary structural elements. In fact, on average 30% of the residues are found as helices, 30% as sheets, and 30% as loops and turns. Proteins can be monomeric or multimeric, with subunits that fold in a dependent or independent manner with respect to other domains.

The different polypeptide domains can be connected by disulfide bonds, hydrogen bonds, or the weaker van der Waals interactions. Tertiary structure is also affected by the environment. Hydrogen bonding with solvent water molecules can stabilize the native conformation, and the salt concentration can affect the compact arrangement of the folded chain.

Molecular graphics packages often display the secondary structural motifs clearly by using ribbon diagrams in which helices are depicted as coils and sheets are shown as twisting planes with arrows (see Figures 4.2, 4.3, 4.4, and 4.5, for example).

4.5.2 Classes in Protein Architecture

Based on the known protein structures at atomic resolution, four major *classes* can be used to describe the arrangement in space of the various secondary structural elements (or domains) of polypeptides:

- α-proteins — proteins which form compact aggregates by packing mainly α-helices, often in a symmetric arrangement around a central hydrophobic core;
- β-proteins — proteins which pack together mainly β-sheets, with adjacent strands linked by turns and loops and various hydrogen-bonding networks

formed among the individual strands, often resulting in layered or barrel structures;

- α/β proteins — proteins that are folded with alternating α-helices and β-strands, often forming layered or barrel-like structures; and
- $\alpha + \beta$ proteins — proteins that combine largely-separated (i.e., non-alternating) helical and strand regions, often by hairpins.

Figures 4.2–4.7 illustrate members of each such class.

Recent statistics for PDB protein structures reveal that approximately 23% belong to the all-α class, 15% to all-β, 17% to α/β, and 30% to $\alpha + \beta$. The remaining 15% includes multidomain proteins, membrane and cell-surface proteins and peptides, and small proteins. For updated statistical information, check scop.mrc-lmb.cam.ac.uk/scop/, click on Statistics 'here'. (See last section of this chapter for SCOP description).

Other classes are defined for proteins found on membrane and cell surfaces, small and/or irregular proteins with multiple disulfide bridges, proteins with multiple domains or with bound ligands, and more. Included, for example are small proteins like **rubredoxin** (PDB entry 1rb9), various zinc-finger and metal-binding proteins like the cysteine-rich domain of **protein kinase** (PDB entry 1ptq), disulphide-rich proteins like **sea anemone toxin k** (PDB entry 1roo), and **proteinase inhibitor PMP-C** (PDB entry 1pmc).

4.5.3 Classes are Further Divided into Folds

The protein *classes* are further divided into observed *folds* for protein structures. Folds describe the arrangement of secondary structural elements and/or chain topology. Each protein class has common folds, as described in turn in the next three sections.

4.6 α-Class Folds

In the α class of proteins (Figures 4.2 and 4.3), bundles, folded-leafs, and hairpin arrays are major fold groups.

4.6.1 Bundles

Bundles occur when α-helices pack together to produce a hydrophobic core. Typically, an array of α-helices is roughly aligned around a central axis. The bundle can be right or left-handed depending on the twist that each helix makes with respect to this axis. A *coiled coil* (two intertwined helices) can be a building block of these bundles. A simple example of a coiled coil is seen in the **DNA-binding leucine zipper protein** shown in Figure 6.5 of Chapter 6.

Among the α-protein bundles, the four-helix bundle motif (often written as α_4) is common, as in **myohemerythrin**, Figure 4.2, and **Rop** (a small RNA-binding protein involved in replication), Figure 3.10. Other α_4 proteins are **ferritin** (a storage molecule for iron in eukaryotes), **cytochrome c′** (heme-containing electron carrier), the **coat protein of tobacco mosaic virus**, and **human growth hormone**.

Multi-helical bundles are also observed in α-proteins; 3–6 and 8-helix aggregates are more frequent than others. Figure 4.2 shows a 5-helix bundle for the transport protein **pix**.

4.6.2 *Folded Leafs*

Complex packing patterns involving layered arrangements are often features of long α-proteins. For example, in folded-leaf folds, a layer of α-helices wraps around a central hydrophobic core. Like bundles, such multihelical assemblies (usually 3 or more) pack together as well as form layers. The longest helices are usually in the center, and often the arrangement contains internal pseudosymmetry.

The globin fold of **myoglobin** (Figure 4.3) shows such a compact arrangement of a folded-leaf arrangement formed by 8 helices, leaving a pocket for heme binding. **Cytochrome C6** in Figure 3.12 also displays a folded leaf.

A more complex layered topology is the two-layered ring structure of one α-helical domain in the N-terminal region of the enzyme **muramidase** in bacterial cell walls, **soluble lytic transglycosylate** (Figure 4.3). It is built from 27 α-helices, arranged in a two-layered superhelix, leaving a large central hole, thought to be important in its catalytic activity.

4.6.3 *Hairpin Arrays*

Other α-helix assemblies that cannot be described by bundle or folded-leaf motifs are often described as hairpin arrays (arrays of α-helix /loop /α-helix motifs). The calcium binding protein **calmodulin**, for example, has a helix/loop/helix motif where the loop region between two helices binds calcium (see Figure 3.11). Figure 4.2 also shows **cellulase cela**, a toroid-like circular array composed from 6 hairpins.

An irregular α protein from an all-α subdomain of the regulator of G-protein signaling 4, namely **guanine nucleotide-binding protein**, is also shown in Figure 4.3. This protein's motif contains a 4-helical bundle with left-handed twist and up-and-down topology.

4.7 β-Class Folds

Proteins in the β-class display a flexible and rich array of folds, as seen in Figures 4.4 and 4.5. Various connectivity topologies can exist within networks of *parallel*, *anti-parallel*, or *mixed* β-sheets that twist, coil and bend in various ways. Indeed, note the much wider regions of the Ramachandran plot associated with β-sheets than with α-helices (Figs. 3.18 and 3.19).

4.7.1 Anti-Parallel β Domains

To describe these intriguing folds, it is simpler to begin with folds associated with the large subclass of β-proteins made exclusively of *anti-parallel β* domains. Such proteins tend to form distorted barrel structures. They can be described in terms of building blocks of two-strand, four-strand, eight-strand units, etc., as follows.

Two-Strand Units

The basic two-strand unit, the hairpin (denoted β_2), involves a β/loop/β motif. It has adjacent anti-parallel β-strands linked head-to-tail by a turn or loop; see the β-strands connected as $1 \to 2$ or $4 \to 5$ for the head-to-tail direction in **fibronectin** in Figure 4.4.

Four-Strand Units

Proceeding to connections of four β-strands, there are 24 ways to combine two β-hairpin units to form a 4-stranded anti-parallel β-sheet unit. The most common topology is the Greek key (or β_4). The four strands of a Greek key produce a β-sandwich through the head-to-tail connectivity of $3 \to 4 \to 5 \to 6$, as shown in the diagrams for **fibronectin** and **satellite panicum mosaic virus** in Figure 4.4. The β-strands in these illustrations are labeled according to their connectivity in the protein.

Eight-Strand Units

Correspondingly, there are many more ways to combine a larger number of β-strands from motifs of smaller systems. The two most common folds for 8 anti-parallel β-strands are jellyrolls (β_8) and up-and-down β-sheet.

- The appetizing jellyroll is illustrated in Figure 4.4 for the β-sandwich coat protein of **satellite panicum mosaic virus**. It is a network of 8 anti-parallel β-sheets with the connectivity $1 \to 2 \to 3 \to 4 \to 5 \to 6 \to 7 \to 8$, where strands are *shuffled* when viewed in the diagram left to right. Note the Greek key submotif in the $4 \to 5 \to 6 \to 7$ subunit of the jellyroll.
- In the up-and-down β-sheet, each β-strand is connected to the next by a short loop. It has the simpler connectivity $1 \to 2 \to 3 \to 4 \to 5 \to 6 \to$

$7 \to 8$, where strands 1 through 8 are written left to right (no shuffling required). Figure 4.4 shows this fold for **fatty acid binding protein** ($1 \to 2 \to \cdots \to 9 \to 10$).

4.7.2 *Parallel and Antiparallel Combinations*

More generally, β-protein topologies made of composites of parallel and anti-parallel strands usually form layered or barrel structures. The sandwich, barrel, and β-propeller are three general reference fold groups.

Sandwiches and Barrels

In sandwiches, β-sheets twist and pack with aligned strands, whereas in barrels the sheets twist and coil so that often the first strand is hydrogen-bonded to the last strand to produce closed structures. See the sandwich protein **fibronectin** and barrel in **fatty acid binding protein** in Figure 4.4. The immunoglobulins in Figure 3.12(d) are also β-sandwiches where seven strands form two sheets.

Propellers

In β-propeller folds, 6 to 8 β-sheets, each with 4 anti-parallel and twisted strands, arrange radially to resemble a propeller. The 7-bladed propeller of **galactose oxidase** is shown in Figure 4.5.

Other β-Folds

Other β-folds include β-prisms (3 sheets that pack around an approximate 3-fold axis), barrel/sandwich hybrids (2 β-sheets, each shaped as a half barrel and packing like a sandwich), and β-clips (3 two-stranded β-sheets, forming a long hairpin folded upon itself in two locations). **Agglutinin** in Figure 4.5 shows a β-prism fold.

Recently, β-helix structures have been identified [243]. The polypeptides contain up to 16 helical turns, each of which contains 2 or 3 β-sheet strands. Unlike the β-sandwiches, the β-sheet strands of a β-helix have little or no twist. Most such β-helix folds known to date are right-handed, as seen in **pectin lyase A** in Figure 4.5. The β-helix motif has been suggested to occur in the infectious scrapie prion protein [197].

4.8 α/β and α+β-Class Folds

Even more diverse fold patterns are known for the α/β class of proteins depending on the sheet types (parallel, anti-parallel, or mixed network) and the location of the helices (exterior, interior, or on both faces) with respect to the sheet assembly.

We can broadly classify three fold motifs in this class (see Figure 4.6): barrels — closely packed β-strands (usually 8) with α-helices on the exterior, open structures made of twisted β-sheets (parallel or mixed) surrounded by α-helices on both the exterior and interior, and leucine-rich motifs of curved β-sheets with exterior α-helices in leucine-rich regions.

4.8.1 α/β Barrels

A classic example of a barrel core is the barrel structure of **triosephosphate isomerase** (TIM), an $(\alpha/\beta)_8$ topology (see Figure 4.6). The TIM barrel is one of the most common polypeptide-chain folds known today. TIM's 8 parallel β-strands coil to form a central core, and its 8 α-helices pack along the exterior. The central barrel 'mouth' is the active site of the protein.

4.8.2 Open Twisted α/β Folds

An example from the highly-variable class of open twisted α/β structures is **flavodoxin** (Figure 4.6). Note that its helices lie on opposite sides of the β-sheet. Typically, the active sites of proteins in this fold class are near the loop regions that connect β-strands to α-helices. Another member of this class is **maltate dehydrogenase**, characterized by the *Rossmann* fold (named after its discoverer Michael Rossmann). This $(\beta\alpha\beta\alpha\beta)_2$ topology has a central, parallel twisted β-sheet surrounded by α-helices and/or loops. It is an important motif in proteins that bind to nucleic acids.

4.8.3 Leucine-Rich α/β Folds

Ribonuclease inhibitor is an example in the leucine-rich class of α/β folds. Its *horseshoe* structure is formed by homologous repeats of right-handed β-loop-α units (see Figure 4.6). The 17 parallel β-strands lie on the inside of this horseshoe, with the 16 α-helices clustering on the outside. The leucine residues present in all three segments of the repeating unit — the β-strand, the loop, and the α-helix — pack snuggly together to form a hydrophobic core between the β-strand and α-helix regions.

4.8.4 $\alpha+\beta$ Folds

Yet more complex fold patterns have been observed for the $\alpha+\beta$ class of proteins (see Figure 4.7). This diversity reflects the various topologies of the subdomains (or layers) as well as the richness of connectivity patterns among them.

4.9 Number of Folds

It has been postulated that the number of folding motifs is finite and that the entire catalog of folds will eventually be known with the rapidly-increasing number of solved globular proteins [144, 244, 245]. Such postulates come from stereochemical considerations — for example, there is a small number of ways to link compactly α-helices and β-strands — database analyses, and statistical sampling approaches.

4.9.1 *Finite Number?*

The exact number of folds has not been determined. Some studies estimate this number to be several thousand, while other yield only several hundred [246, 247] (around 850 total folds in the latter work), so a minimal estimate of around 1000 [248] and the range of 1000-5000 seem reasonable [249]. Only time will tell how many folds Nature has produced, but progress is rapid on this front [248, 250].

Since many computational folding-prediction schemes use known folds for closely-related sequences or closely-related functions of proteins, a finite number of folds suggests that *eventually* we will be able to describe 3D structures from sequence quite successfully!

Zhang and DeLisi estimated in 1998 [247], however, that with the technology available at that time, 95% of the folds will only be determined only in 90 years. They argued that, aside from technological improvements, we should carefully select new sequences for structure determination so as to maximize new fold detection and thereby reduce that time substantially. This is important since the annual number of new folds discovered during 1995–2000 has only averaged around 10%, with even less during 2000-2002. The structural genomics initiatives are expected to change this trend, but the effect of these projects will take time to assess. For updated fold information, see www.rcsb.org/pdb/holdings.html.

4.9.2 *Concerted Target Selection: Structural Genomics*

This new-fold driven solution approach has been a widely-recognized goal in structural genomics [251, 133, 134]. As mentioned in the beginning of Chapter 2, a concerted structural genomics initiative was launched in late 2000, with seven research groups aiming to solve collectively structures of 10,000 representative proteins over the next decade (see November 2000 supplement issue of *Nature Structural Biology*, volume 7). Such structural studies might lead to critical insights into protein-folding mechanisms besides advancing the studies of human diseases and rational drug design. In many cases, protein structure — which can reveal evolutionary relationships better than sequence — can lead to characterization of function. Hence structural genomics is closely related to functional genomics [134] and structural enzymology [252].

To maximize the structural coverage return on effort, Vitkup *et al.* [253] (see also introduction in [254]) estimate that a concerted, optimized effort can be seven times more efficient (i.e., require determination of one seventh of the structures) compared to random target selection. An extrapolation suggests that about 16,000 carefully selected structures will have to be determined to construct representative atomic models for 90% of the proteins, and that this can be done within a decade; a loosely coordinated effort would require a factor of three more structures. Of course, the idea behind structural genomics has limitations, since certain proteins or regions (e.g., membrane proteins, coiled-coil regions, unstructured regions, multi-domain proteins) cannot be easily characterized by typical approaches.

The payoffs of the structural genomics initiatives are already evident from reports at the Keystone symposium on Human Genetics and Genomics held in March/April 2001 [255, 137]: The proteins solved not only have novel folds (or superfamilies) as predicted, but can be used to model by homology functionally-related enzymes. While some structures unexpectedly produce known, common folds like TIM barrels, despite expectations, it is also observed that structures thought to be similar to known ones can lead to different, unanticipated folds. The latter situation — indicating independent structure evolution of two protein active sites — also highlights the notion that function does not always predict structure. Thus, it is important to couple structural genomics initiatives to functional genomics studies.

4.10 Quaternary Structure

Quaternary structures describe complex interactions for multiple polypeptide chains, each independently folded, with possibly other molecules (nucleic acids, lipids, ions, etc.). The interactions are stabilized by hydrogen bonds, salt bridges, and various other complex inter and intramolecular associations in space. The classic example for a quaternary structure is that of the protein **hemoglobin**, which consists of four polypeptide chains. The four subunits, each of which contains an oxygen-binding heme group, are arranged symmetrically.

4.10.1 Viruses

Virus coats are often comprised of many protein molecules and have intriguing quaternary structures. These protein coats envelope the inner domain which consists of infectious nucleic acids. For example, the **poliovirus** — a spherical complex of 310 Å in diameter — has a shell of 60 copies of each of four proteins. The coat of **tobacco mosaic virus** combines 2130 identical protein units, each of 158 residues, arranged in a helix around a coiled RNA structure of 6400 nucleotides. This results in a rod-shaped complex 3000 Å long and 18 Å in diameter.

Figure 4.8 illustrates the structure of the 180-chain **tomato bushy stunt virus** that infects many plants, including tomatoes and cherry trees. Interestingly, virus coats are assemblies of similar proteins rather than one huge protein or combi-

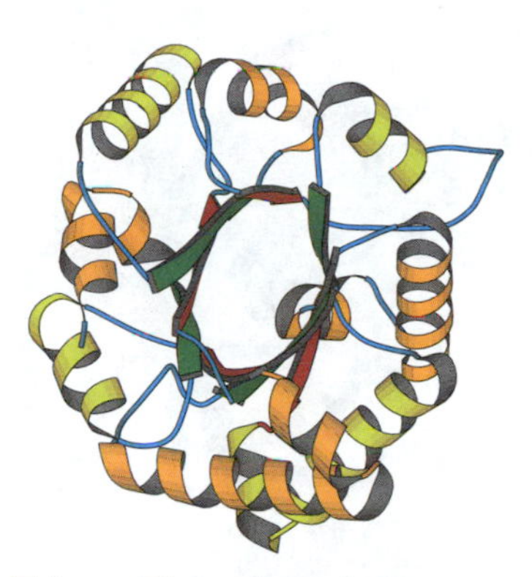

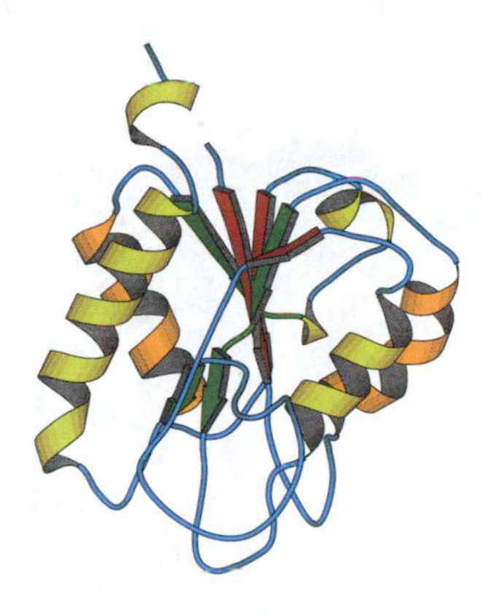

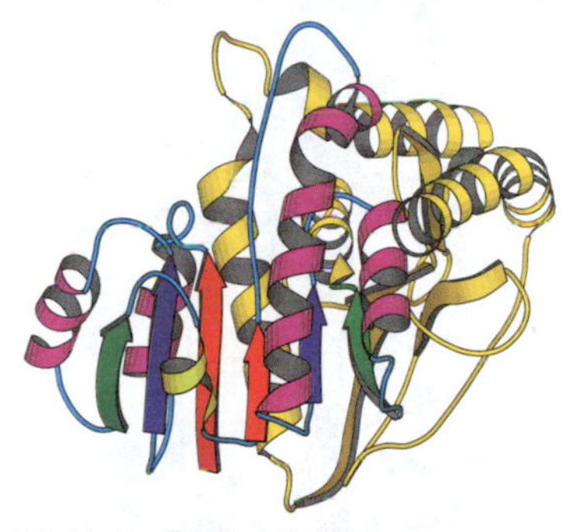

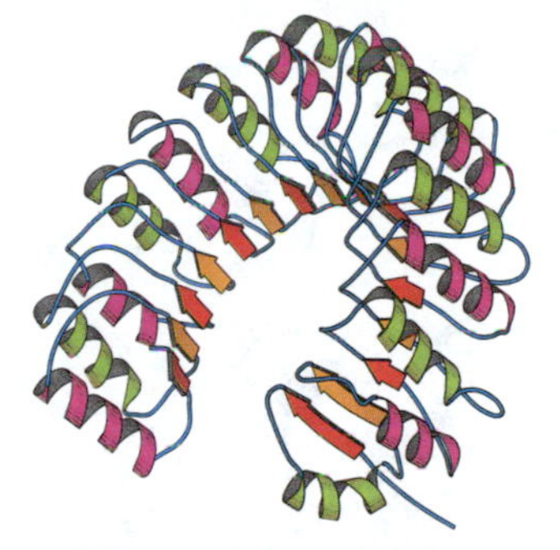

Figure 4.6. Examples of α/β proteins. **TIM** (triosephosphate isomerase) displays an architecture of 8 twisted parallel β-strands which form a barrel surrounded by α-helices. **Flavodoxin**, an electron transport protein that binds to a flavin mononucleotide prosthetic group, displays an open twisted α/β fold made of three layers (2 helices at left, 5 β-strands in the middle, and 2 helices at right). **Maltate dehydrogenase** contains the $(\beta\alpha\beta\alpha\beta)_2$ *Rossmann* fold in the subunit shown. **Ribonuclease inhibitor**, in the leucine-rich class of α/β folds, displays a *horseshoe* structure.

nations of different proteins, because the relatively small amount of viral nucleic acids must encode this protein coat; at the same time, the nucleic acids must be covered completely. Hence a large protein shell consisting of repetitive motifs satisfies both of these criteria.

Among the larger molecular structures determined by X-ray crystallography at moderate resolution (i.e., approaching 3.5 Å) is the core particle of **bluetongue**

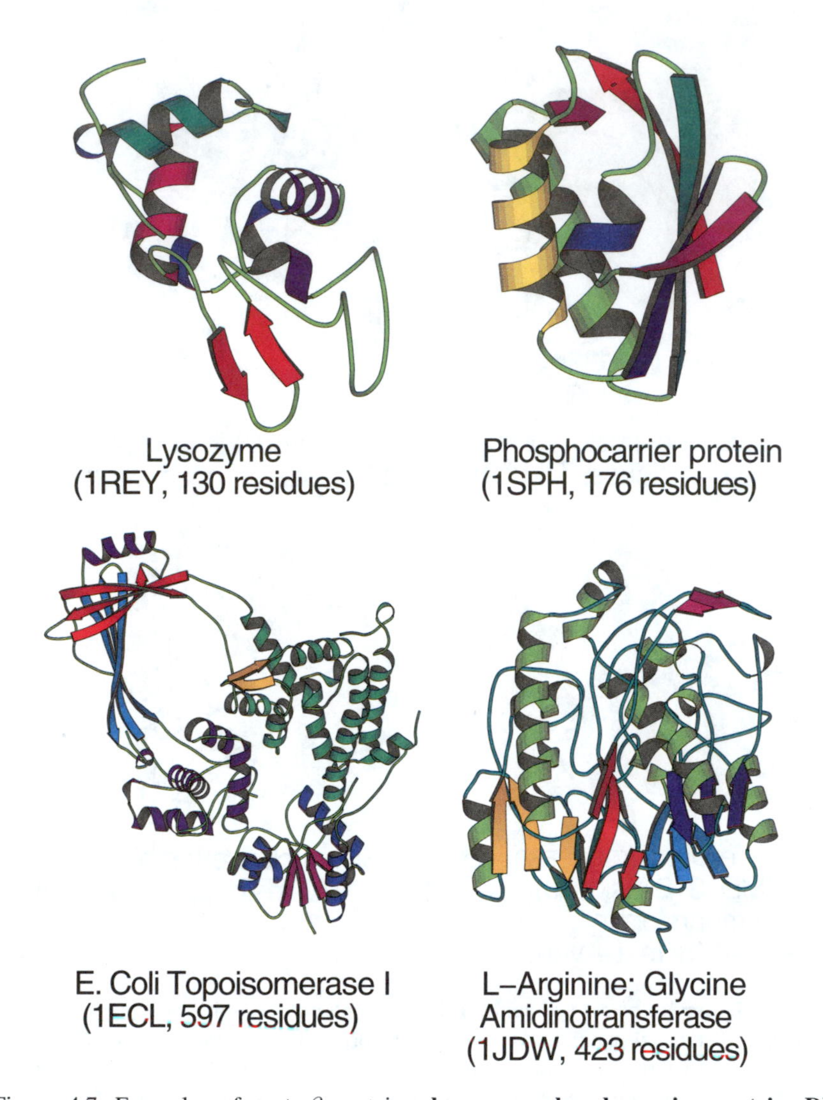

Figure 4.7. Examples of $\alpha + \beta$ proteins: **lysozyme**, **phosphocarrier protein**, **DNA topoisomerase I**, and **glycine amidinotransferase**.

virus, an agent of disease in both plants and mammals. Its transcriptionally active compartment measures 700 Å in diameter and is composed of two principal structural proteins that assemble in two layers, a core and a subcore, together encapsulating the RNA genome (10 segments of doubled-stranded RNA, ~19,000 base pairs total). The crystal structure revealed how these approximately 1000 protein components self-assemble through a complex mixture of packing mech-

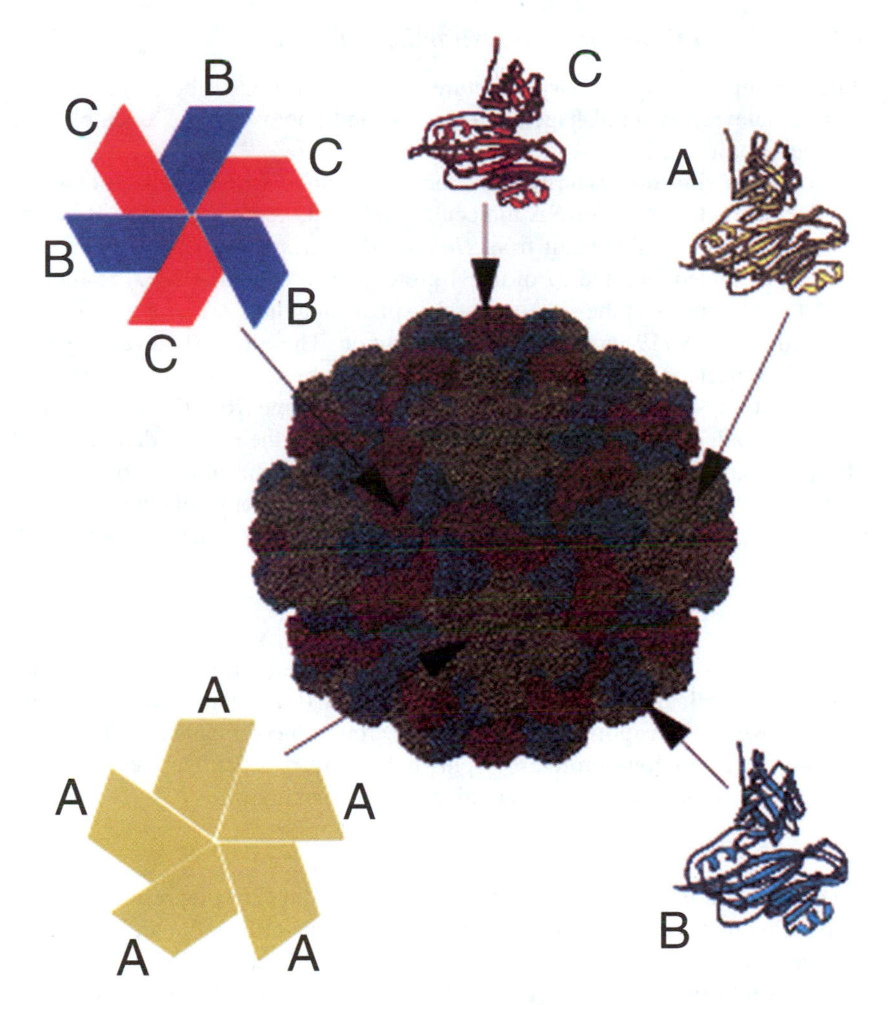

Figure 4.8. The structure of **tomato bushy stunt virus**, a spherical arrangement of 180 polypeptide chains, each of 387 amino acids, with every 3 chains making up an asymmetric unit (the subunits are colored blue, green, and red).

anisms involved in each of the two layers, using triangulation and geometrical quasi-equivalence packing motifs [256].

4.10.2 From Ribosomes to Dynamic Networks

Other examples of quaternary structure are noted for the ribosome, muscle-fiber complexes, bacterial flagellar filaments, and photosynthetic assemblies of membrane proteins.

The *E. Coli* **ribosome** is a ribonucleoprotein complex with a diameter of about 200 Å constructed from 3 RNA molecules and 55 protein chains. The structure of the large ribosomal subunit from *Haloarcula marismortui* (2833 of the subunit's 3045 nucleotides and 27 of its 31 proteins) was solved at 2.4 Å resolution in 2000 [14]; reports of the structure of the small subunit of *T. thermophilus* soon followed at 3.3 Å [19] and 3.0 Å [17] resolution. These eagerly awaited structures of the bacterial ribosome were made possible by cryo-electron microscopy reconstructions — first reported in 1995 for the ribosome from *E. Coli* — which helped crystallographers estimate the initial phasing of their X-ray data (see [20] for a perspective). The combined structural characterizations of the ribosome provided clear evidence that the ribosome is a ribozyme — that is, that the ribosome RNA's component likely catalyzes peptide bond formation (see Chapter 6, RNA sections).

Muscle cells contain parallel myofibrils composed of two kinds of filaments, each with the following proteins: **myosin** (thick filament), and **actin**, **tropomyosin**, and **troponin** (thin filament); around these filaments, **titin** — itself two extremely long proteins — plus nebulin form a flexible mesh. Muscle contraction is produced by the interaction of actin and myosin.

The bacterial flagellar motor of the protein **flagellin** [257] represents another challenging motor complex solved recently. Filaments of flagellin are formed by an arrangement of stacked flagellin proteins ('protofilaments') lined up side by side; an arrangement like loosely rolled sheets of paper results. The remarkable cooperativity among the different filaments leads to conversions between a macroscopic left-handed form — used for swimming — and a right-handed form — used for reorientation of motion. The high-resolution flagellin crystal suggests how this possible structural switch (between left and right-handed supercoiled forms) might occur to direct function.

Insights into the solar energy converters in the membranes of bacteria and plants were provided by the crystal structure of photosystem I, a large photosynthetic assembly of membrane proteins and other cofactors from the thermophilic cyanobacterium *S. elogatus* [258]. The detailed atomic picture (at 2.5 Å resolution) of the network of 12 proteins subunits and 127 cofactors (chlorophylls, lipids, ions, waters, others) shows the beautiful coordination of all components for efficient absorption and conversion of solar energy into chemical energy.

4.11 Protein Structure Classification

Many groups worldwide are working on classifying known protein structures; see [248, 259] for a perspective of protein structure and function evolution. Several classification schemes and associated software products exist. A popular program is SCOP: "Structural Classification of Proteins" [260]. (See scop.mrc-lmb.cam.ac.uk/scop/ or connect to SCOP through links available in many mirror sites such as PDB) [261]. These classifications are currently done manually, by visual inspection, but some automated tools are being used for assistance.

Also noteworthy is the PROSITE (www.expasy.ch/prosite/) database of protein families and domains intended to help researchers associate new sequences with known protein families. Other databases of patterns and sequences of protein families are PFAM and PRODOM; see [142] for a comprehensive list.

The SCOP levels (top-to-bottom) are: class, fold, superfamily, family, and domain. The sequence, or reference PDB structure, can be considered at the very bottom of this tree.

The top level of the SCOP hierarchy is the *class*, several examples of which were described above (all-α, all-β, α/β, and $\alpha + \beta$ proteins). Each *class* denotes common, global topologies of secondary structure.

Next comes the *fold*, which clusters proteins that have the same global structure, that is, similar packing and connectivity schemes for the secondary structural elements. Folds are often also called *supersecondary structure*. More than 50 known folds are currently known for each class, with the repertoire increasing steadily. An example mentioned above, the α/β barrel fold, groups **TIM** with other proteins like **RuBisCo(C)**, **Trp biosynthesis**, and **glycosyltransferase** into a *superfamily*, the next level of the classification hierarchy.

The *superfamily* groups proteins with low sequence identity but likely evolutionary similarity, as judged by similar overall folds and/or related functions. Members of the same superfamily are thus thought to evolve from a common ancestor. Another superfamily, for example, contains **actin**, the ATPase domain of the **heat shock protein**, and **hexokinase**. Superfamilies often pose the greatest challenge in the task of protein classification.

Superfamilies are further divided into *families*, which cluster proteins with substantial sequence, structure, and function similarity. Generally, this requirement implies a sequence identity of at least 30%, but there are instances of low sequence identity (e.g., 15%) but definitive structural and functional similarities, as in the case of globin proteins. For example, families of glycosyltransferase include β-galactosidases, β-glucanase, α-amylase, and β-amylase.

Finally, at the bottom of the tree of the SCOP classification lies the *domain* category, to distinguish further structurally-independent regions that may be found in larger proteins.

For updated information on the number of identified folds, superfamilies, and domains, check scop.mrc-lmb.cam.ac.uk/scop/count.html.

As our knowledge of protein structure increases, our classification schemes and software tools will evolve quickly. Automation of the classification is important for rapid structural analysis and ultimately for relating the sequence and structure to biological function.

The reader is encouraged to re-read at this point the sections in Chapter 2 on protein folding/misfolding (Sections 2.2 and 2.3).

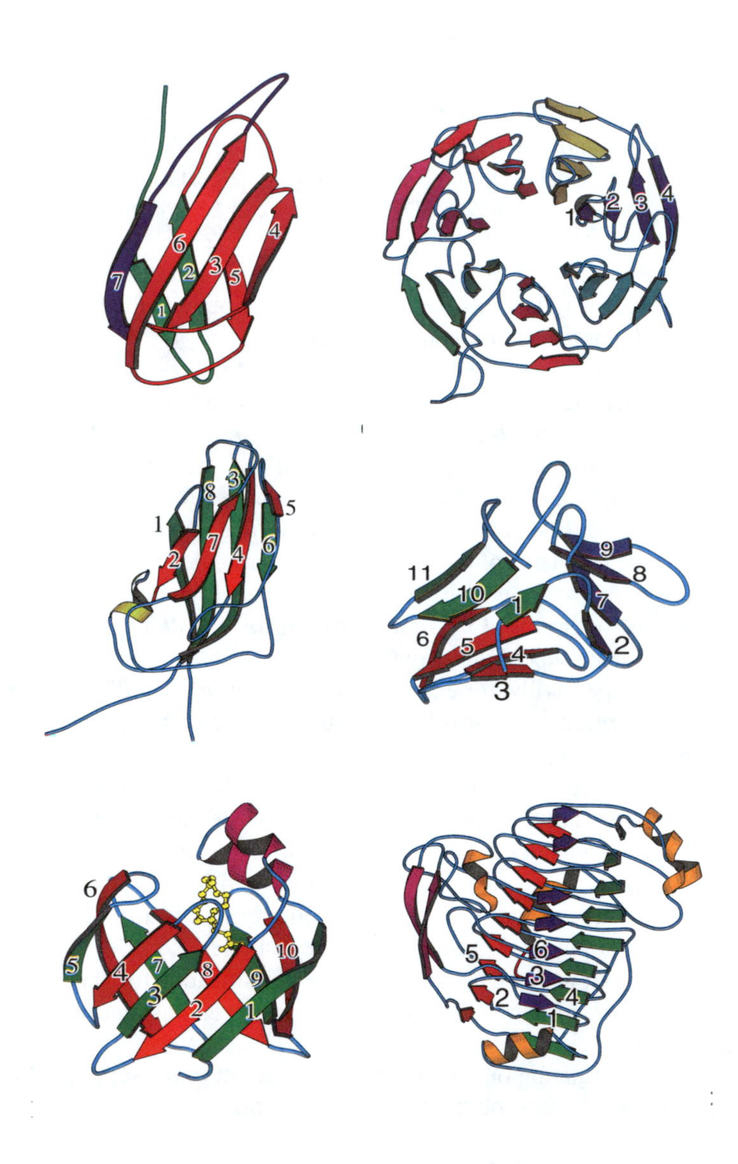

5

Nucleic Acids Structure Minitutorial

Chapter 5 Notation

SYMBOL	DEFINITION
Vectors	
$\{\mathbf{e}_1, \mathbf{e}_2, \mathbf{e}_3\}$	local base-pair coordinate system
Scalars & Terms	
endo/exo	sugar labeling (e.g., C3′-endo)
dx, dy	x and y-displacements (helical parameters)
h	helical rise
q	wave amplitude (pseudorotation description)
n	number of DNA base pairs
n_b	number of base pairs per turn
z_0–z_4	perpendicular displacements of sugar ring atoms
Dx, Dy, Dz	shift, slide, and rise translations (between successive base pairs)
P	phase angle of sugar pseudorotation
P_h	helix pitch
$\alpha, \beta, \gamma, \delta, \epsilon, \zeta$	polynucleotide backbone torsion angles
η	inclination angle (helical parameter)
θ	tip angle (helical parameter)
ρ	roll angle (base-pair step parameter)
τ	tilt angle (base-pair step parameter)
τ_0–τ_4	internal sugar ring torsion angles
$\tau_{\max}$	puckering amplitude of sugar pseudorotation
ϕ	phase-shift angle (pseudorotation description)
χ	glycosyl (sugar/base) torsion angle
ω	propeller twist (base pair parameter)
Ω	twist angle

> DNA plays a role in life rather like that played by the telephone directory in the social life of London: you can't do anything much without it, but, having it, you need a lot of other things — telephones, wires, and so on — as well.
>
> Review of *The Double Helix* in *The Sunday Times, London* (1968).

5.1 DNA, Life's Blueprint

With the master molecule of heredity, deoxyribonucleic acid (or DNA), so frequently mentioned in the media — in connections ranging from PCR (polymerase chain reaction) applications (e.g., criminology, medical diagnoses) to cloning and beyond — it is difficult to imagine today that it was only about 50 years ago when Watson and Crick reported their description of the DNA double helix [262, 263, 264]! Based on analysis of DNA fiber diffraction patterns and Chargaff's rules, they described a spiral image of an orderly helix — two intertwined polynucleotide chains, with a sugar/phosphate backbone on the exterior and pairs of hydrogen-bonded nitrogenous bases in the center. See [78] for a historical perspective of this discovery, including the contribution of all key players.

Though the model was imperfect in several respects (e.g., sugar conformations and base positioning with respect to the helical axis were inexact), Francis Crick, James Watson, and Maurice Wilkins received the 1962 Nobel Prize in Chemistry for this monumental discovery. Indeed, soon after the description of the DNA double helix came a burst of scientific excitement and discovery regarding the mechanism of heredity. The field of *molecular biology* gained extraordinary momentum and, soon after, its tentacles extended to the disciplines of cellular biology, molecular genetics, genomics, and proteomics, all buttressed by impressive advances in technology and computing.

5.1.1 The Kindled Field of Molecular Biology

The description of DNA's three-dimensional (3D) structure provides a classic example of the structure/function relationship. The two key functions of DNA — replication and transcription — were suggested and later proven based on the ordered spatial arrangement of the DNA. *Replication*, or the accurate transmittal of genetic information from parent to progeny, was explained by base pairing specificity. *Transcription*, the transformation of DNA's genetic information into a form that directs protein synthesis, was later understood with the deciphering of the genetic code relating triplet nucleotide codons (see Section 5.2 and Figure 5.2 for a description of the basic building blocks, including the pyrimidine bases T and C and purine bases A and G) to amino acids.

This genetic code (Table 5.1) reveals interesting patterns:

1. **It is highly redundant**: as 61 of the 64 triplet codons represent one of the 20 amino acids, and 3 signal 'stop' messages for protein synthesis (i.e., translation). In particular, Arg, Leu, and Ser are specified by 6 codons; Met and Trp are each represented by only one; and the rest are specified by 2–4 triplets.

2. **Pyrimidine-end commonalities exist**: Codons B_1B_2T and B_1B_2C (where B_1 and B_2 denote any of the 4 bases) always code for the same amino acid.

3. **Purine-end commonalities exist**: Codons B_1B_2A and B_1B_2G code for the same amino acid except in the cases of AT (ATA = Ile, ATG = Met) and TG (TGA = 'STOP', TGG = Trp).

Though the genetic code was initially thought to be 'universal', DNA sequencing in the early 1980s revealed code variations for certain organisms. Now it is clear that the genetic codes for mammalian and plant mitochondria and certain protozoa, for example, exhibit some variations from the code described in Table 5.1.

Recent advances in experimental techniques that allow detailed study of the components of the translation apparatus are also making possible the study of how the modern code has evolved from a simpler form [265].

Thus the model for the DNA double helix led the way to establishment of the first iteration of the central dogma of biology: DNA is self-replicating, DNA makes messenger ribonucleic acid (mRNA) through transcription,[1] and mRNA makes protein through translation (DNA → RNA → protein).[2] Although other arrows in the dogma have been suggested or verified (e.g., RNA → RNA in certain viruses, RNA → DNA in retroviruses), the main components appear known.

Much attention has now expanded to new areas of molecular biology, such as structural molecular biology, computational molecular biology, and offspring of genomics, such as structural genomics and functional genomics. The functional relationships among genes and the interaction of genes within the context of the complete organism are also of great interest.

---

[1]In *transcription*, an RNA polymerase glides along one strand of the DNA double helix and builds an RNA complement by coupling ribonucleotides through dehydration synthesis. The faithful replicate of one DNA strand then functions as the messenger RNA.

[2]In *translation*, the genetic code of the messenger RNA is read by transfer RNA molecules on cellular structures called ribosomes. Every transfer RNA molecule carries a specific sequence of three nucleotides on one end of an L-shaped structure and the corresponding amino acid on the other. The transfer RNA's main task is to deposit its amino acids on the ribosomes in proper sequence. In this process of matching each messenger-RNA codon with the complementary transfer RNA molecule, the polypeptide chain is assembled. As the amino acids link to one another on the ribosomes, polypeptide folding is thought to begin.

Table 5.1. The genetic code in terms of the parental DNA; the mRNA transcript has uracil (U) instead of thymine (T).

Amino Acid (or Instruction)	Encoding Triplets in Parental DNA[a]
Arginine (Arg)	CG(*), AG(A,G)
Leucine (Leu)	CT(*), TT(A,G),
Serine (Ser)	TC(*), AG(T,C)
Alanine (Ala)	GC(*)
Glycine (Gly)	GG(*)
Proline (Pro)	CC(*)
Threonine (Thr)	AC(*)
Valine (Val)	GT(*)[b]
Isoleucine (Ile)	AT(T,C,A)
Asparagine (Asp)	GA(T,C)
Aspartic acid (Asn)	AA(T,C)
Cysteine (Cys)	TG(T,C)
Histidine (His)	CA(T,C)
Phenylalanine (Phe)	TT(T,C)
Tyrosine (Tyr)	TA(T,C)
Glutamine (Gln)	CA(A,G)
Glutamic acid (Glu)	GA(A,G)
Lysine (Lys)	AA(A,G)
Methionine (Met)	ATG[b]
Tryptophan (Trp)	TGG
STOP	TA(A,G), TGA

[a]Short-hand notation is used for the third base of the codon when the first and second bases are the same: GA(T,C) specifies both GAT and GAC, while GC(*) denotes all four possibilities, namely GCT, GCC, GCA, and GCG.

[b]The codons ATG and (less frequently) GTG form part of the initiation signal in addition to their coding for Met and Val, respectively.

5.1.2 Fundamental DNA Processes

As the genetic material, DNA carries structural information in the primary sequence that not only controls faithful duplication but also regulates expression of the hereditary information [266]. Genetic variability can result from errors (or *mutations*), such as insertions, deletions, or substitutions in the daughters compared to the parental DNA, during the template-copying process; these changes can lead to altered triplet codes and hence different polypeptide composition.

Since many basic biological processes rely on protein/nucleic-acid interactions, the base sequence of polynucleotides also affects profoundly the characteristic 3D structure of DNA and RNA and hence the nature of fundamental biological processes. On a higher level of structure (hundreds and thousands of base pairs), compact forms of long DNA — such as supercoils and knots — are central to fundamental mechanisms for replication, transcription, and recombination [267, 268]. DNA supercoiling, namely the coiling in space of the double helix axis itself, can condense the DNA by several orders of magnitude and readily store energy for various activities. The wide range of characteristic timescales associated with the configurational rearrangements and hydrodynamic properties of supercoiled (or superhelical) DNA represents another area of intense study.

5.1.3 Challenges in Nucleic Acid Structure

As described above, there are several levels of nucleic acid structure, from the base-pair level to the cellular level of organization in the chromosomes (see end of next chapter and [269] for example). This study of DNA's rich configurational levels is particularly challenging to modelers.

Much focus has been placed on *protein structure* (including protein/DNA complexes) and the *protein folding problem* — predicting the 3D architecture of a system from the primary sequence. Yet an analogous folding problem is also relevant to DNA and RNA polymers. In fact, the nucleic-acid analogue of the 'protein folding problem' might be viewed as more challenging than protein structure prediction in the sense that it extends over much larger spatial scales (thousands of Ångstroms) as well as temporal scales (picoseconds to minutes and longer) than typically associated with proteins. Biologically, elucidating the folding of DNA in the cell — supercoiling and chromosome condensation, for example — is important for understanding the regulatory role of DNA in fundamental biological processes.

For small segments of nucleic acids, X-ray crystallography and NMR data have been invaluable for providing detailed, atomic-resolution data on single, double, triple and other forms of nucleic acid oligonucleotides, as well as their complexes with proteins, other biomolecules, and ligands [270]. The Nucleic Acid Database (NDB, ndbserver.rutgers.edu/) [271] has been beautifully cataloging these structures and offers many utilities for their analysis.

All-atom molecular dynamics (MD) simulations of nucleic acids have also shed important insights on DNA sequence/structure relationships, the nature of the hydration geometries surrounding nucleic acids, and nucleic acid mobility [270, 272, 273, 68, 274, for example, and see Chapter 6]. However, it is only fairly recent that stable, fully solvated nucleic acid MD trajectories, with representative ionic atmospheres, have been possible [275, 276, 277], as well as ultra-high resolution nucleic-acid crystal structures [278, 279, 280].

The theoretical advances resulted from improvements in long-range electrostatic modeling (e.g., [61]) as well as increases in computer memory and speed. The experimental advances reflect improved methods for crystallization and phase determination, the increased availability of very strong X-ray sources, and improvements in algorithms for model refinement.

Further advances are on the horizon with improvements in crystallographic techniques for solving RNA structures, in simulation protocols for propagating stably solvated systems of DNA, and in the growing appreciation of DNA's higher levels of structural hierarchy. The possibilities of using the strands of DNA *in vitro* for practical applications (e.g., to produce electronic devices like nanowires, or as parallel computers to solve very difficult, combinatorial optimization problems that have non-polynomial complexity; see [281, 282, 283], for example)[3] are also generating much interdisciplinary interest in DNA.

Undoubtedly, progress is expected in the bridging between all-atom and macroscopic representations of nucleic acids and between experiment and theory. This unity will enhance our understanding of DNA structure and DNA/protein interactions and, in turn, will likely have important biomedical applications, for example, in the design of new anti-viral drugs, antibiotics, and anti-cancer agents.

5.1.4 Chapter Overview

In this chapter, the basic elements of nucleic acids and DNA structure on the base-pair and helical level will be introduced: the fundamental building blocks and how they are linked to form polynucleotides, aspects of nucleic acid conformational flexibility (sugar pseudorotation, torsion angle preferences, and global and local base-pair parameters), and the three canonical DNA helices (A, B, and Z-DNA). The next chapter presents selected topics regarding the structural diversity of DNA and RNA, and DNA folding on a higher level, namely supercoiling; many other exciting subjects are omitted.

For basic books on DNA structure, see [268, 284, 285, 286, 287, 288, 289, 290, 291]. Lively, less technical introductions can be enjoyed from [292, 293, 294]. The reader is also well advised to examine some of the rich information on nucleic acid structure available in the public domain through the NDB [271].

Throughout this chapter and the next, we abbreviate a *base pair* as bp and *base pairs* as bps. Note that though we mention the three canonical helices before introducing them formally (using all notation we systematically develop), novices should be able to follow the material. Students are encouraged to re-read the chapter after they are more familiar with all the presented material.

5.2 The Basic Building Blocks of Nucleic Acids

In the classic DNA double helix described by Watson and Crick, a flexible ladder-like structure is formed with the polymer wrapped around an imaginary central axis (Figure 5.1). The two rails of the ladder consist of alternating sugar (deoxyri-

[3] See also the large bibliography for the emerging interdisciplinary field of DNA Computing that unites molecular biology with computer science assembled at the URL www.wi.LeidenUniv.nl/~pier/dna.html.

bose) and phosphate units; the rungs of the ladder consist of nitrogenous bps held together by hydrogen bonds (see Box 3.2 of Chapter 3 for a definition).

5.2.1 *Nitrogenous Bases*

Four nitrogenous bases can be found in DNA: the pyrimidines cytosine (C) and thymine (T) which are 6-membered rings, and the purines guanine (G) and adenine (A), each of which is a fused system with a 5 and 6-membered ring. In the single-stranded RNA polynucleotide, ribose replaces deoxyribose and uracil (U) replaces thymine (Figure 5.2). Some DNAs contain bases that are chemically modified (e.g., substitutions of a hydrogen by a methyl group).[4] Alternative but rare *tautomeric* forms due to proton shifts in aromatic molecules (as introduced in Chapter 1, subsection on the discovery of DNA structure),[5] are possible but may not be biologically significant.

5.2.2 *Hydrogen Bonds*

In the Watson-Crick base pairing scheme (termed WC herewith), cytosine pairs with guanine by forming three hydrogen bonds, and thymine pairs with adenine by forming two hydrogen bonds (Figure 5.3). This arrangement produces CG and TA bps whose widths are nearly identical. The approximately uniform width is significant because any pyrimidine/purine or purine/pyrimidine sequence can be accumulated on one strand, with a pyrimidine opposing a purine, without much alteration in overall structure. The discovery of this orderly 3D structure of DNA helices explained for the first time how a regular polymer made of repeating phosphate/sugar/base units (*nucleotides*) could replicate and encode hereditary information.

5.2.3 *Nucleotides*

The basic building block of nucleic acid polymers, the *nucleotide*, consists of a 5-membered sugar ring — deoxyribose in DNA and ribose in RNA — a phosphate, and a purine or pyrimidine base (see Figure 5.4). The unit containing just the sugar and base is called the *nucleoside*. Nucleotides are linked together through the phosphate group to form the polynucleotide chain.

---

[4]For example, *N6-methyl-dA* is a modified adenine base with the $N6H_2$ (attachment to the ring carbon C6) moiety replaced by $N6HCH_3$; *5-methyl-dC* is a modified cytosine where the C5H becomes $C5CH_3$.

[5]Two classic tautomerization reactions are *keto/enol* and *amino/imino*; the keto and amino forms are typically favored and are shown in Figure 5.2. Keto/enol tautomerization involves alteration of the carbonyl group (–C=O) to a hydroxyl group (=C–O–H), shifting the double bond from the carbonyl group to the nitrogen-carbon bond in the ring (e.g., C6=O of G becomes C6–OH, accompanied by the change of H–N1–C6 to N1=C6). Similarly, an amino/imino tautomerization involves a change in an amino nitrogen $–NH_2$ to an imino form, =NH (e.g., $C6–NH_2$ of A becomes C6=NH, accompanied by the change of N1=C6 to H–N1–C6).

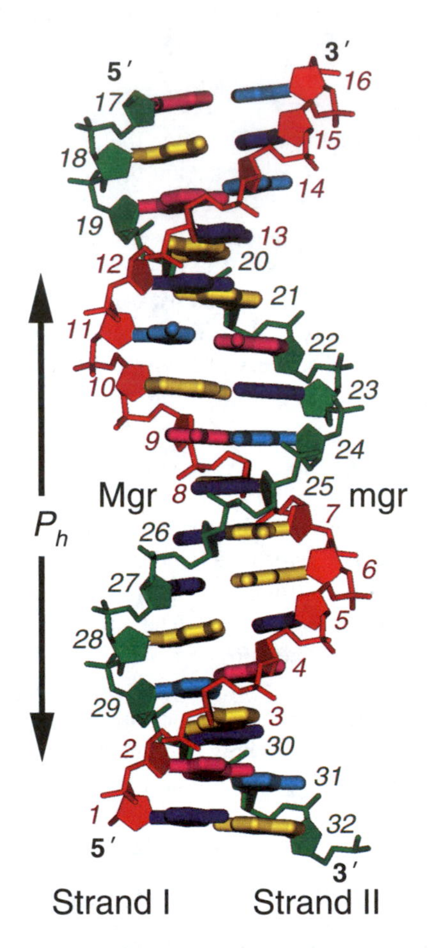

Figure 5.1. The two strands of the classic DNA double helix: alternating subunits of phosphates and deoxyriboses held together by hydrogen-bonded bps, normally with adenine (A) pairing with thymine (T) and guanine (G) pairing with cytosine (C). The two chains of the DNA double helix run in opposite directions: The chain ending in the lower left terminates with a free hydroxyl (–OH) group linked to the sugar carbon denoted as C5′, while the chain ending in the lower right terminates with a free hydroxyl group linked to the sugar carbon denoted as C3′. Individual bases are numbered 1–16 (or 1 to n where n is the number of bps) for the sequence strand (Strand I) and 17–32 ($n + 1$ to $2n$ in general) for Strand II, both in the $5' \rightarrow 3'$ direction; the n base pairs 1–16 are numbered so that they coincide with the bases of Strand I. The *pitch* of the helix P_h is the length along the helix axis for one complete turn, and n_b is the number of bps per turn (around 10.5 for DNA in solution). The unit *twist* is defined as $\Omega = 360°/n_b$, and the helical rise is $h = P_h/n_b$. Mgr and mgr denote the major and minor grooves, respectively.

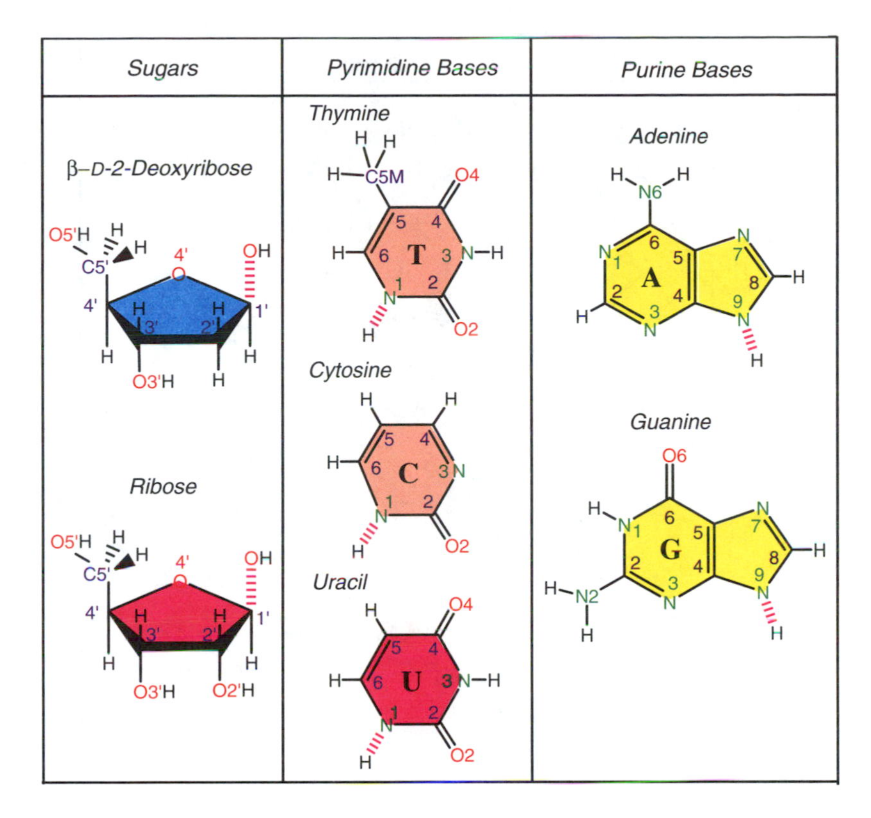

Figure 5.2. Chemical structures and atomic labels for nucleic acid sugars (deoxyribose in DNA and ribose in RNA) and nitrogenous bases: T, C, A, G in DNA and U, C, A, G in RNA. The broken-line pattern in each compound indicates the bond where a link to another building block is made (see Figure 5.4).

5.2.4 *Polynucleotides*

When nucleotides are polymerized into nucleic acid chains by the chemical removal of water molecules, a sugar/phosphate backbone is formed. The C3′–hydroxyl group of the nth nucleotide sugar is joined to the C5′–hydroxyl group of the $(n+1)$th nucleotide by a *phosphodiester* bridge.[6] This negative phosphate group, located on the exterior of the helix, is readily available for physical and chemical interactions with solvent water molecules and metal ions present in the cell (see next chapter). Hydration effects are of great importance for DNA molecules in solution, since water plays a central role in stabilizing a particular helical conformation.

[6] An ester is an –OR group where R represents an organic chemical group.

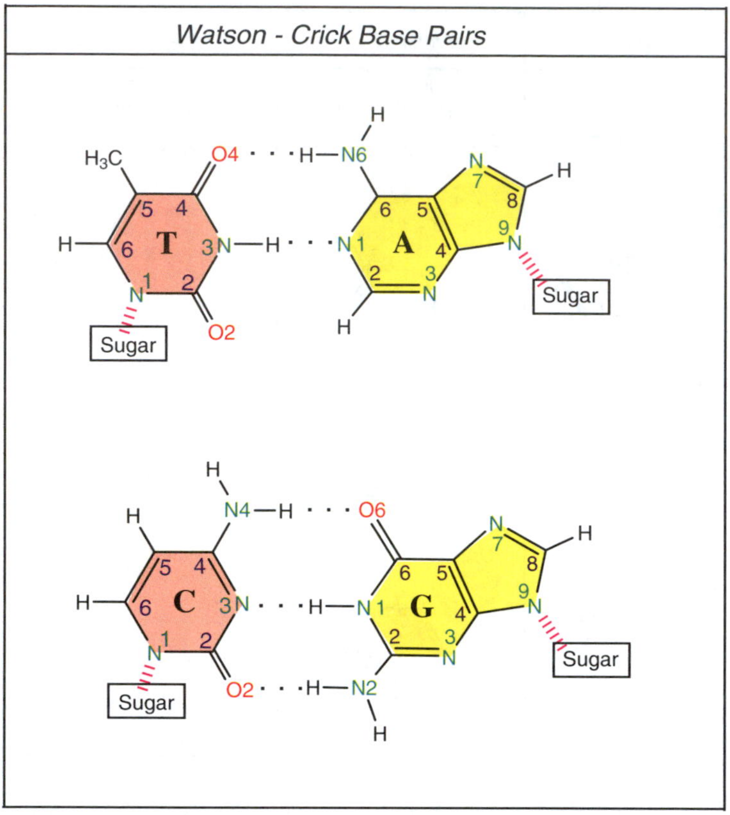

Watson - Crick Base Pairs

Figure 5.3. In the classic Watson-Crick base pairing scheme of DNA double helices, thymine (T) pairs with adenine (A) by forming two hydrogen bonds, and cytosine (C) pairs with guanine (G) by forming three hydrogen bonds. The hydrogen bonds are often represented by dots (· · ·). The structural fit (e.g., as measured by C1′–C1′ distances) is such that the difference in width between the two base pairs is less than 3%, allowing formation of a double helix with a nearly constant diameter.

5.2.5 Stabilizing Polynucleotide Interactions

Both *hydrogen bonding* and *base stacking* are considered to be intrinsic factors for helix stability. The estimates given for intrinsic energies of hydrogen-bonding and base-pair stacking in Chapter 6 (Subsection 6.5.1) suggest that the latter is at least as important for overall helix stability.

The bases in DNA double helices are normally linked by *hydrogen bonds* in the WC bps as depicted in Figure 5.3. Recall that in this weak noncovalent electrostatic interaction a hydrogen atom is shared between negatively-charged donor and acceptor atoms, typically nitrogens or oxygens in biological polymers. Besides WC, other hydrogen-bonding patterns involve variations in the positioning

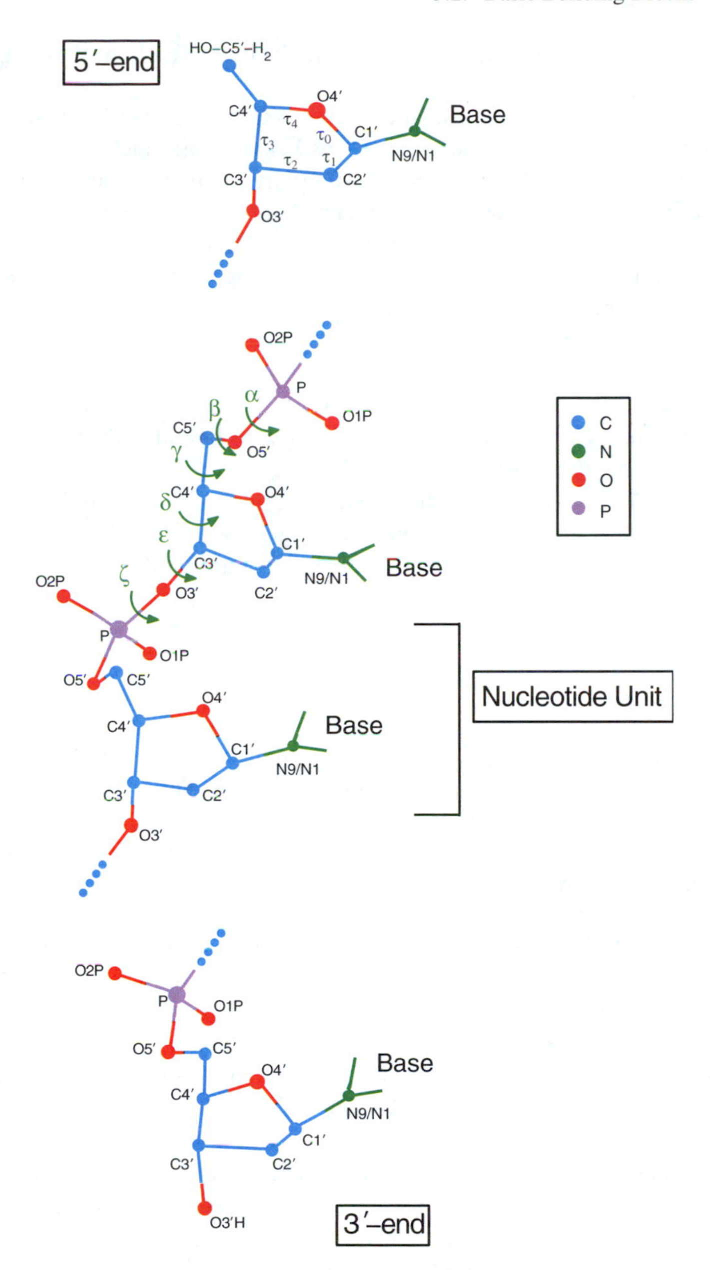

Figure 5.4. The polynucleotide chain, with standard atom labeling shown, runs from the 5′ end (of the sugar atom C5′) to the 3′ end (of the sugar atom C3′). Nucleotide linkage is via the 3′ to 5′ phosphodiester bonds (i.e., C3′–hydroxyl group of one nucleotide sugar to C5′–hydroxyl group of another). Nucleic acid torsion angles are labeled as $\alpha, \beta, \gamma, \delta, \epsilon$ and ζ along the polynucleotide chain, τ_0 through τ_4 about the endocyclic bonds of the sugar, and χ about the glycosyl bond, connecting the sugar and base. See Table 5.2 for full quadruplet sequences of these torsion angles.

of the two bases relative to one another and in the hydrogen-bond defining atoms.

Base-pair stacking refers to favorable interactions between hydrogen-bonded neighboring bps (imagine weak interactions between the stacked ladder rungs). These arise from favorable van der Waals and hydrophobic contacts, which can optimize the water-insoluble areas of contact. Another factor in stability is electrostatics, which leads to preferred distributions of the partial electric charges spread over the aromatic rings. In addition, specific hydrogen bonds, other than classic WC bps, can form between adjacent bps if the bps are properly oriented. One example is *propeller-twisting* of base pairs (introduced below).

In fact, we are growing to appreciate the many hydrogen-bonding and base-pair stacking arrangements that one or more polynucleotide strands can form, including residues with modified bases, bases with bound adducts, and peptide nucleic acid structures [285, 295, 296, 297].

As will be further discussed, the notion of a perfectly ordered double-helical molecules is thus just an ideal reference. Many of the biological functions of DNA and DNA/protein complexes require appreciation of the sequence-specific, atomic-level variations.

5.2.6 Chain Notation

A DNA chain has *polarity*. Each strand of the chain has a terminal C5′–OH group on one end and a terminal C3′–OH group on the other. Thus, the two intertwined strands run in anti-parallel directions. Conventionally, the base sequence in a polynucleotide is specified for the $5' \rightarrow 3'$ direction; the complementary strand is automatic for ideal WC bps.

To number the $2n$ bases and n base pairs in a DNA duplex, Strand I — the sequence strand — and Strand II — its antiparallel companion — are specified, as shown in Figure 5.1. Bases along Strand I are numbered 1 to n in the $5' \rightarrow 3'$ direction, and bases along Strand II are numbered from $n + 1$ to $2n$, also in the $5' \rightarrow 3'$ direction. Base pairs 1 to n are numbered so that they coincide with the bases of Strand I. Thus, bp 1 involves base 1 of Strand I with base $2n$ of Strand II; bp 2 is the hydrogen-bonded pair of Strand I's base 2 with Strand II's base $2n - 1$; and so on.

Nucleosides are abbreviated by a pair of letters: a lower-case Roman letter denotes the sugar type ("r" or "d"), and an upper-case Roman letter represents the base type (C, G, A, T, U). For example, rA is *adenosine* — ribose plus adenine, and dC is *deoxycytidine* — deoxyribose plus cytosine. The base prefixes are often dropped when the type of sugar is obvious. Nucleotides are abbreviated by adding a lower-case "p" (for phosphate) to the nucleoside symbol.

Thus, the sequence dGpdCpdApdC is a *tetramer* in a DNA double helix where G is at the C5′–OH end and C is at the C3′–OH end. For brevity, these lower-case "p"s are often omitted when a specific nucleic-acid sequence is specified. In the nucleic acid literature, duplex units of DNA for 2–12 base pairs are commonly termed as dimer, trimer, tetramer, pentamer, hexamer, heptamer, octamer, nonamer, decamer, undecamer, and dodecamer, respectively.

5.2.7 *Atomic Labeling*

Standard atomic numbering schemes have been recommended for nucleic acids [284], as follows (see Figures 5.3 and 5.4):

- Sugar atoms are distinguished from base atoms by a prime suffix, and within the sugar the numbering sequence is counted clockwise from the ring oxygen in the direction of the carbon attached to the base nitrogen: $O4' \rightarrow C1' \rightarrow C2' \rightarrow C3' \rightarrow C4' \rightarrow O4'$.

- In the polynucleotide backbone, the counting direction for torsion angles ($\alpha, \beta, \gamma, \delta, \epsilon$ and ζ) is specified by the sequence: $P \rightarrow O5' \rightarrow C5' \rightarrow C4' \rightarrow C3' \rightarrow O3' \rightarrow P$.

- Base atoms are numbered systematically as shown in Figure 5.3. The procedure for labeling base atoms is different than that used for the sugar: Whereas the sugar atoms are numbered by atom *types* (O4′, C1′ through C4′), the base atoms are numbered according to *position* in the ring (in cytosine, for example, N1 and N3 are the two ring nitrogens, and C2, C4, C5 and C6 are the four ring carbons).
  One nitrogen of the base is always connected to the C1′ of the sugar by the *glycosyl* bond (N to C1′). Thus, according to the base numbering scheme, a pyrimidine base attached the glycosyl nitrogen has N1–C1′, and a purine base has N9–C1′.

- When more than one oxygen or hydrogen are attached to the same atom, the two substituents are generally distinguished by the numbers 1 and 2. This applies to the two C5′ hydrogens, H1 and H2, and the two oxygens at P, O1P and O2P. (Note that O2 is defined also for atoms of the pyrimidine bases). The order of these substituents is such that when we look along the counting direction of the main chain or the counting direction of the sugar ring, a clockwise direction gives substituent 1 and then 2. For example, when we look along the $O5' \rightarrow C5'$ bond, a clockwise counting gives substituents C4′, H1, and H2 of atom C5′ (the hydrogens are generally labeled H5′1 and H5′2).

5.2.8 *Torsion Angle Labeling*

Recall that a torsion angle describes the relative orientation of a bonded 4-atom sequence i–j–k–l as the angle between the two planes defined by i–j–k and j–k–l (see Box 3.4 and Figure 3.14 in Chapter 3). As with polypeptides, the torsion angles in polynucleotide chains are denoted by a systematic set of Greek letters,

Table 5.2. Nucleic acid torsion angle definitions.

Angle	Sequence	Angle	Sequence
α	O3′–P–O5′–C5′	τ_0	C4′–O4′–C1′–C2′
β	P–O5′–C5′–C4′	τ_1	O4′–C1′–C2′–C3′
γ	O5′–C5′–C4′–C3′	τ_2	C1′–C2′–C3′–C4′
δ	C5′–C4′–C3′–O3′	τ_3	C2′–C3′–C4′–O4′
ϵ	C4′–C3′–O3′–P	τ_4	C3′–C4′–O4′–C1′
ζ	C3′–O3′–P–O5′	χ	O4′–C1′–N1–C2 (pyrimidine)
			O4′–C1′–N9–C4 (purine)

as shown in Figure 5.4 and defined in Table 5.2. For the backbone rotations, the letters α, β, γ, δ, ϵ and ζ are used, in the order along the chain backbone direction.

For the sugar molecule, the endocyclic (i.e., ring) torsion angles are designated by τ_0 through τ_4 along the ring direction noted above (clockwise) so that τ_j is the torsion angle for the sequence $C_{j'-1}$–$C_{j'}$–$C_{j'+1}$–$C_{j'+2}$ where C_0 is O4′ and the integer j' is equal to j modulo 5 (hence C_5 is O4′ and C_{-1} is C4′, for example; see Table 5.2) [298].

For describing the orientation of the nitrogenous base to the sugar, the torsion angle χ is defined about the glycosyl bond; χ is the torsion angle of the O4′–C1′–N1–C2 sequence for a pyrimidine base and the torsion angle of O4′–C1′–N9–C4 for a purine base. Rotational sequences for the nitrogenous bases are not given special symbols, because the bases are approximately planar.

5.3 Nucleic Acid Conformational Flexibility

As in polypeptides, where conformational flexibility is limited due to steric hindrance (as described, for example, in the Ramachandran plots introduced in Chapter 3). the observed set of conformations in nucleic acids is limited by energetic and chemical considerations. A description of these conformations is more complex than those for the backbone ϕ and ψ pairs and secondary-structure elements as used for polypeptides, since a greater conformational variability exists in nucleic acid residues, with additional variation in helical parameters. Sequence and environmental factors (e.g., cations, bound ligands) affect sensitively the local structural fluctuations.

5.3.1 The Furanose Ring

The 5-membered furanose sugar ring is generally nonplanar in nucleic acids. One or two atoms may *pucker* out of the plane defined by the remaining skeletal ring atoms. When four ring atoms are planar, the pucker form is called *envelope*; when

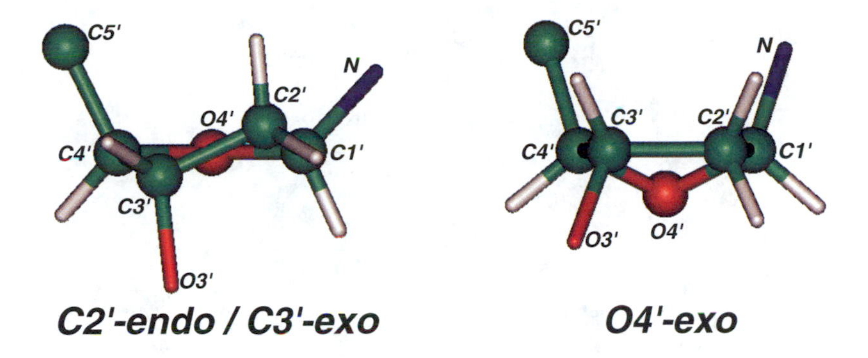

Figure 5.5. *Envelope* and *twist* puckering forms for a 5-membered ring: C2′–endo/C3′–exo symmetric twist (left) and O4′-exo envelope (right); for clarity, hydrogens attached to exocyclic atoms are not shown.

two atoms pucker at opposite sides of the plane defined by the remaining three ring atoms, the conformation is known as a *twist* (see Figure 5.5).

The sugar pucker type is described by the atom or atoms that deviate from that three or four-atom ring plane. Atoms displaced on the same side of C5′ are designated as *endo*, and atoms displaced on the opposite side of C5′ are called *exo*.

To describe nucleic acid sugar conformations, it is convenient to use a *pseudorotation path* developed on the basis of the 5-membered carbon ring cyclopentane [299, 298]. In cyclopentane, a wavelike motion between conformations of equal energy can be imagined with respect to a mean plane. This mean ring plane can be defined (in various ways) by positions of the five skeletal ring atoms. In both the Altona/Sundaralingam (AS) [298] and Cremer/Pople (CP) [299] descriptions, this sinusoidal motion is described by a wave amplitude and phase shift: (q, ϕ) and $(\tau_{\max}, P)$, respectively, as follows.

In the AS description, the five endocyclic torsion angles are restricted to the values:

$$\tau_j = \tau_{\max} \cos \left[\, P + 4\pi/5\,(j-2) \,\right], \quad j = 0, 1, 2, 3, 4. \tag{5.1}$$

In CP, the five perpendicular displacements of the ring carbons from a mean plane are described by the cosine series:

$$z_j = (2/5)^{1/2}\; q \,\cos \left[\, \phi + 4\pi/5\,(j-2) \,\right], \quad j = 0, 1, 2, 3, 4. \tag{5.2}$$

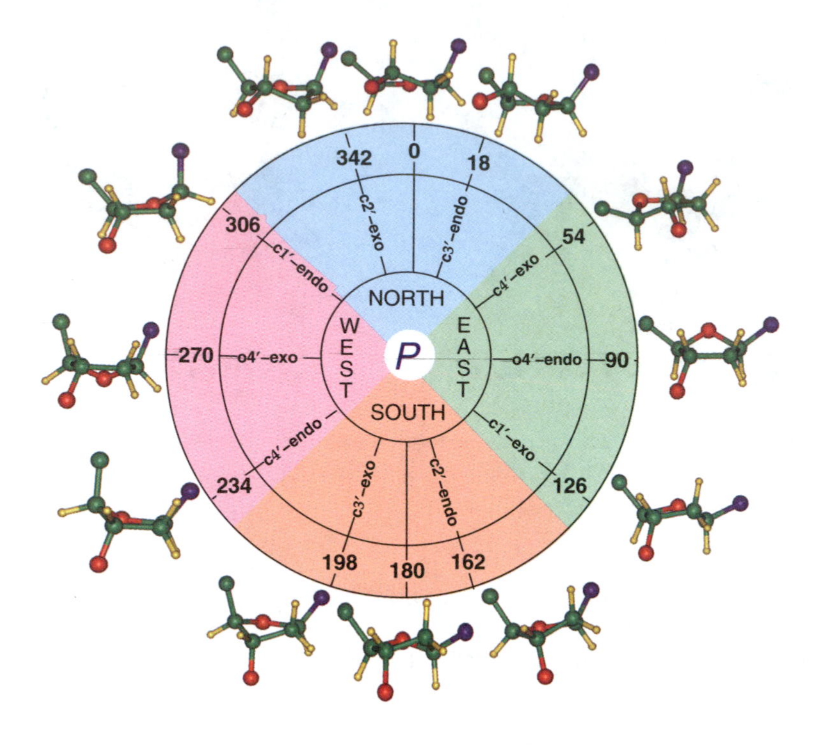

Figure 5.6. Sugar pseudorotation cycle.

The two formalisms are only equivalent under the assumption of infinitesimal displacements from a regular planar pentagon. A more general relation based on a simple analysis of model riboses was derived in [300].

The wave-like pseudorotation path described by eq. (5.1) defines 10 *envelope* conformations for $P = 18^o$, 54°, $90^o, \ldots, 342^o$ and 10 *twist* conformations for $P = 0^o, 36^o, 72^o, \ldots, 324^o$. Figure 5.6 shows the 10 envelopes and two of the 10 twists. The quadrant terminology of N, S, E, and W is often used to describe North, South, East, and West sugar-pucker regions.

Using this description for nucleic-acid sugars, we see from Figure 5.5 that O4′–exo, for example, is an *envelope* conformation with O4′ puckering out of the C4′–C3′–C2′–C1′ plane on the opposite side of C5′. Similarly, C2′–endo/C3′–exo is a *twist* conformation with C2′ puckering out of the C4′–O4′–C1′ plane on the same side of C5′ and C3′ puckering out of the same plane on the opposite side of C5′.

Two major types of puckering modes are observed in nucleic acid sugars. They cluster around the North C3′–endo ($0^o < P < 36^o$) and South C2′–endo ($144^o < P < 188^o$) puckers; see Figure 5.7. The puckering amplitude $\tau_{\max}$ averages around $40^o \pm 5^o$. Still, significant deviations of sugar pseudorotation

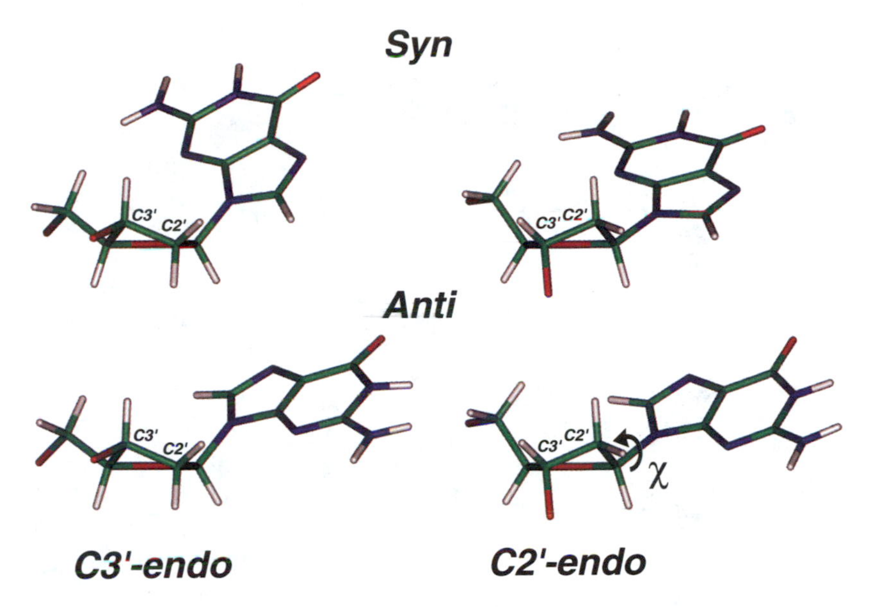

Figure 5.7. The common C3′–endo (left) and C2′–endo (right) sugar puckers for a deoxyguanosine in combination with the glycosyl torsion angle in the *syn* (top) and *anti* (bottom) conformations. Note that the 6-membered ring of G lies over the sugar in *syn* and away in *anti*.

parameters are commonly observed [284], especially as new forms of DNA are being discovered.

Figure 5.8 shows sugar puckering patterns of crystallographically determined nucleosides and nucleotides. Such analyses are possible using the above (Fourier) formalism extrapolated from cyclopentane motions. The numerical values for the two Fourier parameters are estimated by the Cartesian coordinates of atomic-resolution models.

The study of nucleic acid sugar conformations and conformational interchanges has been an active area of research, since the sugar conformation strongly affects the overall helical structure.

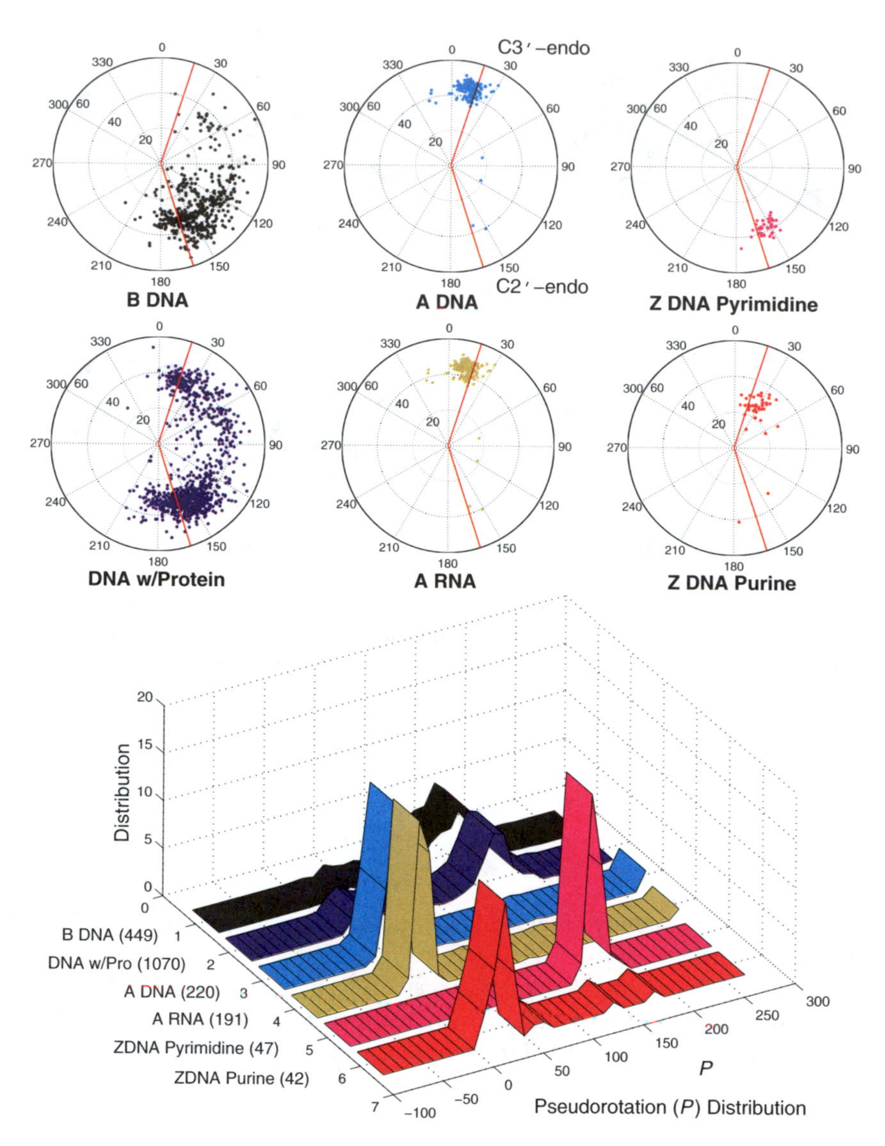

Figure 5.8. Sugar pucker conformations analyzed from the NDB for higher-resolution structures (< 2 Å) as averaged over all inner residues (excluding two at each end). Full details of the analyzed sequences are given in the book's website. The total number of residues analyzed to obtain sugar parameters for A-DNA, B-DNA, Z-DNA pyrimidines, Z-DNA purines, A-RNA, and DNA/Protein complexes is indicated at bottom (in parentheses). In the sugar wheels, the radial coordinate is the phase amplitude $\tau_{\max}$ and the angle is the phase angle of pseudorotation (P).

5.3.2 *Backbone Torsional Flexibility*

In addition to the five internal sugar torsions described above, the six phosphodiester backbone torsion angles $\alpha, \beta, \gamma, \delta, \epsilon$, and ζ are flexible but restricted to sterically allowable regions. Different values are also characteristic of various helical structures, as shown in Table 5.4 (discussed later).

Recall that a torsion angle range can be described by the three exhaustive regions of the conformational space — *gauche*$^+$ (g$^+$), *gauche*$^-$ (g$^-$), and *trans* (t) (see also Figure 3.13 in Chapter 3). In terms of these classifications, both α (about P–O5′) and γ (about C5′–C4′) exhibit relatively large flexibility, with the g$^+$, g$^-$, and t positions observed. The angles β (about O5′–C5′) and ϵ (about C3′–O3′) tend to cluster around the *trans* state, with ϵ occupying a broad t/g$^-$ range. The backbone torsion angle δ (about C4′–C3′) is correlated to the sugar pseudorotation puckering state, since the sugar torsion τ_3 is defined about the same C–C bond. This correlation can be described roughly by $\delta = \tau_3 + 120°$. Finally, ζ (about O3′–P) is rather flexible, with all three ranges observed.

See Figure 5.9 for distributions of these backbone torsion angles, as analyzed for the same high-resolution crystal structures used to generate Figure 5.8.

5.3.3 *The Glycosyl Rotation*

Relative to the sugar moiety, the base can assume two major orientations about the glycosyl C1′–N bond: *syn* and *anti* (torsion angles of 0 and 180°, respectively) [284]. Roughly speaking, four major conformations are favored. They correspond to the combinations of C3′–endo and C2′–endo sugar puckers with *syn* and *anti* values for χ. Favored combinations of $\{P, \chi\}$ pairs vary for the different nucleosides or nucleotides. They depend on the chemical structure of the sugar, the size of the base, and the nature of the nucleoside substituents (chemical derivatives). For example, deoxyribose nucleosides and nucleotides prefer the C2′–endo conformation over C3′–endo [301]. In pyrimidine nucleotides, the *anti* orientation of χ about the sugar ring is finely tuned by the sugar pucker [302].

In Figure 5.7 the two orientations of *syn* and *anti* bases are illustrated for deoxyguanosine in combination with the two common sugar puckers. The {C3′–endo, *syn*} combination of this figure (top left) is that observed in the purine of Z-DNA helices while the {C2′–endo, *anti*} combination (bottom right) is typically observed in the B-DNA varieties (and in Z-DNA pyrimidines). Figure 5.9 also shows the distributions of χ in various nucleic-acid structures.

5.3.4 *Sugar/Glycosyl Combinations*

To further illustrate conformational tendencies in polynucleotides, we generated *adiabatic maps*[7] for two models of deoxyadenosine in the $\{P, \chi\}$ space (Fig-

[7] An adiabatic map is a simple way to examine molecular motion by characteristic low-energy paths along a prescribed reaction coordinate (i.e., variations in specific conformational variables). For each

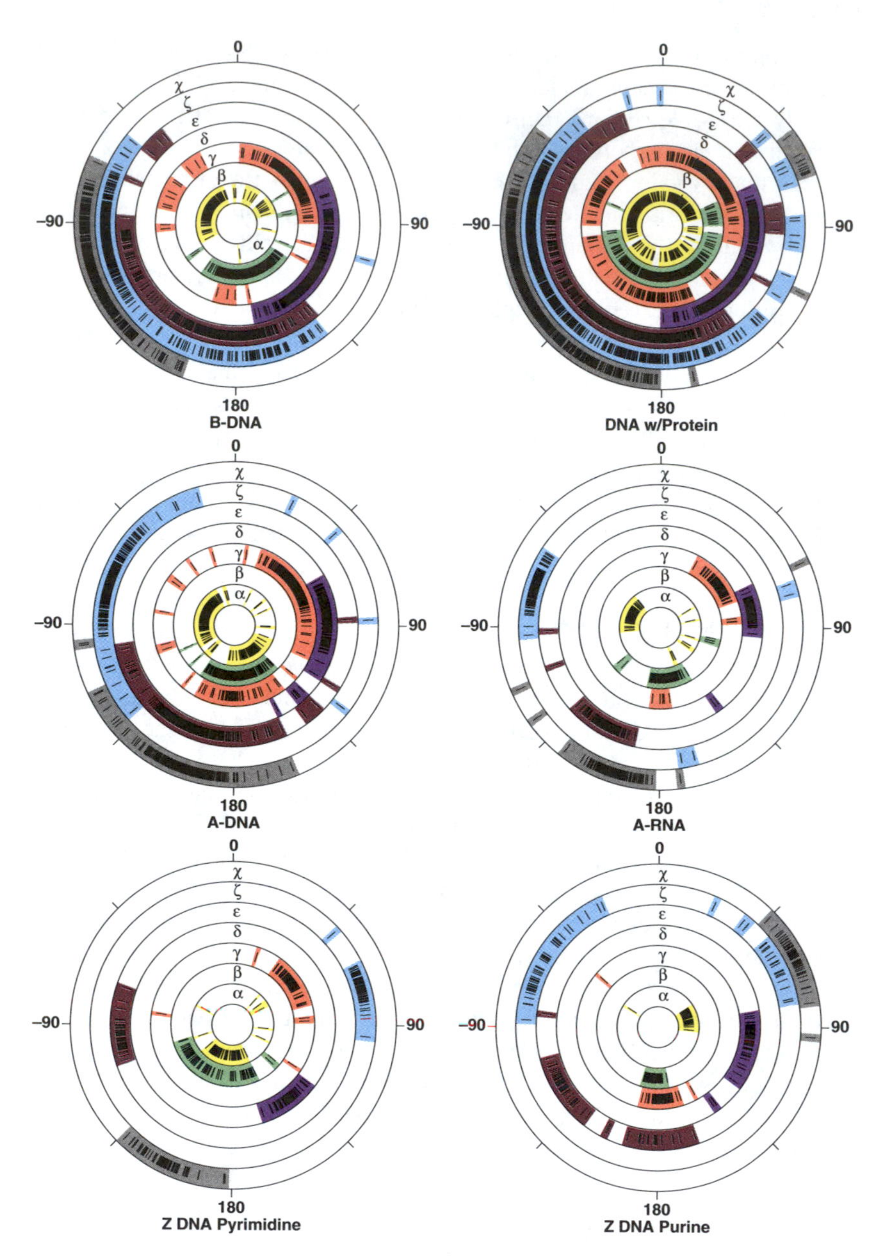

Figure 5.9. Observed ranges of nucleic-acid torsion angles for higher-resolution structures (< 2 Å) as averaged over all inner residues (excluding two at each end) for the same residues analyzed for Figure 5.8. Full details of the analyzed sequences are given in the book's website.

ure 5.10) based on the CHARMM force field [303, 304, 305]. The first model approximates solvation simply with a distance-dependent dielectric function. The latter uses explicit representation of water molecules.[8]

The adiabatic maps were calculated by dividing the $\{P, \chi\}$ grid to 3600 points, using 6° intervals for each angle. For each selected P, the values of the 5 endocyclic sugar torsion angles were determined from eq. (5.1) using $\tau_{\max} = 40°$. Starting structures were then generated from the set of variables $\{\tau_0, \tau_1, \tau_2, \tau_3, \tau_4, \chi\}$ and minimized over the remaining degrees of freedom. Of course, such a map only provides a reference for conformational flexibility, since constraining (or freezing) the angles does not allow complete energy relaxation over all the available degrees of freedom.

We note from Figure 5.10 essentially the same trends for the solvated (top) and more approximate (bottom) models of four minima corresponding to the $\{P, \chi\}$ combinations of C2′–endo/*anti*, C3′–endo/*anti*, C2′–endo/*syn*, and C3′–endo/ *syn*; two maxima are also apparent. See also [306] for a recent quantum-mechanical study of the glycosyl torsion energetics in DNA.

5.3.5 Basic Helical Descriptors

Besides the sugar, backbone, and glycosyl conformational variables introduced above, additional helical parameters are defined to describe the global arrangement of a base pair (bp) in a double helix [308, 284]. See Dickerson [309] and the text [291] for complete definitions of all translational and rotational variables.

Names, symbols, and sign conventions were decided at an international workshop [308]. Before we introduce some of these (as well as other) conformational variables, we define basic features of helix descriptors that are relevant for model helices, whose geometries can be described by characteristic values, as shown in Tables 5.3 and 5.4.

- *Helix sense* refers to the handedness of the double helix.[9]
- *Helix pitch* (per turn), P_h, measures the distance along the helix axis for one complete turn (see Figure 5.1).

---

combination of these conformational coordinates, the entire potential energy of the system is minimized to approximate behavior for the motion under study. Though simple in principle, specification of the reaction coordinate is difficult in general, and the neglect of other degrees of freedom in the process is clearly an approximation whose validity depends on the motion in question.

[8]Namely, each nucleoside is enveloped in a water sphere of radius 11 Å, and the nonbonded interactions are truncated at 12 Å using a 2 Å buffer region, a potential shift function for the electrostatic terms and a potential switch function for the van der Waals terms; see Chapter 9 for details of such procedures.

[9]In a right-handed form, a right hand held with the thumb pointing upward in the direction of the helix axis will wrap right (counterclockwise) and around the axis to follow the chain; a left hand with an upward-pointing thumb will wrap to the left (clockwise) to follow the chain direction of a left-handed helix.

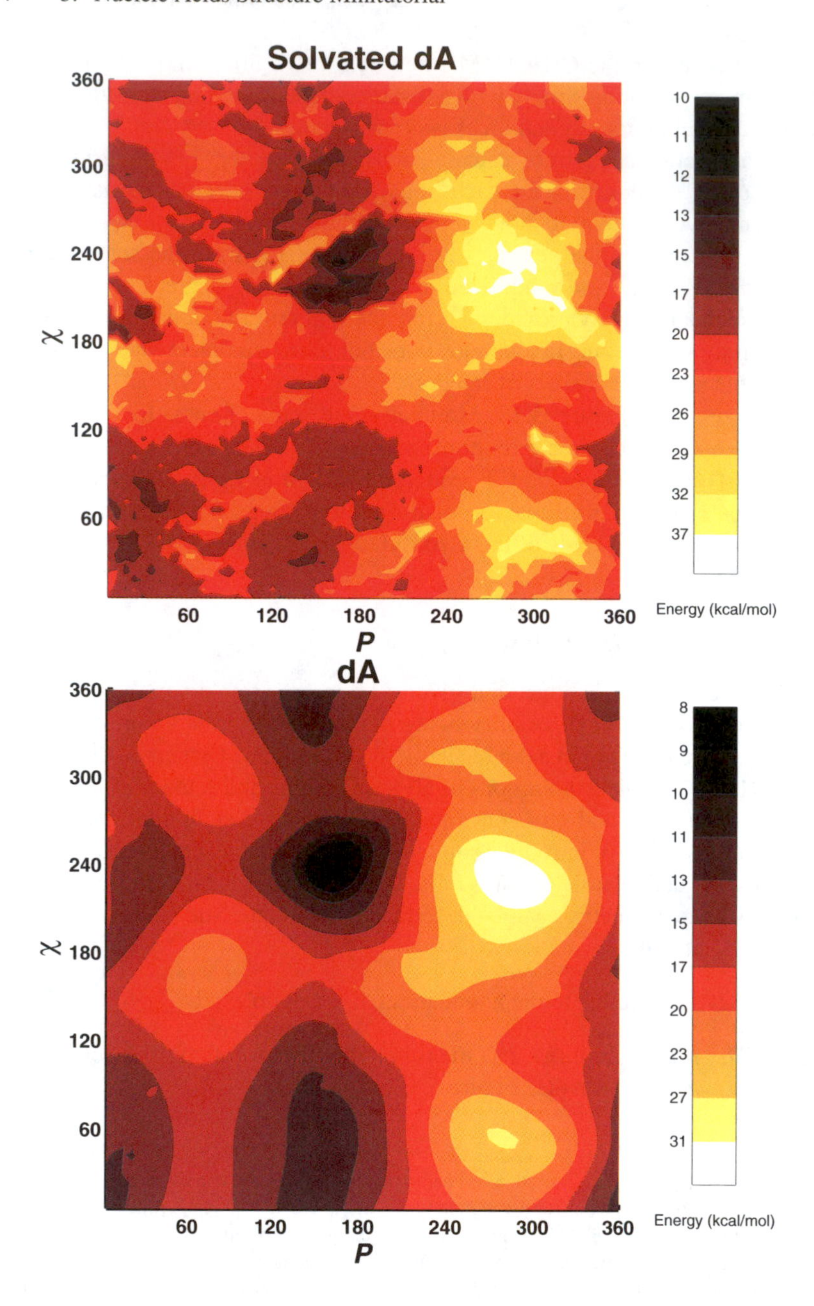

Figure 5.10. Adiabatic maps for two models of deoxyadenosine in the $\{P, \chi\}$ space, the sugar pseudorotation parameter and the sugar-to-base torsion angle, respectively, as calculated with the all-atom CHARMM 27 force field for nucleic acids [307, 305]. Hydration is modeled with approximately 140 water molecules to produce the top figure, and approximated with a distance-dependent dielectric function to generate the bottom figure, as described in the text. Only the internal energy of the dA system (i.e., excluding water/dA and water/water energies) is used for plot generation in both cases.

- *Number of residues per turn*, n_b, is the number of bps for every complete helix turn.

- *Axial rise*, h, is the characteristic vertical distance along the double helix axis between adjacent bps.

- *Unit twist* or *rotation per residue* for a repetitious helix is $\Omega = 360^{\circ}/n_b$ and describes the characteristic rotation about the global helix axis between two neighboring base pairs.

- *Helix diameter* refers to the geometric diameter of the helical cylinder (around 20 Å for DNA).

- *Major* and *minor grooves* (Mgr and mgr in Figure 5.1) refer to the spaces generated by the asymmetry of the DNA. The two different-sized grooves are most apparent from a side view of the double helix (Fig. 5.13). The *minor groove side* is defined as the space generated along the edge *closer* to the two glycosyl linkages of a bp, and the *major groove side* is generated by the other edge, *farther* from those links (see also Figure 5.11).

  In the classic DNA described by Watson and Crick, the minor groove is narrower (around 6 Å wide) compared to the major groove (which is doubly wide) and slightly deeper (8.5 vs. 7.5 Å). For the A-model of DNA discussed below, the 'minor' groove is as large, or larger, and also more shallow, than the 'major' groove according to the above definition.

  Characterizing helical grooves is important for describing interactions of nucleic acids with solvent molecules and with proteins. A larger accessible area of a groove can facilitate nonspecific, as well as sequence-directed, protein recognition and binding. The edges of the bps, which form the bottom of the grooves, contain nitrogen and oxygen atoms available for contacting protein side chains via hydrogen bonds. The hydrogen bonds form the basis of sequence-specific recognition of DNA by proteins.

5.3.6 *Base-Pair Parameters*

The geometric variables above and others are necessary to describe the *global* and *local* arrangements of base pairs in nucleic acid helices. The global parameters describe the overall arrangement of the bps in double-stranded helices, while the local variables specify the orientation between successive bps. The global helical parameters are thus measured for a particular bp with respect to the overall (global) helix axis, while the local variables are defined in the local framework of two successive bps. *The global and local helical parameters can be entirely different quantities.* See [311] for related transformations between global and local variables.

Table 5.3. Mean properties of representative DNA forms.

Property	A-DNA	B-DNA	Z-DNA
Handedness	Right	Right	Left
Representative Structures	GGCCGGCC CGTATACC	CGCGAATTCGCG	CGCGCG
Bps/turn	11	10	12 (6 dimers)
Rise/base pair	2.6 Å	3.4 Å	3.8 Å (ave.)
Helix diameter	≈ 26 Å	≈ 20 Å	≈ 18 Å
Helix pitch	≈ 28 Å	≈ 34 Å	≈ 45 Å
Twist/residue	33°	36°	−60°/dinuc.
Bp inclination	20°	0°	−7°
Sugar pucker	C3′-endo	C2′-endo	C2′-endo (C)/ C3′-endo (G)
Glycosyl rotation	*anti*	*anti* (higher)	*anti* (C)/ *syn* (G)
Major groove	narrow & deep	wide & deep	convex
Minor groove	very wide & shallow	narrow & deep	very narrow & deep

In addition to these two groups of conformational variables, other parameters describe the orientation of the *two bases* in a hydrogen-bonded base pair.

Below, we define some of the global parameters for bp orientations (like *tip* and *inclination*), local variables (associated with successive bps) like *roll*, *tilt*, and *twist*, and a variable that describes orientations within a bp (*propeller twist*). See Dickerson [309] for complete definitions of all rotational and translational variables.

The description of these quantities requires definition of a reference coordinate frame, which we introduce next.

Reference Frame

The commonly used reference frame shown in Figure 5.11 [308] defines a coordinate system with unit vectors $\{\mathbf{e1}, \mathbf{e2}, \mathbf{e3}\}$ so that **e1** represents the short bp axis, **e2** represents the long bp axis, and **e3** is the normal to **e1** and **e2** which completes the right-handed coordinate system (i.e., defined as the cross product $\mathbf{e3} = \mathbf{e1} \times \mathbf{e2}$). The direction of the long-bp axis can be defined by connecting the two $C1'$ atoms of the pyrimidine and purine atoms. The short-bp axis (also called the *dyad*) can be constructed by passing a perpendicular vector from the midpoint of this $C1'$ (purine) to $C1'$ (pyrimidine) line. The intersection of the dyad with the C8 (purine) to C6 (pyrimidine) line is considered the origin of this plane.

An alternative standard reference frame describing nucleic-acid bp geometry was proposed at an international workshop held in 1999 in Tsukuba, Japan [312].

Table 5.4. Selected parameters for constructing model DNA from nucleotide geometric variables (dinucleotide for Z-DNA) according to Figure 5.13, as developed in [310] and used to generate the structures in Figure 5.13. See also Figure 5.11 for definitions of the translational variables dx and dy and the rotational variables tip (θ), inclination (η), and twist (Ω). The parameter h is the helical rise.

Helix	α	β	γ	δ	$P/\tau_{\max}$	χ
A-DNA	−62	173	52	88	3/38	−160
B-DNA	−63	171	54	123	131/36	−117
Z-DNA (dG)	47	179	−165	9	−1/23	68
Z-DNA (dC)	−137	−139	56	138	152/35	−159

Helix	dx	dy	h	θ	η	Ω
A-DNA	4.0	0	2.87	0	13.5	32.2
B-DNA	0	0	3.33	0	0	37.3
Z-DNA (dG)	−3.0	2.5	−3.72	0	−7	52 (G→C)
Z-DNA (dC)	−3.0	−2.5	−3.72	0	−7	8 (C→G)

Global Variables (Base Pair Orientations With Respect to Helical Axis)

The reference frame defined above can be used to define deviations of the bp as a whole with respect to the overall helical axis. These include the rotational variables *tip* and *inclination*, and translational variables like dx and dy.

- Tip (θ in standard conventions) measures the rotational deformation of the bp as a whole about the long bp axis **e2**. It is considered positive when the rotation is clockwise as shown in Figure 5.11, moving the far side of the bases, or the major groove side, as viewed along **e1**, below the paper plane.

- Inclination (η in standard conventions) measures the rotational deformation of the bp as a whole about the short bp axis **e1**. This angle is considered positive when it is clockwise as shown, moving the far base when viewed along **e2** (G in Figure 5.11) down below the paper plane.

- The displacement parameters dx and dy denote the translations of the mean bp plane from the global helical axis, along **e1** and **e2**, respectively (see Figure 5.11). They indicate the shift of the bp origin (the point through which **e3** passes for a particular helical model). A positive dx indicates translation towards the major groove direction, and a positive dy denotes displacement toward the first nucleic acid strand of the duplex (see Strand I in Figure 5.12).

  For A and B-DNA, the helix axis lies approximately on the dyad, but for Z-DNA the helix axis is displaced from the dyad toward a pyrimidine atom and points down rather than up.

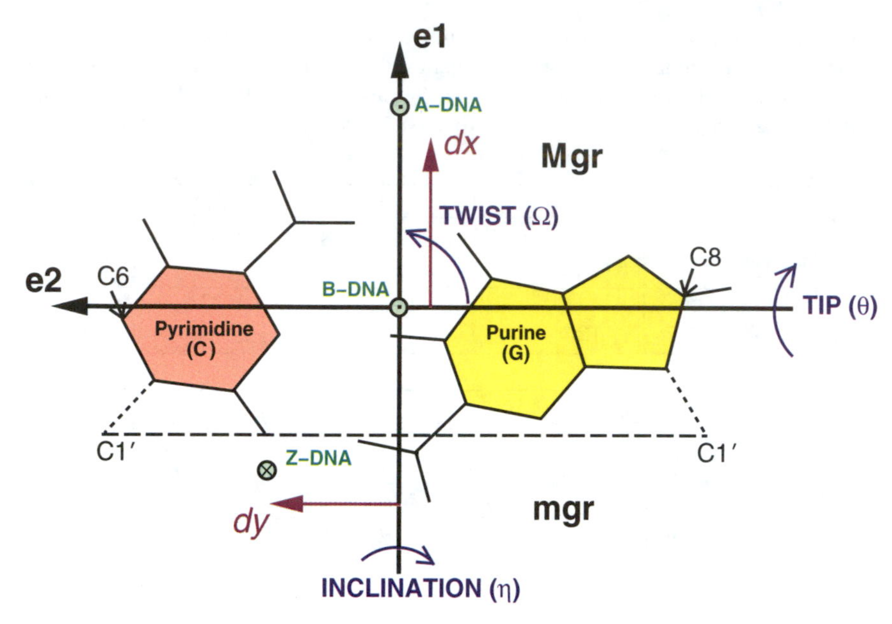

Figure 5.11. The local base-pair coordinate system $\{\mathbf{e1}, \mathbf{e2}, \mathbf{e3}\}$ representing the short base-pair axis, the long base-pair axis, and the normal to both which completes a right-handed coordinate system. Associated translational and rotational parameters are indicated as detailed in the text. The symbols Mgr and mgr define major and minor groove sides of the base pair, respectively. The positions of A-DNA and Z-DNA helix centers are only illustrated for perspective relative to B-DNA. The global helix direction for Z-DNA would point in the opposite direction (down from the paper plane) according to standard definitions [308].

The A-DNA double helix lies on the major groove side ($dx > 0$ as shown in Figure 5.11 with respect to the B-DNA helical axis), while the Z-DNA helix lies on the minor groove side ($dx < 0$, as shown in Figure 5.11). Note that the signs of dx and dy should be reversed when meaning the displacements of the mean A-DNA and Z-DNA bp planes from their respective global helical axes.

Local Variables (Base-Pair Step Orientations)

The *roll*, *tilt*, and *twist* angles (see Figure 5.12) define the rotational deformations that relate the local coordinate frames of two successive bps. The three local translational variables are *slide*, *shift*, and *rise*.

- Roll (ρ) defines the deformation along the long axis of the bp plane and describes DNA groove bending: a positive roll angle opens up a bp-step towards the minor groove, while a negative roll angle opens up a bp-step towards the major groove.

- Tilt (τ) is the deformation defined with respect to the short axis of the bp plane. A positive tilt angle opens the bps on the side of the sequence strand (Strand I in Figure 5.12).
- Twist (Ω) is the helical rotation between successive base pairs, as measured by the change in orientation of the C1′–C1′ vectors between two successive bps, projected down the helix axis, as shown in Figure 5.12.
- The translational *slide* (Dy) motion describes the relative displacement of successive bps along their long axes as measured between the midpoint of each pyrimidine–purine long-bp axis. It is considered positive when the direction is toward the first nucleic acid strand (i.e., positive dy direction), as shown in Figure 5.12. The other local translational variables are the shift (Dx) and rise (Dz).

Deviations Within a Base Pair

- The propeller twist (ω in standard conventions) measures the angle between the normal vectors associated with the planes of the two bases in a hydrogen-bonded pair (from the torsion angle between the individual base planes). Imagine the motion of a helicopter propeller where the two bases twist in opposite directions about the long bp axis **e2** (one up and one down), as shown in Figure 5.12.

  According to nucleic-acid conventions [308], when the angle is viewed from the minor groove edge of a bp, it is positive when the base on the left moves its minor groove edge up while the base on the right moves down. Alternatively, we define a positive angle when each base rotates in a counterclockwise manner when viewed from its attached sugar/phosphate backbone. The propeller twist has a *negative* sign under normal conditions.

- The two other rotational variables that describe deformations within a hydrogen-bonded bp are *buckle* (κ), about the short bp axis **e1**, and *opening* (σ), about the **e3** [308].

5.4 Canonical DNA Forms

Small changes in the global and local parameters introduced above can lead to large overall changes in helix geometries. The double helix described by Watson and Crick in 1953 — now known as B-DNA — was deduced by adjusting wire models so as to fit the X-ray diffraction patterns recorded in the 1950s from calf thymus DNA fibers, first manually and later by various model-building and refinement analyses (summarized in [313]). The fiber diffraction patterns provide an excellent reference for describing features of canonical B-DNA forms since they are generally devoid of the end effects evident in analysis of crystal structures of

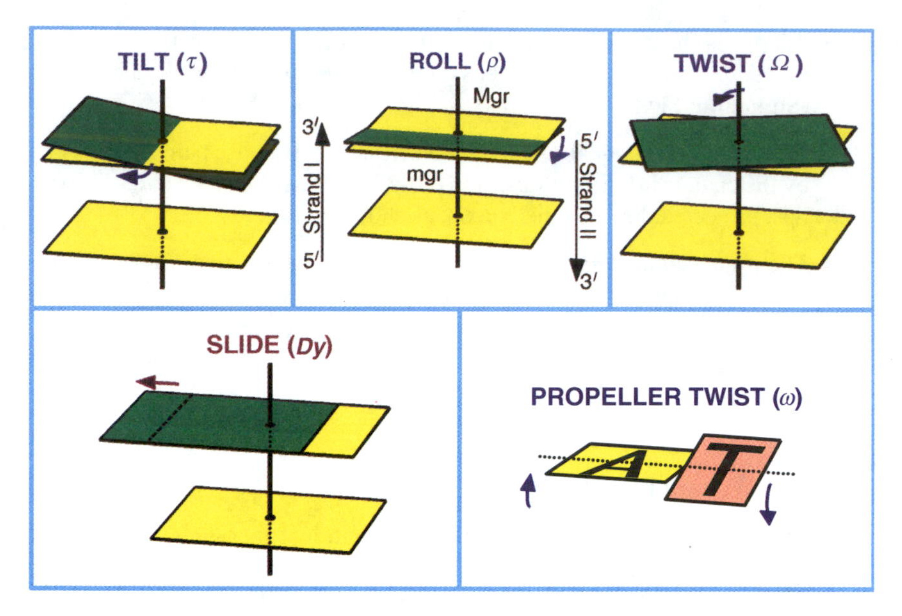

Figure 5.12. Base-pair step (tilt τ, roll ρ, twist Ω, slide Dy) and base pair (propeller twist ω) orientation parameters, with positive displacements shown; see text for details. Mgr and mgr denote the major and minor-groove sides of the bps, respectively. The sequence strand (I) and the complementary strand (II) are indicated in the roll illustration.

DNA oligomers. They also represent average structures over all sequences in the fiber.

The right-handed B-form is the dominant form under physiological conditions. One possibility for its prevalence is that the B-DNA helix can be smoothly bent about itself to form a (left-handed) superhelical (*plectoneme* or toroid-like form; see next chapter) with minimal changes in the local structure. This was first suggested by Levitt by early molecular simulations [314]. This deformability property facilitates the packaging of long stretches of the hereditary material in the cell (especially of circular and topologically-constrained DNA) by promoting volume condensation as well as protein wrapping.

Yet we now recognize numerous variations in polynucleotide structures — both helical and nonhelical forms — that depend profoundly on the nucleotide sequence composition and the environment (counterions, relative humidity, and bound ligands or other biomolecules).

The canonical B-DNA was deduced from X-ray diffraction analyses of the sodium salt of DNA fibers at 92% relative humidity. Another form of DNA — now termed A-DNA — emerged from early X-ray diffraction studies of various forms of nucleic acid fibers at the much lower value of 75% relative humidity. This alternative helical geometry is prevalent in double-helical RNA structures and in duplex DNA under extreme solvation conditions in certain sequences (such as runs of guanines).

Though both these early diffraction-based models were inherently low in resolution and contained several incorrect structural details, later analyses of single DNA crystals concurred with these basic fiber diffraction findings. The DNA fiber structure analyses also served as a reference by which to analyze the sequence-dependent trends that emerged from oligomer crystallography [313].

Both the A and B-DNA forms are right handed. A rather surprising finding, first discovered by single crystal X-ray diffraction and rediscovered 25 years after Watson and Crick's description of DNA, was a peculiar left-handed helix with a zigzag pattern. Andrew Wang, Alexander Rich, and their collaborators observed this form in crystals of cytosine/guanine polymers (dCGCGCG) at high salt concentrations and dubbed it Z-DNA (for its zigzag pattern) [315]. This high ionic environment stabilizes Z-DNA relative to B-DNA by shielding the closer phosphate groups on opposite strands and hence minimizing the otherwise increased repulsive interactions.

The biological function of Z-DNA remains in the forefront of research, but recent evidence suggests that the conversion of helical segments from B to Z-like acts as a genetic regulator.

Below, these three families of DNA helices are detailed; see Figures 5.13, 5.14, and 5.15 for comparative illustrations.

5.4.1 B-DNA

B-DNA can be distinguished by the following characteristics:

- The helix axis runs through the center of each bp ($dx, dy \approx 0$).
- The bps stack nearly perpendicular to the helix axis (small inclination of the bps and very small roll and tilt values). This implies that bases in adjacent steps of the same strand overlap vertically (*stack*), and bases on the opposite strands do not stack.
- The mean helical twist (Ω) is about 34–36°.
- There are about 10–10.5 bps per turn.
- Deoxyriboses favor a C2′-endo sugar conformations (S region of pseudorotation cycle); see Figure 5.8.
- The glycosyl bond orientation is typically higher *anti*; see Figure 5.9.
- The major groove is wider (12 Å) and deeper (8.5 Å) than the minor groove, which is 6 Å wide and only slightly less deep.

Overall, these features produce a model helix of the form shown in Figures 5.13, 5.14, and 5.15. Note that the top view in the space-filling stereo figure (5.15) reveals no hole in the helix cylinder since the global helix axis intersects the bps. Ordered water spines in DNA structures have been reported along the minor groove and around the phosphate groups (see next chapter) [316].

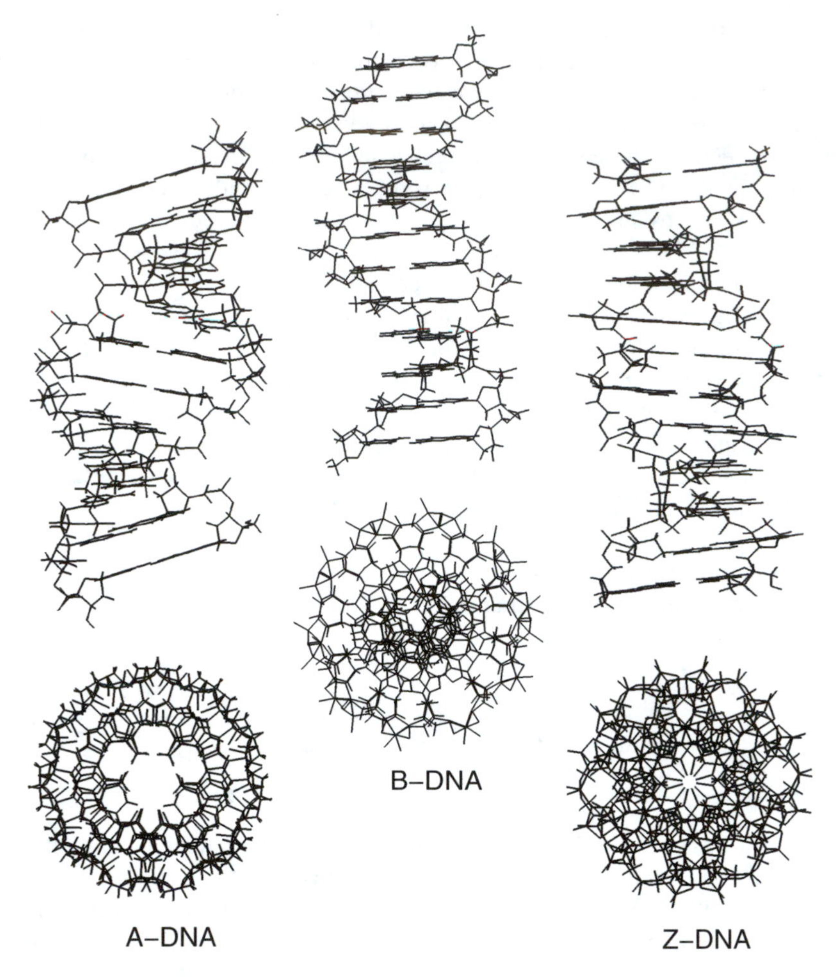

Figure 5.13. Model A, B, and Z-DNA (top and side views) as generated by the polynucleotide building program developed in [310] based on parameters listed in Table 5.4.

5.4.2 A-DNA

The A-DNA helix is very different in overall appearance than B-DNA. Specifically:

- The bp center is shifted from the global helix axis ($dx \approx 4$ Å, $dy \approx 0$ in Figure 5.11).
- A prominent inclination is noted for the bp planes, as large as 20° on average. This implies a combination of both intrastrand and interstrand base stacking for most sequences (the exception being pyrimidine/purine steps, which overlap in an interstrand manner only).

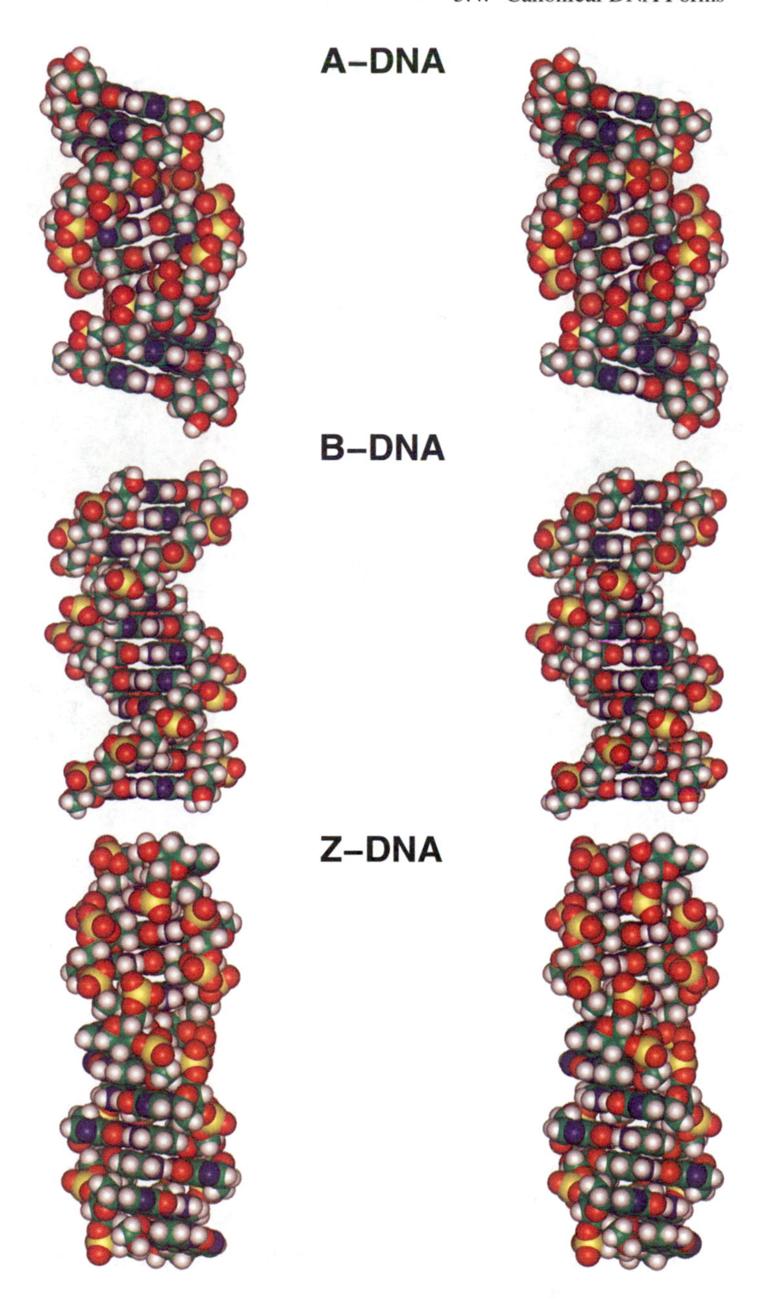

Figure 5.14. Space-filling stereo figures of model A, B, and Z-DNA, as generated by the polynucleotide building program developed in [310] based on parameters listed in Table 5.4. The stereo side-view images are rotated about 8° relative to one another, with the center of images separated by about 6.5 cm, the average distance between two human eyes. Images are prepared for cross-eyed viewing from about 46–51 cm. See Figure 5.15 for corresponding top views.

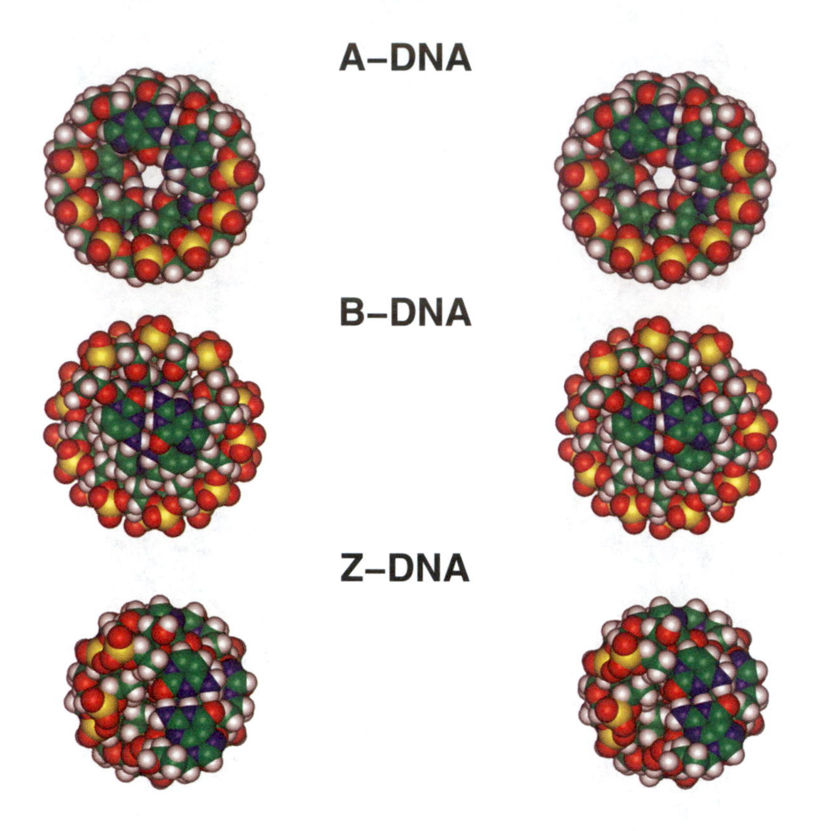

Figure 5.15. Space-filling stereo figures of model A, B, and Z-DNA, as generated by the polynucleotide building program developed in [310] based on parameters listed in Table 5.4. The stereo top-view images are rotated about 8° relative to one another, with the center of images separated by about 6.5 cm, the average distance between two human eyes. Images are prepared for cross-eyed viewing from about 46–51 cm. See also Figure 5.14 for corresponding side views.

- The mean helical twist is less than for B-DNA, about 33°.
- There are thus 11 bps per turn, producing a shorter helix length than B-DNA for the same number of bps.
- The sugars favor a C3′-endo pucker (N region of the pseudorotation cycle) rather than C2′-endo as in B-DNA; see Figure 5.8.
- The glycosyl bond orientation is typically *anti*, as in B-DNA; see Figure 5.9.
- The minor groove is not as deep as in B-DNA, but the major groove is narrower and deeper than the minor groove.

From Figure 5.13, note the dramatic inclination of the bps and the hollow top view of the helix, a consequence of the bps being pulled closer to the sugar/phosphate backbone.

A-DNA regions might exist within a generally B-DNA helix (e.g., in runs of poly(dG)·poly(dC)) under extremes conditions only. Certain RNA molecules that adopt partially double-helical forms, such as tRNA, rRNA, and parts of mRNA, tend to be A-like in the duplex regions, with characteristic C3′-endo sugar puckers, because the B-DNA conformation leads to steric clashes between the two sugar hydroxyl groups.

5.4.3 Z-DNA

In contrast to the A and B-DNA models, specific sequences are required for Z-DNA (e.g., alternating GC). The repeating unit for Z-DNA is a dinucleotide rather than a mononucleotide. The regular composition of alternating pyrimidines and purines allows alternating features in geometry, for example in the sugar and glycosyl orientations. Besides its distinguishing left-handedness, the following are properties of the slender Z-DNA helix (modeled for GC polymers):

- The bp center is shifted slightly from the helix axis along **e1** and **e2** ($dx \approx -3$ Å, $dy \approx \pm 2.5$ Å in Figure 5.11).
- The bps are inclined slightly ($\eta \approx -7^{\circ}$).
- The mean helical twist is about 60° *per dimer*, with about 52° for the G $\rightarrow$ C step and 6° for the C $\rightarrow$ G step [310].
- There are 12 bps per turn.
- The sugars adopt C2′-endo conformers at C but C3′-endo forms at G; see Figure 5.8.
- The glycosyl link is *anti* for C but *syn* for G in alternating CG sequences; see Figure 5.9.
- The major groove bulges out and the minor groove is narrow and deep.

Like the A-form, a stretch of Z-DNA might occur within B-DNA, but direct experimental evidence is lacking. The negative supercoiling of naturally occurring DNA (see next chapter) may also promote Z-DNA, or other left-handed, helix formation.

Biological Significance

The biological role of Z-DNA forms continues to be an active area of research. It has been recently found that when DNA adopts the Z conformation, an editor RNA molecule — the double-stranded RNA enzyme adenosine deaminase (ADAR1) — can bind to left-handed DNA and alter the base in a codon (for example from adenine to guanine) and thereby produce alternative forms of the translated proteins. The Z-DNA binding domain of ADAR1, namely Zα, was recently co-crystallized with a 6-bp DNA fragment [317]. This newly-identified role for Z-DNA — a regulator of the nucleotide template which directs protein synthesis — may have practical benefits as another element of biological control.

5.4.4 Comparative Features

Table 5.3 summarizes basic features of model A, B, and Z-DNA forms as compiled and reconciled from several textbooks. See also Figure 5.8 for sugar analysis and Figure 5.9 for torsion-angle analysis.

Such model helices can also be constructed from building blocks of nucleotides (for A and B-DNA) and dinucleotides (for Z-DNA), for example using the geometric variables shown in Table 5.4 [310].

In practice, departures from model-helix variables occur frequently and the observed ranges in some of these quantities are quite large and differ markedly from one residue to the next.

For example, analyses of high-resolution DNA structures collected since the early 1980s indicate a mean twist angle (Ω) for the right-handed B-DNA of around 34° (rather than 36°), though a broad sequence-dependent range (roughly 20–50°) is observed. The helical twist of 34° corresponds to about $n_b = 10.5$ bps per turn (rather than 10), 36 Å for the helical pitch, and $h = 3.4$ Å for the axial rise (or rise per residue).

A large field of research is devoted to unraveling the sequence-dependent features of DNA. These sensitive patterns are important for numerous biological functions involving DNA, such as protein binding and transcription regulation. The next chapter introduces basic elements of DNA sequence effects, DNA hydration and ion interactions, DNA/protein interactions, RNA structure, cellular aspects of DNA organization, and DNA supercoiling.

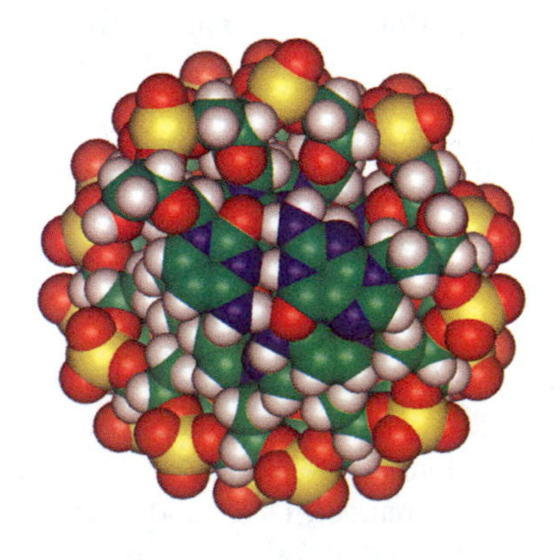

6

Topics in Nucleic Acids Structure

Chapter 6 Notation

Symbol	Definition
h	helical rise, or base-pair step separation (e.g., 3.4 Å)
k_B	Boltzmann's constant
n	number of DNA base pairs
n_b	number of base pairs per DNA turn
p_b	DNA bending persistence length
p_{tw}	DNA twisting persistence length
r	elastic ratio of bending to twisting elastic constants, A/C
s	arclength
A	DNA bending rigidity
C	DNA torsional-rigidity constant
Dy	slide (between successive base pairs)
$\mathcal{L}$	DNA contour length
Lk	linking number (topological invariant)
Lk_0	reference linking number
$\langle R^2 \rangle$	mean square displacement
T	temperature
$\langle T_c \rangle$	site juxtaposition time
Tw	total twist (geometric variable)
Wr	writhing number (geometric variable)
θ_b	bend angle (for isotropic polymer models)
θ_R	net roll angle
θ_T	net tilt angle
κ	curvature (also Debye screening parameter)
ρ	roll angle (between successive base pairs)
σ	$\Delta Lk/Lk_0$ (superhelical density)
σ_e	Poisson's ratio (for an elastic material)

Chapter 6 Notation Table (continued)

SYMBOL	DEFINITION
τ	tilt angle (between successive base pairs)
ω	DNA twist rate; also propeller twist (within a base pair)
ω_0	intrinsic twist rate of DNA (e.g., $2\pi/10.5$ radians)
ΔLk	linking number difference with respect to Lk_0
Ω	twist angle

It has not escaped our notice that the specific pairing we have postulated immediately suggests a possible copying mechanism for the genetic material.

Molecular Structure of Nucleic Acids, J.D. Watson and F.H.C. Crick (1953).

6.1 Introduction

This chapter introduces further topics in nucleic acid structure, building upon the minitutorial of the previous chapter. These topics include DNA sequence effects, DNA hydration, DNA/protein interactions, alternative hydrogen bonding schemes, non-canonical helical and hybrid structures, DNA mimics, overstretched and understretched DNA, RNA structure and folding, and DNA supercoiling by computer simulations.

If I dare write something on any of these topics — given the excellent books and book sections [284, 285, 318, 268, 287, 288, 289, 290, 28, 291] and numerous articles already written on these subjects — it is only because these important areas in nucleic acid structure cannot be omitted.

The sequence-induced variations of DNA, for example, profoundly affect the local structure of DNA and that of DNA complexed with other molecules (water, ions, ligands, and proteins). Such local variations translate into global structural effects when considered on the molecular and cellular levels. The associated energetic and dynamic effects, in turn, affect key functional processes, from the action of regulatory proteins (replication, repair, transcription, etc.) to genome packaging and processing. This functional influence of DNA sequence on biological activity has been shown by the works of Olson, Zhurkin, Dickerson, and many others (e.g., [319, 320]).

Until recently, nucleic acids might have been considered 'the silent partner' in regulatory complex formation with proteins. This view is beginning to change as our appreciation grows for the subtle, sequence-dependent effects of nucleic acids.

Though presenting these topics in this text contrasts with the very brief and basic protein chapters, I hope to encourage mathematical and physical scientists who may not read structural biology texts to pursue research in the area of nucleic acids. Aspects of DNA and RNA structure that may interest mathematicians, computer scientists, or engineers, include DNA supercoiling; RNA folding prediction; application of elasticity theory, topology, and knot theory to DNA structure prediction and functional mechanisms; and using DNA as a parallel computer (e.g., [281, 282, 283]). The presentations here might introduce readers to some of the exciting areas of research on DNA.

Besides the excellent texts [284, 285, 268, 291], the reader is invited to explore the wealth of related structural data on the nucleic acid database NDB (ndbserver.rutgers.edu/) and the protein data bank/research collaboratory for structural bioinformatics (PDB/RCS) resource (www.rcsb.org/pdb/) [321].

As in the previous chapter, we abbreviate *base pair* and *base pairs* as bp and bps, respectively.

6.2 DNA Sequence Effects

6.2.1 Local Deformations

What makes DNA such a rich resource of structural and functional information? The subtle geometric, electrostatic, and mechanical differences inherent in each nucleotide (e.g., C), bp (e.g., G–C), or bp step (e.g., AT in the 5′ GCG**AT**CGC 3′ octamer) combine to affect patterns in local hydration and local helical parameters, such as *twist* (Ω), *tilt* (τ), *roll* (ρ), and *slide* (Dy) between two successive bps (one bp-step), and propeller twist (ω) within each bp, as introduced in the previous chapter and Figure 5.12.

Recall that *twist* is the relative rotation between two successive bps about the helical axis as measured by the change in orientation between the corresponding C1′–C1′ vectors, projected down the helix axis.

Roll and *tilt* define the deformations between two bp planes along the long and short bp axes, respectively, perpendicular to the helical axis. The sign conventions are such that a positive roll angle opens up the minor groove side, while a positive tilt angle opens up bps in the direction of the phosphate backbone of the first strand.

Slide describes the translational displacement of neighboring bps along their long axes, as measured between the midpoint of each long-bp axis. It is considered positive when the direction is toward the first nucleic acid strand (i.e., positive **e2** direction).

The *propeller twist* defines the angle between the planes of the two *bases* in a hydrogen-bonded pair. A positive value corresponds to rotation of each base in a counterclockwise manner when viewed from its attached sugar/phosphate backbone [308].

Normal propeller twists are negative in DNA. This is a consequence of base-pair stacking interactions along polynucleotide strands, as first discussed in [314,

322]. (Note: some of the earlier sign conventions were opposite than the current standard).

Propeller twisting can improve bp stacking interactions through enhanced van der Waals, electrostatic, and hydrophobic contacts. It can also promote stability through additional hydrogen-bonding contacts. Such contacts involve diagonally-distant bases in two adjacent residues. Stability is enhanced since the distance that would result from a perfectly stacked arrangement (perpendicular to the global helix axis) is decreased due to propeller twisting. Regions of repeating AT bps exhibit greater propensity than GC for propeller twisting [268, 323].

The *bending anisotropy* in DNA was established over twenty years ago [324, 325]. Namely, DNA prefers to bend more easily into the minor and major grooves than other directions, and the preference to major or minor-groove depends on the nucleotide sequence; such bending preferences already emerge for dinucleotide steps, as described below.

6.2.2 Orientation Preferences in Dinucleotide Steps

There are 10 unique dimer (dinucleotide) steps in DNA, conventionally specified on the $5' \rightarrow 3'$ strand. Thus, associated with each dimer step is the complementary bp step on the opposite strand (e.g., $5'$ [AC] $3'$ with $5'$ [GT] $3'$); when the usual orientation conventions (for tilt, roll, etc.) are applied to the complementary bp step, the same deformations defined for the main strand result

Local geometric and energetic trends can be associated with each bp step. Trends can also be clustered into three broad classes: pyrimidine/purine (Pyr/Pur), purine/purine (Pur/Pur), and purine/pyrimidine (Pur/Pyr) steps. (Often pyrimidines and purines are also denoted by Y and R for short). This is because common properties such as residue size and electrostatics produce similar geometric and energetic characteristics with these classes of dinucleotide steps.

The three *Pyr/Pur* steps are:

TA (TA) CA (TG) CG (CG);

the four *Pur/Pur* (Pyr/Pyr) steps are:

AA (TT) AG (CT) GA (TC) GG (CC);

and the three *Pur/Pyr* steps are:

AT (AT) GT (AC) GC (GC).

Let us first examine in Table 6.1 the published average roll (ρ), tilt (τ), and twist (Ω) values for these 10 dinucleotide steps as collected over a set of crystallographically-determined for free (unbound) DNA systems [326] (first col-

Table 6.1. Base-pair step parameters for roll (ρ), tilt (τ), and twist (Ω) angles (in degrees) for free [326] and protein-bound DNA [319] (denoted as DNA and DNA$^+$, respectively), as analyzed from crystal structures of DNA and DNA complexes from the NDB. The overline and underline symbols indicate the largest and lowest values, respectively, associated with each angular value in each class (free and complexed DNA). They help distinguish features among the three dinucleotide classes as well as highlight similarities in patterns between free and bound DNA of the same bp step. Note a tie for the minimal value of 0.7° roll for AA and GT steps.

Nuc.		*Pur* **A**		*Pur* **G**		*Pyr* **T**		*Pyr* **C**	
		DNA	DNA$^+$	DNA	DNA$^+$	DNA	DNA$^+$	DNA	DNA$^+$
Pyr **T**	ρ	2.6	3.3						
	τ	0.0	0.0						
	Ω	$\overline{40.0}$	$\overline{37.8}$						
Pyr **C**	ρ	1.1	4.7	$\overline{6.6}$	$\overline{5.4}$				
	τ	$\overline{0.6}$	$\overline{0.5}$	0.0	0.0				
	Ω	36.9	37.3	31.1	36.1				
Pur **A**	ρ	0.5	$\underline{0.7}$	2.9	4.5	−0.6	1.1		
	τ	−0.4	−1.4	$\underline{-2.0}$	$\underline{-1.7}$	0.0	0.0		
	Ω	35.8	35.1	$\underline{30.5}$	31.9	33.4	$\underline{29.3}$		
Pur **G**	ρ	−0.1	1.9	6.5	3.6	0.4	$\underline{0.7}$	$\underline{-7.0}$	0.3
	τ	−0.4	−1.5	−1.1	−0.1	−0.9	−0.1	0.0	0.0
	Ω	39.3	36.3	33.4	32.9	35.8	31.5	38.3	33.6

umn in each dinucleotide entry) and protein-bound DNA systems [319] (second column). For updated values, consult newer works by the authors.

This 4×4 table is symmetric as presented, so the upper triangle is not filled. As ordered, the three top matrix entries (upper triangle cluster: TA, CA, CG) correspond to Pyr/Pur bp steps; the four lower square entries (for AA, AG, GA, GG) correspond to the Pur/Pur steps; and the three entries in the lower triangle cluster (AT, GT, GC) correspond to Pur/Pyr steps. Note that the six maxima (overline symbols) associated with roll, tilt, and twist values for both free and bound DNA occur in the Pyr/Pur cluster; four extremes (lower bounds) are noted in the Pur/Pur cluster, and three in the Pur/Pyr group. (The minimal value of 0.7° average roll displacement occurs for both AA and GT steps).

This observation is related to the decreasing overall flexibility associated with these dinucleotide steps, with Pyr/Pur steps most easily deformed. Note also the similarity in trends for the free and protein-bound DNA entries.

The observed trends in each dinucleotide step can be explained in part by the tendency of DNA to adjust local geometry so as to improve bp *stacking* (overlap) interactions. Since the bp-plane areas are water insoluble, stacking energies can be strengthened by reducing the area between successive bps via local variations, and/or enhancing longer-range interactions through formation of bifurcated hydrogen bonds (between successive bps on opposite strands).

See Box 6.1 for a discussion of the inherent flexibility of Pyr/Pur and Pur/Pur steps, including the contribution of some examples of biological systems where flexibility in Pyr/Pur steps has been noted.

Box 6.1: Inherent Flexibility of Pyr/Pur and Pur/Pur Steps

Pyr/Pur Favorable Configurations. Pyr/Pur steps (i.e., TA, CA, and CG) can produce favorable stacking in one of two general configurations: a combination of *large positive roll* and *small negative slide*, or *near-zero roll* and *small positive slide*, both in association with *propeller twisting* in the two bps [268] (see Figure 5.12 of Chapter 5).

In the former arrangement (large positive roll and small negative slide), the purine bases slide toward each other to improve the cross-chain stacking between them. The concomitant large positive roll inclines the smaller pyrimidine partners to maintain the favored propeller twist in both bps. In the latter (near-zero roll and small positive slide), the large purine bases slide away from each other to avoid a steric clash between them since the propeller twisting of the two bps brings them closer together. The roll is zero since, in the same strand, stacked pyrimidines and purines remain parallel to one another in this propeller-twisted arrangement. (See illustrations in [268]).

Given at least these two options for favorable stacking, the Pyr/Pur class of dinucleotide steps generally displays a wide range of roll values when many structures are analyzed.

CG Example in E2. The high positive roll associated with a CG dinucleotide step has been used to explain the importance of the central CG step in the DNA-binding sequence of the bovine papillomavirus E2 protein [327], whose sequence is ACCGA**CG**TCGGT. This step contributes to the needed deformation of the DNA-binding region — a large overall bending toward the protein.

TG Example in CAP. The characteristic flexibility of another Pyr/Pur step, TG (equivalent to the CA step discussed above), has also been used to explain the importance of the central TG dinucleotide step in the DNA-binding sequence of the catabolite gene-activating protein (CAP) in *E. coli*, a highly conserved dinucleotide segment in both monomers of this dimer protein in different CAP-binding sites. The binding of this complex requires substantial distortion of the DNA, largely modulated by a 40° kink at this central TG step [328, 28].

TA-Rich Regions in Functional Sites. Furthermore, the combination of an observed small energetic barrier to unwinding in Pyr/Pur steps (though not fully understood) and the intrinsic curvature associated with consecutive AT bps (equivalently, AA or TT dinucleotide steps) has been used to explain the prevalence of TA-rich regions in key functional sites. Examples of such sites are TATATATA, also known as the adenovirus E4 promoter, and TATAAAAG, in the adenovirus major late promoter, both of which binds to transcription proteins. These proteins (like the TATA-binding protein TBP in eukaryotes) must unwind DNA, and hence the relatively low barrier to twisting is well exploited. Note also that the standard, Watson-Crick (WC) AT bp has one fewer hydrogen bond than a WC GC bp, which has three, and thus the AT bp should be easier to deform.

Pur/Pur Tendencies. It has also long been noted that the Pur/Pur AA dinucleotide step prefers an orientation where the two successive AT bps are propeller-twisted (i.e., they deviate from planar bp orientations) with near zero roll values (as bases remain parallel to their neighbors on the same strand) [268]. This out-of-plane bending shortens cross-residue distances between the adenines on one strand and the thymines on the opposite strand and allows offset, cross-strand hydrogen bonds to form between an oxygen of thymine and a nitrogen of adenine on the major groove side. This interaction also implies close contact between an oxygen of thymine and carbon of adenine on the minor groove side.

The large positive roll characteristic of CG dinucleotide steps and low roll associated with AA (or TT) steps are particularly consistent trends noted through various structural analyses [329].

6.2.3 Intrinsic DNA Bending in A-Tracts

Now consider all these sequence effects taken together. For heterogeneous, or mixed sequence DNA, all these differing trends will very roughly average out on a global scale for free DNA. In other words, these trends will accumulate slowly in the sense of causing large global distortions. However, this would not be the case if the sequence composition was very regular, as in poly(AT), or contains a motif (such as A_n) that recurs every about 10 bps (i.e., within the repeating unit of the DNA helix); such 'phased' sequences may lead to substantial global effects as a sum of local variations.

A dramatic sequence-dependent pattern of this type from the sum of small net curvature was noted in in the early 1980s in Paul Englund's laboratory [330] in segments of kinetoplast DNA, which displayed anomalously slow gel migration rates.[1]

Studies of this intriguing phenomena continued in Don Crothers' laboratory [331, 332]. Upon close analysis, this behavior was explained by the AT-bp rich content of these DNA sequences. Specifically, kinetoplast DNA contains stretches of consecutive groups of adenines residues (4 to 6), separated by other sequences, *phased* with the helical repeat. This composition of blocks of adenine repeated every ≈10 bps is known as *phased A-tracts*. For example, a representative part of the sequence is:

CCC**AAAAA**TG|TC**AAAAAA**TA|GGC**AAAAAA**T|GCC**AAAAA**TC|,

where the vertical bars mark each 10 residues. Thus, the 5 or 6-membered A-tracts in this sequence occupy positions 3–8, 4–8, or 4–9 within each 10 residues.

[1]The anomaly is a deviation from the usual linear relationship between DNA length and the logarithm of the distance migrated on the gel.

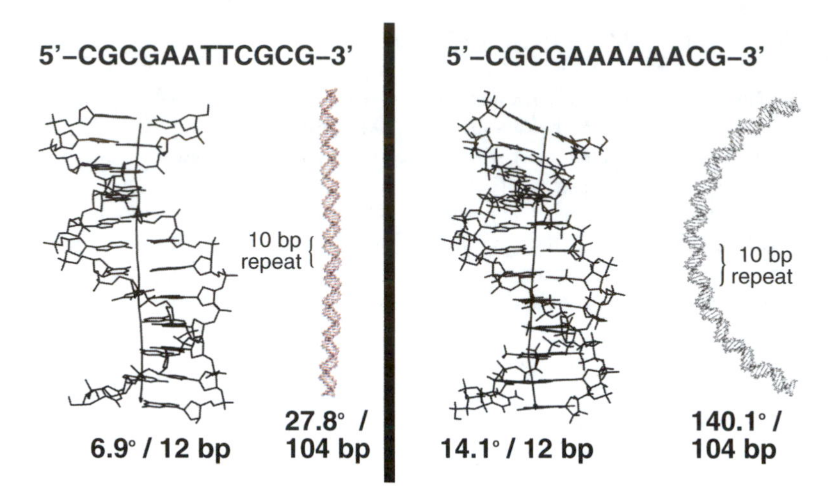

Figure 6.1. Bending on the dodecamer and longer-DNA levels: models of 110 bp DNA were constructed from dodecamer systems with differing bending trends: a gentle degree of helix bending (7° per dodecamer) with no preferred direction of bending, and more substantial bending (14° per dodecamer) with preferred bending realized as minor groove compression, for an A-tract system (1D89 in Figure 6.2). Data are based on molecular dynamics simulations [334]. Scales are different for the long and short DNAs.

The significant curvature observed in association with each A-tract (roughly 15°) [332, 333] thus translates into a highly-curved helix when viewed over hundreds of residues, as shown in Figure 6.1 (right).

The models in Figure 6.1 were built from dodecamer curvature as deduced from MD simulations of two solvated systems [334]: the A-tract system CGCG**AAAAAA**CG solved by crystallography (1D89) [335] and a control sequence CGCG**AA**TTCGCG (known as the Dickerson/Drew dodecamer) [336, 337]. The control sequence — though containing two adenines and thus exhibiting limited aspects of intrinsic bending — has a small overall helix bend. The A-tract oligomer, in contrast, exhibits an overall helix bend of about 14° per dodecamer. This curvature has profound implications on the plasmid level, as seen in the figure, and explains the anomalous migration rates observed in kinetoplast DNA.

Though we now recognize the bent helical axes in DNAs containing phased A-tracts, a clear understanding of the origin and details of intrinsic bending has not yet been achieved despite numerous experimental and theoretical studies [333, 338, 339, 340, 341, 342, for example]. Part of the dilemma in explaining bending involves reconciling experimental data obtained by crystallography [339] versus solution studies since environmental effects can be critical; another component is the consideration of static versus dynamic structures; finally, there is a technical difficulty in quantifying large helical bending in DNA; see Box 6.2.

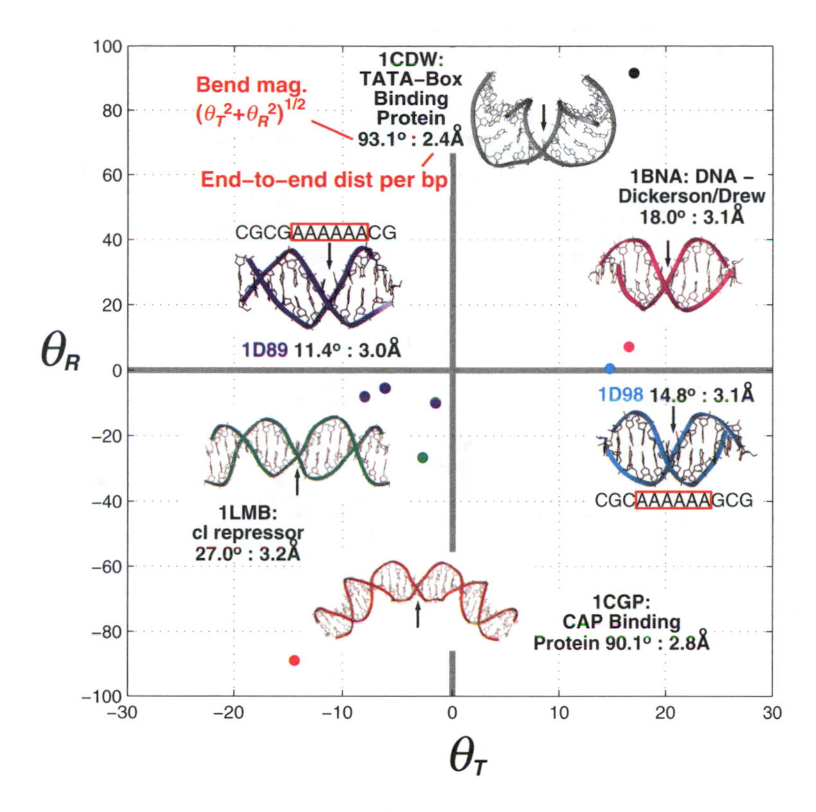

Figure 6.2. Macromolecular examples of *net tilt* (θ_T) and *net roll* (θ_R) combinations, as calculated by eqs. (6.1) and (6.2), for each system in the reference plane indicated by arrows. The bending angle magnitude ($\sqrt[+]{\theta_T^2 + \theta_R^2}$) and the end-to-end distance (normalized per bp) are also shown. Regions of positive and negative θ_R are often termed *"major-groove compression"* or *"minor-groove compression"*, respectively. Co-crystals of dimeric major-groove binding proteins (like CAP and cI repressor) tend to wrap the DNA around the protein with the minor groove at the center of curvature ($\theta_R < 0$). Structures with bends toward the major groove, such as the Dickerson/Drew dodecamer or co-crystals of minor-groove binding proteins like the TATA-box binding protein (TBP), have $\theta_R > 0$. The net bend angles for the two A-tract crystal structures (1D89 and 1D98) correspond to near-zero θ_R (and opposite magnitudes of θ_T) rather than bending toward the minor groove ($\theta_R < 0$), as suggested by simulations and solution models. For 1D89, three crystal forms have been solved (three filled blue circles).

Indeed, various bending models have been proposed for adenine-rich sequences [285]. At the two extremes, the *junction-type model* (developed from models of junctions between two types of helices [343]) suggests that a localized bending between two bps at regions separating A-tract regions from others largely explains the large degree of bending; the *wedge-type model* proposes that bending is smooth, a cumulative effect from the small bending associated with each AA

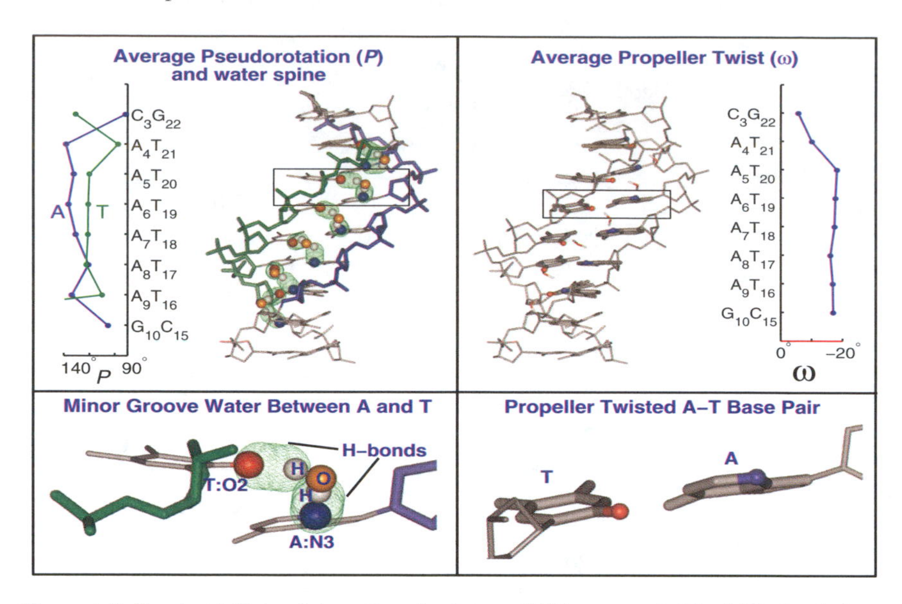

Figure 6.3. Bend-stabilizing interactions in A-tract DNA as revealed by MD simulations [334] of two A-tract DNA dodecamers, 1D89 and 1D98, also shown in Figure 6.2: sugar puckering differences between adenines and thymines on opposite strands (left plot), ordered minor-groove water spine (left dodecamer), and propeller twisting in the A-tract region (right dodecamer and right plot).

dinucleotide step. Clearly, the realistic model involves some combination of sharp and smooth bending models and depends on the residues that flank both the $5'$ and $3'$ sides of the A-tract [334]. A growing body of simulations and experiments are suggesting that the A-tract moieties themselves are relatively straight and that the overall curvature results from substantial rolls at the A-tract junctions, between A-tracts and external sequences, the magnitude of which is sequence modulated [341, 334, for example].

Moreover, the factors that induce and stabilize bending include the tendency of AT bps to be propeller-twisted and exhibit systematic differences in sugar conformations between the adenine and thymine sugars ($\sim 15^\circ$ pucker angle) [344, 68, 345, 346, 334]. These factors combine to compress the DNA minor groove and stabilize it by bifurcated hydrogen bonds (between thymines on one strand and adenines on the opposite strand at successive steps), as illustrated in Figure 6.3. This geometry also leads to a stabilizing, ordered spine of hydration in the minor groove of the A-tracts, with counterion coordination playing a role [68, 278]. Research continues on this front.

6.2.4 Sequence Deformability Analysis Continues

While Crothers' sentiment [329] is now well-accepted: "*the days are over when one dinucleotide step was good as another in determining the structure and mechanical properties of DNA*", our understanding of sequence-dependent deformability in DNA and DNA complexes is far from complete. While clearly much effort has focused on finding a 'set of rules' relating nucleic-acid sequence to preferred structure, it is probably fair to say that we have not yet been illuminated by a useful set of predictive rules.

Consideration of longer bp-step regions, such as trinucleotide and tetranucleotide steps, is one obvious route for establishing clearer predictive rules (and exceptions to the rules); the increasing resolution of DNA crystal structures will undoubtedly facilitate such analyses. Still, even if a set of rules cannot be identified, the local tendencies of DNA are important, as they are known to work along with the deformations induced by proteins.

6.3 DNA Hydration and Ion Interactions

The binding of water molecules (and thus ions) to DNA atoms in the cellular environment is not only important for explaining DNA's conformational stability and energetic properties; it can also help interpret the dependence of DNA conformation on the environment. For example, differences in hydration patterns (specifically with regard to accessible surface area in the grooves) are believed to be a factor in DNA's preference for the canonical A-form at low humidity, higher salt concentrations, and larger amounts of organic solvents, and that of Z-DNA at increased ionic strengths for certain sequences. Both bases and phosphate groups have organized hydration shells [347].

Protein/DNA and drug/DNA interactions are also dependent on water contacts. Proteins can displace water molecules that surround certain DNA regions and replace them with interactions involving amino acid side chains. Thus, hydration sites represent preferential, energetically favored binding sites for hydrophilic DNA binders, like proteins and drugs [348].

Positive counterions are also fundamentally important in shielding the negatively charged DNA backbone and allowing DNA segments to come closer together. Dramatic effects are manifested in long supercoiled DNA molecules, as observed experimentally and studied by simulation (e.g., [349, 350]).

Box 6.2: Net DNA Bending and A-Tract Analysis

A quantitative description of helical DNA bending is far from trivial. Indeed, depending on the choice of the analysis program used for computing DNA parameters, differences in measurements frequently result when the DNA is significantly curved. Part of these differences can be explained by the various protocols used to define reference frames for each base and base pair [351]. This can be resolved by using standard definitions [351]. Still, helical bending is often measured as the angle between two *extremal* vectors connecting two separated bp steps [352, 353, 354, 355, 356, 357].

As an alternative, introduced in [334] are two net-bend angles termed *net tilt* and *net roll* (θ_T, θ_R). These angles are calculated by considering the cumulative projection of tilt and roll at each dinucleotide step, adjusted for twist, into the plane of a reference 'average' bp (typically at the oligomer center). From the tilt, roll, and twist at each bp-step j, namely τ_j, ρ_j, and Ω_j, we compute the net tilt and roll angles as:

$$\theta_T = \sum_j \left[\tau_j \cos\left(\sum_{i=N_c}^{j} \Omega_i \right) + \rho_j \sin\left(\sum_{i=N_c}^{j} \Omega_i \right) \right], \tag{6.1}$$

$$\theta_R = \sum_j \left[-\tau_j \sin\left(\sum_{i=N_c}^{j} \Omega_i \right) + \rho_j \cos\left(\sum_{i=N_c}^{j} \Omega_i \right) \right], \tag{6.2}$$

where $\sum_{i=N_c}^{j} \Omega_i$ is the cumulative twist over i bp-steps from the plane of the reference bp, N_c, to bp-step j [334]. Bends in the helical axis defined by $\theta_R < 0$ are equivalent to 'minor groove compression', a property associated with A-tracts [332] or with the center oligonucleotide of the CAP/DNA complex [328, 28]; bends defined by $\theta_R > 0$ are equivalent to 'major groove compression', as noted for the TATA-box binding protein TBP [28, 207] (see Figure 6.2).

The crystal versus solution data of A-tract systems differ in their characterization of the degree of bending. Crystallographic studies have indicated various bend directions for solved A-tract systems: negative [335, for example] versus positive [358, for example] net tilt, as shown in Figure 6.2. Solution models of A-tract oligomers, in contrast, have suggested a unique bend direction along negative net roll. Such data come from gel electrophoresis (mobility data), NMR (minor-groove water life times), and hydroxyl radical footprinting (compressed minor groove, as inferred from the reduced accessible surface area in the A-tract minor groove to hydroxyl radical [OH^-] cleavage of the DNA backbone). See [359, 332, 335, 360, 361, 362, 363, 339, 364, 365] and other citations collected in [334]. Simulations have shown, however, that small changes from those initially dissimilar bends (from crystallographic models) can yield similar bend directions, equivalent to minor groove compression, in accord with solution data [334]. The crystal models may reflect the inherent conformational disorder (multiple conformations) in A-tract systems, as well as effects by the organic solvent needed for crystallization (2-methyl-2,4-pentanediol or MPD). See also discussion in [340] emphasizing the importance of considering the role of thermal fluctuations and the local chemical environment on DNA curvature and its interpretation.

6.3.1 *Resolution Difficulties*

Though fundamentally important, it has been a challenge in the past to establish precise hydration and ion patterns for biomolecules (see [366], for example, concerning the latter).

First, the number of water molecules, captured in crystal structures of DNA and DNA complexes, is highly dependent on structure resolution, though the high-resolution (e.g., better than 2 Å) structures now appearing should alleviate this problem (e.g., [279]). Counterions, which in the past could not be located in the electron density maps, can also be detected in ultra-high resolution DNA structures [278, 279, 280]. See [366], for example, for a perspective.

Second, the crystal environment of the lattice — periodicity or organic solvent — can also introduce artifacts, so interpretations must be cautious (see also [340]).

Third, NMR spectroscopy measurements for locating protons and other techniques are complicated by the relatively fast, reorganization component of water dynamics (0.1 ps and longer) that can lead to overlap of spectra signals; such experiments can yield more detailed information on the less-transient, and more structured, water molecules, such as the minor groove hydration patterns in DNA A-tracts.

Fourth, computer simulations are vulnerable to the usual approximations (force fields, protocols, etc.), but are nonetheless becoming important in deducing biomolecular hydration and ion patterns [367], especially in identifying counterion distribution patterns for DNA [272, 273] and RNA [368, 369, 370, 371, 372]. Continuum solvent models are also beginning to provide information on the relative stabilities of different sequences and different helical forms (e.g., [101, 373]).

6.3.2 *Basic Patterns*

From all available techniques, the following facts have now been established regarding DNA hydration and ion patterns:

1. **Multiple Layers.** Hydration patterns are inhomogeneous and multilayered, from the first layer closest to the DNA (including the nucleotide-specific hydration patterns and minor-groove 'spine of hydration' [374, 316, 270]) to the outermost layer of the highly transient bulk water molecules. (Intermediate layers are characterized by fast exchange of water molecules and ions with bulk solvent). Water interactions around nucleic acids are important for stabilizing secondary and tertiary structure.

2. **Local Patterns.** Hydration patterns are mostly local, i.e., short range and largely dependent on the neighboring atoms [375]. For example, the water patterns around guanines and adenines are very similar and there are clear differences in the distributions of hydration sites around guanines and cytosines that canonical A, B, and Z-DNA helices share [374]. (Conclusions for adenine and thymine base-pairs are not available due to their strong preference for B-form DNA). The strong local patterns generated near in-

dividual nucleotides permit canonical helices to be reconstructed, complete with preferential hydration sites [374].

To analyze hydration patterns and ions around nucleic acids, thermodynamic, spectroscopic, and theoretical calculations have been used. A useful concept for quantifying hydration patterns is the solvent-accessible surface — introduced to describe the proportion of buried to accessible atomic groups [376]. The three-dimensional quantification of hydration is typically computed as a volume-dependent probability [377, 374, 375].

3. **Structured First Layer.** The first hydration shell is generally more structured. It contains roughly 15–25 water molecules per nucleotide, with varying associated life times. The accessible electronegative atoms of DNA are water bound in this model, including the backbone phosphate region (around phosphate oxygens), sugar oxygen atoms (especially O4′), and most suitable base atoms. In particular, hydrogen bonds form in roughly one quarter of these molecules, bridging base and backbone atoms or two base sites. Interestingly, the accessible, hydrophilic O3′ and guanine N2 groups are statistically un-hydrated [374, 347].

4. **Cation Interactions and Counterion Condensation Theory.** Debye-Hückel theory, along with the counterion condensation theory (see Box 6.3), are invaluable frameworks for understanding how counterions distribute in electrolyte solutions.

   Counterions tend to cluster around a central ion of opposite charge. The *Debye length* κ^{-1} (or Debye radius) is the salt-dependent distance at which the probability of finding a counterion from a central ion is greatest. Debye-Hückel theory predicts the *screening factor* $\exp(-\kappa r)$ (see Chapter 9, Subsection 9.6.4 on continuum electrostatics) by which the ordinary Coulomb electrostatic potential is reduced (see Figure 9.10 in Chapter 9). For DNA, the negatively charged (and thus mutually repulsive) phosphate residues on different strands — separated from each other by about 10 Å — are effectively screened by added salts, thereby stabilizing the helix and allowing the phosphates to come closer together. At physiological ionic concentrations, the Debye length for DNA is about 8 Å. Along with Debye-Hückel theory, counterion condensation theory can be used to analyze the ionic environment around polyelectrolyte DNA and its effect on thermodynamic processes [378, 379, 380, 381].

   As a result of screened Coulomb effects, the association of ions with DNA spreads over a broad range, from very loose associations to tight binding, and the timescale ranges are also broad. Experimental methods tend to capture only average effects, but some insights are emerging from simulation studies. For example, the counterion distribution around DNA that emerged from Monte Carlo simulations [272] revealed two distinct layers of sodium ions, the innermost of which corresponds to the condensed counterion layer, and the outer layer more diffusive [272, Figure 1].[2]

It is well appreciated that cations interact in a nonuniform and sequence-specific manner with the DNA grooves and the phosphate backbone [68, 280]. Major groove cation binding appears to be a more general motif than the more sequence-specific minor-groove cation binding, as observed in A-tracts [382, 383, 384]. Furthermore, divalent cations bind to duplex DNA in a more sequence-specific manner than univalent ions since the former tend to be fully hydrated and thus can donate or accept hydrogen bonds to base atoms through their water ligands [280].

Such details of cation binding to DNA have come from nucleic acid simulations and are beginning to emerge from ultra-high resolution crystal structures of DNA [278, 279, 280]. The facilitation of DNA bending motions by proper cation shielding was demonstrated by classic studies of Mirzabekov, Rich, and Manning on the role of asymmetrically neutralized phosphates in promoting bending around nucleosomes [385, 386]. These findings were highlighted by Maher and co-workers, who examined bending promoted by neutral phosphate analogs [387], and by Williams and co-workers, who examined ion distribution and binding in crystallographic nucleic acid structures [278, 384]. See [366] for a perspective.

5. **Second Hydration Layer.** The second water shell mainly stabilizes the first shell, allowing it to have a favorable tetrahedral structure (see box on water structure in Chapter 3). The outermost, bulk solvent hydration shell is generally more disordered, with its network of interactions rapidly changing.

6. **Hydrated Grooves.** The DNA major and minor grooves are well hydrated. Though the wider major groove can accommodate more water molecules, the minor groove hydration patterns tend to be more structured. The different groove shapes in canonical A, B, and Z-DNA lead to systematic differences in hydration patterns [388, 374, 278]. For example, hydration sites are better defined and form an extensive network in the very deep major groove surface of A-DNA than those in B-DNA (less deep major groove) and Z-DNA (very flat major groove surface); hydration sites have higher densities in the narrower minor groove of B and Z-DNA than in corresponding A-DNA minor-groove sites.

7. **Sequence-Dependent Hydration.** Hydration pattern details depend on the sequence and helical context [388, 374]. The compressed minor groove waters of adenine-rich sequences (A-tracts) are particularly associated with

[2] In nucleic acid simulations, typically one sodium ion per base is incorporated into the first hydration shell for charge neutralization; additional ions (positive and negative) are placed, with positions adjusted by minimization or Monte Carlo procedures so as to produce ions in the most probable regions compatible with the salt concentration of the solution (e.g., [273]).

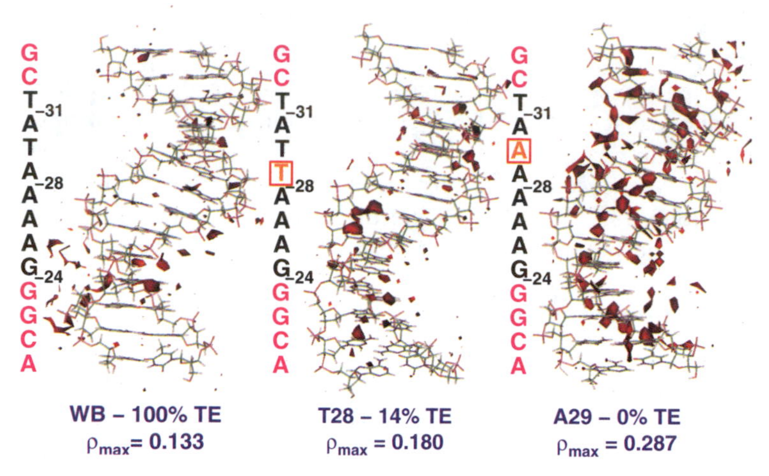

Figure 6.4. Local hydration patterns in three DNA systems (TATA elements differing by one bp) as deduced from molecular dynamics simulations [274] based on crystallographic studies [207]. The sequences are variants of the promoter octamers called TATA elements (for their rich A and T content) [28]. The experimental studies [207] examined the structures of resulting DNA/TBP complexes (TBP is the eukaryotic transcription 'TATA-box Binding Protein') and their relative transcriptional activities (TE values given in figure). The simulations [274] discerned sequence-dependent structural and flexibility patterns, like the maximal water density around the DNA, as shown here for three TATA sequences. The transcriptionally inactive variant shown at right, an A-tract (GCTA$_6$GGGCA), has a much denser local water environment than the wildtype sequence (WB, GCTATA$_4$GGGCA, left). The maximal water densities (number of oxygens) were computed on cubic grids of 1 Å.

long lifetimes. This results from well-ordered water interactions ('spine of hydration') stabilized by cross-strand, bifurcated hydrogen bonds (e.g., N3 of A at step $n+1$ with O2 of T on the opposite strand at step n), where the adenine and thymines are on opposite strands and offset by one step), as shown in Figure 6.3. Such patterns have been extensively studied in association with the central AT-rich region of the Dickerson-Drew dodecamer (CGCG**AA**TTCGCG) [336, 337, 389]. An example of sequence-dependent local water is shown in Figure 6.4 for three DNA sequences that differ by one bp from one another.

8. **Varying Lifetimes.** Water lifetimes vary, from residence times in the several-hundred picosecond range for major groove waters associated with thymine methyl groups to life times of order nanosecond in the minor groove of A-tracts [389, 390, 334].

9. **Phosphate Localization.** Hydration patterns around the phosphate groups exhibit characteristic hydration 'cones' with three tetrahedrally-arranged water molecules in the first hydration shell of each phosphate-charged oxygen [347]. Furthermore, the arrangements depend on the DNA he-

lix type. For example, the shorter distances between successive phosphate groups in A-DNA (with respect to B-DNA) encourage water molecules to form bridges between them. In B-DNA, these waters favor forming hydrogen bonds with the anionic phosphate oxygens O1P and O2P; O5′ is inaccessible and O3′ is surprisingly under-hydrated [348].

Given this multilayered organization of hydration shells in DNA and the large costs associated with fully-solvated simulations of nucleic acids, practitioners have attempted to economize on system size by modeling a limited number of water molecules to represent only the thin, ordered first solvation shell around the DNA [391, 392].

Box 6.3: Counterion Condensation Theory

Counterion condensation theory essentially predicts formation of a condensed layer of mobile and hydrated counterions at a critical polyionic charge density in the vicinity of the DNA surface ($\sim$ 7 Å). For monovalent cations, the concentration of this cloud is relatively independent of the bulk cation concentration. This concentrated cloud of ions effectively neutralizes $\sim$76% of the DNA charge, thereby reducing the negative charge of each phosphate to about one quarter its magnitude (hence the charge scaling done in early nucleic-acid simulations [58, 59, 60]). Divalent and trivalent counterions reduce the residual charge further according to this model (to about one tenth of its magnitude). The resulting phosphate screening reduces the electrostatic stiffness of DNA and also explains the favorable entropy of binding by cation ligands to DNA, since they lead to release of counterions from the condensed ion layer to the bulk solvent.

Various experimental and theoretical models of DNA electrostatics support counterion condensation of polyion charge above a critical threshold of polyion charge density. Theories differ mainly in the structure of the condensed layer [272]. Work on assessment of models in light of experimental observations continues; see [384, 393], for example, for reviews.

6.4 DNA/Protein Interactions

Fundamental biological processes involve the interaction of nucleic acids and proteins. These complexes may involve various types of proteins: regulatory, packaging, replication, repair, recombination, and more. Many questions are now being addressed in this large area of research involving protein/DNA complexes, including:

- How do these proteins recognize DNA? That is, how do the protein and the DNA accommodate each other? What geometric, energetic mechanisms are involved?

Table 6.2. Representative protein/DNA and drug/DNA complexes.

Complex	Binding Motif[a]	Binding Groove[b]	Details of Complex
λ repressor	HTH	Mgr	Canonical HTH; homodimers; 2 helices of Cro dimer cradle Mgr, stabilized by direct H-bond and vdW contacts; little DNA distortion.
CAP repressor	HTH	Mgr	About 90° bend.
trp repressor	HTH	Mgr	Indirect, water-mediated base contacts.
Purine rep.	HTH	Mm	α-helices inserted in mgr.
Yeast MATα2	HTH	Mgr	Homeobox domains bind as monomers.
Zif268	Zn	Mgr	Zinc finger subfamily; each Zn finger recognizes 3 bps.
GATA-1	Zn	Mm	Transcription factors subfamily; single domain coordinated by 4 cysteines.
GAL4	Zn	Mgr	Metal binding subfamily; each of two Zn ions, coordinated by 6 cysteines, recognizes 3 bps.
GCN4	Leu/Zip	Mgr	Canonical; basic region/leucine zipper (α helices) motif; slight DNA bending.
fos/jun	Leu/Zip	Mgr	α-helices resemble GCN4; unstructured basic region folds upon DNA binding.
fos/jun/NFAT	Leu/Zip	Mgr	α-helices bend to interact with NFAT.
MetJ	β-ribbon	Mgr	Two anti-parallel β-strands in Mgr; bends each DNA end by 25°.
papillomavirus E2 DNA target	β-barrel	Mgr	Domed β-sheets form an 8-strand β-barrel dimer interface with 2 α-helices in Mgr; strong tailored fit for every base of the recognition element; bent DNA; compressed mgr; DNA target crystallized without protein.
TBP	β-saddle	mgr	Ten-β-strand saddle binds in Mgr; significant distortion, $\approx$ 90° bend.
p53 tumor supp.	Loop/other	Mm	Binds to DNA via protruding loop and helix anchored to anti-parallel β-barrel.
SRY	Loop/other	mgr	Isoleucine intercalated into mgr.
NFAT	Loop/other	Mm	Flexible binding loop stabilized by DNA.
histones	Loop/other	Mm	Nonspecific PO_4 interactions.
distamycin (drug)		mgr	Selective to AT bps; binds in mgr without distortion.

[a]HTH: helix/turn/helix; Zn: Zinc-binding; β: β-strand motif; Leu/Zip: Leucine zipper/bZIP; Loop/other: motifs with few representative members.

[b]Mgr: binding mainly to major groove; mgr: binding mainly to minor groove; Mm: binding to both grooves.

- How are the mutual interactions stabilized? What are the specific and nonspecific interactions involved?

- How are nucleotide or/and amino acid substitutions tolerated in the target DNA (e.g., [207]) and bound protein regions (e.g., [394, 395])? Numerous mutation studies (experimental and theoretical) are shedding insights into this question and highlighting the residues that are essential to structure and function and those that are more variable, as evidenced by evolutionary trends. Enhancing interactions by design is also an exciting application of such studies.

- What is the relation between structural stability of the complex and biological function? For example, Burley and co-workers [207] showed that TBP can successfully bind to several single-bp mutation variants of the wild type TATA-box octamer element of adenovirus major late promoter (AdMLP), 5′ TATAAAAG 3′, but that the biological transcriptional activity can be sensitively compromised by these mutations.

These are complex questions, but we are addressing them steadily as our collection of complexes between proteins and DNA and between DNA and other molecules (drugs, various polymers) is rapidly growing.

There are now hundreds of solved protein/DNA complexes known, a representation of which is collected in Table 6.2 and illustrated in Figure 6.5. What has certainly been established is that the observed protein/DNA interactions span the whole gamut of possibilities — regarding binding specificity for sequence or grooves, stabilizing motifs, protein topology, overall deformation, tolerance to mutations, and more.

The supplemental section posted on the book's website elaborates on some of these interesting aspects of protein/DNA binding and interactions.

6.5 Variations on a Theme

As evident in the previous sections, nucleic acids are extremely versatile. The classic right-handed double helical structure described by Watson and Crick is an excellent textbook model. In Nature, however, polynucleotide helices have developed an enormous repertoire of sequence-dependent structures, helical arrangements, and folding motifs on both the bp and higher level of folding.

6.5.1 Hydrogen Bonding Patterns in Polynucleotides

Hydrogen-bonding variations are important for DNA's adaptability to base modifications (e.g., methylation, drug or carcinogen binding), bp mutations and mismatches, and various interactions between DNA and proteins and among polynucleotide helices.

Classic Watson-Crick (WC)

The *classic WC* hydrogen-bonding arrangement as shown in Figure 5.3 of Chapter 5 is a particularly beautiful arrangement. It can be appreciated by noting that the C1′ (pyrimidine) to C1′ (purine) distance is around 10.6 Å in both the AT and GC pairs, in good preparation for helix stacking. The intrinsic energy of hydrogen bonds — that is, relative to vacuum — is generally in the 3–7 kcal/mol

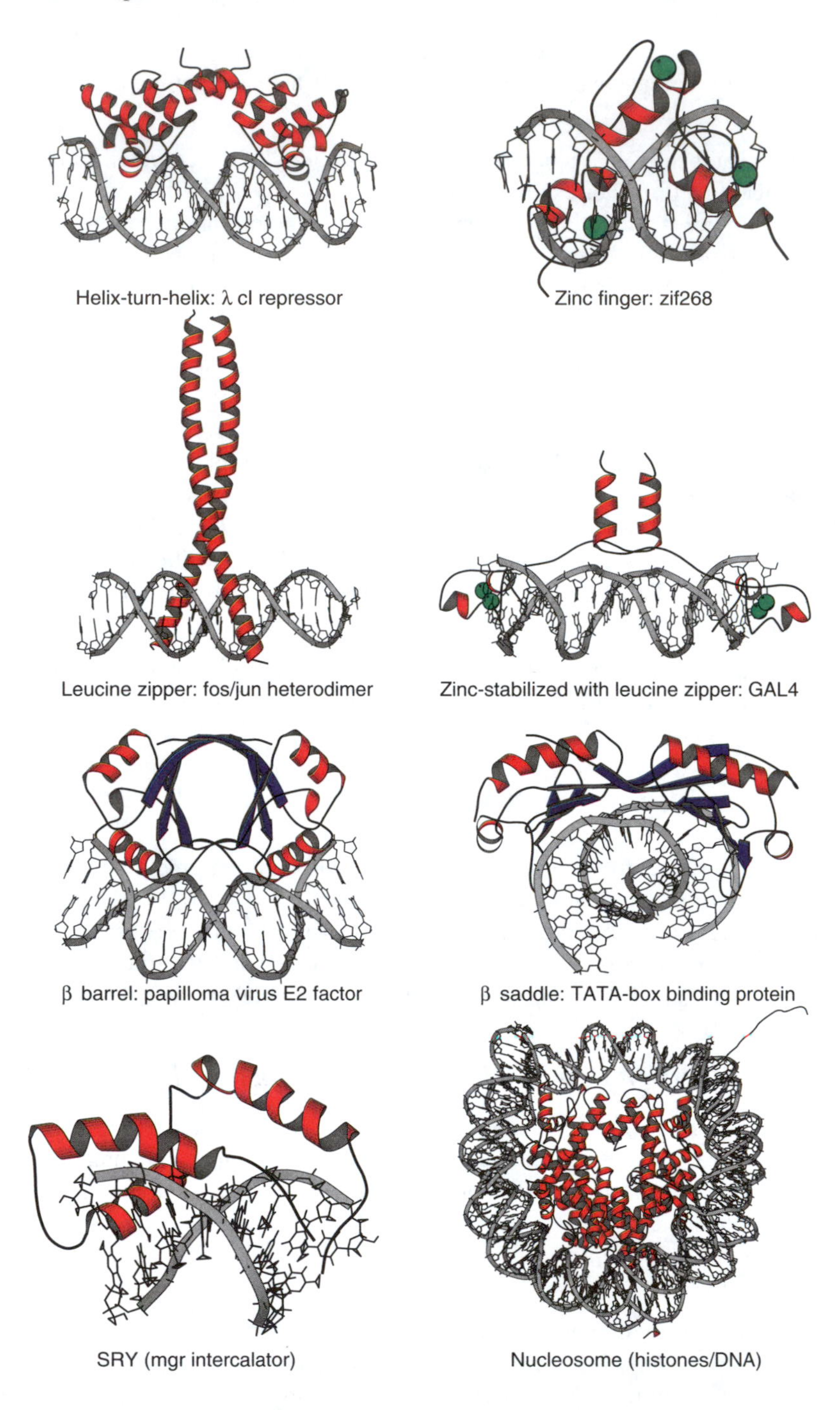

Figure 6.5. DNA/protein binding motifs for various complexes, with secondary structural elements such as α-helices (red), β-strands (blue), metal ions (green), and DNA (grey) shown.

range but only up to 3 kcal/mol in nucleic acids due to geometric constraints.[3] Though much weaker than covalent bonds (80–100 kcal/mol) [285, p. 12–14], the cumulative effect in the double helix is substantial due to the collective impact of hydrogen-bonding interactions resulting from cooperative effects [227]. Based on theoretical studies, the strength of the hydrogen-bond energy depends on base composition but is very similar for the two WC bps. The energy of bp stacking, 4–15 kcal/mol, substantially contributes to overall helix stability and depends on the nucleotide sequence [397].

Yet, hydrogen-bonding patterns in polynucleotides are extremely versatile. There are several nitrogens and oxygens on the bases, allowing various donor and acceptor combinations involving different interface portions of the aromatic rings.

To appreciate this versatility, we first examine the classic WC pattern in detail.

For an AT pair, two hydrogen bonds form between the C4–N3 face of T and the C6=N1 face of A (see Figure 5.3 of Chapter 5 and Figure 6.6 here). One hydrogen bond involves the sequence O4 (T)· · ·H–N6 (A), i.e., between the thymine carbonyl oxygen O4 attached to ring atom C4 and the adenine N6 atom, which is attached to the ring C6 atom. The other hydrogen bond is N3–H (T)· · ·N1 (A), between the thymine ring nitrogen N3 and the adenine N1.

For the GC pair, three hydrogen bonds form between the C4=N3–C2 face of C and the C6–N1–C2 face of G: N4–H (C)· · ·O6 (G), N3 (C)· · ·H–N1 (G), and O2 (C)· · ·H–N2 (G). All hydrogen-bond lengths in both the AT and GC bps are very similar, with individual lengths ranging from about 2.85 to 2.95 Å between the heavy atoms.

Reverse WC

Reverse WC base-pairing schemes can result when one nucleotide rotates 180° with respect to its partner nucleotide, as shown in Figure 6.6. Thus, one hydrogen bond from the original WC scheme might be retained, but another can involve a different atom combination. This flip of one bp changes the position of the glycosyl linkage. Hence, an AT pair can more easily accommodate this flip due to near symmetry about the N3–C6 axis of T. That is, since the ring face C4–N3–C2 of T has carbonyl attachments at both C4 and C2, the hydrogen bonding face can be changed from C4–N3 to C2–N3 (see Figure 6.6). This type of reverse WC base pairing exists in parallel DNA [284].

Hoogsteen

Hoogsteen bps utilize a different part of the aromatic ring for hydrogen bonding. Specifically, the N7–C5–C6 face of A and G, where N7 is in the 5-membered

[3] Estimates of base-pairing energetics in solvated nucleic acids come mainly from theoretical calculations [396]. This is because the resolution of macromolecular thermodynamic measurements into subcontributions (hydrogen bonding, stacking, and electrostatic forces) remains a challenge, despite numerous thermodynamic studies on nucleic acid systems [397, 398, 399].

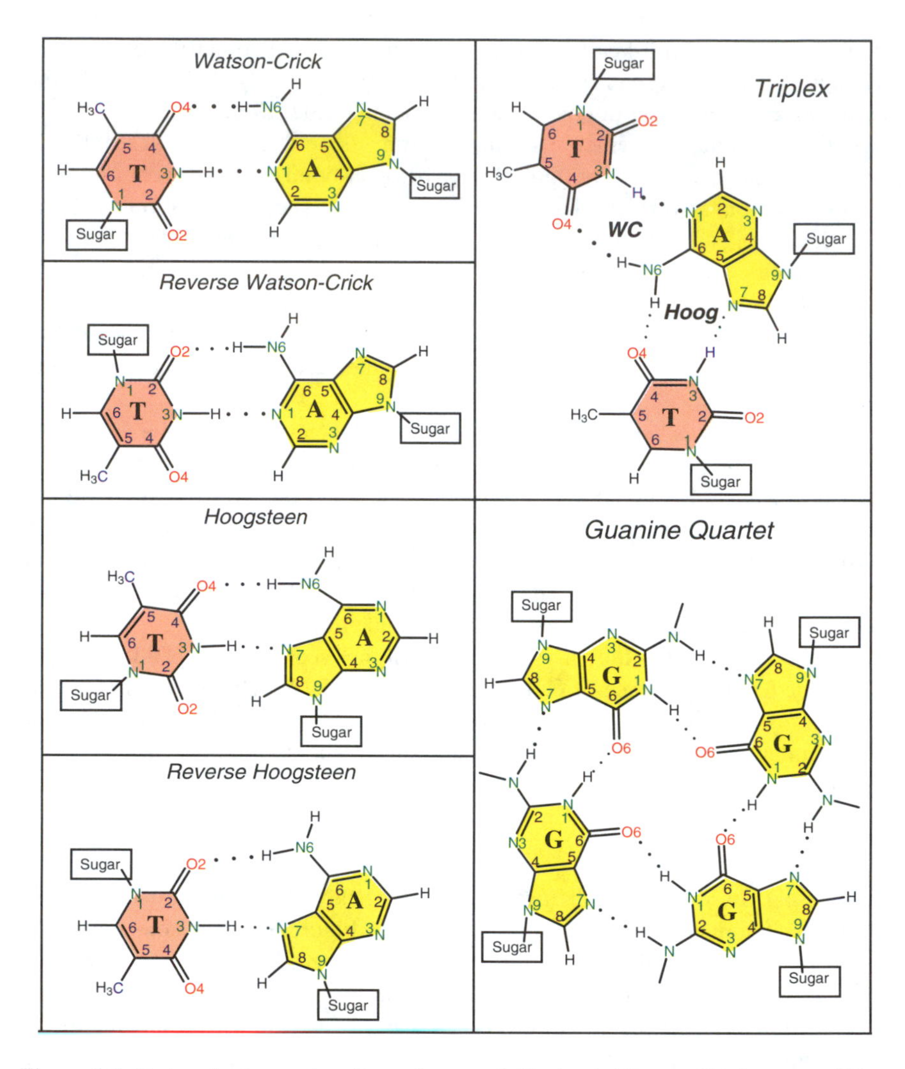

Figure 6.6. Various hydrogen-bonding schemes: (left) classic Watson-Crick, reverse Watson-Crick, Hoogsteen, and reverse Hoogsteen in base pairs; (right) WC/Hoogsteen in a triplex (from an H-DNA structure, which forms in sequences that have stretches of purine followed by stretches of pyrimidines, PDB code 1B4Y) [400], and patterns in a guanine-quartet quadruplex (or guanine tetraplex, PDB code 1JB7) [401].

ring and C6 is in the 6-membered ring of the purine, is used in the Hoogsteen arrangement (see Figure 6.6). This is instead of the C6=N1 face of A and G — both atoms of which are in the 6-membered ring of the fused-ring compound — in combination with the C4–N3 face of T and C — as in WC interactions.

The Hoogsteen arrangement also implies a change in the glycosyl torsion angle χ, from *anti* to *syn*, and a shortening of the C1′–C1′ distance, from about 10.5 Å in WC to about 8.5 Å in Hoogsteen base pairs (e.g., [402]).

Hoogsteen base pairing helps stabilize structures of highly bent DNA (e.g., TATA elements [207]), tRNAs, and DNA triplexes where two strands are held by WC hydrogen bonds and where two strands are stabilized by Hoogsteen-type pairing (see Figure 6.6). They also appear occasionally in complexes of DNA with anti-cancer drugs. Recently, an unusual Hoogsteen A–T bp was discovered in a high-resolution complex of unbent DNA bound to 4 MATα2 homeodomains (see Figure 6.7) [402, 403].

Reverse Hoogsteen

Similarly, a *reverse Hoogsteen* arrangement involves the flipping of one nucleotide by 180° with respect to its partner, by analogy to reverse WC, as shown in Figure 6.6.

Mismatches and Wobbles

Other bp arrangements involve *anti/syn* bps between two purines (a *mismatch*), in which one base adopts an atypical *syn* glycosyl configuration. This type of pairing can occur due to protonation of one base (favored at acidic conditions) or keto-enol tautomerizations. (See introduction to the term in Chapter 1, subsection on the discovery of DNA structure, and Chapter 5, subsection on nitrogenous bases).

Such ionized bases reverse the polarity of the hydrogen-bonding sites and hence lead to various hydrogen-bonding schemes, such as between G^+·C (Hoogsteen) or the ionized mispair A^+·C (*wobble*). Wobble pairs are formed when one base is shifted in the bp plane relative to its partner due to steric misalignment of the bases. Experimental evidence for bp wobbling has come from tRNA structural analysis.

Thus, various non-WC base-pairing schemes can occur when bases are chemically modified, found in the rare enol and imino tautomeric forms, or mismatched, for example. The alternative arrangements for mispaired bases, in particular, can in turn lead to mutations upon DNA replication if not corrected by the elaborate repair machinery of the cell.

Other Patterns

For further details regarding these and many other base-pairing schemes in nucleic acids, see [285] and [296]. The latter work provides an encyclopedic compendium of the "bewildering" variety of observed canonical and non-canonical nucleic acid arrangements, as compiled from RNA base-pairing interactions. This variety led to a proposal for new nomenclature for RNAs [404].

In general, non WC bps are generally lower in stability because they typically incur a greater distortion in the sugar/phosphate backbone (assuming that the rest of the DNA is in a canonical A, B, Z-type conformation). Though they are more

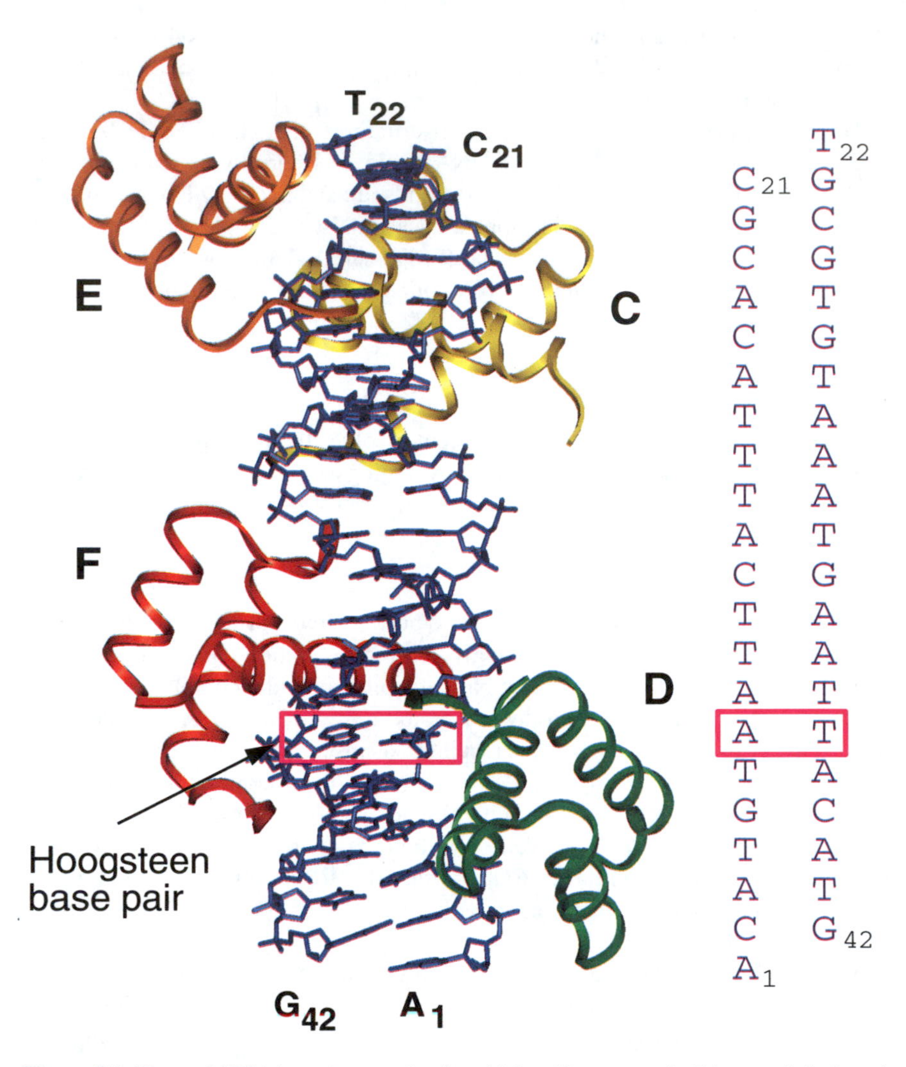

Figure 6.7. Unusual DNA/protein complex in which a Hoogsteen A–T base pair is found in unbent DNA. The complex is a high-resolution crystal of four MATα2 homeodomains bound to DNA [403, 402].

difficult to accommodate in general, many local alterations in sequence composition, environmental factors, and conformational patterns favor these alternative base-pairing arrangements.

6.5.2 *Hybrid Helical/Nonhelical Forms*

Alternative Helical Geometries

Numerous variations are noted in the helical structures of polynucleotides (see [284, 289, 291], for example). For example, C-DNA, D-DNA, and T-DNA helices have been defined with 9.3, 8.5, and 8.0 bps/turn, respectively. The C-DNA motif forms in fibers of low humidity (57–66%), D-DNA is observed in poly(dA)·poly(dT) strands, and T-DNA has been noted in bacteriophages that contain modified C residues and glucose derivatives for the sugar attachments.

Within polynucleotide structures, local distortions can also produce *hairpins* (DNA and RNA strands that fold back on themselves), *cruciforms* (intrastrand hydrogen bonds between complementary bases, often leaving a few unpaired bases at the hairpin tip), and *bulges* and *loops* (unpaired extrusions) (Figure 6.9). Such motifs are especially important in stabilizing the folding of single-stranded RNA.

DNA Triplexes and Quadruplexes

Though Linus Pauling was ridiculed for his early suggestion of a triple helix structure for DNA [405], we now recognize the existence or the possibility for formation of various DNA triplexes (involving WC and Hoogsteen bps) [406, 407]. Other existing or possible forms are quadruplexes (stabilized by hydrogen bonding and cations as in guanine-quartets; see Figure 6.6) [408, 409]; parallel DNA (requiring reverse WC base pairing); DNA analogues; and hybrids of RNA, DNA, and other polymers such as oligonucleotide mimics (see below) [410].

Triplex forming oligonucleotides are promising pharmaceutical targets since they modulate gene activity in vivo through recognition of duplex DNA [411]. Triplexes and other unusual oligonucleotides are good subjects for simulation techniques since the factors that govern sequence-dependent properties can be explored [412].

Triple-helical DNA gained wide attention in the mid-1980s when various groups demonstrated that homopyrimidine (polynucleotide chains containing pyrimidines, such as poly(dT) or poly(dCT)) and some purine-rich oligomers (such as poly(dA) or poly(dAG)) can form stable and sequence-specific triple-stranded complexes with corresponding sites on duplex DNA; the third strand associates with the duplex via Hoogsteen pairing (see Figure 6.6) [406]. Homopurine/homopyrimidine stretches in supercoiled DNA were also found to adopt an unusual structure which includes a triplex called H-DNA.

DNA Mimics

A growing number of DNA analogues (or 'DNA mimics') is also appearing (see Figure 6.8) [411]. Analogues can be sought in several ways (see Figure 6.8 and Box 6.4):

- by modifying the *phosphate backbone* segment (e.g., one P–O link changed to P–S or P–CH_3);

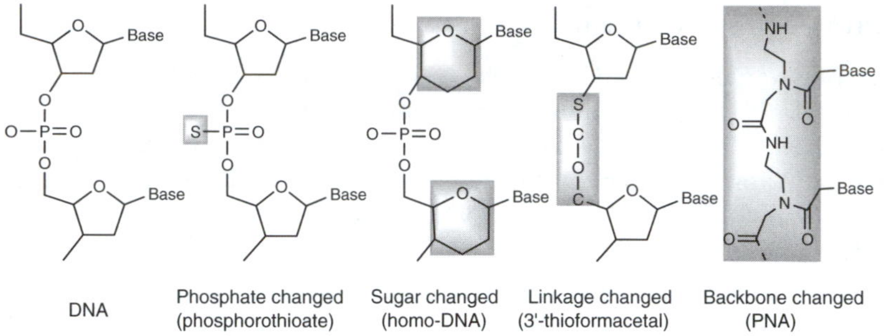

DNA
Phosphate changed (phosphorothioate)
Sugar changed (homo-DNA)
Linkage changed (3'-thioformacetal)
Backbone changed (PNA)

Figure 6.8. Various oligonucleotide analogues that involve modifications of the DNA phosphate unit, sugar, entire phosphodiester link, or the entire sugar/phosphate backbone.

- by modifying the *sugar* (e.g., 6-membered instead of 5-membered ring);
- by modifying the entire *phosphodiester linkage* of 4 atoms, O3′ through C5′ (to increase DNA hybrid stability through charge neutrality); or
- by replacing the entire *sugar/phosphate backbone*, retaining just the bases as in PNA, peptide (or polyamide) nucleic acids.

Such close analogues of oligonucleotides are designed to bind selectively to DNA bps through triple strands or duplexes. Various duplex DNA/RNA hybrids are key elements in transcription and replication and thus are potential therapeutic agents in anti-sense techniques. Triplex structures are also potential agents for sequence-specific cutting of DNA and drugs (by attacking various disease targets at the genetic level). Non-complementary or 'anti-sense' binding to a sequence of complementary messenger RNA or single-stranded DNA can also be exploited in antisense technology to suppress gene expression.

To be suitable, DNA analogues must also be biologically stable (i.e., resistant to nuclease digestion) and have favorable cellular absorption properties. All issues of geometry, flexibility, hydrophobicity, and triplex stability must be weighed in designing DNA mimics with practical utility.

Box 6.4: PNA (Peptide Nucleic Acid)

PNA is an especially interesting oligonucleotide analogue that forms very stable complexes with double-stranded DNA [297, 413, 414, 415, 411, 416]. In studies of evolution, PNA has been proposed as a candidate for the backbone of the first genetic material that predates the RNA world. While RNA is unstable and difficult to synthesize, the polymeric constituent of PNA, N-(2-aminoethyl)glycine (or AEG), may spontaneously polymerize and is accessible in prebiotic syntheses. No firm evidence, however, exists regarding this hypothesis [265].

The sugar/phosphate backbone of oligonucleotides is replaced in PNA by a pseudopeptide chain of AEG units with the N-acetic acids of the bases (N9 for purines and N1 for pyrimidines) linked via amide bonds (Figure 6.8). This substitution of the DNA backbone makes the PNA backbone electrically neutral rather than negatively charged; though the same number of chemical bonds as found in DNA and RNA exist in PNA (albeit different types), the replacement of the sugar ring by a linear bond sequence introduces additional torsional flexibility.

Stable hybrids between PNA and complementary DNA or RNA oligonucleotides form in a sequence-selective manner. Binding of PNA to double-stranded DNA leads not only to triple helices but also to P-loops strand-displacement complexes [411]. Model building, molecular mechanics and dynamics calculations, as well as experimental studies have shown how the replacement of DNA by PNA disrupts the canonical DNA triple helix, by displacing the hydrogen-bonded bases away from the global helical axis with respect to their positions in B-DNA helices, and changes the groove structure substantially [295]. This makes PNA/DNA hybrids (involving both single and double helical DNA) A-like rather than an intermediate between A and B-DNA forms, as is typical for canonical DNA triplexes. Helix stabilization in PNA/DNA hybrids stems from favorable base pairing, stacking, and (reduced) electrostatic interactions.

PNA molecules have already found practical applications as probes — an alternative to conventional DNA probes — for the detection of bacteria like *Salmonella enterica* or *Staphylococcus aureus*; chemical advantages stem from PNA's strong resistance to degradation by proteases and nucleases in the cell and its salt-independent hybridization kinetics. Triplexes and DNA analogues like PNA also have potential as drugs in gene therapy [411]. In particular, PNA is quite attractive for drug development, given its high triplex stability. However, given the A-like form of PNA/DNA hybrids, the design of stable B-like hybrids requires a different chemical approach than that used to construct PNA.

6.5.3 Unusual Forms: Overstretched and Understretched DNA

Single-Molecule Manipulations

Recently, a new experimental achievement of single-molecule observation and manipulation experiments using laser tweezers or single glass fibers [417] has been applied to DNA [418], re-invigorating interest in DNA's structural versatility [419, 420, 421, 422, 21]. It has been found that DNA subjected to previously unattainable forces of 10–150 picoNewtons (pN) displays a highly cooperative, sharp, and reversible transition to a new DNA structure at around 70 pN. This new overstretched helical structure, termed the *S-DNA ladder* (related to P-DNA), has a rise per residue of around 5.8 Å (compare to values in Table 5.3 of Chapter 5) and a notable inclination (possibly as large as 70°) [423]. The transition under a force of extension is also affected by changes in the ionic strength of the medium, sequence, and addition of intercalators. Interestingly, DNA can sustain extensions to roughly *twice* its original length without severe distortions to the base pairing.

Modeling studies soon followed these experiments to investigate global conformational features of overstretched DNA duplexes [423, 424] and the local morphological features associated with the deformation process [311] (see Box 6.5).

Similar experiments have also been applied to RNA [425].

Biological Relevance

The extreme deformations of DNA are of great interest because flexible DNA molecules undergo a variety of distortions in biological environments. Notable examples occur during DNA recombination and cell division. It has been suggested, for example, that a helix to ladder transition in DNA near the chromosomal centromere region may occur during cell division and thus play a regulatory role [423]. Intriguingly, the overstretched DNA in recombination filaments of 5.1 Å per bp is close to that associated with the phase transition observed in single-molecular micro-manipulation experiments.

It has also been suggested that longitudinal deformations of DNA might be associated with many DNA/protein binding events, such as DNA binding to the TATA-box binding protein TBP (where DNA is locally compressed, strongly bent, and unwound) and DNA binding to nucleosomes, where variable lengths of DNA associated with nucleosome wrapping might accommodate sequence and ionic variations [311].

Box 6.5: Modeling Overstretched DNA

Results of modeling studies of overstretched DNA — by energy minimization or molecular dynamics — are protocol dependent. For example, results and interpretations depend on which end of the DNA is pulled, what minimization method is used, how minimization is implemented (e.g., how fine the helical rise increments are), what force field is employed, and how solvent is treated.

The minimization work of Lavery and coworkers [424] suggested that the stretched conformation may be a flat, unwound duplex or a narrow fiber with substantial bp inclination [424]. The molecular dynamics simulations of Konrad and Bolonick [423] reproduced the helix to ladder transition and analyzed the geometric and energetic properties stabilizing the *S-ladder*. The constrained minimization studies of Olson, Zhurkin and coworkers [311] produced a wealth of structural analyses for both compressed and stretched DNA duplexes (from 2 to 7 Å per bp) of poly(dA)·poly(dT) and poly(dG)·poly(dC) homopolymers under high and low salt conditions. It was found that DNA can stretch to about double, and compress to half, its length before the internal energy rises sharply. Energy profiles spanning four families of right-handed structures revealed that DNA extension/compression deformations can be related to concerted changes in rise, twist, roll, and slide parameters. The lowest energy configurations correspond to canonical A and B-DNA. These models may be relevant to nucleoprotein filaments between bacterial Rec-A-like proteins and overstretched, undertwisted DNA [426].

Such fascinating single-molecule biochemistry manipulations have also shown that under much smaller forces (< 3 pN), a new DNA phase is achieved, with about 2.6 bps/turn and thus about 75% more extended than B-DNA [427]. This new DNA conformation, termed *P-DNA* (for Pauling, see below), occurs at moderate positive supercoiling for molecules that cannot relieve the torsional stress via writhing (nonplanar bending). (For negative supercoiling, DNA denatures in this force range). Intriguingly, some modeling suggests that such a structure is 'inside-out', with bases on the helical exterior and the sugar/phosphate backbone at the center, as Linus Pauling once suggested for the structure of the double helix based on a model with un-ionized bases [405]. Such a conformational transition arrives from major rotation of torsion angles (notably α and γ, toward the *trans* conformation) and is compatible with both C2′-endo and C3′-endo sugar puckers. The intrastrand P–P distances in P-DNA (around 7.5 Å) are larger than the corresponding values in canonical A and B-DNA (5.8 and 6.6 Å, respectively). Evidence shows that this P-DNA conformation is found in the packed DNA inside some virus complexes, where the DNA is constrained by the helical coat protein [428, 429]. *Still, the interpretation of positively-supercoiled, overstretched DNA as an 'inside-out' model remains controversial.*

Polymer statistical-mechanics calculations suggest that the experimental data involved in these force-versus-extension measurements of DNA can be fit to the elastic theory of the entropic force required to extend a worm-like chain [430]. However, such studies of entropic force are only relevant to small fluctuations about the B-DNA equilibrium conformation.

6.6 RNA Structure

6.6.1 RNA Chains Fold Upon Themselves

Many of the variations described above are common for single-stranded RNA molecules, which can form double-helical segments interspersed with loops. Such patterns are governed by the primary sequence and can accommodate a variety of hydrogen bonding patterns. The double stranded regions (stems) can be imperfect with bulges, mismatched pairs, and unusual hydrogen bond schemes, as shown in Figure 6.9. The stem/loop structures themselves can fold further, seeking favorable stacking, hydrogen bonding, and other interactions between distant partners.

The multitude of RNA secondary structures and tertiary interactions are stabilized by various double-stranded regions, interior and terminal loops, bulges, K-turns, U-turns, S-turns, A-platforms, tetraloops, and other motifs (e.g., [431, 432, 433]).

For example, *pseudoknot* motifs are produced from a special intertwining of secondary structural elements by Watson-Crick base pairing (see Figure 6.10). Namely, pseudoknots form when a consecutive single-stranded domain with segments **a**, a′, **b**, b′, **c**, c′ and **d** (a′, b′, and c′ are connectors) folds to form two hydrogen bonds: **a** with **c**, and **b** with **d**. For short RNA segments this is a pseudoknot rather than a knot since the strands do not actually pass through one another [434]. Pseudoknots are 3D structures and cannot be represented easily in conventional 2D trees describing RNA secondary structure. See [435], for example, and end of Box 6.6 for an overview of the motifs that stabilize RNA and contribute to the activity of large RNAs.

6.6.2 RNA's Diversity

The wonderful capacity of RNA to form complex, stable tertiary structures has been exploited by evolution. RNA molecules are integral components of the cellular machinery for protein synthesis and transport, RNA editing, chromosome replication and regulation, catalysis and many other functions[4] [437]. RNA/DNA hybrid double helices also occur in transcription of RNA onto DNA templates and in the initiation of DNA replications by short RNA segments. As mentioned in Chapter 1, a highlight of RNA's functional diversity is its catalytic capability, as discovered in the 1980s and recognized in the 1989 Nobel Prize in Chemistry to Thomas Cech and Sidney Altman. Such catalysis is performed by RNA molecules termed *ribozymes*; more than 500 ribozyme types have been found in a diverse range of organisms, and some have been designed (e.g., [438, 439, 440].

RNA's conformational flexibility and versatility [441] are also important for recognition by, and interactions with, ligands (e.g., as in the RNA of the human immunodeficiency virus, HIV), and other molecules (e.g., [442]). These features are also of practical importance, for the design of therapeutic agents that exploit RNA's functional sites as potential drug targets [443]. By designing ligands that bind to RNA and interfere with protein synthesis, transcription, or viral replication, we can produce new therapeutic agents, such as antibiotic/antiviral drugs. Such design poses a major challenge at present given the resurgence of previously known viral and bacterial diseases and the emergence of resistant mutants of antibiotic and antiviral drugs. These applications together with novel RNA design have important potential as therapeutic agents, molecular switches, and molecular sensors (e.g., [439, 442, 444]).

6.6.3 RNA at Atomic Resolution

Until fairly recently, less was known at atomic resolution on RNA structure in comparison to DNA, but this has changed as the 'RNA era' has begun. Progress in RNA structure elucidation can be attributed to vast improvements in crystallization procedures (e.g., RNA structure determination through crystallization with a

[4]For example, tRNA molecules carry amino acids and deposit them in correct order, mRNAs translate hereditary information from DNA into protein, rRNA are involved in protein biosynthesis (within a complex of ribosomal RNA and numerous proteins), gRNAs edit RNA messages (by 'guide' sequences), cRNAs play roles in catalysis and autocatalysis, and snRNAs are splicing agents (by 'small nuclear' components of the mRNA splicing machinery known as the spliceosome).

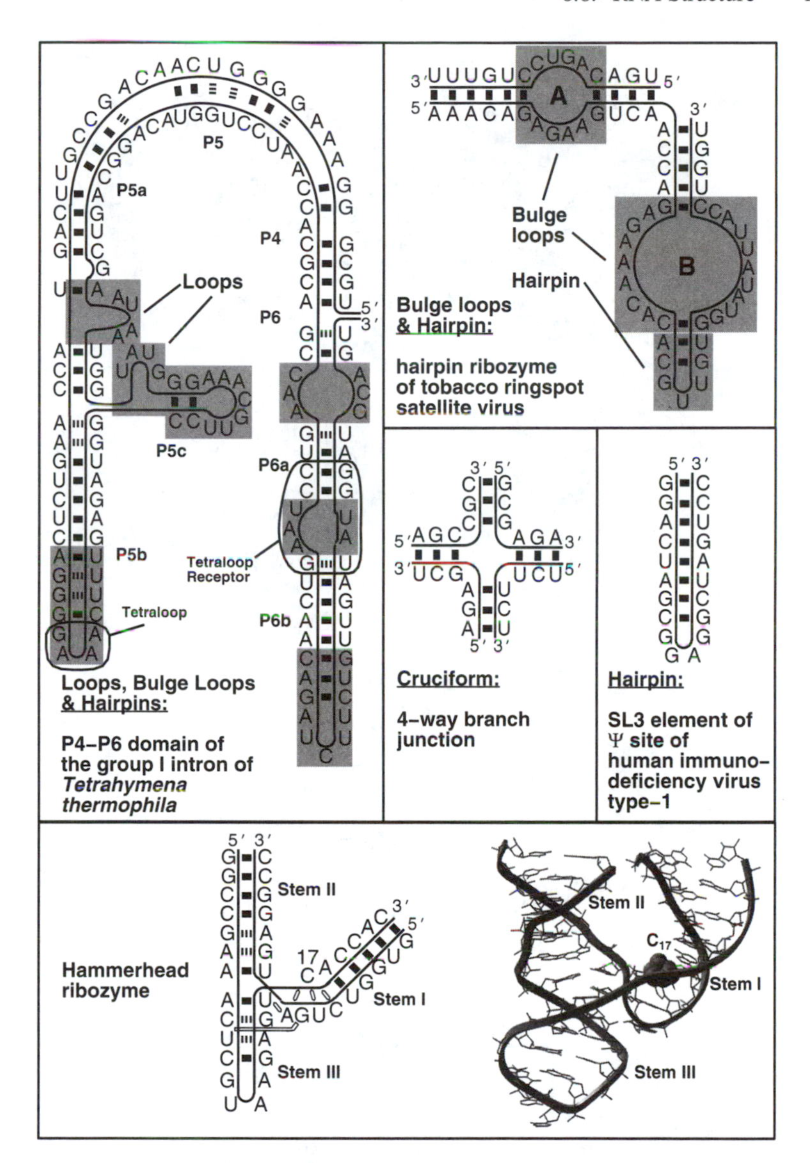

Figure 6.9. Various nucleotide-chain folding motifs from solved RNA systems: loops, bulge loops, hairpins, hairpin loops, cruciforms (four-way junctions, found in recombination intermediates and synthetic designs), and the complex tertiary contacts found in RNA (illustrated on the hammerhead ribozyme), which compact the secondary-structure elements. Broken connections are used for non-WC hydrogen, including mismatches. The PDB files for the structures (clockwise from left) are as follows: tetrahymena intron, 1GID; tobacco ringspot ribozyme, 1B36; HIV-1 Ψ-stem loop 3, 1A1T; and hammerhead ribozyme, 1MME. The tertiary interactions shown for the hammerhead drawing (ellipses or connecting line segments) are based on [436]. The nucleotide C17 shown in space-filling form is the strand cleavage site of the ribozyme.

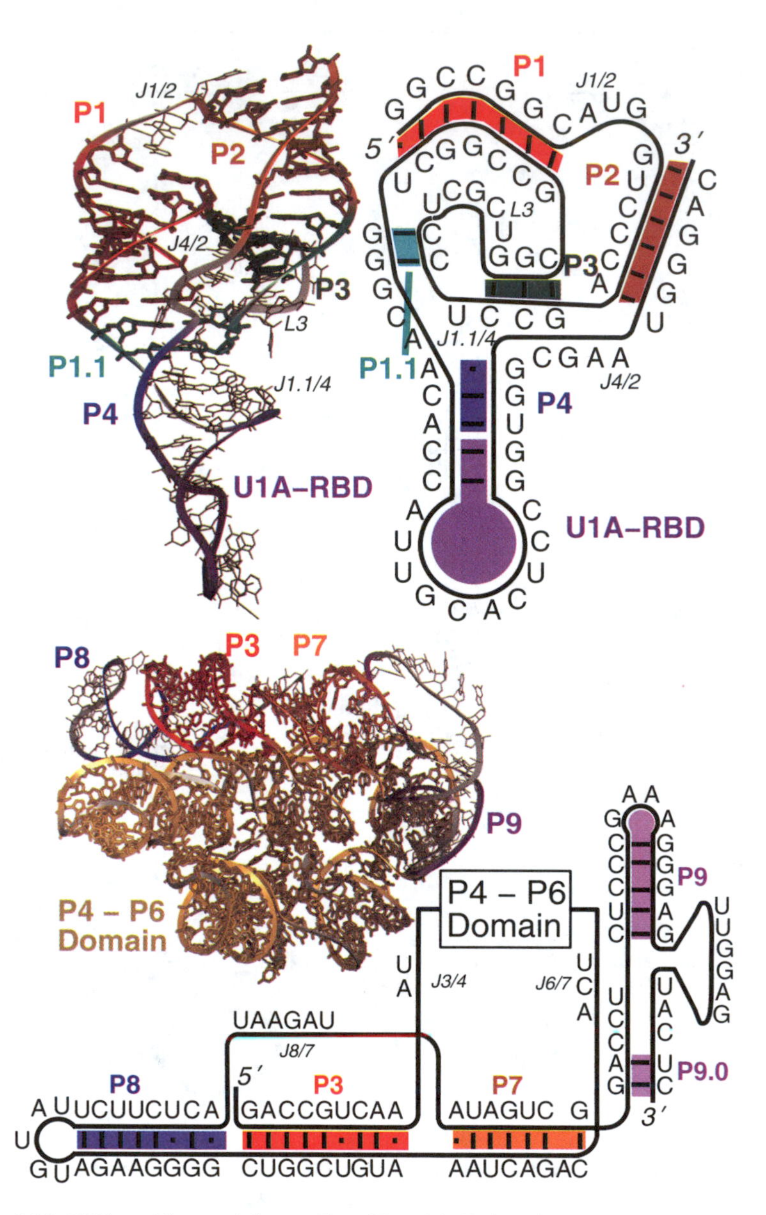

Figure 6.10. RNAs with pseudoknots. Top: Hepatitis Delta virus (HDV) ribozyme (PDB code 1DRZ) shows two pseudoknots: a main one formed by regions P1 (red) and P2 (orange-red), and a minor one formed by regions P1.1 (cyan) and P3 (green). A hairpin formed by the P4 helix (blue) and the U1A ribonucleoprotein binding domain (purple) is also present. Bottom: P3-P9 domain of the group I intron *Tetrahymena thermophila* ribozyme (PDB code 1GRZ) shows one pseudoknot formed by regions P3 (red) and P7 (orange-red). The P4-P6 domain (yellow) folds independently. See Box 6.6 for a discussion of some of those structures.

protein that would not interfere with the enzyme's activity [445]), as well as alternative approaches for studying RNAs such as high-resolution NMR, spectroscopy, crosslinking reactions, and phylogenetic data analysis. Our knowledge of RNA structures has also increased dramatically with recent solutions of ribosomes [11, 13, 14, 15, 16, 17, 18, 19, 12].

The clover-leaf structure of the tRNA molecule has been known for over 25 years, and for a long time was the only well-characterized major structure of an RNA molecule;[5] see [446] for a perspective following the high-resolution determination of yeast phenylalanine tRNA in 2000. Its structure whet our appetite for RNA appreciation by revealing the long-distance tertiary interactions.

RNA folds characterized by X-ray crystallography include the tRNA, hammerhead ribozyme, *Tetrahymena* group I intron, and hepatitis delta virus ribozyme [447]. See Box 6.6 for details of some solved catalytic RNAs.

6.6.4 Emerging Themes in RNA Structure and Folding

Deducing the functional structure of RNA molecules from the primary sequence has been called the *RNA folding problem* [370, 448, 425]. The challenge is to understand how strong electrostatic repulsions between closely packed phosphates in RNA are alleviated. Indeed, the stability of compact RNA forms is strongly maintained through interactions with both monovalent and divalent cations and by pseudoknotting, namely WC base pairing between two unpaired regions of the secondary structure (e.g., loop/loop interactions when two hairpin loops are involved).

Predicting the secondary and tertiary folding of RNA is a difficult and ongoing enterprise [449, 450, 370]. Secondary structural elements are easier to identify through modeling combined with evolutionary and database relationships [451, 452]. The new structures, however, provide opportunities for learning what works, as well as what fails, in structure prediction.

Emerging themes in RNA structure include the importance of metal ions and loops for structural stability, various groove binding motifs, architectural motifs tailored for intermolecular interactions [442], hierarchical folding, fast establishment of secondary structural elements, and extreme flexibility of the molecule as a whole [441, 370, 447, 448].

At present, relatively successful algorithms are available to predict secondary structure of RNA molecules by calculating the most energetically favorable base-pairing schemes. However, discriminating among the possible tertiary interactions to obtain the final folded state remains a challenge [448]. Still, findings concerning the folding kinetics of *Tetrahymena* ribozyme [453] have suggested that, as thermodynamic data on tertiary structure interactions become available [448], the RNA folding problem might be easier to solve than protein folding [454, 448].

[5]Structures of a few small RNA oligonucleotides, both single and double-stranded helices, some complexed with proteins, have also become available.

Therefore, with advances in RNA synthesis and structure determination [455] and the availability of thermodynamic data on tertiary interactions [448], it is likely that our understanding of RNA structure, RNA folding, and RNA's role in enzyme evolution will dramatically increase in the coming decade. These developments are propelling studies in RNA structural genomics (or *ribonomics*) [456] and RNA design (e.g., [440, 438, 457, 458]).

Box 6.6: Recent RNAs at Atomic Resolution

The *hammerhead ribozyme* (in the family of self-cleaving catalytic RNA) was solved in 1994. This Y, or wishbone-shaped, structure has three base-paired stems resembling the head and handle of a carpenter's hammer (see Figure 6.9) [459, 460]. The structure is unpaired in its U-turn core and stabilized by non-WC, non-wobble bps in the stems. Visualizing its structure led to further analysis of the mechanism of RNA self cleavage through trapping of intermediates in the ribozyme reaction pathway [461, 462].

The crystal structure of one self-folding, *P4–P6 domain* of the *Tetrahymena thermophila* RNA intron (see Figure 6.9) [463, 464], in the Group-I intron family of catalytic RNA, was solved in 1996. This 160-nucleotide domain consists of a helical *tetraloop* region connected to another helical segment, the *tetraloop receptor*, by a large bend ($\approx 150^o$). Base paired regions (both WC and noncanonical) are interspersed with internal loops, and an *adenosine-rich bulge* region mediates the long-range tertiary contacts in the RNA. The adenosine platform motif emerges from this structure as an important general architectural component of RNA that might have arisen early in evolution. These residues are well suited for RNA structure and function as they improve base stacking interactions and minimize steric clashes. As found in other RNA structures, metal-phosphate coordination is also important in stabilizing tertiary contacts in this intron RNA domain.

The *Hepatitis Delta Helper virus* (HDV) [445] — a member of the self-cleaving ribozymes like the hammerhead — and the *Group I intron* from *Tetrahymena thermophila* [465] were solved in 1998 (see Figure 6.10). The latter contains the other half of the Group I ribozyme active site (domain P3–P9), which wraps around the previously solved P4–P6 structure. This completion provides a satisfactory structural explanation to kinetic-folding results which revealed faster folding of the P4–P6 domain [453]. In both ribozymes, the catalytic site is hidden, in marked contrast to the hammerhead ribozyme; the sheltered active site instead resembles the catalytic organization in globular proteins. These ribozymes also exemplify the complexity of RNA structure. In both cases, an intertwining of secondary structural elements results in *pseudoknot* motifs (see main text for a definition and [434]).

Recent works on both crystal structures [466, 467] and statistical frequencies of RNA sequences [468] have delineated specific examples of conformational motifs involving the interaction of single-stranded adenines with minor-groove RNA helices. This motif stabilizes tertiary contacts in RNAs and may be an important global theme in RNA structure.

6.7 Cellular Organization of DNA

Thus far we have discussed the structure of DNA at the atomic and molecular levels. Understanding the organization of DNA in the cell is important for appreciating DNA's role as the hereditary material and its versatility in structure and function.

6.7.1 Compaction of Genomic DNA

DNA's cellular organization is critical because of the enormous content of genomic DNA. The genome size — in terms of nucleotide bps per chromosome haploid (eukaryotic chromosomes are each made of two haploids) — varies from organism to organism. Though the number of bps that specifies our makeup basically increases with the number of different cell types present in each organism, it ranges greatly within an organism. Some organisms also have more genomic content than mammals. For example, bacterial genomes have about 10^6–10^7 bps per haploid; for algae, the range is 10^8 to 10^{11}; for mammals it is around 10^9; yet salamanders reach the large number of 10^{10}–10^{11} residues.

Table 6.3 shows representative examples of the genomic content of different organisms, along with the corresponding total length of DNA (assuming the DNA is fully stretched). We see that this total DNA length isolated for bacterial chromosomes is only of order 1 mm, but for human DNA the comparative length is 3 orders of magnitude greater. In fact, if the genomic content in each human chromosome would be stretched, 4 cm of DNA would result; stretching out all the DNA (two haploids for each of the 23 diploid human chromosomes) would produce *2 meters of DNA*! Yet, the eukaryotic nucleus size (or the cell size in prokaryotes) is much smaller, around 5 μm, thus more than 5 orders of magnitude smaller. *This necessitates extreme condensation of the DNA.*

This condensation is largely achieved by *supercoiling*, the coiling imposed on top of the double-helical coiling, involving the twisting and bending of the DNA about the global double helix axis itself. The right-handed B-DNA form is most suitable for this coiling, as first shown via modeling by Levitt [469]: B-DNA can be bent smoothly about itself to form a (left-handed) superhelical structure with minimal changes in the local structure. This property facilitates the cellular packaging of DNA, not only by reducing the overall volume DNA occupies but also by promoting protein wrapping in the chromatin fiber complex.

Different types of DNA compaction can be distinguished for prokaryotes and eukaryotes. The former have supercoiled closed circular genomes (leading to the solenoids introduced below), while the latter have linear DNA duplexes and nucleosomes (leading to the toroids discussed below). Multivalent cations, such as polyamines, are known to be important in the spontaneous condensation of DNA to form compact, orderly toroids [470] and are likely to have a significant role in chromosomal condensation.

Table 6.3. DNA content of representative genomes.

Organism[a]	kb[b]	Total DNA (# haploids)[c]
Bacteriophage λ virus	49	17 μm
Haemophilus influenzae bacterium	1800	0.6 mm
E. coli bacterium	5000	1.6 mm
Yeast (*S. cerevisiae*)	13,000	4.6 mm (16)
Roundworm (*C. elegans*)	100,000	3.4 cm (6)
Mustard plant (*Arabidopsis thaliana*)	135,000	4.6 cm (5)
Fruitfly (*Drosophila*)	137,000	4.7 cm (4)
Mouse (*M. musculus*)	3,100,000	1.1 m (21)
Human (*H. sapiens*)	3,300,000	1.1 m (23)
Salamander (*A. tigrinum*, axolotl)	42,000,000	14.3 m (14)

[a]All listed are eukaryotes except the virus and bacteria; a bacteriophage is a virus that invades bacteria.

[b]One kb = 1000 bps.

[c]Contour lengths of stretched DNA are calculated based on 3.4 Å per bp. The number of haploids reflected in this total DNA length is given in parentheses.

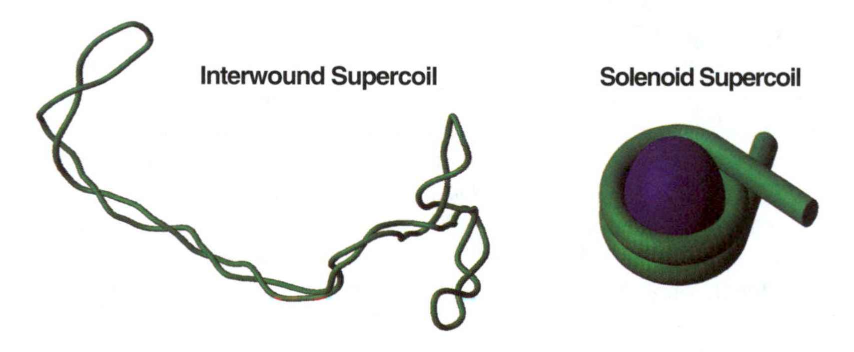

Figure 6.11. Left: Interwound supercoiling with writhing number $Wr \approx -13$. Right: a solenoidal supercoil, drawn wrapped around a protein core to mimic the nucleosome.

6.7.2 *Coiling of the DNA Helix Itself*

Supercoiling is a property of both free and protein-bound DNA. When this wrapping of the global DNA helix itself involves self-interaction, like a braid, a *plectoneme* or *interwound* configuration results. If instead, the winding occurs around the imaginary axis of a torus, a *toroidal* or *solenoidal* supercoil (or superhelix) is formed (see Figure 6.11).

Interwound supercoiling is common for circular DNA, such as found in the genomes of many viruses and bacteria, as well as for topologically constrained DNA in higher organisms. In addition, eukaryotic DNA is circular in certain energy-producing organelles, the mitochondria. Toroidal-type supercoiling is characteristic of the packaged form of chromosomal DNA in the chromatin fiber (see below).

The orderly packaging of the DNA in the cell has two major roles. It contributes to the flexibility of the DNA fiber and to the accessibility of DNA for performing vital biological processes — replication, transcription, and translation — all of which require the DNA to unwind. The packaging also compacts the chromosomal material by orders of magnitude, as required.

6.7.3 Chromosomal Packaging of Coiled DNA

Cellular DNA is organized in the chromosomes. Each chromosome in eukaryotes (two haploids) is made up of a fiber called the *chromatin* that contains DNA wound around proteins. Specifically, in this packaged form of the DNA, DNA wraps around many large *histone* protein aggregates like a long piece of yarn around many spools (see Figure 6.13). The diameter of this DNA/protein fiber in its compact form is around 300 Å, an order of magnitude greater than the diameter of the double helix [267].

Several levels of folding are recognized for the chromatin fiber, but only the basic units of the chromatin fiber and the associated low-level packaging are well characterized. This view has been deduced from various experimental studies of the individual globular histone proteins (class types H1, H2A, H2B, H3, H4) as well as of the chromatin fiber. Techniques used for the fiber analysis include electron microscopy, X-ray diffraction, neutron diffraction, nuclease digestion combined with gel electrophoresis,[6] and chromatin reconstitution *in vitro*.[7]

The Nucleosome: DNA + Histones

A key fact established in 1974 by Roger Kornberg and Jean Thomas was that the repeating unit of the chromatin is the *nucleosome* [471]. This unit consists of about 200 bps of DNA, most of which is wound around the outside core of histones; the remainder, *linker DNA*, joins adjacent nucleosomes (Figures 6.11 and 6.13).[8]

---

[6] In this procedure, the phosphodiester bond of DNA in solution is cleaved. This leaves chromatin protected and therefore reveals overall chromatin organization when analyzed by gel electrophoresis.

[7] Reconstitution can involve the construction of a chromatin-like fiber by adding histones to specific sequences of DNA.

[8] The lengths associated with the total nucleosomal DNA and with the linker component vary from organism to organism and tissue to tissue. Specifically, the total length ranges from around 160 to 260 bps; the length of linker DNA ranges broadly from about 10 to 110 bps, though it is usually around 55 bps.

The histone proteins have a large proportion of the positively-charged residues Arg and Lys. Both residues make up between 20 and 30% of all residues: the percentages of Lys/Arg residues for H1, H2A, H2B, H3, and H4 are 29/1, 11/9, 16/6, 10/13, and 11/14, respectively. (These proteins have 215, 129, 125, 135, and 102 amino acids, respectively). Therefore, electrostatic interactions between the negatively charged DNA backbone and the positively-charged histone side chains are thought to stabilize this protein/DNA complex.

Nucleosome Structure

The earlier works, combined with more recent chemical, enzymatic, and structural studies (e.g., [472, 473, 5]) suggest detailed organization of the nucleosome units and reveal some aspects of stabilizing electrostatic interactions. For example, based on a nucleosome crystal structure without the wound DNA [472], Moudrianakis and collaborators have shown that the nucleosome has a tripartite organization — assembly of two dimers of H2A–H2B, one on each side of a centrally located H3–H4 tetramer [472]. The nucleosome was later shown to be surrounded by a positive ion cloud with an average local density exceeding the bulk ion concentration significantly.

The 1997 11-nanometer nucleosome core particle with the DNA [5] elucidated by crystallography (see Figure 6.12) revealed further details. Namely, 146 bps of core DNA are wound on the outside of an octamer core of the histone proteins (two dimers each of proteins H2A, H2B, H3, and H4) to form 1.75 turns of a left-handed *supercoil*; the histone component H1 is thought to be a key player in regulation through binding to the *outside* of this core particle and contacting the linker DNA. (H1 is not present in all eukaryotes; it is absent in yeast chromatin, for example).

The nucleosome core structure shows that each histone contains unstructured end regions which make important points of contact between the protein and the DNA. Specifically, an underwinding (10.2 vs. 10.25 bp/turn) of the nucleosome-bound DNA superhelix lines up neighboring DNA grooves to form a channel through which the histone ends can pass. These tails may, in fact, play key roles in regulating biological processes, such as transcription, that require a conformational change of the complex for initiation.

The current challenge is to crystallize higher-order structures of the nucleosome to see how the tails are involved in this mechanism. Elucidating details of the transition between the more open and more compact nucleosome structure will help us to understand better transcriptional regulation and DNA packaging. Determining whether all DNA sequences have the same conformation at each binding site on the nucleosome also remains to be confirmed.

Figure 6.12 illustrates the electrostatic view that emerges from Poisson-Boltzmann calculations (see discussion of theory and methodology at the end of Chapter 9) for a nucleosome core particle. Note the positively charged H3 tails (blue) and the positively-charged regions inside the complex.

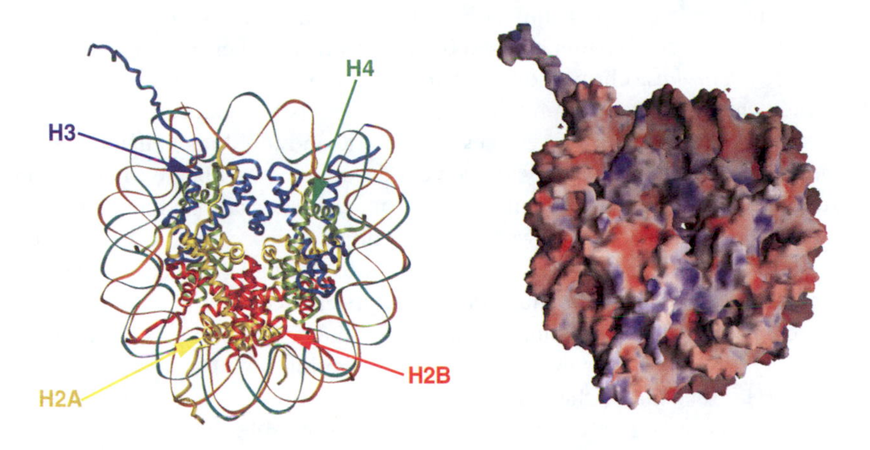

Figure 6.12. The nucleosome core particle (left) [5] and its corresponding electrostatic surface potential (right) as computed from the Poisson Boltzmann (PB) equation with program DelPhi (see Chapter 9 for theoretical framework and algorithms). The grid resolution for solving the nonlinear PB equation was 2.5 Å; the salt concentration was set to 0.04 M, and partial charges from the AMBER program were used as input. As customarily done for macromolecules [474], the unitless (i.e., relative) dielectric constant ϵ_r is set to the numerical values of 2 and 80 inside and outside the macromolecule (i.e., 2 and 80 times the vacuum dielectric constant ϵ_0).

Polynucleosome Assembly

The nucleosomal packing described above — superhelical DNA around a histone core — represents only a tenfold compression of the DNA. This is because $\approx$ 166 bps of DNA (or $\approx$560 Å of contour length if stretched) are organized as a core cylindrical particle of dimensions 110 $\times$ 110 $\times$ 55 Å, where 110 Å is the diameter and 55 Å represents the height of this unit. It is believed that at physiological ionic strengths, a next level of chromosomal organization emerges to produce the 300 Å chromatin filament observed by electron microscopy.

Several models have been suggested for this polynucleosomal level of folding.[9] For example, a crossed linker/zigzag model has been proposed, leading to a rather flat ribbon pattern. Alternatively, John Finch, Aaron Klug, and collaborators have deduced [476], based on electron micrographs, that the fiber is a helical form with 6 nucleosomes per turn wound as solenoids and stabilized by contacting H1 molecules. This compact solenoid view would look like a *beads-on-a-string* model if unraveled. It compacts the DNA further by a factor of about 40.

---

[9] A recent computational study (described in Box 6.10) predicted a chromatin model organized as a compact helical zigzag with about 4 nucleosomes per 100 Å and a cross-sectional radius of gyration of 87 Å, in close agreement with corresponding values for rat and chicken chromatin [475] (see also Figure 6.17, bottom).

Still, the condensation at this polynucleosomal level does not achieve the 5 orders of spatial compaction realized by the chromatin fiber near the end of the cell cycle (*metaphase* chromosomes). Various looping, scaffolding, wrapping, and specific contacts with other proteins and possibly RNA have been suggested for this higher folding to occur. Yet, these geometric and topological details remain rather mysterious. Undoubtedly, in the next decade, we will begin to understand better how this nucleoprotein chromosomal material is stabilized and packed in the cells, and how packaging works with biological processes, such as replication and transcription, that require full access to the DNA. Much work is also in progress on understanding the biochemical mechanisms by which the electrostatic charge density of polynucleosomes is modulated (e.g., by acetylation and phosphorylation mechanisms) to regulate transcription.

Interested readers in the flurry of research focusing on chromatin structure and function are referred to the compilation of relevant 'hot papers' and other information on chromatin research collected at the site www.cstone.net/~jrb7q/chrom.html.

6.8 Mathematical Characterization of DNA Supercoiling

Now that we have covered many local features of nucleic acids as well as basic global characteristics, I will introduce quantitative tools to describe how DNA is condensed in the cell.

6.8.1 DNA Topology and Geometry

Topological methods are important for analyzing some reactions of supercoiled DNA. For example, successful collaborations between biologists and mathematicians have led to techniques that establish mechanisms for various reactions that produce knots and catenanes (linked DNA rings) from supercoiled DNA substrates by recombination [477].

Basic Topological Identity

The topological method, as well as the many computational methods used to study DNA supercoiling, rely on the fundamental identity attributed to Grigore Calugareanu, Jim White, and F. Brock Fuller [478, 479, 480]. This equation relates the topological invariant Lk, or *linking number*, to the geometric quantities *twist*, Tw, and *writhe*, Wr as:

$$Lk = Tw + Wr\,. \tag{6.3}$$

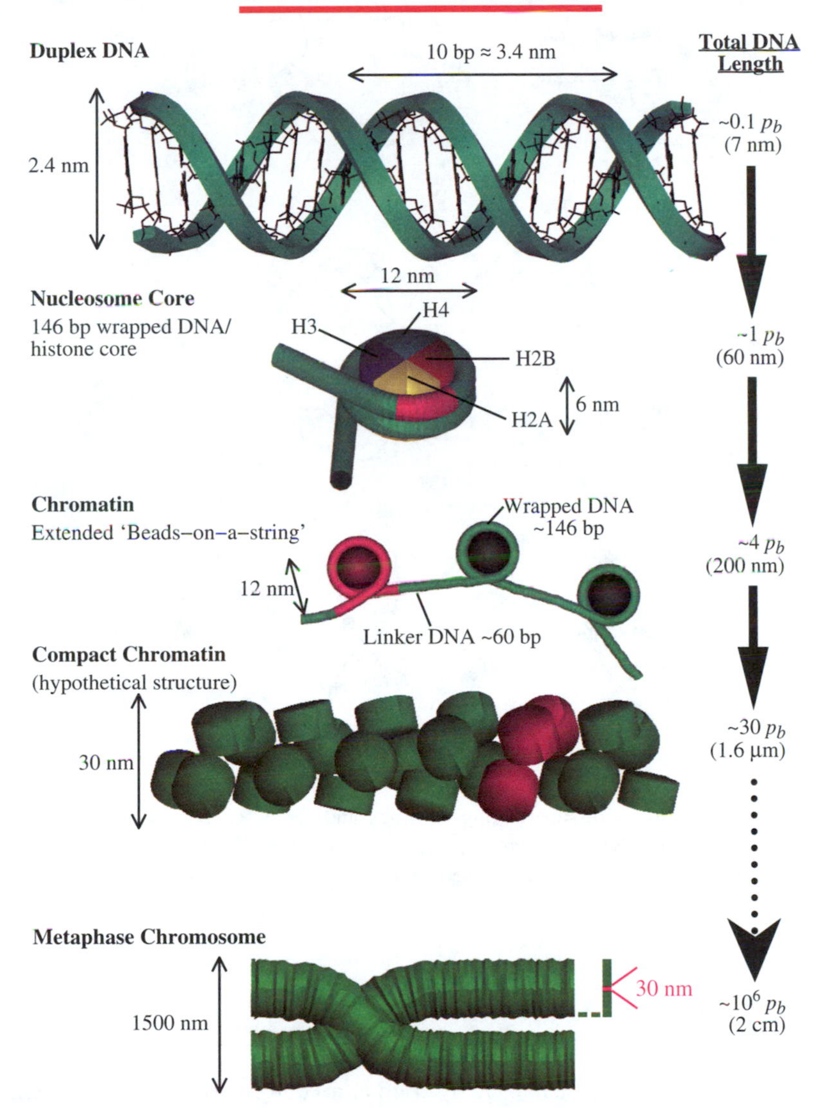

Figure 6.13. Schematic view of DNA's many levels of folding. On length scales much smaller than the persistence length, p_b, DNA can be considered straight. In eukaryotic cells, DNA wraps around a core of histone proteins (see also Figure 6.12) to form the chromatin fiber. The fiber is shown in both the extended view and a hypothetical compact view (the '30-nm fiber') deduced from a modeling study [475]; the compact structure of the chromatin fiber is unknown. Chromosomes are made up of a dense chromatin fiber, shown here in the metaphase stage. For reference, we highlight with pink in all the DNA/protein views the hierarchical organizational unit preceding it. The length scale at right indicates the level of compaction involved.

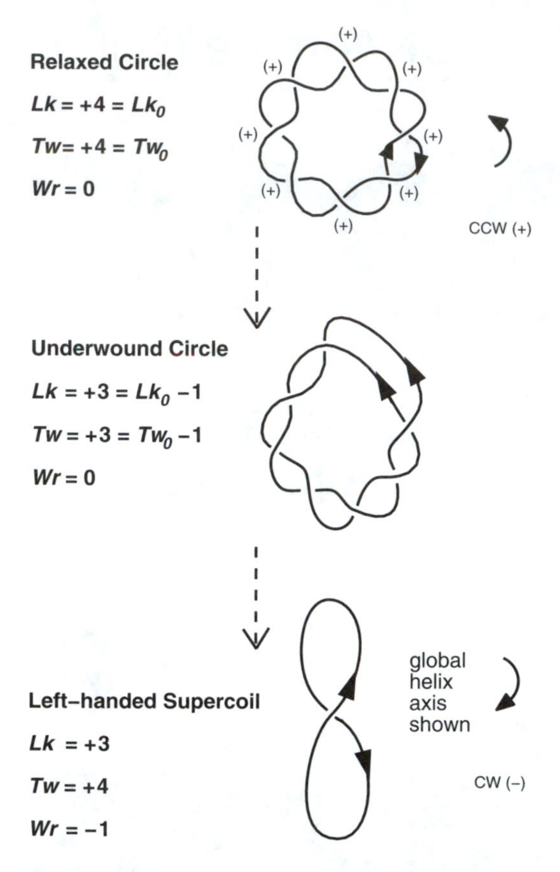

Figure 6.14. Descriptors of supercoiling topology and geometry. For the relaxed and underwound circles of DNA, each line represents one polynucleotide strand of the double helix. For the supercoil, the curve represents the *double helical* axis of the DNA.

Linking Number

Essentially, Lk characterizes the order of linkage of two closed-oriented curves in space (Figure 6.14). It is unchanged by continuous deformations but altered when strand breaks occur. By convention, for right-handed DNA, the linking number of the relaxed state, Lk_0, is the number of primary turns: $Lk_0 = n/n_b$ where n is the number of bps and n_b is the number of bps/turn (e.g., 10.5).

Lk can be rigorously computed from any planar projection of these curves onto the plane by summing all signed crossings (ignoring self interaction) and dividing the result by two, as shown in Figure 6.14. By convention, a crossover index is considered *positive* if the tangent vector of the top vector is rotated *counterclockwise* to coincide with the tangent of the bottom curve; the sign is *negative* if this rotation is *clockwise*. For precise mathematical definitions, refer to [480, 481].

Twist

The variable Tw reflects the familiar concept of the helical twist angle introduced earlier, namely the winding angle per residue. For the case of a helix spinning about a global helical axis, Tw measures the number of times the helix revolves about that axis. As conventional for Lk, Tw is positive if the helix is right-handed and negative if it is left-handed (see Figure 6.14).

When the global helix axis is a closed circle and the DNA winds n times around this planar axis, $Tw = Lk \equiv Lk_0 = n/n_b$. From eq. (6.3), this implies that $Wr = 0$. This case is shown for the relaxed circle in Figure 6.14.

A nonzero writhing number is introduced by the freeing of the two ends of the closed ring and overwinding or underwinding the duplex before resealing the ends. Topoisomerase enzymes perform this magical act in the cell. The unwrithed (planar) state is unfavored because there is substantial local disruption to the natural twist of the DNA (Figure 6.14). The induced torsional stress can be alleviated through nonplanar writhing, or supercoiling, namely the twisting of the global helix axis itself. This is illustrated in Figure 6.14 where a figure-8 interwound forms the path of the double helical axis.

Writhe

The writhing number describes the geometry of this global helix axis in space. Essentially, it is the average number of crossings in space of this curve. Like the linking number, Wr can be obtained for each planar projection and averaged over all these projections. The same sign convention used for Lk is used for Wr (Figure 6.14).

Rigorously, the writhe can be computed by the *Gauss double integral* [481] (mathematically speaking, a map that associates each pair of curve points with the unit sphere). For example, the simplest interwound — a figure-8 form — has $Wr = \pm 1$; the interwound with two self-crossings, separating the chain into three regions, has $Wr = \pm 2$. The figure-8 interwound in Figure 6.14 (bottom) has $Wr = -1$, and the interwound supercoil in Figure 6.11 (left) has $Wr \approx -13$. See [482, for example] for computational determination of Wr.

Typically, DNA is subjected to a *linking number deficit*, or *underwinding*, with respect to its relaxed state: $\Delta Lk = Lk - Lk_0 < 0$. This state is often described by the normalized quantity $\sigma = \Delta Lk / Lk_0$. The value -0.06 is typical of DNA *in vivo*. In the absence of loop-constraining agents like histone proteins, DNA tends to absorb most of this stress into writhing, by forming left-handed supercoils (negative Wr).

6.9 Computational Treatments of DNA Supercoiling

While the mathematical concepts of Lk, Tw, and Wr have been invaluable for interpreting supercoiling, the energetic and geometric details of the process remain unknown and require further computational treatments. The processes of interest involve fundamental reactions such as replication, recombination, and transcription, that often require, or are facilitated by, a supercoiled DNA substrate [483, 484]. Experimental techniques such as gel electrophoresis and electron microscopy (atomic, scanning, and cryo) have been crucial for studying properties associated with large supercoiled DNA molecules. The former allows separation of DNA molecules into topological isomers, and the latter offers views of overall shapes of supercoiled DNA. However, the experimental resolution is limited due to the large size (thousands of bps) of the DNA subjects and their extreme floppiness in solution. Hence, theoretical tools based on numerical analysis and computational modeling have been useful.

Computational treatments of the energetics and dynamics of supercoiled DNA have largely relied on the theories of polymer statistical mechanics and elasticity in the framework of elastic rod mechanics and dynamics. For realistic results, non-elastic contributions — electrostatics and hydrodynamics — must be included. These can be easily incorporated into computational models and investigated using various configurational sampling techniques (Metropolis/Monte Carlo) [485] and dynamic techniques (by molecular, Langevin, and Brownian dynamics) [486].

6.9.1 DNA as a Flexible Polymer

Very long DNA can be described by concepts from polymer statistical mechanics [487]. The DNA polymer is characterized by two key quantities: a contour length $\mathcal{L}$, and a bending rigidity A. This view of a random coil slithering through space is reasonable only for naturally occurring DNA of mixed sequences that are not intrinsically bent at lengths much greater than DNA's *persistence length*, p_b.

The persistence length is essentially the *length scale on which the polymer directionality is maintained*. It can be computed as the average projection angle between the end-to-end distance vector on the first bond vector of a polymer, in the infinite-length limit [487]. The main result from polymer statistical mechanics is that the length of this vector is proportional to the square root of the contour length; that is, the mean square displacement, $\langle R^2 \rangle$, satisfies:

$$\langle R^2 \rangle = 2 p_b \mathcal{L} \tag{6.4}$$

with $2p_b$ as the proportionality constant.

Hence, for lengths $\ll p_b$, the DNA can be considered straight, but for lengths $\gg p_b$, a better description is a bent random coil undergoing Brownian motion. This is evident from the views of DNA on different length scales, as shown in Figure 6.13. For DNA, the persistence length *in vivo* is around 500 Å or about 150 bps at physiological monovalent salt concentrations [488]. A salt dependence of the persistence length is recognized for DNA [489], but the magnitude of this effect is not currently understood.

The persistence length is also related to the bending force constant of a long flexible polymer of length $\mathcal{L}$ and bending rigidity A via

$$A = p_b \, k_B \mathrm{T} \tag{6.5}$$

(k_B is Boltzmann's constant and T is the temperature). In this description, the floppy polymer writhes through space as a worm-like chain. The bending rigidity — which tries to keep the DNA straight — is balanced by thermal forces — which tend to bend it in all directions.

A decomposition of the persistence length into static and dynamic components has been developed [490, 491].

6.9.2 *Elasticity Theory Framework*

Since the pioneering applications of Fuller [480, 492], the elastic energy approximation has proven valuable for studying global (i.e., long range and time flexibility) features of superhelical DNA. In this approximation, the DNA polymer is idealized as a long, thin and naturally straight (i.e., no intrinsic curvature) elastic isotropic rod with a circular cross section. Homogeneous bending (i.e., equal flexibility in all directions) is often assumed as a first approximation, though current computational models follow twist locally, for example by using body-centered coordinate frames along the helical axis with associated Euler transformations.

The elastic deformation energy can then be written as a sum of bending and twisting potentials, with bending and torsional-rigidity constants (A and C, respectively) deduced from experimental measurements of DNA bending and twisting [488]. The bending term is proportional to the square of the curvature κ, and the twisting energy is proportional to the twist deformation:

$$E = E_B + E_T = \frac{A}{2} \oint \kappa^2(s) \, \mathrm{d}s \; + \frac{C}{2} \oint (\omega - \omega_0)^2 \, \mathrm{d}s. \tag{6.6}$$

In these equations, s denotes arclength, and the integrals are computed over the entire closed DNA curve of length $\mathcal{L}_0$. The intrinsic twist rate of the DNA is ω_0 (e.g., $2\pi/10.5$ radians between successive bps). The parameters A and C, denoting the bending rigidity and the torsional rigidity constants, respectively, can be estimated from various experimental measurements of DNA, such as the persistence length. These force constants are key characteristics of an elastic material.[10] See Box 6.7 for the relationship between the force constants A and C and the bending and twisting persistence lengths, and Box 6.8 for the relationship between A and measured variations in roll and tilt angles in solved DNA structures.

[10]For example, a rubber band, for which bending is facile, has a small ratio $r = A/C$, while a stiff material with strong bending resistance has large r. This bending to torsional-rigidity ratio is a key descriptor of an elastic material. It is related to both Poisson's ratio (σ_e) — a characteristic of a homogeneous isotropic material — and the geometry of the rod cross section. Specifically, for a homogeneous elastic rod of circular cross section, $r = 1 + \sigma_e$. **Note**: the ratio A/C is usually designated by the symbol ρ, but here we reserve ρ for the roll angle.

Box 6.7: Elastic Constants and Persistence Length

As discussed in the text (eq. (6.5)), the elastic bending constant of DNA, A, can be linearly related to the bending persistence length of DNA, p_b, as: $A = k_B \mathrm{T} p_b$, where k_B is Boltzmann's constant and T is the temperature. In addition, A can be related to the root-mean-square (RMS) bending angle $\langle \theta_b^2 \rangle^{1/2}$ of a semiflexible chain with a preferred axis of bending (perpendicular to the helix axis), for which the angular fluctuations are independent from one another [493]:

$$A = (2k_B \mathrm{T} h)/\langle \theta_b^2 \rangle \,. \tag{6.7}$$

Here h is the bp-separation distance in DNA (around 3.4 Å) [494, 488]. (Box 6.8 describes how θ_b is related to the roll and tilt variables introduced earlier).

Similarly, the torsional rigidity constant can be related to the persistence length of twisting p_{tw} and the RMS twist angle Ω as [495]:

$$C = (p_{tw}\, k_B \mathrm{T})/2 \tag{6.8}$$

and

$$C = (k_B \mathrm{T} h)/\langle \Omega^2 \rangle \,. \tag{6.9}$$

For the bending to torsional rigidity constant ratio $r = A/C$, the relationship

$$\langle \Omega^2 \rangle^{1/2} = \langle \theta_b^2 \rangle^{1/2} \sqrt{\frac{r}{2}} \tag{6.10}$$

follows.

The typical values for DNA ($A = 2.0 \times 10^{-19}$ erg cm and $C = 3.0 \times 10^{-19}$ erg cm) correspond to persistence lengths for bending and twisting of $p_b = 500$ Å [488], $p_{tw} = 750$ Å, and $r = 2/3$. These values correspond to RMS bend and twist-angle values of $\langle \theta_b^2 \rangle^{1/2} = 6.7^{\circ}$ and $\langle \Omega^2 \rangle^{1/2} = 3.9^{\circ}$ at room temperature, in the context of isotropic bending. The bending constant A can also be related to measured local angular variations in solved DNA structures, as detailed in Box 6.8.

6.9.3 Simulations of DNA Supercoiling

Many groups are studying the geometric, thermodynamic, statistical, and dynamical properties associated with supercoiling using a variety of models both for representing the DNA and for simulating conformations. See reviews in [485, 486, 496, 497] and the articles [498, 269, 499] for details of the computational models that include stretching, bending, twisting, electrostatic, and hydrodynamic terms.

A few of the many topics that have been studied using such models are the behavior of supercoiled DNA as a function of salt [349, 350], solvent [500], superhelical density [501], and length [495, 502, 503]; extended theoretical treatments based on elastic rod mechanics and dynamics [504, 505, 506]; the response of supercoiled DNA to constraints imposed by proteins [507]; theory of elastic rods with applications to DNA [508, 509]; and the *site juxtaposition* time (the bringing together in space of linearly-distant DNA segments due to supercoiling) [510, 511, 512]. See Boxes 6.9 and 6.10 for examples of site juxtaposition and nucleosome folding studies, respectively. Static and dynamic models for studying nucleosome particles are also emerging [513, 475] and should offer further insights into the nature of the complex condensation of the chromatin fiber.

Box 6.8: Relationship between the DNA Bending Constant and Local Angular Measurements in Solved Nucleic Acid Structures

To relate the elastic bending constant A and *measured* standard deviations of DNA bending angles, the two components of bending — namely roll (ρ) and tilt (τ), each with associated stiffness constants A_ρ and A_τ [514] — can be related to A via:

$$A = \frac{2k_B \mathrm{T} h}{\langle \rho^2 \rangle + \langle \tau^2 \rangle}\,. \tag{6.11}$$

As above, h is the bp-separation distance in DNA (around 3.4 Å). By comparing the above to eq. (6.7), we see that the bending persistence length of $p_b = 500$ Å corresponds to an isotropic model where $\langle \theta_b^2 \rangle^{1/2} = 6.7^\circ$, as well as to an anisotropic bending model where $\langle \rho^2 \rangle^{1/2} = 5.7^\circ$ and $\langle \tau^2 \rangle^{1/2} = 3.6^\circ$. Values from analysis of B-form crystal structures yield somewhat smaller values for these roll and tilt fluctuations (e.g., $\langle \rho^2 \rangle^{1/2} = 5^\circ$ and $\langle \tau^2 \rangle^{1/2} = 3^\circ$) and thus a larger effective bending persistence length and rigidity constant (by about 1.3, or $A = 2.6 \times 10^{-19}$ erg cm and $p_b = 500$ Å). However, the static fluctuations in the crystal structures do not directly correspond to the dynamic range of DNA flexibility.

For sufficiently small angular deflections, we can partition the total bending magnitude as

$$\langle \theta_b^2 \rangle = \langle \rho^2 \rangle + \langle \tau^2 \rangle = [k_B \mathrm{T} h (A_\rho + A_\tau)]/(A_\rho A_\tau)\,. \tag{6.12}$$

This first-order model does not account for the well-recognized preferential directions of bending of DNA into the major and minor grooves with respect to other directions [324, 355, 325, 515, 326]. A nonuniform bending chain view is required for better representations, especially for sequences with intrinsic curvature, like A-tracts.

Box 6.9: Simulations of DNA Site Juxtaposition

Illustrative results from computer simulations of supercoiled DNA are shown in Figure 6.15. These snapshots are taken from a Brownian dynamics simulation of a 3000-bp DNA system at the superhelical density of $\sigma = -0.06$ at high salt using a homogeneous bead/wormlike model [499]. Simulations have tabulated the times for distant sites to *juxtapose* due to the ambient floppiness of DNA as a function of the superhelical density, DNA length, site separation, and the salt concentration [512].

In the study of juxtaposition, two linearly-distant sites are considered *juxtaposed* when the distance between them is 100 Å or less. Interestingly, we find (Figure 6.16) that at low salt [516] juxtaposition times are accelerated by a factor of 10 or more due to supercoiling (i.e., juxtaposition times for relaxed DNA are much slower), but that supercoiling beyond $|\sigma| = 0.03$ does not accelerate such site synapsis times [516]. Such critical behavior at moderate $|\sigma|$ agrees with experimental findings regarding supercoiling-dependent synapsis rates for the resolvase recombination system [517], as also displayed in Figure 6.16. The explanation for the fact that further supercoiling does not enhance juxtaposition times significantly emerges from our simulations: the balance between flexibility and tight supercoiling is optimized at this mean value of superhelicity.

At high salt, the effect of superhelicity is far less dramatic. Further analysis of the mechanism for juxtaposition shows that large conformational rearrangements due to Brownian motion operate at low superhelical density and low salt, whereas systematic *slithering* motions (reptation, or bidirectional conveyer-like motion) along the polymer contour [518] are more prominent at high salt, as shown in Figure 6.15. At high salt, site juxtaposition can be described by a combination of slithering, branch rearrangements, and interbranch collisions, as also shown in the figure. The figure, prepared for studying site juxtaposition kinetics [499], shows at top a plot in time of the bp i_2 that juxtaposes with the first bp (i.e., is within 100 Å of it). The continuous pattern in i_2 in time indicates ordered slithering; at low salt, the patterns are much more random.

Box 6.10: Simulations of Polynucleosome Folding

As mentioned in the text, little is known about the detailed organization of the chromatin fiber — the DNA/protein complex in eukaryotes made of DNA wound around histone proteins (see Figure 6.13) — both in terms of the linker DNA geometry and the nucleosome packing arrangement in the fiber. However, the crystallographic triumphs that produced an atomic-level view of the basic building block of the chromatin — the nucleosome core [472, 5] — provide a firm foundation for complementary modeling work.

To explore how this nucleoprotein complex might be organized, the geometry of DNA in small minichromosome systems was examined by modeling conformations of closed DNA loops containing phantom histone octamers [519] (where portions of the DNA loop are modeled wound as supercoils around two phantom protein complexes). Stable conformations were found by minimizing the elastic energy of the DNA chain. An interesting result is that the observed linking number difference induced by the nucleosomes is heavily influenced by the exact number of superhelical turns around each core and by the overall length of the DNA chain [519].

A more recent macroscopic computer model has been developed to describe the mechanics of the chromatin fiber on the polymer level [475]. The core particles are treated as electrostatically charged disks linked via elastic DNA segments. Each nucleosome unit is represented by several hundred charges, as shown in the top of Figure 6.17, optimized so that the effective Debye-Hückel electrostatic field matches that predicted by the nonlinear Poisson-Boltzmann equation [498]. This electrostatic representation is important, since properties of chromatin are sensitively dependent on the internal charges as well as on the ionic concentration of the medium. The chromatin model has been parameterized and tested on dinucleosomes, for which experimental translational-diffusion data are available from chicken and rat erythrocytes.

Figure 6.17 shows snapshots from a dynamic simulation of a dinucleosome model for chromatin at high salt [475]. The nucleosome cores are represented as rigid, charged cylinders, and the 200-Å linker DNA is modeled with the bead/wormlike formulation as described for the site juxtaposition studies.

From the condensed structural motif of the dinucleosome system, a model for oligonucleosome systems was constructed and refined by Monte Carlo simulations (Figure 6.17). The refined structure brings to close proximity each core particle to its third neighbor in both directions along the chain. The favorable electrostatic interactions between pairs of nucleosome cores form a chain that wraps around the fiber as a compact, left-handed helical zigzag pattern. The resulting geometry of the chromatin model — 4 nucleosomes per 100 Å and a cross-sectional radius of gyration of 87 Å — is strikingly close to experimental measurements of rat and chicken chromatin [475]. Substantial expansion of the structure at low salt was also observed.

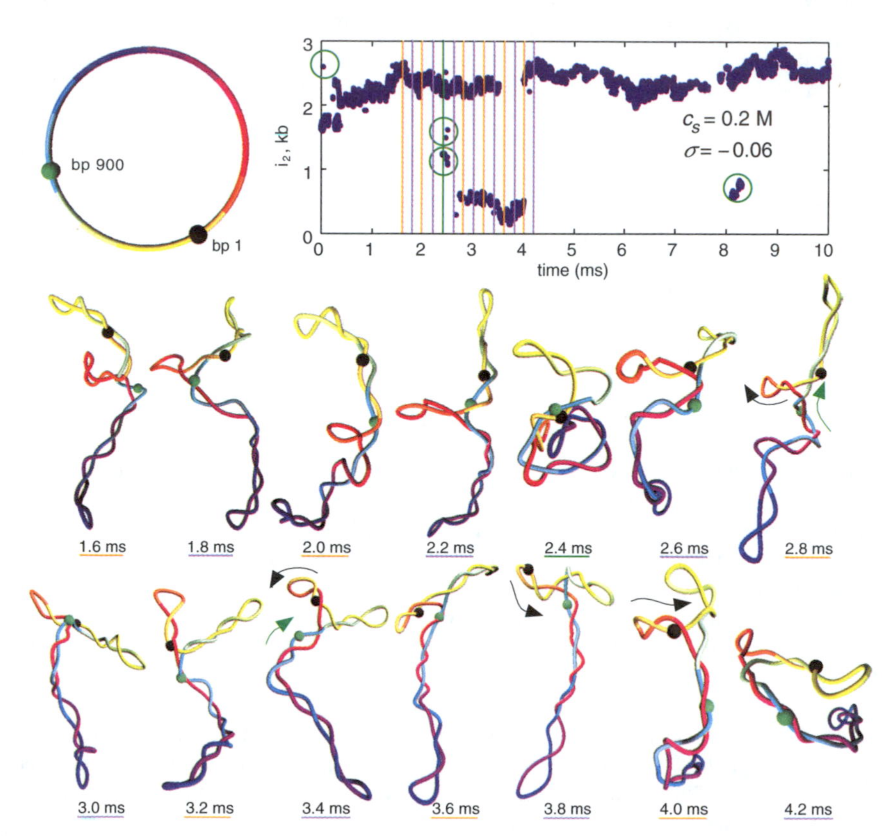

Figure 6.15. Juxtaposition kinetics, as analyzed for two beads separated by 900 bps in circular DNA of length 3000 bps and $\sigma = -0.06$ by Brownian dynamics simulations at high salt ($c_s = 0.2$ M sodium ions) [499]. (Top) Juxtaposition event plot, showing bps i_2 that juxtapose with bp 1; the circled values correspond to more random juxtaposition events, as illustrated for the snapshot at 2.4 ms. (Bottom) BD snapshots, showing ordered motions along the DNA contour and rarer random collision events.

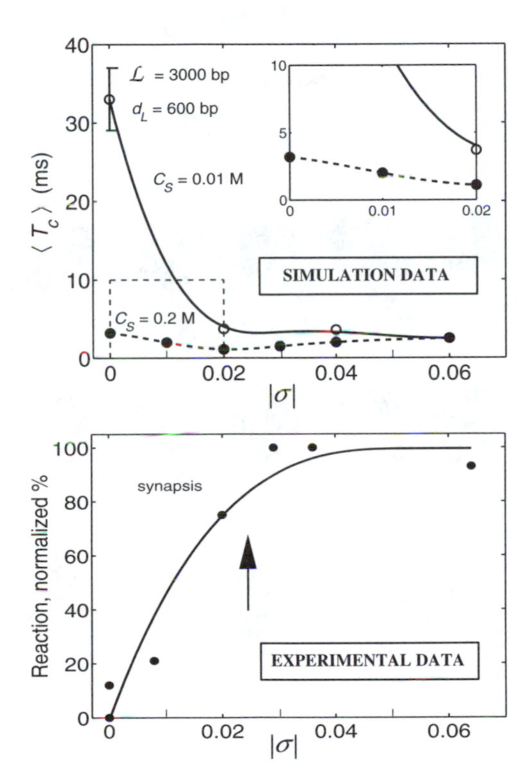

Figure 6.16. (*Top*): Site juxtaposition time measurements, $\langle T_c \rangle$, defined as the mean time for two sites d_L bps apart to come within 100 Å, as averaged over all such separated pairs, as a function of $|\sigma|$ for $d_L = 600$ bps for a 3000 bp system and two salt concentrations [499]. The inset focuses on the large salt effect at low $|\sigma|$. Note also the plateauing effect of $\langle T_c \rangle$ around $|\sigma| = 0.03$ for both salt concentrations. This is also the experimental observation (*bottom*) [517] for the synapsis-time dependence on the superhelical density.

Discrete Charge Parameterization:

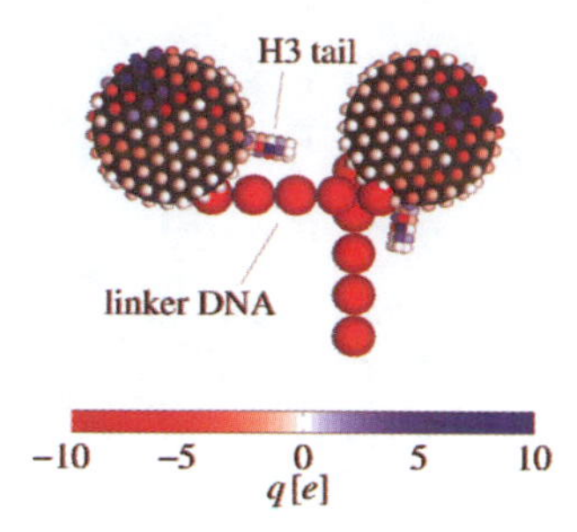

Dinucleosome Folding Trajectory:

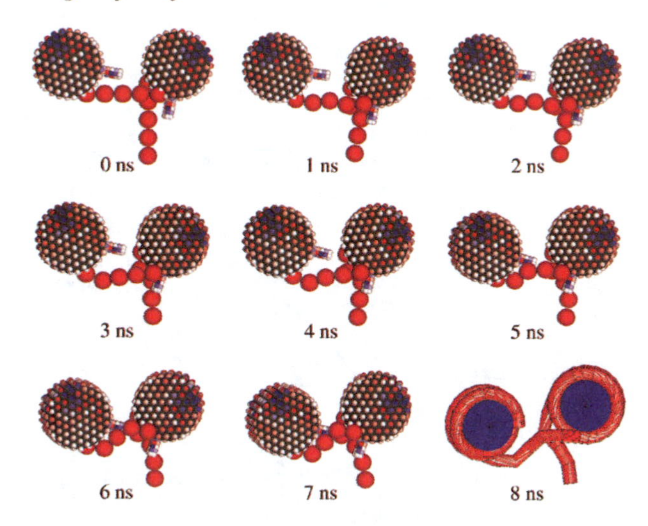

48-Core Fiber Reconstructed from Basic Nucleosome Repeat:

Figure 6.17. The dinucleosome model (upper panel) consists of two electrostatically-charged core particles connected by a 200-Å linker DNA modeled as an elastic wormlike chain. Each core particle is represented as 277 charges, as modeled in [498]. The Brownian dynamics snapshots of a dinucleosome model at high salt, 0.05 M sodium ions (middle), reveal spontaneous folding into a structure where the linker DNA gently bent. The '30-nm fiber' shown in the lower panel (48 nucleosome units) was constructed based on Monte Carlo refinements of a model built from the dinucleosome system and shows a compact helical zigzag pattern.

7

Theoretical and Computational Approaches to Biomolecular Structure

Chapter 7 Notation

SYMBOL	DEFINITION
Vectors	
$\mathbf{a}, \mathbf{b}, \mathbf{c}$	interatomic distance vectors (in definition of τ)
$\mathbf{n}_{ab}, \mathbf{n}_{bc}$	unit normals
$\mathbf{r}_{ij}$	interatomic distance vector $(\mathbf{x}_j - \mathbf{x}_i)$
$\mathbf{x}_i$	position vector of vector i, components x_{i1}, x_{i2}, x_{i3}
$\widetilde{P}$	collective momentum (for nuclei and electrons)
$\widetilde{X}$	collective position (for nuclei and electrons)
X	collective position (nuclei only), components $\mathbf{x}_1, \ldots, \mathbf{x}_N \in \mathcal{R}^{3N}$
V	collective velocity (nuclei only)
ψ_n	eigenfunctions of the Hamiltonian operator
Scalars & Functions	
b_i	bond length i
q_i	Coulomb partial charge for atom i
r_{ij}	interatomic distance (between atoms i and j)
t_0	initial time
A_{ij}, B_{ij}	Lennard-Jones coefficients
$\widehat{H}$	Hamiltonian operator $(E_k + E_p)$
E_{bond}	bond length energy
E_{bang}	bond angle energy
E_{coul}	Coulomb energy
E_k	kinetic energy
E_{local}	local (short-range) energy
E_n	eigenvalues of the Hamiltonian operator (quantum states)
E_{nonlocal}	nonlocal (long-range) energy

Chapter 7 Notation Table (continued)

SYMBOL	DEFINITION
E_p	potential energy (also E)
E_{tor}	torsional (dihedral) angle energy
E_{LJ}	Lennard-Jones energy
K_{ijk}	bond angle force constant
M_F	dimension of the Fock matrix
N	number of atoms
N_e	number of electrons
$\mathcal{R}^n$	Euclidean space of dimension n
S_{ij}	bond length force constant
S_B	set of bonds
S_{BA}	set of bond angles
S_{DA}	set of dihedral angles
S_{NB}	set of atom pairs computed in the nonbonded energy
T	temperature
V_n	torsional barrier height of periodicity n
ϵ	dielectric function
θ_i or θ_{ijk}	bond angle (e.g., θ_{dha} for donor–hydrogen$\cdots$acceptor sequence in a hydrogen bond)
τ_i or τ_{ijkl}	dihedral or torsion angle
Δt	timestep

Science says the first word on everything, and the last word on nothing.

Victor Hugo, 1907 (1802–1885).

7.1 The Merging of Theory and Experiment

7.1.1 *Exciting Times for Computationalists!*

Computational techniques for exploring three-dimensional (3D) structures of nucleic acids and proteins are now well recognized as invaluable tools for revealing details of molecular conformation, motion, and associated biological function. Over a decade ago, the theoretical chemist Henry Schaefer declared: "*It is clear that theoretical chemistry has entered a new stage ... with the goal of being no less than full partner with experiment*" [520].

Commenting on the two 1998 Chemistry Nobel Prize awardees in quantum chemistry — Walter Kohn of the University of California, Santa Barbara, and John Pople of Northwestern University — a reporter in *The Economist* wrote: "*In the real world, this could eventually mean that most chemical experiments are conducted inside the silicon of chips instead of in the glassware of laboratories. Turn off that Bunsen burner; it will not be wanted these ten years*" (17 October 1998)!

It remains to be seen how powerful "in silico biology" will be, but clearly a new enthusiasm is stirring in the molecular biophysics community in the wake of many methodological and technological improvements. The following categories reflect improvements that lend computation a stronger basis than ever before.

1. **Improvements in instrumentational and experimental techniques** [521]. Rapid advances in sequencing and mapping genomes are being made. Many procedures in biomolecular NMR spectroscopy [98] and X-ray crystallography [89] are accelerating the rate and improving both the accuracy and scope (e.g., time-resolved crystallography [85]) of structure determination. And newer microscopic techniques, such as scanning tunneling and cryo [522, 523, 524, 525, 526, 109], and pulsed gel electrophoresis and optical mapping for single molecules of DNA [527, 528], are being applied. With such advances, structures that were considered unconquerable a decade ago are being resolved — RNAs, nucleosomes, ribosomes, ion channels, and membrane proteins, for example. Theoreticians can use these structures with other experimental data as solid bases for simulations and computer analyses.

2. **New models and algorithms for molecular simulations**. Improved force fields are being developed, some simpler and others more complex. Innovative protein folding models are shedding insights into important biological processes. Faster and more advanced dynamic simulations of complex systems with full account of long-range solvation and ionic effects are being used. And quantum-mechanical studies of biomolecules are becoming possible.

3. **The increasing speed and availability of supercomputers, parallel processors, and distributed computing**. Faster, cheaper, and smaller computing platforms and graphics workstations are entering the laboratory at reduced costs, though computer memory requirements are exploding.

4. **Successful multidisciplinary collaborations**. Notable examples are collaborations between mathematicians and biologists in relating knot theory with DNA topology and geometry [529, 530, 481]; between computer scientists/engineers and biologists regarding DNA computing [531, pp. 26–38], [532, 533, 534, 283] and nanotechnology [535, 536];[1] and between

---

[1]Nanotechnology is an emerging science of creating functional materials, devices, and systems on the basis of matter at the nanometer scale, including macromolecules, and the exploitation of novel properties and phenomena on this scale for biomedicine, technology, and more. See itri.loyola.edu/nano/IWGN.Research.Directions/ for a nanotechnology vision report [536], nano.gov/nsetmem.htm for a more general link to President Clinton's proposed U.S. National Nanotechnology Initiative, www.nano.gov, the 24 November 2000 issue of *Science*, volume 290,

biological and mathematical/physical scientists in propelling biology-information technology, or bioinformatics. (See [136], for example, for a description of some mathematical challenges in genomics and molecular biology).

5. **The wealth of readily available Internet and web resources, sequence and structure databases, and highly-automated analysis tools.**[2]

7.1.2 *The Future of Biocomputations*

"*One of these days*," believes *Nature's* former editor John Maddox, "*somebody will begin a paper by saying, in effect, 'Here is the Hamiltonian of the DNA molecule' and will then, after a little algebra, explain just why it is that the start and stop codons of the genetic code have their precise functions, or how polymerase molecules work in the process of transcription*" [537].

Some of us might be somewhat skeptical that "a little algebra" will suffice to explain DNA's functional secrets, but many remain confident that *a large amount of computation* with carefully developed models will bring us closer to that goal in the not-too-distant future.

7.1.3 *Chapter Overview*

In this chapter, we introduce molecular mechanics from its quantum-mechanical roots via the Born-Oppenheimer approximation. Following a brief overview of current quantum mechanical approaches, we discuss the three underlying principles of molecular mechanics: the thermodynamic hypothesis, additivity, and transferability. We then describe the choices that must be made in formulating the potential energy function: configuration space, functional form, and energy parameters. We end by mentioning some of the current limitations in force fields. The next chapter discusses details of the force field form and origin.

7.2 Quantum Mechanics (QM) Foundations of Molecular Mechanics (MM)

Quantum mechanical methods are based on the solution of the Schrödinger equation [538, 539]. This fundamental approach is attractive since 3D structures,

---

highlighting issues in nanotechnology, and the September 2001 special issue of *Scientific American* devoted to 'The Science of the Small', for example.

[2]Caution is certainly warranted regarding the quality of some unreviewed online information. As stated in the New York Academy of Sciences Newsletter of Oct./Nov. 1996, "There may be debate about whether the explosive growth in electronic communication has made life better or worse, but there's no question that it has made life faster".

molecular energies, and many associated properties can be calculated on the basis of fundamental physical principles, namely electronic and nuclear structures of atoms and molecules. Indeed, the quantum mechanics pioneer Paul Dirac is believed to have expressed the sentiment that the Schrödinger equation reduces theoretical chemistry to applied mathematics [540]!

Although historically quantum calculations were practical only for very small systems, exciting developments in both software and hardware (computer speed as well as memory) have made quantum-mechanical calculations feasible for larger systems, even biomolecules, with various approximations (see [541], for example; more below). See the Nobel Lecture address by John Pople for field perspectives [542].

7.2.1 The Schrödinger Wave Equation

The Schrödinger wave equation describes the motions of the electrons and nuclei in a molecular system from first principles. This equation can be written as

$$\widehat{H}\psi_n = E_n\psi_n \,, \tag{7.1}$$

where the Hamiltonian operator $\widehat{H}$ is the sum of the kinetic (E_k) and potential (E_p) energy of the system:

$$\widehat{H}(\widetilde{P},\widetilde{X}) = E_k(\widetilde{P}) + E_p(\widetilde{X}), \tag{7.2}$$

where $\widetilde{P}$ and $\widetilde{X}$ denote the collective momentum and position vectors for all the nuclei and electrons in the molecule. The potential energy $E_p(\widetilde{X})$ originates from electrostatic interactions among all the variables.

According to this description, the quantum states E_n (eigenvalues) form a discrete set, corresponding to the eigenfunctions ψ_n, for the system of electrons and nuclei. The Schrödinger equation thus describes the spatial *probability distributions* corresponding to the energy states in a *stationary* quantum system.

Traditional electronic structure methods — important more from a historical perspective — calculate these eigenstates associated with the discrete energy levels by diagonalization of the Hamiltonian matrix, of order of the number of basis functions. This cubic scaling is prohibitive for large systems.

7.2.2 The Born-Oppenheimer Approximation

In the Born-Oppenheimer approximation to the Schrödinger equation, the motions of the molecule are separated into two levels: electrons and nuclei.

First, only the electrons are considered as independent variables of $\widehat{H}$, while positions of the nuclei are assumed fixed. This is generally a good approximation because the nuclei — much heavier than the electrons — are typically fixed on the timescale of electronic vibration. The resulting eigenvalues from this analysis of electron motion represent electronic energy levels of a molecule as a function of atomic coordinates. These energy states are known as Born-Oppenheimer energy surfaces (BOES).

In the second level of the Born-Oppenheimer approximation, the BOES of the electronic ground state ($E_p(X)$, where X is the collective position vector of the nuclei in the system), are used as the *potential* energy of the Hamiltonian (eq. (7.2)) instead of the Coulombic potential. The quantum mechanical behavior of the nuclei is then investigated.

In theory, quantum mechanical treatments should be the tool of choice for reliable description of complex chemical processes, since no experimental information is needed as input. Yet, an accurate analytic solution of the Schrödinger equation is not feasible except for small molecules. Thus the equation must be generally solved through standard approximations, which naturally reduce the accuracy obtained. The range of application is further limited by the computational complexity required by these techniques, but advances are rapidly occurring on this front.

Two basic quantum-mechanical approaches are used in practice: *ab initio* and *semi-empirical.* Both rely on the Born-Oppenheimer approximation that the nuclei remain fixed on the timescale of electronic motion, but different types of approximations are used. The former is more rigorous than the latter and hence more computationally demanding. For example, in Hartree-Fock type calculations, the computational requirements, including computer time and memory, scale as M_F^2 to M_F^4 where M_F is the dimension of the Fock matrix. This dimension corresponds to the size of the computational basis set used to approximate the wave functions, related to the number of electron orbitals, N_e.

7.2.3 Ab Initio QM

The name *ab initio* implies *non-empirical* solution of the time-independent Schrödinger equation, or solutions based on genuine theory. However, besides the Born-Oppenheimer approximation, relativistic effects are ignored, and the concept of molecular orbitals (or wave functions) is introduced.

In *ab initio* methods, molecular orbitals are approximated by a linear combination of atomic orbitals. These are defined for a certain basis set, often Gaussian functions. The coefficients describing this linear combination are calculated by a variational principle, that is, by minimizing the electronic energy of the molecular system for a given set of chosen orbitals.

This energy (known as Hartree Fock energy) and the associated coefficients are calculated iteratively by the Self Consistent Field (SCF) procedure with positions of the nuclei fixed. This calculation is expensive for large systems — as computation of many integrals is involved — and makes *ab initio* methods computationally demanding.

In practice, the basis set for the molecular wave functions is represented in computer programs by stored sets of exponents and coefficients. The associated calculated integrals are then used to formulate the Hamiltonian matrix on the basis of interactions between the wave function of pairs of atoms (off-diagonal elements) and each atom with itself (diagonal elements) via some potential that varies from method to method. An initial guess of molecular orbitals is then obtained, and the Schrödinger equation is solved explicitly for a minimum state of electronic energy.

The quality of the molecular orbitals used, and hence the accuracy of the calculated molecular properties, depends on the number of atomic orbitals and quality of the basis set. The electronic energy often ignores correlation between the motion of the electrons, but inclusion of some correlation effects can improve the quality of the *ab initio* results.

Density Functional Theory (DFT)

DFT can be formulated as a variant of *ab initio* methods where exchange/ correlation functionals are used to represent electron correlation energy [543]. DFT methods are based on the use of the electron density function as a basic descriptor of the electronic system. In the DFT Kohn-Sham formulation, the electronic wave function is represented by a single ground-state wave function in which the electron density is represented as the sum of squares of orbital densities.

DFT schemes differ by their treatment of exchange/correlation energy, but in general this class of methods offers a good combination of accuracy and computational requirements, especially for large systems. DFT methods are computationally more efficient than conventional *ab initio* methods that correct for electron correlations, but have a similar scaling complexity due to the N_e^3 diagonalization cost of the Hamiltonian matrix. With linear-algebra advances, however, this standard diagonalization expense can be circumvented with approaches that yield *linear scaling* by localization of the electronic degrees of freedom (i.e., electron density, density matrix, and orbital calculations) [544, 545, 546, 541, for example] (see below).

7.2.4 Semi-Empirical QM

In semi-empirical methods, the matrix elements associated with the wave-function interactions are not explicitly calculated via integrals but are instead constructed from a set of predetermined parameters. These parameters define the forms and energies of the atomic orbitals so as to yield reasonable agreement with experimental data. Thus, most integrals are neglected in semi-empirical methods, and empirical parameters are used as compensation. Although good parameterization is a challenging task in these approaches, and parameters are not automatically transferable from system to system, semi-empirical methods retain the flavor of quantum approaches (solution of the Schrödinger equation) and are less memory intensive than *ab initio* methods. As in *ab initio* methods, the quality of the results depends critically on the quality of the approximations made.

7.2.5 Recent Advances in Quantum Mechanics

Traditional quantum calculations are dominated by the cost of solving the electronic wave function expressed in terms of the Hartree-Fock operator. Solution of the wave functions via inversion of the Fock matrix for a minimal basis set (roughly the number of electrons, N_e) requires $\mathcal{O}(M_F^4)$ work. While still much better than classical orbital calculations that include correlations, this scaling limits its applications to large systems. It has been noted, however, that by exploiting the sparsity of the Fock matrix — zero elements due to the rapid decay with distance of orbital overlapping — algorithms developed in linear algebra for banded systems can reduce the cost to linear scaling, i.e., $\mathcal{O}(M_F)$, or to the near-linear cost of $\mathcal{O}(M_F \log M_F)$ when Ewald forces are included [547, 548, 549, 550]. See also [551] for a physics-community perspective of electronic structure calculations (rather than the chemistry community).

Linear Scaling

Linear scaling algorithms are also possible by various divide-and-conquer algorithms that localize electronic calculations (for both *ab initio* and semi-empirical methods) [544, 541] and by reformulating the Hamiltonian diagonalization as a minimization problem. The localization is achieved by using different treatments for the strong intramolecular interactions and the weak intermolecular interactions [552]. The reformulation as an optimization problem entails construction of an energy functional, which is minimized with respect to some variational parameters that represent the electronic ground state [546].

The nonlinear conjugate gradient method (see Chapter 10) exploits the matrix sparsity pattern and is very modest in both computational and storage requirements. See Daniels and Scuseria [546] for a comparison among several linear-scaling approaches for semi-empirical calculations, including pseudo-diagonalization of the Fock matrix via orthogonal rotations of the molecular orbitals, conjugate-gradient density matrix search, and purification of the density matrix via transformations applied to a function expanded in terms of Chebyshev polynomials.

Biomolecular Applications

These advances in complexity reduction are making possible quantum calculations on biomolecules under various simplifications. Semi-empirical methods have been applied to protein and DNA systems of up to 10,000 atoms [553, 554, 555], and *ab initio* approaches have been applied to systems of fewer than 1000 atoms (e.g., DNA bases and base pairs [552, 556] and cyclic peptides [557]).

Exciting developments for treating the boundary in QM/MM methods (quantum mechanics/molecular mechanics) also make possible the application of a combined *ab initio* QM/MM approach to study enzyme mechanisms [558] and protein dynamics [559]. See Figures 7.1 and 7.2 and Box 7.1 for examples.

Box 7.1: Illustrative QM and QM/MM Applications

Semi-empirical quantum-mechanical applications have been used, for example, to study aqueous polarization effects on biological macromolecules, by comparing free energies in the solvated versus gas-phase states [553]. Figure 7.1 from [554] illustrates the quantum-mechanically derived electrostatic potential of the polarized electron density of A, B, and Z-DNA relative to the gas phase density. The maps indicate how the electrostatic potential changes when the electron density of the DNA is polarized by the reaction field of the solvent. Another example of semi-empirical applications is the study of the active site of an enzyme (cytidine deaminase) to delineate a mechanism for ligand attack [555].

In *ab-initio* applications, electronic and vibrational properties have been determined from optimized geometries. *Ab initio* methodologies have also been applied to many nucleic-acid base systems and their complexes, for example, to study stacking interactions [560, 561, 556], hydrogen-bonding [562], and basepair planarity properties.

Combined *ab initio* QM/MM approaches can be used to study enzyme catalysis [558], for example to deduce the reaction paths and free energy barriers for the two steps of the reaction catalyzed by enolase [563, 564]. This two-step reaction involves abstraction of a proton from the substrate to produce an intermediate, followed by departure of a hydroxyl group with the assistance of a general acid.

The calculations have identified catalytically important residues and water molecules at the active site of the enzyme (see Figure 7.2) and have shown that the electrostatic interactions driven by metal cations at the active site strongly favor the first, but strongly disfavor the second, step of the reaction. This dilemma appears to be resolved by a tailored organization of polar and charged groups at the enolase active site, exploiting the two different orientations of charge redistribution involved in each of the two reaction steps. Thus, the enzyme environment might provide an essential platform for the reaction mechanism. This finding may ultimately be tested experimentally.

7.2.6 *From Quantum to Molecular Mechanics*

An alternative approach is known as *molecular mechanics*, also referred to as the *force-field* or *potential energy* method [566, 567, 568, 569, 570, 571, 572, 573, 53]. The Born-Oppenheimer approximation to the potential energy with respect to the nuclei, $E_p(X)$, can be imagined as the target function in molecular mechanics (X represents the collective position vector for the nuclei). The electrons can be regarded as implicit variables of this potential. However, unlike quantum mechanics, this potential function must be evaluated empirically.

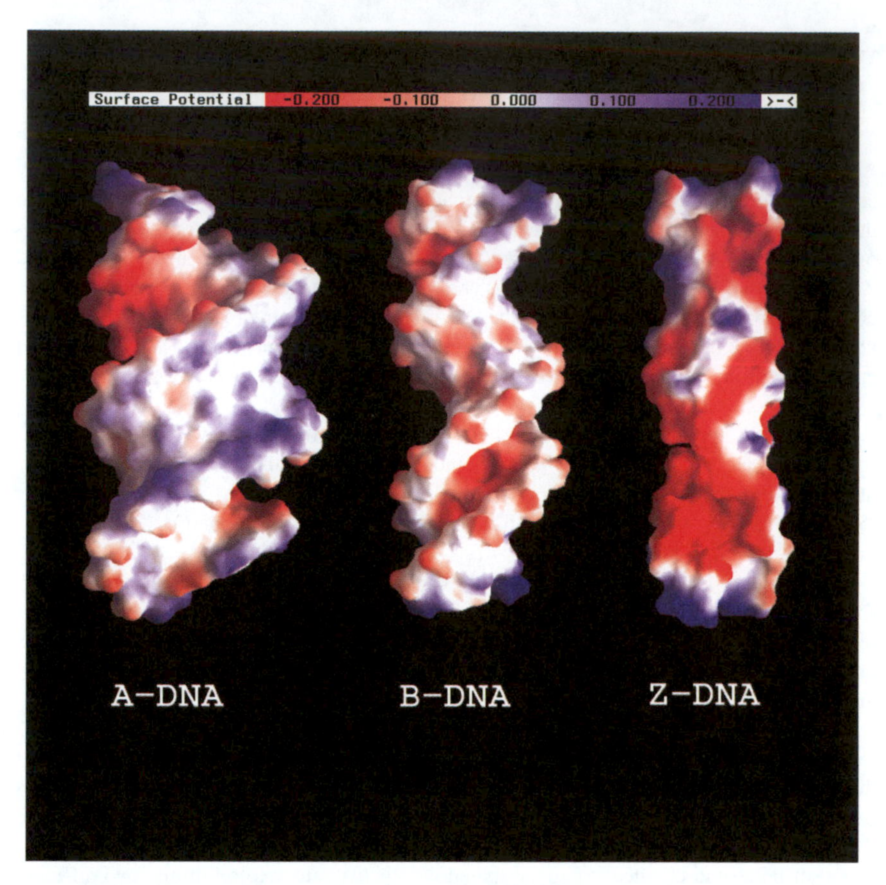

Figure 7.1. The electrostatic potential surface of the electronic response density of A, B and Z-DNA calculated by the linear-scaling electronic structure and solvation methods of York and coworkers [554]. The electronic response density is defined as $\Delta\rho(\mathbf{r}) = \rho_{\text{sol}}(\mathbf{r}) - \rho_{\text{gas}}(\mathbf{r})$ where $\rho_{\text{gas}}(\mathbf{r})$ is the relaxed electron density in the gas phase and $\rho_{\text{sol}}(\mathbf{r})$ is the relaxed electron density in solution (approximated by a continuum dielectric model).

Mechanical Molecular Representation

An underlying principle in molecular mechanics is that *cumulative* physical forces can be used to describe molecular geometries and energies. The spatial conformation ultimately obtained is then a natural adjustment of geometry to minimize the total internal energy. A molecule is considered as a collection of masses centered at the nuclei (atoms) connected by springs (bonds); in response to inter and intramolecular forces, the molecule stretches, bends, and rotates about those bonds (Fig. 7.3). This simple description of a molecular system as a mechanical body is usually associated with a "classical" system. Remarkably, this classical mechanics description — an appropriate characterization even as the amount of quantum-

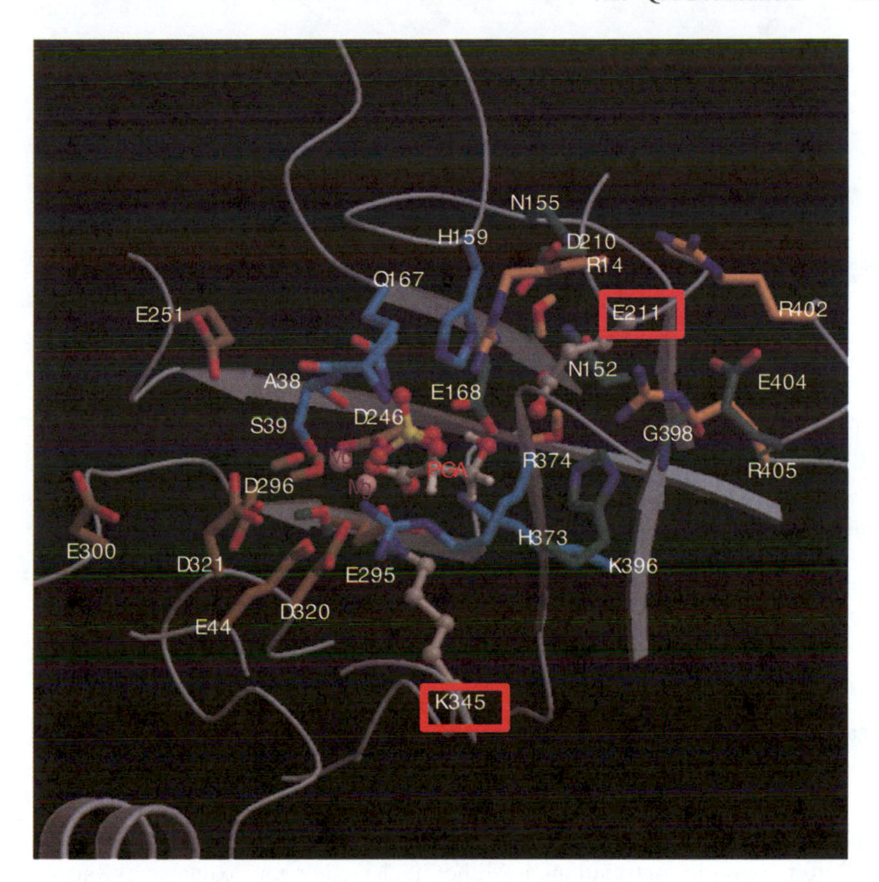

Figure 7.2. Catalytically important protein residues and water molecules at the active site of enolase as identified by *ab initio* QM/MM calculations by Weitao Yang and coworkers [565]. The color coding is according to categories of energetic contribution to stabilization of the transition state in either the first or second step of the reaction (e.g., sky blue, dark golden, green, orange); oxygens are colored red, nitrogens are dark blue, and hydrogen atoms of amino acid residues are omitted. Lys345 (K345, bottom center) is the general base that captures a proton from the substrate (PGA, 2-phospho-D-glycerate) in the first reaction step. Glu211 (E211, top right) is the general acid that assists departure of a hydroxyl group in the second reaction step. Residues in sky blue are hydrogen bond donors interacting with the substrate. Those in dark golden and orange are positioned to counteract the effects of the two magnesium ions (labeled) in the second reaction step but have small energetic effects on the first reaction step. Residues colored green are others found to have energetic effects on the reaction.

mechanical information used to derive force fields increases — works generally well for describing molecular structures and processes, with the exception of bond-breaking events.

In practical terms, molecular mechanics involves construction of a potential energy, a function of atomic positions, from a large body of molecular data (crystal

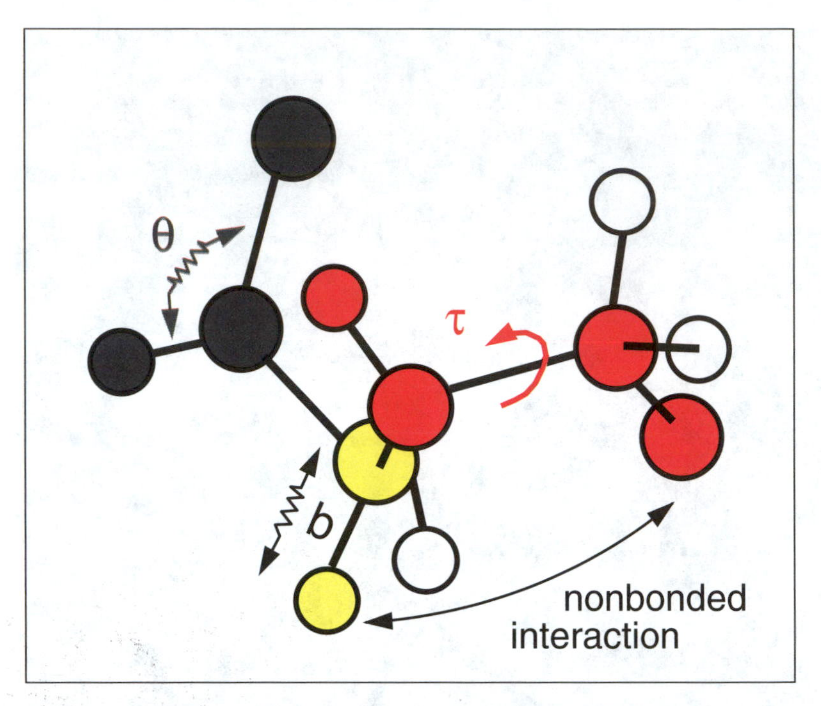

Figure 7.3. A molecule is considered a mechanical system in which particles are connected by springs, and where simple physical forces dictate its structure and dynamics.

structure geometries, vibrational and microwave spectroscopy, heats of formation, etc.). Entropic contributions are either neglected or approximated by various techniques. Minimization of this function can then be used to compute favorable regions in the multidimensional configuration space, and molecular dynamics simulations can further explain the system's thermally accessible states.

Early Days

As introduced in Chapter 1, the first molecular mechanics implementations date to the 1940s, but only in the late 1960s/early 1970s did the availability of digital computers make such calculations tractable. One of the force-field pioneers, the late Shneior Lifson, recalled that "*the empirical [force field] method did not always enjoy a high prestige value among theoretical chemists, particularly those engaged in quantum chemistry*" [540]. But Lifson went on to suggest that in the early days of the field the notions of force-field *consistency* and *transferability* were neither always carefully treated nor well understood. The power of the empirical force-field approach to describe collectively properties of related molecules — a description not possible by quantum mechanics — was also not fully appreciated.

A classic example illustrating the early success of molecular modeling computations is the correct interpretation of peculiar experimental results. In 1967, consistent force-field calculations predicted two stable conformations for a cycloalkane (1,1,5,5–tetramethylcyclodecane). Oddly, neither structure resembled the crystal form [574]. However, the *average* of the two computed molecular mechanics structures matched the experimental data *exactly*! Thus, empirical calculations revealed that the regular crystal contained a distribution of two conformers. This was indeed confirmed later by experimentation. Today, this theme emerges often from molecular dynamics simulations, where the spatial and temporal conformational-ensemble average corresponds to the experimental structure [575].

7.3 Molecular Mechanics: Underlying Principles

In theory, the overall effectiveness of molecular mechanics depends on the validity of its three underlying principles: (1) basic thermodynamic assumption, (2) additivity of the effective energy potentials, and (3) transferability of these potentials.

7.3.1 The Thermodynamic Hypothesis

Does Sequence Imply Structure?

The thermodynamic hypothesis assumes that many macromolecules are driven naturally to their folded native structure (i.e., with minimal or no intervention of catalysts) largely on the grounds of thermodynamics. This behavior is in contrast to other processes in the central dogma that require the crucial interplay of enzyme machinery to lower activation barriers. Though it is important to consider these complex factors when modeling macromolecules, the basic thermodynamic hypothesis of the empirical approach appears reasonable for many biomolecules under study by molecular mechanics and dynamics.

Indeed, experiments have supported the intrinsic folding/entropy connection for many small globular proteins [576]. Thus, a strong thermodynamic force drives 'scrambled' conformations of high free energy to the native state, of low free energy. Folding in many cases is reversible (denatured $\Longleftrightarrow$ native state) and attainable from many different configurations.

Theoretical analyses based on statistical studies of spin glasses further suggest that the probability that a randomly synthesized protein will exhibit a thermodynamically dominant fold (i.e, the global minimum of the free energy) increases rapidly as temperature decreases [577].

As discussed in Chapter 3, the resilience in general of overall protein structures to small mutations — in some cases even at critical regions, as in the Rop turn ([206] and Figure 3.10) — is another indicator of the strong thermodynamic tendency of proteins to adopt and retain native structure/folds. For higher organizational forms of DNA, the intrinsic topology also appears to induce a folded configuration determined in large part by base composition and the ionic medium [286].

The kinetics involved in such folding processes is yet another subject of intense interest (see [166, 183], the discussion in Chapter 2, and Box 7.2).

Box 7.2: The Thermodynamic Hypothesis as Exemplified by The Levinthal Paradox

The concept introduced by Cyrus Levinthal in 1969 [154], which became termed the "Levinthal paradox", is that a protein cannot find its native state by a random search through all its possible conformations. This is because such an exhaustive enumeration would in theory require eons of years, while real proteins fold on the sub-second and second timeframe. Levinthal thus hypothesized that well defined pathways must exist for protein folding in nature [153].

Recently, protein-folding kinetics have been analyzed and interpreted on the basis of simple simulations of polypeptide models. The "old view" that emerged from Levinthal's work is now being merged with the "new view" [158, 159, 23], which promotes many competing folding pathways rather than a unique pathway with well-defined intermediates. The two views can be merged when the folding pathways all share a unifying pattern of native contacts and can be described within a common multidimensional energy landscape [160]. (See also discussion in Chapter 2).

While this protein folding problem is generally formulated purely in terms of "sequence dictates structure", it is clear that in many cases the folding process *in vivo* is enhanced or facilitated through the external intervention of additional molecules such as chaperones. Such agents may stabilize intermediates, prevent aggregates of misfolded proteins, and/or reduce the activation energies effectively [578, 579]. Members of the chaperone class that includes the heat-shock protein hsp70 bind to newly-synthesized proteins and short linear peptides, possibly preventing aggregates. Other chaperones assist in protein translocation across membranes. Members of the *chaperonin* class include the well-studied GroEL/GroES complex from bacteria [182] that bind to partially-folded polypeptides and assist in the folding. Many questions are now being investigated regarding the folding kinetics in all these systems (see [180, 181], for example).

7.3.2 Additivity

The molecular mechanics principle of additivity assumes that the effective molecular energy can be expressed as a sum of potentials derived from simple physical forces: van der Waals, electrostatic, mechanical-like strains arising from "ideal" bond length and angle deviations, and internal torsion flexibility (rotation of two chemical groups about the bond joining them). The forces can be separated into local (bonded) and nonbonded (nonlocal) terms.

Local Terms

The local components for macromolecules can typically be written as:

$$E_{\text{local}}(X) = \sum_{\substack{\text{bonds} \\ i}} E_{\text{bond}}(b_i) + \sum_{\substack{\text{bond angles} \\ i}} E_{\text{bang}}(\theta_i) + \sum_{\substack{\text{dihedral angles} \\ i}} E_{\text{tor}}(\tau_i)\,, \tag{7.3}$$

where summations extend over the sets of bonds $\{b_i\}$, bond angles $\{\theta_i\}$, and dihedral angles $\{\tau_i\}$ (see Fig. 7.3). Functional forms are typically harmonic (e.g., $S[b - \bar{b}]^2$) and trigonometric (e.g., functions of $\cos(\theta)$), as discussed in the next chapter. Note that all internal variables are functions of interatomic distances, r_{ij}. Cross terms (like $E(b, b')$ or $E(b, \theta)$) are not commonly used for proteins and nucleic acids because the associated force constants ('off-diagonals' in the force-constant matrix) usually have much smaller values than those associated with the 'diagonal' terms. Such force constants, for example $S_{bb'}$ for bond/bond interactions or $K_{b\theta}$ for bond/bond-angle terms, are associated with potentials of the form $S_{bb'}(b - \bar{b})(b' - \bar{b'})$ and $K_{b\theta}(b - \bar{b})(\theta - \bar{\theta})$, respectively, which are common in small-molecule force fields (e.g., [580]).

Nonlocal Terms

The nonlocal components can be written as:

$$E_{\text{nonlocal}}(X) = \sum_{\substack{\text{nonbonded pairs} \\ (i,j),\ i<j}} E_r(r_{ij})\,, \tag{7.4}$$

where the functions are often rational (e.g., $\sum_{i=1}^{N_i} r^{-m}$). These energies are usually comprised of van der Waals and Coulombic contributions between non-bonded atom pairs. (Nearby atom pairs counted in the local energy may be omitted to avoid double counting).

Benefits of Separability

The natural separability into bonded (local) and nonbonded (nonlocal) terms can be exploited in the design of minimization algorithms that use function structure to accelerate convergence (see Chapter 10). This separability is also the basis for multiple-timestep protocols for molecular dynamics that update local and nonlocal forces at different frequencies (see Chapters 12 and 13). This difference in force updating is reasonable because the local terms generally change rapidly, while the nonlocal terms change more slowly, with distance and time.

While the number of nonbonded terms grows quadratically $\mathcal{O}(N^2)$ with the number of atoms, the number of local terms — involving pairs, triplets, and quadruplets of bonded atomic sequences — grows only linearly with size. This computational complexity of the nonbonded terms is especially a burden in simulation protocols for large systems that require updating for many iterations, as in macromolecular dynamics. Fortunately, some work can be reduced since the nonbonded Coulomb forces change more slowly with distance than the bonded terms, and hence can be updated less often than the local forces. Chapter 9 is devoted to the nonbonded forces.

Multibody Potentials

In this typical energy formulation, nonbonded interactions are "effective pair potentials" (i.e., additive two-body forces). For electrostatic interactions, for example, these effective pair potentials only reflect in some average sense the charge distribution due to molecular polarizability. It is clear that many-body contributions are important for accurate reproduction of certain molecular properties. For example, accurate account of dispersion forces for polar molecules (i.e., molecules in which the charges are nonuniformly distributed), such as water, require three-body interaction potentials:

$$E_{ijk}(r_{ij}, r_{jk}, r_{ik}).$$

These potentials can be expressed as a composite of trigonometric and rational functions in terms of three bond lengths and three bond angles [581]. Indeed, the importance of water as a solvent for proteins and nucleic acids has prompted development of 'polarized' water potentials [582, for example]. In practice, these models increase the number of *interaction sites* per water molecule.

7.3.3 Transferability

The principle of transferability assumes that potentials can be developed to incorporate all experimental data for *representative* structures and then applied successfully to the prediction of large biological molecules composed of the same chemical subgroups. This is basically a reasonable assumption since bond lengths and bond angles tend to adopt similar values in different molecular species under normal conditions. However, under special straining forces, such as in cycloalkanes, these values may vary significantly from 'ideal' or average values. Modeling the structures and properties of complex aromatic or conjugated systems[3] also requires special treatment, such as using sophisticated quantum-mechanical calculations to obtain bond orders of conjugated bonds, which can be related to bond lengths and stretching constants, and in this way reducing the problem to molecular mechanics ([583, 584, 585, 586], for example).

Functional Variations in Geometry

In small molecules, the environment-dependence of geometric trends can be modeled by a *function* rather than a constant. For example, Allinger and coworkers devised functions for bond lengths in small-molecule force fields (MM3/MM4) to account for the *electronegativity* of attached substituents [587] and for *hyperconjugation* [588]. The former parameterization [587] can, for example, accurately

[3] Aromatic compounds are benzene-like in structure and properties. Conjugation is a structural feature caused by the overlap of the π orbital with other orbitals in the molecule.

model the *shorter* C–C bond in fluoroethane (C_2H_5F) relative to ethane (C_2H_6), since the fluorine is electronegative; similarly, it can also account for a *longer* C–O bond in an alcohol where the electropositive hydrogen is attached to an oxygen (like ethyl alcohol, CH_3CH_2OH), relative to an analogous molecule in which a carbon rather than hydrogen is attached (as in dimethyl either, $C_2H_5OC_2H_5$). A functional dependence for reference bond lengths used for hyperconjugated systems [588] can account for bond-length trends in molecular species in which bond orbitals overlap and lead to resonance, like longer C–H or C–C bonds in carbonyl compounds (e.g., H–C–C=O $\leftrightarrow$ H^+ + C=C–O).

Proliferation of Atom Types

An alternative approach for incorporating the environment dependence of geometric tendencies is to increase the number of 'atom types'. Thus, atom types reflect the molecular environment (e.g., aromatic carbon in a nitrogenous base) and hybridization (e.g., sp^2 or sp^3) [566, 568, 589].

There are around 57 atom types, for example, in the CHARMM 22 program for proteins (version 22) [589] (21 carbons, 12 hydrogens, 11 nitrogens, 7 oxygens, 3 sulfurs, one heme iron, one calcium cation, and one zinc ion); see Table 7.1 for some examples of these atom types and Figure 7.4 for illustrations for selected residues [589].[4] The proliferation of atom types in modern force fields thus attempts to improve compatibility with experiment. Alternatively, force fields may be restricted to certain families of molecules such as alkanes, amides, and carboxylic acids [571, 569].

I emphasize that *individual* functions *should not be transferred* from one force field to another, since the entire potential is parameterized as a whole to reproduce consistently experimental data.

Overall, while the transferability assumption is inherent in this empirical science, molecular mechanics has steadily gained recognition through many important contributions. Today's force fields are excellent for deducing structures and properties of many molecular systems, especially for small molecules using very accurate force fields. One advantage of theoretical calculations is that the thermal motions and lattice effects that influence crystallographically-determined structures may not be a problem when accurate force fields are used to predict molecular structures and properties.

[4]CHARMM 22 has 51 atom types if only carbon, nitrogen, oxygen, hydrogen, and phosphorus atoms are considered, versus 69 types for these atoms in the CHARMM 27 force field.

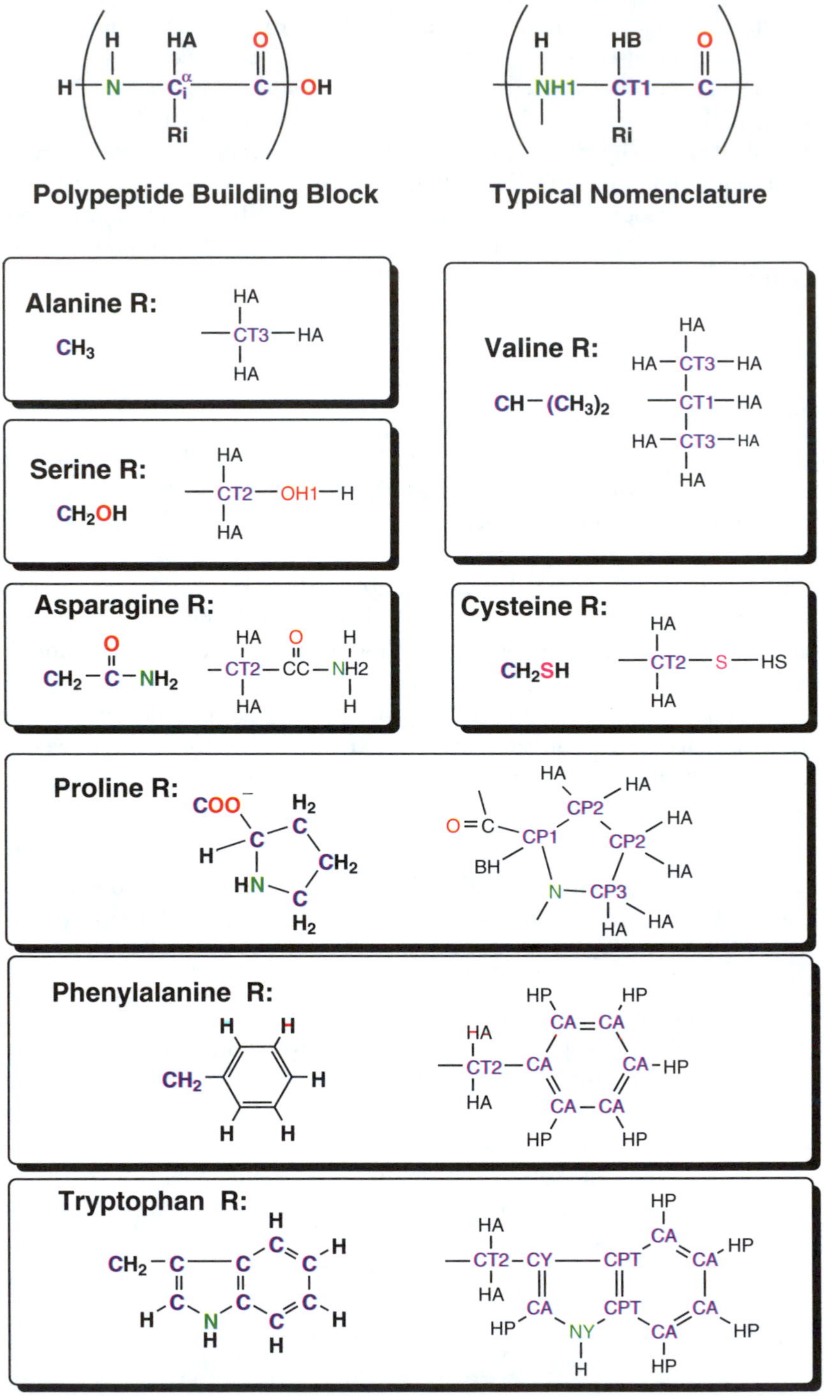

Figure 7.4. Examples of atom types as used in polypeptides in the CHARMM program [589].

Table 7.1. Examples of atom types defined in CHARMM 22 [589].

Atom	Symbol	Atom modifier
Carbon	C	polar (carbonyl, peptide backbone)
	CA	aromatic
	CC	carbonyl (Asn, Asp, Gln, Glu)
	CPT	inter-ring in tryptophan
	CP1, CP2, CP3	special tetrahedral, in proline
	CT1	aliphatic sp3 in CH
	CT2	aliphatic sp3 in CH_2
	CT3	aliphatic sp3 in CH_3
Oxygen	O	carbonyl
	OC	carboxylate
	OH1	hydroxyl
Nitrogen	N	proline
	NH1	peptide
	NH2	amide
Hydrogen	H	polar
	HA	nonpolar
	HP	aromatic
	HS	thiol

7.4 Molecular Mechanics: Model and Energy Formulation

In practice, the challenge and success of molecular mechanics rely on both effective formulation of the potential energy function and the application of suitable search algorithms (e.g., multivariate minimization, sampling, dynamic simulations). With the well known difficulties of large-scale minimization (see Chapter 10) and the timestep problem in dynamics (Chapters 12 and 13), the structural outcome — and hence biological implications of any calculation — depends on the combination of modeling and algorithmic techniques employed. Indeed, some strengths and weaknesses of molecular mechanics have been debated openly in a series of papers [590, 591, 592, 593]; see also Homework Assignment 7. Some valid questions are:

- How accurate are quantum-mechanically derived partial charges?
- How do Cartesian and dihedral-angle representations affect results?
- Is it appropriate to introduce arbitrary scale factors in energy coefficients to enhance agreement with experiment?

The cautions and possible pitfalls of force field derivations and parameter choices are thus important to emphasize, even for state-of-the-art force fields.

In formulation of the energy, three basic and important decisions are involved: (1) representative configuration space, (2) functional form, and (3) numerical values for the parameters. We discuss each in turn.

7.4.1 *Configuration Space*

A Question of Size

The atomic configuration[5] space for a molecular system consists of $3N - 6$ degrees of freedom, where N is the number of atoms. (Six degrees of freedom are removed for rigid-body translation and rotation invariance of the energy). Thus, the number of degrees of freedom for proteins and nucleic acids typically ranges in order from 10^3 to 10^4. With explicit representation of water molecules, this number rises by at least an order of magnitude (see, for example, solvated-macromolecular system sizes in Table 1.2 of Chapter 1).

The complete set of $3N - 6$ degrees of freedom can be described directly by lists of Cartesian coordinates for all the atoms in the system. Alternatively, internal variables such as bond lengths, bond angles, and dihedral angles may be used; this internal representation has been quite successful for the study of proteins [590, for example].

Other representations have been used for biomolecules to reduce this number of variables. Reductions are possible by fixing bond lengths and angles and working in dihedral angle space, or by restricting energetic pathways to approximate formulations. However, these representations do not usually lead to enhanced efficiency unless the work involved in the nonbonded computations — the real computational 'bottleneck', see Chapter 9 — is reduced significantly.

The Pseudorotation Description

An example of a parameter reduction approach is the pseudorotation path used earlier for nucleic acids [299, 298]. This energetic path, initially developed for the hydrocarbon cyclopentane and later extended to ribose and deoxyribose sugars, constrains the energy of five-membered sugar rings to a wavelike motion from a mean plane. This plane can be defined in various ways by positions of the five skeletal ring atoms. See Chapter 5 for a discussion of these descriptions in the section on furanose conformations.

While conceptually simple, the reduction in degrees of freedom from 9 $(3{\cdot}5-6)$ to 2 is clearly approximate. The pseudorotation approximation was noted to produce anomalies in overall nucleic acid structures [594, 301], inconsistencies with

---

[5] *Configuration*, a more general term than *conformation*, is often used by mathematicians to describe the shapes of objects in space. The more chemical term *conformation* often refers to configurations that differ from one another through rotations of groups of atoms about the bond connecting them (i.e., dihedral angles). Note, however, that in organic chemistry a *configuration* refers to stereoisomers, molecules with different connectivity arrangements which cannot be converted into one another through rotations about bonds.

ring closure, and mathematical difficulties in expressing the energy derivatives in terms of the independent conformational variables. More generally, while for some systems simplifications in the representative conformation space may be acceptable, it is usually advantageous to avoid constraints altogether [595]. (The pseudorotation concept remains a useful analysis tool, however).

Cartesian Space

Cartesian coordinate space is most convenient for direct differentiation of the energy in terms of the independent parameters and therefore application of powerful second-derivative Newton minimization methods (see Chapter 10). Cartesian space is also most natural for implementation of molecular dynamics, since the generated molecular trajectories

$$\{X(t_0),\ X(t_0+\Delta t),\ \ldots,\ X(t_0+n\,\Delta t),\ \ldots\}\,,$$

where t_0 is the initial time reference and Δt is the timestep (see molecular dynamics chapters), generally rely on recursive expressions. That is, the new positions and velocities

$$X(t_0+n\,\Delta t) \quad \text{and} \quad V(t_0+n\,\Delta t)$$

are explicit functions of the previous positions and velocities

$$X(t_0+(n-1)\,\Delta t) \quad \text{and} \quad V(t_0+(n-1)\,\Delta t)\,.$$

With all these considerations, it is usually advantageous to enforce constraints — if necessary — by means of penalty functions (i.e., *soft* constraints). Harmonic penalty functions can be used, for example, to keep bond lengths and angles close to their observed values.

7.4.2 *Functional Form*

Composition

The potential energy E of a molecular model is typically constructed as the sum of contributions from the following types of terms: bond length and bond angle strain terms (E_{bond} and E_{bang}), a torsional potential (E_{tor}), a Lennard-Jones potential to model repulsion at short interatomic separations and attraction at long distances (E_{LJ}), and a Coulombic potential among the pairs of charged particles in the system (E_{coul}):

$$E = E_{\text{bond}} + E_{\text{bang}} + E_{\text{tor}} + E_{\text{LJ}} + E_{\text{coul}} \tag{7.5}$$

$$E_{\text{bond}} = \sum_{i,j\in S_{\text{B}}} S_{ij}\big(r_{ij}-\bar{r}_{ij}\big)^2$$

$$E_{\text{bang}} = \sum_{i,j,k\in S_{\text{BA}}} K_{ijk}(\cos\theta_{ijk}-\cos\bar{\theta}_{ijk})^2$$

$$E_{\text{tor}} = \sum_{ijk\ell \in S_{\text{DA}}} \sum_n \left(\frac{V n_{ijk\ell}}{2} \left[1 \pm \cos(n\tau_{ijk\ell}) \right] \right)$$

$$E_{\text{LJ}} = \sum_{i,j \in S_{\text{NB}}} \left(\frac{-A_{ij}}{r_{ij}^6} + \frac{B_{ij}}{r_{ij}^{12}} \right)$$

$$E_{\text{coul}} = \sum_{i,j \in S_{\text{NB}}} \left(\frac{q_i q_j}{\epsilon(r_{ij})\, r_{ij}} \right)$$

These equations represent first approximations to the physical potentials but are nonetheless reasonable for biomolecules. In these general expressions, the symbols S_{B}, S_{BA}, and S_{DA} denote the sets of all bonds, bond angles, and dihedral angles. The nonbonded set, S_{NB}, typically includes all (i, j), $i < j$, atom pairs separated by three bonds or more. Bond and angle variables capped by bar symbols denote reference values associated with these quantities.

The 6/12 Lennard Jones potential above (i.e., attraction of form $-A/r^6$ and repulsion of form B/r^{12}) is typical for large-molecule force fields because of its mathematical convenience; a Buckingham potential (with the same functional form for attraction but an exponential repulsion term of form $B\exp(-B'r)$) [596] is in principle closer to the electronic structure of the atom and thus provides a better potential fit over a broader range of interparticle distances [580].

Molecular Geometry

To be more precise, we define the analytic expressions for the geometric quantities in the potential energy.

Positions and distances. For a molecular system of N atoms (possibly including atom groups) in Cartesian coordinate space, let

$$\mathbf{x}_i = (x_{i1}, x_{i2}, x_{i3}), \qquad i = 1, \ldots, N,$$

denote the position vector of atom i, and

$$\mathbf{r}_{ij} \equiv \mathbf{x}_j - \mathbf{x}_i$$

denote the distance vector from atom i to j. Our potential energy function E then depends on all Cartesian variables of the atoms:

$$E = E(X) \equiv E(\mathbf{x}_1, \mathbf{x}_2, \ldots, \mathbf{x}_N)$$

where X is the collective vector in the Euclidean space $\mathcal{R}^n$ of dimension $n = 3N$.

For any vector $\mathbf{a}$ we denote its standard Euclidean magnitude by $\|\mathbf{a}\|$. For convenience later in writing the potential energy function, we also denote an interatomic vector magnitude $\|\mathbf{r}_{ij}\|$ as r_{ij}, in nonbold type.

Bond angles. A bond angle θ_{ijk} formed by a bonded triplet of atoms i–j–k is expressed as an inner (or dot) product:

$$\cos\theta_{ijk} = \frac{(\mathbf{x}_k - \mathbf{x}_j) \bullet (\mathbf{x}_i - \mathbf{x}_j)}{r_{ij}\; r_{kj}} \,. \tag{7.6}$$

Dihedral angles. A dihedral angle $\tau_{ijk\ell}$, defining the rotation of bond i–j around j–k with respect to k–l, is expressed as (see Fig. 3.14 in Chapter 3 and Fig.7.3):

$$\cos\tau_{ijk\ell} = \mathbf{n}_{ab} \bullet \mathbf{n}_{bc}. \tag{7.7}$$

The vectors $\mathbf{n}_{ab}$ and $\mathbf{n}_{bc}$ denote unit normals to planes spanned by vectors $\{\mathbf{a},\ \mathbf{b}\}$ and $\{\mathbf{b},\ \mathbf{c}\}$, respectively, where $\mathbf{a} = \mathbf{r}_{ij}$, $\mathbf{b} = \mathbf{r}_{jk}$, and $\mathbf{c} = \mathbf{r}_{k\ell}$. Denoting θ_{ab} and θ_{bc} as angles θ_{ijk} and $\theta_{jk\ell}$, respectively, we write:

$$\cos\tau_{ijk\ell} = \frac{\mathbf{a}\times\mathbf{b}}{\|\mathbf{a}\|\,\|\mathbf{b}\|\sin\theta_{ab}} \bullet \frac{\mathbf{b}\times\mathbf{c}}{\|\mathbf{b}\|\,\|\mathbf{c}\|\sin\theta_{bc}} \,. \tag{7.8}$$

The sign of $\tau_{ijk\ell}$ is set by the sign of the triple scalar product $\mathbf{a} \bullet (\mathbf{b} \times \mathbf{c})$.

Lagrange's identity. To simplify potential energy equations and differentiation [597], it is convenient to work with inner product expressions and use Lagrange's identity:

$$(\mathbf{a}\times\mathbf{b}) \bullet (\mathbf{c}\times\mathbf{d}) = (\mathbf{a}\bullet\mathbf{c})(\mathbf{b}\bullet\mathbf{d}) - (\mathbf{b}\bullet\mathbf{c})(\mathbf{a}\bullet\mathbf{d})\,. \tag{7.9}$$

This produces the alternative expression for $\cos\tau$:

$$\begin{aligned}\cos\tau_{ijk\ell} &= \frac{(\mathbf{a}\times\mathbf{b})\bullet(\mathbf{b}\times\mathbf{c})}{[\,(\mathbf{a}\times\mathbf{b})\bullet(\mathbf{a}\times\mathbf{b})\ \ (\mathbf{b}\times\mathbf{c})\bullet(\mathbf{b}\times\mathbf{c})\,]^{1/2}} \\ &= \frac{(\mathbf{a}\bullet\mathbf{b})(\mathbf{b}\bullet\mathbf{c}) - (\mathbf{a}\bullet\mathbf{c})(\mathbf{b}\bullet\mathbf{b})}{\{\,[\,(\mathbf{a}\bullet\mathbf{a})(\mathbf{b}\bullet\mathbf{b}) - (\mathbf{a}\bullet\mathbf{b})^2\,]\ [\,(\mathbf{b}\bullet\mathbf{b})(\mathbf{c}\bullet\mathbf{c}) - (\mathbf{b}\bullet\mathbf{c})^2\,]\,\}^{1/2}}\end{aligned} \tag{7.10}$$

According to this convention, $\tau = 0^\circ$ defines a *cis* coplanar orientation for atoms i–j–k–l, $\tau = 180^\circ$ defines a *trans* coplanar orientation, and a positive sign corresponds to a clockwise rotation of the far bond with respect to the near bond (when viewed along the j–k bond).

Derivatives. First and second derivatives of these expressions are often needed for structure minimization, molecular dynamics, and various conformational analyses (e.g., normal modes). The derivative expressions are tedious to derive (see Homework Assignment 8) and care must be used to avoid singularities, but various algorithmic procedures have been developed to simplify the task in practice [597, for example].

7.4.3 Some Current Limitations

Parameterization details for each potential energy term, including functional variations, are described in the next chapter. We conclude this chapter by mentioning some general limitations of current molecular force fields:

1. **Many Force Field Choices**. At present, there is no "universal force field", nor are the many force fields in use close to converging to one another in some sense. Essentially, users around the world develop a preference for one force field over another on the basis of their target application and practical factors such as cost and convenience.

   For example, the MM2/3/4 force-field family developed by Norman Allinger and coworkers is a popular choice for small molecular systems. Protein modelers often use the CHARMM package developed by Martin Karplus and coworkers. Nucleic acid modelers might prefer the AMBER program developed by Peter Kollman and coworkers. Many from the Scheraga school prefer the dihedral-angle ECEPP program for proteins. Other programs are available, such as GROMOS, OPLS, CFF, SIGMA, and NAMD; see also [598, p. 6109] and [599, p. 491] for lists of available programs, and Table 10.1 of Chapter 10 for contact information for some of these programs.

   A current effort in force field development focuses on covering a wider range of chemical systems within one program. The MMFF force field, for example, developed at Merck is especially designed to yield moderately high accuracy for a wide range of organic molecules and proteins (see next chapter). However, it may also be argued that specialized, rather than general-purpose, force fields may be appropriate, to allow specialization in terms of applications (e.g., drug/protein energetics on one hand and long-time, approximate behavior or macromolecules on the other) [600].

2. **Variability in Functional Form and Numerical Parameters**. As detailed in the next chapter, a great deal of variability exists in the functional forms used for each potential energy term as well as in the numerical values for the associated parameters.

3. **Inclusion of Explicit Hydrogen Bonding**. Some macromolecular force fields use explicit hydrogen-bonding potentials, though these potentials were more common in older versions that did not consider hydrogens explicitly. Note that the strength of a hydrogen bond is determined by its geometry, which depends on the distances associated with the Donor-Hydrogen $\cdots$ Acceptor sequence and the Donor-Hydrogen $\cdots$ Acceptor angle (θ_{dha}) formed about the hydrogen atom. For a linear hydrogen bond (usually strongest), θ_{dha} has the ideal value of 180°. In the current biomolecular AMBER and CHARMM force fields, for example, the proper dependence of hydrogen bonding on distance and angular orientations is adequately treated with the electrostatic and Lennard-Jones terms.

In theory, classical electrostatic and (quantum) bonding forces can account for hydrogen-bonding interactions. Thus much work has gone into formulating appropriate hydrogen-bond potentials for small-molecule force fields.

Hydrogen-bond potentials can be derived from crystal packing studies or quantum-mechanical calculations, or introduced to correct weak interactions [567]. A re-optimization of hydrogen-bond potentials for MM3 [601] based on *ab initio* calculations concluded that hydrogen-bond interactions are more complex than previously thought. In that work, the hydrogen bond potential was formulated as a complex function of the Hydrogen $\cdots$ Acceptor distance, the Donor-Hydrogen bond length, and the cosine of the θ_{dha} angle. While this [601] and earlier works treated the overall nuclear charges, more recent work has shown that a better representation involves placing the center of the electron density where the lone pairs are and not where the nucleus is [602].

4. **Electrostatic Approximations**. In the simple Coulomb potential described above, interactions between pairs of atoms often consider only point charges for each atom. However, the charge distribution about each nucleus is clearly more complex, and in some cases induced dipole effects (i.e., distortion of the electron distribution) are important to consider. This notion of modeling electronic polarizability (roughly a measure of the distortion of an electronic charge cloud) is now a current focus in force field design (see [53, 603] for example).

5. **Limited Use of Quantum-Mechanical Information**. In theory, quantum-mechanical theory from molecular orbital techniques can be applied to small molecules to improve functional form and assign more accurate parameters (e.g., to fit parameters to *ab initio* relative energies or to the quantum-mechanical energy surface curvature). In practice, this fitting is not easy to perform and depends on the quality of the quantum calculations (e.g., basis set). Yet, quantum-based calculations have been used to assign the electrostatic partial charges, and the newer generation of force fields relies on quantum calculations wherever possible (e.g., [604, 599, 589, 601]), in combination with increasingly accurate experimental measurements. Thus, using both experimental and quantum calculations — despite each technique's limitations — can provide the best overall results.

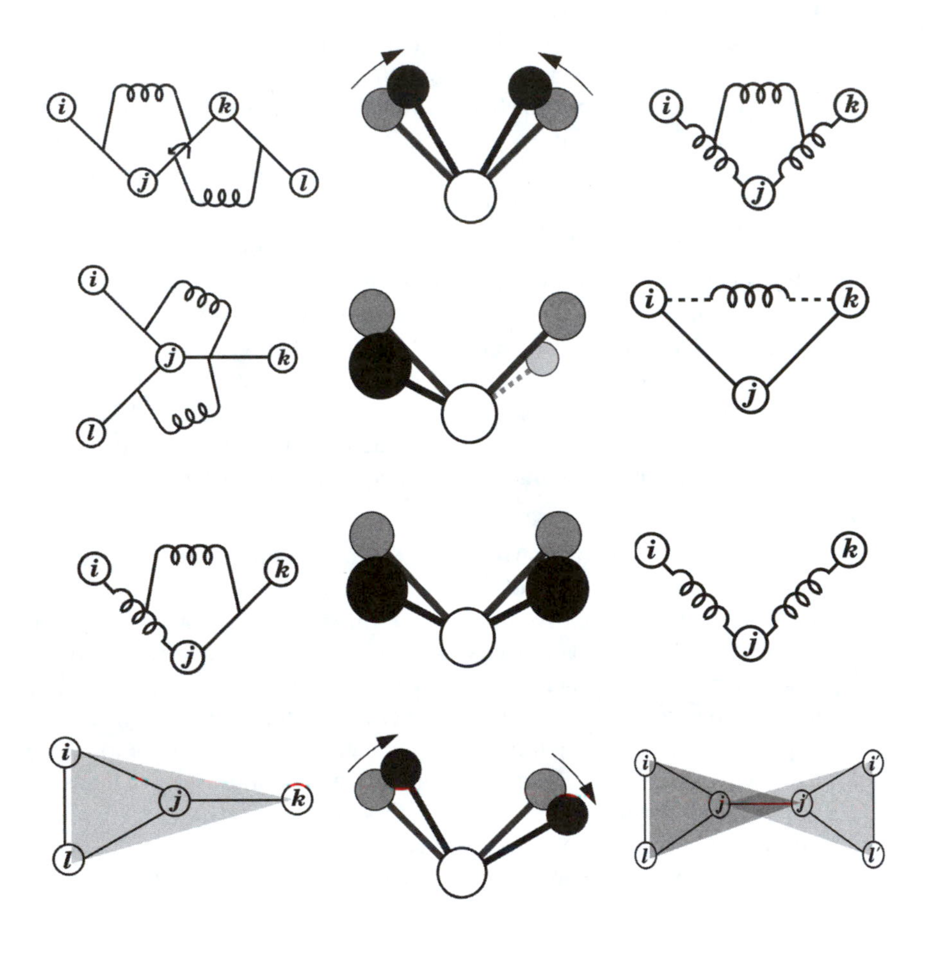
i
j
k
l
i
j
k
i
j
k
l
i
j
k
i
j
k
i
j
k
i
j
k
l
i
j
l
j'
i'
l'

8

Force Fields

Chapter 8 Notation

SYMBOL	DEFINITION
Scalars & Functions	
c	speed of light
c_s	ionic concentration
e	electron charge
$\hbar$	Planck's constant over 2π
k	harmonic spring constant
m	particle mass
m_e	electron rest mass
n	periodicity of rotational barrier (torsional potential)
q_i	Coulomb partial charge of atom i
r	bond length
$\bar{r}$	reference bond length
r_{ij}	interatomic distance (between atoms i and j)
x	displacement
A_{ij}, B_{ij}	Lennard-Jones coefficients for atom pair i, j (attraction, repulsion)
D	Morse well depth parameter
E	potential energy
E_{coul}	Coulomb energy
E_{LJ}	Lennard-Jones energy
E^r	bond length energy
$E^{rr'}$	stretch/stretch energy
$E^{r\theta}$	stretch/bend energy
$E^{r\theta r'}$	stretch/bend/stretch energy
E^{θ}	bond angle energy

Chapter 8 Notation Table (continued)

SYMBOL	DEFINITION
$E^{\theta\theta'}$	bend/bend energy
E^{ρ}	Urey-Bradley energy
E^{τ}	dihedral (torsional) angle energy
$E^{\tau\theta\theta'}$	torsion/bend/bend energy
$E^{\tau\theta}$	torsion/bend energy
E^{χ}	improper torsion energy
$E^{\chi\chi'}$	improper/improper torsion energy
F	force
F_{coul}	Coulomb force
K_{coul}	Coulomb potential constant
K_h	harmonic bending constant
K_t	trigonometric bending constant
N	number of atoms
N_e	number of outer shell electrons
S_h	harmonic stretching constant
S_m	Morse stretching constant (well width parameter)
S_q	stretching force constant for special quartic potential
V_{ij}, r_{ij}^0	Lennard-Jones coefficients for atom pair i, j (energy minimum/interaction distance)
V_n	barrier height of torsional potential associated with periodicity n
α	atomic polarizability
ϵ	dielectric constant
ϵ_0	permittivity of vacuum
θ	bond angle
$\bar{\theta}$	reference bond angle
θ_{tet}	tetrahedral bond angle, 109.47°, or $\cos^{-1}(-1/3)$
κ	Coulomb screening parameter
λ	wave number (wavelength of absorption)
μ	reduced mass
ν	characteristic frequency
τ	torsion (or dihedral) angle
τ_0	reference torsion (or dihedral) angle
χ	Wilson angle (for improper torsion potential)
ω	angular frequency

The purpose of models is not to fit the data but to sharpen the questions.

Samuel Karlin, 1983 (1923–).

8.1 Formulation of the Model and Energy

In this chapter, we discuss only basic functional expressions of the potential energy function, emphasizing the simple forms typically used for biomolecules. For biomolecular systems, computational speed is premium, and the use of more complex terms (higher-order expansions, cross terms, etc.), as employed for accurate modeling of smaller systems, is not practical. The next chapter discusses important topics related to this computational complexity of the nonbonded terms: spherical cutoff techniques, fast electrostatic evaluation techniques (Ewald and fast multipoles), and implicit solvation alternatives.

While improvement of potential energy functions — both in terms of functional form and parameters — has been an ongoing enterprise, the current, "*second-generation*" molecular mechanics and dynamics force fields are more sophisticated than those originating from pioneering works in the 1960s and 1970s. Specifically, parameterization depends quite significantly now on quantum mechanical calculations. "*Third generation*" force fields that account more accurately for electronic polarizabilities are already emerging (see [603] for example).

Parameterization of force fields ensures that calculations produce appropriate molecular geometries and interaction energies for a set of model compounds that are appropriate for the force field. For proteins, test compounds are peptides or model peptides [589]. For nucleic acids, deoxyribonucleosides [605, 606] and compounds containing the furanose ring and oligonucleotide crystals [303] are appropriate.

Force field parameters are optimized in a '*self consistent*' fashion [607, 596, 569] so as to reproduce the increasing body of experimental information on molecular geometries (from crystal and solution studies) to many other properties: measurements of vibrational frequencies, heats of formation, intermolecular energies and geometries, torsional barriers, and more. The use of dynamic simulations to assess the quality of the force field and to refine parameters — so as to better reproduce structural properties and molecular interaction energies — is also an improvement over procedures that utilize energy minimization alone [589, 606, 304].

Before describing each potential-energy term in turn, we review the fundamental molecular motions, called *normal modes*, that form the basis for parameterization of bonded deformations.

8.2 Normal Modes

8.2.1 Quantifying Characteristic Motions

Molecular vibrational spectra of small molecules form the basis for deriving various force constants, internuclear distances, and bond dissociation energies for bonded nuclei. Such bond motions describe vibrations about equilibrium states.

Specifically, all possible vibrations of a molecule can be described as a superposition of the fundamental oscillations (termed *normal modes*) for that molecule. Each molecule of N atoms has $3N - 6$ normal modes: 3 degrees of freedom per atom (giving $3N$) minus 3 translational and 3 rotational degrees of freedom for the molecule as a whole.

Experimental Determination

The vibrational energy levels of molecules can be detected experimentally by spectroscopic techniques, for example through the vibrational absorption of infrared radiation (IR) and by Raman scattering.

IR spectroscopy is a powerful technique that captures information on the transitions between vibrational quantum states, since these transitions lead to absorption and emission of infrared radiation. The IR wavelength range is 1–100 μm; this spectral range can be compared with the shorter wavelength of the visible spectrum, which has the range 400–750 nm. IR transitions are 'allowed' if there is a change in the dipole moment of the molecule during the transition.

The complementary technique of Raman spectroscopy captures transitions between vibrational levels, but the selection rules are different compared to IR: Raman bands appear only if there is a change in the polarizability of a system during the transition.

IR absorption and Raman spectroscopy are complementary techniques since some transitions that have changing dipole moments absorb light, whereas others have changing polarizability and scatter light. Thus, certain light-induced transitions can be weak, or even absent, in one technique, but of high intensity in the other. For small symmetric molecules, the observed transitions are complementary. For larger and asymmetric molecules, however, the selection rules are not rigidly obeyed, and Raman and IR spectra are essentially the same.

Frequency Units

Rather than expressing these frequencies in inverse seconds (or hertz, Hz, units), these fundamental frequencies are typically reported in *wavenumbers* of inverse centimeters, that is, the number of waves per centimeter. The higher the frequency, the more difficult (i.e., energetically costly) the deformation. For this reason, bond-stretching modes generally have higher frequencies than angle-bending modes, which in turn have higher frequencies than torsion-angle modes.

Note that stretching a bond significantly amounts to breaking it, so the energy is high; angle bending can be considered to have smaller effects on bonding (hence energy barriers are lower); torsion barriers are small for rotations about single bonds since the effects on bonding are smaller still. For double bonds, torsional motion corresponds to bond breaking and the frequencies are higher than for torsional barriers about single bonds.

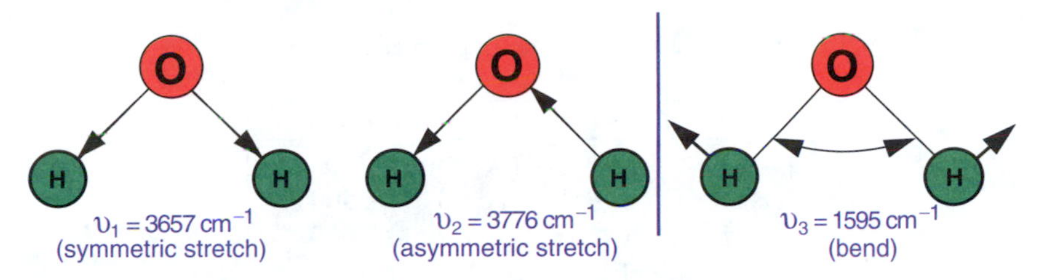

Figure 8.1. Normal modes of a water molecule.

Illustration

For larger molecules and complex mixtures, vibrational spectroscopy is a powerful analytical technique for determining which chemical groups are present. To illustrate, Figure 8.1 shows the three fundamental vibrations of a water molecule: an *asymmetric stretch* (around 3750 cm^{-1}), a *symmetric stretch* (around 3650 cm^{-1}), and an *angle-bending* mode around 1600 cm^{-1}. The symmetric stretch describes the contraction or elongation of both O–H bonds in concert, while the asymmetric mode involves this stretching motion in alternate fashion. The latter has a higher vibrational frequency than the symmetric mode (by about 100 wavenumbers) since the asymmetric vibration is slightly more energetically costly. The more facile angle-bending deformation has the lowest frequency in water among these three modes.

8.2.2 *Complex Biomolecular Spectra*

As the number of atoms in a molecule increases, so does the number of modes, as well as the associated complexity of the vibrational spectrum. Assigning normal modes to observed peaks in the experimental spectra becomes more difficult. Help is available from the characteristic modes of small molecules, which serve as excellent references for interpretation. In addition, the intensities in the vibrational spectrum can be calculated, at least to a rough approximation, by theoretical techniques [608].

While vibrational frequencies for the same bond type (e.g., O–H) vary depending on the molecular context, general values can be assigned to basic two-atom and three-atom sequences separated into distinct bond types (single, double, hydrogen-bonded, etc.).

For example, the symmetric O–H stretch in one water molecule (water vapor) is about 50 wavenumbers higher than that in the H–O–Cl molecule, but 300 wavenumbers higher than the O–H stretching frequency of a hydrogen-bonded water molecule in liquid water and ice (O–H $\cdots$ O), where $\nu = 3400\ \text{cm}^{-1}$. This reduction is due to the attractive force acting on the hydrogen atom in hydrogen-bonded species, since such attraction reduces the energy and hence the frequency of the O–H stretching motion.

Table 8.1. Characteristic **stretching** vibrational frequencies.

Vibrational Mode	Frequency [cm^{-1}]
H–O stretch	3600–3700
H–N stretch	3400–3500
H–C stretch	2900–3000
H–Br stretch	2650
C≡C, C≡N stretch	2200
C=C, C=O stretch	1700-1800
C–N stretch	1250
C–C stretch	1000
C–S stretch	700
S–S stretch	500

Table 8.2. Characteristic **bending and torsional** vibrational frequencies.

Vibrational Mode	Frequency [cm^{-1}]
H–O–H, H–N–H bend	1600
H–C–H bend	1500
H–C–H scissor	1400
H–C–H rock	1250
H–C–H wag	1200
H–S–H bend	1200
O–C=O bend	600
C–C=O bend	500
S–S–C bend	300
C=C torsion	1000
C–O torsion	300–600
C–C torsion	300
C–S torsion	200

8.2.3 Spectra As Force Constant Sources

Such spectroscopic measurements and analyses are used to derive appropriate force constants for biomolecular force fields. Tables 8.1 and 8.2 display examples of approximate stretching (Table 8.1) and bending and torsional (Table 8.2) frequencies.

Vibrational spectra of alkane molecules are a good source of parameters for C–C and C–H vibrational modes in proteins and nucleic acids. The spectrum of a butane molecule (CH_3–CH_2–CH_2–CH_3), for example, reflects both methyl (CH_3) and methylene (CH_2) stretching and bending modes.

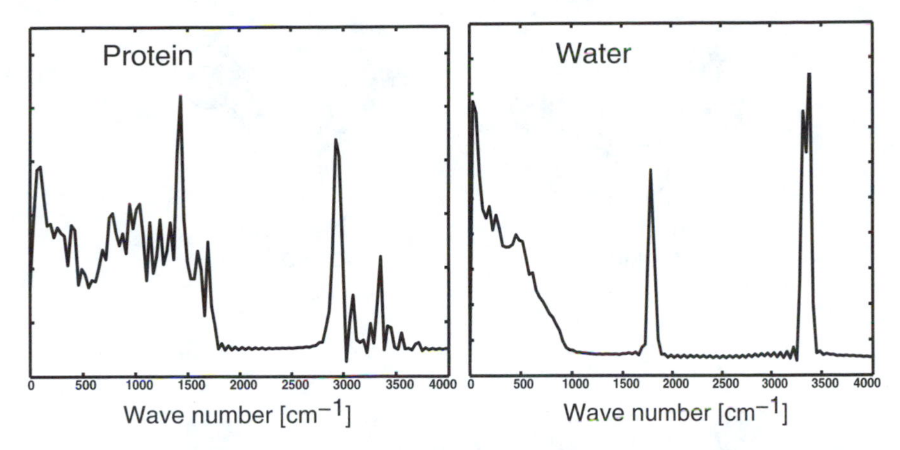

Figure 8.2. Characteristic frequencies calculated over 5 ps molecular dynamics simulations of solvated BPTI for the protein (left) and water (right) atoms. Data are from [609].

The alkane frequencies can be grouped into two strong stretching modes slightly below 3000 cm^{-1}, symmetric and asymmetric bending deformations within the range 1350–1500 cm^{-1}, and a C–C stretching mode around 1000 cm^{-1}. Indeed, these ranges of modes are evident in simulation-computed spectra corresponding to a small solvated protein, bovine pancreatic trypsin inhibitor (BPTI) as seen in Figure 8.2 [609].[1]

8.2.4 *In-Plane and Out-of-Plane Bending*

Bending modes include two types of in-plane deformations: *scissoring* and *rocking*, and two out-of-plane deformations: *wagging* and *twisting* (see Figure 8.3).

The in-plane *scissoring* deformation of an X–Y–Z sequence makes atoms X and Z move closer together. The *rocking* deformation moves both atoms in one direction while keeping their distance about the same.

The *out-of-plane wagging* bending deformation moves these atoms in the same direction with respect to the reference plane. *Twisting* moves one atom in one direction and the other in the opposite direction.

Figure 8.2 shows the power spectrum of a solvated protein system (BPTI) as computed from Fourier transforms of velocity autocorrelation functions [609]. The calculated peak *locations* depend sensitively on the force field, assuming the simulation protocol and frequency calculation procedure are sound.[2]

---

[1]Essentially, vibrational spectra can be computed from molecular dynamics simulations by transforming time-dependent properties, such as velocity autocorrelation functions, into the frequency domain using Fourier transforms. See [609], for example, for the precise procedure.

[2]The simulation-derived peak *heights* can, at best, reproduce frequency values corresponding to the force field parameters, not the experimental values, though the latter often serve as a reference. For

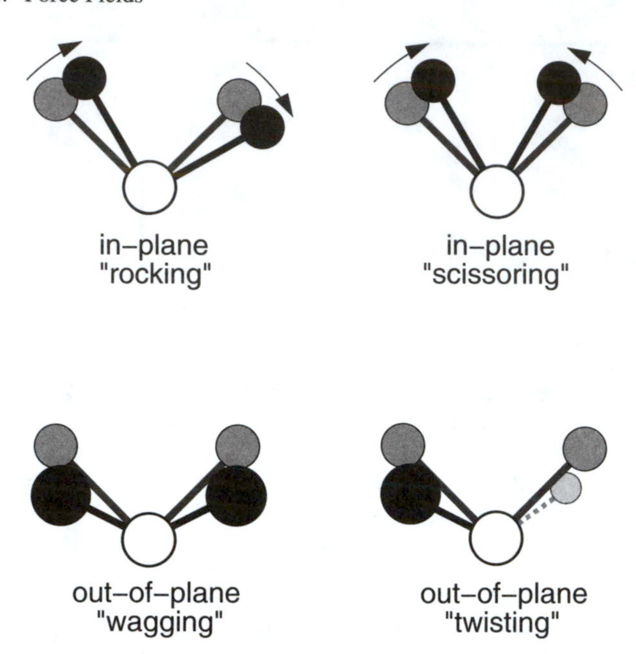

Figure 8.3. Various in-plane and out-of-plane bending vibrational modes. The reference molecular position is shown with grey-shaded atoms, and the position after the move is shown in black (and very-light grey for twisting). For the *wag*, the two nonbonded atoms move out of the paper plane toward us, while for *twist* one atom moves up (toward us) and the other down (away from us, below the paper plane).

For water, we note in Figure 8.2 characteristic peaks for vibrational modes of stretching around 3500 cm^{-1} and bending around 1700 cm^{-1}, as well as the slower *tumbling* or rocking (*librational*) motion for the water molecules as a whole in liquid water. For the protein, a wide range of vibrations is captured, from the fastest stretching modes for bonds involving hydrogens (O–H and N–H) around 3300 cm^{-1} to various angle-bending modes below 1700 cm^{-1}, to much slower deformations.

8.3 Bond Length Potentials

Bond length potentials can be considered as "strain" terms that model small-scale deviations about reference values. The reference values for different chemical bonds can be obtained from solved X-ray crystal structures as well as from quantum mechanical solutions to equilibrium structures of small molecules. For

---

example, the bond stretching force constant used for O–H water bonds may be physically unrealistic, so an unnatural spectral peak may emerge from simulations using unconstrained O–H bonds. (The peak is absent if these bonds are constrained).

accepted values, see organic chemistry textbooks such as [610] and Handbooks of Chemistry and Physics (e.g., CRC Press Handbook of Chemistry and Physics).

Note that *ab initio* methods calculate *equilibrium* bond lengths whereas experimentally measured values are usually vibrationally averaged bond lengths; the averaging depends on the experiment and hence there are many bond length values in theory. See [611, 612, 613] for a discussion of these different procedures for determining bond lengths and for the interconversion among the bond length values. For macromolecules, these small differences among the reference values are not usually important.

8.3.1 Harmonic Term

The harmonic potential modeled after Hooke's law is the simplest molecular-mechanics formulation for bond deformations. According to Hooke's law, the force F is proportional to the displacement, x, and the acceleration, $\ddot{x}$, as follows:

$$F(x) = -kx = m\frac{d^2x}{dt^2}\,, \qquad k = m\omega^2 > 0\,. \tag{8.1}$$

Thus, the *angular frequency* ω (number of radians per unit time), or 2π times the *circular frequency* ν ($= c/\lambda$ where c is the speed of light and λ is the wavelength), is related to the spring constant k as:

$$\omega \equiv 2\pi\nu = \sqrt{k/m}\,. \tag{8.2}$$

The corresponding potential energy E is:

$$E(x) = \frac{k}{2}\,x^2\,. \tag{8.3}$$

More generally, we write this harmonic bond potential as:

$$E^r_{\text{harmonic}}(r) = S_h\,[r - \bar{r}]^2\,, \tag{8.4}$$

where r is the bond length, $\bar{r}$ is the reference bond-length value, and S_h is a constant. From the measured mass and frequency for a particular bond vibration, force constants can be derived accordingly ($k = m\,\omega^2$). For atomic pairs of different species (with masses m_1, m_2), the general guideline is

$$k = \mu\,\omega^2\,, \tag{8.5}$$

where μ is the *reduced mass* defined as:

$$\mu = (m_1 m_2)/(m_1 + m_2)\,. \tag{8.6}$$

The harmonic potential is only adequate for small deviations from reference values, around one vibrational level above the ground state, or on length deformations of the order of 0.1 Å or less. It is not valid for larger deviations from equilibrium since the atoms dissociate and no longer interact; thus the energy levels off, rather than increases, rapidly as the distance increases beyond $\bar{r}$. For very small interaction distances, however, the deformation energy is very large.

This physical picture is described by a parabolic potential-well shape for small distance separations and a leveling curve for separations greater than the reference value. Rather than a harmonic function, the precise shape of $E(r)$ is more adequately represented by higher-order functions that approximate better the experimentally-measured energy trend as a function of distance.

Electronic spectroscopy is necessary to measure this precise distance dependency since frequencies and energies are higher for changes in electronic states. Specifically, the potential energy as a function of internuclear separation for both the ground and excited states is obtained from visible and ultraviolet (UV) spectroscopic techniques.[3] The commonly used ultraviolet and visible spectrometers measure absorption of light in the range 200–750 nm. Visible and UV absorption bands are broad since several vibrational states are contained in each electronic state, and each vibrational state is further decomposed into many rotational states. Thus, a rigorous assignment of bands to specific transitions is difficult, but the overall shape of the energy well can be deduced.

8.3.2 Morse Term

To model bond deformations that exceed very small fluctuations about equilibrium states, the empirical bond potential due to P.M. Morse [614] has proven very successful for reproducing vibrational levels of small molecules [615, 571]. The Morse function has the form:

$$E^r_{\text{Morse}}(r) = D\{1 - \exp[-S_m(r - \bar{r})]\}^2 , \tag{8.7}$$

where the adjustable parameters S_m and D characterize the well width and well depth respectively. As seen in Figure 8.4, the Morse potential correctly rises steeply for contraction of the bond length ($E(r) \to \infty$ as $r \to 0$) but levels off to the *dissociation energy* D at large r: $E(r) \to D$ as $r \to \infty$.

The empirical Morse potential can be written as an infinite series in the powers of the bond displacement $(r - \bar{r})$. Using Taylor series, we expand eq. (8.7) as:

$$\begin{aligned} E^r_{\text{Morse}}(r) &= D\{1 - 2\exp[-S_m(r - \bar{r})] + \exp[-2S_m(r - \bar{r})]\} \\ &= D\Bigg\{1 - 2\left[1 - S_m(r - \bar{r}) + \frac{S_m^2(r - \bar{r})^2}{2} - \frac{S_m^3(r - \bar{r})^3}{3!}\right. \\ &\qquad\qquad \left. + \frac{S_m^4(r - \bar{r})^4}{4!} \quad + \quad \cdots \right] \\ &\qquad\qquad + \left[1 - 2S_m(r - \bar{r}) + \frac{4S_m^2(r - \bar{r})^2}{2} - \frac{8S_m^3(r - \bar{r})^3}{3!}\right. \end{aligned}$$

[3] UV spectroscopy measures wavelengths just beyond the violet end of the visible spectrum, that is, with $\lambda < 400$ nm.

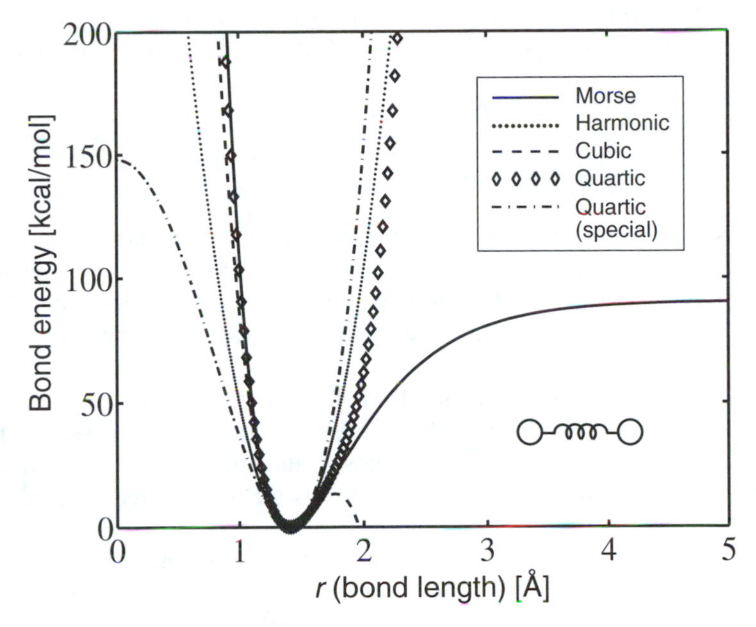

Figure 8.4. Morse, harmonic, cubic, and two quartic bond potentials for H–Br. The Morse, harmonic, and special quartic potentials are given in eqs. (8.7), (8.4), and (8.11), respectively. The cubic and non-degenerate quartic polynomials are given polynomial coefficients to match the Taylor-series expansion of the Morse potential given in eq. (8.8) up to the desired order. Note that the cubic potential causes a problem for significant bond stretches because of the change in curvature.

$$
\left. \left. + \frac{16 S_m^4 (r-\bar{r})^4}{4!} \quad + \quad \cdots \quad \right] \right\}
$$

$$
\begin{aligned}
&= D S_m^2 (r-\bar{r})^2 - D S_m^3 (r-\bar{r})^3 + \frac{7}{12} D S_m^4 (r-\bar{r})^4 \\
&\quad + \mathcal{O}(r-\bar{r})^5 .
\end{aligned}
\tag{8.8}
$$

Hence we can relate the harmonic stiffness constant S_h in eq. (8.4) to the Morse constant S_m of eq. (8.7) as:

$$
S_h \approx D (S_m)^2 . \tag{8.9}
$$

Put another way, from the relationship between S_h and the spring constant of a harmonic oscillator, namely $S_h = k/2 = (\mu\omega^2)/2$, the value of the Morse well-depth parameter S_m can be reasonably approximated as $\sqrt{S_h/D}$ or

$$
S_m = \omega \sqrt{\frac{\mu}{2D}} = \pi\nu \sqrt{\frac{2\mu}{D}} . \tag{8.10}
$$

Figure 8.4 shows the harmonic and Morse potentials for a hydrogen bromide molecule. The parameters $D = 90.5$ kcal/mol, S_m = 1.814 Å^{-1}, $\bar{r} = 1.41$ Å, and

$S_h = 297.8$ (kcal/mol)/Å^2 (from eq. (8.9)), are used. We note that the harmonic potential is a good approximation to the energy surface only for small displacements from equilibrium.

8.3.3 Cubic and Quartic Terms

To reproduce Morse potentials better than possible with the harmonic potential, cubic and quartic polynomials can be used (through terms added to the quadratic potential) to match the Taylor series expansion of eq. (8.8) up to a desired order, as shown in Figure 8.4. The MM3 force field, for example, uses cubic and quartic bond potentials, which work well for most molecules [580]; a sextic bond potential works even better and is used in MM4. The Merck force field, MMFF [599], uses a quartic bond function. Note that for better optimization of the shape of the bond length potential function, coefficients of the polynomial can be adjusted; thus they need not coincide with those given by the Morse Taylor expansion of eq. (8.8).

Note that a quartic is preferable to a cubic bond potential because the cubic function has an inflection point at some value $r > \bar{r}$; thus, significant bond stretches lead to negative rather than positive energy ($E \to -\infty$ as $r \to \infty$) (see Figure 8.4). This can cause the molecular energy to have large negative values and the computation (energy minimization, for example) to become nonsensical. Series that end in even powers (like quartic rather than cubic polynomials) can provide better approximations and drive the molecule more rapidly toward the energy minimum.

For large-molecule force fields where computational time is important, a special quartic has been suggested to avoid square root computations [595, 597]; it measures the square of the squared bond differences as:

$$E^r_{\text{quartic}} = S_q[r^2 - \bar{r}^2]^2 \,. \tag{8.11}$$

This quartic is special (degenerate) since it does not contain a cubic term. Its shape is therefore similar to the harmonic potential, as shown in Figure 8.4.

At small displacements from equilibrium, we have the relation

$$[r^2 - \bar{r}^2]^2 \equiv [r - \bar{r}]^2 \, [r + \bar{r}]^2 \approx [r - \bar{r}]^2 \, (2\bar{r})^2 \,.$$

Hence, comparing the series expansion in eq. (8.11) with the harmonic potential of eq. (8.4), we can relate the quartic-potential force constant to that of the harmonic potential by:

$$S_q \approx S_h / 4\bar{r}^2 \,. \tag{8.12}$$

Figure 8.4 displays this quartic potential with S_q calculated from S_h as above. As for the quadratic potential, the energy approximation is good only for very small deviations from equilibrium.

8.4 Bond Angle Potentials

The bond angle arrangement around each atom in a molecule is governed by the hybridization of the orbitals around the atom. For example, when an atom has two identical hybrid orbitals (*sp*) around it (e.g., Be in $BeCl_2$), the bond angle is 180°. When three identical orbitals surround an atom (e.g., B in BF_3, sp^2), the arrangement is trigonal and coplanar with bond angles all 120°. When four identical orbitals surround an atom (e.g., C in CH_4, sp^3), the arrangement is tetrahedral — all angles are 109.47° $\left(\theta_{\rm tet} = \cos^{-1}[-1/3]\right)$.

This simple rule serves as a first approximation for bond angle geometries. However, small deviations from these estimates generally occur, and large deviations sometimes occur. Even small differences of 1–2° between different bond angles in a molecule can have important global influence on molecular structure, as in riboses.

Indeed, it is important to realize that exact sp^3 orbits exist only for tetrahedrally-symmetric compounds like methane. Ordinary alkanes already have their orbits deformed: propane, for example, has the C–C–C bond angle of about 112.5° and H–C–H bond angles around 107.5°; its C–C–H bond angles are approximately tetrahedral. Another common example of bond angle deviations involves ring molecules, like cycloalkanes and riboses; ring-closure constraints can alter the geometry significantly. Finally, electron lone pairs about atoms influence the geometry: in water, the oxygen lone pair forms bond-like orbitals to produce the liquid water bond angle θ(H–O–H) $\approx 105^{\circ}$. See Figure 8.5 for such illustrations. As for bond potentials, stiffness constants for bond angle bending are determined from measured vibrational frequencies.

8.4.1 *Harmonic and Trigonometric Terms*

Commonly used bond-angle potentials are harmonic functions that involve the difference between angles and angle cosines:

$$E^{\theta}_{\rm harmonic}(\theta) = K_h\,[\theta - \bar{\theta}]^2\,, \tag{8.13}$$

$$E^{\theta}_{\rm trig.}(\theta) = K_t\,[\cos\theta - \cos\bar{\theta}]^2\,. \tag{8.14}$$

As shown above for bond potentials, we can expand the trigonometric function above by a Taylor series in powers of $\theta - \bar{\theta}$ to relate K_t to K_h:

$$E^{\theta}_{\rm trig.}(\theta) = K_t\left\{-\sin\bar{\theta}\,(\theta-\bar{\theta}) \;-\; \frac{\cos\bar{\theta}}{2}\,(\theta-\bar{\theta})^2 \;+\;\cdots\right\}^2 \qquad \Longrightarrow$$

$$K_t \approx K_h\,\sin^2\bar{\theta}\,. \tag{8.15}$$

The advantage of the trigonometric potential is its boundedness and its ease of implementation and differentiation. This is because no inverse trigonometric functions need to be calculated, and singularity problems for linear bond angles can

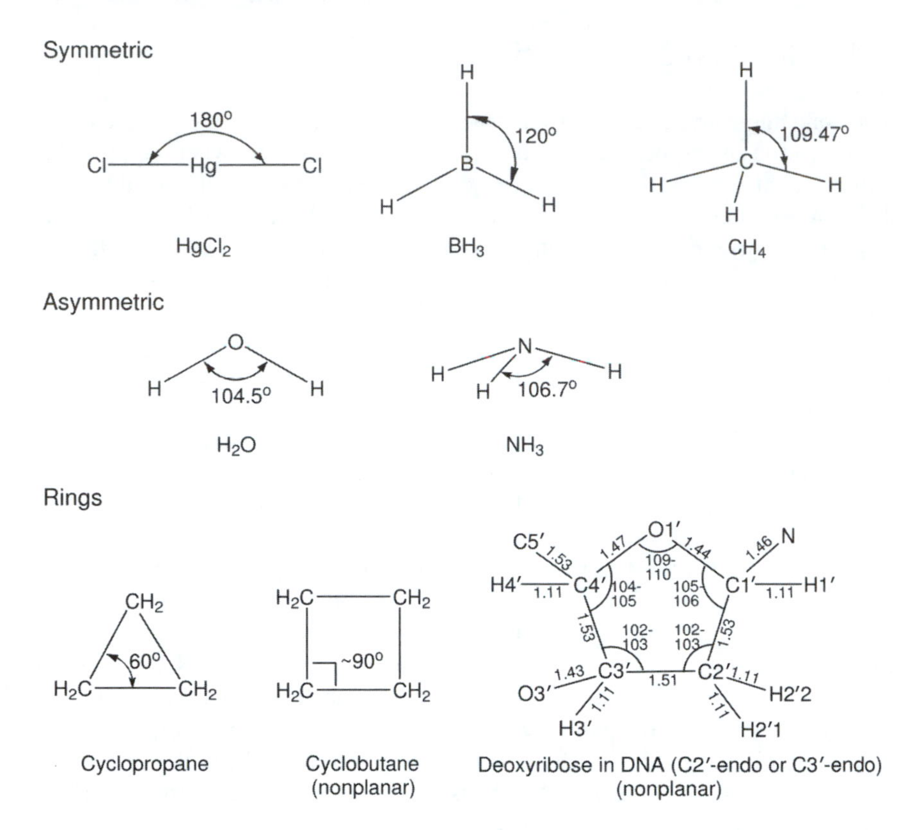

Figure 8.5. Bond angle geometries for simple chain and cyclic molecules. Note that cyclobutane and deoxyribose are nonplanar. The former has bond angles less than 90° by only 1-2°, but the dihedral angle is substantial, around 30°, which relieves the eclipsing of the hydrogens substantially. The geometry shown for deoxyribose (bond lengths in Å and bond angles in degrees) corresponds to observed C3′-endo and C2′-endo conformations in B-DNA. The five endocyclic deoxyribose dihedral angles ν_0 through ν_5 have values (as computed from solvated dodecamers in CHARMM) of approximately −24, 38, −38, 24, and 0 degrees for C2′-endo, and 0, −24, 38, −38, and 24 degrees for the C3′-endo sugar pucker.

be avoided [310, 597]. As a compromise between a quadratic and infinite series in $\theta - \bar{\theta}$, the MMFF force field uses a cubic bond-angle function of form [599]

$$E^{\theta}_{\text{cubic}} = K_1(\theta - \bar{\theta})^2 + K_2(\theta - \bar{\theta})^3 \tag{8.16}$$

for non-colinear atom orientations. For linear (or near linear) reference angles, the function

$$E^{\theta}_{\text{trig.}'} = K_3(1 + \cos\theta) \tag{8.17}$$

is used instead.

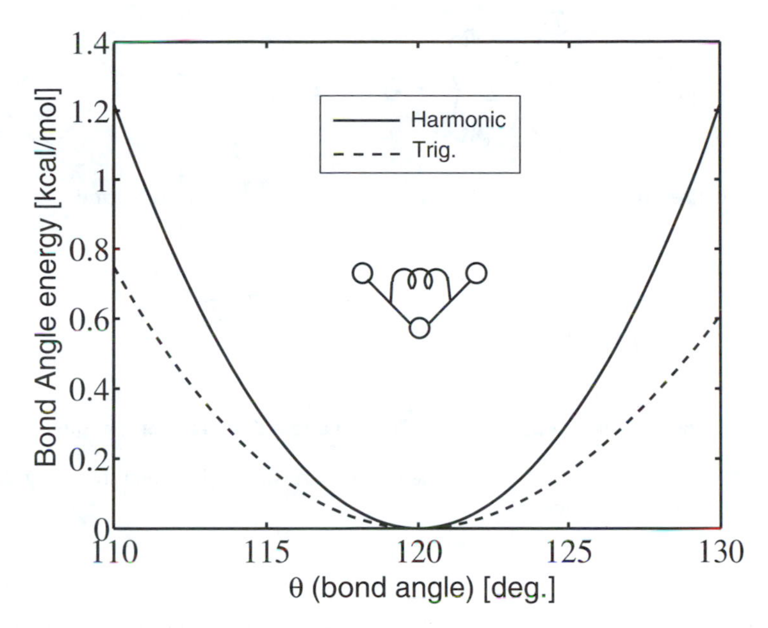

Figure 8.6. Harmonic bond-angle potentials of the form (8.13) and (8.14) for an aromatic C–C–C bond angle (CA–CA–CA atomic sequence in CHARMM) with parameters $K_h = 40$ kcal/(Mol-rad.^{2}) and $\bar{\theta} = 2.1$ rad (120°). The K_t force constant for eq. (8.14) is calculated via eq. (8.15).

Note that, as for the bond potential, it is hazardous to use a function that ends in an odd power of the deformation since these terms have negative coefficients, which dominate for large deviations. Hence, potential functions that end in even powers are preferable.

Figure 8.6 displays harmonic bond angle potentials of the forms given in eqs. (8.13) and (8.14). Note that the harmonic cosine potential is very similar to the harmonic potential for a small range of fluctuations. It should thus be preferred in practice if computational time is an issue.

8.4.2 Cross Bond Stretch / Angle Bend Terms

Cross terms are often used in force fields targeted to small molecular systems (e.g., MM3, MM4) to model more accurately compensatory trends in related bond-length and bond-angle values. These cross terms are considered correction terms to the bond-length and bond-angle potentials.

For example, a stretch/bend term for a bonded-atom sequence ijk allows bond lengths i–j and j–k to increase/decrease as θ_{ijk} decreases/increases. A bend/bend potential couples the bending vibrations of the two angles centered on the same atom appropriately, so the corresponding frequencies can be split apart to match better the experimental vibrational spectra [580].

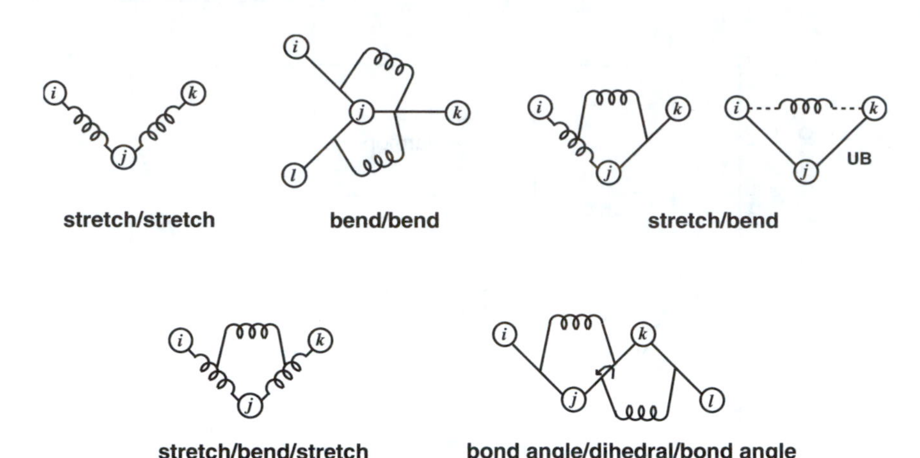

Figure 8.7. Schematic illustrations for various cross terms involving bond stretching, angle bending, and torsional rotations.

These correlations can be modeled via stretch/stretch, bend/bend, and stretch/ bend potentials for such ijk sequences (see Figure 8.7), where the distances r and r' are associated with bonds ij and jk, and θ is the ijk bond angle:

$$E^{rr'}(r, r') = S\,[r - \bar{r}]\,[r' - \bar{r'}]\,, \tag{8.18}$$

$$E^{\theta\theta'}(\theta, \theta') = K\,[\theta - \bar{\theta}]\,[\theta' - \bar{\theta'}]\,, \tag{8.19}$$

$$E^{r\theta}(r, \theta) = S\,K\,[r - \bar{r}]\,[\theta - \bar{\theta}]\,. \tag{8.20}$$

Through addition, a stretch/bend/stretch term of form

$$E^{r\theta r'}(r, \theta, r') = K\,[S(r - \bar{r}) + S'(r' - \bar{r'})]\,[\theta - \bar{\theta}]\,, \tag{8.21}$$

can be mimicked, as in the Merck force field [599].

A variation of a stretch/bend term, devised to obtain better agreement of calculated with experimental vibrational frequencies for a bonded atom sequence ijk and associated bond angle θ [616], is a potential known as Urey-Bradley. It is commonly used in molecular mechanics force fields, like CHARMM. The potential is a simple harmonic function of the *interatomic* (not bond) distance ρ, between atoms i and k in the bonded ijk sequence (see Figure 8.7):

$$E^{\rho}(\rho) = S\,[\rho - \bar{\rho}]\,. \tag{8.22}$$

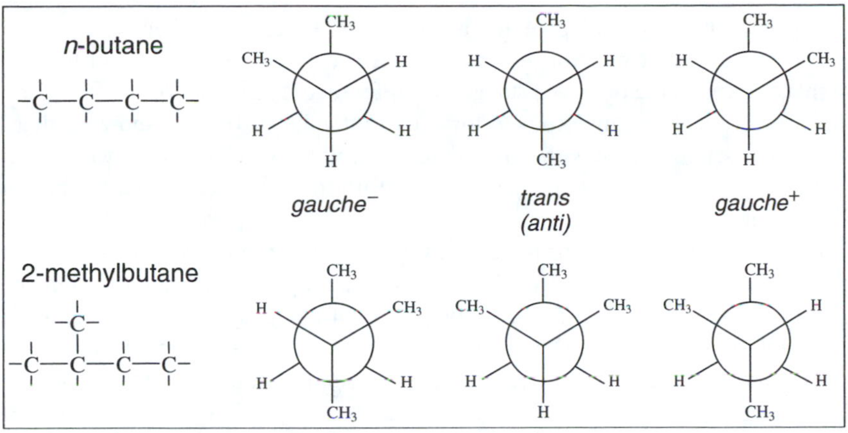

Figure 8.8. Torsional orientations for n-butane and 2-methylbutane, as illustrated by Newman projections viewed along the central C–C bond. Note that while n-butane has the global minimum at the *trans* (or *anti*) configuration, the *gauche* states of 2-methylbutane are lower in energy than the corresponding trans (symmetrical) state since the methyl groups are better separated and hence the molecule is less congested.

8.5 Torsional Potentials

8.5.1 Origin of Rotational Barriers

Torsional potentials are included in potential energy functions since all the previously described energy contributions (nonbonded, stretching, and bending terms) do not adequately predict the torsional energy of even a simple molecule such as ethane. In ethane, the C–C bond provides an axis for *internal rotation* of the two methyl groups.

Although the origin of the barrier to internal rotation has not been resolved [617, 618, 619],[4] explanations involve the relief of steric congestion and optimal achievement of resonance stabilization (another quantum-mechanical effect arising from the interactions of electron orbitals) [619]. This latter consideration represents a newer hypothesis. For a long time, the primary interactions that give rise to rotational barriers have been thought to be repulsive interactions, caused by overlapping of bond orbitals of the two rotating groups [620].

In ethane (C_2H_6), for example, the torsional strain about the C–C bond is highest when the two methyl groups are nearest, as in the *eclipsed* or *cis* state, and lowest when the two groups are optimally separated, as in the *anti* or *trans* state (see Figure 8.8 and the related illustration for n-butane in Figure 3.13 of Chap-

[4]Goodman *et al.* [617] refer to the barrier origin as "the Bermuda Triangle of electronic theory", reflecting a complex interdependence among three factors: electronic repulsion, relaxation mechanisms, and valence forces.

ter 3). Without a torsional potential in the energy function, the eclipsed form for ethane is found to be higher in energy than the staggered form, but not sufficiently higher to explain the observed staggered preference [572]. See [621] for a meticulous theoretical treatment of the torsional potential of butane, showing that a large basis set combined with a treatment of electron correlations in *ab initio* calculations is necessary to obtain accurate equilibrium energies relative to the *anti* conformation.

Accounting accurately for rotational flexibility of atomic sequences is important since rotational states affect biological reactivity. Such torsional parameters are obtained by analyzing the energetic profile of model systems as a function of the flexible torsion angle. The energy differences between rotational isomers are used to parameterize appropriate torsional potentials that model internal rotation.

8.5.2 Fourier Terms

The general function used for each flexible torsion angle τ has the form

$$E^{\tau}(\tau) = \sum_{n} \frac{V_n}{2} \left[1 \pm \cos n\tau\right], \tag{8.23}$$

where n is an integer. For each such rotational sequence described by the torsion angle τ, the integer n denotes the *periodicity* of the rotational barrier, and V_n is the associated *barrier height*. The value of n used for each such torsional degree of freedom depends on the atom sequence and the force field parameterization (see below). Typical values of n are 1, 2, 3 (and sometimes 4). Other values (e.g., $n = 5, 6$) are used in addition by some force fields (e.g., CHARMM [589]). A reference torsion angle τ_0 may also be incorporated in the formula, i.e.,

$$E^{\tau}(\tau) = \sum_{n} \frac{V_n}{2} \left[1 + \cos(n\tau - \tau_0)\right]. \tag{8.24}$$

Often, $\tau_0 = 0$ or π and thus the cosine expression of form (8.23) suffices; this is because eq. (8.24) can be reduced to the form of eq. (8.23) in such special cases, from relations like:

$$1 + \cos(n\tau - \pi) = 1 - \cos(n\tau).$$

Experimental data obtained principally by spectroscopic methods such as NMR, IR (Infrared Radiation), Raman, and microwave, each appropriate for various spectral regions, can be used to estimate barrier heights and periodicities in low molecular weight compounds. According to a theory developed by Pauling [622], potential barriers to internal rotation arise from exchange interactions of electrons in adjacent bonds; these barriers are thus similar for molecules with the same orbital character. This theory has allowed tabulations of barrier heights as class averages [607, 623, 573, 567]. Since barriers for rotations about various single bonds in nucleic acids and proteins are not available experimentally, they must be estimated from analogous chemical sequences in low molecular weight compounds.

8.5.3 Torsional Parameter Assignment

In current force fields, these parameters are typically assigned by selecting several classes of model compounds and computing energies as a function of the torsion angle using *ab initio* quantum-mechanical calculations combined with geometry optimizations. The final value assigned in the force field results from optimization of the combined intramolecular and nonbonded energy terms to given experimental vibrational frequencies and measured energy differences between conformers of model compounds.

This procedure often results in several Fourier terms in the form (8.23), that is, several $\{n, V_n\}$ pairs for the same atomic sequence; see examples in Table 8.3. Moreover, parameters for a given quadruplet of atoms may be deduced from more than one quadruplet entry. For example, the quadruplet C1′–C2′–C3′–O3′ in nucleic acid sugars may correspond to both a C1′–C2′–C3′–O3′ entry and a $\star$–C–C–$\star$ entry, the latter designating a general rotation about the endocyclic sugar bond; here, $\star$ designates any atom. When general rotational sequences are involved (e.g., $\star$–C–N–$\star$), CHARMM may list *a pair* of $\{V_n, \tau_0\}$ values (i.e., different τ_0 for the same rotational term); only one τ_0 is used when a specific quadruplet atom sequence is specified.

Twofold and Threefold Sums

Twofold and threefold potentials are most commonly used (see Figure 8.9 for an illustration). Inclusion of a one-fold torsional term as well is a force-field dependent choice.

A threefold torsional potential exhibits three maxima at 0°, 120°, and 240° and three minima at 60°, 180°, and 300°. Ethane has a simple 3-fold torsional energy profile, as seen in the 3-fold energy curve in Figure 8.9, with all maxima energetically equivalent and corresponding to the eclipsed or *cis* form, and all minima corresponding to the energetically equivalent staggered or *trans* form.

In related hydrocarbon sequences, the local minima may not be of equal energies. For n-butane, for example, the *anti* (or *trans*) conformer is roughly 1 kcal/mol lower in energy than the two *gauche* states. In 2-methylbutane, the *gauche* states are lower in energy than the *anti* state due to steric effects. The torsional conformations of these molecules are illustrated in Figure 8.8.

In general, the *gauche* forms may be higher in energy than the *trans* form, as in n-butane [623], or lower in energy, as in 2-methylbutane due to steric effects. Certain X–C–C–Y atomic sequences where the X or Y atoms are electronegative like O or F (e.g., 1-2-difluoroethane, certain O–C–C–O linkages in nucleic acids) [624] also have lower-in-energy *gauche* conformations, but this is due to a fundamentally different effect than present in hydrocarbons — neither steric, nor dipole/dipole — called the *gauche effect* [624].

Reproduction of *Cis/Trans* and *Trans/Gauche* Energy Differences

A combination of a twofold and a threefold potential can be used to reproduce both the *cis/trans* and *trans/gauche* energy differences. Figure 8.9 illustrates the case for different parameter combinations of the twofold and threefold parameters. One interaction is modeled after the O–C–C–O sequence in nucleic acids (e.g., O3′–C3′–C2′–O2′ in ribose) [594], showing a minimum at the *trans* state. The other is a rotation about the phosphodiester bond (P–O) in nucleic acids, showing a shallow minimum at the *trans* state. Note that for small molecules, torsional potentials with $n = 1, 2, 3, 4$ and 6 are often used to reproduce torsional frequencies accurately.

To see how to combine twofold and threefold torsional potentials, for example, denote the experimental energy barrier by ΔV, the total empirical potential energy function as E, and let E^{τ} represent the following combination of twofold and threefold torsional terms:

$$E^{\tau} = \frac{V_2}{2}\left[1 + \cos 2\tau\right] + \frac{V_3}{2}\left[1 + \cos 3\tau\right]. \tag{8.25}$$

The torsional parameters V_2 and V_3 for a given rotation τ are then computed from the relations:

$$\begin{aligned} \Delta V_{\text{cis/trans}} &= E_{\tau=0^{\circ}} - E_{\tau=180^{\circ}} \\ &= V_3 + [E - E^{\tau}]_{\tau=0^{\circ}} - [E - E^{\tau}]_{\tau=180^{\circ}}\,, \end{aligned} \tag{8.26}$$

$$\begin{aligned} \Delta V_{\text{trans/gauche}} &= E_{\tau=180^{\circ}} - E_{\tau=60^{\circ}} \\ &= (3V_2)/4 + [E - E^{\tau}]_{\tau=180^{\circ}} - [E - E^{\tau}]_{\tau=60^{\circ}}\,, \end{aligned} \tag{8.27}$$

since

$$V_3 = E^{\tau}_{\tau=0^{\circ}} - E^{\tau}_{\tau=180^{\circ}}$$

and

$$\frac{3}{4}V_2 = E^{\tau}_{\tau=180^{\circ}} - E^{\tau}_{\tau=60^{\circ}}\,.$$

Thus, Cartesian coordinates of the molecule must be formulated in terms of τ. A simplification can be made by assuming fixed bond lengths and bond angles and calculating nonbonded energy differences only. In theory, then, every different parameterization of the nonbonded coefficients requires an estimation of the torsional potentials V_2 and V_3 to produce a consistent set.

This general requirement in development of consistent force fields explains why energy parameters are not transferable from one force field to another.

In general, many classes of model compounds must be used to represent the various torsional sequence in proteins and nucleic acids. Since rotational barriers in small compounds have little torsional strain energy compared with large systems, rotational profiles are routinely computed for a series of substituted mol-

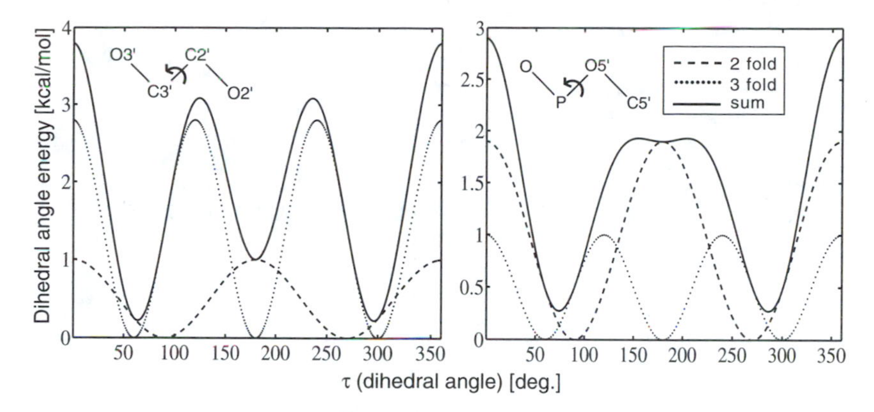

Figure 8.9. Twofold and threefold torsion-angle potentials and their sums for an O–C–C–O rotational sequence in nucleic-acid riboses ($V_2 = 1.0$ and $V_3 = 2.8$ kcal/mol, reproducing the known *trans/gauche* energy difference [594]) and a rotation about the phosphodiester (P–O) bond in nucleic acids ($V_2 = 1.9$ and $V_3 = 1.0$ kcal/mol, from CHARMM [589]).

ecules. Substituted hydrocarbons are used, for example, to determine rotational parameters for torsion angles about single C–C bonds in saturated species.

Model Compounds

Examples of model compounds used for assigning torsional parameters to nucleic acids and proteins include the following (see Figure 8.10).

- Hydrocarbons like **ethane**, **propane**, and n-**butane**, and the more crowded environments of **2-methylbutane** and **cyclohexane**: for rotations about single C–C bonds in saturated species (i.e., each carbon is approximately tetrahedral), such as C–C–C–C, H–C–C–H, and H–C–C–C;

- The **ethylbenzene** ring: for rotations about CA–C bonds in ring systems (CA denotes an aromatic carbon);

- Alcohols like **methanol** and **propanol**: to model H–C–O–H, C–C–O–H, H–C–C–O, and C–C–C–O sequences (e.g., in the amino acids serine and threonine);

- Sulfide molecules (e.g., **methyl propyl sulfide**): to model rotations involving sulfur atoms (e.g., C–C–C–S, C–C–S–C, and H–C–S–C) as in methionine, or **dimethyl disulfide** (C–S–S–C), to model disulfide bonds in proteins;

- Model amine, aldehydes, and amides (e.g., **methylamine**, **ethanal**, N-**methyl formamide**): for sequences involving nitrogens such as H–C–N–H, H–C–C=O, and H–C–N–C.

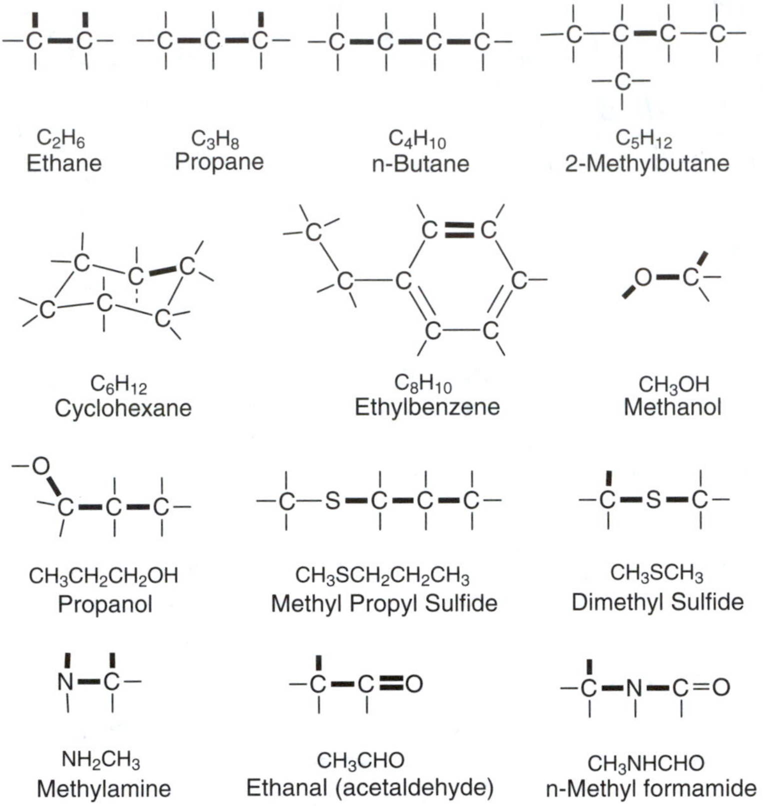

Figure 8.10. Model compounds for determining torsional parameters for various atomic sequences in biomolecules. Illustrated in **bold** is τ: the bond about which the τ rotation occurs, or the complete τ sequence (3 bonds), as needed for an unambiguous definition.

See [604] for a comprehensive description of such a parameterization for peptides based on *ab initio* energy profiles.

In the CHARMM and AMBER force fields, the parameters n and V_n are highly environment dependent. Furthermore, more than one potential form can be used for the same bond about which rotation occurs; these different torsion-angle terms may be weighted proportionately. Some examples of torsion angle parameters from the AMBER [605] and CHARMM [307, 589] force fields are given in Table 8.3.

Table 8.3. Parameters for selected torsional potentials from the AMBER (first row for each sequence) [606] and CHARMM (second row of that sequence) [589, 303] force fields. Barrier heights are in units of kcal/(mol rad^2), angles are in radians, and $\star$ represents any atom.

SEQUENCE	$\frac{V_1}{2}$	τ_0	$\frac{V_2}{2}$	τ_0	$\frac{V_3}{2}$	τ_0	DESCRIPTION
$\star$–C–C–$\star$					1.4	0	alkane C–C (e.g., Lys
					0.2	0	or Leu C^α–C^β–C^γ–C^δ)
O4′–C1′–N9–C8	2.5	0					purine glycosyl
	1.1	0					C1′–N9 rotation (A, G)
C–C–S–C					1.0	0	rotation about C–S
	0.24	π			0.37	0	in Met
O3′–P–O5′–C5′			1.2	0	0.5	0	P–O5′ rotation in
	1.2	π	0.1	π	0.1	π	nucleic-acid backbone

8.5.4 Improper Torsion

Harmonic 'improper torsion' terms, also known as 'out-of-plane bending' potentials, are often used in addition to the terms described above to improve the overall fit of energies and geometries. They can enforce planarity or maintain chirality about certain groups. The potential has the form

$$E^\chi(\chi) = (V'/2)\,\chi^2\,, \tag{8.28}$$

where χ is the improper *Wilson* angle. It is defined for the four atoms i, j, k, l for which the central j is bonded to i, l, and k (see Figure 8.11) as the angle between bond j–l and the plane i–j–k.

The following are examples of atom centers used to define sequences of atoms counted in improper torsion potentials (taken from CHARMM):

- Polypeptide backbone nitrogen and carbonyl carbons attached to the C^α atom (i.e., N and C in the sequence –N–C^α–C=O), for maintaining planarity of the groups around the peptide bond;
- Terminal side-chain carbons in the CO_2 unit of the Asp and Glu amino acids;
- Terminal side-chain carbons in the O=C–NH_2 unit of the Asn and Gln amino acids;
- Terminal side-chain carbon in the NH_2–C–NH_2 unit of the Arg amino acid;
- Various carbons and nitrogens in the side-chain rings of the Phe; Tyr, and Trp amino acids (not used in all CHARMM versions);
- Glycosyl nitrogen (N1 in pyrimidines or N9 in purines) in nucleic acid bases;

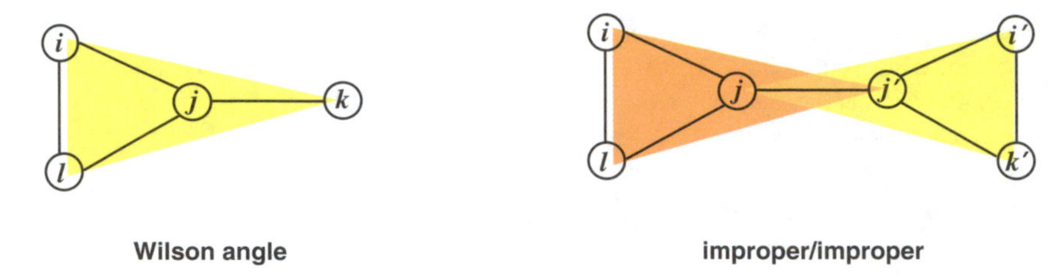

Figure 8.11. Wilson angle definition (left) and geometry for a cross improper/improper torsion term (right).

- Exocyclic nitrogen of the NH_2 unit attached to the aromatic ring of the adenine and cytosine bases in nucleic acids (N6 in adenine and N4 in cytosine).

8.5.5 *Cross Dihedral/Bond Angle and Improper/Improper Dihedral Terms*

Cross terms are used in certain force fields (e.g., CVFF used in the Insight/Discover program) to model various relationships between quadruplet sequences (see Figure 8.11) or to couple torsion angles with bond angles or torsion angles with bond lengths (see Figure 8.7). For example, the association between a torsion angle and related bond angles can take the form

$$E^{\tau\theta\theta'}(\tau,\theta,\theta') = K\,V^{\tau\theta\theta'}\ \cos\tau\,[\theta-\bar{\theta}]\,[\theta'-\bar{\theta}'] \,. \tag{8.29}$$

This kind of potential couples the two bending motions (e.g., symmetric and antisymmetric wagging of the two methyl groups in ethane) and has an important affect on spectra (vibrational frequencies). A cross term that relates the dihedral angle to one bond angle can, however, have instead a large effect on geometry (with a small effect on spectra):

$$E^{\tau\theta}(\tau,\theta) = K\,V^{\tau\theta}\ \cos\tau\,[\theta-\bar{\theta}] \,. \tag{8.30}$$

(More generally, $\cos\tau$ above may be replaced by a Fourier series in the form of eq. (8.23) or eq. (8.24)).

This torsion/bend effect may be important to reproduce fine features like a small C–C–H bond-angle opening in the eclipsed configuration in ethane relative to the staggered configuration.

Torsion/stretch potentials can similarly be used for fine tuning, for example to reproduce a bond-stretching tendency for ethane in the eclipsed torsional configuration (relative to the staggered-configuration bond length) [580].

The association between neighboring Wilson angles, in which the two centers (j and j') are bonded, can be represented as:

$$E^{\chi\chi'}(\chi,\chi') = V^{\chi\chi'}\,\chi\,\chi' \,. \tag{8.31}$$

8.6 The van der Waals Potential

8.6.1 *Rapidly Decaying Potential*

The van der Waals potential for a nonbonded distance r_{ij} has the common 6/12 Lennard-Jones form in macromolecular force fields:

$$E_{\text{LJ}}(r_{ij}) = \frac{-A_{ij}}{r_{ij}^6} + \frac{B_{ij}}{r_{ij}^{12}}, \tag{8.32}$$

where the attractive and repulsive coefficients A and B depend on the type of the two interacting atoms. The attractive r^{-6} term originates from quantum mechanics, and the repulsive r^{-12} term has been chosen mainly for computational convenience. Together, these terms mimic the tendency of atoms to repel one another when they are very close and attract one another as they approach an optimal internuclear distance (Figure 8.12).

As mentioned in the last chapter, the 6/12 Lennard Jones potential represents a compromise between accuracy and computational efficiency; more accurate fits over broader ranges of distances are achieved with the Buckingham potential [596, 580]: $-A_{ij}/r_{ij}^6 + B_{ij}\exp(-B'_{ij}r_{ij})$.

The steep repulsion has a quantum origin in the interaction of the electron clouds with each other, as affected by the Pauli-exclusion principle, and this combines with internuclear repulsions.

The weak bonding attraction is due to London or dispersion force (in quantum mechanics this corresponds to electron correlation contributions). Namely, the rapid fluctuations of the electron density distribution around a nucleus create a transient dipole moment and induce charge reorientations (dipole-induced dipole interactions), or London forces.

Note that as $r \to \infty$, $E_{\text{LJ}} \to 0$ rapidly, so the van der Waals force is short range. For computational convenience, van der Waals energies and forces can be computed for pairs of atoms only within some "cutoff radius".

8.6.2 *Parameter Fitting From Experiment*

Parameters for the van der Waals potential can be derived by fitting parameters to lattice energies and crystal structures, or from liquid simulations so as to reproduce observed liquid properties [605]. From crystal data, minimum *contact distances* $\{r_i\}$ of atom radii (see below) can be obtained. Liquid simulations, such as by Monte Carlo, are typically performed on model liquid systems (e.g., ethane and butane) to adjust empirical minimum contact distances $\{r_i\}$ (see below) so as to reproduce densities and enthalpies of vaporization of the liquids.

8.6.3 *Two Parameter Calculation Protocols*

To determine the attractive and repulsive coefficients for each $\{ij\}$ pair, two main procedures can be used for each different type of pairwise interaction (e.g., C–C, C–O).

Energy Minimum/Distance Procedure (V_{ij}, r^0_{ij})

In the first approach, coefficients can be obtained by requiring that an energy minimum V_{ij} will occur at a certain distance, r^0_{ij}, equal to the sum of van der Waals radii of atoms i and j. This requirement between the minimum energy and distances produces the relations:

$$A_{ij} = 2\,(r^0_{ij})^6\,V_{ij} \tag{8.33}$$

and

$$B_{ij} = (r^0_{ij})^{12}\,V_{ij}\,. \tag{8.34}$$

The latter equation can be written using eq. (8.33) in terms of A_{ij} and r^0_{ij} as

$$B_{ij} = \frac{A_{ij}}{2}\,(r^0_{ij})^6\,. \tag{8.35}$$

The van der Waals radii can be calculated from the measured X-ray *contact distances* — which reflect the distance of closest approach in the crystal. These contact distances are appreciably smaller than the sum of the van der Waals radii of the atoms involved. This relationship — between the X-ray contact distances, which crystallographers refer to as "van der Waals radii", and the molecular mechanics meaning of van der Waals radii — was recognized in the early days of molecular mechanics [625] but still may not be widely appreciated.

The CHARMM and AMBER force fields employ this energy-minimum / distance procedure outlined above, using the following definitions for V_{ij} and r^0_{ij}. For each atom type, the pair $\{\epsilon_i, r_i\}$ is specified as a result of the parameterization procedure (crystal fitting or liquid simulations). The parameters for a given ij interaction are then obtained by setting V_{ij} and r^0_{ij} as the geometric and arithmetic (see below) means, respectively:

$$V_{ij} = \sqrt{\epsilon_i \epsilon_j}\,, \tag{8.36}$$
$$r^0_{ij} = (r_i + r_j)\,, \tag{8.37}$$

and then using eqs. (8.33) and (8.34) to define the attractive and repulsive coefficients, respectively. The presence/absence of the $\frac{1}{2}$ factor in the eq. (8.37) depends on the original definition of $\{r_i\}$ (i.e., given radii may have already been divided by 2).

Slater-Kirkwood Procedure (A_{ij}, B_{ij})

The second parameterization procedure is based on the Slater-Kirkwood equation. The attractive coefficient A_{ij} is determined on the basis of the atomic properties of the interacting atom pair [626, 567, 627]:

$$A_{ij} = \frac{365\,\alpha_i\alpha_j}{(\alpha_i/N_{e_i})^{1/2} + (\alpha_j/N_{e_j})^{1/2}}\,, \tag{8.38}$$

where α represents the experimentally determined atomic polarizability, and N_e denotes the number of outer-shell electrons. The numerical factor in the Slater-Kirkwood equation (that is, 365) is derived from universal constants so that the energies are produced in kcal/mol. This factor is computed from the relation $3e\hbar/2m_e$, where e = electron charge, $\hbar$ = Planck's constant divided by 2π, and m_e = electron rest mass.

Slater and Kirkwood have shown, based on experiments for noble gases, that this derivation for A_{ij} produces a London attraction potential in good agreement with experiment.

When A_{ij} is obtained from the Slater-Kirkwood equation (8.38), the coefficient B_{ij} of the repulsive term can then be obtained from A_{ij} using equation (8.35). This leads to a combined van der Waals potential that produces an energy minimum at r_{ij}^0.

The MMFF force field van der Waals parameterization employs the Slater-Kirkwood equation to define V_{ij} via $V_{ij} = A_{ij}/[2(r_{ij}^0)^6]$ (from eq. (8.33)) where the minimum-energy separation r_{ij}^0 is determined from special combination rules. In addition, a "buffered 7/14 attraction/repulsion" van der Waals term is used, found to better fit rare gas interactions [628]. The form of this potential is:

$$E_{\mathrm{LJ}}^{\mathrm{MMFF}}(r_{ij}) = \frac{-A_{ij}}{(r_{ij} + \gamma_1 r_{ij}^0)^7} + \frac{B_{ij}}{(r_{ij} + \gamma_1 r_{ij}^0)^7\,[r_{ij}^7 + \gamma_2 (r_{ij}^0)^7]}\,, \tag{8.39}$$

where the ij-dependent A and B values are functions of V_{ij} and r_{ij}^0, and γ_1 and γ_2 are constants.

Such more complex nonbonded terms (also the 6/exponential combination used by the MM2/3/4 force field) represent a tradeoff between the accuracy of fitting and the computational expense involved in potential evaluation and differentiation.

8.7 The Coulomb Potential

8.7.1 Coulomb's Law: Slowly Decaying Potential

Ionic interactions between fully or partially charged groups can be approximated by Coulomb's law[5] for each atom pair $\{i, j\}$:

$$F_{\mathrm{coul}}(r_{ij}) \propto q_i q_j / r_{ij}^2\,, \tag{8.40}$$

[5]Charles Augustin de Coulomb (1736–1806) was a French physicist who formulated around 1785 the famous inverse square law, now named after him. This result was anticipated 20 years earlier by Joseph Priestly, one of the discoverers of oxygen and author of a comprehensive book on electricity.

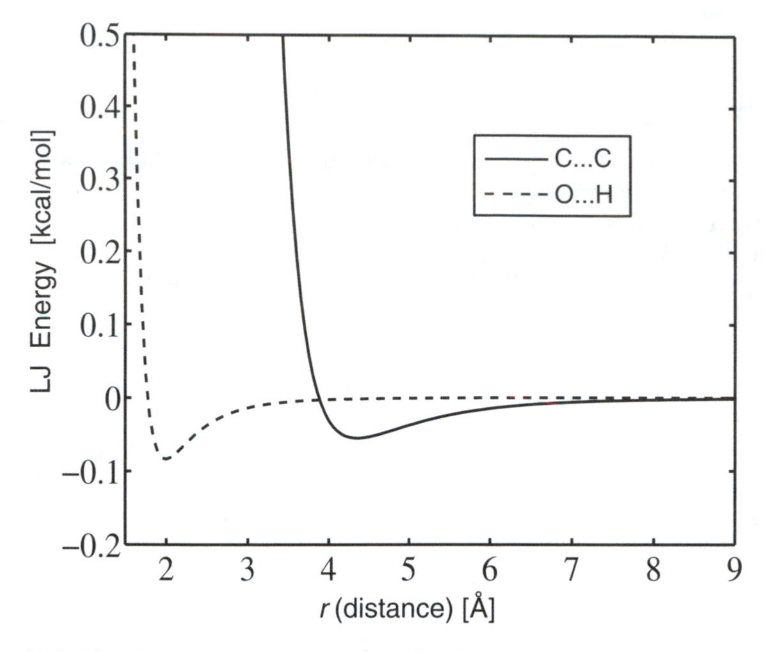

Figure 8.12. Van der Waals nonbonded potentials for C...C and water O...H distances using the CHARMM parameters -0.055, -0.1521, and -0.046 for ϵ_i and 2.1750, 1.7682, and 0.2245 for the distances r_i. The coefficients A and B are computed as: $A_{ij} = 2(r_i + r_j)^6 \sqrt{\epsilon_i \, \epsilon_j}$, $B_{ij} = (r_i + r_j)^{12} \sqrt{\epsilon_i \, \epsilon_j}$, where coefficients produce energies in kcal/mol.

where F is the force and q_i is the effective charge on atom i. The force is positive if particles have the same charge (repulsion) and negative if they have opposite signs (attraction). This leads to the Coulomb potential of form

$$E_{\text{coul}} = K_{\text{coul}} \frac{q_i q_j}{\epsilon \, r_{ij}} \,, \tag{8.41}$$

where ϵ is the dielectric constant and K_{coul} is the conversion factor needed to obtain energies in units of kcal/mol with the charge units used (see Box 8.1).

Unlike the van der Waals potential, the Coulomb interactions decay slowly with distance (see Figure 8.13). In fact, electrostatic interactions are important for stabilizing biomolecular conformations in solvent and associating distant residues in the primary sequence closer in the folded structure. However, because of the $\mathcal{O}(N^2)$ complexity in evaluating all pairwise terms directly, prior calculations have often only counted interactions within a cutoff radius (e.g., 10 Å).

To account for all long-range electrostatic interactions in macromolecules, implementation of fast-electrostatics algorithms, such as the multipole method and Ewald methods, are needed (see next chapter). Such methods can reduce the computational complexity to $\mathcal{O}(N)$. The Ewald and Particle Mesh Ewald treatments have been found to be more amenable to biomolecular dynamics simulations (e.g.,

[629, 67]), especially for parallel platforms; this trend may change as hardware and software evolve. See next chapter for details.

8.7.2 Dielectric Function

The reduction of the force or energy by a dimensionless factor of $1/\epsilon$ is appropriate if the charged particles are immersed in any medium other than a vacuum. That is, a weaker interaction occurs in a polarizable medium (e.g., water) than in vacuum, since then charges become "screened". Very simple approximations to the dielectric function, such as $\epsilon(r) = r$ or $\epsilon(r) = \tilde{D}r$ where $\tilde{D}$ is a constant, have been used in first-generation force fields when water molecules were not represented explicitly in the model [566].

Sigmoidal Function

More sophisticated formulations rely on a large body of theoretical and experimental data, which suggest screening functions in the general *sigmoidal* form [630, 631, 632, 633] that can also account for different ionic concentrations. Such a sigmoidal function uses a screening parameter κ related to the ionic strength of the medium [634]:

$$\epsilon(r) = \tilde{D} \exp(\kappa r) \, ,$$

where κ has dimensions of inverse length. Other forms employ distance-dependent dielectric functions like

$$\epsilon(r) = (D_s + D_0)/(1 + k \, \exp[-\kappa(D_s + D_0)r]) - D_0 \, ,$$

where D_s is the permittivity of water, D_0 and κ are parameters, and k is a constant [635]. These screened Coulomb potentials simultaneously model two effects between two charges in a dielectric medium, like water: a substantially-damped electrostatic field when the charges are separated by moderate distances (e.g., several Ångstroms apart), and the diminished dielectric screening, approaching toward vacuum values, when the charges are in close proximity.

Such screened Coulomb potentials have been used for molecular dynamics simulations and have been incorporated into a procedure to calculate pH-dependent properties in proteins (pK_a shifts, i.e., shifts in hydrogen dissociation constants) with excellent accuracy [635].

The sigmoidal functions are more suitable than the simpler forms above for implicit solvation models of biomolecules. Still, for detailed structural and thermodynamic studies of biomolecules, the trend has been to model water molecules explicitly (i.e., unit value ϵ). Nonetheless, screened Coulomb potentials are of great usage in other applications, such as macroscopic simulations of long DNA in a given monovalent ionic concentration [350, for example]. In such applications, the coefficient $\tilde{D}$ is a function of the salt concentration through a salt-dependent linear charge density for DNA, and the screening parameter κ is the inverse of the

Debye length [24].[6] See end of Chapter 9 (section on continuum solvation) and [498], for example, for further details.

Box 8.1: Coulomb Potential Constant ($K_{\rm coul}$)

Chemists typically use an *electrostatic charge unit* (esu) to define the unit of charge. This unit is defined as the charge repelled by a force of 1 dyne when two equal charges are separated by 1 cm. In these units, the electron charge is 4.80325×10^{-10} esu.

In cgs units, the constant of proportionality in the Coulomb energy is unity; in the SI international system of units, the unit of charge is the *coulomb* (C) = 1 Ampere second (As). In coulomb units, the electron charge is 1.6022×10^{-19} C. The corresponding constant of proportionality in Coulomb's law is:

$$K_{\rm coul} = 1/(4\pi\epsilon_0)\,, \tag{8.42}$$

where the permittivity of a vacuum ϵ_0 is

$$\begin{aligned} \epsilon_0 &= 8.8542 \times 10^{-12}\ {\rm kg}^{-1}\ {\rm m}^{-3}\ {\rm s}^4\ {\rm A}^2 \\ &= 8.8542 \times 10^{-12}\ {\rm J}^{-1}\ {\rm m}^{-1}\ {\rm C}^2\,. \end{aligned}$$

Thus, to obtain energy in kcal/mol using Ångstroms for distance and esu for partial charges, the conversion constant is

$$\begin{aligned} K_{\rm coul} &= \frac{1\ {\rm J\ m}}{4\pi\,(8.8542 \times 10^{-12})\ {\rm C}^2} \\ &= \frac{(6.0221 \times 10^{23}\ {\rm mol}^{-1})\,(10^{10}\ {\rm Å})\,(4184^{-1}\ {\rm kcal})\,(1.6022 \times 10^{-19})^2}{4\pi\,(8.8542 \times 10^{-12})\ {\rm esu}^2} \\ &\approx 332\,\frac{\rm kcal}{\rm mol}\cdot\frac{\rm Å}{{\rm esu}^2}\,. \end{aligned} \tag{8.43}$$

8.7.3 Partial Charges

Selection of partial charges to fit atom *centers* has been a difficult issue in molecular mechanics calculations, since different quantum mechanical approaches yield significantly different values [636]; earlier, determinations of atomic partial charges based on X-ray diffraction data were suggested [637]. However, extensive developments in high-level quantum-derived electrostatic potentials [589, 605, 606] are being applied to determine partial atomic charges for iso-

---

[6]The electrostatic energy is typically expressed as the sum over pairwise interactions between hydrodynamic DNA beads separated by segments of length l_0 as: $[(\nu l_0)^2/\epsilon]\sum_{i<j}[\exp(-\kappa r_{ij})/r_{ij}]$, where ν is an effective linear charge density of the DNA, ϵ is the dielectric constant of water, and $1/\kappa$ is the Debye length at the monovalent salt concentration c_s, given in Molar units ($\kappa \approx 0.33\sqrt{c_s}$ inverse Ångstrom units at room temperature for 1:1 electrolyte solutions).

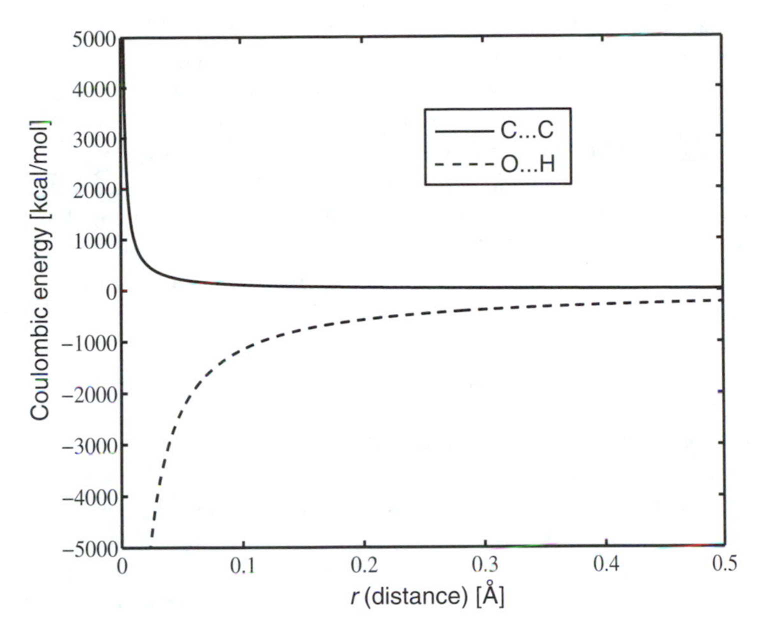

Figure 8.13. Coulomb potentials $K_{\rm coul}\, q_i q_j / r_{ij}$ for C...C and O...H interactions using the CHARMM partial charge parameters (in esu units) of -0.1800 (Arg carbon), -0.8340 (water oxygen), and 0.4170 (water hydrogen).

lated model compounds, with various adjustments to correct for the neglect of many-body polarization effects in liquid water and other factors (e.g., [638]).

For reference, the MMFF force field uses a simple modification of the Coulomb potential of form [628]

$$E_{\rm coul}^{\rm MMFF}(r_{ij}) = K_{\rm coul}\, \frac{q_i q_j}{\epsilon\,(r_{ij} + \delta)^n}\,, \tag{8.44}$$

where ϵ is the dielectric constant (unity by default), $\delta = 0.05$ Å (the *buffering constant*), and n is 1 (default) or 2 (distance-dependent model). This form is required to dampen the attraction between oppositely charged atoms in combination with the buffered van der Waals term used.

8.8 Parameterization

8.8.1 A Package Deal

The general parameterization process for potential energy functions is a difficult task. Several important decisions must be made regarding choices for the functional form and numerical values for the parameters. Even if one is given a specific energy form and a set of structural and energetic data to reproduce, the combinations of parameters that can be used are endless. Unrealistic choices for one group of parameters can be compensated for by adjustment of another.

In theory, the energy terms should have clear physical significance with parameters calibrated by empirical fitting of crystal data, rotational barriers of analogous small molecules, and vibrational frequencies. However, an approximation is inherent in the extension of data from small to large systems. Moreover, interaction with solvent and counterions, reflected in the experimental data, must be interpreted and incorporated in the energy model. In summary, much freedom and manipulation are possible in constructing empirical energy surfaces. Only if constructed and parameterized correctly will the energy model generate reliable structural predictions.

In the case of nucleic acid sugars, the importance of parameter choices in the energy function has already been realized. For example, different potential energy models have produced results that are *qualitatively* different regarding sugar pseudorotation [594, 566, 300, 639, 469]. This is particularly possible when choosing appropriate equilibrium values for endocyclic bond angles [594] or torsion-angle parameters about sugar atoms [606]. Since the unusual puckering geometry and ring closure constraints produce significant deviations from tetrahedral bond angle arrangements, it is not clear what equilibrium values should be used in the harmonic bending terms, more appropriate for small fluctuations.

Some force-field dependent conformations for DNA have been discussed with respect to the AMBER and CHARMM force fields [640, 641, 303, 304, 638, 305]. With some earlier versions, average structures tended to be more B-like with AMBER and intermediate between A and B-DNA with CHARMM in large part due to differences in sugar conformations; newer force fields better balance local geometry with global helical propensities and can reproduce the equilibrium between A and B-DNA in solution (both when Ewald and nonbonded cutoffs are used for electrostatics) [304, 638, 642]. Differences in backbone and base geometries, as well as dynamic properties, are also believed to be force-field dependent. Such disparities, rectifiable with improved parameters, warrant a particularly cautionary note in interpreting structural transitions (such as between A and B-DNA [276]) in molecular dynamics simulations.

8.8.2 *Force Field Performance*

Several discussions of force field performance [590, 591, 592, 593] highlight general issues that remain unresolved and in need of improvement:

- Determination of partial charges;
- Improvement of electrostatic potentials (e.g., use of distributed multipole analysis);
- Methods for solvent representation;
- Interpretation of results in the absence of solvent;

- The approximation reflected by Cartesian vs. torsion space representation; and
- General interpretation of conflicting results by different models and potentials.

Frequent comparisons among force-field results with respect to experimental and high-accuracy *ab initio* data, such as done in [643, 644], reveal the sizable errors made by most force fields and inherent deficiencies. Improvements in the electrostatic formulation that refines the simple atom-centered charged models are repeatedly urged by computational chemists.

As force fields improve, some of these issues may be resolved, but currently force fields are far from "converging" to one another [600].

Still, I emphasize that force fields need not be perfect to be useful! Their overall utility is in generating qualitative and quantitative insights into structural, energetic, and dynamic properties of complex systems through systematic studies (for example by varying critical parameters), especially when trends are compared for a related group of biological systems (like single base-pair variants of DNA).

Ultimately, predictions or interpretations can be tested experimentally.

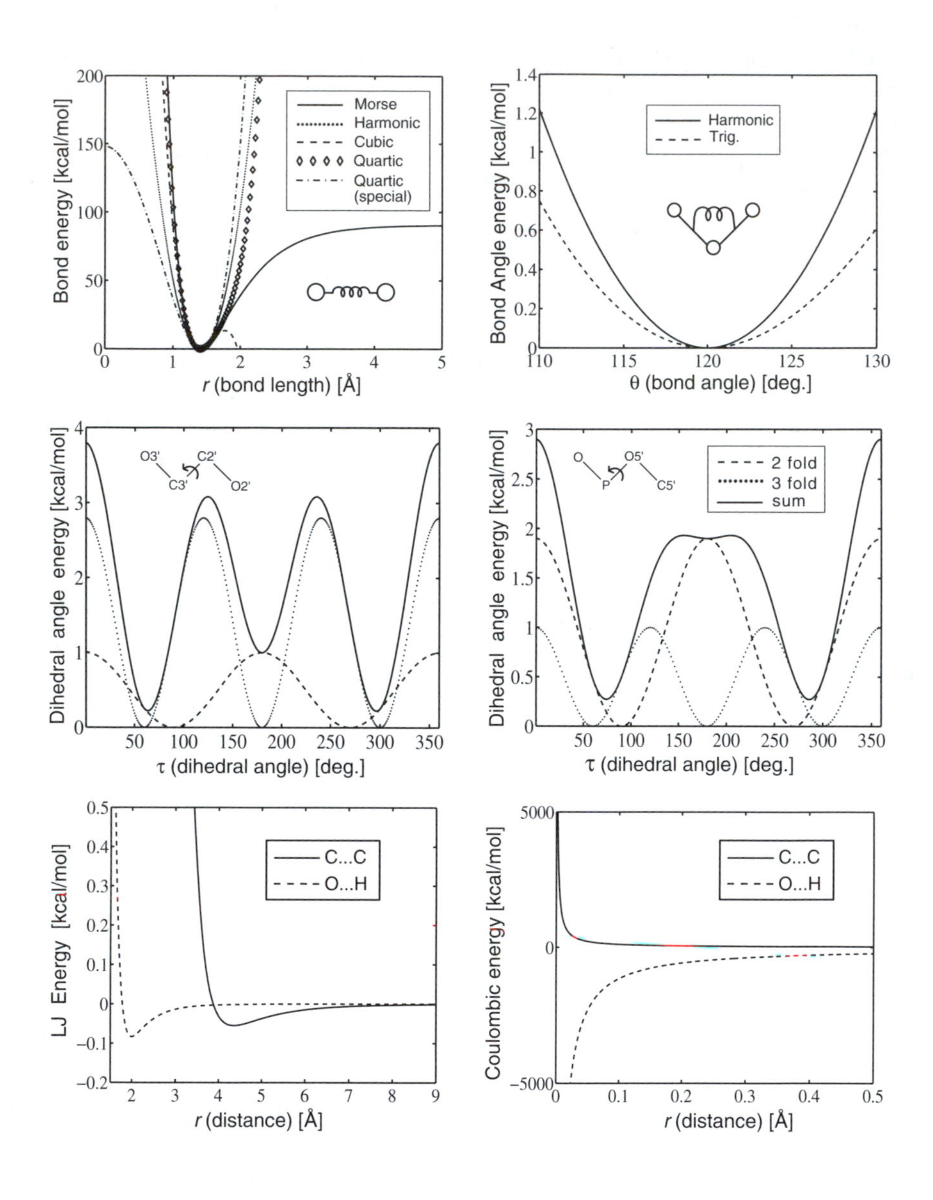
Bond energy [kcal/mol]
r (bond length) [Å]
Morse
Harmonic
Cubic
Quartic
Quartic (special)
Bond Angle energy [kcal/mol]
θ (bond angle) [deg.]
Harmonic
Trig.
Dihedral angle energy [kcal/mol]
τ (dihedral angle) [deg.]
O3'
C2'
C3'
O2'
O
O5'
P
C5'
2 fold
3 fold
sum
LJ Energy [kcal/mol]
r (distance) [Å]
C...C
O...H
Coulombic energy [kcal/mol]
r (distance) [Å]
C...C
O...H

9

Nonbonded Computations

Chapter 9 Notation

Symbol	Definition
Matrices	
$\mathbf{D}$	diffusion tensor (also defined for Coulomb manipulations)
$\mathbf{H}$	Hessian of potential function
$\mathbf{M}$	mass matrix
$\mathbf{T}$	hydrodynamic tensor
Vectors	
$\mathbf{m}$	reciprocal lattice vector, components $\{L/n_x, L/n_y, L/n_z\}$
$\mathbf{n}$	periodic domain vector, components $\{n_x\, L, n_y\, L, n_z\, L\}$
$\mathbf{q}$	arbitrary vector (defined for matrix/vector product)
$\mathbf{r}_{ij}$	interatomic distance vector
$\mathbf{s}$	scattering vector, components $\{h, k, l\}$ (indices of reflection)
$\mathbf{v}(\mathbf{x})$	gravitational field vector at point $\mathbf{x}$
$\mathbf{x}_j$	position vector of atom j, components x_{j1}, x_{j2}, x_{j3}
F_{ij}	force term
$F_{\mathbf{s}}$	complex-valued structure factor, $F_{\mathbf{s}} = A_{\mathbf{s}} + iB_{\mathbf{s}}$
$F_{\mathbf{s}_j}$	scattering vector at atom j
F_{NB}	nonbonded force component, $-\nabla E_{\text{NB}}$
R	Langevin random force
$X, \mathbf{x}$	collective position vector
μ	dipole moment
$\nabla \Phi$	gradient of potential function
Scalars & Functions	
$a, b, c, \tilde{c}$	distance parameters
a_r	hydrodynamic radius
c_s	ionic concentration

Chapter 9 Notation Table (continued)

SYMBOL	DEFINITION
dv, dv^*	volume elements
e	protonic charge
f_j	scattering amplitude for atom j
l	integer, level of refinement in multipole expansion
m	particle mass
m^0	zeroth moment of Φ
m_1^1, m_2^1, m_3^1	components of dipole moment μ of Φ
$m_{11}^2, m_{12}^2, \cdots, m_{33}^2$	components of quadrupole moment of Φ
n_x, n_y, n_z	integers
p	integer (power of expansion)
q_j	Coulomb partial charge for atom j
r, r_{ij}	interatomic distance
$\{r, \theta, \phi\}$	spherical coordinates
$u(\mathbf{x})$	gravitational potential function
A_{ij}, B_{ij}	Lennard-Jones coefficients for atom pair i, j (attraction, repulsion)
C_{ij}, D_{ij}	coefficients for modified Lennard-Jones potential
D_t	translational diffusion constant
E_{coul}	Coulomb potential
E_{LJ}	Lennard-Jones potential
E_{NB}	nonbonded potential
$F_{\mathbf{s}}$	scattered amplitude of whole crystal
G	gravitational constant
K_{coul}	Coulomb potential constant
L	box size dimension
$\{M_n^m\}$	moments of the multipole expansion
N	number of variables (atoms)
N_A	Avogadro's number
$\{P_n^m\}$	Associated Legendre polynomials of degree n
R_e	earth's radius
R_{ij}	Squared interatomic distance (between atoms i and j), $(r_{ij})^2$
$S(r)$	shift/switch function
T	temperature
V	volume of unit cell (associated with volume element dv)
V^*	volume of reciprocal space (associated with volume element dv^*)
$\{Y_n^m(\theta, \phi)\}$	spherical harmonics functions
α, β, γ	angles
β	Gaussian screening parameter
γ	Langevin damping constant
ϵ	dielectric constant
$\epsilon(\mathbf{x})$	position-dependent dielectric function
ϵ_{acc}	desired accuracy
η	solvent viscosity
κ	Debye screening parameter
$\rho(\mathbf{x})$	electron (or charge) density
ρ_G	screening Gaussian
$\phi(\mathbf{s})$	phase angle associated with structure factor $F_{\mathbf{s}}$
ω_{ij}	weight

Chapter 9 Notation Table (continued)

SYMBOL	DEFINITION
Φ	electrostatic potential
$\Phi_{\rm real}$	real (or direct-space) component of Φ
$\Phi_{\rm recip}$	reciprocal-space component of Φ
$\Phi_{\rm cor,ex}$	correction term for excluded nonbonded interactions
$\Phi_{\rm cor,self}$	correction term for self nonbonded interactions
$\Phi_{{\rm cor},\epsilon}$	correction term for finite dielectric

Hofstadter's law: It always takes longer than you expect, even when you take into account Hofstadter's law.

Douglas R. Hofstadter, in *Godel, Escher, Bach*, 1979 (1945–).

9.1 A Computational Bottleneck

Reducing the cost of the nonbonded energy and force computations is of primary importance in molecular mechanics and dynamics simulations of biomolecules. This is because the direct evaluation of these nonbonded interactions involving all atom pairs has the complexity of $\mathcal{O}(N^2)$ where N is the number of atoms. Recall that the bonded terms are local and thus have a linear computational complexity; see homework assignment 8 for a related exercise.

The rapid, quadratic growth in CPU time when all nonbonded interactions are summed directly versus the linear growth associated with the "cutoff" procedures (consideration of interactions within a limited distance range) is shown in Figure 9.1; see CPU scale at left for the evaluation of an energy and force. The implication of these CPU times on total times for 1 ns trajectories (of one million steps) is also shown (see scale at right), explaining the urgency in reducing the nonbonded-term evaluation cost.

The data in Figure 9.1 and Table 9.1 show, for example, that computing all the nonbonded energy and force interactions directly for a hen-eggwhite (HEW) lysozyme protein (2857 atoms or 8571 Cartesian variables) in vacuum requires about one second on a single MIPS R12000/300 MHz processor of an SGI Origin machine. One million such steps to span one nanosecond by molecular dynamics with a 1 femtosecond timestep would require 2 CPU weeks. A system that is five times larger (e.g., size range of a small solvated protein) requires roughly a factor of 45 more computational time, or nearly 2 years of CPU to span a single nanosecond!

Fortunately, techniques have been developed to reduce this cost dramatically without destroying the value of simulations of biomolecules in solvent.

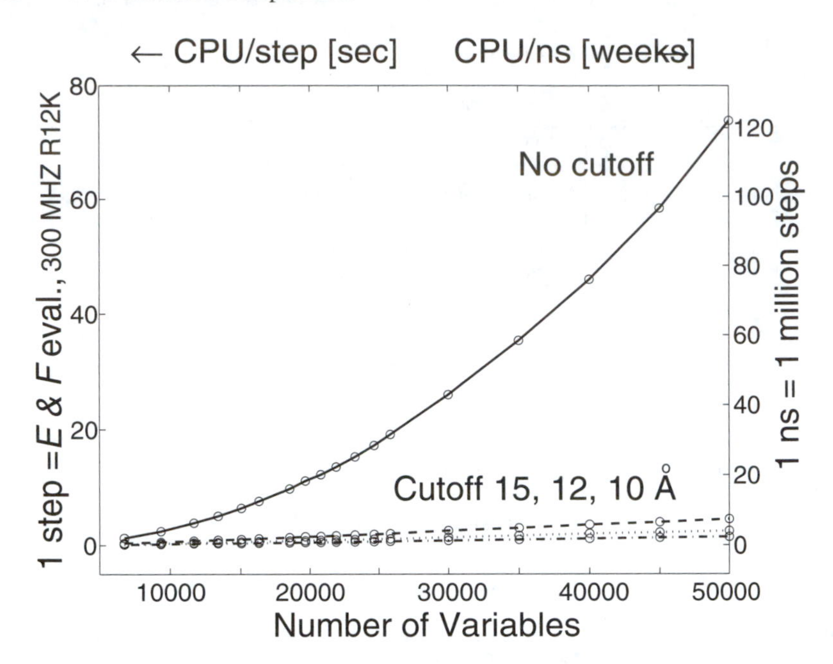

Figure 9.1. CPU time per step (energy plus force evaluation) for water clusters of various sizes modeled in CHARMM when cutoffs are used (at 10, 12, and 15 Å) versus no cutoffs (see left vertical scale) by direct calculation, and corresponding times required for 1 ns trajectories assuming 10^6 steps of 1 fs (see right vertical scale). The number of variables is nine times the number of water molecules. Timings were obtained on a single MIPS R12000/300 MHz processor of an SGI Origin machine.

This chapter introduces novice modelers to three fundamental techniques for handling the nonbonded interactions of large biomolecular systems: spherical cutoffs, particle-mesh Ewald, and fast multipole schemes. We conclude with a brief mention of alternative continuum solvent models, such as Langevin and Brownian dynamics and Poisson-Boltzmann calculations.

9.2 Approaches for Reducing Computational Cost

9.2.1 Simple Cutoff Schemes

The spherical cutoff techniques introduced in the next section are easy to implement as well as computationally cheap ($\mathcal{O}(N)$). More sophisticated than straightforward truncation, these methods can yield reasonable approximations to the energy and force functions up to some threshold separation-distance value. They are particularly suitable for van der Waals interactions which decay rapidly with distance and therefore can be considered zero beyond some interatomic separation distance. Spherical cutoff methods have been used by necessity for simulations of large systems

Table 9.1. Computational requirements of nonbonded calculations on CHARMM version 28a2 by cutoffs, direct nonbonded computations ('all nonbonded'), or all nonbonded by particle-mesh Ewald (PME).

Model[a]	Atoms/ Variables	CPU/step[b] [sec.]	CPU/ns[c] [days]
HEW Lysozyme, 12 Å cutoffs	2857/	0.3	3.0
HEW Lysozyme, all nonbonded	8571	1.2	14.1
Solvated DNA, 12 Å cutoffs	12389/	2.7	31.5
Solvated DNA, PBC, 12 Å cutoffs	37167	6.5	75.3
Solvated DNA, all nonbonded		41.4	479.6
Solvated BPTI, 12 Å cutoffs	14275/	3.1	35.7
Solvated BPTI, PBC, 12 Å cutoffs	42825	7.8	90.3
Solvated BPTI, PBC/PME		8.5	98.8
Solvated BPTI, all nonbonded		54.8	634.0

[a](For lysozyme): nonbonded cutoffs via **group**-based electrostatic switch and **atom**-based Lennard-Jones switch functions, switch buffer 10 to 12 Å, pairlist buffer 12 to 13 Å, and SHAKE used for all bonds involving hydrogens.

(For DNA dodecamer and BPTI): nonbonded cutoffs with periodic boundary conditions (PBC) via **atom**-based electrostatic and Lennard-Jones switch functions, switch buffer 10 to 12 Å, pairlist buffer 12 to 13 Å, and SHAKE used for all bonds involving hydrogens.

[b]Each step entails an energy and gradient evaluation performed in the CHARMM program. The timings were made on a single MIPS R12K/300 MHz processor of an SGI Origin machine.

[c]We assume 1 fs timesteps.

9.2.2 *Ewald and Multipole Schemes*

Of course, cutoff methods neglect long-range interactions beyond some distance and therefore are poor approximations for highly charged systems where large-scale conformational arrangements of linearly-distant residues are involved. The alternative techniques mentioned in the following sections are more suitable for these cases.

Approximating all nonbonded interactions can be accomplished by fast electrostatic techniques based on multipole expansions or Ewald lattice techniques. These schemes have revolutionized the biomolecular simulation field in the past decade since their computational complexity is only $\mathcal{O}(N \log N)$, a dramatic computational saving with respect to the direct value of $\mathcal{O}(N^2)$.

For example, computing the energy and force using spherical cutoffs of range 12 Å for a solvated small protein (BPTI) of size 14275 atoms (42825 variables) modeled in a periodic domain requires about 8 seconds on an R12000/300MHz processor of an SGI machine; one million such steps to span one nanosecond by molecular dynamics with a 1 fs timestep would require about 3 months. If all Coulomb interactions for the same system are computed with the particle-mesh Ewald technique available in CHARMM, the computing time per step (and nanosecond) would increase slightly (but of course accuracy will increase; see Table 9.1). In comparison, direct evaluation of all nonbonded terms would make the project untenable!

An efficient three-dimensional version of the fast multipole method is rather involved to implement; this might explain the preference to date for Ewald techniques in the biological simulation community. The Ewald approach is applied to periodic domains, and this has been known to produce nonphysical long-range correlations for the system [421, 645, 646, 75]. These effects may, however, be considered secondary in general in comparison with truncation artifacts.

Such problems remind practitioners that rarely in the field of biomolecular simulations are pure gains involved due to improving methodologies; there is often a balance between the approximations made and the physical reality of the resulting models. See [647] for an overview of Ewald and multipole methods for computing long-range electrostatic effects in biomolecular dynamics simulations.

Before we introduce the Ewald and fast multipole techniques, we present spherical cutoff methods. Continuum solvation models based on the Poisson-Boltzmann equation are also discussed at the end of this chapter.

The notation used in this chapter (e.g., lattice vectors, scattering factors), though different from some other parts of this text, follows presentations elsewhere on the Ewald summation; these conventions originated in the crystallographic community and have been adopted by the molecular simulation community.

9.3 Spherical Cutoff Techniques

9.3.1 Technique Categories

There are three basic categories of cutoff techniques: *truncation*, *switch*, and *shift* formulations. All approaches set the distance-dependent nonbonded function to zero beyond some distance value $r = b$; however the functional values for $r < b$ are treated differently (see Figure 9.2; the mathematical formulas mentioned in the caption are discussed below):

- The simplest approach, **truncation**, abruptly defines values to be zero at b and does not alter the values of the energies and forces for distances $r < b$.
- **Switching** schemes begin to change values at a nonzero value $a < b$ but leave values for $r < a$ unchanged.
- **Shift** functions alter the function more gradually for all $r < b$.

These three general categories can be applied to either the *energy* or the *force* function of the nonbonded potential (van der Waals or electrostatic). When the force rather than energy function is altered, the energy value is obtained by integration.

In addition, *atom-based* or *group-based* schemes can be used. In the latter, distance thresholds are applied to distances between group *centers*. Group-based cutoffs can better maintain charges associated with entire residues. They can thus avoid potential instabilities in the energy or force that arise when only a subset of atoms of a particular residue is altered.

Besides choosing the particular approach (e.g., atom-based potential switch, group-based force switch), care is required in specifying the distance parameters a and b.

The cutoff techniques described here are also employed when multiple-timestep integration schemes are applied to different force classes (see Chapter 13); in these applications, *force switching* techniques are often used.

9.3.2 Guidelines for Cutoff Functions

In developing nonbonded cutoff functions, we are guided by the following considerations.

1. The short-range energies and forces should be altered as minimally as possible (while satisfying other criteria below).

2. The energies should be altered gradually rather than abruptly to avoid the introduction of artificial minima (where the potential energy and gradient values are suddenly zero).

3. The cutoff approach should avoid introducing large forces around the cutoff region (spikes in right panels of Figure 9.2). This is especially important for molecular dynamics simulations.

4. Also for molecular dynamics, it is important that the cutoff approach alters the functions in a way to approximately conserve the energy.

Truncation schemes satisfy criterion 1 above but violate all others and are removed from further consideration. Switching schemes alter the potential less than shift schemes (since function values for $r < a$ are not altered) but can introduce artificial minima and large sudden forces (Fig. 9.2), violating criterion 3. Energy conservation (criterion 4) can be problematic for certain group-based implementations when polar groups are involved near the cutoff region. Improved (force) shift and force switch functions are often preferred for molecular dynamics applications [648] with a sufficiently wide buffer region $[a, b]$ (e.g., 8 to 12 Å or 11 to 15 Å) and a large enough cutoff value b ($\geq$ 12 Å).

In general, the choice of the best spherical cutoff scheme to use depends on the force field and the system being studied. See [649] and references cited therein

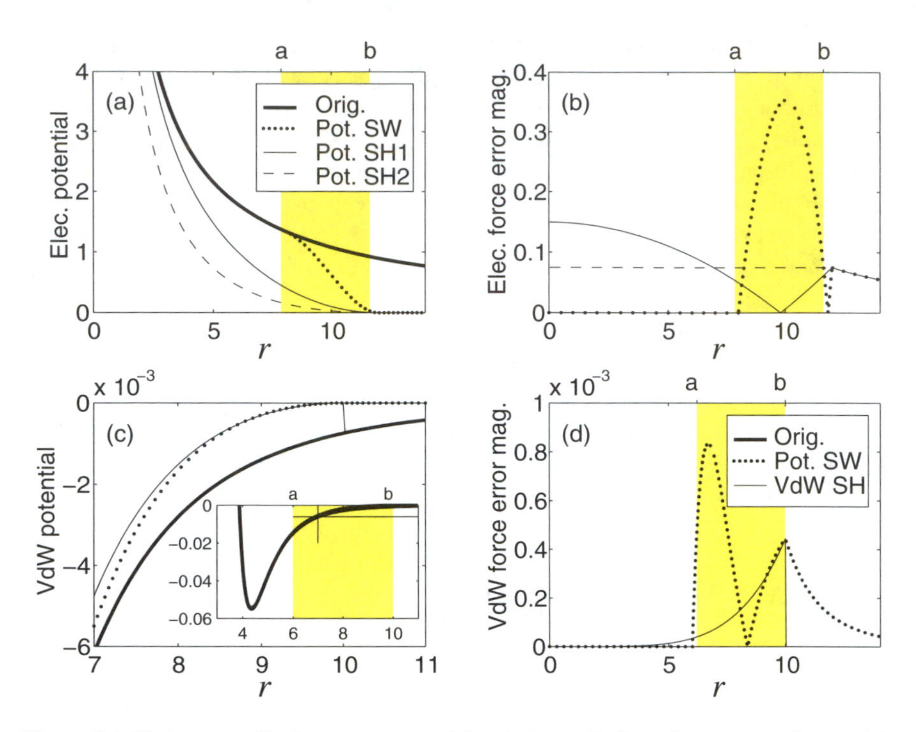

Figure 9.2. Various cutoff schemes — potential switch, eq. (9.4), and two types of potential shift, eqs. (9.10), (9.11) — with buffer regions of 8–12 Å (electrostatic) and 6–10 Å (van der Waals). The altered potentials $E(r)$ are shown on the left panels (a,c), and the corresponding errors in the associated forces ($\Delta|F_{\text{mod}} - F_{\text{orig}}|$), reflecting the magnitude of the difference between the original and modified force, are shown on the right panels (b,d). The potential shift function for the van der Waals interaction corresponds to eq. (9.14). Parameters from the CHARMM program are used, modeling a C^{β}–C^{β} interaction in peptides. The corresponding charge for this atom type is -0.18 esu, and the van der Waals parameters $\{\epsilon_i, r_i\}$ are -0.055 kcal/mol and 2.175 Å, respectively (see eqs. (8.36), (8.37) of Chapter 8 for deriving V_{ij} and r_{ij}^0), producing: $A_{ij} = 745.295$ [kcal/mol] Å^6 and $B_{ij} = 2.525 \times 10^6$ [kcal/mol] Å^{12}; the values of C_{ij} and D_{ij} from eqs. (9.15) and (9.16) are $C_{ij} = -7.403 \times 10^{-10}$ [kcal/mol] / Å^6 and $D_{ij} = -0.0015$ kcal/mol.

for studies on the effect of different cutoff and long-range electrostatic models on the stability and accuracy of biomolecular simulations.

9.3.3 General Cutoff Formulations

Consider the following general modification to the nonbonded energy function E_{NB} by a distance-dependent switch or shift function $S(r)$:

$$E_{\text{NB}}(X) = \sum_{i,j} \omega_{ij}\, S(r_{ij}) \left[\frac{-A_{ij}}{r_{ij}^6} + \frac{B_{ij}}{r_{ij}^{12}} + \frac{q_i q_j}{\epsilon\, r_{ij}} \right] . \tag{9.1}$$

Here X is the collective position vector, r_{ij} represents the distance between atoms i and j, and the parameters $0 \leq \omega_{ij} \leq 1$ are weights. These weights can be used to exclude bonded terms (i.e., 1–2 interactions) or bond-angle interactions (1–3) (i.e., with $\omega_{ij} = 0$), or to scale other interactions such as those involving a sequence of four bonded atoms (1–4).

Truncation

Simple truncation can be expressed by the switch function S defined as

$$S(r) = \begin{cases} 1 & r < b \\ 0 & r \geq b \end{cases} . \tag{9.2}$$

This function introduces a force discontinuity at $r = b$ and fails to conserve energy. In general, $S(r)$ is a distance-dependent distance function which assigns a constant value or a function of r depending on the value of r with respect to the distance parameters a and b. Thus $S(r)$ may be set separately for the three cases: $r \leq a$, $a < r \leq b$, and $r > b$.

Switch/Shift

The switch/shift function can be different for the van der Waals and electrostatic terms; in that case eq. (9.1) should be written as two terms with different functions $S(r)$. The CHARMM program, for example, uses the same switch functions for both van der Waals and Coulomb interactions but different shift functions for these terms when selected.

Atoms/Groups

The above formulation is atom-based. Group-based formulations apply the distance-dependent switch/shift functions based on the separation of *group centers*. For example, if such an intergroup distance is in the buffer region $[a, b]$, the modification that S applies in this range (see below) is used for all atoms in the two groups; similarly, this intergroup distance determines the modifications applied to all atoms in the two groups when the distance falls in the other two regions: $r < a$ and $r > b$.

Energy/Force Modifications

When a modification is applied to the force F_{NB} instead of the potential energy of the nonbonded terms, each r_{ij}-dependent force term, namely $F_{ij}(r_{ij})$, is modified as:

$$\widehat{F}_{ij}(r_{ij}) = \omega_{ij}\, S(r_{ij})\, F_{ij}(r_{ij}) \tag{9.3}$$

where the force rather than the potential is switched by the operator $S(r)$. The corresponding energy must then be obtained by integration.

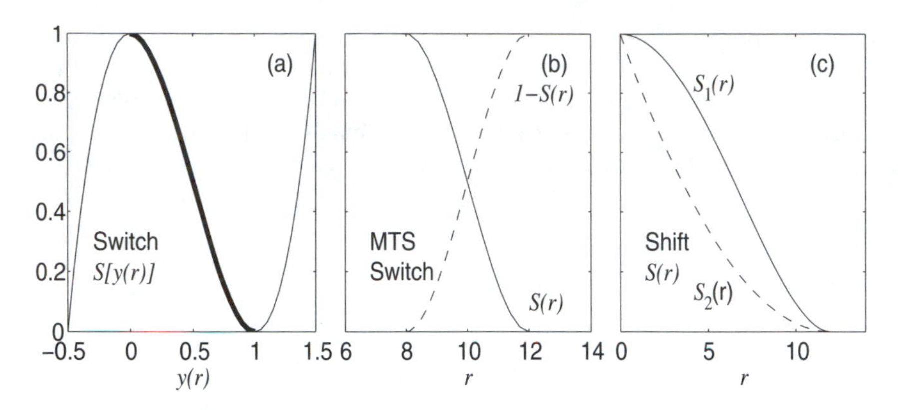

Figure 9.3. Switch and shift functions: (a) the potential switch function of eq. (9.4) as a function of $y(r)$ given in eq. (9.5); only the heavier part of the curve is relevant since $y(r)$ increases from 0 to 1 as r increases from a to b; (b) the switch function $S(r)$ of eq. (9.9) that is applied to the fast force component in multiple-timestep schemes (solid curve), along with $1 - S(r)$ (dashed curve) that is applied to the slow force component (see eq. (9.8)); (c) the shift functions $S_1(r)$ and $S_2(r)$ of eqs. (9.10) and (9.11).

9.3.4 *Potential Switch*

For potential switch functions, $S(r)$ is a polynomial of r that alters the nonbonded energy smoothly and gradually over the buffer region $[a, b]$ so that $E(b) = 0$, while leaving values of the energy function $E(r)$ for $r \leq a$ unchanged. The polynomial degree must be sufficiently high to ensure that both the energy and its gradient are continuous functions. A satisfactory function is the following cubic polynomial of r^2 (see Figure 9.3, part (a)):

$$S(r) = \begin{cases} 1 & r < a \\ 1 + y(r)^2\,[2y(r) - 3] & a \leq r \leq b \\ 0 & r > b \end{cases}, \tag{9.4}$$

where

$$y(r) = (r^2 - a^2)/(b^2 - a^2)\,. \tag{9.5}$$

The $S(r)$ expression above for $a \leq r \leq b$ can also be written as (following algebra):

$$S(r) \;=\; \frac{(b^2 - r^2)^2\,(b^2 + 2r^2 - 3a^2)}{(b^2 - a^2)^3} \qquad \text{for } a \leq r \leq b\,. \tag{9.6}$$

From this form, it is clear that $S(r)$ decreases monotonically from 1 to 0 as r increases from a $[y(a) = 0$ and $S(a) = 1]$ to b $[y(b) = 1$ and $S(b) = 0]$. Note that for $a < r < b$ the derivative from eq. (9.6) is:

$$S'(r) = 12\,r\,y(r)\,[y(r) - 1]/(b^2 - a^2)\,, \tag{9.7}$$

and thus both the right derivative of $S(r)$ at a and the left derivative of $S(r)$ at b (where $y(r) = 0$ and 1, respectively) are zero. Since these derivatives are also zero from the left of a and the right of b (where $S(r)$ is a constant function), both the potential and the gradient functions are continuous. This function $S(r)$ also yields continuous second derivatives when the force, rather than the energy, is switched on or off.

9.3.5 *Force Switch*

Rather than switching the potential, force switching is used in multiple-timestep schemes to gradually separate the short from long-range forces.[1] Often, force classes are defined according to a distance parameter b closely related to the cutoff distance above. For example, interactions within a region of b Å can be considered 'fast' and those beyond that value 'slow' (e.g., $b = 6$ Å). In this case, the short-range forces are turned off gradually by a switch function $S(r)$ that decreases from 1 to zero as r increases from a to b in a manner closely resembling that of S defined in eq. (9.4). Here the region $[a, b]$ is the switching buffer region, and $c = b - a$ is the size of this buffer.

The switching function is applied as follows to the fast and slow force components of $F_{\rm NB}$:

$$\begin{aligned} F_{\rm NB,\,fast} &= S(r)\, F_{\rm NB}(r) \\ F_{\rm NB,\,slow} &= [1 - S(r)]\, F_{\rm NB}(r)\,, \end{aligned} \tag{9.8}$$

where the function $S(r)$ is defined as:

$$S(r) = \begin{cases} 1 & r < a \\ 1 + r^2(2r - 3) & a \le r \le b \\ 0 & r > b \end{cases} . \tag{9.9}$$

See Figure 9.3, part (b), for an illustration.

Buffer Parameters

In addition to the buffer length c (which can range from 1 to 4 Å, for example), another buffer parameter $\widetilde{c}$ is typically used for bookkeeping purposes. Namely, to keep track of the nonbonded atom pairs $\{i, j\}$ used to associate each pair of atoms with a force term in which it is calculated (i.e., fast or slow), the pairlist is monitored up to a distance $b + \widetilde{c}$. Implementations vary from program to program, but the main idea is to use this buffer size $\widetilde{c}$ to monitor changes in interatomic distances that would require a new pairlist generation (see [650] for example).

---

[1]These two components are also termed 'fast' and 'slow' because the short-range terms are rapidly varying in time while the long-range terms change more slowly with time; see Chapter 13.

9.3.6 Shift Functions

Shift-type functions can avoid the sudden changes in force that occur with truncation and switch methods at the cost of underestimating the short-range forces (see Fig. 9.2). This requires alteration of the nonbonded function over the larger region $r \leq b$ (rather than $0 < a \leq r \leq b$).

Shift functions include the following two formulations:

$$S_1(r) = [1 - (r/b)^2\,]^2, \quad \text{for } r \leq b\,, \tag{9.10}$$

or

$$S_2(r) = [1 - r/b\,]^2, \quad \text{for } r \leq b\,. \tag{9.11}$$

Both decrease monotonically from 1 to 0 as r increases from 0 to b, but their curvature is different (Fig. 9.3, part (c)). These two functions work well for Coulomb interactions.

In fact, $S_2(r)$ above augments the true Coulombic force by a constant $(1/b^2)$ for $r \leq b$. This can be seen by writing the modified energy $\widetilde{E}(r)$ of the scalar Coulomb potential $E(r) = 1/r$ as

$$\widetilde{E}(r) \equiv S_2(r) \cdot E(r) = \frac{1}{r} \cdot \left(\frac{b-r}{b}\right)^2 = \frac{1}{b^2} \cdot \frac{(b-r)^2}{r}\,. \tag{9.12}$$

The associated derivative is:

$$\widetilde{E}\,'(r) = \frac{r^2 - b^2}{b^2\, r^2} = \frac{1}{b^2} - \frac{1}{r^2} = E\,'(r) + \frac{1}{b^2}\,, \tag{9.13}$$

showing a derivative augmentation by the constant $1/b^2$.

The potential shift approach by $S_2(r)$ is often termed *force shift* for this reason. Note, however, that this name is misleading since S_2 is applied to the *potential* and not to the force.

CHARMM, for example, employs $S_1(r)$ above for group-based potential shift and $S_2(r)$ for atom-based shifts. For the van der Waals interactions, the potential shift function is additive rather than multiplicative, applied as an auxiliary term so as to dampen the force monotonically to zero:

$$E_{\rm LJ}(X) = \begin{cases} \sum_{i,j} \omega_{ij} \left[\left(\frac{-A_{ij}}{r_{ij}^6} + \frac{B_{ij}}{r_{ij}^{12}} + C_{ij} r_{ij}^6 + D_{ij}\right)\right] & r \leq b \\ 0 & r > b \end{cases}\,. \tag{9.14}$$

Here, C and D are chosen so that both the van der Waals potential and force functions are zero at $r = b$. This leads to the expressions in terms of A_{ij} and B_{ij} as:

$$C_{ij} = -A_{ij}\, b^{-12} + 2\, B_{ij}\, b^{-18} \tag{9.15}$$

$$D_{ij} = 2\, A_{ij}\, b^{-6} - 3\, B_{ij}\, b^{-12}\,. \tag{9.16}$$

See Figure 9.2 (lower panels) for an illustration of this shift function approximation.

9.4 The Ewald Method

In this section, we only sketch the efficient particle-mesh Ewald (PME) method, popular for biomolecular dynamics. See [651, 652, 653] for technical details, and [654, 277, 647], for example, for works which emphasize the impact of PME on nucleic acid simulations. An outline of the Ewald summation technique can also be found in [31].

Throughout this section, we use notation consistent with other works in this subject (coming from crystallography), though different from notation used elsewhere in this text. The symbol X for the collective Cartesian vector used elsewhere is identical to the collective position vector $\mathbf{x}$ used here. The position vector of atom i is $\mathbf{x}_i$, and the interparticle distance vector from atom j to i is $\mathbf{r}_{ij} \equiv \mathbf{x}_i - \mathbf{x}_j$; we express the magnitude of this distance as the Euclidean norm of the vector, or $|\mathbf{r}_{ij}|$.

9.4.1 *Periodic Boundary Conditions*

The Ewald method is a technique for calculating the electrostatic energy of a system on a lattice with periodic boundary conditions. By periodic conditions, we mean that the modeled system (biomolecules and solvent molecules) is placed in the *unit cell* and considered to have infinitely many images in space. This replication forms an infinite lattice in 3D space. Figure 9.4 shows a standard square lattice in 2D of dimension L; this lattice generalizes to a cubic geometry in 3D.

Space-Filling Polyhedra

Non-cubic periodic domains can also be modeled in 3D, such as the truncated octahedron, hexagonal prism, and face-centered cube, as shown in Figure 9.5. These lattice geometries reflect variations in the relationships among the three side lengths (a, b, c) and the three angles between these sides (α, β, γ) of the domain. For some commonly used space-filling lattices we have the following relations:

1. **Cubic**: $a = b = c$ and $\alpha = \beta = \gamma = 90^\circ$.

2. **Hexagonal**: (e.g., hexagonal prism): $a = b$ and $\alpha = \beta = 90^\circ$ and $\gamma = 120^\circ$.

3. **Rhombic Dodecahedron** (e.g., face-centered cube): $a = b = c$ and $\alpha = \gamma = 60^\circ$ and $\beta = 90^\circ$.

More generally, other space-filling polyhedra are shown in Figure 9.6.

Minimum-Image Convention

Only coordinates of the unit cell need to be recorded and propagated. As an atom leaves the unit cell by crossing the boundary, an image enters to replace it, and hence the total number of particles is conserved. In biomolecular simulations, the unit cell is usually set to be large enough to limit such occurrences

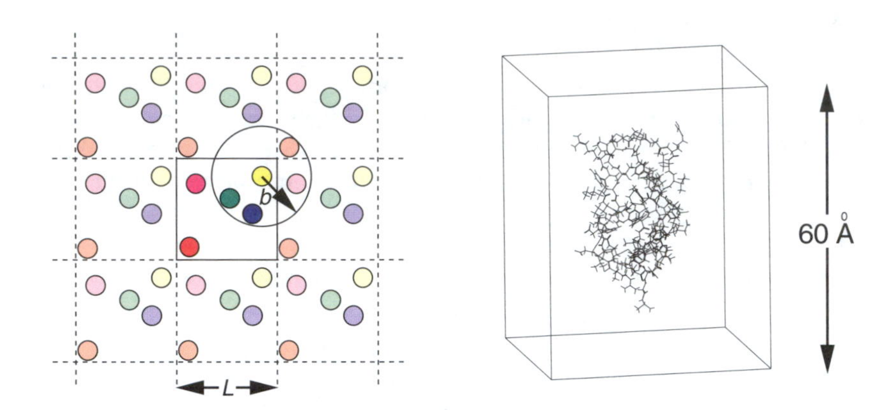

Figure 9.4. Periodic domain in 2D, showing the unit cell (center) and its images (surrounding replicas), and in 3D (rectangular), as used for a solvated protein (BPTI) system [609].

and to avoid artifacts that can be caused by artificial periodic boundary conditions [421, 645, 646, 75]. Such artifacts may be more pronounced in a solvent of low dielectric permittivity (not water), for a solute having a large overall charge, and when the solute cavity is relatively large compared to the size of the unit cell (due to artifactually small solvation of extended solute). Specifically, it has been found that artificial periodicity perturbs the potentials of mean force associated with conformational equilibria of solvated biomolecules and can tend to overstabilize compact folded forms. A reasonable cell size involves a solvated biomolecule with at least a 10 Å layer of water molecules [655].

In practical terms, a *minimum-image convention* is typically used so that each atom i interacts only with the *closest* periodic image of the other $N - 1$ particles. In addition, a spherical cutoff (as described in the last section) is applied to restrict this number of calculated interactions further. For a consistent combination of spherical cutoffs and the minimum-image convention, the cutoff distance (b of last section) is at most $L/2$, where L is the dimension of the side of the box. Note from Table 9.1 that applying periodic boundary conditions only about doubles the computational work with respect to using 12 Å cutoffs (for a finite system). Further, implementing the particle-mesh Ewald fast summation scheme (to approximate all nonbonded interactions in the periodic model) requires only slightly more work than the periodic summation with this cutoff.

Choice of Geometry

The geometry of the periodic domain used also affects the total number of pairwise interactions considered. Note from Figure 9.5, for example, that a truncated octahedron may be more compact than a cubic lattice for a solvated protein, while a hexagonal prism is appropriate for solvated DNA systems.

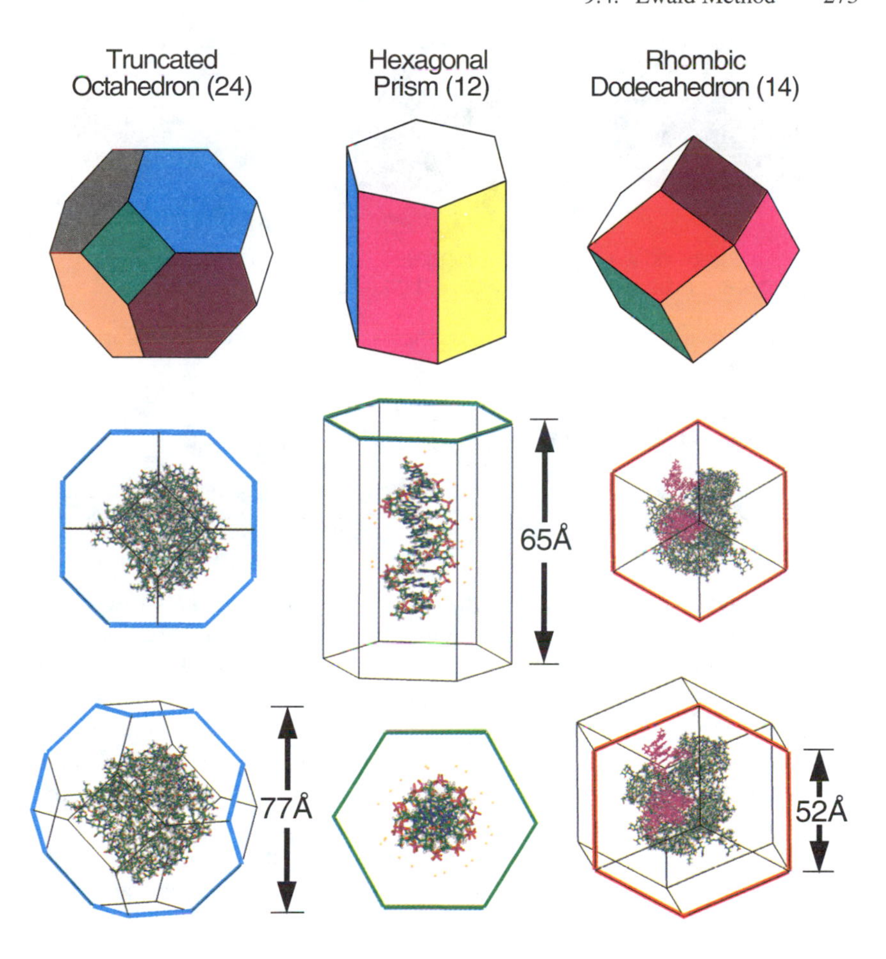

Figure 9.5. Examples of nonrectangular periodic domains in 3D used for biomolecular simulations: *truncated octahedron*, containing a solvated protein (villin) [69]; *hexagonal prism*, containing a solvated DNA dodecamer [334]; and *rhombic dodecahedron* (face-centered cube), containing a polymerase/DNA complex [656]; water molecules (not shown) fill the domain. A side and top view is shown for each geometry, as well as a space-filling rendering (at top) indicating the number of vertices in each case.

Programs to optimize such models, that is, to orient and solvate the solute macromolecule in a given domain subject to a minimal water layer thickness (e.g., 10 Å) so as to yield the lowest number of water molecules, were pioneered by Mihaly Mezei [657] (see references and details for program Simulaid on fulcrum.physbio.mssm.edu/~mezei/) and recently generalized, with enhanced efficiency, for additional space-filling polyhedra in the program PBCAID [658].

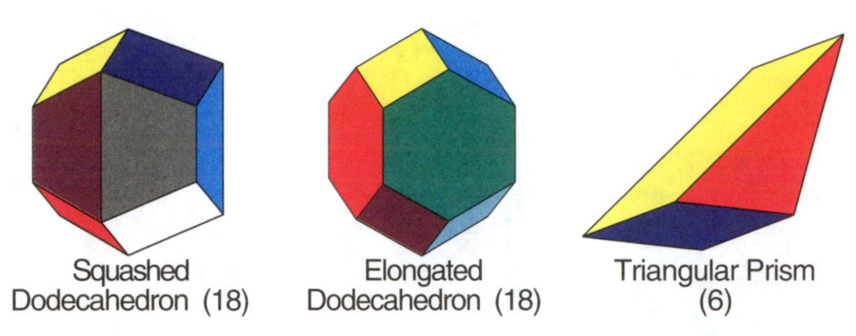

Figure 9.6. Three additional space-filling polyhedra are shown, with the number of vertices for each domain indicated.

9.4.2 Ewald Sum and Crystallography

The Ewald construct has its roots in the crystallographic community. There are three specific commonalities between crystallography and the Ewald summation for fast electrostatics:

1. **Reciprocal Term.** The notion of the *reciprocal space* is used in addition to the direct space. In crystallography, an Ewald sphere describes the diffraction conditions in terms of the reciprocal rather than direct lattice. The reciprocal lattice is an orthogonal system related to the orthogonal system associated with the atoms in the unit cell (termed the real-space lattice). This lattice is used to express the scattering vector in crystals (see Box 9.1). In the Ewald summation, two terms arise for the computation of the electrostatic energy, for atom pairs in the *direct lattice* and for atoms pairs corresponding to interactions with *images* of the unit-cell atoms. The former is evaluated by direct calculation, and the latter (smooth and long-range) by Fourier transforms.

2. **Fast Fourier transforms.** FFTs are key ingredients in both applications. In crystallography, FFTs are used to express the amplitude of diffraction for the whole crystal from the electron density function associated with one unit cell of the crystal (see Box 9.1). In the Ewald sum, FFTs are used to evaluate the electrostatic energy corresponding to the reciprocal lattice.

3. **Electron Density Distributions.** The notion of *electron density distribution* is key in both applications. Crystallographers compute the total electron density to obtain a spatial arrangement of the atoms in space in terms of FFTs of the *structure factors*, which describe the scattering pattern of each atom with respect to an electron at the cell origin (see Box 9.1). Modelers use an analogous concept to describe the electrostatic energy in terms of charge densities rather than point charges to make the infinite sums convergent. Terms analogous to the crystallographic structure factors also arise in the Ewald sum FFTs.

Box 9.1: Some Tricks from X-Ray Crystallography

A crystal is a 3D periodic arrangement of atoms (see [93], for example, for an introduction to crystallography, and Chapter 1 for an introduction to the phase problem). The structural subunit of crystals is the *unit cell*. The X-ray diffraction pattern by the electrons of a molecular crystalline system is interpreted in terms of both the *direct lattice*, the orthogonal system associated with the unit-cell atoms, and the *reciprocal lattice*, an auxiliary orthogonal system used to express the scattering vector in crystals. Specifically, the scattering of each atom j with respect to an electron at the origin of the cell it occupies is:

$$F_{\mathbf{s}_j} \;=\; f_j \, \exp\left(2\pi i \mathbf{x}_j \cdot \mathbf{s}\right), \tag{9.17}$$

where $i = \sqrt{-1}$, f_j is an experimentally-measured amplitude, $\mathbf{x}_j = \{x_{j1}, x_{j2}, x_{j3}\}$ denotes the Cartesian position vector of atom j in the direct lattice, and $\mathbf{s}$ is the scattering vector corresponding to a triplet of indices (*indices of reflection*), such as $\mathbf{s} = \{h, k, l\}$, in the reciprocal lattice. Thus for N atoms in the unit cell, we write the complex-valued *structure factor* $F_{\mathbf{s}}$ as:

$$F_{\mathbf{s}} \;=\; \sum_{j=1}^{N} f_j \, \exp\left(2\pi i \mathbf{x}_j \cdot \mathbf{s}\right) \;=\; A_{\mathbf{s}} + iB_{\mathbf{s}} \,, \tag{9.18}$$

$$A_{\mathbf{s}} \;=\; \sum_{j=1}^{N} f_j \, \cos\,\left(2\pi i \mathbf{x}_j \cdot \mathbf{s}\right), \qquad B_{\mathbf{s}} = \sum_{j=1}^{N} f_j \, \sin\,\left(2\pi i \mathbf{x}_j \cdot \mathbf{s}\right). \tag{9.19}$$

From these quantities, the phase angles associated with $F_{\mathbf{s}}$ are computed as $\phi_{\mathbf{s}} = \tan^{-1}(B_{\mathbf{s}}/A_{\mathbf{s}})$. The intensity of the diffracted X-ray beam is proportional to the square of the structure factor's amplitude and depends only on the interatomic vectors, not on the actual, origin-dependent atomic coordinates:

$$|F_{\mathbf{s}}|^2 = A_{\mathbf{s}}^2 + B_{\mathbf{s}}^2 = \sum_{i=1}^{N} \sum_{j=1}^{N} f_i \, f_j \, \cos\,\left[2\pi \,(\mathbf{x}_i - \mathbf{x}_j) \cdot \mathbf{s}\right]. \tag{9.20}$$

The relationship between crystallography and the electrostatic potential arises by relating the diffracted amplitude to a continuous distribution of electron density, $\rho(\mathbf{x})$, expressed as electrons per unit volume. The scattered amplitude for the whole crystal, $F(\mathbf{s})$, from a small volume element dv, results from an effective point charge of $\rho(\mathbf{x})\,dv$ electrons:

$$F(\mathbf{s}) = \int_V \rho(\mathbf{x}) \, \exp\left(2\pi i \mathbf{s} \cdot \mathbf{x}\right) dv \,, \tag{9.21}$$

where the integration is taken over the volume V of the unit cell. This is the Fourier transform of the electron density in one unit cell of the crystal, sampled at points $\mathbf{s}$ on the reciprocal lattice. The inverse transformation yields the electron density via integration over the entire volume of the reciprocal space in which $\mathbf{s}$ is defined:

$$\rho(\mathbf{x}) = \int_{V^*} F(\mathbf{s}) \, \exp\left(-2\pi i \mathbf{s} \cdot \mathbf{x}\right) dv^* \,. \tag{9.22}$$

Crystallographers compute a discrete analog of this electron density to obtain the atomic arrangement in the crystal in which each point of the reciprocal lattice (where $F(\mathbf{s})$ is defined) has an associated weight of $F_{\mathbf{s}}/V$ where $F_{\mathbf{s}}$ is the structure factor defined in eq. (9.18):

$$\rho(\mathbf{x}) = \frac{1}{V} \sum_{h,k,l=-\infty}^{\infty} F_{\mathbf{s}} \exp\left(-2\pi i \mathbf{s} \cdot \mathbf{x}\right). \tag{9.23}$$

This computation is performed using Fast Fourier transforms. The amplitudes of these structure factors are known from the observed intensities of the X-ray reflections, but the phase angles cannot be measured. This is the heart of the problem of deducing structure from X-ray diffraction patterns.

9.4.3 *Mathematical Morphing of a Conditionally Convergent Sum*

Coulomb Energy in Periodic Domains

The following Ewald sum describes the total Coulomb energy corresponding to a system in an infinite periodic domain; the conversion factor K_{coul} (see eq. (8.41) of Chapter 8) is omitted for simplicity, and the dielectric constant $\epsilon = 1$:

$$E_{\text{coul}} = \frac{1}{2} \sum_{i,j=1}^{N} \sum_{\text{images } |\mathbf{n}|}{}' \frac{q_i q_j}{|\mathbf{r}_{ij} + \mathbf{n}|} = \frac{1}{2} \sum_{j=1}^{N} q_j \, \Phi(\mathbf{x}_j), \tag{9.24}$$

where

$$\Phi(\mathbf{x}_j) = \sum_{i=1}^{N} \sum_{|\mathbf{n}|}{}' \frac{q_i}{|\mathbf{r}_{ij} + \mathbf{n}|}. \tag{9.25}$$

The absolute value signs above denote vector magnitudes, and the summation extends over all images of the unit cell. The vector $\mathbf{n}$ denotes the vector $\mathbf{n} = (n_x L, n_y L, n_z L)$, where n_x, n_y, and n_z are integers and L is the box size, and where the triplet of indices (n_x, n_y, n_z) are related to the magnitude of $\mathbf{n}$, namely $|\mathbf{n}|$, by $|\mathbf{n}| = |n_x| + |n_y| + |n_z|$.

Thus, for example, the 'image' corresponding to $|\mathbf{n}| = 0$ has only one triplet: $\mathbf{n} = (0, 0, 0)$; $|\mathbf{n}| = 1$ has all triplets that can be written as: $\mathbf{n} = (\pm L, 0, 0)$, $(0, \pm L, 0)$, and $(0, 0, \pm L)$; and $|\mathbf{n}| = 2$ is associated with the vectors that can be expressed as $\mathbf{n} = (\pm 2\,L, 0, 0)$, $(0, \pm 2\,L, 0)$, $(0, 0, \pm 2\,L)$, $(\pm L, \pm L, 0)$, $(\pm L, 0, \pm L)$, and $(0, \pm L, \pm L)$. Similarly, the reciprocal lattice vectors $|\mathbf{m}|$ defined below correspond to vectors $(L/n_x, L/n_y, L/n_z)$.

The sum over the images extends from $|\mathbf{n}| = 0$ to ∞. The prime symbol in the sum ($\sum_{\mathbf{n}}'$) indicates that for $\mathbf{n} = 0$ we omit the $i = j$ interaction (so the denominator is well defined).

Conditional Convergence

The sum in eq. (9.24) is only *conditionally convergent* since the terms decay as $1/|\mathbf{n}|$, like the harmonic series.[2] The value of the sum also depends on the nature of the surrounding medium (dielectric constant ϵ) [659]. For a unit cell surrounded by vacuum, a dipolar layer exists on the surface and its contribution must be canceled.

Ewald's Trick

Ewald noted a trick that can be used to convert this sum into an expression involving a sum of two absolutely and rapidly convergent series in direct and reciprocal space. This conversion is accomplished by representing each point charge as a Gaussian charge density, producing an exponentially decaying function. This Gaussian transformation must be counteracted by an analogous subtraction to leave the net result of an effective point charge (delta function charge, see Fig. 9.7). This canceling distribution is summed in the reciprocal space (and transformed back to real space) because it is a smoothly-varying periodic function which can be represented by a rapidly convergent Fourier series.

Essentially, the basic idea can be written for the radial function $\Phi(r)$ as the splitting

$$\Phi(r) \equiv \frac{1}{r} = \Phi_{\text{real}}(r) + \Phi_{\text{recip}}(r)\,, \tag{9.26}$$

where

$$\Phi_{\text{real}}(r) = \frac{1}{r} - \frac{\text{erf}(\beta r)}{r} = \frac{\text{erfc}(\beta r)}{r}\,, \tag{9.27}$$

and

$$\Phi_{\text{recip}}(r) = \frac{\text{erf}(\beta r)}{r}\,. \tag{9.28}$$

The first term (Φ_{real}) is short-range with singularity at the origin, and the second (Φ_{recip}) is a smooth and long-range term (which can be Fourier transformed). The error function erf is defined as

$$\begin{aligned} \text{erf}(x) &= \frac{2}{\sqrt{\pi}} \int_0^x \exp(-t^2)\, dt\,, & (9.29)\\ &= \frac{2}{\sqrt{\pi}} \left(x - \frac{x^3}{3} + \frac{1}{2!}\frac{x^5}{5} - \frac{1}{3!}\frac{x^7}{7} + \cdots \right), & (9.30) \end{aligned}$$

[2] A series $S = \sum_{n=1}^{\infty} a_n$ is conditionally convergent if the infinite sum $\sum_n a_n$ converges but $\sum_n |a_n|$ diverges. It is also known that the sum for a conditionally convergent series depends on the order of summation. For $a_n = (-1)^{n+1}/n$, the 'alternating harmonic series' $1 - \frac{1}{2} + \frac{1}{3} - \frac{1}{4} + \cdots$ converges, but the harmonic series $1 + \frac{1}{2} + \frac{1}{3} + \frac{1}{4} + \cdots$ diverges, though $a_n \to 0$ as $n \to \infty$. To see that the sum for finite n terms can be as large as we please for the harmonic series, note that terms can be grouped so that each subsum is larger than $\frac{1}{2}$: $\sum_{n=1}^{\infty} \frac{1}{n} = 1 + \left(\frac{1}{2} + \frac{1}{3}\right) + \left(\frac{1}{4} + \frac{1}{5} + \frac{1}{6} + \frac{1}{7}\right) +$ (next 8 terms) $+ \cdots > 1 + \frac{1}{2} + \frac{1}{2} + \frac{1}{2} + \cdots$.

and the complementary error function erfc is

$$\operatorname{erfc}(x) = 1 - \operatorname{erf}(x)\,. \tag{9.31}$$

In practice, $\operatorname{erf}(x)$ can also be evaluated from its relationships to other known functions, like incomplete gamma functions [660] or the normal probability function $f(t) = \exp(-t^2/2)/\sqrt{2\pi}$, since

$$\operatorname{erf}(x/\sqrt{2}) = 2 \int_0^x f(t)\,dt\,,$$

and this integral (area under the standard normal curve from 0 to x) is often tabulated (see CRC Standard Mathematical Tables).

The Screening Gaussian ρ_{G_j}

More generally (for the periodic potential), we perform this splitting using a screening spherical Gaussian ρ_G, centered at each point charge, parameterized by β. The parameter β controls the width of the distribution (width = $\sqrt{2}/\beta$) and also the rate of convergence of the sum (see below). For each point charge, the surrounding distribution is given by:

$$\rho_{G_j}(\mathbf{x}) = -\,q_j \left(\frac{\beta}{\sqrt{\pi}}\right)^3 \exp\left[-\beta^2\,|\mathbf{x}|^2\right]. \tag{9.32}$$

Thus, instead of the cumulative point charge density described by a sum of delta functions:

$$\rho(\mathbf{x}) = \sum_{j=1}^{N} q_j\,\delta(\mathbf{x} - \mathbf{x}_j)\,, \tag{9.33}$$

we use a sum of localized densities based on the Gaussian of eq. (9.32) to define the total Gaussian screening charge density $\rho_G(\mathbf{x})$:

$$\rho_G(\mathbf{x}) \equiv \sum_{j=1}^{N} \rho_{G_j}(\mathbf{x}) = -\sum_{j=1}^{N} q_j \left(\frac{\beta}{\sqrt{\pi}}\right)^3 \exp\left[-\beta^2\,|\mathbf{x} - \mathbf{x}_j|^2\right]. \tag{9.34}$$

The real-space sum can be written in terms of the complementary error function using Poisson's equation[3] for the electrostatic potential, due to the Gaussian charge cloud followed by integration [31]. (See Box 9.2 for a definition of the divergence and Laplace operators). Namely, the solution of Poisson's equation in the form $\nabla^2\Phi_{G_j}(r) = -4\pi\rho_{G_j}(r)$ leads to $\Phi_{G_j}(r) = \frac{q_j}{r}\operatorname{erf}(\beta r)$ after two steps

[3]Simeon Denis Poisson (1781–1840) was a brilliant mathematician whose name frequently appears in text books. The Poisson equation (1812) in potential theory is a result of his discovery that Laplace's equation for the gravitational force holds only at points where no mass is located (see also Box 9.2).

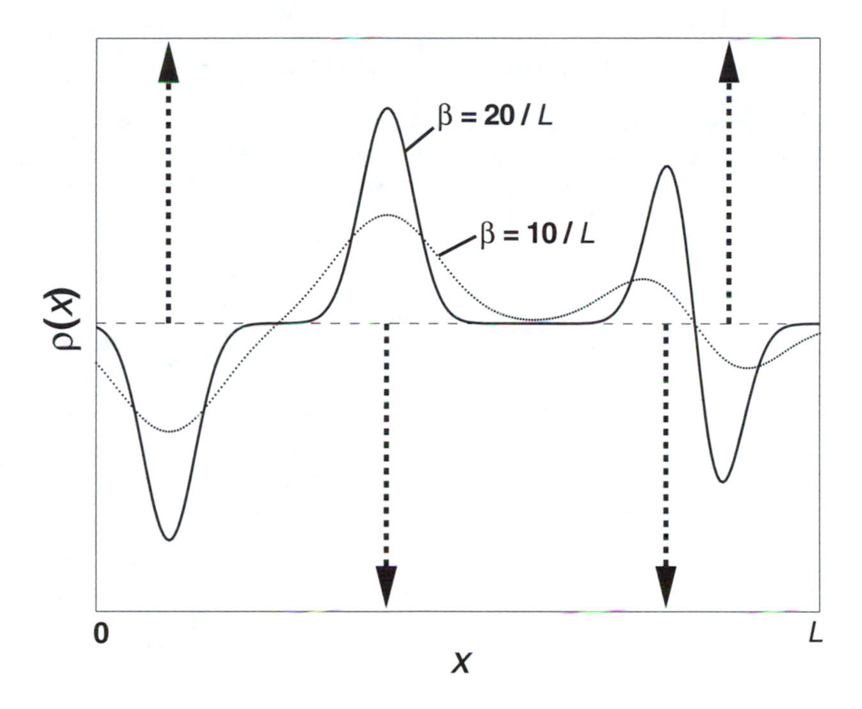

Figure 9.7. Schematic illustration of Ewald's trick for converting a conditionally convergent sum into two terms that are rapidly and absolutely convergent: each point charge (represented by delta functions, dashed spikes) is masked by a Gaussian function; correspondingly, a canceling Gaussian function must be added (not shown). The charge density when point charges are screened by Gaussians is shown for four point charges for two values of β, the parameter controlling the width of the Gaussian distribution.

of integration. Thus, the sum in eq. (9.24) is converted into the pair of terms:

$$E_{\text{coul}} = \frac{1}{2} \sum_{j=1}^{N} \sum_{|\mathbf{n}|}{}' \frac{q_j \operatorname{erfc}(\beta|\mathbf{x}_{ij} + \mathbf{n}|)}{|\mathbf{r}_{ij} + \mathbf{n}|} + \frac{1}{2} \sum_{j=1}^{N} \sum_{|\mathbf{n}|}{}' \frac{q_j \operatorname{erf}(\beta|\mathbf{x}_{ij} + \mathbf{n}|)}{|\mathbf{r}_{ij} + \mathbf{n}|} . \tag{9.35}$$

The first term (E_{real}) is short-range, and the second (E_{recip}) is the smooth long-range component of the Coulomb potential.

Box 9.2: The Divergence and Laplacian Operators

The divergence symbol, as used in the Poisson and Poisson-Boltzmann equations, is a differential operator on a continuously differentiable vector $\mathbf{w} \in \mathcal{R}^n$ whose components $\{w_i\}$ are functions of the coordinate variable $\mathbf{x}$ with components $\{x_i\} \in \mathcal{R}^n$. The divergence of $\mathbf{w}$ is written as div $\mathbf{w}$ or $\nabla \cdot \mathbf{w}$ and defined as:

$$\operatorname{div} \mathbf{w} \equiv \nabla \cdot \mathbf{w} \equiv \frac{\partial w_1(\mathbf{x})}{\partial x_1} + \frac{\partial w_2(\mathbf{x})}{\partial x_2} + \cdots + \frac{\partial w_n(\mathbf{x})}{\partial x_n} .$$

For example, for the gravitational field vector $\mathbf{v}(\mathbf{x})$ at a point $\mathbf{x}$ in space whose distance from the origin is r, we have $\mathbf{v}(\mathbf{x}) = -(GR_e^2/r^3)\,\mathbf{x}$ where R_e is the earth's radius and G is the gravitational constant. For $r > R_e$, it can be shown that the divergence of $\mathbf{v}$ is zero:

$$\begin{aligned} \nabla \cdot \mathbf{v}(\mathbf{x}) &= -\frac{GR_e^2}{r^3}\,\nabla \cdot \mathbf{x} + \nabla \left(\frac{-GR_e^2}{r^3} \right) \cdot \mathbf{x} \\ &= -GR_e^2 \left[\frac{1}{r^3}\,\nabla \cdot \mathbf{x} + \frac{d}{dr}\left(\frac{1}{r^3} \right) \nabla r \cdot \mathbf{x} \right] = 0 , \end{aligned}$$

since $\nabla \cdot \mathbf{x} = 3$ and $\nabla r \cdot \mathbf{x} = (\mathbf{x} \cdot \mathbf{x})/r = r$. The corresponding potential function $u(\mathbf{x}) = GR_e^3/r$, where $\nabla u = \mathbf{v}(\mathbf{x})$, thus satisfies $\nabla \cdot \nabla u = \nabla \cdot \mathbf{v} = 0$ or Laplace's equation, also written as $\nabla^2 u = 0$ or $\Delta u = 0$. The *Laplacian* is a differential operator from scalar to scalar fields whereas the *divergence* is a differential operator from vector to scalar fields.

9.4.4 *Finite-Dielectric Correction*

The decomposition above, while convergent, is strictly correct only for an infinite dielectric ($\epsilon = \infty$) medium or for unit cells with a zero dipole moment. The correction term ($E_{\text{cor},\epsilon}$) for a nonuniform field associated with a macroscopic crystal in a dielectric continuum with external dielectric constant ϵ was only derived 60 years after Ewald's original derivation [659], and yields:

$$E_{\text{cor},\epsilon} = E_{\text{coul}}\,(\epsilon = 1) - E_{\text{coul}}\,(\epsilon = \infty) = \frac{2\pi}{3L^3} \left| \sum_{j=1}^{N} q_j\, \mathbf{x}_j \right|^2 . \quad (9.36)$$

Though reliable, Ewald's algorithm as corrected in [659] was still $\mathcal{O}(N^2)$ in computational complexity. This is because the long-range reciprocal-space Fourier sum requires $\mathcal{O}(N^2)$ to be a sufficiently accurate approximation for large β (see below).

9.4.5 *Ewald Sum Complexity*

The key breakthroughs for reducing the computational complexity of the Ewald sum came in two steps.

Optimization of β

First, by optimizing the parameter β that controls the width of the screening Gaussians (see Figure 9.7), the relative convergence rates of the real and reciprocal-space series are adjusted to optimize the work involved [661]. Namely, as β increases, the real-space sum converges more rapidly and the reciprocal sum more slowly. Both sums are truncated in practice (finite number of terms). Thus, for example, a sufficiently large β yields an accurate direct-space sum with an appropriate truncation, resulting in $\mathcal{O}(N)$ work for the direct-space sum rather than $\mathcal{O}(N^2)$; however, the reciprocal-space sum still requires $\mathcal{O}(N^2)$ work. The optimal work balance between the two components can yield an overall $\mathcal{O}(N^{3/2})$ method by adjusting β [662]. This is much better than $\mathcal{O}(N^2)$ but still considerably expensive for large biomolecular systems.

Mesh Interpolation

The second breakthrough in the Ewald sum came by noting that the trigonometric-function values in the Fourier series used to represent the reciprocal-space term can be evaluated through a smooth interpolation of the potential over a regular grid. The resulting particle-mesh Ewald (PME) method reduces the overall computation to $\mathcal{O}(N \log N)$. The smoothing can be done by Lagrange interpolation [651] or by B-spline interpolation [653].

Variations

Credit for this interpolation idea is due to Hockney and Eastwood who developed several methods in the early 1970s for simulations of hot gas plasmas in fusion machines. Included is a 'particle-particle particle-mesh' (P^3M) scheme, as detailed in their text [663]. The P^3M method for evaluating long-range forces in large systems splits the interparticle force summation into a short-range, rapidly-varying part and a smooth, slowly-varying remainder. A direct 'particle-particle' sum is used to compute the former, while a 'particle-mesh' interpolation procedure is used to approximate the latter on a uniform grid. A 'Q-minimizing' method is also used to optimize the scheme's parameters (given a desired accuracy), such as mesh size, cutoff radius, and charge assignment scheme.

The P^3M method and its cousins are thus closely related to the fast PME schemes used in biomolecular dynamics. Application of the P^3M scheme as described in [663] to molecular dynamics has also demonstrated good performance [664, 665], with guidelines for the scheme's parameters obtained by optimizing an auxiliary function.

9.4.6 *Resulting Ewald Summation*

Following a series of mathematical manipulations (see [666], for example), the resulting composition of the Ewald summation for E_{coul} of eq. (9.24) at $\epsilon = 1$

has five terms:

$$E_{\rm coul} = E_{\rm real} + E_{\rm recip} + E_{\rm cor,self} + E_{\rm cor,ex} + E_{{\rm cor},\epsilon}\,. \qquad (9.37)$$

The first term, $E_{\rm real}$, corresponds to a *real-space* (or direct) sum of the electrostatic energy due to the point charges screened by oppositely-charged Gaussians. The second sum, $E_{\rm recip}$, is the *associated canceling term* (periodic sum of Gaussians) summed in reciprocal space using smooth interpolation of Fourier-series values. The last three terms are correction terms.

The first correction (which is position independent) subtracts the *self-interaction term* (each point charge and its corresponding Gaussian charge cloud) which is included in the first two terms. The second correction subtracts the Coulomb contribution from the *nonbonded pairs excluded* from the Coulomb energy (denoted as pairs $i, j \in$ Ex below) since they are separately accounted for in bonded, bond-angle, and possibly dihedral-angle terms. The third correction accounts for the *non-infinite dielectric medium* (eq. (9.36)).

Following algebraic manipulations, the resulting Ewald sum can be expressed as follows:

$$E_{\rm real} = \frac{1}{2}\sum_{i,j=1}^{N} q_i\,q_j \sum_{|\mathbf{n}|}{}' \frac{{\rm erfc}\,(\beta\,|\mathbf{r}_{ij}+\mathbf{n}|)}{|\mathbf{r}_{ij}+\mathbf{n}|}\,; \qquad (9.38)$$

$$E_{\rm recip} = \frac{1}{2\pi L^3}\sum_{|\mathbf{m}|\neq 0} \frac{\exp(-\pi^2|\mathbf{m}|^2/\beta^2)}{|\mathbf{m}|^2}\,S(\mathbf{m})\,S(-\mathbf{m})\,, \qquad (9.39)$$

$$\text{where} \quad S(\mathbf{m}) = \sum_{j=1}^{N} q_j\,\exp\,[2\pi i\mathbf{m}\cdot\mathbf{x}_j]\,; \qquad (9.40)$$

$$E_{\rm cor,self} = \frac{-\beta}{\sqrt{\pi}}\sum_{j=1}^{N} q_j^2\,; \qquad (9.41)$$

$$E_{\rm cor,ex} = -\frac{1}{2}\sum_{i,j\in{\rm Ex}}^{N} q_i\,q_j\,\frac{{\rm erf}\,(\beta\,|\mathbf{r}_{ij}|)}{|\mathbf{r}_{ij}|}\,; \qquad (9.42)$$

$$E_{{\rm cor},\epsilon} = \frac{2\pi}{3L^3}\left|\sum_{j=1}^{N} q_j\,\mathbf{x}_j\right|^2\,. \qquad (9.43)$$

The erfc function decays to zero with increasing values of the independent variable. If β is sufficiently large, the only term that contributes to the real-space sum is for $|\mathbf{n}| = 0$, and it can be computed in practice with a spherical cutoff of order $b = 10$ Å.

The reciprocal space or Fourier term corresponds to a summation over the reciprocal vectors $\mathbf{m}$, where the Fourier terms $S(\mathbf{m})$ are *charge-weighted structure factors* (see Box 9.1). Eqs. (9.39) and (9.40) can also be written as:

$$E_{\text{recip}} = \frac{1}{2\pi L^3} \sum_{i,j=1}^{N} q_i \, q_j \sum_{|\mathbf{m}| \neq 0} \frac{\exp(-\pi^2 |\mathbf{m}|^2/\beta^2)}{|\mathbf{m}|^2} \exp[2\pi i \mathbf{m} \cdot (\mathbf{x}_j - \mathbf{x}_i)] \,. \tag{9.44}$$

When the charges are interpolated over a uniform set of grid points, these structure factors are easily computed by FFTs.

9.4.7 *Practical Implementation: Parameters, Accuracy, and Optimization*

Gaussian Width

In practice, the Gaussian width parameter β is determined so that the real-space term achieves a desired accuracy tolerance. In typical solvated proteins or DNA simulations, β has an order of magnitude of roughly $10/L$ (L ranges roughly from 60 to 100 Å; values exceeding 70 Å have been implicated with smaller artifacts from the enforced periodicity [646]). The effective cutoff used for the real-space interactions is around 10 to 12 Å. When multiple-timestep schemes are implemented for molecular dynamics, the parameter β (or the cutoff for the direct-space term) may be further optimized to distribute the work for the real and reciprocal terms appropriately, so as to yield the greatest overall speedup; see [667], for example.

Grid Size and Accuracy

The reciprocal-space uses multidimensional piecewise interpolation (e.g., B-splines) for evaluating the Fourier terms, with grid size and number of terms chosen to achieve the desired accuracy. For instance, moderate accuracy (e.g., 10^{-4} relative force error) might be achieved with a coarse interpolation grid (1 to 2 Å). Very high accuracy (e.g., 10^{-10} relative force error) might be obtained with a finer grid ($\sim$ 0.5 Å). The reciprocal-space energy and force terms are expressed as *convolutions* and can thus be evaluated quickly using FFTs.[4] Work has also shown that the finite number of wave vectors used in the discrete approximation gives rise to truncation errors due to the exclusion of intramolecular interactions [668]; this, and the related existence of fast terms in the reciprocal-force component [669, 667, 670, 650], create a problem for development of efficient multiple-timestep integrators for molecular dynamics simulations using PME approximations [671].

[4]For two functions of time $f(t)$ and $g(t)$ and corresponding Fourier transforms $F(f)$ and $F(g)$, we define the *convolution* of the two original functions f and g, $f*g$, as: $f*g = \int_{-\infty}^{\infty} f(\tau)\, g(t-\tau) d\tau$. It can be shown that $F(f*g) = F(f)\,F(g)$. That is, the Fourier transform of the convolution of two functions $(f*g)$ is just the product of the individual Fourier transforms of those functions.

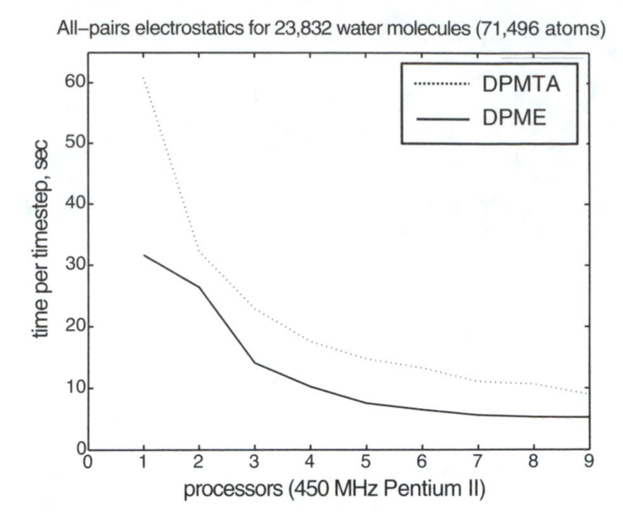

Figure 9.8. CPU time for evaluating the electrostatic energy for 23,832 water molecules (71,496 atoms) by codes developed at Duke by John Board and co-workers for the distributed PME (DPME) and distributed parallel multipole tree algorithm (DPMTA), run at modest accuracy of 10^{-3} relative force accuracy, on a tightly coupled network of 450 MHz Pentium II processors.

Computer Architecture Considerations

In the implementation of the PME method on multiprocessors, the cutoff radius used for the direct sum may be increased to balance the work between the real-space and reciprocal-space components. This accelerates the computations because 3D FFTs are challenging to parallelize well. In contrast, the direct sum parallelizes easily by spatial decomposition. Using a larger cutoff in the real-space sum reduces the number of lattice vectors (Fourier terms) needed for the same accuracy in the reciprocal sum and therefore improves the overall performance.

Experience to date shows that PME implementations for biomolecules — especially when tightly adapted to the computer architecture — are very fast; the best implementations require about the same work as needed to evaluate the same periodic version of the potential but with cutoffs in the range of 10–12 Å [67]. Figure 9.8 shows performance times for an efficiently distributed PME code by John Board and co-workers at Duke University for a huge water system.

9.5 The Multipole Method

In this section, we present a brief introduction to the efficient fast multipole technique. For details, see [672, 673] and other references cited below.

9.5.1 Basic Hierarchical Strategy

Fast multipole techniques are powerful alternatives to the Ewald schemes introduced above for evaluating the pairwise interactions in large molecular systems. They are widely used for many important problems in applied mathematics, engineering, physics, chemistry, and biology [674, 675]. Examples in astrophysics include evaluation of gravitational potentials, and examples in chemistry include electrostatic potentials in molecular dynamics and quantum mechanics (Hartree-Fock calculations).

Series Expansion

Multipole schemes rely on a power-series expansion that describes the interaction between groups of particles (charged bodies in molecular dynamics) [676]. The term *multipole* refers to the *moments* $m^k = \sum_{i=1}^{N} q_i \; x_i^k$ that appear in the expansion, such as dipole and quadrupole. To see this, consider the electrostatic potential Φ at $\mathbf{x}$ for unit dielectric written in terms of the charge distribution about each atom j as (see also eqs. (9.24), (9.25) for the periodic version):

$$\Phi(\mathbf{x}) = \Phi(\mathbf{x}_1, \mathbf{x}_2, \ldots, \mathbf{x}_N) = \frac{1}{2} \sum_{\{i,j\},\, i \neq j} q_i \, q_j / |\mathbf{r}_{ij}| = \frac{1}{2} \sum_j q_j \, \Phi(\mathbf{x}_j) \,, \tag{9.45}$$

where

$$\Phi(\mathbf{x}_j) = \sum_{i \neq j} q_j / |\mathbf{r}_{ij}| \,. \tag{9.46}$$

An expansion of $\Phi(\mathbf{x}_j)$ yields (see Box 9.3):

$$\Phi(\mathbf{x}_j) = \Phi(0) + \nabla \Phi^{\mathrm{T}} \mathbf{x} + \frac{1}{2} \mathbf{x}^{\mathrm{T}} \mathbf{H} \mathbf{x} + \mathcal{O}(|\mathbf{x}|^3) \tag{9.47}$$

where the derivatives are evaluated at $\mathbf{x} = 0$. This expansion produces expressions for $\Phi(0)$ based on the *zeroth moment* ($\Phi(0) = m^0 / |\mathbf{x}_j|$), for $\nabla \Phi(\mathbf{x})$ in terms of the *dipole moment* (vector $\mu = [m_1^1, m_2^1, m_3^1]$), and for the second-derivative term based on the *quadrupole moment* (elements $m_{11}^2, m_{12}^2, m_{13}^2, m_{21}^2, \cdots, m_{33}^2$). See Box 9.3 for details.

Note that multipole expansions can be expressed using Cartesian coordinations, as in Boxes 9.3 and 9.4, or spherical coordinates, as in Subsection 9.5.3. Such power series can save computational work since the target function can be written as a linear combination of moments (see Box 9.4 for a simple example).

Domain Decomposition

Accelerated multipole algorithms use a hierarchy of approximations to represent these interactions on the basis of a spatial particle partitioning in a tree-like structure (typically oct-tree): the original domain (level 0) is subdivided into 8 level-1 domains, leading to 64 regions in level 2, and so on (see also illustration in [677]). Thus, each successive refinement of the computational domain produces 'offspring' corresponding to the 'parents' at the prior level. This recursive partitioning, like FFTs, works together with a systematic formulation and manipulation of power series expansions of the target potential to reduce the evaluation computational complexity from $\mathcal{O}(N^2)$ to the more modest $\mathcal{O}(N \log N)$ and $\mathcal{O}(N)$, depending on the implementation (see Figure 9.9).

Multipole expansions represent one suitable choice of power series, since they converge well when groups of particles (charges) are well-separated, allowing a small number of coefficients in the expansions while maintaining good accuracy; other expansions are possible [678, 677, 679, 680, 681]. These power series expansions are used to compute interactions between well-separated pairs (contained in well-separated clusters). Interactions between nearby particles are computed directly.

Summation Protocol

The various algorithms differ by how the tree structure is generated and by the protocol used to determine which power-series approximation to invoke at every stage (on the basis of distance separations monitored by interaction lists). Figure 9.9, for example, illustrates the definition of interaction lists for multipole expansions by boxes beyond first neighbors with respect to a given particle (shown in red); starting from second neighbors is an alternative definition.

Generally, good efficiency — linear complexity for moderate-sized systems — for 3D implementations requires meticulous programming and algorithmic structure [673]. This programming sophistication is especially important for applications where the particles are distributed *heterogeneously* through the computational domain (usually not the case for molecular dynamics applications). For heterogeneous cases, *adaptive* schemes are needed to distribute the computations in a balanced and efficient manner since regular subdivisions may generate some empty cells and empty cells need not be divided.

The seasoned programming required for fast multiple algorithms partly explains why Ewald codes have been more popular in the molecular dynamics community over the last few years. It is possible that as larger systems (of order 10^5 atoms) become more common in macromolecular simulations, the difference between the $\mathcal{O}(N \log N)$ (best Ewald implementation) and $\mathcal{O}(N)$ (best multipole implementation) may become significant, and the effort required for multipole codes may well be worth the programming investment.

Besides Coulomb interactions, multipole (and Ewald) methods apply more generally to any function that can be approximated by a converging power-series expansion, such as functions of $1/r^n$ or exponentials as in screened Coulomb (Debye-Hückel) expressions $(\exp(-\kappa r)/r)$ [682, 683] (see below).

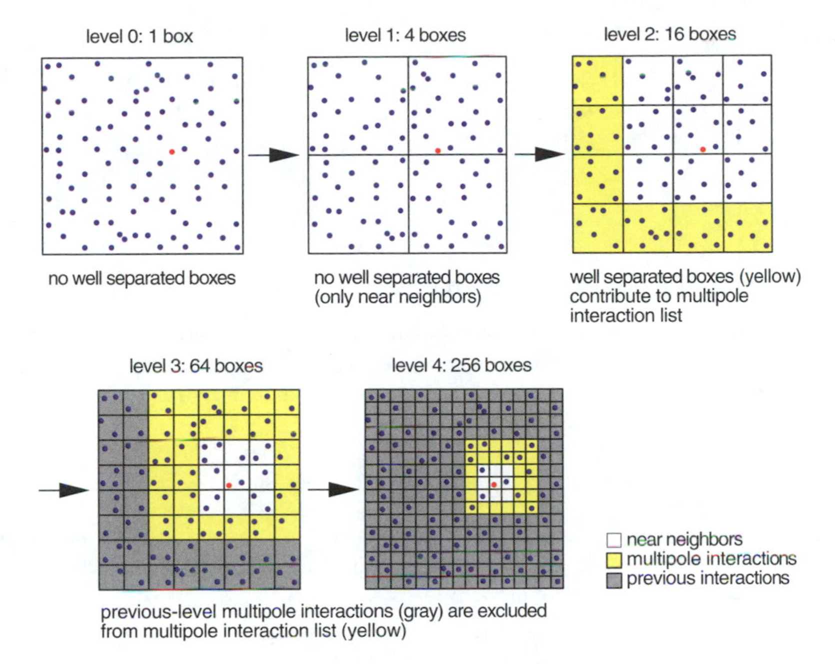

Figure 9.9. Hierarchical domain partitioning approach for fast multipole schemes. The partitioning is illustrated with respect to a central particle (shown in red).

Box 9.3: Multipole Expansion

Consider a charge distribution about each atom j (in a system of N atoms) as defined in eq. (9.46), where $\mathbf{r}_{ij} = \mathbf{x}_i - \mathbf{x}_j = [x_{i1} - x_{j1}, x_{i2} - x_{j2}, x_{i3} - x_{j3}]$, $r_{ij} = |\mathbf{r}_{ij}|$, and $R_{ij} = (r_{ij})^2$. An expansion at position $\mathbf{x}$ about the origin can be written as

$$\begin{aligned}
\Phi(\mathbf{x}_j) &= \sum_{i=1}^{N} \frac{q_i}{[(x_{i1} - x_{j1})^2 + (x_{i2} - x_{j2})^2 + (x_{i3} - x_{j3})^{\,2}\,]^{\,1/2}} \\
&= \Phi(0) + \sum_{i=1}^{N}\sum_{k=1}^{3} \frac{\partial \Phi}{\partial x_{ik}}\, x_{ik} + \frac{1}{2}\sum_{i=1}^{N}\sum_{l=1}^{3}\sum_{k=1}^{3} \frac{\partial^2 \Phi}{\partial x_{ik}\partial x_{il}}\, x_{ik}x_{il} + \text{higher order terms} \\
&\equiv \Phi(0) + \nabla\Phi^{\mathrm{T}}\mathbf{x} + \frac{1}{2}\mathbf{x}^{\mathrm{T}}\mathbf{H}\mathbf{x} + \mathcal{O}(|\mathbf{x}|^3)
\end{aligned}$$

where derivatives are evaluated at $x_{i1} = x_{i2} = x_{i3} = 0$. The gradient vector $\nabla\Phi$ and Hessian matrix $\mathbf{H}$ above contain partial derivatives corresponding to the $3N$ components of the vector $\mathbf{x}$. Namely, $\nabla\Phi$ has the $3N$ components $\{\partial\Phi/\partial x_{11}, \partial\Phi/\partial x_{12}, \cdots, \partial\Phi/\partial x_{N3}\}$, and the Hessian is:

$$\mathbf{H} = \begin{pmatrix} \partial^2\Phi/\partial x_{11}^2 & \partial^2\Phi/\partial x_{11}\partial x_{12} & \cdots\cdots & \partial^2\Phi/\partial x_{11}\partial x_{N3} \\ \partial^2\Phi/\partial x_{12}\partial x_{11} & \partial^2\Phi/\partial x_{12}^2 & \cdots\cdots & \partial^2\Phi/\partial x_{12}\partial x_{N3} \\ \vdots & \vdots & \vdots & \vdots \\ \vdots & \vdots & \vdots & \vdots \\ \vdots & \vdots & \vdots & \vdots \\ \partial^2\Phi/\partial x_{N3}\partial x_{11} & \partial^2\Phi/\partial x_{N3}\partial x_{12} & \cdots\cdots & \partial^2\Phi/\partial x_{N3}^2 \end{pmatrix}.$$

The *moments* $\{m^k\}$ for integers k corresponding to the partial charges $\{q_i\}$ are defined by $m^k = \sum_{i=1}^{3N} q_i \, x_i^k$ where the summation extends over all components of $\mathbf{x}$. The first term in the above expansion is written in terms of the zeroth moment m^0:

$$\Phi(0) = \sum_{i=1}^{N} \frac{q_i}{[x_{j1}^2 + x_{j2}^2 + x_{j3}^2]^{\ 1/2}} = \sum_{i=1}^{N} \frac{q_i}{|\mathbf{x}_j|} = \frac{m^0}{|\mathbf{x}_j|}.$$

The gradient, or the *dipole* potential, has 3 components: m_1^1, m_2^1, m_3^1. They represent the *dipole moment* μ. The expression for the gradient in terms of the dipole moments can be derived following the relation for the derivatives:

$$\frac{\partial R_{ij}^{-1/2}}{\partial x_{ik}} = -\frac{1}{2}\, R_{ij}^{-3/2}\, \frac{\partial R_{ij}}{\partial x_{ik}} = -r_{ij}^{-3}\, (x_{ik} - x_{jk})\Big|_{x_{ik}=0} = x_{jk} \,/\, |\mathbf{x}_j|^3 .$$

We then have:

$$\begin{aligned} \Phi'(\mathbf{x}_j) &= \nabla\Phi^{\mathrm{T}}\mathbf{x} = \sum_{i=1}^{N} q_i \left(\frac{\partial R_{ij}^{-1/2}}{\partial x_{i1}}\, x_{i1} + \frac{\partial R_{ij}^{-1/2}}{\partial x_{i2}}\, x_{i2} + \frac{\partial R_{ij}^{-1/2}}{\partial x_{i3}}\, x_{i3} \right) \\ &= \sum_{i=1}^{N} \left(q_i \, x_{i1} \frac{x_{j1}}{|\mathbf{x}_j|^3} + q_i \, x_{i2} \frac{x_{j2}}{|\mathbf{x}_j|^3} + q_i \, x_{i3} \frac{x_{j3}}{|\mathbf{x}_j|^3} \right) \\ &= \frac{1}{|\mathbf{x}_j|^3} \left(m_1^1 \, x_{j1} + m_2^1 \, x_{j2} + m_3^1 \, x_{j3} \right) = \frac{(\mu^{\mathrm{T}} \cdot \mathbf{x}_j)}{|\mathbf{x}_j|^3} . \end{aligned}$$

The second-derivative term, the *quadrupole potential*, has six terms (because of symmetry), again with all partial derivatives evaluated at $x_{i1} = x_{i2} = x_{i3} = 0$:

$$\begin{aligned} \Phi''(\mathbf{x}_j) &= \frac{1}{2}\mathbf{x}^{\mathrm{T}} H \mathbf{x} = \frac{1}{2}\sum_{i=1}^{N} q_i \left(\frac{\partial^2 R_{ij}^{-1/2}}{\partial x_{i1}^2}\, x_{i1}^2 + \frac{\partial^2 R_{ij}^{-1/2}}{\partial x_{i2}^2}\, x_{i2}^2 + \frac{\partial^2 R_{ij}^{-1/2}}{\partial x_{i3}^2}\, x_{i3}^2 \right. \\ &\quad \left. + \ 2\, \frac{\partial^2 R_{ij}^{-1/2}}{\partial x_{i1}\partial x_{i2}}\, x_{i1}x_{i2} + 2\, \frac{\partial^2 R_{ij}^{-1/2}}{\partial x_{i1}\partial x_{i3}}\, x_{i1}x_{i3} + 2\, \frac{\partial^2 R_{ij}^{-1/2}}{\partial x_{i2}\partial x_{i3}}\, x_{i2}x_{i3} \right) \\ &= \frac{1}{2} \left(m_{11}^2 \frac{\partial^2 R_{ij}^{-1/2}}{\partial x_{i1}^2} + m_{22}^2 \frac{\partial^2 R_{ij}^{-1/2}}{\partial x_{i2}^2} + \ \ldots \ + 2\, m_{23}^2 \frac{\partial^2 R_{ij}^{-1/2}}{\partial x_{i2}\partial x_{i3}} \right) . \end{aligned}$$

Box 9.4: Example of Work Reduction by Multipole Expansion

Let our potential be defined as:

$$\Phi(x) = \sum_{j=1}^{N} \Phi(x_j) = \sum_{j=1}^{N} \sum_{i=1}^{N} q_i \, (x_j - y_i)^3 \, .$$

That is, $\Phi(x)$ is the sum of the following components:

$$\begin{array}{llllll}
\{\Phi(x_1)\} & q_1 \, (x_1 - y_1)^3 + q_2 \, (x_1 - y_2)^3 & + & \ldots & + q_N \, (x_1 - y_N)^3 \\
\{\Phi(x_2)\} & q_1 \, (x_2 - y_1)^3 + q_2 \, (x_2 - y_2)^3 & + & \ldots & + q_N \, (x_2 - y_N)^3 \\
\vdots & \vdots & \vdots & & \vdots \\
\{\Phi(x_N)\} & q_1 \, (x_N - y_1)^3 + q_2 \, (x_N - y_2)^3 & + & \ldots & + q_N \, (x_N - y_N)^3 \, .
\end{array}$$

Evaluation by *straightforward* summation requires $\mathcal{O}(N^2)$ work, but our function of x and y simplifies as:

$$(x - y)^3 = x^3 - 3x^2 y + 3xy^2 - y^3 \, .$$

This is a finite power series, with degree $p = 3$ and moments $m^k = \sum_{i=1}^{N} q_i \, y_i^k$. Therefore we can rewrite each $\Phi(x_j)$ as a linear combination of the moments $\{m^k\}$ for k = 0, 1, 2, 3:

$$\Phi(x_j) \quad = \quad x_j^3 \, m^0 \; - 3x_j^2 \, m^1 \; + 3x_j \, m^2 \; - m^3 \, .$$

The work can thus be substantially reduced: once the moments are computed in $\mathcal{O}(Np)$ work, evaluating each $\Phi(x_j)$ requires only $\mathcal{O}(p)$ work, where p is the power of the expansion, 3 here. *This reduction $\mathcal{O}(N^2)$ to $\mathcal{O}(3N)$ is significant!*

9.5.2 *Historical Perspective*

Hierarchical Refinements

Coming from the astrophysics community, Appel first introduced the multipole approach for solving the N-body problem by a hierarchical, power-series approach [684]. Barnes and Hut accelerated the association scheme between particles at successive refinement levels to produce a method that is asymptotically $\mathcal{O}(N \log N)$ [685]. To see this let $\mathbf{D}_{N \times N}$ be the matrix defined by

$$\mathbf{D}_{ij} = \begin{cases} q_i / |\mathbf{r}_{ij}| & i \neq j \\ 0 & i = j \end{cases} \tag{9.48}$$

and $\mathbf{q}$ be the vector of N partial charges. Hence, the potential for each atom j due to the charges induced by all other atoms is

$$q_j \, \Phi(\mathbf{x}_j) = q_j \sum_{i \neq j} q_i / |\mathbf{r}_{ij}| = \{\mathbf{Dq}\}_j \, ,$$

the jth component of the matrix/vector product $\mathbf{Dq}$. Summing up the values of the components of the resultant product (N potentials evaluated at N points) yields the desired potential. This is clearly $\mathcal{O}(N^2)$ computation by a straightforward matrix/vector multiplication. However, if the potential Φ is sufficiently smooth when distances are sufficiently large, the matrix $\mathbf{D}$ can be approximated by low-rank submatrices; in this case, an $\mathcal{O}(Np)$ scheme results where p is the rank of the approximating matrices. See also the simple example in Box 9.4 and Figure 9.9.

Hierarchical Protocol

This $\mathcal{O}(N \log N)$ method works roughly as follows.

A hierarchy of boxes is introduced in the computational domain so that each refinement level l is a subdivision (e.g., into 8) of the domain in level $l-1$. Boxes at the same refinement level that share a boundary are considered *near neighbors*, and those at the same refinement level that do not share a boundary are considered *well-separated*.

As the algorithm sweeps through refinement levels, a multipole expansion is associated with each box at that level to describe the far-field potential contribution from the particles in that box. (The clusters that contribute to this far field are determined by an *interaction list*). These multipole expansions for the clusters are then used to compute interactions between distant clusters of particles, with nearby interactions computed directly.

This clustering is at the heart of such hierarchical methods: the computation for the interactions between particles is recast as work between clusters containing groups of particles.

$\mathcal{O}(N \log N)$ Work

The recursive computation described above is completed in about $\log_8 N$ steps (levels), leading to total work of $\mathcal{O}(N \log N)$. However, the *constant* associated with ($N \log N$) depends on the number of operations required to form the expansions. This cost of the local expansions depends on the number of coefficients p used (like the number of Fourier terms in FFTs). This number can be specified based on the desired accuracy, $\epsilon_{\rm acc}$, for the resulting potential.

It can be shown that p can be related to $\epsilon_{\rm acc}$ as

$$p = \log_{\sqrt{3}}(1/\epsilon_{\rm acc})$$

at sufficiently large separations. The corresponding total work is approximately $200\, Np^2 \log_8 N$ [673].

Fast Multipole Machinery ($\mathcal{O}(N)$)

Unfortunately, these estimates do not produce substantial speedups in practice with respect to direct calculations (considering also the additional bookkeeping work required for multipole schemes), since $p \approx 20$ for 7 digits of accuracy. Significant speedups are achieved only for very large system sizes.

Independently of the Barnes and Hut scheme, the fast multipole method was also developed in the mid 1980s by Rokhlin [686]. Greengard and Rokhlin then made a seminal contribution [672] by developing further mathematical machinery that exploits the smoothness of the far-field potential and works with the tree codes. Relying on translation operators that act on the distant (multipole) as well as local expansions (harmonics), they lumped and converted expansions associated with several clusters into local expansions. The cost of translating the multipole-to-local expansions is reduced via fast schemes for application of rotation matrices. This combined clever mathematical and numerical machinery yields an asymptotically $\mathcal{O}(N)$ method. The linear constant depends sensitively on the practical algorithmic implementation.

9.5.3 Expansion in Spherical Coordinates

The multipole expansion is most conveniently written in spherical coordinates. A point $\mathbf{x}$ in 3D space can be represented by the triplet $\{r,\ \theta,\ \phi\}$ instead of the Cartesian triplet $\{x,\ y,\ z\}$, where:

$$r = \sqrt{x^2 + y^2 + z^2}, \qquad \theta = \cos^{-1}(z/r), \qquad \phi = \tan^{-1}(y/z).$$

Consider N charges located at points represented in spherical coordinates as $\{\rho_i,\ \alpha_i,\ \beta_i\}$ lying inside a sphere of radius a, i.e., with $|\rho_i| < a$ for all i. Then for any 3D point $\mathbf{x} \equiv \{r,\ \theta,\ \phi\}$ outside that sphere, i.e., with $r > a$, the potential at $\mathbf{x}$ can be written in terms of *spherical harmonics* functions, $\{Y_n^m\}$, solutions of the Laplace equation in spherical coordinates, as follows:

$$\Phi(\mathbf{x}) = \sum_{n=0}^{\infty} \sum_{m=-n}^{+n} \frac{M_n^m}{r^{n+1}}\ Y_n^m(\theta, \phi)\,. \tag{9.49}$$

The functions $Y_n^m(\theta, \phi)$ are called the *spherical harmonics of degree* $-(n+1)$, or *multipoles*, and the coefficients M_n^m are known as *moments of the expansion*:

$$M_n^m = \sum_{i=1}^{k} q_i\ \rho_i^n\ Y_n^{-m}(\alpha_i, \beta_i)\,. \tag{9.50}$$

The spherical harmonics functions can be expressed in terms of partial derivatives of $1/r$ from the following relations [687]:

$$Y_n^m(\theta,\phi) = \begin{cases} r^{n+1}\ A_n^m\ \frac{\partial^n}{\partial z^n}\left(\frac{1}{r}\right) & m = 0 \\ r^{n+1}\ A_n^m\ \left(\frac{\partial}{\partial x} + i\frac{\partial}{\partial y}\right)^m\ \left(\frac{\partial}{\partial z}\right)^{(n-m)}\left(\frac{1}{r}\right) & m = 1, 2, \ldots \\ r^{n+1}\ A_n^{-m}\ \left(\frac{\partial}{\partial x} - i\frac{\partial}{\partial y}\right)^{-m}\ \left(\frac{\partial}{\partial z}\right)^{(n+m)}\left(\frac{1}{r}\right) & m = -1, -2, \ldots \end{cases} \tag{9.51}$$

where

$$A_n^m = \frac{(-1)^n}{\sqrt{(n-m)!(n+m)!}}\,. \tag{9.52}$$

The spherical harmonic functions $\{Y_n^m(\theta, \phi)\}$ are also related to the *associated Legendre polynomials of degree* n, $\{P_n^m\}$, defined as

$$P_n^m(x) = (-1)^m \, (1 - x^2)^{m/2} \, \frac{d^m}{dx^m} P_n(x)$$

by

$$Y_n^m(\theta, \phi) \equiv \sqrt{\frac{(n - |m|)!}{(n + |m|)!}} \, P_n^{|m|} \, (\cos\theta) \, \exp(im\phi) \, .$$

The expansion power p is chosen to achieve the desired accuracy so that the remainder

$$\left| \sum_{n=p+1}^{\infty} \sum_{m=-n}^{+n} \frac{M_n^m}{r^{n+1}} \, Y_n^m(\theta, \phi) \right| \quad \leq \quad \frac{\sum_{i=1}^N |q_i|}{r - a} \, \left(\frac{a}{r}\right)^{p+1} \quad = \quad \epsilon_{\rm acc} \, . \tag{9.53}$$

Thus to achieve greater accuracy, we can increase the expansion power p or decrease the radius a. For example, if $r = 2a$, the remainder above is $(2^{-(p+1)}/a) \sum_i |q_i|$. Setting $p = \log_2(\epsilon_{\rm acc}^{-1})$ yields an accuracy of $\epsilon_{\rm acc}$ relative to $\sum_i |q_i|/a$.

The spherical harmonics functions are believed to provide a more efficient basis than Cartesian expansions, since they involve p^2 coefficients rather than p^3.

9.5.4 *Biomolecular Implementations*

Thorough comparisons of the balance among speed, accuracy, and scalability of Ewald and fast multipole schemes are ongoing areas of research. The fast multipole approach was first implemented for biomolecular dynamics in the early 1990s [688, 689], and later parallelized to achieve good speedup on a large number of processors [690, 691, 67], both on workstation clusters and on supercomputers like Cray T3D/T3E. A periodic version of the fast multipole method relies on machinery similar to that used for the periodic Ewald summation [692], namely obtaining a convergent sum by masking the original sum by Gaussians.

Implementations in 3D of the periodic fast multipole method were adapted for, and applied to, molecular dynamics by Board and co-workers in the late 1990s [693, 67] and by Figueirido *et al.* [694], as well as compared to the PME alternative.

In 3D applications, implementors have found it particularly difficult to achieve good performance with high accuracy. For example, in [694] it has been reported that the multipole approach is three times *slower* than PME for about 6000 atoms and slightly slower than PME for about 22,000 atoms; it is suggested, however, that fast multipoles will be competitive for larger sizes.

In serial test implementations, Board and collaborators found that the periodic version of the fast multipole method is still significantly slower than PME by roughly a factor of two for a large water system of 71,496 atoms [67]; this performance involves modest accuracy (4–5 digits of accuracy in the potential and 3–4 in the corresponding force) and required 8 multipole terms. This performance result may be explained by the very fast, 'custom' version of PME of Board and co-workers which uses a table lookup for the erfc function and yields equivalent run times for calculations involving spherical cutoffs of 10–12 Å. Work for distributed PME versus distributed parallel multipole tree code, run at the modest accuracy, still shows PME to be faster on a small number of processors (Figure 9.8).

The newer version of the fast multipole method extends the 2D scheme of translation operators acting on harmonic functions to produce a new diagonal form for these operators [673]. This approach produces a more modest break-even system size in 3D when compared to direct calculations, such as 2000 particles for single-precision accuracy or 5000 for 10-digit accuracy. Application of this fast multipole method to biomolecular dynamics should be forthcoming.

9.5.5 Other Variants

Besides these commonly-used fast multipole and Ewald methods, scientists have also developed variants as well as new methods. Summations based on multigrid techniques developed for partial differential equations, as implemented for the fast evaluation of integral transforms [695], were extended to molecular electrostatic potentials [696, 697]. A variable-order extension of Appel's algorithm in Cartesian coordinates was also reported, with adaptive tree structure and error control [698, 699].

9.6 Continuum Solvation

9.6.1 Need for Simplification!

The fast electrostatic techniques described above have been designed and are necessary for 'deluxe' models of macromolecular systems — very detailed solvated biomolecular representations involving thousands of explicit water molecules surrounding the biomolecule or biomolecules. This large size mimics the cellular environment and is needed to reproduce the bulk properties of water as well as the effects of local solvation on important associations such as those between two proteins, proteins and ligands, and proteins and nucleic acids.

While such elaborate models can yield invaluable information on the rich complexity of biomolecular environments, they are clearly computationally-intensive and, unfortunately, free neither from approximations nor artifacts. As described previously, artificial periodicity can lead to noncanceling errors associated with the Coulomb and solvation contributions to the electrostatic free energy [646] (see also below). Trajectories are also highly sensitive to various force-field approximations, propagation protocols, and modeling choices (e.g., positions of ions and water molecules); see discussion in the chapters on molecular dynamics.

A thorough introduction to implicit solvation models (with emphasis on quantum-mechanical based approaches) can be found in [700], which also contains a literature survey of relevant works; see also [701, 702] for reviews of Poisson-Boltzmann and generalized Born models.

9.6.2 Potential of Mean Force

Implicit solvent models based on *continuum electrostatics* treatments form an alternative to this deluxe approach to modeling [703, 704, 701, 702]. These methods essentially approximate a *potential of mean force* [705] for the solvated biomolecular environment and can provide a broad, qualitative and quantitative view of overall biomolecular structure, recognition, and functional properties.

Balancing Biophysics with Numerics

The potential of mean force is an effective free energy potential that depends on state variables like temperature and pressure [705]. It augments the potential energy used to describe the intramolecular interactions by indirectly incorporating the effects of rearrangement in the medium on the biomolecule (solute). This effect is important to consider since water molecules and ions in the solute environment perturb, in a cooperative manner, the pairwise forces acting on pairs of solute particles through their own locations. The basic idea is to construct a function that captures in an average sense solute configurations without explicit consideration of the solvent degrees of freedom. Additionally, hydrodynamics can be applied to provide continuum approximations to the effects on the solute of the motions of the solvent molecules. In this way, the model complexity can be compromised with physical reproducibility/reliability.

Electrostatic and Non-Electrostatic Components

When such potentials of mean force are constructed, they are generally separated into two components: *electrostatic* and *non-electrostatic* (or nonpolar) interactions [704, 646]. This decomposition is useful for dissecting the different microscopic factors that influence molecular conformations and solvation. The electrostatic component can also be related to various continuum electrostatic approximations (see below).

Variations

There are many ways to construct the effective potential of mean force for this broader structural and functional description, both empirically and theoretically.

Empirical constructs lump all effects into an 'information based' or 'statistical' potential, derived, for example, from observed protein structures in solution [706].

Other approaches involve approximations of integral equations [707], free energy simulations, solvent-accessible surface models, various combinations of implicit/explicit solvent representations [708, 709, 710], generalized Born models

[711, 373], solutions based on classical continuum electrostatics (i.e., Poisson-Boltzmann equation) [712] and quantum mechanics [700], and phenomenological dynamic models such as Langevin and Brownian dynamics (BD) [703]. Generalized Born models, in particular, construct a dielectric cavity surrounding the molecular charges to approximate solvation and salt-screening effects. In all cases, the dielectric constant must be chosen with care [713].

A good overview of the state-of-the-art in methods that implicitly account for solvation effects and their applications is available in *Biophysical Chemistry*, volume 78 (5 April 1999), guest edited by Benoît Roux and Thomas Simonson, with updates in [701, 702]. The study in [714] shows that implicit solvent models yield reasonable agreement in terms of protein/ligand binding free energies when compared to their much more computational-costly explicit-solvent models. Applications to membrane systems [702] reveal the progress of modeling such heterogeneous materials by implicit solvation models but also some computational problems (e.g., artificial periodicity imposed by Ewald sums and related methods) that require further work to characterize and resolve.

Below, we sketch two approaches, based on Langevin dynamics and continuum electrostatics.

9.6.3 Stochastic Dynamics

The theory of hydrodynamics has a long history in molecular modeling, originating from the study of liquids, polymers, and simple molecular reactions [715, 716, 717, 718, 719]. These theories have been applied to macromolecular dynamics via Langevin and Brownian dynamics simulations that generate *stochastic trajectories*, so called because the governing dynamic equations include stochastic forces that mimic solvent effects, in addition to the systematic force (negative gradient of the potential energy).

See classic statistical mechanics texts such as [29] for extensive background, and Chapter 13 of this text for a brief discussion of generating stochastic trajectories by Langevin and Brownian dynamics algorithms; a thesis by Hongmei Jian [512] nicely summarizes the theory and numerical applications of Langevin and Brownian dynamics to long DNA molecules.

In the stochastic treatment, the influence of solvent particles on the solute is incorporated through additional frictional and random terms in a manner consistent with physical laws regarding equilibrium and nonequilibrium processes (e.g., equilibrium conformation distributions, fluctuation/dissipation theorem) [29]. Applications of Langevin and Brownian dynamics simulations have been particularly successful for macroscopic models of biomolecules, such as long DNA of thousands of base pairs [720, 721, 516, 722]. In such applications, polymer theory has been used to guide parameterization (for hydrodynamic radii, timesteps, etc.) through governing macroscopic polymer properties [487] such as persistence length, radius of gyration, and diffusion constants. Protein applications include long-timescale enzyme catalysis events such as the loop opening/closing motion in triosephosphate isomerase (TIM) [723, 724, 725, 726].

The Langevin Equation

In the simplest form of the Langevin equation, the friction kernel is taken to be space and time independent for each particle, and the influence of the environment on the systematic, internal force is represented in an average sense. Thus, explicit hydrodynamic interactions are ignored, and the internal force is augmented by a frictional term, proportional to the velocity, and a random force R, which crudely mimics molecular collisions and viscosity in the realistic cellular environment:

$$\mathbf{M}\frac{d^2\mathbf{x}}{dt^2} = -\nabla E(\mathbf{x}) - \mathbf{M}\gamma\frac{d\mathbf{x}}{dt} + R(t). \tag{9.54}$$

Here E is the potential energy governing the solute, γ is the damping constant (or collision frequency), and $R(t)$ is the 'white noise' vector with mean, $\langle R(t)\rangle$, of zero.

From classic theories of Brownian motion, it can be shown that although molecular collisions are random, the ensemble of these collisions produces a systematic effect. In other words, random motions exist at thermal equilibrium as a fluctuation. It follows that the frictional force and the random force are related by the *fluctuation/dissipation* theorem [727]. This relation can be expressed by the γ-dependence of the covariance of R:

$$\langle R(t)R(t')^T\rangle = 2\gamma k_B \mathrm{T}\mathbf{M}\delta(t-t') \tag{9.55}$$

(k_B is Boltzmann's constant and T is the temperature). Here the covariance matrix is diagonal since hydrodynamic interactions between particles have been discounted. The damping constant γ controls both the magnitude of the frictional force and the variance of the random forces. It thus ensures that the system converges to a Boltzmann distribution characterized by the temperature T. The larger the value of γ, the greater the influence of the surrounding fluctuating force (solvent).

In the limit of small γ, the motion is termed *inertial*, and in the limit of large γ it is *diffusive* or *Brownian*; see [500] for vivid illustrations of those regimes for models of supercoiled DNA. Different integration algorithms are generally applied depending on the relevant regime (e.g., [728, 729, 730]). See also Chapter 13 for a simple discretization [728] of the above Langevin equation and other discretizations in [729, 730].

The stochastically modeled system reaches the same equilibrium as the original system obeying $\mathbf{M}\ (d^2\mathbf{x}/dt^2) = -\nabla E(\mathbf{x})$, but the *rate* at which equilibrium is reached depends on the viscous coupling to the environment. Since the number of degrees of freedom in the Langevin model is the number corresponding to the solute particles, the model is computationally much cheaper than the corresponding all-atom representation which includes explicit solvent.

Langevin Parameters from Hydrodynamic and Other Considerations

Guidelines for choosing numerical values for γ come from hydrodynamic theory. Stokes' law describes how the frictional resistance of a spherical particle in solution varies linearly with its radius: the effective force magnitude is $6\pi\eta a_r$ times the particle's velocity, where a_r is the hydrodynamic radius of each spherical particle and η is the solvent viscosity. Stokes' law is frequently applied to particles of molecular size. Hence, γ in the Langevin equation can be set to

$$\gamma = 6\pi\eta a_r/m\,,$$

where m is the particle's mass.

From a modeling point of view, it is also possible to choose γ so that the resulting translational diffusion constants D_t match the the experimental values in the *diffusive limit* (large γ). This follows the relationship

$$D_t = k_B\mathrm{T}/\sum_i m_i\gamma\,,$$

where the summation extends over all masses m_i in the system. Note that this expression for D_t reduces to

$$D_t = k_B\mathrm{T}/6\pi\eta a_r$$

when Stokes' law for setting γ is used. This is the Stokes-Einstein law of diffusion for a Brownian particle; see homework assignment 12 for an example.

A *computationally efficient* value for γ can also be selected from practical considerations, since an optimal coupling of the system to the thermal reservoir can accelerate configurational sampling during finite-length simulations [731, 500, 732]. This choice of γ implies relinquishing the idea of approximating the 'true' dynamics in favor of efficient sampling.

The Brownian Limit

In the diffusive limit of the Langevin equation, the motion is more random or *Brownian* in character. Such motion characterizes in a global sense a dense system in which the solute collides often with the surrounding fluid particles, and is thus continuously and significantly reoriented by the solvent molecules. Specifically, the Brownian regime assumes that the velocity relaxation time is much more rapid than position relaxation time.

Theories for Brownian motion, dating from Einstein's work (circa 1905), have been based on the generalized Langevin equation and on the Fokker-Planck equation, a partial differential equation that describes the evolution of a system in a probabilistic sense [29].

Practical BD algorithms [733, 703, 734] are similar to Monte Carlo procedures (see Chapter 11) in that the current position is perturbed by a random displacement vector. However, this random displacement is more complicated to formulate, as it depends on the forces and the diffusion tensor. See the brief section in Chapter 13 on Brownian dynamics algorithms and illustrations at the end of Chapter 6 for long DNA and polynucleosomes.

When very large biomolecular polymers are modeled, such as long DNA of thousands of base pairs, a more detailed account of hydrodynamic interactions is required to accurately represent the changes in solute forces induced by the flow of the surrounding fluid particles. This is done by formulating a position-dependent hydrodynamic tensor $\mathbf{T}$ related to the diffusion tensor $\mathbf{D}$ instead of the γ-dependent friction term in the simple Langevin equation.

The Brownian methods that incorporate hydrodynamic effects via a tensor that is configuration dependent yield better descriptions of large polymers in solution and can reach longer time frames such as milliseconds [516, 722]. Still, the computational complexity grows rapidly, as $\mathcal{O}(N^3)$, with the number of modeled beads (N particles) when hydrodynamic effects are included [512].

Various approximations can be used to make such BD simulations with hydrodynamics applicable to long-time macroscopic models of biological polymers like DNA [722]. An idea of Marshall Fixman [735] to use Chebyshev polynomial approximations for matrix/vector products was developed and applied in [269], demonstrating that an $\mathcal{O}(N^2)$ complexity is possible for large systems (see Chapter 13).

9.6.4 *Continuum Electrostatics*

For introductions into the study of polyelectrolyte solutions and related theories, such as the Poisson-Boltzmann equation and Debye-Hückel theory, readers may wish to consult the book [736] for a good treatment of statistical thermodynamics and kinetics, the text [737] for statistical thermodynamics, and the volume [738] for an early overview of how Poisson-Boltzmann and Debye-Hückel theories have been applied to polyelectrolyte solutions. Classic references on electrolyte solutions are [739, 740].

Gauss' Law for the Electrostatic Potential

Continuum electrostatic approximations are based on numerical solutions to the Poisson-Boltzmann (PB) equation (see [737, 736], for example, for introductions). This second-order differential equation combines theories from statistical mechanics — the Boltzmann distribution for a charge density ρ — with an equation from electrostatics — Gauss' law (or the Poisson equation),[5] which relates the second derivative of the electrostatic potential Φ to the charge density.

Gauss' law describes the electrostatic potential Φ at position $\mathbf{x}$ in terms of the fixed charged density of the solute, $\rho_{\rm solute}(\mathbf{x})$ and the position-dependent dielectric function $\epsilon(\mathbf{x})$ as:

$$\nabla \cdot [\epsilon(\mathbf{x}) \, \nabla\Phi(\mathbf{x})] = -4\pi \, \rho_{\rm solute}(\mathbf{x}) \,. \tag{9.56}$$

(See Box 9.2 for the definition of the divergence operator ∇ used above). This equation reduces to Coulomb's law when the dielectric constant is uniform throughout space and the charges are modeled as point charges.

[5] Physicists often refer to this equation as Gauss' law while mathematicians tend to favor the term Poisson's equation.

For a polar solvent like water, the effective dielectric constant increases as the point $\mathbf{x}$ moves farther away from the solute. This spatially dependent dielectric function — low ϵ near the solute and high ϵ for the bulk solvent — allows a better description than Coulomb's law: it takes into account the difference in electric polarizability between the macromolecule and the solvent. (See [632], for example, for estimates of local dielectric constants in the environment of B-DNA in solution). Poisson's equation cannot be solved analytically for arbitrary geometries and must be solved numerically by finite-difference or other methods (see below).

When mobile ions are also present in the solution, the charge density is delocalized from the solute/solvent boundary. The charge atmosphere is thus position-dependent, since the ions redistribute in the solution in response to the electric potential. A better approximation is obtained by considering a charge density around the solute resulting from the *distribution of charges* in the medium.

Assume that we have an electrolyte solution occupying a volume V and containing N_i ions of corresponding charge q_i for n_i ion species i. Let $c_i = N_i/V$ denote the bulk concentration of the ionic species i. The total *charge density* resulting from the sum of all charge densities of the ions can be described by a Boltzmann distribution as:

$$\rho(\mathbf{x}) = \sum_{i=1}^{n_i} q_i\, c_i\, \exp[-\widetilde{E}_i(\mathbf{x})/k_B \mathrm{T}]\,, \tag{9.57}$$

where $\widetilde{E}_i(\mathbf{x})$ is the 'effective potential of mean force' for electrolytes of type i at position $\mathbf{x}$ for a *given* solute configuration, or the energy change required to bring the ion from infinity to the position $\mathbf{x}$.

In practice, it is assumed that $\widetilde{E}_i(\mathbf{x})$ is approximated by the product charge times the potential, that is, the distribution of ions is determined by the electrostatic field:

$$\widetilde{E}_i(\mathbf{x}) \approx q_i\, \Phi(\mathbf{x})\,.$$

Thus, the Boltzmann distribution leads to an exponential relation between the charge density ρ and the electrostatic potential.

The Poisson-Boltzmann Equation

Combining Gauss' law (eq. (9.56)) with the Boltzmann charge density described by eq. (9.57) yields the *Poisson-Boltzmann (PB) equation*:

$$\nabla \cdot [\epsilon(\mathbf{x})\; \nabla\Phi(\mathbf{x})] = -4\pi\rho_{\mathrm{solute}}(\mathbf{x}) - 4\pi \sum_{i=1}^{n_i} q_i\, c_i\, \exp[-q_i\, \Phi(\mathbf{x})/k_B \mathrm{T}]\,. \tag{9.58}$$

The PB equation is the basis for modern electrolyte theory. The solution of this nonlinear equation for $\Phi(\mathbf{x})$ yields thermodynamic properties in the electrolyte solution. Since the electrostatic potential does not vary linearly in space with the source charges, properties of linear systems do not generally apply. As for

Poisson's equation, analytic solutions are not available in general and various numerical methods are used in practice.

Linear Approximations to the PB Equation; Debye-Hückel Theory

For the case of dilute or moderate solutions (e.g., molar concentrations $c_s < 10^{-3}$ M) and low fixed charge, it is possible to approximate the PB equation using results from *Debye-Hückel theory* (see [736, 737, 581, 25], for example, for introductions). Specifically, a linearized approximation to the PB equation can be used to represent the ionic atmosphere of a solute immersed in aqueous solution and counterions.

A linearized version of the PB equation (eq. (9.58)) can be obtained by a Taylor-series expansion of the Boltzmann factor with a truncation beyond the first-order terms:

$$\exp[-q_i\,\Phi(\mathbf{x})/k_B\mathrm{T}] \approx 1 - [q_i\,\Phi(\mathbf{x})/k_B\mathrm{T}]\,. \tag{9.59}$$

The linearized version is justified when

$$q_i\,\Phi(\mathbf{x}) \ll k_B\mathrm{T}\,,$$

that is, when energies are much smaller than the thermal energy. This often holds for monovalent electrolytes and weak source charges (dilute solution), as mentioned above.

In the special case of spherical symmetry about the origin for the distribution of charges, the potential Φ depends on the distance r of the point from the origin. Linearization of the PB equation yields the following *linear* second-order differential equation for $\Phi(r)$ [736]:[6]

$$\nabla^2\Phi(r) = \kappa^2\,\Phi(r)\,, \tag{9.60}$$

where

$$\kappa^2 = \frac{8\pi\,N_A\,e^2\,\rho_A}{1000\,\epsilon\,k_B\mathrm{T}}\,c_s\,. \tag{9.61}$$

Here N_A is Avogadro's number, ρ_A is the solvent density, e is the protonic charge (4.803×10^{-10} esu), and ϵ is the solvent dielectric constant. The ionic concentration c_s is measured in molar units as a sum over all molar concentrations per liter of solution, c_i, associated with charges (or valences) q_i: $c_s = \frac{1}{2}\sum_{i=1}^{n_i} c_i q_i^2$.

The linearized PB equation as expressed in eq. (9.60) with (9.61) can be solved to determine the effective potential Φ at a distance r from a central ion. Recall that the Coulomb potential $q/(\epsilon r)$ is produced at a point separated by distance r from an isolated central ion, which is represented as a uniform charged sphere of radius a_r in a medium of dielectric constant ϵ. The Debye-Hückel solution for the modified electrostatic potential representing the influence of the ionic atmosphere

[6] We can also write $\frac{1}{r^2}\frac{d}{dr}\left(r^2\frac{d}{dr}\right)\Phi(r) = \kappa^2\,\Phi(r)$.

is (e.g., [736]):

$$\Phi(r) = B(\kappa)\,\frac{q\exp(-\kappa r)}{\epsilon r}\,, \tag{9.62}$$

where $B(\kappa)$ is the salt-dependent coefficient

$$B(\kappa) = \exp(\kappa a_r)/(1 + \kappa a_r)\,. \tag{9.63}$$

Thus, Debye-Hückel theory produces an effective electrostatic potential in which the Coulomb interactions are *screened* by ions. The theory predicts the range of electrostatic influence of a central ion to be the Debye screening length κ^{-1}. In other words, κ^{-1} is the characteristic distance of exponential screening.

From eq. (9.61), we see that the *screening parameter* κ is proportional to the square root of the ionic concentration c_s. For 1:1 electrolytes (monovalent:monovalent salts like NaCl),

$$\kappa \approx 0.33\sqrt{c_s}\ \text{Å}^{-1} \tag{9.64}$$

at room temperature (25°), with $\epsilon = 78.5$.

For dilute solutions, or $\kappa a_r \ll 1$, we have $B(\kappa) \approx 1$; it follows that the Coulomb screening — reduction of the unscreened potential by the factor $\exp(-\kappa r)$ — reflects the reduced effective charge ρ on the central ion due to counterion accumulation. For physiological ionic strengths, such as 0.15M, the Debye length is approximately 8 Å; this distance represents considerable damping of Coulomb interactions.

Figure 9.10 compares the Coulomb potential $(1/\epsilon r)$ to the screened Coulomb potential $(B(\kappa)\ \exp[-\kappa r]/\epsilon r)$ for two values of κ corresponding to c_s = 0.15M and 0.015M monovalent salt, where the Debye lengths are about 8 and 25 Å, respectively.

The DH approximation has been applied to long DNA for exploring conformational stability and mobility as a function of monovalent ionic concentrations in the natural cellular environment (e.g., [350, 349]). Models are based on the pioneering work of Stigter, who modeled DNA as charged cylinders [741] and reproduced the experimentally-observed dependence of the effective diameter of DNA on salt concentration by a tail approximation to the PB equation. Extensions of such DH approximations to macroscopic models of protein/DNA systems have been described [475] and applied to model chromatin [498].

General Solutions to the Poisson-Boltzmann Equation

Many practical procedures have been developed to solve the PB equation numerically, including solutions of its various approximations (e.g., [474, 703, 742, 743, 744, 146, 745]); see also references cited in [746], and [747, 748] for applications to calculate charge distributions. The linearized approximations are useful in many cases, as mentioned above. For high charge density (such as for polyelectrolyte DNA) and high salt concentrations, the nonlinear PB version is preferred.

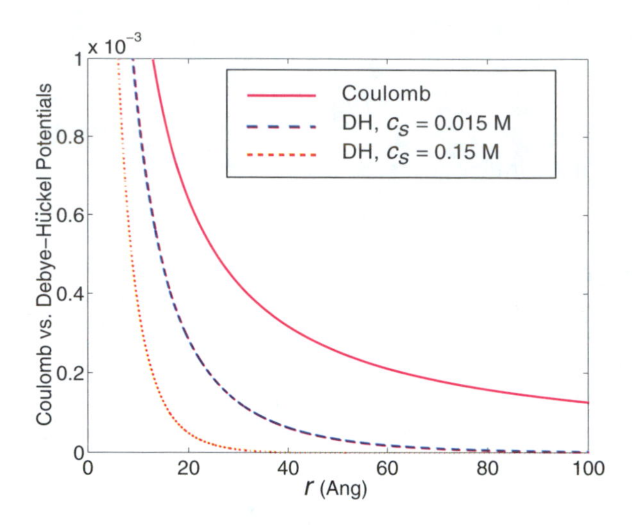

Figure 9.10. Screened Coulomb potentials at two values of the Debye length. The Coulomb potential $(1/\epsilon r)$ is compared to the screened Coulomb potential of form $B(\kappa)\ \exp[-\kappa r]/\epsilon r$ as a function of distance r for two values of κ. These two values are set according to eq. (9.64) for c_s = 0.15M and 0.015M monovalent salt concentrations, with coefficients $B(\kappa)$ computed according to eq. (9.63) (the coefficients are very close to 1). The two κ values (0.128 and 0.040 Å, respectively) correspond to Debye lengths of about 8 and 25 Å.

Numerical solutions are typically obtained through finite-difference or finite-element (or boundary element) methods. Both involve a discretization of the (irregular) biomolecular domain in 3D, so that the potential, charge density, and dielectric constant are defined at grid points (ϵ is usually defined over broader domain regions). The dielectric constant is assigned appropriate values depending on its proximity to the solute, though a two-dielectric model is typically used — to distinguish the inside region, near the solute, from the outside region, far from the solute. The finite-difference solution can be obtained iteratively by various linear algebra solvers (e.g., linear conjugate gradient method, Gauss-Seidel, or Successive Over Relaxation methods) [474, 146, 745].

The resulting quality of the numerical solutions depends on the various assumptions made and settings used. The convergence of the solvers also depends on such parameters as the grid size, initial charge and ϵ assignments, and algorithmic parameters.

Recently, an analytical gradient minimization method based on a finite-element discretization for solving the PB equation has also been presented [749]. While the DelPhi program uses a regular cubic grid, this gradient-based approach uses an adaptive grid so as to include more grid points at the solvent-accessible surface, where the dielectric value is changing rapidly. Overall, preliminary comparisons show that the computational efficiency of both approaches is comparable [749]; the gradient-based method is also used for geometry optimization applications, serving as an improvement to gas-phase molecular mechanics minimizations since solvation effects are included.

Popular packages used in the biomolecular community are those developed at research groups at Columbia University (DelPhi and GRASP) and the University of Houston (UHBD, now at UCSD, which includes segments for Brownian dynamics).[7] The latter has been used to study huge macromolecular systems [746] (see also Figure 9.11).

Classical electrostatics solutions can provide useful information on charge distributions around macromolecules (e.g., [747, 748]), specific localization of charged and polar groups in biomolecular systems (e.g., [5, 750]), the shape of molecular surfaces, and the relation between salt effects and conformational changes (e.g., [350]). Effects of the ionic atmosphere on intermolecular binding associations, such as between proteins and ligands, can also be analyzed [712, 751].

Effective numerical packages such as DelPhi and UHBD, combined with advanced graphical rendering programs, offer valuable modeling tools for exploring the effects of electrostatics and solvation on molecular structure and function. The electrostatic analysis of a ribosomal unit is shown in Figure 9.11. (See also analysis of the nucleosome core particle in Chapter 6).

Algorithmic Challenges

Recent algorithmic advances — by adaptive finite element methods [742, 743] tailored to multiprocessor systems [746] — allow solutions of the linear PBE for very large systems of order one million atoms. Figure 9.11 illustrates the PB solution for the 30S ribosome subunit.

A future goal in the field is repeated solution of the PBE, for example in the course of a dynamic simulation.

[7] At present, information for DelPhi can be obtained at the URL trantor.bioc.columbia.edu/delphi/ (Honig's lab homepage is honiglab.cpmc.columbia.edu), and for UHBD at the URL mccammon.ucsd.edu/uhbd.html (mccammon.ucsd.edu/ for McCammon's lab homepage).

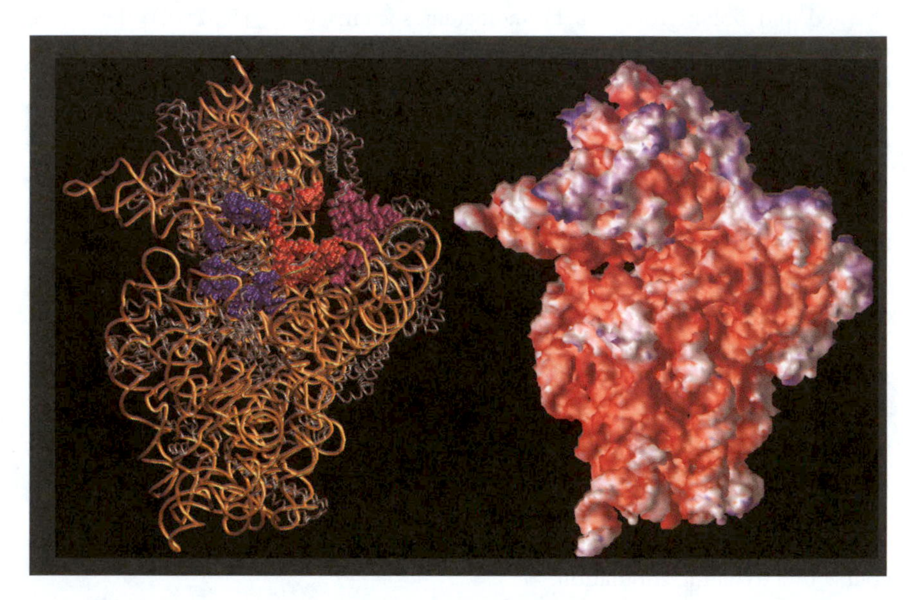

Figure 9.11. The electrostatic environment of the 30S ribosomal unit as computed by the linear PB equation at 150 mM ionic strength with a solute dielectric of 2 and solvent dielectric of 78.5 [746]. The protein (silver) and nucleic acid (gold) atoms are distinguished on the left, with selected components of A, P, and E sites shown in blue, red, and purple respectively. At right, the electrostatic potential mapped on the 30S surface is rendered in blue (positive) and red (negative) to illustrate regions with values greater than 2.6 k_BT/e and less than -2.6 k_BT/e, respectively.

10

Multivariate Minimization in Computational Chemistry

Chapter 10 Notation

SYMBOL	DEFINITION
Matrices	
$\mathbf{A}$	symmetric matrix, components $\{A_{ij}\}$
$\widehat{\mathbf{B}}_k$	approximation to Hessian inverse at $\mathbf{x}_k$ (QN methods)
$\mathbf{D}_k$	scaling matrix at step k of minimization method (trust region approach)
$\mathbf{H}$	Hessian matrix, components $H_{ij}(\mathbf{x}) \equiv \partial^2 f(\mathbf{x})/\partial x_i \partial x_j$
$\mathbf{I}$	identity matrix
$\mathbf{M}$	preconditioning matrix, related to $\mathbf{H}$ (TN methods)
$\mathbf{M}_k$	preconditioning matrix at step k of TN method
$\mathbf{U}_k$	QN low-rank update matrix at step k
Vectors	
$\mathbf{b}, \mathbf{y}$	constant vectors
$\mathbf{e}_j$	unit vectors
$\mathbf{g}$	gradient vector of f, components $g_i(\mathbf{x}) \equiv \partial f(\mathbf{x})/\partial x_i$
$\mathbf{g}_k$	gradient vector at $\mathbf{x}_k$ (short hand for $\mathbf{g}(\mathbf{x}_k)$)
$\mathbf{p}_k$	search vector at step k of minimization method
$\mathbf{p}_k^j$	inner-loop CG iterate j for outer-loop search vector $\mathbf{p}_k$ (TN method)
$\mathbf{r}$	residual vector defined in TN methods ($\mathbf{Mz} = \mathbf{r}$)
$\mathbf{s}_k$	displacement vector at step k, $\mathbf{x}_{k+1} - \mathbf{x}_k$ (QN method)
$\mathbf{x}$	vector of n components $\{x_i\}$
$\mathbf{x}_0$	starting point vector for minimization
$\mathbf{x}_k$	minimization iterate at step k of method
$\mathbf{x}^*$	local minimum point of objective function
$\mathbf{y}_k$	gradient difference vector at QN step k, $\mathbf{g}_{k+1} - \mathbf{g}_k$
$\mathbf{z}$	solution vector defined in TN methods ($\mathbf{Mz} = \mathbf{r}$)

Chapter 10 Notation Table (continued)

SYMBOL	DEFINITION
Scalars & Functions	
a, b	numbers
$c_i(\mathbf{x})$	constraint function i
c_r	small positive number (in TN methods)
f_0	constant (function value)
$f(\mathbf{x})$	objective function, dependent on vector $\mathbf{x}$
h	small number (finite difference interval)
n	problem dimension
p	convergence order
$q(\mathbf{x})$	quadratic function
$q_k(\mathbf{s})$	quadratic model of objective function
r_k	residual norm at step k of TN methods
α	line search parameter for sufficient decrease condition
β	line search parameter for sufficient decrease of curvature (also convergence ratio)
β_k	scheme-dependent scale parameter of search vector $\mathbf{p}_k$ (CG and QN methods)
ϵ_f, ϵ_g	small positive numbers
ϵ_m	small positive number, machine precision
η_k	forcing sequence in TN methods
λ	line search steplength
λ_t	trial line search steplength
ξ	variable in the neighborhood of x for a univariate function $f(x)$
$\phi(\lambda)$	polynomial of steplength λ
Δ_k	size bound in QN methods at step k

'Pon my word Watson, you are coming along wonderfully. We have really done very well indeed. It is true that you have missed everything of importance, but you have hit upon the method.

Arthur Conan Doyle (1859–1930), in *A Case of Identity (1891).*

10.1 Ubiquitous Optimization: From Enzymes to Weather to Economics

Optimization is a fundamental component of molecular modeling. The determination of a low-energy conformation for a given force field can be the final objective of the computation. It can also serve as a starting point for subsequent calculations, such as molecular dynamics simulations or normal-mode analyses.

Both local and global optimization problems lie at the heart of numerous scientific and engineering problems — from the biological and chemical disciplines to architectural and industrial design to economics. Optimization is part of our everyday life — responsible for our weather forecasts, flight planning, telephone routing, microprocessor design, and the functioning of enzymes in our bodies.

10.1.1 *Algorithmic Sophistication Demands Basic Understanding*

The mathematical techniques developed to address these optimization problems are just as robust and varied as the target problems themselves. The algorithmic complexity of such techniques has led to many available computer programs that require minimal input from the user (e.g., the starting point and a routine for function evaluation).

However, the prudent user of these canned software modules — even within standard molecular mechanics and dynamics packages — should understand the fundamental structure of the optimization algorithms and associated performance issues to make their application both efficient and correct, in terms of the physical interpretations.

This chapter introduces key optimization concepts for this purpose. We also highlight the fundamentals of local optimizers for *large-scale nonlinear unconstrained problems*, an important optimization subfield relevant to biological macromolecules. We describe the most promising approaches among them, and discuss practical issues, such as parameter variations and termination criteria. Of course, the latter are best learned by experimentation in the context of real problems. To illustrate behavior for complex problems, some comparisons among three competitive minimizers are also included, for molecular models minimized in the molecular mechanics and dynamics program CHARMM.

10.1.2 *Chapter Overview*

Specifically, Section 10.2 introduces optimization fundamentals such as problem formulation and terminology. Section 10.3 describes the basic algorithmic framework of iterative minimization protocols (based on line search and trust region methods); it also discusses convergence criteria and line search procedures and introduces the key concept of descent directions.

In Section 10.4, we present the Newton method, including a historical perspective, and one-dimensional implementations for nonlinear equations as well as optimization. This presentation familiarizes readers with the Newton method framework — the basis for formulating many other optimization methods — and with performance and convergence issues relevant to minimization of multivariate functions.

In Section 10.5, we mention effective methods for large-scale nonlinear optimization, namely quasi-Newton (QN), nonlinear conjugate gradient (CG), and truncated Newton (TN) schemes. Section 10.6 outlines available software and presents comparative performance in CHARMM for two molecular models (a small model system and a protein). Finally, in Sections 10.7 and 10.8, we summarize recommendations to optimization practitioners and offer a future perspective to field developments.

For details, as well as for other categories of the rich and exciting field of optimization, I refer readers to classic texts [752, 753, 754, 755, 756], some reviews [757, 758, 759, 760], and a perspective [761].

10.2 Optimization Fundamentals

The methods for solving an optimization task depend on the problem classification. Since the value of the independent variable that maximizes a function f also minimizes the function $-f$, it suffices to deal with minimization.

The optimization problem is classified according to the *type* of independent variables involved (real, integer, mixed), the *number* of variables (one, few, many), the *functional characteristics* (linear, least squares, nonlinear, nondifferentiable, separable, etc.), and the problem *statement* (unconstrained, subject to equality constraints, subject to simple bounds, linearly constrained, nonlinearly constrained, etc.). For each category, suitable algorithms exist that exploit the problem's structure and formulation.

10.2.1 Problem Formulation

For a vector $\mathbf{x}$ of n components $\{x_i\}$, we write the minimization problem as:

$$\min_{\mathbf{x}} \{f(\mathbf{x})\}, \quad \mathbf{x} \in \mathcal{D}, \tag{10.1}$$

where f is the objective function and $\mathcal{D}$ is a given region (which can be the entire Euclidean space $\Re^n$). The problem can be subject to m constraints, which can be written more generally as a combination of equality and inequality constraints:

$$\begin{aligned} c_i(\mathbf{x}) &= 0 \quad \text{for } i = 1, \ldots, m', \\ c_i(\mathbf{x}) &\leq 0 \quad \text{for } i = m'+1, \ldots, m. \end{aligned} \tag{10.2}$$

This general formulation can be obtained for problems with bound constraints in the form

$$c_i(\mathbf{x}) = x_i \,,$$

where x_i is the ith component of the vector $\mathbf{x}$, or for problems with two-sided constraints such as

$$l_i \leq c_i(\mathbf{x}) \leq u_i \,.$$

10.2.2 Independent Variables

In most computational chemistry problems, $\mathbf{x}$ is a real vector in Euclidean space, i.e., $\mathbf{x} \in \Re^n$, and f defines a transformation to a real number, i.e., $f(\mathbf{x})$:

$\Re^n \rightarrow \Re$. When the components of $\mathbf{x}$ are integers, the optimization problem is classified as *integer-programming*. When $\mathbf{x}$ is a mixture of real and integer variables, the problem is of *mixed-integer programming* type. Common examples of integer-programming are network optimization and the 'traveling salesman problem',[1] also classified as combinatorial optimization. See [761], for example, and references cited therein.

10.2.3 Function Characteristics

The nature of the function f is the next step in problem classification. Many application areas such as finance and management-planning tackle *linear* or *quadratic* objective functions.

Linear and Quadratic Functions

Linear objectives can be written in vector form as

$$f(\mathbf{x}) = \mathbf{b}^T \mathbf{x} + f_0 \,, \tag{10.3}$$

where $\mathbf{b}$ is a column vector of dimension n, and f_0 is a scalar. Quadratic objective functions can be expressed as

$$f(\mathbf{x}) = \mathbf{x}^T \mathbf{A} \mathbf{x} + \mathbf{b}^T \mathbf{x} + f_0 \,, \tag{10.4}$$

where $\mathbf{A}$ is a constant *symmetric* matrix of dimension $n \times n$. (By definition, the n^2 entries of a symmetric matrix $\mathbf{A}$ satisfy $A_{i,j} = A_{j,i}$). The superscripts ^{T} above refer to a vector transpose; thus $\mathbf{x}^T\mathbf{y}$ is an inner product.

Linear programming problems refer to linear objective functions subject to linear constraints (i.e., a system of linear equations and inequalities), and *quadratic programming* problems have quadratic objective functions and linear constraints.

Least-Squares Functions

Nonlinear functions can be classified further. *Least-squares* functions have the form

$$f(\mathbf{x}) = \frac{1}{2} \sum_{i=1}^{m} f_i(\mathbf{x})^2 \,. \tag{10.5}$$

---

[1] The notorious 'traveling salesman' problem seeks to find the optimal travel route that covers a given number of cities, each one only once, and returning to the home town. Visually, imagine drawing such a route on a map, where each city k for $k = 0, \ldots, n$ is designated by coordinates $\{x_k, y_k\}$. The connected route started at $\{x_0, y_0\}$ covers each city and returns to the original point. Though simple to envision, there are clearly many such routes, and the number of combinations that connect all these cities grows steeply with n. This problem in fact belongs to a class of very difficult problems (known as *NP-complete*) for which no polynomial-complexity algorithm is known (i.e., the computational time for an *exact* solution of this problem increases exponentially with n).

Separable Functions

Separable functions can be expressed as a sum of subfunctions, namely

$$f(\mathbf{x}) = \sum_{i=1}^{m} f_i(\mathbf{x}) \,, \tag{10.6}$$

where each subfunction f_i depends only on a subset of the independent variables. That is, for each subfunction f_i there are many unit vectors $\mathbf{e}_j$ (with 1 in component j and 0 elsewhere) for which $f_i(\mathbf{x} + \mathbf{e}_j) = f_i(\mathbf{x})$. All molecular mechanics potential functions arising from the local, bonded interactions can be written this way.

Nonsmooth Functions

Because most optimization algorithms exploit derivative information to locate optima, nonsmooth functions pose special difficulties, and very different algorithmic approaches must be used. See [762] and [754, Chapter 14] for a general introduction to nonsmooth optimization, and the two-volume set [763, 764] for the special case of nonsmooth convex problems. Optimization of nonsmooth functions requires new mathematical machinery (e.g., *subdifferentials*) that extends ordinary differentiation and leads to counterparts of most results in differential calculus (Taylor expansions, mean value theorem, etc.).

Potential Energy Functions

Geometry optimization problems for molecular potential functions in the context of standard all-atom force fields in computational chemistry are typically of the multivariate, continuous, and nonlinear type [760]. They can be formulated as constrained (as in adiabatic relaxation, an example of which was shown in Chapter 5) or unconstrained. Discontinuities in the derivatives may be a problem in certain formulations involving truncation, such as of the nonbonded terms (see Section 10.6).

The large number of independent variables for biomolecules, in particular, warrants their classification as *large-scale* and rules out the use of many algorithms that are effective for a small number of variables. However, as we will discuss, effective techniques are available today that achieve rapid convergence even for large systems. In practice, for macromolecular applications these optimization algorithms must be modest in storage requirements and economical in computations, which are dominated by the function and derivative evaluations.

10.2.4 Local and Global Minima

Definitions

The *local* unconstrained optimization problem in the Euclidean space $\Re^n$ can be stated as in eq. (10.1) for $\mathbf{x} \in \mathcal{D} \subset \Re^n$ where $\mathcal{D}$ denotes a neighborhood of

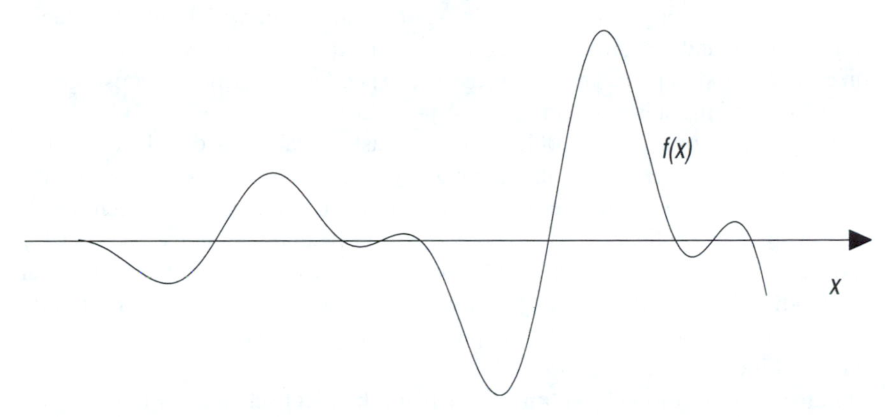

Figure 10.1. A one-dimensional function with several minima. This function was constructed from the actual univariate function at one line search step of the truncated Newton algorithm (see later in chapter) applied to minimization of a small protein's potential energy function.

the starting point, $\mathbf{x}_0$. The *global* optimization problem is much more difficult because it requires finding the global minimum among all the local minima, and the number of minima can be exponentially large.

A (strong) *local minimum* $\mathbf{x}^*$ of $f(\mathbf{x})$ satisfies

$$f(\mathbf{x}^*) < f(\mathbf{y}) \qquad \text{for all } \mathbf{y} \in \mathcal{D}, \;\; \mathbf{y} \neq \mathbf{x}^*. \tag{10.7}$$

The point $\mathbf{x}^*$ is a *weak local minimum* if $f(\mathbf{x}^*) \leq f(\mathbf{y})$.

A *global minimum* $\mathbf{x}^*$ satisfies the stringent requirement that

$$f(\mathbf{x}^*) < f(\mathbf{y}) \qquad \text{for all } \mathbf{y} \neq \mathbf{x}^*. \tag{10.8}$$

See Figure 10.1 for an illustration of a one-dimensional function with several minima. The function corresponds to the actual univariate function minimized in the line search substep of the TN method (see later in chapter for details).

Convergence

Finding a *local minimum* is a challenging task for a large biological system governed by a nonlinear potential energy function. This is because the optimization scheme must find a minimum from any point along the potential surface, even one associated with a very high-energy, and should not get trapped at local maxima or saddle points. Finite-precision arithmetic and various errors that accumulate over many operations also degrade practical performance in comparison to theoretical expectations (which can be described as *convergence order*; see Box 10.1). Nonetheless, the local optimization problem is solved in a mathematical sense: convergence to a local minimum can be achieved on modern computers.

In the mathematical literature, this is referred to as *global convergence* to a local minimum. Still, though many algorithms are available in widely-used molecular mechanics and dynamics packages, performance and solution quality vary considerably and depend greatly on the user-specified algorithmic convergence parameters and the starting point.

The global optimization problem, by contrast, remains unsolved in general. This is because the exponentially-growing number of minima with system size cannot be exhaustively surveyed. Certainly, effective strategies have been developed in specific application contexts (e.g., for polypeptides) and work well for moderately-sized systems. See [765], for example, for a review, the website at www.mat.univie.ac.at/~neum/glopt.html for general information, and homework 13 for the deterministic global optimization approach based on the *diffusion equation* [766].

Global minimization algorithms differ from the local schemes in that they do not necessarily require the energy to decrease systematically, making possible escape from local potential wells and entry into others. Global optimization methods can be stochastic or deterministic, or a combination thereof; they often rely on local optimization components.

Box 10.1: Convergence Definitions

A sequence $\{x_k\}$ *converging* to x^* has *order p* if p is the largest number such that a finite limit β (the "convergence ratio", not to be confused with the line search parameter β) exists, where:

$$0 \leq \lim_{k \to \infty} \frac{|x_{k+1} - x^*|}{|x_k - x^*|^p} = \beta < \infty . \tag{10.9}$$

When $p = 2$, we have *quadratic convergence*. When $p = 1$, we refer to the convergence as *superlinear* if $\beta = 0$ and as *linear* if the nonzero β is less than 1.

For example, the reader can verify that the sequences $\{2^{-2^k}\}$, $\{k^{-k}\}$, and $\{2^{-k}\}$ converge, respectively, quadratically, superlinearly, and linearly. Quadratic convergence is faster than superlinear, which in turn is faster than linear.

10.2.5 Derivatives of Multivariate Functions

Gradient

When f is a smooth function with continuous first and second derivatives, we define its *gradient vector* of first derivatives by $\mathbf{g}(\mathbf{x})$, where each component of $\mathbf{g}$ is

$$g_i(\mathbf{x}) = \partial f(\mathbf{x}) / \partial x_i . \tag{10.10}$$

Hessian and Curvature

The $n \times n$ symmetric matrix of second derivatives, $\mathbf{H}(\mathbf{x})$, is called the *Hessian*. Its components are defined as:

$$H_{i,j}(\mathbf{x}) = \partial^2 f(\mathbf{x}) / \partial x_i \partial x_j . \tag{10.11}$$

At a *stationary* point, the gradient is zero. At a minimum point $\mathbf{x}^*$, in addition to stationarity, the curvature is positive. For higher dimensions, convexity is expressed as *positive-definiteness* of the Hessian. A multivariate function is positive-definite at a point $\mathbf{x}^*$ if

$$\mathbf{y}^T \mathbf{H}(\mathbf{x}^*) \mathbf{y} > 0 \qquad \text{for all nonzero } \mathbf{y}. \tag{10.12}$$

In particular, positive definiteness guarantees that all the eigenvalues are positive at $\mathbf{x}^*$. A *positive semi-definite* matrix has nonnegative eigenvalues; a *negative semi-definite* matrix has nonpositive eigenvalues; and a *negative-definite* matrix has only negative eigenvalues. Otherwise, the matrix is *indefinite*. The utilization of curvature information is important for formulating effective multivariate optimization algorithms.

Figure 10.2 illustrates this notion of curvature for quadratic functions of two variables:

$$q(\mathbf{x}) = \mathbf{x}^T \mathbf{A} \mathbf{x} + \mathbf{b}^T \mathbf{x} .$$

Namely, it displays the *contours* of these functions — curves on which the function is constant — in four cases. These cases are defined by different properties of the matrix $\mathbf{A}$: (a) indefinite, (b) positive definite, (c) negative definite, and (d) singular (i.e., not invertible). Figure 10.3 displays corresponding three-dimensional views of the functions, with circles and a line indicating stationary points. We use similar contour plots later (Figure 10.10) to illustrate paths of different minimization algorithms.

10.2.6 *The Hessian of Potential Energy Functions*

Sparsity

A matrix is termed *sparse* if it has a large percentage of zero entries; otherwise it is *dense*. (There is no specific threshold percentage of zero elements below which a matrix is considered 'sparse'). A sparse matrix can be *structured*, as in a banded matrix of bandwidth p where there are zeros for $|i-j| > p$. Alternatively, a sparse matrix can be *unstructured*, as shown in Figures 10.4 and 10.5.

In these figures, the matrix indices are the independent variables (three times the number of atoms) of the potential energy function for molecular systems. A point in the matrix position $\{i, j\}$ indicates a nonzero Hessian element for the second-derivative term of the potential energy objective function. Examples are shown for various molecular systems. The left-column matrices correspond to the Hessian pattern resulting when 8 Å cutoffs are used for the nonbonded terms. The right-column patterns correspond to only the local, bonded second-derivative

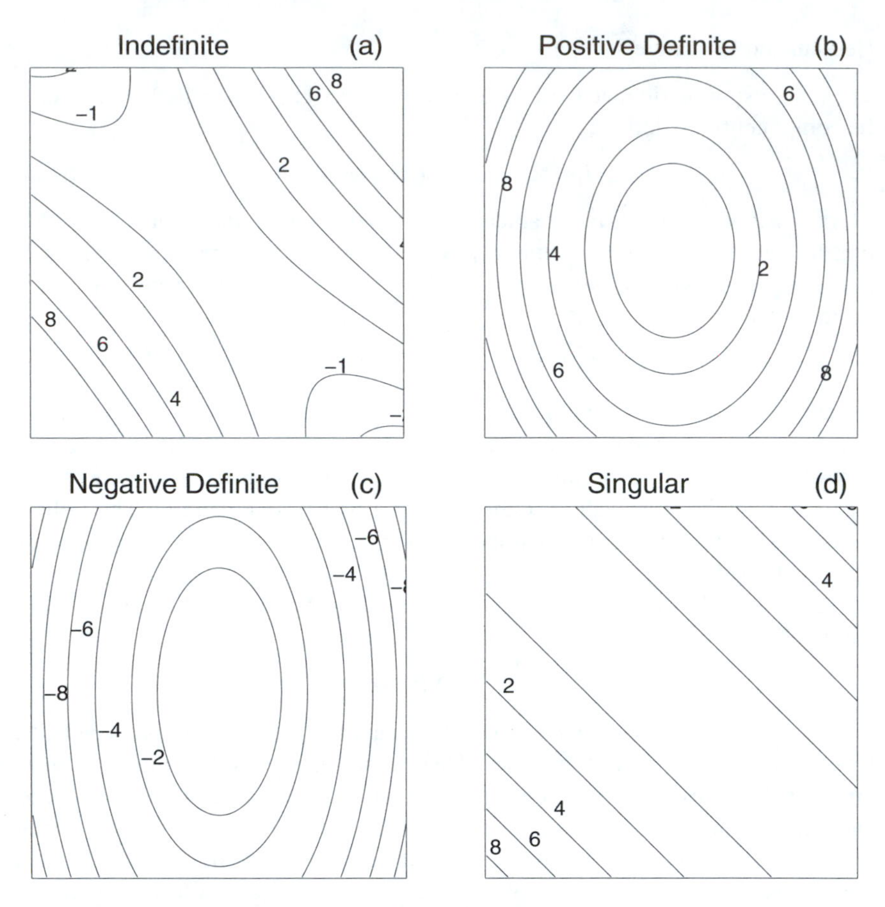

Figure 10.2. Two-dimensional contour curves for the quadratic function $q(\mathbf{x}) = \mathbf{x}^T\mathbf{A}\mathbf{x} + \mathbf{b}^T\mathbf{x}$ of two variables, where $\mathbf{A}$ is: (a) indefinite, with entries by row 1,2,2,2; (b) positive definite, entries 4,0,0,2; (c) negative definite, entries -1,0,0,-4; and (d) singular, entries 1,1,1,1. See also Figure 10.3.

terms (bond-length, bond angle, and dihedral-angle). The insets zoom on two submatrices and illustrate how the sparsity pattern repeats in triplets (for the x, y, and z components), and how nearly banded the local Hessian structure is due to the finite range of the bonded interactions.

We also see that although the matrices corresponding to 8 Å cutoffs are sparse for the larger systems, the atom ordering used determines the resulting pattern. For example, the X pattern for the DNA system results from the consecutive ordering of atoms down one strand and up the complementary strand; the water atoms are numbered following the DNA atoms.

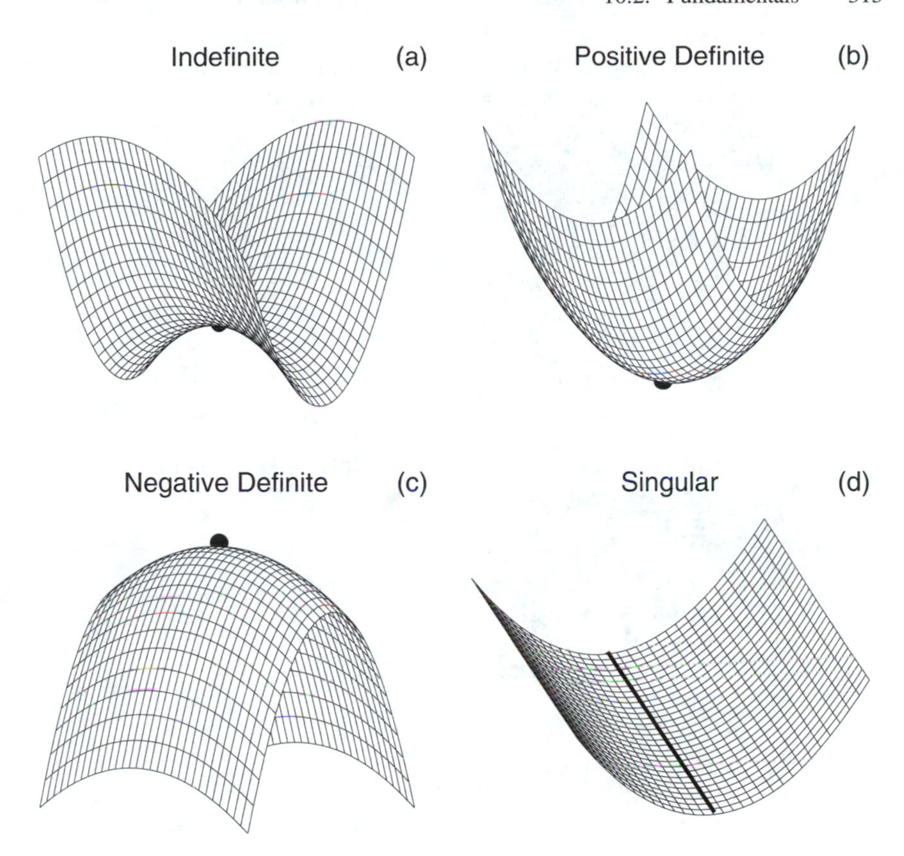

Figure 10.3. Three-dimensional curves for the quadratic functions as described for Figure 10.2. Critical points are shown by thick circles (a–c) and a line (d).

Memory Intensity

Because the formulation of a dense Hessian (n^2 entries) is both memory and computation intensive, many Newton techniques for minimization approximate curvature information implicitly and often progressively, i.e., as the algorithm proceeds. Limited-memory versions reduce computational and storage requirements so that they can be applied to very large problems and/or to problems where second derivatives are not available.

Exploitation of Derivatives

In most molecular mechanics packages, the second derivatives are programmed, though sparsity (when relevant) is not often exploited in the storage techniques for large molecular systems. The optimizer should utilize some of this second-derivative information to make the algorithm more efficient. Truncated Newton methods, for example, are designed with this philosophy.

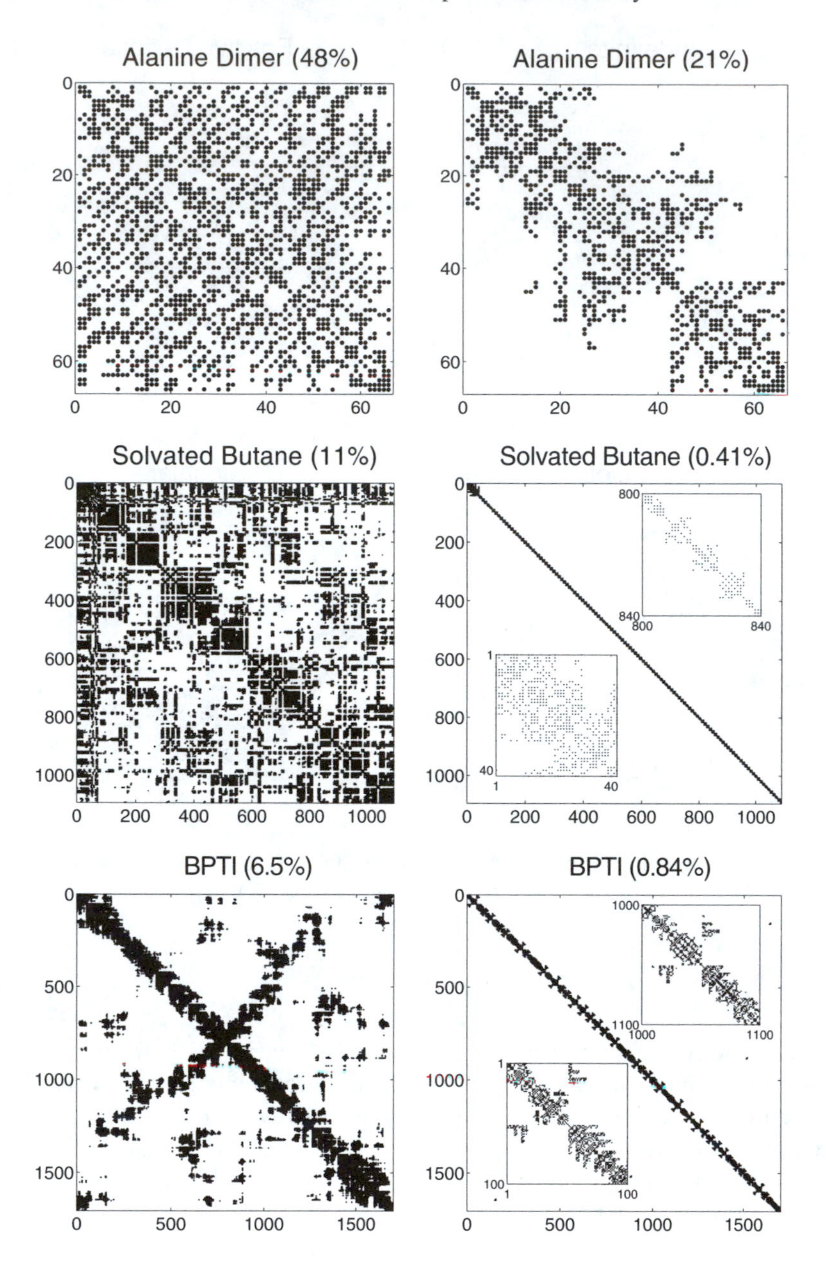

Figure 10.4. Hessian patterns from the potential energy functions of various molecular systems corresponding to 8-Å cutoffs (matrices at left column) or to local terms (right column; bond-length, bond-angle, and dihedral-angle components). The percentage sparsity is shown for each case, and insets show enlargements of some Hessian submatrices. The matrix axes label Cartesian coordinates, i.e., the x, y, z coordinates of each atom in turn; the atom ordering comes from the molecular mechanics package (CHARMM used here).

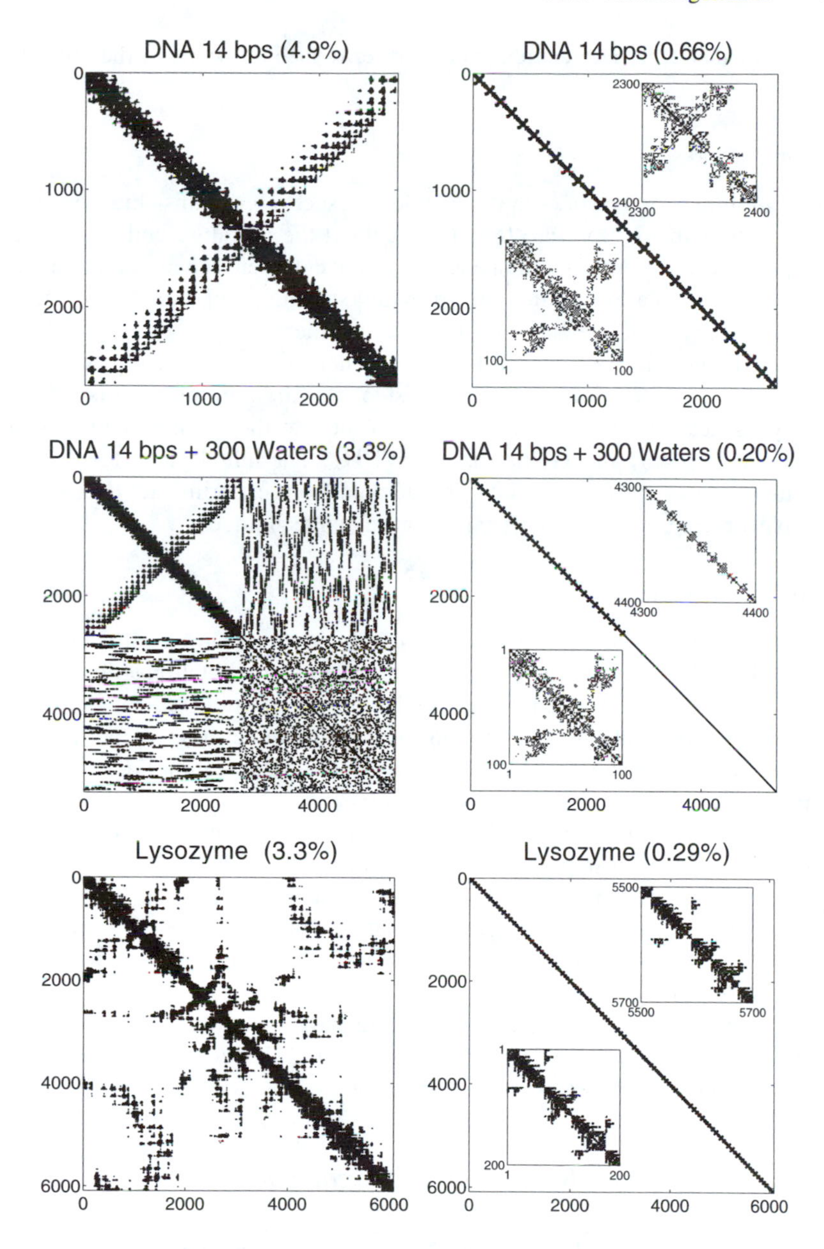

Figure 10.5. Sparse Hessian patterns, continued (see caption to Figure 10.4).

10.3 Basic Algorithmic Components

10.3.1 Greedy Descent

The basic structure of an iterative local optimization algorithm is one of "greedy descent". Namely, a sequence $\{\mathbf{x}_k\}$ is generated from a starting point $\mathbf{x}_0$ in such a way that each iterate attempts to further reduce the value of the objective function.[2]

Two Frameworks

Two algorithmic frameworks are available for such algorithms: line-search or trust-region methods. Both are found throughout the literature and in software packages and are essential components of effective descent schemes that guarantee convergence to a local minimum from any starting point. No clear evidence has emerged to render one class superior over another.

In describing iterative minimization techniques, it is convenient to use short hand notation for quantities used at each step k of the minimization algorithm. Namely, associated with each iterate $\mathbf{x}_k$, we denote the gradient and Hessian at $\mathbf{x}_k$, namely $\mathbf{g}(\mathbf{x}_k)$ and $\mathbf{H}(\mathbf{x}_k)$, as $\mathbf{g}_k$ and $\mathbf{H}_k$. The initial guess for the iterative minimization process ($\mathbf{x}_0$) can be derived from experimental data, where available, or from results of conformational search techniques.

Algorithmic Parameters

The final stopping criteria must be chosen with care to ensure a sufficiently accurate solution and, at the same time, avoid wasting computational effort when further progress is not realized. For example, the norm of the gradient alone (i.e., $||\mathbf{g}_k||$) may not be a satisfactory stopping criterion in unconstrained optimization, as it often exhibits oscillations in the course of the optimization [767]; see also Figure 10.11.

The line search framework requires careful implementation of convergence criteria of its own at each step, for a one-dimensional optimization procedure. This segment is a tricky part of minimization methods and requires well tested software with safeguards against many undesirable situations that can occur in practice, like very small steplengths and failure to bracket the univariate minimum (see [753, 756, 768], for example).

We now describe in turn the line search and trust-region frameworks for minimization (Subsections 10.3.2 and 10.3.3); this is followed by a discussion of convergence criteria for the minimization process (Subsection 10.3.4).

10.3.2 Line-Search-Based Descent Algorithm

Algorithm [A1]: Basic Descent Using Line Search

From a given point $\mathbf{x}_0$, perform for $k = 0, 1, 2, \ldots$ *until convergence*:

1. Test $\mathbf{x}_k$ for convergence (see subsection 10.3.4).

---

[2]This does not imply that the reduction in the gradient norm *is monotonic*; see Figure 10.11 for example.

2. Calculate a *descent* direction $\mathbf{p}_k$ (method dependent).

3. Determine a steplength λ_k by a one-dimensional line search so that the new position vector, $\mathbf{x}_{k+1} = \mathbf{x}_k + \lambda_k \mathbf{p}_k$, and corresponding gradient $\mathbf{g}_{k+1}$, satisfy:

$$f(\mathbf{x}_{k+1}) \leq f(\mathbf{x}_k) + \alpha \, \lambda \, \mathbf{g}_k^T \mathbf{p}_k \qquad [\text{“sufficient decrease”}] \tag{10.13}$$

and

$$|\mathbf{g}_{k+1}^T \mathbf{p}_k| \leq \beta \, |\mathbf{g}_k^T \mathbf{p}_k| \qquad [\text{“sufficient directional derivative reduction”}] \tag{10.14}$$

where $0 < \alpha < \beta < 1$
(e.g., $\alpha = 10^{-4}$, $\beta = 0.9$ in Newton methods).

4. Set $\mathbf{x}_{k+1}$ to $\mathbf{x}_k + \lambda_k \mathbf{p}_k$ and k to $k + 1$ and go to step 1.

Step 2: Descent Direction

A *descent direction* $\mathbf{p}_k$ is one along which the function must decrease locally. Formally, we define such a vector as one for which the directional derivative is negative:

$$\mathbf{g}_k^T \mathbf{p}_k < 0 \, . \tag{10.15}$$

To see why this property implies that f can be reduced, approximate the nonlinear objective function f at $\mathbf{x}$ by a linear model along the descent direction $\mathbf{p}$, assuming that higher-order terms are smaller than the gradient term. Then we see that the difference in function values is negative:

$$\begin{aligned} f(\mathbf{x} + \lambda \mathbf{p}) - f(\mathbf{x}) &= \lambda \, \mathbf{g}(\mathbf{x})^T \mathbf{p} + \frac{\lambda^2}{2} \, \mathbf{p}^T \mathbf{H}(\mathbf{x}) \, \mathbf{p} \\ &\approx \lambda \, \mathbf{g}(\mathbf{x})^T \mathbf{p} < 0 \, , \end{aligned} \tag{10.16}$$

for sufficiently small positive λ.

Steepest Descent

The descent condition is used to define the algorithmic sequence that generates $\mathbf{p}_k$. The simplest way to specify a descent direction is to set

$$\mathbf{p}_k = -\mathbf{g}_k \tag{10.17}$$

at each step. This “steepest descent” direction defines the steepest descent (SD) method. SD methods generally lead to improvements quickly but then exhibit slow progress toward a solution. Though it has become customary to recommend the use of SD for initial minimization iterations when the starting function and gradient-norm values are very large, this approach is not necessary when a more robust minimization method is available.

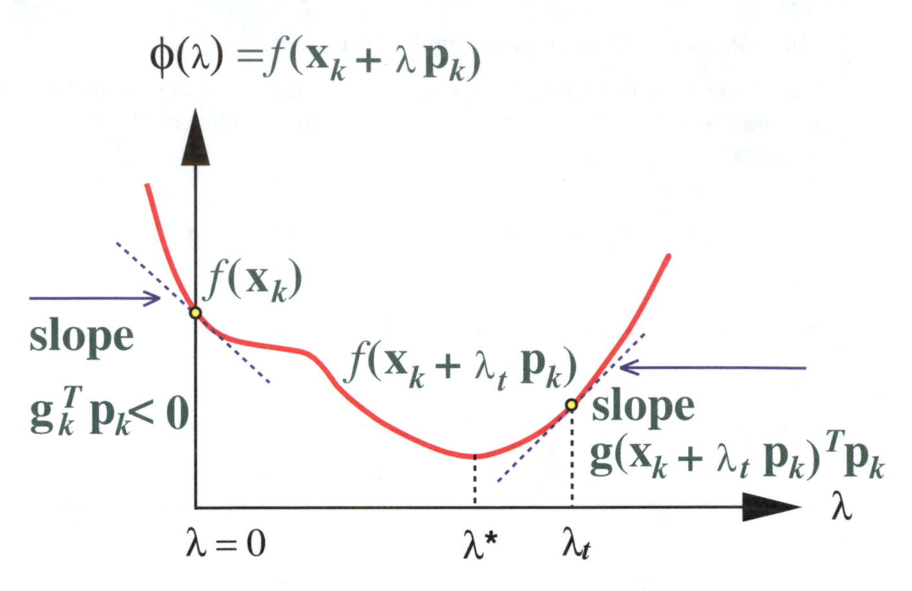

Figure 10.6. One-dimensional line search minimization at step k of the multivariate method, using polynomial approximation to estimate the optimal steplength λ^* in the region $[0, \lambda_t]$.

Step 3: The One-Dimensional Optimization Subproblem (Line Search)

The line search procedure, a univariate minimization problem, is typically performed via approximate minimization of a quadratic or cubic polynomial interpolant of the one-dimensional function of λ (given $\mathbf{x}_k$ and $\mathbf{p}_k$):

$$\phi(\lambda) \equiv f(\mathbf{x}_k + \lambda\mathbf{p}_k)\,.$$

See Figure 10.6 for an illustration of the first step of this univariate minimization process, where the minimum (λ^* in figure) is sought in the initial interval $[0, \lambda_t]$ where $f(\mathbf{x}_k) = \phi(0)$ and $f(\mathbf{x}_k + \lambda_t\mathbf{p}_k) = \phi(\lambda_t)$. For details, consult standard texts (e.g., [756]). This iteration process is generally continued until the λ value that minimizes the polynomial interpolant of $\phi(\lambda)$ satisfies the line search criteria (eqs. (10.13) and (10.14)). The resulting steplength λ_k defines the next iterate for minimization of the multivariate function by $\mathbf{x}_{k+1} = \mathbf{x}_k + \lambda_k\,\mathbf{p}_k$.

The line search criteria in Step 3 of Algorithm [A1] have been formulated to ensure sufficient decrease of f relative to the size of step (λ) taken. The first condition (eq. (10.13)) prescribes an upper limit on acceptable new function values; recall that the second term on the right is negative by the descent property. The second criterion, eq. (10.14), imposes a lower bound on λ. The control parameters α and β determine the balance between the computational work performed in the line search and the reduction in function achieved. (See [758] for illustrations and further discussion). The work in the line search (number of polynomial interpolations) should be balanced with the overall progress realized in the minimization algorithm.

10.3.3 Trust-Region-Based Descent Algorithm

Algorithm [A2]: Basic Descent By A Trust Region Subsearch

From a given point $\mathbf{x}_0$, perform for $k = 0, 1, 2, \ldots$ *until convergence*:

1. Test $\mathbf{x}_k$ for convergence (see subsection 10.3.4).
2. Calculate a step $\mathbf{s}_k$ by solving the subproblem

$$\min_{\mathbf{s}} \{q_k(\mathbf{s})\}\,, \tag{10.18}$$

   where q_k is the quadratic model of the objective function:

$$q_k(\mathbf{s}) \;=\; f(\mathbf{x}_k) \;+\; \mathbf{g}_k^T\mathbf{s} \;+\; \frac{1}{2}\,\mathbf{s}^T\,\mathbf{H}_k\,\mathbf{s}\,, \tag{10.19}$$

   subject to a size bound, Δ_k (a positive value), on s. This bound involves a scaling matrix, $\mathbf{D}_k$, and requires

$$\|\,\mathbf{D}_k\mathbf{s}\,\| \;<\Delta_k\,, \tag{10.20}$$

   where $\|\cdot\|$ denotes the standard Euclidean norm.
3. Set $\mathbf{x}_{k+1}$ to $\mathbf{x}_k + \mathbf{s}_k$ and k to $k+1$ and go to step 1.

Basic Idea

The idea in trust-region methods — the origin of the quadratic optimization subproblem in step 2 above — is to determine the vector $\mathbf{s}_k$ on the basis of the size of region within which the quadratic functional approximation can be "trusted" (i.e., is reasonable). The quality of the quadratic approximation can be assessed from the following ratio:

$$\rho_k \;=\; \frac{f(\mathbf{x}_k) \;-\; f(\mathbf{x}_k + \mathbf{s}_k)}{f(\mathbf{x}_k) \;-\; q_k(\mathbf{s}_k)}\,. \tag{10.21}$$

A value near unity implied that the bound Δ_k imposed on s can be increased; in contrast, a negative or a small positive value for ρ_k implies that the quadratic model is poor, requiring a decrease in Δ_k.

Many Newton methods (see next section for details) based on trust-region approaches determine a candidate $\mathbf{s}_k$ by solving the linear system

$$\mathbf{H}_k\,\mathbf{s} \;=\; -\mathbf{g}_k \tag{10.22}$$

that results from minimizing $q_k(\mathbf{s})$. (A related system may also be formulated). The scaling of this vector s is determined according to the quality of the quadratic model at the region of approximation. A good source of a trust-region Newton method is the program LANCELOT [769].

10.3.4 Convergence Criteria

The criteria used to define convergence of the minimization algorithm (in step 1 of Algorithms [A1] or [A2]) must be chosen with care. The desire to obtain as accurate a result as possible should be balanced with the amount of computation involved. In other words, it is wasteful to continue a loop when the answer can no longer be improved. A well-structured algorithm should halt the iteration process when progress is poor.

The gradient norm value, together with measures of progress in the function values and the independent variables (e.g., eqs. (10.25) and (10.26)) are used to assess performance. Upper limits for the total number of allowable function and/or gradient evaluations are important safeguards against wasteful computing cycles.

Specifically, reasonable tests for the size of the gradient norm are:

$$\|\mathbf{g}_k\| \ \leq \epsilon_g \, (1 + |\ f(\mathbf{x}_k) \ |) \tag{10.23}$$

or

$$\|\mathbf{g}_k\| \ \leq \epsilon_g \ \max\,(1, \ \|\ \mathbf{x}_k \ \|)\,, \tag{10.24}$$

where the gradient norm may be set to the Euclidean norm divided by $\sqrt{n}$ (this introduces a dependency on the number of variables). The parameter ϵ_g is a small positive number such as 10^{-6} that might depend on the *machine precision*, ϵ_m; ϵ_m is roughly the largest number for which $1 + \epsilon_m = 1$ in computer representation. The student is encouraged to code a routine for determining ϵ_m.[3]

For example, our truncated-Newton package TNPACK [771, 772, 773] checks the following four conditions at each iteration:

$$\begin{aligned}
f(\mathbf{x}_{k-1}) - f(\mathbf{x}_k) \quad &< \quad \epsilon_f \, (1 + |f(\mathbf{x}_k)|)\,, && (10.25)\\
\|\ \mathbf{x}_{k-1} - \mathbf{x}_k \ \| \quad &< \quad (\epsilon_f)^{1/2} \, (1 + \|\ \mathbf{x}_k \ \|)\,, && (10.26)\\
\|\ \mathbf{g}_k \ \| \quad &< \quad (\epsilon_f)^{1/3} \, (1 + |f(\mathbf{x}_k)|)\,, && (10.27)\\
\|\ \mathbf{g}_k \ \| \quad &< \quad \epsilon_g \, (1 + |f(\mathbf{x}_k)|)\,. && (10.28)
\end{aligned}$$

Here, all norms are the Euclidean norm divided by $\sqrt{n}$, and ϵ_f and ϵ_g are small numbers (like 10^{-10} and 10^{-8}, respectively). If the first three conditions above are satisfied, or the fourth condition alone is satisfied, convergence is considered to have been satisfied and the minimization process is halted; otherwise, the loop continues. Note that the first and second conditions test for convergence the sequences of the function values and iterates of the independent variables, respectively, while the third and fourth conditions test the size of the gradient norm.

---

[3]Typically, ϵ_m is 10^{-15} and 10^{-7}, respectively, for double and single-precision IEEE arithmetic [770].

Box 10.2: Historical Perspective of 'Newton's' Method

The method's credit to Sir Isaac Newton is a partial one. Although many references also credit Joseph Raphson, the contributions of mathematicians Thomas Simpson and Jean-Baptiste-Joseph Fourier are also noteworthy. Furthermore, Newton's description of an algebraic procedure for solving for the zeros of a polynomial in 1664 had its roots in the work of the 16th-century French algebraist François Viète. Viète's work itself had precursors in the 11th-century works of Arabic algebraists.

In 1687, three years after Newton described a root finder for a polynomial, Newton described in *Principia Mathematica* an application of his procedure to a nonpolynomial equation. That equation originated from the problem of solving Kepler's equation: determining the position of a planet moving in an elliptical orbit around the sun, given the time elapsed since it was nearest the sun. Newton's procedure was nonetheless purely *algebraic* and not even iterative.

In 1690, Raphson turned Newton's method into an *iterative* one, applying it to the solution of polynomial equations of degree up to ten. His formulation still did not use calculus; instead he derived explicit polynomial expressions for $f(x)$ and $f'(x)$.

Simpson in 1740 was first to formulate the Newton-Raphson method on the basis of calculus. He applied this iterative scheme to solve general systems of nonlinear equations. In addition to this important extension of the method to nonlinear systems, Simpson extended the iterative solver to multivariate minimization, noting that the nonlinear solver can be applied to optimization by setting the gradient to zero.

Finally, Fourier in 1831 published the modern version of the method as we know it today in his celebrated book *Analyse des Équations Determinées*. The method for solving $f(x) = 0$ was simply written as: $x_{k+1} = x_k - f(x_k)/f'(x_k)$. Unfortunately, Fourier omitted credits to either Raphson or Simpson, possibly explaining the method's name.

Thus, strictly speaking, it is appropriate to title the method as the Newton-Raphson-Simpson-Fourier method.

10.4 The Newton-Raphson-Simpson-Fourier Method

Newton's method is a classic iterative scheme for solving a nonlinear system $f(\mathbf{x}) = 0$ or for minimizing the multivariate function $f(\mathbf{x})$. These root-finding and minimization problems are closely related.

Though the method is credited to Newton or Newton and Raphson, key contributions were made also by Fourier and Simpson; see Box 10.2 for a historical perspective. For brevity, we refer to the "Newton-Raphson-Simpson-Fourier method" as Newton's method.

A Fundamental Optimization Tool

Many effective methods for nonlinear, multivariate minimization can be related to Newton's method. Hence, a good understanding of the Newton solver, including performance and convergence behavior, is invaluable for applying optimization techniques in general.

We first discuss the univariate case of Newton's method for obtaining the zeros of a function $f(x)$. In one dimension, instructive diagrams easily illustrate the method's strengths and weaknesses. We then discuss the general multivariate formulations. The section that follows continues by describing the effective variants known as quasi-Newton, nonlinear conjugate gradient, and truncated-Newton methods.

10.4.1 The One-Dimensional Version of Newton's Method

Iterative Recipe

The modern version of Newton's method (see Box 10.2) for solving $f(x) = 0$ is:

$$x_{k+1} \; = \; x_k \; - \; f(x_k)/f'(x_k) \, . \tag{10.29}$$

This iterative scheme can be derived easily by using a Taylor expansion to approximate a twice-differentiable function f locally by a quadratic function about x_k:

$$f(x_{k+1}) \; = \; f(x_k) \, + \, (x_{k+1} \, - \, x_k) \, f'(x_k) \, + \, \frac{1}{2}(x_{k+1} \, - \, x_k)^2 \, f''(\xi) \tag{10.30}$$

where $x_k \; \leq \; \xi \; \leq \; x_{k+1}$. Omitting the second-derivative term, the solution of $f(x_{k+1}) \; = \; 0$ yields the iteration process of eq. (10.29). The related *discrete-Newton* and *quasi-Newton* methods [756] correspond to approximating $f'(x)$ by *finite-differences*, as

$$f'(x_k) \; \approx \; [f(x_k + h) \, - \, f(x_k)] \, / \, h \, , \tag{10.31}$$

or by the method of *secants*

$$f'(x_k) \; \approx \; [f(x_k) \, - \, f(x_{k-1})] \, / \, (x_k \, - \, x_{k-1}) \, , \tag{10.32}$$

where h is a suitably-chosen small number.

Geometric Interpretation

Newton's method in one dimension has a simple geometric interpretation: at each step, approximate $f(x)$ by its tangent at point $\{x_k \, , \; f(x_k)\}$ and take x_{k+1} as the abscissa of the intersection of this line with the x-axis (see Figure 10.7).

Performance

The method works well in the ideal case (Figure 10.7a), when x_0 is near the solution (x^*) and $|f'(\xi)| \geq M > 0$ nearby.

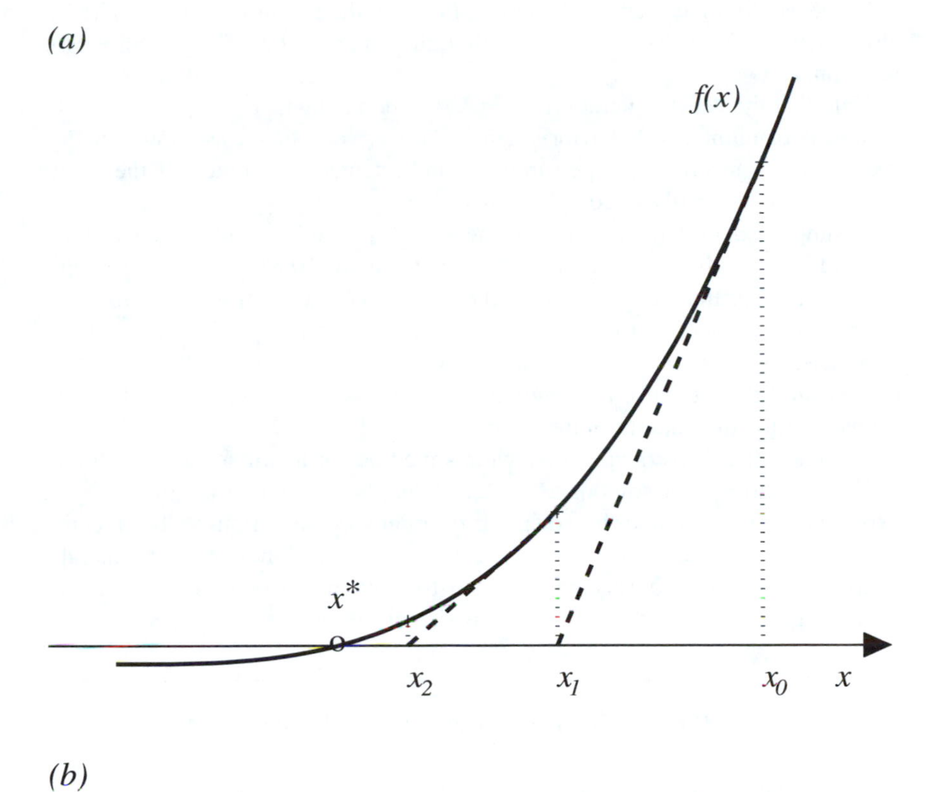

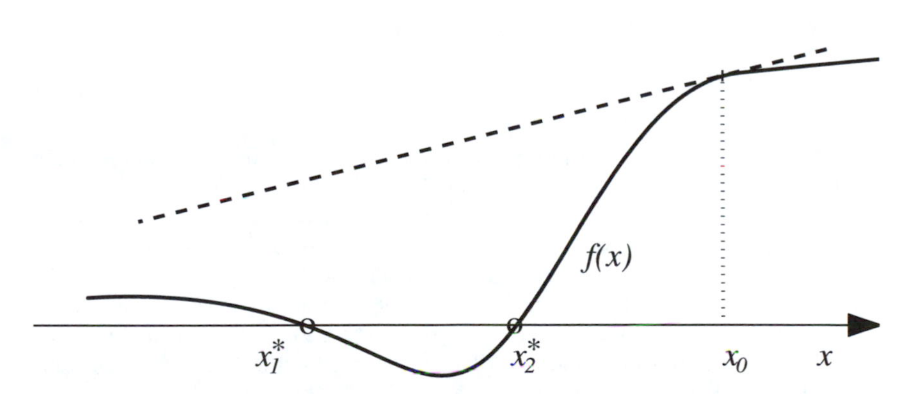

Figure 10.7. Newton's method in one dimension: (a) geometric interpretation and behavior in the ideal case (rapid convergence), and (b) divergent behavior near point where $f'(x) = 0$.

However, difficulties arise when x_0 is far from the solution, or when $f'(x)$ is close to zero (Figure 10.7b). Further difficulties emerge when $f'(x)$ is zero at the solution.

Note that the Newton iteration process is undefined when $f'(x) = 0$ and can exhibit poor numerical behavior when $|f'(x)|$ is very small, as shown in Figure 10.7b. In general, both performance and attainable accuracy of the solver worsen if any of the above complications arise.

A simple example for solving for the square root of a number by Newton's method in Box 10.3 (with associated data in Figure 10.8) illustrates the rapid convergence for the ideal case when the root of $f(x)$ is simple and reasonably separated from the other root. In the non-ideal case (e.g., x_0 far from the solution or $f'(x)$ close to zero), convergence is slow at the beginning but then improves rapidly until the region of *quadratic convergence* is approached. (See Box 10.1 for a definition of quadratic convergence).

The quadratic convergence of Newton's method for a simple root and for x_0 sufficiently close to the solution x^* can easily be shown on the basis of the Taylor expansion. Tensor methods based on fourth-order approximations to the objective function can achieve more rapid convergence [774], but they are not generally applicable. The attainable accuracy for Newton's method depends on the function characteristics, and on whether the root is simple or not.

Box 10.3: Newton's Method: Simple Examples

We can apply Newton's method to solve for the square root of a number a by defining

$$f(x) = x^2 - a = 0\,;$$

the resulting iterative scheme for computing $x = \sqrt{a}$ is:

$$x_{k+1} = \frac{1}{2}\left[x_k + \frac{a}{x_k}\right], \qquad (10.33)$$

defined for $x_k \neq 0$. A computer result from a double-precision program is shown in Figure 10.8 for $a = 0.01$ with four starting points: 5, -100, 1000, and 10^{-6}.

The rapid, *quadratic* convergence (see Box 10.1) can be noted in all cases at the last 3–4 steps. In these steps, the number of correct digits for the solution is approximately doubled from one step to the next! Note the larger number of iterations for convergence when x_0 is near zero ($x_0 = 10^{-6}$ shown). Since the derivative of the objective function is zero at $x = 0$, the tangent takes the iterates very far, as illustrated in Fig. 10.7b, but then works back systematically toward the solution.

Figure 10.9 further illustrates the notions of of accuracy, convergence, and problem conditioning for solving

$$f(x) = x^3 - bx = 0, \quad b > 0\,,$$

Figure 10.8. Computer output from the application of Newton's method to solve the simple quadratic $x^2 = a, a = 0.01$, from four different starting points.

	Newton iterate, x	Error: $\lvert x - x^* \rvert / x^*$ (for nonzero x^*)
	$x_0 = 5$	
0	5.000000000000000	49.00000000000000
1	2.501000000000000	24.01000000000000
2	1.252499200319872	11.52499200319872
3	0.6302416186767726	5.302416186767727
4	0.3230542746187304	2.230542746187304
5	0.1770044127792565	0.7700441277925646
6	0.1167500897135064	0.1675008971350636
7	0.1012015644103529	1.2015644103528789E-02
8	0.1000071330766507	7.1330766507338161E-05
9	0.1000000002543858	2.5438576245484512E-09
10	0.1000000000000000	0.0000000000000000
	$x_0 = -100$	
0	−100.0000000000000	999.0000000000000
1	−50.00005000000000	499.0005000000000
2	−25.00012499990000	249.0012499990000
3	−12.50026249895001	124.0026249895001
4	−6.250531241075214	61.50531241075214
5	−3.126065552544528	30.26065552544528
6	−1.564632230895322	14.64632230895322
7	−0.7855117545897353	6.855117545897353
8	−0.3991211544161648	2.991211544161648
9	−0.2120881016068179	1.120881016068179
10	−0.1296191592706879	0.2961915927068787
11	−0.1033841239244204	3.3841239244203625E-02
12	−0.1000553871053945	5.5387105394461011E-04
13	−0.1000000153301663	1.5330166275306922E-07
14	−0.1000000000000012	1.1657341758564144E-14
15	−0.1000000000000000	0.0000000000000000
	$x_0 = 1000$	
0	1000.000000000000	9999.000000000000
1	500.0000050000000	4999.000050000000
2	250.0000124999999	2499.000124999999
3	125.0000262499989	1249.000262499989
4	62.50005312499107	624.0005312499106
5	31.25010656242754	311.5010656242753
6	15.62521328066817	155.2521328066817
7	7.812926635966156	77.12926635966156
8	3.907103283034968	38.07103283034968
9	1.954831361974763	18.54831361974762
10	0.9799734463768975	8.799734463768974
11	0.4950889022510404	3.950889022510404
12	0.2576436474057566	1.576436474057566
13	0.1482284733538422	0.4822847335384220
14	0.1078459475072978	7.8459475072977791E-02
15	0.1002854019724900	2.8540197248999588E-03
16	0.1000004061123768	4.0611237676901890E-06
17	0.1000000000008246	8.2461815154033502E-12
18	0.1000000000000000	0.0000000000000000
	$x_0 = 10^{-6}$	
0	9.9999999999999995E-07	0.9999900000000000
1	5000.000000500000	49999.00000499999
2	2500.000001250000	24999.00001250000
3	1250.000002625000	12499.00002625000
4	625.0000053125000	6249.000053124999
5	312.5000106562499	3124.000106562499
6	156.2500213281244	1561.500213281244
7	78.12504266405783	780.2504266405783
8	39.06258533199397	389.6258533199396
9	19.53142066571737	194.3142066571736
10	9.765966330621753	96.65966330621752
11	4.883495147415947	47.83495147415947
12	2.442771430566446	23.42771430566446
13	1.223432570726772	11.23432570726771
14	0.6158031472956539	5.158031472956539
15	0.3160210514743984	2.160210514743984
16	0.1738322565259314	0.7383225652593136
17	0.1156794895626801	0.1567948956268007
18	0.1010626187661944	1.0626187661944286E-02
19	0.1000055864307498	5.5864307498265653E-05
20	0.1000000001560323	1.5603232594862959E-09
21	0.1000000000000000	0.0000000000000000

by Newton's method. The three roots are $-\sqrt{b}$, 0, and $+\sqrt{b}$. The corresponding iterative scheme for solving this cubic polynomial of x becomes:

$$x_{k+1} = 2x_k^3/(3x_k^2 - b)\,. \tag{10.34}$$

Near the point where the Newton iteration is undefined ($x_k = \sqrt{b/3}$), the iterative process converges very slowly. When b is small, as in our example ($b = 0.0001$), the three roots are relatively close. The Newton iterates in Figure 10.9 started from -50, $\sqrt{b/3} + 10^{-10}$, -1, 10^{-10}, 0.009, and 0.011 show that the solution obtained depends on the starting point.

10.4.2 Newton's Method for Minimization

To derive the iteration process of Newton's method for minimization of the one-dimensional $f(x)$, we use a quadratic, rather than linear, approximation:

$$f(x_{k+1}) \approx f(x_k) + (x_{k+1} - x_k)\, f'(x_k) + \frac{1}{2}(x_{k+1} - x_k)^2\, f''(x_k)\,. \tag{10.35}$$

Figure 10.9. Computer output from the application of Newton's method to solve $x^3 - bx = 0, b = 10^{-4}$, from various starting points.

	Newton iterate, x	Error: $\lvert x - x^* \rvert / x^*$ (for nonzero x^*)
	$x_0 = -50$	
0	−50.00000000000000	4999.000000000000
1	−33.33333377777778	3332.333377777778
2	−22.22222318518520	2221.222318518520
3	−14.81481645679016	1480.481645679016
4	−9.876545804526833	986.6545804526833
5	−6.584366119684733	657.4366119684732
6	−4.389580788123710	437.9580788123710
7	−2.926392254584279	291.6392254584279
8	−1.950935763478845	194.0935763478845
9	−1.300635232964303	129.0635232964302
10	−0.8671072413145484	85.71072413145484
11	−0.5780971233434421	56.80971233434420
12	−0.3854365263554411	37.54365263554411
13	−0.2570153518629270	24.70153518629269
14	−0.1714300741869435	16.14300741869435
15	−0.1144164918146208	10.44164918146208
16	−7.6472379206287938E-02	6.647237920628794
17	−5.1273843468759718E-02	4.127384346875972
18	−3.4621530718982767E-02	2.462153071898276
19	−2.3741241957549244E-02	1.374124195754924
20	−1.6822346583303217E-02	0.6822346583303216
21	−1.2712265750664345E-02	0.2712265750664345
22	−1.0677216879173919E-02	6.7721687917391901E-02
23	−1.0059418708361139E-02	5.9418708361139161E-03
24	−1.0000522346453959E-02	5.2234645395859980E-05
25	−1.0000000040921884E-02	4.0921883937006243E-09
26	−1.0000000000000000	0.0000000000000000
	$x_0 = -1$	
0	−1.000000000000000	99.00000000000000
1	−0.6666888896296543	65.66888896296543
2	−0.4444925944752626	43.44925944752625
3	−0.2963783993367317	28.63783993367316
4	−0.1976606072449507	18.76606072449507
5	−0.1318862603202570	12.18862603202570
6	−8.8092992421240537E-02	7.809299242124054
7	−5.8982008504053302E-02	4.898200850405330
8	−3.9701746654566113E-02	2.970174665456611
9	−2.7039652798163824E-02	1.703965279816382
10	−1.8887531503436573E-02	0.8887531503436572
11	−1.3889510079148154E-02	0.3889510079148153
12	−1.1193786717951709E-02	0.1193786717951709
13	−1.0167292368239794E-02	1.6729236823979400E-02
14	−1.0004040357696271E-02	4.0403576962704663E-04
15	−1.0000002446367255E-02	2.4463672546048976E-07
16	−1.0000000000000899E-02	8.9858676055598607E-14
17	−1.0000000000000002E-02	1.7347234759768071E-16
	$x_0 = 10^{-10}$	
0	1.0000000000000000E-10	
1	−2.0000000000000006E-26	
2	1.6000000000000012E-73	
3	−8.1920000000000178E-215	
4	0.0000000000000000	
	$x_0 = .009$	
0	8.9999999999999993E-03	0.1000000000000001
1	1.0195804195804197E-02	1.9580419580419658E-02
2	1.0005499738015171E-02	5.4997380151706327E-04
3	1.0000004531252979E-02	4.5312529787372435E-07
4	1.0000000000003079E-02	3.0791341698588326E-13
5	9.9999999999999985E-03	1.7347234759768071E-16
	$x_0 = .0058 + 10^{-10}$	
0	5.7735027918962576E-03	0.4226497208103743
1	111111.1159104916	11111110.59104916
2	74074.07727366130	7407406.727366131
3	49382.71818244116	4938270.818244115
4	32921.81212162789	3292180.212162788
5	21947.87474775260	2194786.474775260
6	14631.91649850275	1463190.649850275
7	9754.610999003351	975460.0999003351
8	6503.073999337846	650306.3999337845
9	4335.382666228647	433537.2666228647
10	2890.255110824224	289024.5110824224
11	1926.836740557172	192682.6740557172
12	1284.557827049647	128454.7827049647
13	856.3718847170644	85636.18847170644
14	570.9145898373255	57090.45898373255
15	380.6097265971409	38059.97265971409
16	253.7398177898131	25372.98177898131
17	169.1598786141209	16914.98786141209
18	112.7732525407821	11276.32525407821
19	75.18216855757360	7517.216855757360
20	50.12144600062744	5011.144600062744
21	33.41429777711917	3340.429777711918
22	22.27619918313077	2226.619918313077
23	14.85080045299750	1484.080045299750
24	9.900535131697179	989.0535131697179
25	6.600358999013134	659.0358999013134
26	4.400242699498232	439.0242699498232
27	2.933500183234001	292.3500183234001
28	1.955674364178697	194.5674364178697
29	1.303794272497512	129.3794272497512
30	0.8692132262697133	85.92132262697133
31	0.5795010512121747	56.95010512121746
32	0.3863723851127763	37.63723851127763
33	0.2576391179580148	24.76391179580147
34	0.1718457086006079	16.18457086006078
35	0.1146932668316612	10.46932668316612
36	7.6656423605242732E-02	6.665642360524274
37	5.1395830038388754E-02	4.139583003838875
38	3.4701786556898713E-02	2.470178655689871
39	2.3793131896999858E-02	1.379313189699986
40	1.6854498439733485E-02	0.6854498439733484
41	1.2730083741629223E-02	0.2730083741629223
42	1.0684415288974207E-02	6.8441528897420639E-02
43	1.0060600942778943E-02	6.0600942778942651E-03
44	1.0000543191397791E-02	5.4319139779039627E-05
45	1.0000000044252924E-02	4.4252924241705571E-09
46	1.0000000000000000E-02	0.0000000000000000
	$x_0 = .011$	
0	1.0999999999999999E-02	9.9999999999999908E-02
1	1.0121673003802283E-02	1.2167300380228237E-02
2	1.0002159360555317E-02	2.1593605553163475E-04
3	1.0000000699073473E-02	6.9907347272080145E-08
4	1.0000000000000073E-02	7.2858385991025898E-15
5	9.9999999999999985E-03	1.7347234759768071E-16

Since $f(x_k)$ is constant, minimization of the second and third terms on the right-hand-side in eq. (10.35) yields the iteration process:

$$x_{k+1} = x_k - f'(x_k)/f''(x_k). \tag{10.36}$$

Thus, we have replaced f and f' of eq. (10.29) by f' and f'', respectively. This Newton scheme for minimizing $f(x)$ is defined as long as the second derivative at x_k is nonzero.

10.4.3 The Multivariate Version of Newton's Method

We generalize Newton's method for minimization in eq. (10.36) to multivariate functions by expanding $f(\mathbf{x})$ locally along a search vector $\mathbf{p}$ (in analogy to eq. (10.35)):

$$f(\mathbf{x}_k + \mathbf{p}_k) \;\approx\; f(\mathbf{x}_k) \;+\; \mathbf{g}(\mathbf{x}_k)^T \mathbf{p}_k \;+\frac{1}{2}\, \mathbf{p}_k^T \mathbf{H}(\mathbf{x}_k)\, \mathbf{p}_k \,. \tag{10.37}$$

Minimizing the right-hand side leads to solving the linear system of equations, known as the *Newton equations*, for $\mathbf{p}_k$, as long as $\mathbf{H}_k$ is positive definite:

$$\mathbf{H}_k\, \mathbf{p}_k \;=\; -\mathbf{g}_k \,. \tag{10.38}$$

Performing this approximation at each step k to obtain $\mathbf{p}_k$ leads to the iteration process

$$\mathbf{x}_{x+1} \;=\; \mathbf{x}_k \;-\; \mathbf{H}_k^{-1}\, \mathbf{g}_k \,. \tag{10.39}$$

Thus, the search vector

$$\mathbf{p}_k = -\mathbf{H}_k^{-1}\, \mathbf{g}_k \tag{10.40}$$

is used at each step of the *classic* Newton method for minimization. This requires repeated solutions of a linear system involving the Hessian. Not only is this an expensive, order n^3 process for general dense matrices; for multivariate functions with many minima and maxima, the Hessian may be *ill-conditioned* (i.e., have large maximal-to-minimal eigenvalue ratio $\lambda_{\max}/\lambda_{\min}$) or *singular* (zero eigenvalues) for certain $\mathbf{x}_k$.

Thus, in addition to the line-search or trust-region modifications that essentially *dampen* the Newton step (by scaling $\mathbf{p}_k$ by a positive scalar less than unity), effective strategies must be devised to ensure that $\mathbf{p}_k$ is well defined at each step. Such effective strategies are described in the next section. These include quasi-Newton (QN), nonlinear conjugate gradient (CG), and truncated Newton (TN) methods.

10.5 Effective Large-Scale Minimization Algorithms

The popular methods that fit the descent framework outlined in subsections 10.3.2 and 10.3.3 require gradient information. In addition, the truncated-Newton (TN) method may require more input to be effective, such as second-derivative information from components of the objective function that can be computed cheaply.

The steepest descent method (recall $\mathbf{p}_k = -\mathbf{g}_k$) can be viewed as a simple version of the $\mathbf{p}_k$ definition in eq. (10.40) in which the Hessian replaced by the identity matrix.

Nonlinear CG methods improve upon (the generally poor) convergence of SD methods by using better search directions than SD that are still cheap to compute.

QN methods, which are closely related to nonlinear CG methods, can also be presented as robust alternatives to the classic Newton method ($\mathbf{p}_k$ by eq. (10.40)) which update curvature information as the algorithm proceeds.

TN methods are another clever and robust alternative to the classic Newton framework that introduce curvature information only when locally warranted, so as to balance computation with realized convergence. Hybrid schemes have also been devised, e.g., limited-memory QN with TN [775].

The methods described in turn in this section — QN, nonlinear CG, and TN methods — render SD obsolete as a general method.

10.5.1 Quasi-Newton (QN)

Basic Idea

QN methods avoid using the actual Hessian and instead build-up curvature information as the algorithm proceeds [756, 776]. Actually, it is often the Hessian inverse ($\widehat{\mathbf{B}}$) that is updated in practice so that a term $\widehat{\mathbf{B}}_k \, \mathbf{g}_k$ replaces $\mathbf{H}_k^{-1} \, \mathbf{g}_k$ in eq. (10.39). Here $\widehat{\mathbf{B}}_k$ is short hand for $\widehat{\mathbf{B}}(\mathbf{x}_k)$.

The Hessian approximation $\mathbf{B}_k$ is derived to satisfy the *quasi-Newton condition* (see below). QN variants define different formulas that satisfy this condition.

Because memory is considered premium for large-scale applications, the matrix $\mathbf{B}_k$ or $\widehat{\mathbf{B}}$ is formulated through several vector operations, avoiding explicit storage of an $n \times n$ matrix. In practice, $\mathbf{B}_k$ is updated by adding a *low rank* update matrix $\mathbf{U}_k$.

Recent Advances

Two important developments have emerged in modern optimization research in connection with QN methodology. The first is the development of *limited-memory* versions, in which the inverse Hessian approximation at step k only incorporates curvature information generated at the last few m steps (e.g., $m = 5$) [777, 778, 779, 756]. The second is the emergence of insightful analyses that explain the relationship between QN and nonlinear CG methods.

QN Condition

The QN condition specifies the property that the new approximation $\mathbf{B}_{k+1}$ must satisfy:

$$\mathbf{B}_{k+1} \, \mathbf{s}_k = \mathbf{y}_k \, . \tag{10.41}$$

Here

$$\mathbf{s}_k = \mathbf{x}_{k+1} - \mathbf{x}_k \tag{10.42}$$

and

$$\mathbf{y}_k = \mathbf{g}_{k+1} - \mathbf{g}_k \,. \tag{10.43}$$

(Note that the 'step vector' $\mathbf{s}_k$ can be equated with the displacement from $\mathbf{x}_k$, namely $\lambda_k \, \mathbf{p}_k$, used in the basic Algorithm [A1]). If $f(\mathbf{x})$ were a quadratic function, its Hessian $\mathbf{H}$ would be a constant and would satisfy (from the Taylor expansion of the gradient) the following relation:

$$\mathbf{g}_{k+1} - \mathbf{g}_k = \mathbf{H} \, (\mathbf{x}_{k+1} - \mathbf{x}_k) \,. \tag{10.44}$$

This equation makes clear the origin of the QN condition of eq. (10.41).

Updating Formula

The updating QN formula can be written symbolically as:

$$\mathbf{B}_{k+1} = \mathbf{B}_k + \mathbf{U}_k(\mathbf{s}_k, \mathbf{y}_k, \mathbf{B}_k) \tag{10.45}$$

where $\mathbf{U}_k$ is a matrix of low rank (typically 1 or 2). Note that a rank 1 matrix can be written as the outer product of two vectors: $\mathbf{u}\mathbf{v}^T$. In addition to rank, imposed symmetry and positive-definiteness are used in the formulation of $\mathbf{U}_k$.

BFGS Method

One of the most successful QN formulas in practice is associated with the BFGS method (for its developers Broyden, Fletcher, Goldfarb, and Shanno). The BFGS update matrix has rank 2 and inherent positive definiteness (i.e., if $\mathbf{B}_k$ is positive definite then $\mathbf{B}_{k+1}$ is positive definite) as long as $\mathbf{y}_k^T \mathbf{s}_k > 0$. This condition is satisfied automatically for convex functions but may not hold in general without the sufficient reduction of curvature criteria (eq. (10.14)) in the line search. In practice, the line search must check for the descent property; updates that do not satisfy this condition may be skipped.

The BFGS update formula is given by

$$\mathbf{B}_{k+1} = \mathbf{B}_k - \frac{\mathbf{B}_k \mathbf{s}_k \mathbf{s}_k^T \mathbf{B}_k^T}{\mathbf{s}_k^T \mathbf{B}_k \mathbf{s}_k} + \frac{\mathbf{y}_k \mathbf{y}_k^T}{\mathbf{y}_k^T \mathbf{s}_k} \,. \tag{10.46}$$

The corresponding formula used in practice to update the inverse of $\mathbf{B}$, namely $\widehat{\mathbf{B}}$, is:

$$\widehat{\mathbf{B}}_{k+1} = \left(\mathbf{I} - \frac{\mathbf{s}_k \mathbf{y}_k^T}{\mathbf{y}_k^T \mathbf{s}_k} \right) \widehat{\mathbf{B}}_k \left(\mathbf{I} - \frac{\mathbf{y}_k \mathbf{s}_k^T}{\mathbf{y}_k^T \mathbf{s}_k} \right) + \frac{\mathbf{s}_k \mathbf{s}_k^T}{\mathbf{y}_k^T \mathbf{s}_k} \,. \tag{10.47}$$

From this $\widehat{\mathbf{B}}$, the BFGS search vector is defined as

$$\mathbf{p}_k = -\widehat{\mathbf{B}}_k \, \mathbf{g}_k \,; \tag{10.48}$$

(compare to the Newton search vector defined in eq. (10.40)).

Practical Implementation

Because we only require the product of $\widehat{\mathbf{B}}$ with the gradient (and not $\mathbf{B}$ or $\widehat{\mathbf{B}}$ *per se*), effective matrix/vector products have been developed to minimize storage requirements by using low-rank QN updates. This requires $\mathcal{O}(n)$ memory to store the successive pairs of update vectors ($\mathbf{s}_k$ and $\mathbf{y}_k$) and the respective inner products $\mathbf{y}_k^T \mathbf{s}_k$.

Limited-memory QN methods reduce storage requirements further by only retaining the $\{\mathbf{s}, \mathbf{y}\}$ pairs from the previous few iterates (3–7). The identity matrix, $\mathbf{I}$, or a multiple of it, is typically used for the initial Hessian approximation $\mathbf{B}_0$. Updating this scaling at each iteration enhances overall efficiency [778, 779].

The limited-memory BFGS code of Nocedal and co-workers [780] is one of the most effective methods in this class. The combination of modest memory, requiring only gradient information, and good performance in practice makes it an excellent choice for large-scale multivariate minimization [781]. The method has been extended to constrained optimization [782, 783, 784], used to propose preconditioners for CG methods [785], and combined with TN methods in a QN/TN cyclic fashion [775]. The text of Nocedal and Wright [756] presents a comprehensive description of the limited-memory BFGS method.

10.5.2 Conjugate Gradient (CG)

Nonlinear CG methods form another popular type of optimization scheme for large-scale problems where memory and computational performance are important considerations. These methods were first developed in the 1960s by combining the linear CG method (an iterative technique for solving linear systems $\mathbf{A}\mathbf{x} = \mathbf{b}$ where $\mathbf{A}$ is an $n \times n$ matrix [786]) with line-search techniques. The basic idea is that if f were a convex quadratic function, the resulting nonlinear CG method would reduce to solving the Newton equations (eq. (10.38)) for the search vector $\mathbf{p}$ when $\mathbf{H}$ is a constant positive-definite matrix.

CG Search Vector

In each step of the nonlinear CG method, a search vector $\mathbf{p}_k$ is defined by a recursive formula. A line search is then used as outlined in Algorithm [A1]. The iteration process that defines the search vectors $\{\mathbf{p}_k\}$ is given by:

$$\mathbf{p}_{k+1} = -\mathbf{g}_{k+1} + \beta_{k+1} \mathbf{p}_k , \tag{10.49}$$

where $\mathbf{p}_0 = -\mathbf{g}_0$. The scheme-dependent parameter β_k that defines the search vectors is chosen so that if f were a convex quadratic and the line search exact (i.e., $\mathbf{x}_k + \lambda_k \mathbf{p}_k$ minimizes f exactly along $\mathbf{p}_k$), then the *linear* CG process would result. The reduction to the linear CG method in this special case is important because linear CG is known to terminate in at most n steps of exact arithmetic. This finite-termination property relies on the fundamental notion that two sets of vectors ($\{\mathbf{g}\}$ and $\{\mathbf{p}\}$) generated in the CG method satisfy

$$\mathbf{g}_k^T \mathbf{p}_j = 0 \quad \text{for all} \quad j < k \,.$$

This orthogonality condition implies that the search vectors span the entire n-dimensional space after n steps, so that $\mathbf{g}_{n+1} = 0$ in finite arithmetic.

CG Variants

Different formulas for β_k (not to be confused with the line search parameter introduced earlier) have been developed for the nonlinear CG case, though they all reduce to the same expressions for convex quadratic functions. These variants exhibit different behavior in practice.

Three of the best known algorithms are due to Fletcher-Reeves (FR), Polak-Ribière (PR), and Hestenes-Stiefel (HS). They are defined by the parameter β (for eq. (10.49)) as:

$$\beta_{k+1}^{\mathrm{FR}} = \mathbf{g}_{k+1}^T \mathbf{g}_{k+1} \,/\, \mathbf{g}_k^{\mathrm{T}} \mathbf{g}_k \,, \tag{10.50}$$

$$\beta_{k+1}^{\mathrm{PR}} = \mathbf{g}_{k+1}^T \mathbf{y}_k \,/\, \mathbf{g}_k^{\mathrm{T}} \mathbf{g}_k \,, \tag{10.51}$$

$$\beta_{k+1}^{\mathrm{HS}} = \mathbf{g}_{k+1}^T \mathbf{y}_k \,/\, \mathbf{p}_k^{\mathrm{T}} \mathbf{y}_k \,. \tag{10.52}$$

(Recall $\mathbf{y}_k = \mathbf{g}_{k+1} - \mathbf{g}_k$).

The PR version is often found in software packages. Still, to be effective PR restarts the iteration process, i.e., sets β_k to zero occasionally, for example when β_k becomes negative.

Some important modifications of this version are due to Powell [787], available in the IMSL library, and to Shanno & Phua [788], available in the NAG library. These modifications have slightly more memory requirements but fewer function evaluations. An interesting CG–PR–FR hybrid algorithm might also be an effective alternative [789].

A careful line search is important for nonlinear CG methods.

CG/QN Connection

Key connections between CG and QN-Newton algorithms for minimization began to emerge in the late 1970s. Essentially, it was found that the CG conjugacy property can be closely related to the QN condition, and thus an appropriate formula for β_k could be obtained from both viewpoints.

The many developments in the 1980s have shown that the limited-memory QN class of algorithms balances the extremely modest storage requirements of nonlinear CG with good convergence properties in practice. The fact that the unit steplength in QN methods is often acceptable leads to greater efficiency in terms of function evaluations and hence less computational time overall.

Still, the linear and nonlinear CG methods play important theoretical roles in the numerical analysis literature, as well as practical roles in many numerical techniques; see the research monograph of [790] for a modern perspective. The linear CG method, in particular, proves ideal for solving the linear subproblem in the truncated Newton method for minimization (discussed next), especially with convergence-accelerating techniques known as *preconditioning*.

10.5.3 Truncated-Newton (TN)

Approximate Solution of the Newton Equations

In the early 1980s a very simple but important idea emerged in connection with the Newton equations: why solve this linear system for the search vector $\mathbf{p}_k$ exactly [791]? In the context of large-scale nonlinear optimization, an accurate solution of eq. (10.38) is not warranted! Far away from a minimum, any descent direction that can be computed cheaply may still produce progress toward a minimum. Only near the solution, where the quadratic model is good, should the system be solved more accurately.

In practice, truncated-Newton (TN) methods allow a *nonzero residual*, r_k, for the Newton equations. For example, we can require

$$r_k \equiv \|\mathbf{H}_k \mathbf{p}_k + \mathbf{g}_k\| \leq \eta_k \|\mathbf{g}_k\| , \tag{10.53}$$

where η_k is the *forcing sequence*.

This condition on the size of the residual r_k at step k of the minimization scheme becomes stricter as the gradient norm becomes smaller. Thus, near the solution we solve for $\mathbf{p}_k$ more accurately, whereas far away we permit a cruder approximation.

Theoretical work further showed that asymptotic quadratic convergence of the method can be realized for a well chosen η_k as $\|\mathbf{g}_k\| \to 0$ [791]. For example, an effective setting is:

$$\eta_k = \min\{c_r/k , \|\mathbf{g}_k\|\} , \qquad 0 < c_r \leq 1 . \tag{10.54}$$

This choice forces the residuals to be progressively smaller as the number of iterations (k) increases and as the gradient becomes smaller. Another termination criterion based on the quality of the quadratic approximation has also been suggested [781].

Truncated Outer Iteration; Effective Residual

To implement an upper bound on the residual norm in practice, an iterative, rather than direct, procedure that can be "truncated" is required for approximating $\mathbf{p}_k$ from eq. (10.38) at each outer step k.

The linear CG method is an excellent candidate since it is simple and very modest in memory. The linear CG algorithm mirrors in structure the general descent method of Algorithm [A1]. That is, it generates search vectors $\{\mathbf{p}_k^1, \mathbf{p}_k^2, \cdots\}$ at each step recursively (as the nonlinear conjugate gradient method of the previous subsection) until the residual (eq. 10.54), or another suitable truncation criterion, is satisfied for the jth iterate $\mathbf{p}_k^j$. However, in place of the line search, an explicit formula for the steplength is used. This expression is derived analytically by minimizing the quadratic model at the current point along $\mathbf{p}_k^j$ and then using the conjugacy condition to simplify the formula.

Preconditioning

To accelerate convergence of this inner iteration process, *preconditioning* is essential in practice. This technique involves modification of eq. (10.38) through application of a closely-related matrix to $\mathbf{H}_k$, $\mathbf{M}_k$ (effectively, multiplication of both sides by the inverse of $\mathbf{M}_k$).

The preconditioner $\mathbf{M}$ is typically chosen as a sparse symmetric matrix that is rapid to assemble and factor. Theoretically, convergence improves if $\mathbf{M}_k^{-1}\mathbf{H}_k$, the coefficient matrix of the new linear system, has clustered eigenvalues or approximates the identity matrix.

The TN code in CHARMM [792, 793] uses a preconditioner from the local chemical interactions (bond length, bond angle, and dihedral-angle terms). This sparse matrix is rapid to compute and was found to be effective in practice, whether the matrix is indefinite or not, with an appropriate (unusual) modified Cholesky factorization [793]. Other possibilities of preconditioners in general contexts have also been developed, such as a matrix derived from the BFGS update (defined in the QN subsection) [785].

Overall Work

Although more complex to implement than QN or nonlinear CG methods, TN algorithms can be very efficient overall in terms of total function and gradient evaluations, convergence behavior, and solution accuracy, as long as the many components of the algorithm are carefully formulated (truncation, solution process for the inner loop, preconditioning, etc.).

In terms of the computational work per outer Newton step (k), TN methods based on preconditioned CG require a Hessian/vector product ($\mathbf{Hp}$) at each inner loop iteration, and one solution of a linear system $\mathbf{Mz} = \mathbf{r}$ where $\mathbf{M}$ is the preconditioner. Because $\mathbf{M}$ may be sparse, this linear solution often takes a very small percentage of the total CPU time (e.g., $< 3\%$ [793]). The benefits of faster convergence generally far outweigh these additional costs associated with the preconditioner.

Hessian/Vector Products

The Hessian/vector products in each linear CG step ($\mathbf{H}_k\, \mathbf{p}_k^j$) are more significant in terms of computer time. For a Hessian formulated with a nonbonded cutoff radius (e.g., 8 Å), many zeros result for the Hessian (see Figures 10.4 and 10.5); when this sparsity is exploited in the multiplication routine, performance is fast compared to a dense matrix/vector product. However, when the Hessian is dense and large in size, the following forward-difference formula of *two gradients* often works faster (we omit subscripts k from $\mathbf{H}$, $\mathbf{p}$, and $\mathbf{x}$ for clarity):

$$\mathbf{H}\mathbf{p} \approx [\mathbf{g}(\mathbf{x} + h\mathbf{p}) - \mathbf{g}(\mathbf{x})] / h\,, \tag{10.55}$$

where h is a suitably-chosen small number. The central difference approximation,

$$\mathbf{H}\,\mathbf{p} \approx [\mathbf{g}(\mathbf{x} + h\mathbf{p}) - \mathbf{g}(\mathbf{x} - h\mathbf{p})]\,/\,2h\,, \tag{10.56}$$

may alternatively be used for greater accuracy at the cost of one more gradient evaluation with respect to the one-sided difference formula.

In either case, finding an appropriate value for the finite-difference stepsize h is nontrivial, and the accuracy of the product near the solution (where the gradient components are small) can be problematic.

Performance

Thus, TN methods require more care in implementation details and user interface, but their performance is typically at least as good overall as limited-memory QN Newton methods. If simplicity is at a premium, the latter is a better choice. If partial second-derivative information is available, the objective function has many quadratic-like regions, and the user is interested in repeated minimization applications, TN algorithms may be worth the effort.

In general, though Newton methods may not always perform best in terms of function calls and CPU time, they are the most reliable of methods for multivariate minimization and have the greatest potential for achieving very small final-gradient norms. This can be especially important if normal-mode analysis is performed following minimization.

10.5.4 Simple Example

To illustrate performance of the methods described in this section, we have constructed a nonlinear minimization problem with an objective function dependent on two variables, whose contour lines are shown in Figure 10.10. Though the original problem has more variables, this construct represents a 'slice' of the real problem. Illustrations on more realistic, multivariate functions (potential functions of molecular systems) are presented in the next section.

The two-variable problem is derived from our charge optimization procedure [475] that determines electrostatic charge parameters for particles distributed on a virtual surface enclosing a macromolecular system; the electrostatic energy is modeled by a Debye-Hückel potential as an approximation (in the far zone) to a continuum, Poisson-Boltzmann solution to the electrostatic field surrounding the system. The objective function thus reflects the error in electric field (or potential) between the discrete and continuum approximations to the electrostatic potential of the complex macromolecular system.

Specifically, our constructed two-dimensional example seeks to optimize two charge values on the surface of the nucleosome, with the remaining 275 charges fixed.

The contour plots of our function, most readily seen from the darker illustration in Figure 10.10 (bottom right), show the unique minimum lying inside a shallow valley in the function surface.

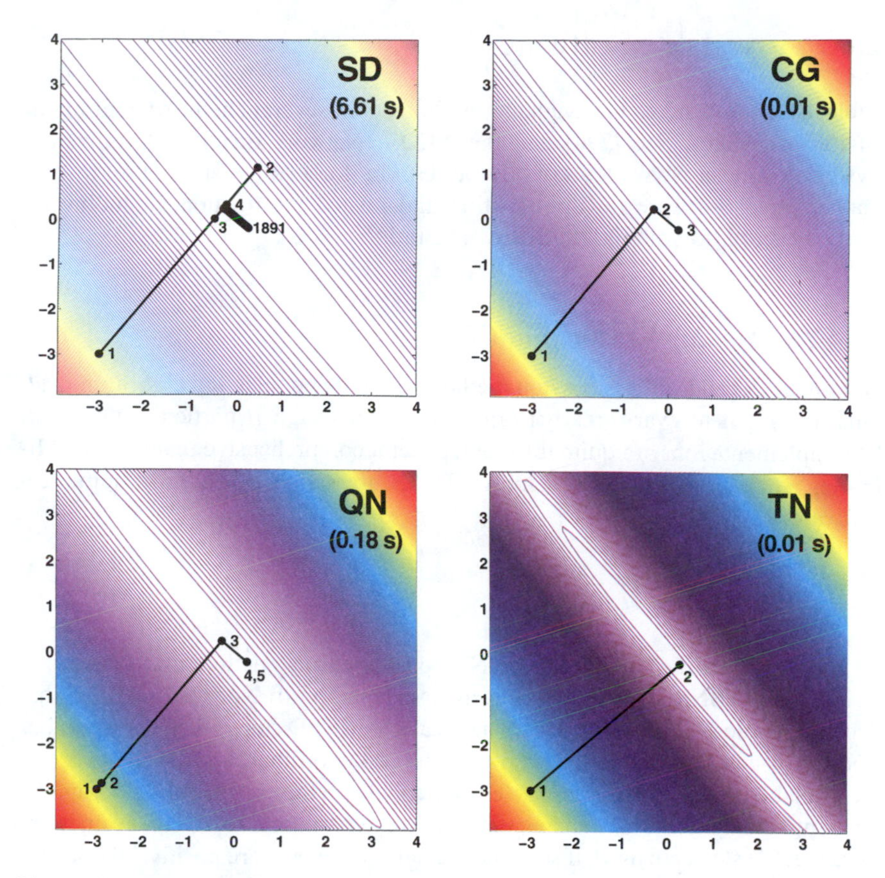

Figure 10.10. Minimization paths (with corresponding CPU times) for a function of two variables shown on top of function contour plots, for steepest descent (SD), nonlinear conjugate gradient (CG), BFGS quasi-Newton (QN), and truncated-Newton (TN) algorithms. See text for functional construction details. All contours are the same, but different levels of resolution are used in each plot to discern both the region near the minimum (darkest contour plot) and the higher-energy regions (lighter plots).

The steepest descent (SD) path (top left) first overshoots the minimum and then slowly approaches it, reaching the high desired gradient accuracy of order 10^{-12} after nearly 2000 iterations. It also requires two orders of magnitude more CPU time that the other methods.

The CG path (top right) is direct and efficient for this problem, likely because of the quadratic nature of the function. Equally direct and efficient are the Newton minimizers: QN BFGS in Matlab and the TN package TNPACK (bottom illustrations). TNPACK, in particular, achieves a low gradient norm in one step.

10.6 Available Software

Table 10.1 summarizes the available minimizers in several chemistry and mathematics packages. See [794] and the NEOS (Network-Enabled Optimization System) Guide at www.mcs.anl.gov/otc/Guide/ for a compilation of up-to-date mathematical software for optimization and related information, including on www.mcs.anl.gov/otc/Guide/SoftwareGuide/.

10.6.1 Popular Newton and CG

Nonlinear CG and various Newton methods are quite popular, but algorithmic details and parameters vary greatly from package to package. In particular, nonlinear CG implementations are quite different. Several comprehensive mathematical libraries, such as IMSL, NAG, and MATLAB are sources of quality numerical software.

10.6.2 CHARMM's ABNR

Of special note is the "adopted-basis Newton-Raphason" method implemented in CHARMM, ABNR. It is a memory-saving adaptation of Newton's method that avoids analytic second derivatives. The idea is to use SD steps for a given number of iterations, m (e.g., 5), after which a set of $m+1$ coordinate and gradient vectors are available. A Hessian is constructed numerically in this $m \times m$ subspace, and all corresponding eigenvalues and eigenvectors are computed. If all eigenvalues are negative, SD steps are used; if some are negative and some are positive, the search direction is modified by a Newton direction constructed from the eigenvectors corresponding to the positive eigenvalues only. In all cases, the n-dimensional search vector $\mathbf{p}_k$ is determined via projection onto the full space. The ABNR algorithm is similar in strategy to limited-memory QN methods in that it uses only recent curvature information and exploits this information to make steady progress toward a solution.

10.6.3 CHARMM's TN

The TN method in CHARMM [792] is detailed elsewhere [771, 772, 773, 793, 797]. It uses a preconditioner constructed from the local chemical interactions (see Figures 10.4 and 10.5, right panels) and determines $\mathbf{p}_k$ from a truncated preconditioned CG loop. When negative curvature is detected, the preconditioned CG loop is halted with a guaranteed direction of descent. Interestingly, numerical analysis and experiments have shown that the method can produce quadratic convergence near a solution regardless of whether the preconditioner is indefinite or not [793]. As implemented, the method is applicable only to moderate system sizes (due to Hessian memory limitations).

10.6.4 Comparative Performance on Molecular Systems

In Table 10.2 we illustrate the minimization performance of three methods in CHARMM — nonlinear CG, ABNR, and TNPACK — for several molecular systems; see [793, 797] for details.

Note that the same minimum is obtained for the small systems (butane and *n*-methyl-alanyl-acetamide) but that different minima typically result for the larger systems.

Considerable differences in CPU times can also be noted. The CG method performs much slower and can fail to produce very small gradient norms. Both Newton methods perform well for these problems, though ABNR is relatively expensive for the small system. TNPACK displays much faster convergence overall and yields smaller final gradient norms.

Note also that CG requires about two function evaluations per iteration (in the line search), while ABNR employs only one on average. TNPACK uses more than one function evaluation per outer iteration, since the (unscaled) magnitude of the produced search vector often leads to small steplengths at some iterations of the line search. The quadratic convergence of TNPACK is evident from Figure 10.11, where the gradient norm per iteration is shown.

10.7 Practical Recommendations

In general, geometry optimization in the context of molecular potential energy functions has many possible caveats. Hence, a novice user especially should take the following precautions to generate as much confidence as possible in a minimization result.

1. *Use many starting points.* There is always the possibility that the method will fail to converge from a certain starting point, or converge to a nearby stationary point that is not a minimum.

   A case in point is minimization of biphenyl from a planar geometry [798]; many minimizers will produce the flat ring geometry, but this actually corresponds to a maximum! Different starting points will produce the correct nonplanar structure. See the homework assignment on minimization (number 10).

2. *Compare results from different algorithms.* Many packages offer more than one minimizer, and thus experimenting with more than one algorithm is an excellent way to check a computational result. Often, one method fails to achieve the desired resolution or converges very slowly. Another reference calculation under the same potential energy surface should help assess the results.

   Minimization of the DNA in vacuum system in Figure 10.11 and Table 10.2 by three algorithms also reveals very different final energies. This is because the DNA strands have separated! Proper solvation and ions remedy

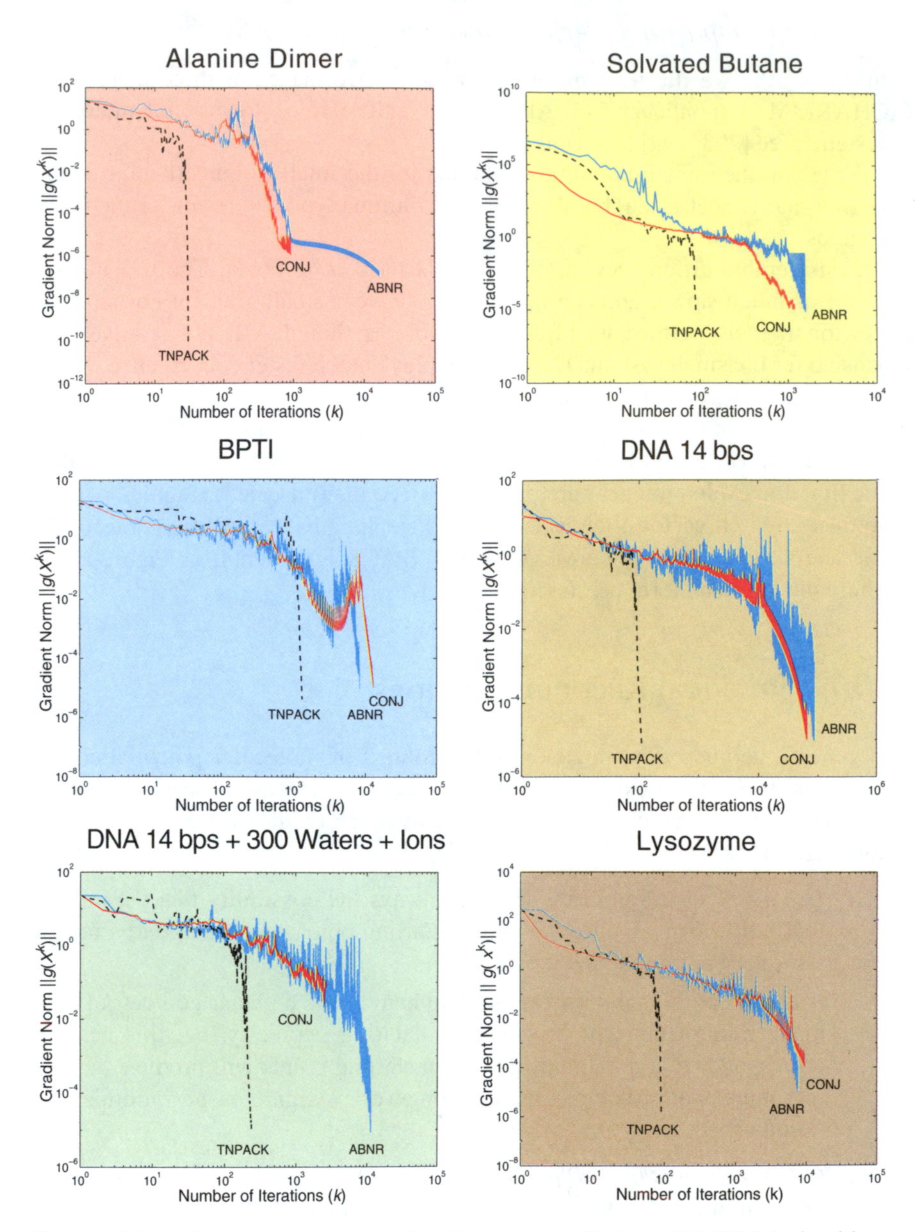

Figure 10.11. Minimization progress (gradient norm) of three CHARMM algorithms (CONJ: nonlinear conjugate gradient, ABNR: adopted-basis Newton-Raphson, and TNPACK: truncated Newton) for various molecular systems. See Figures 10.4 and 10.5 for corresponding sparse matrix patterns; the local Hessians are used as preconditioners for TNPACK. See also Table 10.2 for final function and gradient-norm values and required CPU time.

this physical/chemical problem. Interestingly, adding only water keeps the strands nearby but untwists the strands; only added ions and water maintain the proper DNA chemistry.

3. *Compare results from different force fields whenever possible.* Putting aside the quality of the minimizer, the local minimum produced by any package is only as good as the force field itself. Since force fields for macromolecules today are far from converging to one another — in fact there are very large differences both in parameters and in functional forms — a better understanding of the energetic properties of various conformations can be obtained by comparing the relative energies of the different configurations as obtained by different force fields. Differences are expected, but the results should help identify the lowest-energy configuration. If significant differences are observed, the researcher could further investigate both the associated force fields (e.g., a larger partial charge, an additional torsional term) and the minimization algorithms for explanations.

4. *Check eigenvalues at the solution when possible.* If the significance of the computed minima is unclear, the corresponding eigenvalues may help diagnose a problem. Near a true minimum, the eigenvalues should all be positive (except for the six zero components corresponding to translation and rotation invariance). In finite-precision arithmetic, "zero" will correspond to numbers that are small in absolute value (e.g., 10^{-6}). Values larger than this tolerance might indicate deviations from a true minimum, perhaps even a maximum or saddle point. In this case, the corresponding structure should be perturbed substantially and another trial of minimization attempted.

5. *Be aware of artificial minima caused by nonbonded cutoffs or improper physical models!* When cutoffs are used for the nonbonded interactions, especially in naive implementations involving sudden truncation or potential-switching methods, the energy and/or gradient can exhibit numerical artifacts: deep energy minima and correspondingly-large gradient value near the cutoff region. Good minimization algorithms can find these minima, which are correct as far as the numerical formulation is involved, but unfortunately not relevant physically.

   One way to recognize these artifacts is to note their large energy difference with respect to other minima computed for the same structure (as obtained from different starting points or minima). These artificial minima should disappear when all the nonbonded interactions are considered, or improved spherical-cutoff treatments (such as force shifting and switching methods) are implemented instead.

   Besides artifacts caused by nonbonded cutoffs, improper physical models — such as that for DNA lacking solvent and ions, as discussed above — also produce artificial minima. For this example of DNA in vacuum, parallel strands rather than intertwined polynucleotide strands are produced as a result of minimization.

10.8 Looking Ahead

Only a small subset of topics was covered here in the challenging and ever-evolving field of nonlinear large-scale optimization. Interested readers are referred to the comprehensive treatments in the texts cited at the beginning of this chapter. The increase in computer memory and speed, and the growing availability of parallel computing platforms will undoubtedly influence the development of optimization algorithms in the next decade. Parallel architectures can be exploited in many ways: for performing minimization simulations concurrently from different starting points; for evaluating function and derivatives in tandem; for greater efficiency in the line search or finite-difference approximations; or for performing matrix decompositions in parallel for structured, separable systems.

The increase in computing speed is also making *automatic differentiation* a powerful resource for nonlinear optimization. In this technique, automatic routines are available to construct program codes for function derivatives. The construction is based on the chain-rule application to the elementary constituents of a function [799]. It is foreseeable that such codes will introduce greater versatility in Newton methods [780]. The cost of differentiation is not reduced, but the convenience and accuracy may increase.

Function separability is a more general notion than sparsity, since problems associated with sparse Hessians are separable but the reverse is not true. It is also another area where algorithmic growth can be expected [780]. (Recall that separable functions are composites of subfunctions, each of which depends only on a small subset of the independent variables; see eq. (10.6)). Therefore, efficient schemes can be devised in this case to compute the search vector, function curvature, etc., much more cheaply by exploiting the invariant subspaces of the objective function.

Such advances in local optimization will certainly lead to further progress in solving the global optimization problem as well; see [800, 765, 801] for examples. Scientists from all disciplines will anxiously await all these developments.

Table 10.1. Available optimization algorithms.

Package	Contact	Minimizers
AMBER	www.amber.ucsf.edu/amber/amber.html	SD, nonlinear CG from the IMSL library (due to Powell), and Newton.
CHARMM	yuri.harvard.edu	SD, nonlinear CG (FR, and modified PR version, the latter from the IMSL library),[a] Adopted-Basis Newton (ABNR), Newton, truncated-Newton (TNPACK).
DISCOVER	Biosym Technologies, San Diego, CA	SD, nonlinear CG (PR, FR versions),[a] quasi-Newton, truncated Newton.
DUPLEX	Brian E. Hingerty hingertybe@ornl.gov	Powell's coordinate descent method (no derivatives).[b]
ECEPP/2	QCPE 454 qcpe5.chem.indiana.edu/	Calls SUMSL, a quasi-Newton method based on a trust-region approach (by Gay).
GROMOS	igc.ethz.ch/gromos	SD and nonlinear CG (FR version),[a] both with and without SHAKE constraints.
IMSL Lib.	IMSL, Inc., Sugar Land, TX. www.vni.com/products/imsl/	Many routines for constrained and unconstrained minimization (nonsmooth, no derivatives, quadratic and linear programming, least-squares, nonlinear, etc.), including a nonlinear CG method of Powell (modified PR version with restarts).[a]
LANCELOT	Philippe Toint www.cse.clrc.ac.uk/ Activity/LANCELOT	Various Newton methods for constrained and unconstrained nonlinear optimization, specializing in large-scale problems and including a trust-region Newton method and an algorithm for nonlinear least squares that exploits partial separability.
MATLAB	The Math Works, Inc., Natick, MA info@mathworks.com, www.mathworks.com	SD, DFP[c] and BFGS quasi-Newton, simplex algorithm, and others for linear and quadratic programming, least squares, etc..
MMFF94 /94s	www.ccl.net/cca/data /MMFF94/	Calls OPTIMOL which uses a BFGS quasi-Newton method, with variable-metric updating scheme, but for Cartesian optimization (there is also a torsion-only optimizer) initiates the initial inverse Hessian approximated from the inverse of a 3×3 block-diagonal Hessian.
MM3	europa.chem.uga.edu	3×3 block-diagonal Newton and full Newton.
MM2	europa.chem.uga.edu	3×3 block-diagonal Newton.
NAG Lib.	NAG, Inc., Downers Grove, IL www.nag.com	Quasi-Newton, modified Newton and nonlinear CG (CONMIN by Shanno & Phua, modified PR version); also quadratic programming, least squares minimization, and many service routines.
SIGMA	femto.med.unc.edu /SIGMA/	Nonlinear CG (FR version).[a]
X-PLOR	atb.csb.yale.edu/xplor/	Nonlinear CG (from IMSL library).

[a] FR and PR refer to the Fletcher-Reeves and Polak-Ribière nonlinear CG versions.

[b] See [795, 796] for details on DUPLEX.

[c] DFP is a rank-1 QN method, credited to Davidon, Fletcher, and Powell.

Table 10.2. Performance of three CHARMM minimizers on various molecular systems. See Figures 10.4 and 10.5 for patterns of the preconditioner used in TNPACK and Figure 10.11 for minimization progress. *Note*: For the DNA system in vacuum, though minimization produces a local minimum for each method, the structures are physically incorrect: without proper solvation, the DNA strands intertwine and separate. This also explains the very different values of final energies. For the DNA system with water and ions, the CG method terminates prematurely with an error message; the final gradient is relatively large.

Method[a]	Final f	Final $\|\|g\|\|$	Itns.[b]	f&g Evals	CPU[c]
N-Methyl-Alanyl-Acetamide ($n = 66$)					
CG	−15.245	9.83×10^{-7}	882	2507	2.34 s
ABNR	−15.245	9.96×10^{-8}	16466	16467	7.47 s
TNPACK	−15.245	7.67×10^{-11}	29 (210)	44	1.32 s
Solvated Butane ($n = 1125$)					
CG	−2374.00	1.27×10^{-5}	1152	3175	49.48 m
ABNR	−2398.22	7.0×10^{-6}	1574	1575	48.52 m
TNPACK	−2381.04	7.7×10^{-6}	90 (1717)	263	59.44 m
BPTI ($n = 1704$)					
CG	−2792.93	9.9×10^{-6}	12469	32661	97.8 m
ABNR	−2792.96	8.9×10^{-6}	8329	8330	25.17 m
TNPACK	−2773.70	4.2×10^{-6}	65 (1335)	240	5.21 m
DNA 14 Bps ($n = 2664$)					
CG	−538.41	9.72×10^{-6}	62669	62670	20.42 h
ABNR	−1633.90	9.23×10^{-6}	86496	86497	6.69 h
TNPACK	−560.68	7.62×10^{-6}	111 (3724)	268	0.54 h
DNA 14 Bps + 300 Waters + Ions ($n = 5364$)					
CG	−11921.00	7.52×10^{-2}	2580	6616	1.62 h
ABNR	−11774.64	8.19×10^{-6}	11306	11307	2.78 h
TNPACK	−11928.50	9.84×10^{-6}	236 (6555)	687	1.75 h
Lysozyme ($n = 6090$)					
CG	−4628.362	9.89×10^{-5}	9231	24064	19.63 h
ABNR	−4631.584	9.97×10^{-6}	7637	7638	6.11 h
TNPACK	−4631.380	1.45×10^{-6}	78 (1848)	218	1.49 h

[a]CG: nonlinear conjugate gradient, ABNR: adopted basis Newton Raphson, TNPACK: truncated Newton based on the TNPACK package.

[b]For TNPACK, the total number of inner (preconditioned CG) iterations is indicated in parentheses, following the number of outer iterations.

[c]s: seconds, m: minutes, h: hours.

11

Monte Carlo Techniques

Chapter 11 Notation

SYMBOL	DEFINITION
Matrices	
M	mass matrix (components m_i)
Vectors	
p (or P)	collective momentum vector
q (or X)	collective position vector
R	collective random force vector
Scalars & Functions	
a	multiplier (of random number generator, a prime)
c	increment (of random number generator)
i, j, k, m, q, r	integers
t	time
u, v	real numbers
v_i	velocity component i
$\{x_i\}, \{\tilde{x}_i\}$	sequences of numbers
B_i	data batch (for MC mean)
E_k	kinetic energy
E_p	potential energy
F	probability distribution function
H (or E)	total energy
M	modulus (of random number generator, usually of order of computer word size); also used for number of batches in a sample
N	number of particles
N	sample size
Rg	radius of gyration
S, T	polynomial functions

Chapter 11 Notation Table (continued)

SYMBOL	DEFINITION
T	temperature
Wr	DNA writhing number
β	Boltzmann factor $1/(k_B \mathrm{T})$
γ	Langevin damping constant
μ	mean (also chemical potential in MC Carlo sampling section)
ρ	probability density function
σ^2	variance (σ is standard deviation)
τ	period (of random number generator)
$\langle A(x) \rangle_{\mathcal{D}}$	mean of property $A(x)$ over domain $\mathcal{D}$
$\lfloor y \rfloor$	largest integer smaller than or equal to y

It is a pollster's maxim that the truth lies not in any one poll but at the center of gravity of several polls.

Michael R. Kagay, *New York Times (Week in Review)*, 19 October 1998.

11.1 MC Popularity

From Washington D.C. to Wall Street to Los Alamos, statistical techniques termed collectively as Monte Carlo (MC) are powerful problem solvers. Indeed, disciplines as disparate as politics, economics, biology, and high-energy physics rely on MC tools for handling daily tasks.

Many problems that can be formulated as stochastic phenomena and studied by random sampling can be solved through MC simulations. Essentially, a game of chance is played, but with theoretical and practical rules from probability theory, stochastic processes, and statistical physics (Markov chains, Brownian motion, ergodic hypothesis) that lend the 'sport' practical utility.

11.1.1 A Winning Combination

MC methods are used for numerical integration, global optimization, queuing theory, structural mechanics, and solution of large systems of linear, partial differential or integral equations. MC methods are employed widely in statistical physics and chemistry, where the behavior of complex systems of thousands or more atoms in space and time is studied. Their appeal can be explained by a winning combination of simplicity, efficiency, and theoretical grounding.

11.1.2 From Needles to Bombs

Early records of random sampling to solve quantitative problems can be found in the 18th and 19th centuries with needle throwing experiments to calculate geometrical probabilities (George Louis Leclerc, a.k.a. Comte de Buffon, 1777)[1] or to determine π (Simon de Laplace, 1886). In 1901, Lord Kelvin also described an important application to the evaluation of time integrals in the kinetic theory of gases. Yet a novel class of MC methods (using Markov chains) provides the modern roots of MC theory, and is largely credited to the Los Alamos pioneers (Von Neumann, Fermi, Ulam, Metropolis, Teller, and others).

These brilliant scholars studied properties of the newly discovered neutron particles in the middle of the 20th century by formulating mathematical problems in terms of probability and solving analogues by stochastic sampling. Their work led to a surge of publications in the late 1940s and early 1950s on solving problems in statistical mechanics, radiation transport, and other fields by carefully-designed sampling experiments.

Most notable among these works was the famous algorithm of Metropolis *et al.* in 1953 [802]. With the rapid growth of computer speed and the development of many techniques to improve sampling, reduce errors, and enhance efficiency, MC methods have become a powerful utility in many areas of science and engineering.

11.1.3 Chapter Overview

In this chapter, only the most elementary aspects of MC simulations are described, including the generation of uniform and normal random variables, basic probability theory background (see also Box 11.1), and the Metropolis algorithm (due also to A.W. Rosenbluth, M.N. Rosenbluth, A.H. Teller and E. Teller).

Such methods can be used in molecular simulations to generate efficiently conformational ensembles that obey *Boltzmann statistics*, that is, the probability of a configuration X with energy $E(X)$ is proportional to $\exp(-E(X)/k_B\,\mathrm{T})$ (k_B is Boltzmann's constant and T is the temperature). From such ensembles, various geometric and energetic means estimated. Low-energy regions can also be identified by decreasing the temperature in the sampling protocol (this is termed "simulated annealing"). Method extensions that are of particular interest to the biomolecular community, such as hybrid MC, are also mentioned.

The codes illustrated in this chapter are provided in Fortran, still the language of choice in some of the popular molecular mechanics and dynamics packages

[1] Buffon used a Monte Carlo integration procedure to solve the following problem: a needle of length L is thrown at a horizontal plane ruled with parallel straight lines separated by $d > L$; what is the probability that the needle will intersect one of these lines? Buffon derived the probability as an integral and attempted an experimental verification by throwing the needle many times and observing the fraction of needle/line intersections. It was Laplace who in the early 1800s generalized Buffon's probability problem and recognized it as a method for calculating π.

like CHARMM and AMBER. Analogous routines written in the C language can be obtained from the course website.

Since MC methods rely strongly on random number generators, the first section of this chapter is devoted to the subject. Students can skip Sections 11.2 and 11.3 if they wish to read material related to other aspects of MC simulations. To see immediately why random number generators are important and how they are used in MC simulations, students may wish to read at the onset Subsection 11.5.4 and see the example there (subroutine `monte`) for calculating π by MC sampling.

Good general introductions to MC simulations can be found in the texts by Kalos and Whitlock [803], Bratley, Fox and Schrage [33], and Frenkel and Smit [31]. There are many web resources for Monte Carlo tutorials, for example, obtained through the Molecular Monte Carlo Home Page of www.cooper.edu/engineering/chemechem/monte.html.

11.1.4 Importance of Error Bars

A point which cannot be overstressed in any introduction to MC methods is the fundamental importance of error bars in any MC estimate. Unlike in politics, perhaps, the reliability of any conclusion (e.g., estimate) in science depends on the associated accuracy. Scientists would no doubt have discarded the results of an election whose "margin of error ... is far greater than the margin of victory, no matter who wins", an assessment by mathematician John Allen Paulos of the rocky 2000 U.S. Presidential race, between Texas governor George W. Bush (who became President) and former President Clinton's Vice President Albert Gore.

11.2 Random Number Generators

11.2.1 What is Random*?*

The computer sampling performed in MC simulations of stochastic processes relies on generation of "random" numbers. Actually, those numbers are typically *pseudorandom* since a deterministic recursion rule is used to generate a sequence of numbers given an *initial seed* x_0: $x_{i+1} = f(x_i, x_{i-1}, x_{i-2}, \ldots)$, where f is a function.[2] This reproducibility of the sequence is an essential requirement for debugging computer programs. Even sequences obtained via chaos theory (see [804], for example, and references cited therein) are deterministic.

It is essential to use 'good' random number generators in MC applications to avoid artifacts in the results. (The statement by Dilbert's cartoon character, a horned accounting troll, that "you can never be sure" [of randomness] is not an option for scientists! See cartoon posting on the dilbert.com archives for 10/25/01).

---

[2] We use the term *random* for brevity in most of this chapter, though the terms *pseudorandom* or *quasi-random* are technically correct.

The quality of a generator is determined not only by subjecting the generating algorithm to a large number of established tests (both empirical and theoretical). It is also important to test the *combination* of generator and application. The two examples described in the Artifacts subsection below (following the introduction of generator algorithms) illustrate how the performance of generators is application specific.

Much work has gone into developing random number generators on both serial and parallel computer platforms, as well as associated criteria for testing them. Concurrently, work has focused on the careful implementation of the mathematical expressions to ensure good numerical performance (e.g., avoid overflow or systematic loss in accuracy) and efficiency, on both general and special-purpose hardware.

Novices are well advised to use a routine from a reputable library of programs rather than programming a simple procedure reported in the literature, since many such procedures have not been actually tested comprehensively. Still, caveats are warranted even for some library routines; see below.

The reader is referred to classic texts by Kalos and Whitlock [803], Knuth [805], and Law and Kelton [806] for general introductions into random number generators. Some of these books also review basic probability theory. Good reviews by L'Ecuyer can be found in [807, 808] (see also www.iro.umontreal.ca/~lecuyer) and [809].

11.2.2 Properties of Generators

Let

$$\{x_1, x_2, \ldots, \ldots\}$$

be a sequence of numbers. In theory, we aim for sequences of numbers that exhibit independence, uniformity, and a long period τ. In addition, it is important that such generators be as portable and efficient as possible.

Uniformity and Subtle Correlations

Most MC algorithms manipulate hypothetical *independent uniformly distributed* random variables (*variates*). That is, the independent variables are assumed to have a *probability density function* ρ (see Box 11.1) that satisfies $\rho_u(x) = 1$ for x in the interval $[0, 1]$ and $\rho_u(x) = 0$ elsewhere.[3] From such uniform variates, we can obtain other probability distributions than the uniform distribution, such as the normal, exponential, Gamma, or Poisson distributions; see subsection below on normal variates and [806] for generating continuous and discrete random variates from many distributions.

[3]We say that x lies in $[a, b]$ if $a \leq x \leq b$ and that x lies in $[a, b)$ if $a \leq x < b$; similarly, x in (a, b) means $a < x < b$.

Roughly speaking, independence of two random variables means that knowledge of one random variate reveals no information about the distribution of the other variate. In the strict sense of probability theory,[4] it is impossible to obtain true independence for random numbers. Generating *uncorrelated* random variates is a weaker goal than independence. (This is because independent random variables are uncorrelated but uncorrelated variables are not independent in general). Though correlations exist even in the best known random number generators, quality random number generators can defer correlations to high-order and high-complexity relations.

Long Period

The period τ associated with a sequence of random numbers is the number of sequential random values before the series repeats itself, that is,

$$x_{i+\tau} = x_i \qquad \text{for all integers } i \geq 0.$$

We require the sequence to have as *long a period as possible* to allow long simulations of independent measures.

The period length is an important consideration for modern large-scale simulations.[5] For a 32-bit computer, if the generator's state uses only 32 bits, the maximum period is usually $2^{30} \sim 10^9$ (assuming 2 bits are lost). This number is not a large number by today's standards. More than one million iterations may be performed in dynamics simulations and far more in MC sampling simulations. Moreover, each iteration may require large random *vectors* (e.g., in Langevin dynamics). Thus, the random number generators that might have been adequate only a decade ago on 32-bit machines quickly exhaust their values for the complex applications at present. Unfortunately, many such generators, which experts deem *unacceptable* [808], are often the default methods for many operating systems and software packages.

State-of-the-art generators use more bits for their state than the computer type and employ combinations of methods to defer correlations to high-order and high-complexity relations. This makes possible formulation of sequences with very long periods. For example, the codes given in [810] produce sequences with period lengths of up to order 2^{200} on 32-bit computers and 2^{400} on 64-bit machines!

Portability

Portability and efficiency are also important criteria of generators.

---

[4]The random variables x_1 and x_2 are independent if the joint probability density function $\rho(x_1, x_2)$ is equal to the product of the individual probability density functions: $\rho(x_1, x_2) = \rho_1(x_1)\rho_2(x_2)$

[5]Though for complex systems, the state descriptors (e.g., coordinates) are unlikely to be repeated in phase with the cycle of a (short) random number generator, subtle problems may occur in some applications, making the goal of long period generally desirable.

Portable generators are those that produce the same sequence across standard compilers and machines, within machine accuracy. Portability permits code comparison and repeatability on different platforms. This requirement is nontrivial because even if the mathematical recipe is identical certain floating-point calculations may involve hardware-wired instructions and branched directives for the sub-operations.

Efficiency

The issue of *speed* of random number generators can be important for some problems that involve a large number of computationally-intensive iterations. (See Table 11.1 for CPU data on different generators). Even if the relative computational cost of random number generators in large-scale applications is small, it is important to use quality compiler optimization utilities to reduce most of the overhead associated with *calling* the random number generator function itself. For this reason, it is also important to use a subroutine that returns a *vector* of random variates if an array of such numbers is desired, rather than calling the function multiple times for each vector component.

Box 11.1: The Probability Density and Distribution Functions

Let X be a random variable that takes on values x. We say that X is a *discrete* random variable if it takes on a countable number of values and *continuous* if it takes on an uncountably-infinite number of values.

The distribution function $F(x)$ (also termed the *cumulative distribution function*) of a random variable X defined as the probability that X takes on a values no larger than x (a real number), that is

$$F(x) = P(X \leq x)\,, \qquad -\infty < x < \infty\,. \tag{11.1}$$

A continuous random variable X has the closely related *probability density function* $\rho(x)$. (For discrete random variables, analogous definitions are formulated using a probability function $p(x)$). This relation is given by:

$$F(x) = P(X \leq x) = \int_{-\infty}^{x} \rho(y)\,dy\,, \qquad -\infty < x < \infty\,. \tag{11.2}$$

Thus, $\rho(x)$ is closely related to the derivative of $F(x)$ (under some additional assumptions of regularity, we have $\rho(x) = F'(x)$).

For example, a uniform random variable on $[0, 1]$ has the probability density function

$$\rho(x) = \begin{cases} 1 & 0 \leq x \leq 1 \\ 0 & \text{otherwise} \end{cases}\,, \tag{11.3}$$

and the corresponding density function F is defined by:

$$F(x) = \int_0^x \rho(y)\,dy = \int_0^1 1\,dy = x\,. \tag{11.4}$$

The reader can verify that the mean μ (or *expected value*) of this continuous uniform random variable (by definition, $\mu \equiv E(X) = \int_{-\infty}^{\infty} x\, \rho(x)\, dx$) is $\mu = \int_0^1 x\, \rho(x)\, dx = \frac{1}{2}$ and that the variance σ^2 (by definition, $\sigma^2 \equiv E(X-\mu)^2 = E(X^2) - \mu^2$) is $\sigma^2 = \int_0^1 x^2\, \rho(x)\, dx - (\frac{1}{2})^2 = \frac{1}{3} - \frac{1}{4} = \frac{1}{12}$.

A Gaussian (or *normal*) random variable with mean μ and variance σ^2 has the density function

$$\rho(x) = \frac{1}{\sigma\sqrt{2\pi}} \exp\left(\frac{-(x-\mu)^2}{2\sigma^2}\right), \qquad -\infty < x < \infty \,. \tag{11.5}$$

The associated distribution function is often denoted as $\mathcal{N}(\mu, \sigma^2)$. The probability density function for a *standard normal* random variable (with $\mathcal{N}(0,1)$) is

$$\rho(x) = \frac{1}{\sqrt{2\pi}} \exp(-x^2/2)\,, \qquad -\infty < x < \infty \,. \tag{11.6}$$

11.2.3 Linear Congruential Generators (LCG)

The simplest type of random number method is a *linear congruential generator* (LCG), first used in 1948 by D. H. Lehmer.

Basic Recipe

LCGs compute successive iterates by multiplying the previous iterate by a constant, a, adding this product to another constant, c, and then taking the modulus of this result with respect to another large number, M.

Specifically, the LCG recipe relies on three integers. M (the *modulus*) is a large positive number; a (the *multiplier*) is a positive integer less than M and shares no divisors with M; and c (the *increment*) is less than M.

We then generate a sequence of variates from an initial integer seed $\tilde{x}_0$ less than M, namely $\{\tilde{x}_1, \tilde{x}_2, \ldots\}$, according to the recursion relation:

$$\tilde{x}_{i+1} = (a\tilde{x}_i + c) \bmod M, \qquad i = 0, 1, \ldots \,. \tag{11.7}$$

The uniform variates for this LCG are then obtained by division as:

$$x_i = \tilde{x}_i / M \,.$$

If $c = 0$, these real numbers $\{x_i\}$ are in the open unit interval $(0,1)$, and if $c \neq 0$ they are contained in the interval $[0,1)$. When $c = 0$, this LCG is called *multiplicative linear congruential generator* (MLCG).

The recurrence relation of eq. (11.7) has a period no greater than M. If the integers are properly chosen, the period will have the maximal length M. Judicious choice of a, c, and M must be made, as well as thorough tests for randomness of the resulting sequences. See [805, pp. 170–171] for specific recommendations.

Simple Example

As a simple illustration, consider the MLCG sequence with $M = 11$ and $a = 8$ $(c = 0)$. From $\tilde{x_0} = 1$, we generate the sequence

$$\begin{aligned} \tilde{x}_{i+1} &= (8\tilde{x}_i) \bmod 11 \;:\Longrightarrow \\ &\{ \quad 1, 8, 9, 6, 4, 10, 3, 2, 5, 7, \quad 1, 8, 9, \ldots \}. \end{aligned} \tag{11.8}$$

We see that this sequence has the maximal period length of $M - 1 = 10$ and that each integer in the interval $[1, 10]$ is generated exactly once per cycle. The reader can verify that, for this choice of M (with $c = 0$), the values $a = 2, 6, 7, 8$ also have these favorable properties; the other values generate sequences with only two or five elements and hence violate the uniformity criteria strongly. However, as will also be discussed below, even the full-length sequences exhibit unacceptable correlations.

Of course, we are interested in much longer sequence lengths in real applications. Often, M is taken to be the *word size* of the machine and a is a prime number. However, M and a must be chosen with care, and the resulting algorithm carefully programmed, to avoid an integer *overflow* for the product $a\,\tilde{x}_i$; this is explained further below.

IBM's SURAND and Unix's rand and drand48

One old and still widely used MLCG method (possibly because its modulus M is the largest prime that fits in the 32-bit signed integer word used by many computers [811]) is SURAND, though it is considered poor by experts [808] (see discussion under Lattice Structure below and Figure 11.2). Developed by IBM for its system/360 series, SURAND has the values:

SURAND MLCG:

$$a = 7^5 = 16807; \qquad M = 2^{31} - 1 = 2147483647; \qquad c = 0\,.$$

A 'naive' FORTRAN implementation of this generator might be the simple implementation above, that is, include the two statements:

```
seed = mod (a * seed, m)
    ranu = seed / m.
```

However, this would not produce the right sequence because of overflow.

To avoid the overflow in the product $a\,\tilde{x}_i$ (or `a * seed` in the code), it is necessary to ensure that all intermediate integers are bounded by $M - 1$. The basic idea, based on [812], is outlined in Box 11.2.

Box 11.2: Avoiding Overflow in Linear Congruential Generator Implementation

To avoid overflow in the computation of the product $a\,x$ in the implementation of eq. (11.7) (we suppress subscripts for clarity), let us *assume* for the moment that we could factor M as

$$M = a\,q\,, \qquad q = \text{integer}\,. \tag{11.9}$$

Then, we could write the MLCG recursion relation as

$$f(x) = a\,x \bmod M = a\,x \bmod (a\,q) = a\,(x \bmod q)\,.$$

Of course M is a prime, and no such factorization $M = aq$ exists. However, instead of eq. (11.9), we can *approximately factor* M as:

$$M = a\,q + r, \qquad 1 \le r \le a-1\,, \tag{11.10}$$

where

$$q = M \operatorname{div} a \equiv \lfloor M/a \rfloor, \qquad r = M \bmod a\,. \tag{11.11}$$

Here $\lfloor y \rfloor$ denotes the largest integer smaller than, or equal to, y; in other words, $\lfloor M/a \rfloor$ is the integer division of M by a. If $r < q$, this approximate factorization is useful since then the magnitude of the intermediate product is not greater than $M-1$.

For the SURAND MLCG ($a = 16807$ and $M = 2^{31} - 1$), we obtain $q = 127773 > r = 2836$.

This better implementation leads to the following correct implementation of SURAND (see [812, 33] for further details):

```
c***********************************************************************
      double precision function ranu ()
c Good implementation of SURAND. See Park & Miller,
c Comm. ACM 31:1192, 1988.  Subroutine ranset should be called
c (once) before the first function call.
      integer a, m, q, r, seed
      double precision rm
      parameter (a=16807, m=2147483647, q=127773, r=2836, rm=1d0/m)
      common /random/ seed
      save /random/
      data seed /1/
      seed = a * mod(seed, q) - r * (seed/q)
      if (seed .le. 0) seed = seed + m
      ranu = seed * rm
      return
      end
c***********************************************************************
```

However, this LCG is not recommended since there are far better procedures today. There are also faster and simpler ways to implement this recursion [810].

Other known MLCG combinations are the default random number generators available at the time of this writing on our SGI's Unix System Library, rand and drand48, using 32-bit and 48-bit integer arithmetic, respectively. Their parameters are as follows.

rand MLCG:

$$a = 1103515245; \qquad M = 2^{31} = 2147483648; \qquad c = 12345\,.$$

drand48 MLCG:

$$a = 25214903917; \qquad M = 2^{48}; \qquad c = 11\,.$$

Note the somewhat confusing online documentation for drand48, which reports a and c in base 8 rather than 10 (273673163155_8 and 13_8, respectively).

We discuss some of the defects of rand and drand48 below (see also Figure 11.2).

Lattice Structure in Linear Congruential Generators

Many statistical tests have been formulated to assess the suitability of random number generators. Linear congruential methods, for example, are known to exhibit correlations in certain hyperspaces; this basic defect is termed *coarse lattice structures*. Essentially, this means that when subsets of such sequences are represented in Euclidean space (two dimensions or higher), a lattice structure emerges; in other words, points lie on a number of hyperplanes rather than cover the space in a random-like manner. This pattern indicates that the sequence is not truly as random and uniform as sought. One way to visualize lattice structure is to plot k-lag pairs of numbers of the sequence, namely $\{x_i,\, x_{i+k}\}$ in the unit-square plane for fixed k. Often, we plot pairs of consecutive numbers in the sequence on the unit square and triplets of consecutive numbers in the sequence on the unit cube.

This defect of LCG methods has been credited to G. Marsaglia in 1968; see also [813]. *Spectral tests* have since been developed to measure such k-dimensional uniformities. Such tests essentially determine the maximum distance between adjacent hyperplanes; the larger this value, the worse the generator.

To illustrate, consider $k = 1$-lag pairs for our simple MLCG above with $M = 11$ and $a = 8$ (see expression in (11.8)). If we plot in two dimensions all consecutive pairs of points, that is:

$$\{1, 8\}, \{8, 9\}, \{9, 6\}, \{6, 4\}, \; \ldots \;, \{7, 1\}\,, \tag{11.12}$$

we see alarmingly that these points lie on four parallel lines with either positive and negative slopes (see Fig. 11.1). The spectral test would determine the maximum distance between these parallel lines.

Figure 11.1 shows the lattice structure generated from the four a values that yield full periods for this generator ($\tilde{x}_{i+1} = (a\,\tilde{x}_i) \bmod 11$). Clearly, defects emerge.

You might think that this toy problem is especially misleading. Unfortunately, even long LCG sequences are known to display such uniform patterns or structures that indicate imperfect uniform sampling.

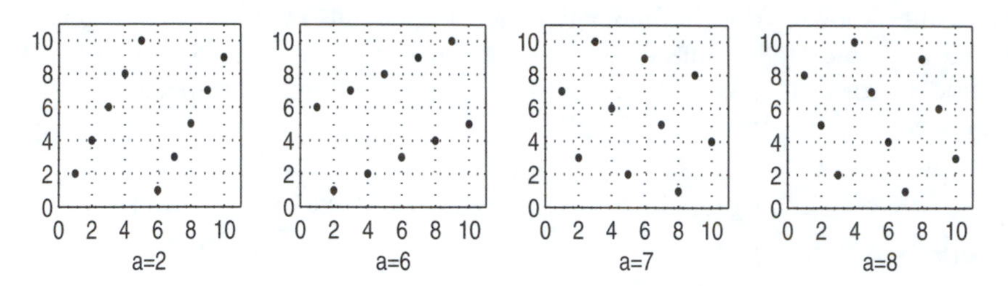

Figure 11.1. Lattice structure in two dimensional space for the multiplicative linear congruential generator (MLCG) generator $y_{i+1} = (a\, y_i) \bmod 11$ for various values of a.

Figure 11.2 shows the structure obtained for SURAND resulting from generating 5 billion random numbers and plotting pairs (in two dimensions) and triplets (in three dimensions) of consecutive (lag $k = 1$) numbers that appear on a subregion of the unit square. Clearly, defects are evident: a regular pattern emerges, indicating limited coverage. These defects would not have been apparent from a similar plot using far fewer numbers in the sequence — say 50,000 — as done in [811, Figure 3.3].

Figure 11.2 also shows patterns obtained from the Unix default generators rand and drand48 discussed above. For both rand and drand48, we also generated 5 billion consecutive numbers in the sequence. The corresponding (lag $k = 1$) plots on subregions of the unit square reveal a lattice pattern for rand but not drand48.

Most texts and review articles on the subject illustrate such patterns (e.g., [811, Figure 3]). For vivid color illustrations of the artifacts introduced by poor random number generators on the lattice structure of a polycrystalline Lennard-Jones spline, see [814], for example.

See also the related Monte Carlo exercise which involves generating 2D and 3D plots to search for structure of a particular (faulty!) random number generator termed RANDU. The LCG RANDU defined in that exercise can already exhibit a high degree of correlation when a relatively small number of sequence points (e.g., 2500) is generated!

11.2.4 Other Generators

To overcome some of these deficiencies of linear congruential generators, other methods have been designed. Two alternative popular classes are *lagged Fibonacci* and *shift-register* generators.

Fibonacci Series

A *Fibonacci series* is one in which each element is the sum of the two preceding values, e.g., $\{1, 2, 3, 5, 8, 13, 21, 34, 55, 89, 144, 233, 377, \ldots\}$. The series is

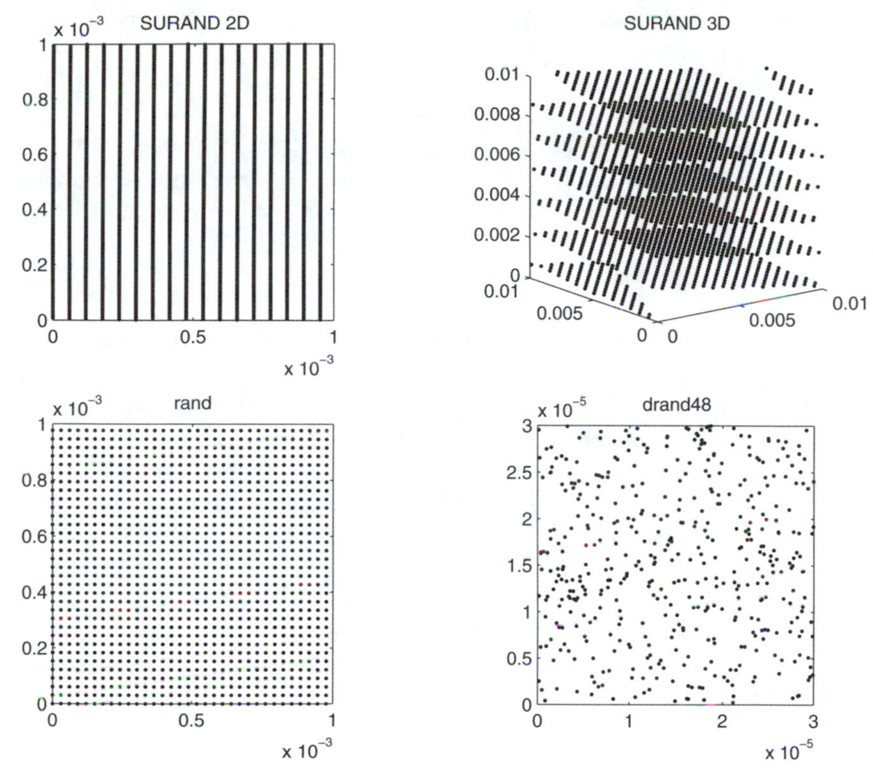

Figure 11.2. Structure corresponding to three linear congruential generators, with basic formula $y_{i+1} = (a\,y_i + c) \bmod M$, and displayed on subregions of the unit square or cube: SURAND ($a = 16807$, $M = 2147483647$, $c = 0$); rand ($a = 1103515245$, $M = 2^{31}$, $c = 12345$); and drand48 ($a = 25214903917$, $M = 2^{48}$, $c = 11$). In all cases, 5 billion numbers in the sequence were generated; fewer numbers correspond to the points displayed on the subregions in view (pairs in 2D and triplets in 3D). The code for SURAND was based on the one given in this text, and the codes for rand and drand48 were these available internally on our SGI's Unix System Library. The programs used for plotting the data are available on the text's website. See also Tables 11.1 and Figure 11.3 for Monte Carlo averages using rand and drand48.

named after the master of unit fractions, made famous for his rabbit breeding question;[6] the answer is 377, the 13th element of the Fibonacci series above.

[6]How many pairs of rabbits can be produced in a year from one rabbit pair? Assume that every month each pair produces a new offspring couple, which from the second month also becomes productive.

A Fibonacci random number generator computes each variate by performing some operation on the previous two iterates. Schematically, we write:

$$\tilde{x}_{i+1} = (\tilde{x}_{i-j} \odot \tilde{x}_{i-k}) \bmod M \,, \qquad j < k, \tag{11.13}$$

where $\odot$ is an arithmetic or logical operation and j and k are *integer lags*.

Multiplicative or additive lagged Fibonacci generators are common, and their periods can be quite long. Popular, easy-to-implement additive generators in this class have the form

$$\tilde{x}_{i+1} = (\tilde{x}_{i-j} + \tilde{x}_{i-k}) \bmod 2^m \,, \qquad j < k, \tag{11.14}$$

where k is sufficiently large (e.g., 1279). The maximal period of such a generator is $(2^k - 1)\, 2^{m-1}$ [805], and the procedure has the advantage of working directly in floating point, without the usual integer-to-floating point conversion.

An example of an additive lagged Fibonacci generator used by the Thinking Machines Library has $j = 5, k = 17, m = 32$ (period around 2^{48}). A multiplicative lagged Fibonacci generator considered better has the form

$$\tilde{x}_{i+1} = (\tilde{x}_{i-j} \times \tilde{x}_{i-k}) \bmod 2^m \,, \qquad j < k, \tag{11.15}$$

though its period, $(2^k - 1)\, 2^{m-3}$, is smaller by a factor of 4. This class of generators is the recommended choice by Knuth [805] and Marsaglia [815], though it was noted [805] that little theory exists to demonstrate their desirable randomness properties. L'Ecuyer later cautioned against the use of these lagged Fibonacci generators, as they display highly unfavorable properties when subjected to certain spectral tests [816].

Shift-Register Generators

Shift-register (or *Tausworthe*) random number generators have a similar form to lagged Fibonacci series generators but employ $M = 2$ in eq. (11.13). This means that only binary bits of a variate are generated and then collected into words by relying on a shift register. The operation $\odot$ is the 'exclusive or'.

An example of a shift-register generator is a k-step method which generates a sequence of k random numbers by splitting this sequence into consecutive blocks and then taking each block as the digit expansion in base p (typically 2). We can thus write this family of generators as:

$$\tilde{x}_{i+k} = \left[\sum_{j=0}^{k-1} a_j \, \tilde{x}_{i+j} \right] \bmod 2 \,, \tag{11.16}$$

where the $\{\tilde{x}_i\}$ and $\{a_j\}$ are either 0 or 1. The output values $\{x_i\}$ are constructed from these bits.

Shift register methods with carefully chosen parameters are very fast, have strong uniformity properties, and possess period lengths that are not bounded by the word size of the machine. Though they may not exhibit lattice structure in the same space as LCGs, quality parameters must be selected based on analysis of lattice structure in a different space (of formal series); see [817, 818] for example. The maximal length of series (11.16) is 2^{k-1}.

Combination Generators

Many other methods exist, including linear matrix generators, nonlinear recurrence generators such as inversive congruential and quadratic, and various combination of these generators. Combining output of good basic generators to create new random sequences can improve the quality of the sequence and increase the length. Schematically, we form $\{\tilde{z}_i\}$ from $\{\tilde{x}_i\}$ and $\{\tilde{y}_i\}$ by defining

$$\tilde{z}_i = \tilde{x}_i \odot \tilde{y}_i ,$$

where $\odot$ is typically a logical (e.g., exclusive-or operator) or addition modulo M. If the associated periods of each sequence, τ_x and τ_y, are relatively prime, the cycle length of $\{\tilde{z}_i\}$ can be as large as the product $\tau_x \, \tau_y$. The properties of the combination sequence will be no worse than those of either sequence and typically better.

L'Ecuyer proposes good combined generators based on the addition of linear congruential sequences of higher order [810]. Three such highly efficient methods are offered programmed in the C language. The two sequences for 32-bit machines have lengths of $2^{191}(\sim 10^{57})$ and $2^{319}(\sim 10^{96})$, and the 64-bit version has the impressive length of $2^{377}(\sim 10^{113})$. All combine two sequences defined by

$$\tilde{x}_{1,i} = (\, a_{1,1}\, \tilde{x}_{1,i-1} + a_{1,2}\, \tilde{x}_{1,i-2} + \cdots a_{1,k}\, \tilde{x}_{1,i-k}\,) \bmod M_1 \,, \qquad (11.17)$$

$$\tilde{x}_{2,i} = (\, a_{2,1}\, \tilde{x}_{2,i-1} + a_{2,2}\, \tilde{x}_{2,i-2} + \cdots a_{2,k}\, \tilde{x}_{2,i-k}\,) \bmod M_2 \,, \qquad (11.18)$$

for $i = 0, 1, \ldots$, where M_1 and M_2 are distinct primes and the two sequences have period lengths $M_1^k - 1$ and $M_2^k - 1$, respectively.

A combined multiplicative linear congruential generator can be formed by adding δ multiples of the variates from the two series, or by forming

$$X_i = (\, \delta_1\, \tilde{x}_{1,i}/M_1 \; + \delta_2\, \tilde{x}_{2,i}/M_2\,) \,, \qquad (11.19)$$

where δ_1 and δ_2 are integers, each relatively prime to its associated sequence modulus (M_1 and M_2). With properly chosen parameters, the period length of $\{X_i\}$ will be $(M_1^k - 1)(M_2^k - 1)/2$. These formulas can be generalized to more than two sequences.

L'Ecuyer's sequence of length 2^{191} [810] combines two sequences by using $k = 3$ (the number of prior sequence iterates), $M_1 = 2^{32} - 209$, $M_2 = 2^{32} - 22853$, and $a_{1,1} = a_{2,2} = 0$, $a_{1,2} = 1403580, a_{1,3} = -810728$, $a_{2,1} = 527612, a_{2,3} = -1370589$. The second 32-bit-machine generator uses more terms with $k = 5$, and the long, 64-bit generator has $k = 3$ but two larger moduli, of order 2^{63}. See the programs for these generators (in the C language) in [810].

FORTRAN and C codes for another generator of length $\sim 2^{121}$ with good statistical properties [819] are available on the course website. This generator combines four MLCGs defined by

$$\tilde{x}_{j,i} = (a_j \, \tilde{x}_{j,i-1}) \bmod M_j \, , \qquad j = 1, 2, 3, 4, \tag{11.20}$$

via

$$w_i = \sum_{j=1}^{4} (\delta_j \, \tilde{x}_{j,i}/M_j) \bmod 1, \qquad \delta_j = (-1)^{j+1} \, . \tag{11.21}$$

The respective multipliers and moduli are set to $a_1 = 45991$, $a_2 = 207707$, $a_3 = 138556$, $a_4 = 49689$, and $M_1 = 2147483647$, $M_2 = 2147483543$, $M_3 = 2147483423$, $M_4 = 2147483323$, respectively [819].

11.2.5 Artifacts

Two interesting examples that illustrate the importance of quality random number generators and their appropriate testing with the application at hand are summarized in Boxes 11.3 and 11.4.

The first, from a real situation in 1989, reflects a coincidental relationship between the problem (matrix) size and the period of the generator (the matrix dimension divides the period τ), as well as short τ and limited accuracy (single-precision computer arithmetic). These problems could have been avoided by averting this matrix-dimension/generator-period relationship, increasing the generator period, and using double-precision arithmetic.

The second instructive example of "hidden errors" stemming from apparently good random number generators was reported in 1992 [820]. Essentially, the researchers showed that incorrect results can be produced under certain circumstances by random number generators that have a long period and have passed certain tests for randomness. Thus, a careful testing of the combination of random number generator and application is generally warranted. Though thought to be generators of high-quality, the generators used in [820] are known to have unfavorable lattice structure [816]. Of course, these examples also argue for using generators with as-long-a-period as possible.

In addition to possible systematic errors with high-quality random number generators for some algorithms due to subtle correlations, researchers showed that even good generators can yield inconsistent results regardless of the algorithm [821]. Though researchers believed that the two used generators had passed all known statistical tests, the generators produced (with the same algorithm) critical temperature estimates for the continuous clock model that differed by 2%, much higher than intrinsic errors. This model has a second-order phase transition, and the exact answer is not known. Thus, it cannot be determined which result is more reliable.

The best advice, therefore, appears to be not only to use reliable generators with as long periods as possible, but also to experiment with several generators as well as algorithms to the extent possible.

Box 11.3: Accidental Relationship Between Problem Size and Generator Length (1989 Linpack Benchmarks)

In 1989, David Hough, a numerical analyst working at Sun Microsystems, noticed very peculiar behavior in benchmark testing of the package Linpack. (The original posting can be found on the archived NA Digest, Volume 89, Issue 1, 1989). The single-precision factorization of 512×512 randomly-generated matrices produced a perplexing underflow (roughly around 10^{-40}) in the diagonal pivot values. (*Sherlocks: note that* $512 = 2^9$). However, matrices composed of random data from a uniform distribution are known to be remarkably well conditioned! So how can this seemingly well conditioned matrix be nearly singular?

The answer came upon examination of the random number generator and how it was used to set the matrix elements. Specifically, the matrix elements a_{ij} were set to be in the range $[-2, 2]$ according to the following subprogram, which relies on a simple MLCG with $a = 3125$ and $M = 65536$. (*Holmes fans: note that* $M = 2^{16}$).

```
c*********************************************************************
      subroutine matgen(amat,Lda,n,b,norma)
      real amat(Lda,1), b(1), norma, halfm, quartm
      integer a, m
      parameter  (a = 3125, m = 65536)
      parameter  (halfm = 32768.0, quartm = 16384.0)
      iseed = 1325
      norma = 0.0
      do 30 j = 1, n
         do 20 i = 1, n
            iseed = mod (a * iseed, m)
            amat(i,j) = (iseed - halfm) / quartm
            norma   = max (amat(i,j), norma)
   20    continue
   30 continue
      return
      end
c*********************************************************************
```

A quick examination first showed that the period of this MLCG is only $\tau = M/4 = 16384$; the full period of M could have easily been generated by changing one line above, to incorporate the nonzero $c = -1$ increment value. Still, this would have only delayed the underflow problem to a 1024×1024 matrix.

The main problem here lies in the fact that the matrix size chosen, $512 = 2^9$, *divides* the modulus, also a power of 2 ($M = 2^{16}$). Hence, the period $\tau = 2^{14}$ factors $\tau = 512 \times 32$. *This means that the first 32 columns of the "random" matrix are repeated 16 times!* The matrix is subsequently singular, and after each 32 steps of Gaussian elimination the zeros in the lower triangular part of the matrix are reduced by a factor of order (10^{-7}). Clearly, after six rounds of such transformations (each treating 32 columns), the element size in the lower triangle would drop to order $\mathcal{O}(10^{-42})$, explaining the underflow. This problem could have been removed by using matrix sizes that do not divide the period, resorting to double precision arithmetic (underflow threshold of 10^{-300}), and by increasing the period of the generator substantially.

Note also that besides underflow, the generator could have experienced other problems for matrices smaller than 512×512, since $512 \times 512 = 2^{18}$, and 2^{18} is greater than 2^{14}, the generator's period.

11.2.6 Recommendations

In sum, high-quality, long-sequence random number generators are easy to find in the literature, but they are not necessarily available on default system implementations.

For the best results, an MC simulator is well advised to consult the resident mathematical expert for the most suitable generator and computing platforms with respect to the application at hand.

Certainly, the user should compare application results as obtained for several random number generators. Indeed, L'Ecuyer likens generators to cars [810]: no single model nor size of an automobile can possibly be a universal choice. Good expert advice can be obtained by examining Pierre L'Ecuyer's website (presently at www.iro.umontreal.ca/~lecuyer). Certainly, given today's computationally-intensive biomolecular simulations, it is advisable to use sequences with long periods. Combined multiplicative linear congruential generators are good choices. See [810] for good recommendations. See also [512, pp. 76–78] for an efficient FORTRAN implementation of a good combination MLCG used for DNA work, though with a small period by today's standards (10^{18}).

Generators particularly suitable for parallel machines are also available [822, 814], characterized by different streams of variates produced by good seeding algorithms and variations in the parameters of the underlying recursion formulas. See the SPRNG scalable library package (Scalable Parallel Random Number Generation) targeted for large-scale parallel Monte Carlo applications: sprng.cs.fsu.edu, for software. The package can be used in C, C++, and Fortran and has been ported to most major computer platforms.

Box 11.4: Accidental Relationship Between Simulation Protocol and Generator Structure (Ising Model)

Ferrenberg *et al.* [820] used different generators in the context of simulating an Ising model, a model characterized by an abrupt, temperature-dependent transition from an ordered to a disordered state. The states, characterized by the spin directionality of the particles, were generated by an algorithm termed Wolff that determines the flips of a cluster on the basis of a random number generator. Surprisingly, the researchers found that the correct answer was approximated far better by the 32-bit multiplicative linear congruential generator SURAND, well recognized to have lattice-structure defects. So why does the apparently-superior shift register generator produce *systematically incorrect results* — energies that are too low and specific heats that are too high?

An explanation to these observations came upon inspection of the Wolff algorithm. Namely, subtle correlations in the random number sequence affect the Wolff algorithm in a specific manner! If the high order bits are zero, they will remain zero according to the spin generation algorithm. This in turn leads to a bias in the cluster size generated and hence the type of equilibrium structures generated.

The main message from this work was a note of caution on the effect of subtle correlations within random number generators on the system generation algorithms used for simulating the physical system. This suggests that not only should a generator be tested on its own; it should be tested together with the algorithm used in the MC simulation to reduce the possibility of artifacts.

11.3 Gaussian Random Variates

11.3.1 Manipulation of Uniform Random Variables

Our uniform random variates computed in the last section, U, can be used to generate variates X that correspond to a more general, given probability distribution by a simple transformation. To generate a continuous variate X with distribution function $F(x)$ (see Box 11.1) which is continuous and strictly increasing on $(0, 1)$ (i.e., $0 < F(x) < 1$), we set x to $F^{-1}(u)$ where F^{-1} is the inverse of the function F. The challenge in practice is to establish good algorithms for evaluating $F^{-1}(u)$ to the desired accuracy.

Below we only describe generating variates from a Gaussian (normal) distribution, commonly needed in molecular simulations. For information on generating variates from many other distributions, see [806] and the web page of Luc Devroye (cgm.cs.mcgill.ca/~luc/), for example.

11.3.2 Normal Variates in Molecular Simulations

A vector of normally-distributed random variates satisfying a given mean (μ) and variance (σ^2) (see Box 11.1, eq. (11.5)) is often required in molecular simulations. One example is the initial velocity vector (of components $\{v_i\}$) in a molecular dynamics simulation corresponding to the target temperature of an n-atom system,

$$\sum_{i=1}^{3n} m_i \, v_i^2 = 3 \, n \, k_B \, \mathrm{T} \, .$$

Another example is the Gaussian random force vector R in Langevin dynamics with zero mean and variance chosen so as to satisfy:

$$\langle R(t)R(t')\rangle = 2\gamma k_B \mathrm{T}\mathbf{M}\delta(t-t')\,, \tag{11.22}$$

where $\mathbf{M}$ is the mass matrix and γ is the damping constant.

In the molecular and Langevin dynamics cases above, we first set each component of the vector vec from a standard normal distribution (zero mean and unit variance) so that the sum is also a normal distribution with the additive means and variances (see Central Limit Theorem below); to obtain the desired variance σ^2 rather than unit variance, we then modify each component according to the vector update relation

$$\mathsf{vec} \leftarrow \sigma\ \mathsf{vec} + \mu\,.$$

Since, by this modification, it is easy to generate normal variates from the normal distribution with mean μ and variance σ^2 ($\mathcal{N}(\mu,\sigma^2)$) from variates sampled from a standard normal distribution ($\mathcal{N}(0,1)$), it suffices to generate standard normal variates.

There are several techniques to set a variate X_u from a Gaussian or normal distribution on the basis of a uniformly distributed variate U (for which procedures were discussed above). Two are described below.

11.3.3 Odeh/Evans Method

One efficient approach was described by Odeh and Evans [823]. For a given u value in the range $0 < u < 1$, the corresponding normal variable x_u is computed to satisfy:

$$u \;=\; \frac{1}{\sqrt{2\pi}}\int_{-\infty}^{x_u} \exp(-t^2/2)\,dt\,. \tag{11.23}$$

This is accomplished by approximating x_u as the sum of two terms:

$$x_u \;=\; y + S(y)/T(y)\,, \qquad y = \sqrt{\{\ln(1/u^2)\}}\,,$$

where S and T are polynomials of degree 4 chosen to yield the minimal degree rational approximation to $x_u - y$ with maximum error less than 10^{-7}.

In practice, a vector of random variates is first formed ($U \equiv \{u_1, u_2, \cdots, u_n\}$ in the notation above; see subroutine ranuv below, based on function ranu), from which a standard normal distribution is formulated ($X_U \equiv \{x_{u_1}, x_{u_2}, \cdots, x_{u_n}\}$); each component is then adjusted to yield the desired mean and standard deviation (see subroutine rannv1 below).

```
c*****************************************************************
      subroutine rannv1 (n, vec, mean, var)
c A vector of n pseudorandom numbers is generated, each from a
c standard  normal distribution (mean zero, variance one), based on
c Odeh and Evans, App. Stat. 23:96 (1974).
c For a nonzero mean MU and/or non unity variance,
```

```
c set  vec(i) = mu + sqrt(sigma(i))*vec(i).
c Subroutine ranset should be called before the first subroutine
c call.
      integer n
      double precision vec(n),mean,var(n),
     *  temp,p0,p1,p2,p3,p4,q0,q1,q2,q3,q4
      parameter (p0=-.322232431088d0, p1=-1d0, p2=-.342242088547d0,
     *  p3=-.204231210245d-1, p4=-.453642210148d-4,
     *  q0=.99348462606d-1, q1=.588581570495d0, q2=.531103462366d0,
     *  q3=.10353775285d0, q4=.38560700634d-2)
      if (n .lt. 1) return
      call ranuv(n, vec)
      do 10 i = 1, n
         temp = vec(i)
         if (temp .gt. 0.5d0) vec(i) = 1d0 - vec(i)
         vec(i) = sqrt(log(1d0/vec(i)**2))
         vec(i) = vec(i) +
     *           ((((vec(i) * p4 + p3) * vec(i) + p2) *
     *           vec(i) + p1) * vec(i) + p0) /
     *           ((((vec(i) * q4 + q3) * vec(i) + q2) *
     *           vec(i) + q1) * vec(i) + q0)
         if (temp .lt. 0.5d0) vec(i) = -vec(i)
   10 continue
      do 20 i = 1, n
         vec(i) = sqrt(var(i)) * vec(i) + mean
   20 continue
      return
      end
c***********************************************************************
      subroutine ranuv (n, vec)
c Generate a vector of n pseudorandom uniform variates
      integer n, a, m, q, r, seed
      double precision vec(n), rm
      parameter (a=16807, m=2147483647, q=127773, r=2836, rm=1d0/m)
      common /random/ seed
      save /random/
      if (n .lt. 1) return
      do 10 i = 1, n
         seed = a * mod(seed, q) - r * (seed/q)
         if (seed .le. 0) seed = seed + m
         vec(i) = seed * rm
   10 continue
      return
      end
c***********************************************************************
```

For example, to set the initial velocity vector according to the target temperature using the equipartition theorem (each degree of freedom has $k_B\mathrm{T}/2$ energy at thermal equilibrium), the routines above are used for the velocity vector V of $3n$ components $\{v_i\}$ with $\mu = 0$ and var(i) = $(k_B\mathrm{T})/m_i$. For the Langevin random force vector, the variance for each vector coordinate i is: $(2\gamma k_B m_i \mathrm{T})/\Delta t$, where the delta function δ in eq. (11.22) is discretized on the basis of the timestep Δt (see also Chapter 12 on molecular dynamics).

11.3.4 Box/Muller/Marsaglia Method

Another popular algorithm to form normal variates x_1 and x_2 is the Box/Muller/Marsaglia method [805, pp. 117–118]. It involves generating two *uniformly* distributed random variates u_1 and u_2, setting v_1 and v_2 as uniform variates between -1 and $+1$ ($v_1 \leftarrow 2u_1 - 1, v_2 \leftarrow 2u_2 - 1$), checking that $s = v_1^2 + v_2^2$ is less than 1 (if $s \geq 1$, the procedure is repeated), and then setting the two *normal* variates x_1 and x_2 as:

$$x_1 = v_1 \sqrt{-2 \ln s/s}\,, \qquad x_2 = v_2 \sqrt{-2 \ln s/s}\,. \tag{11.24}$$

Essentially, we are using the polar-coordinate representation of x_1 and x_2 by v_1 and v_2 ($x_1 = \tilde{r}\cos\tilde{\theta}\,, x_2 = \tilde{r}\sin\tilde{\theta}\,, \tilde{r} = \sqrt{-2\ln s}\,, \tilde{\theta} = \tan^{-1}(v_2/v_1)$) to construct the joint probability distribution of the two normal variates in polar coordinates:

$$\begin{aligned} &\left(\frac{1}{\sqrt{2\pi}} \int_{-\infty}^{x_1} e^{-x^2/2}\, dx \right) \left(\frac{1}{\sqrt{2\pi}} \int_{-\infty}^{x_2} e^{-y^2/2}\, dy \right) \\ &\qquad = \frac{1}{2\pi} \int_{\substack{\{(r,\theta)\,| \\ r\cos\theta \leq x_1 \\ r\sin\theta \leq x_2\}}} e^{-r^2/2}\, r\, dr\, d\theta\,. \end{aligned} \tag{11.25}$$

11.4 Means for Monte Carlo Sampling

11.4.1 Expected Values

Armed with a uniform random variate generator, we can now address the important task of estimating a mean property of interest. In molecular simulations, we might seek the average geometric and energetic properties associated with an equilibrium distribution of conformations.

MC Estimate

In its simplest form, we write such a mean, or expected value, as an integral

$$I = \int_{\mathcal{D}} A(x)\, dx = \langle A(x) \rangle_{\mathcal{D}} \tag{11.26}$$

where the average is computed over the uniformly distributed elements $x \in \mathcal{D}$. For example, assume that the function $A(x)$ is defined on [0, 1]. Choose a sequence of N random variates for large N,

$$x_1, x_2, \ldots, x_N\,,$$

and generate corresponding function values $y_i = A(x_i)$:

$$y_1, y_2, \ldots, y_N \, .$$

Then we compute the average, termed the Monte Carlo estimate of I, by:

$$\bar{y}_N = \frac{1}{N} \sum_{i=1}^{N} y_i \, . \tag{11.27}$$

Simple Example: Calculate π by MC

As a simple example, consider calculating π by Monte-Carlo integration of the area of a quarter-circle of radius 1 circumscribed inside the unit square in the plane (with center at the origin of the plane). The integral to be evaluated is:

$$\int_0^1 \int_0^1 \rho(x, y) \, dx \, dy$$

where

$$\rho(x, y) = \begin{cases} 1 & x^2 + y^2 \leq 1 \\ 0 & \text{else} \end{cases} \, .$$

This integral's value is $\pi/4$. A simple Fortran program to perform this integration by Monte Carlo sampling consists of the following:

```
c***********************************************************************
      subroutine monte(nstep)
      implicit none
      integer nstep, i, nin, iseed
      double precision  x, y, tmp, rand

      nin = 0
      iseed = 12345
      call srand(iseed)

      do 30 i = 1, nstep                    *
         x = rand()                        /|\
         y = rand()                         |   ^
         tmp = sqrt(x*x + y*y)              |      ^
         if (tmp .lt. 1.d0) then            |
            nin = nin + 1                   |         ^
         endif                              |          \
   30 continue                              +-----------*
                                                       /
      print *, nstep, (4.d0 * nin)/nstep
      return
      end
c***********************************************************************
```

Results as a function of the sample size (nstep) are presented in Table 11.1 and Figure 11.3 using the Unix rand and drand48 random number generators and also more sophisticated methods. Since drand48 is the fastest of the generators,

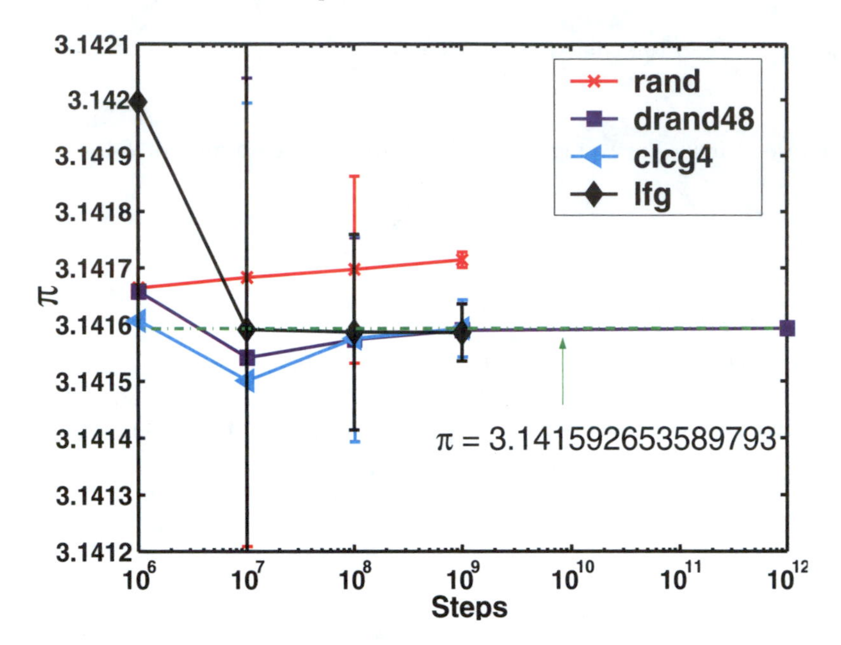

Figure 11.3. Results (means and error bars) of MC estimates for π based on different random number generators, as tabulated in Table 11.1.

we also record the estimated value corresponding to 10^{12} steps (only up to 10^9 steps for the rest).

Note that, unfortunately, this procedure for calculating π is not very accurate. At best, the first six decimal places of $\pi = 3.14159265358979323846\ldots$ are obtained. The accuracy is limited not only by the sample size — statistical error — but also by any possible defects of the random number generator (e.g., lattice structure and limited coverage; see Figure 11.2). Here we see that the accuracy of the means starts to deteriorate after the number of steps exceeds the period length. We also learn from this example that the longer-period generators have greater resolution (another order of magnitude of two).

11.4.2 Error Bars

Law of Large Numbers

According to the *Law of Large Numbers* in probability theory, the average of N sampled random variables converges (in probability) to its expected value. Stated more formally, if the uniform variates are independent and drawn from the same distribution so that the expected value of each y_i is μ, then as $N \to \infty$ the average value $\bar{y}_N$ converges to μ asymptotically:

$$P\{\lim_{N\to\infty} (\bar{y}_N) = \mu\} = 1\,.$$

However, the rate of convergence to the expected value is a different matter and requires stronger assumptions.

Variance

As stressed in this chapter's introduction, it is essential to provide *error bars* when reporting an MC average. The variance of $\bar{y}_N$ is defined as

$$\sigma_{\bar{y}}^2 = \mathrm{var}(\bar{y}) = \frac{1}{N}\sum_{i=1}^{N}(y_i - \bar{y}_N)^2 \,. \tag{11.28}$$

The variance measures the distribution of $\bar{y}$ about its mean μ; the larger N is, the narrower the interval about I where $\bar{y}_N$ can be found.

Variance Relation to Central Limit Theorem

This interval can be determined as a probability of deviation in units of σ on the basis of the *Central Limit Theorem*. This beautiful and powerful result states that as $N \to \infty$, the *limiting distribution* for a *sum of random variates* is the *normal distribution*.

Specifically, if $\{y_1, y_2, \ldots\}$ is a sequence of independent, identically distributed random variates having mean μ and finite nonzero variance σ^2, then the random variable

$$S_N = y_1 + y_2 + \cdots + y_N$$

has the normal density with mean $N\mu$ and variance $N\sigma^2$, $\mathcal{N}(N\mu, N\sigma^2)$. In other words, the normalized random variable

$$\frac{S_N - N\mu}{\sqrt{\mathrm{var}(S_N)}} = \frac{S_N - N\mu}{\sigma\sqrt{N}}$$

has the standard normal distribution:

$$\lim_{N\to\infty} P\left(\frac{S_N - N\mu}{\sigma\sqrt{N}} \leq x\right) = \frac{1}{\sqrt{2\pi}}\int_{-\infty}^{x} \exp\left[-t^2/2\right] dt \,.$$

Thus, in reporting an MC average, we say that we have estimated I within one standard error (i.e., $\sigma/\sqrt{N}$) of $\bar{y}_N$ as:

$$I = \bar{y}_N \pm \sigma_{\bar{y}}/\sqrt{N} \,. \tag{11.29}$$

For example, $N = 10,000$ yields a result that is at best roughly 1% accurate; for correlated data, such as from molecular dynamics simulations, a much larger N is required for that accuracy. Note that this $1/\sqrt{N}$ scaling of errors is a general feature of MC methods and is independent of the space dimension involved; that is why MC is frequently the method of choice for multidimensional integrals.

Since we know that the limiting distribution for $\bar{y}_N$ is the normal distribution $\mathcal{N}$, we say that 68.3% of the time this estimate is within one standard error of

I [803]. The above integral estimates can be generalized to independent random samples from other probability densities, as described in the next section.

Note that the above errors are statistical *and can be controlled. The more serious errors in MC algorithms are the* systematic *errors, such as discussed above in connection with random-number-generator artifacts. Both errors should be monitored to the extent possible.*

11.4.3 Batch Means

When the MC data are highly correlated, the error bars may decrease much more slowly with N as in the above idealized case. The effective number of samples is then N/τ where τ is the decorrelation time (number of steps) for the data. This value can be determined by examining auto and cross-correlation functions for the most slowly-varying properties of the system or by using the method of *batch means*. Extensive mathematical/statistical tests for independence are also available to estimate *confidence intervals* of independent means [806, Chapter 4].

Essentially, we divide the sample size N into M batches $\{B_1, B_2, \ldots, B_M\}$ each of $b = N/m$ elements where M should be significantly greater than τ; we then obtain a mean over each batch sample:

$$\bar{y}_{B_i} = \frac{1}{b} \sum_{y_i \in B_i} y_i \,, \qquad i = 1, \ldots, M \,, \tag{11.30}$$

and then set the $\bar{y}_N$ estimate as the average over all these means:

$$\bar{y}_M = \frac{1}{M} \sum_{i=1}^{M} \bar{y}_{B_i} \,. \tag{11.31}$$

In reporting the estimator of form (11.29), the relevant sample size (M rather than N) and variance to is determined from the above mean, that is:

$$\sigma^2_{\bar{y}_M} = \mathrm{var}(\bar{y}_M) = \frac{1}{M} \sum_{i=1}^{M} (\bar{y}_{B_i} - \bar{y}_M)^2 \,. \tag{11.32}$$

It can be shown that if the batch size M is sufficiently large, the means of the batches are approximately uncorrelated. In practice, variations on the basic batch means method sketched above, and additional tests, are needed to yield good statistics [806, pages 528–530].

11.5 Monte Carlo Sampling

11.5.1 Density Function

The properties of many molecular systems can be described by a separable Hamiltonian of general form

$$H(q,p) = E_k(p) + E_p(q) \; = \; \frac{1}{2} p^{\mathrm{T}} \mathbf{M}^{-1} p + E_p(q) \,, \qquad (11.33)$$

where E_k and E_p are the kinetic and potential energy components, respectively, and q and p are the collective position and momentum vectors of the system. This Hamiltonian function forms the basis for MC simulations applied to estimate various properties of large molecular systems, such as geometric and thermodynamic functions. However, the MC estimates must emulate a probability density function $\rho(q,t)$ or $\rho(q,p,t)$ appropriate for the statistical ensemble (t denotes time). This probability density ρ may or may not be known.

11.5.2 Dynamic and Equilibrium MC: Ergodicity, Detailed Balance

MC simulations can be used to mimic a *dynamic process* (ρ depends on time t), as in Brownian dynamics. They can also generate an ensemble around a *statistical equilibrium*, as in some conformational sampling studies.

Dynamic Process

In the former case, a deterministic rule (such as based on Newton's equations of motion in the diffusive limit) is used to generate each configuration from the previous configuration given initial conditions $\rho(q,p,t_0)$, and that rule determines the resulting $\rho(q,p,t)$ for $t > t_0$.

Below we use the notation X to represent the collective phase-space vector (q,p); *when discussing the Metropolis algorithm later, the momentum component drops out*. (This variable X should not be confused with the random variable X defined in Box 11.1).

Equilibrium Process

The equilibrium ensemble regime is appropriate when $\rho(X,t) = \rho_0(X)$ for some $t > t_0$, as in the Metropolis algorithm (see below). The *ensemble average* is then considered as an estimate for the *time average*, which may be much more complex to follow. This assumption, though very difficult to prove in practice, is known as the *ergodic hypothesis*.

In this statistical equilibrium case, the rule that generates X_{n+1} from X_n need not have a clear physical interpretation. However, to be useful for sampling, the rule must ensure that any starting distribution $\rho(X,t)$ should tend to the stationary density $\rho_0(X)$ and that the system be *ergodic* (i.e., as $t \to \infty$, the system spends equal times in equal volumes of phase space); see [737] for a rigorous definition. When the rule also obeys *detailed balance* (i.e., moving from state X to Y is as likely as returning to X from Y), an equilibrium process is approached (though biased techniques may violate detailed balance and still approach the right answer through correcting for violations).

These criteria are crucial for constructing practical sampling algorithms for physical systems; efficient sampling of configurational space is another important aspect of computer simulations, especially for large systems where configuration space cannot be sampled exhaustively.

11.5.3 Statistical Ensembles

Common statistical ensembles used in biomolecular simulations are the *canonical* or constant–NVT (N = number of particles, V = volume, T = temperature), *micro-canonical* or constant–NVE (E = energy), *isothermal-isobaric* or constant–NPT (P = pressure), and *grand canonical* or constant–μVT (μ = chemical potential) [30].

Canonical Ensemble and Boltzmann Factor

The probability density function for the *canonical* ensemble is proportional to the Boltzmann factor:

$$\rho_{\mathrm{NVT}}(X) \propto \exp\left(-\beta E(X)\right), \tag{11.34}$$

where E is the total energy of the system and $\beta = (1/k_B\mathrm{T})$.

(In the Metropolis algorithm, it is sufficient to work with the *potential energy*, since the potential energy is independent of momenta; see below).

Hence for two system states X and X', the corresponding probability ratio is:

$$\frac{\rho_{\mathrm{NVT}}(X)}{\rho_{\mathrm{NVT}}(X')} = \exp\left(-\beta \Delta E\right), \tag{11.35}$$

where

$$\Delta E = E(X) - E(X').$$

See the sketch of Figure 11.4. The normalizing factor in the proportionality relation (eq. (11.34)) is the total partition function for all of phase space. That is:

$$\rho_{\mathrm{NVT}}(X) = \frac{1}{(h^{3\mathrm{N}})\,\mathrm{N}!}\,\frac{\exp\left(-\beta E(X)\right)}{Q_{\mathrm{NVT}}} \tag{11.36}$$

where h is Planck's constant, the factor N! accounts for the indistinguishability of the N particles, and Q_{NVT} is the canonical partition function:

$$Q_{\mathrm{NVT}} = \frac{1}{(h^{3N})\,N!}\int \exp\left(-\beta E(x)\right) dx\,, \tag{11.37}$$

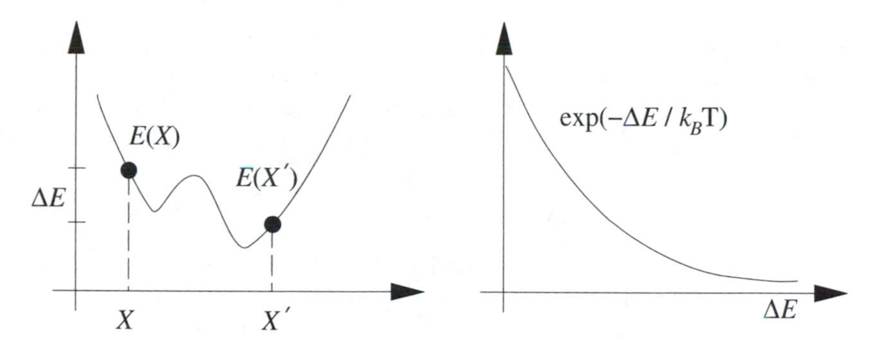

Figure 11.4. The Boltzmann probabilities for two system states X and X' with energies $E(X)$ and $E(X')$ are related at equilibrium by the probability ratio $\exp\left[-\Delta E/k_B\mathrm{T}\right]$.

where x is a point in phase space.

The corresponding means for the system for a function A can then be written as:

$$\langle A(x)\rangle_{\mathrm{NVT}} = \int \rho_{\mathrm{NVT}}(x)\, A(x)\, dx\,. \tag{11.38}$$

The Metropolis algorithm described below is used to generate an appropriate Markov chain (see below) from which the expected value of A is calculated as:

$$\langle A(x)\rangle_{\mathrm{NVT}} = \lim_{M\to\infty} \frac{1}{M} \sum_{i=1}^{M} (A(x_i))\,. \tag{11.39}$$

In other words, the density function is already built into the generation algorithm through an appropriate acceptance probability.

11.5.4 *Importance Sampling: Metropolis Algorithm and Markov Chains*

To obtain reliable statistical averages, it is essential to use computer time efficiently to concentrate calculations of the configurational-dependent functions in regions that make important contributions. This concept is known as *importance sampling*.

In many applications, it is possible to focus sampling on the *configurational* part of phase space (i.e., E is the potential energy component) since the projection of the corresponding trajectory on the momentum subspace is essentially independent from that projection on the coordinate subspace. *Hence X below is the collective position vector only.*

Markov Chain

Metropolis *et al.* [802] described such an efficient and elegantly simple procedure for the canonical ensemble. In mathematical terms, we generate a *Markov chain* of molecular states $\{X_1, X_2, X_3, \ldots\}$ constructed to have the limiting distribution $\rho_{\rm NVT}(X)$. In a Markov chain, each state belongs to a finite set of states contained in the state space $\mathcal{D}_0 \in \mathcal{D}$, and the conditional distribution of each state relative to all the preceding states is equivalent to that conditional distribution relative to the last state:

$$P\{X_{n+1} \in \mathcal{D}_0 \mid X_0, \ldots, X_n\} = P\{X_{n+1} \in \mathcal{D}_0 \mid X_n\}\,;$$

in other words, the outcome X_{n+1} depends only on X_n. The Metropolis algorithm constructs a transition matrix for the Markov chain that is stochastic and ergodic so that the limiting distribution for each state X_i is $\rho_i = \rho_{\rm NVT}(X_i)$ and thereby generates a phase space trajectory in the canonical ensemble.

This transition matrix is defined by specifying a transitional probability π_{ij} for X_i to X_j so that *microscopic reversibility* is satisfied:

$$\rho_i \pi_{ij} = \rho_j \pi_{ji}\,. \tag{11.40}$$

In other words, the ratio of transitional probabilities depends only on the energy change between states i and j:

$$\frac{\rho_i}{\rho_j} = \frac{\pi_{ji}}{\pi_{ij}} \propto \exp\left(-\beta \Delta E_{ij}\right), \tag{11.41}$$

where $\Delta E_{ij} = E(X_i) - E(X_j)$. See subsection below on MC Moves with examples of biased sampling.

Metropolis Algorithm

Briefly, the Metropolis algorithm generates a trial $\tilde{X}_{i+1}$ from X_i by a system-appropriate random perturbation (satisfying detailed balance) and accepts that state if the corresponding energy is lower. If, however, $E(\tilde{X}_{i+1}) > E(X_i)$, then the new state is accepted with probability $p = \exp\left(-\beta \Delta E\right)$, where $\Delta E = E(\tilde{X}_{i+1}) - E(X_i) > 0$, by comparing p to a uniformly-generated number on (0,1): if $p > \mathsf{ran}$, accept $\tilde{X}_{i+1}$, and if $p \geq \mathsf{ran}$, generate another trial $\tilde{X}_{i+1}$ but recount X_i in the Markov chain (see Fig. 11.4).

The result of this procedure is the acceptance probability at step i of:

$$\begin{aligned} p_{\rm acc,\,MC} &= \min\left[1, \exp\left(-\beta \Delta E\right)\right] \\ &= \min\left[1, \frac{\rho_{\rm NVT}(\tilde{X}_{i+1})}{\rho_{\rm NVT}(X_i)}\right]. \end{aligned} \tag{11.42}$$

In this manner, states with lower energies are always accepted but states with higher energies have a nonzero probability of acceptance too. Consequently, the sequence tends to regions of configuration space with low energies, but the system can always escape to other energy basins.

Simulated Annealing

An extension of the Metropolis algorithm is often employed as a global-minimization technique known as *simulated annealing* where the temperature is lowered as the simulation evolves in an attempt to locate the global energy basin without getting trapped in local wells.

Simulated annealing can be considered an extension of either MC or molecular/Langevin dynamics. Simulated annealing is often used for refinement of experimental models (NMR or crystallography) with added nonphysical, constraint terms that direct the search to target experimental quantities (e.g., interproton distances or crystallography R factors; see [50], for example, for a review of the application of simulated annealing in such contexts.

Noteworthy is also a Monte Carlo sampling/Minimization (MCM) hybrid method developed by Scheraga and coworkers [824] to search for the global energy minimum of a protein. It generates a large random conformational change followed by local minimization of the potential energy and applies the Metropolis criterion for acceptance or rejection of the new conformation. This procedure is then iterated. Friesner and coworkers have found this MC method to perform well in a variety of applications [825]. Other stochastic global optimization methods have been developed and successfully applied by the Scheraga group [801].

In this connection, see homework 13 for the *deterministic* global optimization approach based on the *diffusion equation* as suggested and implemented for molecular systems by Scheraga and colleagues [766].

Metropolis Algorithm Implementation

The Metropolis algorithm for the canonical ensemble can be implemented with the potential energy E_p rather than the total energy when the target measurement A for MC averaging is velocity independent. This is because the momentum integral can be factored and canceled. From eq. (11.38) combined with eq. (11.34), we expand the state variable x to represent both the momentum (p) and position (q) variables, both over which integration must be performed:

$$\begin{aligned} \langle A(x) \rangle &= \frac{\int \exp\left[-\beta E_k\right] dp \int A(q) \exp\left[-\beta E_p\right] dq}{\int \exp\left[-\beta E_k\right] dp \int \exp\left[-\beta E_p\right] dq} \\ &= \frac{\int A(q) \exp\left[-\beta E_p\right] dq}{\int \exp\left[-\beta E_p\right] dq} . \end{aligned} \tag{11.43}$$

The Metropolis algorithm is summarized below.

Metropolis Algorithm (Canonical Ensemble)

For $i = 0, 1, 2, \ldots$, given X_0:

1. Generate $\tilde{X}_{i+1}$ from X_i by a perturbation technique that satisfies *detailed balance* (i.e., the probability to obtain $\tilde{X}_{i+1}$ from X_i is identical to that going to X_i from $\tilde{X}_{i+1}$).
2. Compute $\Delta E = E(\tilde{X}_{i+1}) - E(X_i)$.
3. **If** $\Delta E \leq 0$ (downhill move), accept X_{i+1} : $\quad X_{i+1} = \tilde{X}_{i+1}$;

Else, set $p = \exp(-\beta\Delta E)$. Then
If $p >$ ran, accept $\tilde{X}_{i+1}: \quad X_{i+1} = \tilde{X}_{i+1}$.
Else, reject $\tilde{X}_{i+1}: \quad X_{i+1} = X_i$.

4. Continue the i loop.

MC Moves

Specifying appropriate MC moves for step 1 is an art by itself. Ideally, this could be done by perturbing all atoms by independent (symmetric) Gaussian variates with zero mean and variance σ^2, where σ^2 is parameterized to yield a certain acceptance ratio (e.g., 50%). However, in biomolecular simulations, moving all atoms is highly inefficient (that is, leads to a large percentage of rejections) [31], and it has been more common to perturb one or few atoms at each step.

The type of perturbations depends on the system and the energy representation (e.g., rigid or nonrigid molecules, a biomolecule or pure liquid system, Cartesian or internal degrees of freedom). The perturbation can be set as translational, rotational, local, and/or global moves. For example, in the atomistic CHARMM molecular mechanics and dynamics program, protein moves are prescribed from a list of possibilities including rigid-residue translation/rotation, single-atom translation, and single or multiple torsional motions.

For MC simulations of a bead/wormlike chain model of long DNA, we use local translational moves of one bead at a time combined with a rotational move of a chain segment [512]; we must also ensure that no move changes the system's topology (e.g., linking number of a closed chain) for simulating the correct equilibrium ensemble.

Figure 11.5 illustrates such moves for long DNA. Figure 11.6 illustrates corresponding MC (versus Brownian dynamics) distributions of the DNA writhing number (Wr) and the associated mean, as a function of length for two salt environments. Figure 11.7 demonstrates how a faulty move (like moving only a subset of the DNA beads instead of all) can corrupt the probability distributions of Wr and the radius of gyration (Rg). Not only do we note a corruption of the distributions when incorrect MC protocols are used, but a large sensitivity to the initial configuration (sharp distributions around starting configurations).

The rule of thumb usually employed in MC simulations is to aim for a perturbation in Step 1 (e.g., displacement magnitude or the variance σ^2 associated with the random Gaussian variate) that yields about 50% acceptance. Thus, we seek to balance too small a perturbation that moves the system in state space slowly with too large a perturbation that yields high trial-energy configurations, most of which are rejected. *However, the appropriate percentage depends on the application and is best determined by experimentation guided by known outcomes of the statistical means sought.* Much smaller acceptance probabilities may be perfectly adequate for some systems.

Figure 11.5. Translational and rotational MC moves for a bead model of DNA.

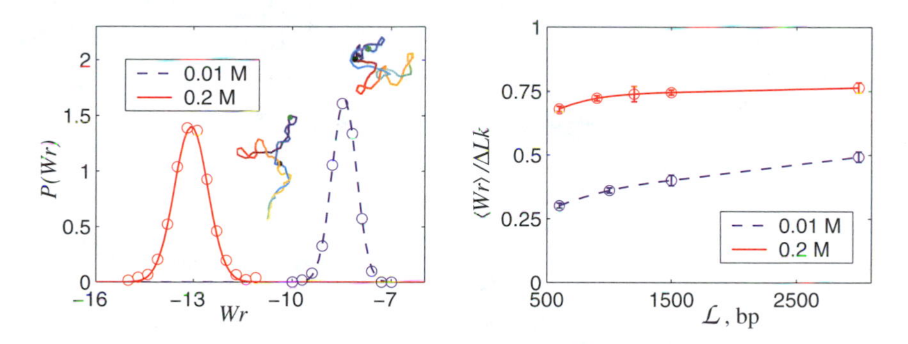

Figure 11.6. MC distributions (curves) versus Brownian dynamics data (circles) of the writhing number (Wr) distribution of supercoiled DNA (left), and mean values of Wr (normalized by the linking number difference) as a function of the DNA length (right), at two salt concentrations. The DNA superhelical density is $\sigma = -0.06$ in both panels, and the left panel involves DNA of length 3000 base pairs. Error bars for BD are shown only if they are larger than the circle symbol.

11.6 Hybrid Monte Carlo

11.6.1 Exploiting Strengths of MC and MD

To enhance the efficiency of MC simulations, a simple idea emerged that attempts to combine the favorable properties of molecular dynamics (MD) simulations — sampling phase space in a directed manner guided by the shape of the energy gradient — with that of MC — sampling phase space more globally. Ideally, following conformation space by MD would generate a correct Boltzmann distribution of states, but the relatively short simulation lengths that are possible (see MD chapters) imply local rather than global sampling.

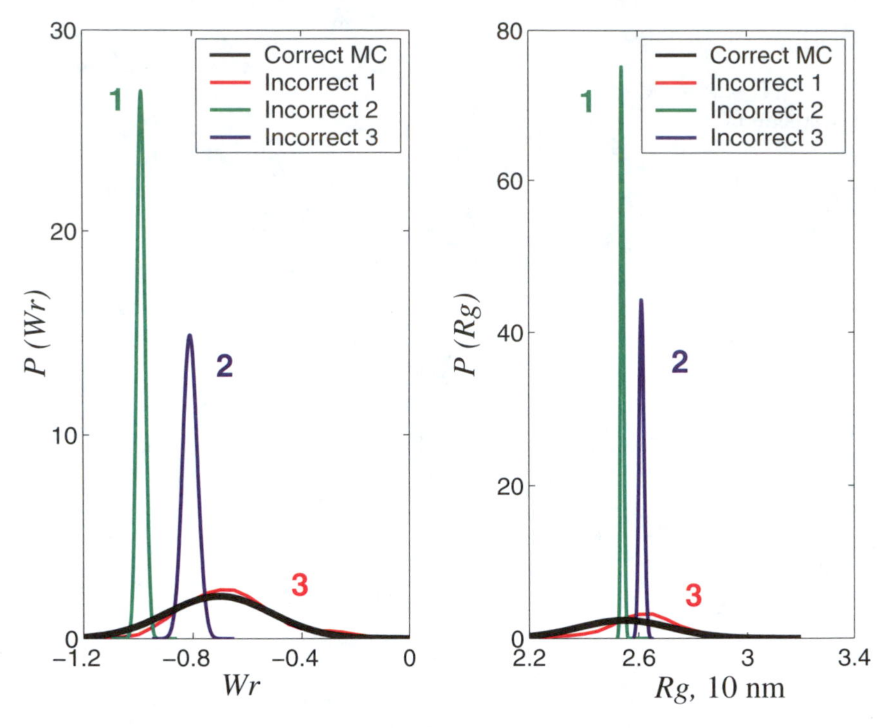

Figure 11.7. DNA writhing number Wr (left) and radius of gyration Rg distributions as generated by one correct versus three incorrect MC procedures. The incorrect MC protocols allow only a subset of the beads to move: 25 out of 30 ("Incorrect 1"), or 5 out of 30 ("Incorrect 2, 3"); the last two schemes are started from different initial points. The DNA modeled has 600 base pairs, with superhelical density of $\sigma = -0.04$, and the MC simulations consist of ten million steps.

Though in theory a good MC protocol would sample configuration space exhaustively, this becomes more difficult and inefficient in practice as the system size increases. When the cost of evaluating the energy function for large biomolecular systems is also a factor, millions of MC steps (with possibly many rejections) can become quite expensive.

11.6.2 Overall Idea

The idea to overcome these MC difficulties with *hybrid Monte Carlo* (HMC) [826] is to combine global updates in position space via MD with reasonable acceptance criteria by MC.

The first step of a hybrid MC method uses a molecular dynamics framework to specify the system's candidate move: $X[i\Delta t] \longrightarrow X[(i+1)\Delta t]$ where Δt is the timestep. The MD algorithm must be *time reversible* and volume preserving so as to ensure detailed balance. The commonly used symplectic Verlet method satisfies this requirement (see Chapters 12 and 13). Since the recursive MD recipe relies on the velocities in addition to positions, the required velocity vectors $V = \mathbf{M}^{-1}P$ (P here designates momentum, not to be confused with the symbol used earlier for *probability*) are generated from a Gaussian distribution so as to obtain the target kinetic energy at the temperature T assuming energy equipartition. That is, the velocity components are drawn from a Gaussian distribution proportional to the Boltzmann factor applied to the kinetic energy:

$$\rho(P) \propto \exp\left[-\beta\, E_k(P)\right]. \tag{11.44}$$

Following this MD-guided step, the second step of the hybrid MC method applies the standard Metropolis acceptance criterion where the energy in the Boltzmann factor is the *Hamiltonian* (potential *plus* kinetic energy). Namely, the Metropolis criterion is applied to accept the new candidate $\tilde{X}_{i+1}$ with probability

$$\begin{aligned} p_{\text{acc, HMC, }\rho} &= \min\left[1, \exp\left[-\beta\left(H(\tilde{X}_{i+1}, \tilde{P}_{i+1}) - H(X_i, P_i)\right)\right]\right] \\ &= \min\left[1, \frac{\exp\left[-\beta E_p(\tilde{X}_{i+1})\right]\, \exp\left[-\beta E_k(\tilde{P}_{i+1})\right]}{\exp\left[-\beta E_p(X_i)\right]\, \exp\left[-\beta E_k(P_i)\right]}\right] \\ &= \min\left[1, \frac{\rho_{\text{NVT}}(\tilde{X}_{i+1}) \exp\left[-\beta E_k(\tilde{P}_{i+1})\right]}{\rho_{\text{NVT}}(X_i) \exp\left[-\beta E_k(P_i)\right]}\right]. \end{aligned} \tag{11.45}$$

HMC methods have been quite successful for biomolecular sampling, and many other extensions to general ensembles and hybrid methods have been described.

Note that *generalized ensembles* can be emulated by HMC methods by applying weighting factors to the Metropolis-generated configurations. That is, to sample from a general ensemble with $\mu(x)$ rather than $\rho(x)$, we adjust the acceptance criteria of eq. (11.45) to be:

$$p_{\text{acc, HMC, }\mu} = \min\left[1, \frac{\mu(\tilde{X}_{i+1}) \exp\left[-\beta E_k(\tilde{P}_{i+1})\right]}{\mu(X_i) \exp\left[-\beta E_k(P_i)\right]}\right]. \tag{11.46}$$

The weighting must be accomplished to maintain *detailed balance*.

11.6.3 Variants and Other Hybrid Approaches

MC variants include *Smart MC*, *J-walking*, stochastic dynamics, umbrella sampling, various methods to enhance barrier crossing events (e.g., the transition path sampling method of Chandler and coworkers [827]), and potential-modification approaches such as smoothing techniques. See [828], for example, for a perspective. Interesting also are papers describing an adaptive-temperature HMC method for a mixed-canonical ensemble that enhances sampling [829] and an efficient Monte Carlo method based on known molecular minima for medium-sized flexible molecules [830]. The simple MC/MD combination [831] — moving some particles by MC rules and others by MD — may be more effective for solvated biomolecular systems than either MC or MD alone.

There has been continued discussion on the relative merits of MC and MD for biomolecules (e.g., [832, 833]). The limits of MD for simulating slow molecular events is clearly recognized, as is its high computational cost, though only MD can ultimately yield detailed dynamic information such as folding pathways and rates of conformational changes. At present, a practitioner is well-advised to use a combination of MC, MD, and local and global minimization algorithms, as appropriate, for the problems at hand.

Table 11.1. Results of computing π by MC integration with various random number generators and sample sizes by the procedure described in the text, subroutine monte (on an SGI R12K/300 MHz Octane processor). The value σ is computed from 100 runs for each NSTEP value, except for 10^{12}, for which it is based on one run. rand has $\tau \approx 2^{31}$; 10^9 calls require 771 seconds (13 min.). drand48 has $\tau \approx 2^{48}$; 10^9 calls require 656 seconds (11 min.). clg4 has $\tau \approx 2^{31}$; 10^9 calls require 5205 seconds (87 min.). lfg has $\tau \approx 2^{64}$; 10^9 calls require 2106 seconds (35 min.). See also Figure 11.3.

Nstep	Estimate	Error	σ, s.d.
rand, IRIX 6.5 system library			
10^2	3.1604000000000E+00	1.8807346410208E − 02	1.59E−01
10^3	3.1346800000000E+00	−6.9126535897936E − 03	4.75E−02
10^4	3.1383680000000E+00	−3.2246535897933E − 03	1.79E−02
10^5	3.1416736000000E+00	8.0946410208504E − 05	5.22E−03
10^6	3.1416646400000E+00	7.1986410206115E − 05	1.61E−03
10^7	3.1416831800000E+00	9.0526410207570E − 05	4.74E−04
10^8	3.1416971676000E+00	1.0451401020672E − 04	1.65E−04
10^9	3.1417139545200E+00	1.2130093020657E − 04	1.37E−05
drand48, IRIX 6.5 system library			
10^2	3.1408000000000E+00	−7.9265358979264E − 04	1.72E−01
10^3	3.1456400000000E+00	4.0473464102098E − 03	5.62E−02
10^4	3.1431880000000E+00	1.5953464102063E − 03	1.53E−02
10^5	3.1418572000000E+00	2.6454641020690E − 04	5.92E−03
10^6	3.1416580000000E+00	6.5346410206946E − 05	1.61E−03
10^7	3.1415415040000E+00	−5.1149589793908E − 05	4.97E−04
10^8	3.1415733104000E+00	−1.9343189793464E − 05	1.79E−04
10^9	3.1415892614400E+00	−3.3921497935019E − 06	4.70E−05
10^{12}	3.1415928451280E+00	1.9153820707274E − 07	1.00E−06
clg4, based on four linear congruential generators [819]			
10^2	3.1600000000000E+00	1.8407346410206E − 02	1.51E−01
10^3	3.1413200000000E+00	−2.7265358979189E − 04	4.98E−02
10^4	3.1405440000000E+00	−1.0486535897933E − 03	1.74E−02
10^5	3.1416476000000E+00	5.4946410205758E − 05	4.94E−03
10^6	3.1416064800000E+00	1.3826410206530E − 05	1.59E−03
10^7	3.1415008760000E+00	−9.1777589792841E − 05	4.92E−04
10^8	3.1415751016000E+00	−1.7551989793141E − 05	1.82E−04
10^9	3.1415925054000E+00	−1.4818979332532E − 07	5.05E−05
lfg, SPRNG package, modified lagged-Fibonacci generator			
10^2	3.1372000000000E+00	−4.3926535897922E − 03	1.70E−01
10^3	3.1414400000000E+00	−1.5265358979244E − 04	5.20E−02
10^4	3.1442160000000E+00	2.6233464102083E − 03	1.55E−02
10^5	3.1431008000000E+00	1.5081464102074E − 03	5.23E−03
10^6	3.1419958000000E+00	4.0314641020611E − 04	1.64E−03
10^7	3.1415909920000E+00	−1.6615897910910E − 06	5.46E−04
10^8	3.1415866680000E+00	−5.9855897953653E − 06	1.72E−04
10^9	3.1415854581200E+00	−7.1954697937748E − 06	5.03E−05

12

Molecular Dynamics: Basics

Chapter 12 Notation

SYMBOL	DEFINITION
Matrices	
$\mathbf{M}$	mass matrix
$\mathbf{P}$	pressure tensor (stress)
Vectors	
$g(X)$	constrained dynamics vector, components $g_i = r_{jk}^2 - \overline{r_{jk}}^2$
$\mathbf{r}_i$	position vector for atom i
$\mathbf{v}_i$	velocity vector component corresponding to atom i
F	force
$\widetilde{F}$	acceleration ($\ddot{X}$), or $\mathbf{M}^{-1}\,F(X(t))$
F_i	force on particle i due to all particles
$V, \dot{V}, \ddot{V}$	velocity $= \dot{X}$ (components $\{v_i\}$), and its first and second time derivatives
V_0	initial velocity
$X, \dot{X}, \ddot{X}$	position (components $\{x_i\}$), and its first and second time derivatives
X_0	initial position
Λ	Lagrange multiplier vector for constrained dynamics (components λ_i)
∇E	energy gradient
Scalars & Functions	
c	speed of light
c_p, c_t	coordinate and velocity scaling parameters (for constant pressure and temperature simulations)
h	Planck's constant
m	particle mass
m_p, m_t	masses for fictitious thermostat and barostat dynamic

Chapter 12 Notation Table (continued)

SYMBOL	DEFINITION
	variables
r_{ij}, $\overline{r_{ij}}$	interatomic distance and associated equilibrium value
$v_{\rm cm}$	velocity of the center of mass
x_p, x_t	dynamic variables for fictitious thermostat and barostat
E_k	kinetic energy
E_p	potential energy
H, $\widetilde{H}$	Hamiltonian and nearby Hamiltonian
N	number of atoms
N_F	number of degrees of freedom
P, P_0	pressure, target pressure
S_h	harmonic bond stretching constant
T, T_0	temperature, target temperature
β	isothermal compressibility
γ_t	thermostat coupling parameter
$\delta(t)$	Lyapunov exponent
ϵ	trajectory perturbation
ζ_p, ζ_t	thermodynamic coefficients for constant pressure and temperature simulations
λ	wavelength ($1/\lambda$ is wave number)
ν, ω	characteristic frequencies
ν_l	volume variable
τ	thermostat coupling parameter
Δt, Δt_m, $\Delta\tau$	timesteps

Time is defined so that motion looks simple.

John Archibald Wheeler, American theoretical physicist (1911–).

12.1 Introduction: Statistical Mechanics by Numbers

12.1.1 Why Molecular Dynamics?

Molecular dynamics (MD) simulations represent the computer approach to statistical mechanics. As a counterpart to experiment, MD simulations are used to estimate equilibrium and dynamic properties of complex systems that cannot be calculated analytically. Representing the exciting interface between theory and experiment, MD simulations occupy a venerable position at the crossroads of mathematics, biology, chemistry, physics, and computer science.

The static view of a biomolecule, as obtained from X-ray crystallography for example — while extremely valuable — is still insufficient for understanding a wide range of biological activity. It only provides an average, frozen view of a complex system. Certainly, molecules are live entities, with their constituent atoms continuously interacting among themselves and with their environment. Their dynamic motions can explain the wide range of thermally-accessible states of a system and thereby connect structure and function. By following the dynamics of a molecular system in space and time, we can obtain a rich amount of information concerning structural and dynamic properties.

Such information includes molecular geometries and energies; mean atomic fluctuations (see Figure 12.1 for an example of MD snapshots of a protein/DNA system compared to the initial structure); local fluctuations (like formation/breakage of hydrogen bonds, water/solute/ion interaction patterns, or nucleic-acid backbone torsion motions); rates of configurational changes (like ring flips, nucleic-acid sugar repuckering, diffusion), enzyme/substrate binding; free energies; and the nature of various types of concerted motions. Ultimately perhaps, large-scale deformations of macromolecules such as protein folding might be simulated. This formidable aspect, however, is more likely to be an outgrowth of hand-in-hand advances in both experiment and theory.

Though the MD approach remains popular because of its essential simplicity and physical appeal (see below), it complements many other computational tools for exploring molecular structures and properties, such as Monte Carlo simulations, Poisson-Boltzmann analyses, and energy minimization, as discussed in preceding chapters, and Brownian dynamics (Section 13.5); see examples in Table 12.1 and Figure 12.1. Each technique is appropriate for a different class of problems.

12.1.2 Background

A solid grounding in classical statistical mechanics, thermodynamic ensembles, time-correlation functions, and basic simulation protocols is essential for MD practitioners. Such a background can be found, for example, in the books by McQuarrie [29], Allen and Tildesley [30], and Frenkel and Smit [31]. Basic elements of simulation can be learned from the book of Ross [32] and Bratley, Fox and Schrage [33]. MD fundamentals, as well as advanced topics, are also available in the Allen and Tildesley [30] and the newer Frenkel and Smit [31] texts, with basic introductions and useful examples in Gould and Tobochnik [34], Rapaport [35], and Haile [36]. Some of these texts also describe analysis tools for MD trajectories, a topic not considered here.

An advanced text for Hamiltonian dynamics and integration schemes is that by Sanz-Serna and Calvo [837]. Of particular interest are two books that aim at biomolecular simulations: by McCammon and Harvey [37] and by Brooks, Pettitt and Karplus [38].

Given these many sources for the theory and application of MD simulations, the focus in this chapter and the next is on providing a broad introduction into the computational difficulties of biomolecular MD simulations and some long-time integration approaches to these important problems. The reader is advised to complement these aspects with the more standard topics available elsewhere.

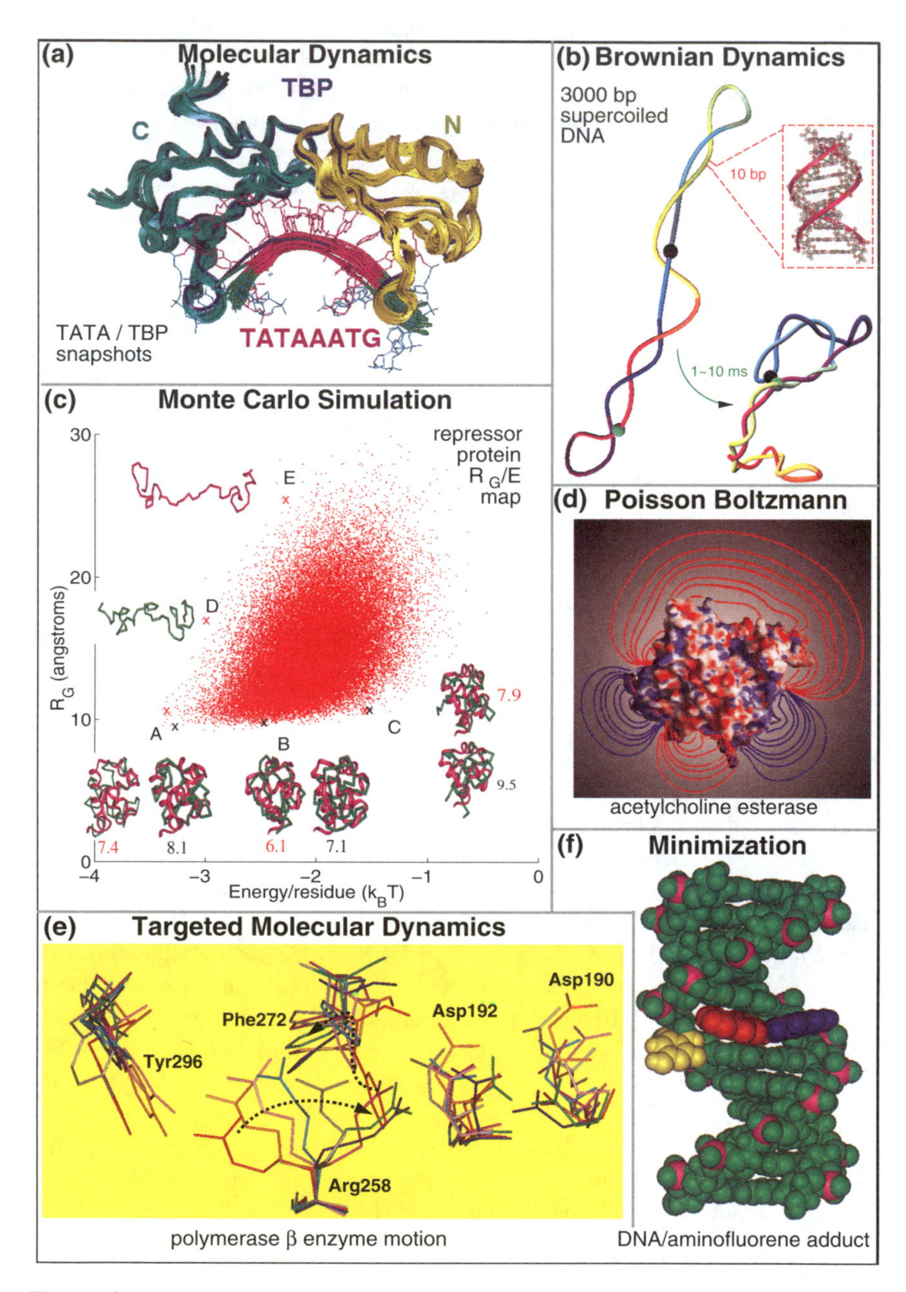

Figure 12.1. Illustrative images for selected biomolecular sampling methods. See caption on the following page.

Table 12.1. Selected biomolecular sampling methods. Continuum solvation includes empirical constructs, generalized Born models, stochastic dynamics, or Poisson Boltzmann solutions, as discussed in Chapter 9. See Figure 12.1 and Figure 13.17 for illustrations of these techniques.

Method	Pros	Cons	CPU
• *Molecular Dynamics* (MD) [37, 38, 834]	continuous motion, experimental bridge between structures and macroscopic kinetic data	expensive; short timespan	high
• *Targeted MD* (TMD) [835]	connection between two states; useful for ruling out steric clashes and suggesting high barriers	not necessarily physical	moderate
• *Stochastic Path Approach* [836]	high-frequency motion filtering; approximate long-time trajectories	expensive (global optimization of entire trajectory)	high
• *Continuum Solvation* [703, 712, 704, 711, 701, 702]	mean-force potential approximates environment and reduces model's cost; useful information on ionic atmosphere and intermolecular associations	approximate	high (if repeated in time)
• *Brownian Dynamics* (BD) [703, 269]	large-scale and long-time motion	approximate hydrodynamics; limited to systems with small inertia	moderate
• *Monte Carlo* (MC) [833]	large-scale sampling; useful statistics	move definitions are difficult; unphysical paths	low
• *Minimization* [758]	valuable equilibria information; experimental constraints can be incorporated	no dynamic information	low

Caption to Figure 12.1: (a) MD snapshots of a DNA TATA-element bound to the TBP protein show flexibility with respect to the initial crystal structure (blue) [274, 838]. (b) BD snapshots of long DNA capture large-scale motions, such as site-juxtaposition kinetics (see Box 6.9 of Chapter 6) of two segments located 900-bp along the DNA contour, in the millisecond timeframe [499]. (c) MC configurations and ensemble radius of gyration/energy plot from a folding simulation of 434-repressor protein show RMS values for predicted structures, superimposed with native conformations [839]. (d) PB contours illustrate the electrostatic potential of mouse acetylcholinesterase (D. Sept, K. Tai, and J.A. McCammon, personal communication, and see [750, 840]). (e) TMD snapshots of polymerase β capture the large-scale conformational change of the enzyme from closed (red) to open (green) forms [656], shown here as reflected by motion of several residues near the active site. (f) The minimized adduct of a DNA duplex with a carcinogen (2-aminofluorene) was computed with NMR constraints [841].

12.1.3 Outline of MD Chapters

We begin by describing in Section 12.2 the roots of molecular dynamics in Laplace's 19th-century vision. We then describe two basic limitations of Newtonian (classical) mechanics: determinism, and failure to capture quantum effects (or electronic motion). We elaborate on the latter by noting that the classical physics approximation is especially poor at lower temperatures and/or higher frequencies. Section 12.3 discusses general issues in molecular and biomolecular dynamics simulations, such as setting up the system (initial conditions, solvent and ions, equilibration), estimating temperature, and other simple aspects of the simulation protocol (equilibration, nonbonded interaction handling). We also elaborate upon the inherent chaotic (but bounded) behavior of biomolecules first mentioned in Section 12.2 by demonstrating sensitivity to initial conditions. We motivate various methods for reducing the computational time of dynamic simulations by emphasizing the large computational times required for biomolecules, as well as extensive data-analysis and graphical requirements associated with the voluminous data that are generated.

Given the large computational requirements, we introduce in the same overview section the concepts of accuracy, stability, symplecticness, and multiple-timesteps (MTS) that are key to developing numerical integration techniques for biomolecular dynamics simulations. In this context, we also mention constrained dynamics and rigid-body approximations.

Though we refer to several of the methods in this introductory section (Section 12.3) that we only define later, a detailed understanding is not needed for grasping the essentials that are communicated here. I recommend that students re-read Section 12.3 after completing the two MD chapters.

Section 12.4 describes the symplectic Verlet/Störmer method and develops its variants known as leapfrog, velocity Verlet, and position Verlet. This material is followed by a sketch of the extension of Verlet to constrained dynamics formulations (Section 12.5) and a reference to various MD ensembles (Section 12.6).

The next chapter introduces more advanced topics on various integration approaches for biomolecular dynamics; the material is suitable for students with a good mathematical background. Specifically, we discuss symplectic integration methods, MTS (or force splitting) methods — via *extrapolation* and *impulse* splitting techniques — and introduce the notion of resonance artifacts. Enhancing our understanding of resonance artifacts in recent years has been important for developing large-timestep methods for MD. We continue to describe methods for Langevin dynamics, Brownian dynamics, and implicit integration.

12.2 Laplace's Vision of Newtonian Mechanics

12.2.1 The Dream

Besides "statistical mechanics by numbers", I like to term MD as "Laplace's vision of Newtonian mechanics on supercomputers".

Sir Isaac Newton (1642–1727) described in 1687 in his *Principia* masterpiece (*Philosophiae Naturalis Principia Mathematica*) the grand synthesis of basic concepts involving force and motion. This achievement was made possible by the large pillars of empirical data laid earlier by scientists who studied the motion of celestial and earthly bodies: Galileo Galilei (1564–1642), Nicolas Copernicus (1473–1543), Tycho Brahe (1546–1601), and Johannes Kepler (1571–1630). Newton's second law of motion — that a body's acceleration equals the net force divided by its mass — is the foundation for molecular dynamics.

The celebrated French mathematician Pierre Simon de Laplace (1749–1827) recognized the far-reaching implications of Newtonian physics almost two centuries ago. In this now-classic piece, Laplace dreams about predicting the future by animating Nature's forces (see Box 12.1) [842].

While stated in an era when the effect of high-speed computers on modern life and science could hardly be imagined, Laplace's vision already contains the essential ingredients of present-day biomolecular simulations: mathematical construction of the suitable force field (*forces*), design of appropriate numerical integration tools (*analysis*), and long-time propagation of the equations of motion (*far-reaching intelligence*).

12.2.2 Deterministic Mechanics

As Laplace stated, the capability to analyze and predict motion — be it of the solar system or a biological system — provides the link between the past and the future. Still, even Newtonian mechanics taken to its extreme cannot predict with certainty the future motion of all bodies. As became evident by the work of Poincaré less than a century after Laplace's statement, the solar system is chaotic even though the underlying laws can be clearly expressed. This understanding, however, should not deter us from pursuing Laplace's dream; rather, it should stimulate us to explore as deeply as possible the consequences of Newtonian physics.

12.2.3 Neglect of Electronic Motion

In addition to the limitation of deterministic mechanics, only the *nuclear* motion of many-body systems is typically followed in molecular dynamics. Thus, electronic motion is not considered, and quantum effects are generally ignored. The classical approximation is excellent for a wide range of systems and materials but is unsuitable for reactions involving electronic rearrangements such as bond formation and cleavage, polarization, and chemical bonding of metal ions. Quantum dynamical approaches are used for this purpose. They are, however, at a relatively early stage with respect to macromolecular applications and are not covered in this text.

Box 12.1: Laplace's 'Far-Reaching Intelligence'

Laplace, the son of a Normandy farmer, is famous for his masterpieces on Celestial Mechanics and the Theory of Probability [842]. The Laplace equation (though written by Euler in 1752 in a hydrodynamic context) is Laplace's chief contribution to potential theory. In his 1820 *oeuvre* [842], Laplace states:

> *Une intelligence qui, pour un instant donné, connaîtrait toutes les forces dont la nature est animée et la situation respective des êtres qui la composent, si d'ailleurs elle était assez vaste pour soumettre ces données à l'Analyse, embrasserait dans la même formule les mouvements des plus grands corps de l'univers et ceux du plus léger atome: rien ne serait incertain pour elle, et l'avenir, comme le passé, serait présent à ses yeux. L'esprit humain offre, dans la perfection qu'il a su donner à l'Astronomie, une faible esquisse de cette intelligence.*

> *An intelligence which could, at any moment, comprehend all the forces by which nature is animated and the respective positions of the beings of which it is composed, and moreover, if this intelligence were far-reaching enough to subject these data to analysis, it would encompass in that formula both the movements of the largest bodies in the universe and those of the lightest atom: to it nothing would be uncertain, and the future, as well as the past, would be present to its eyes. The human mind offers us, in the perfection which it has given to astronomy, a faint sketch of this intelligence.*

12.2.4 Critical Frequencies

Classical MD simulations are also unsuitable for low temperatures, where the energy gaps among the discrete levels of energy dictated by quantum physics are much larger than thermal energy available to the system. This is because the system is confined to one or a few of the low-energy states under such conditions. This discrete description of energy states becomes less important as the temperature is increased and/or the frequencies associated with motion are decreased (i.e., have longer timescales). Under those conditions, more energy states become thermally accessible.

Rough estimates for the characteristic motions for which Newtonian physics is reasonable can be made on the basis of harmonic analysis. For a harmonic oscillator, the quantized energies are separated by $h\nu$ where h is Planck's constant and ν is the vibrational frequency.

Clearly, the classical approach is unsuitable for capturing motions with relatively high frequencies ν, that is with $\nu \gg k_B\,\mathrm{T}/h$, or

$$\frac{h\,\nu}{k_B\,\mathrm{T}} \gg 1\,, \tag{12.1}$$

where k_B is Boltzmann's constant, and T is the temperature. This is because the probability of finding the system with this mode at the ground state energy is high. The larger this ratio, the greater this probability.

Conversely, classical behavior is approached for frequency/temperature combinations for which

$$\frac{h\,\nu}{k_B\,\mathrm{T}} \ll 1\,. \tag{12.2}$$

Table 12.2. Ratios for some high-frequency vibrational modes at T = 300 K.

Vibrational mode	Wave number $(1/\lambda)$ [cm^{-1}]	Frequency $\nu = c/\lambda$ [s^{-1}]	Ratio $h\nu/(k_B\mathrm{T})$
O–H stretch	3600	1.1×10^{14}	17
C–H stretch	3000	9.0×10^{13}	14
O–C–O asym. stretch	2400	7.2×10^{13}	12
C=O (carbonyl) stretch	1700	5.1×10^{13}	8
C–N stretch (amines)	1250	3.8×10^{13}	6
O–C–O bend	700	2.1×10^{13}	3

Around the room temperature of 300 K, $k_B\,\mathrm{T}$ = 0.6 kcal/mol. As we see from Table 12.2, the high-frequency vibrational modes present in biomolecules have ratios larger than unity. They are thus not well treated by classical physics. Specifically, Newtonian physics, which distributes the energy equally among all vibrational modes according to the equipartition theorem, overestimates the energy/motions associated with these high-frequency modes.

The critical frequency ν for which equality of the above ratio holds is around 6.25×10^{12} s^{-1} or 6 ps^{-1}, corresponding to an absorption wavelength of 208 cm^{-1} (or period of 160 fs). Thus, modes with characteristic timescales in the picosecond and longer timeframes are reasonably treated by classical, Newtonian physics. The second and third classes of motions identified for biomolecules in Table 12.3 fall in this range.

12.2.5 *Hybrid Electronic/Nuclear Treatments*

Electronic motions which have much higher characteristic frequencies must be treated by alternative approaches such as quantum-mechanical and hybrid quantum/classical approximations [558, 843, 844]; good progress has been made in

Table 12.3. The broad spectrum of characteristic timescales in biomolecules.

Internal Motion	Timescale [seconds]
Light-atom bond stretch	10^{-14}
Double-bond stretch	2×10^{-14}
Light-atom angle bend	2×10^{-14}
Heavy-atom bond stretch	3×10^{-14}
Heavy-atom angle bend	5×10^{-14}
Global DNA twisting	10^{-12}
Sugar puckering (nucleic acids)	10^{-12}–10^{-9}
Collective subgroup motion (e.g., hinge bending, allosteric transitions)	10^{-11}–10^{-7}
Surface-sidechain rotation (proteins)	10^{-11}–10^{-10}
Global DNA bending	10^{-10}–10^{-7}
Site-juxtaposition (superhelical DNA)	10^{-6}–1
Interior-sidechain rotation (proteins)	10^{-4}–1
Protein folding	10^{-5}–10

recent years on these techniques. Essentially, in these approaches, the reacting part of the system (e.g., active site of an enzyme) is treated quantum mechanically, while the other components (e.g., remaining amino acids and solvent) are modeled classically, by molecular mechanics. Such treatments are critical for calculating reaction rates for organic compounds in solution, or free energies of hydration. They are also necessary for describing chemical reactions that involve bond breaking and forming, localized enzymatic activity entailing charge transfer, or solvent polarization effects. Important recent progress in this area was mentioned in Chapter 7. Still, many technical details must be perfected regarding these more complex simulation protocols. For example, challenging are the proper definition of the quantum-mechanical treatment and the merging between the quantum and classical approximations.

12.3 The Basics: An Overview

12.3.1 Following the Equations of Motion

The molecular dynamics approach is simple in principle. We simulate motion of a system under the influence of a specified force field by following molecular configurations in time according to Newton's equation of motion. We write these equations for a system of N atoms as the following pair of first-order differential equations:

$$\begin{aligned} \mathbf{M}\dot{V}(t) &= F(X) = -\nabla E(X(t)) + \ldots, \\ \dot{X}(t) &= V(t). \end{aligned} \tag{12.3}$$

In these equations, $X \in \boldsymbol{R}^{3N}$ denotes the collective Cartesian vector of the system (i.e., the x, y, and z components of each atom are listed in turn); V is the corresponding collective velocity vector; $\mathbf{M}$ is the diagonal mass matrix (i.e., the masses of each atom are repeated three times in the diagonal array of length $3N$); and the dot superscripts denote differentiation with respect to time, t.

The total force F in the right-hand-side of eq. (12.3) is composed of the systematic force, which is the negative gradient (vector of first partial derivatives) of the potential energy E and, possibly, additional terms that mimic the environment. (See section 13.4 on stochastic dynamics for an example of these additional terms). Each gradient component i, $i = 1, \ldots 3N$, is given by:

$$\nabla E(X)_i = \partial E(X)/\partial \alpha_i \,,$$

where α_i denotes an x, y, or z component of an atom. These equations must be integrated numerically since analytic (closed-form) solutions are only known for the simplest systems. Such numerical integration generates a sequence of positions and velocity pairs, $\{X^n, V^n\}$, for integers n that represent discrete times $t = n\Delta t$ at intervals (timesteps) Δt.

12.3.2 *Perspective on MD Trajectories*

Force Field Dependency

Results of a molecular dynamics simulation can only be as good as governing force field. Essentially, the mechanical representation of a system — particles connected by springs — assumes simple, pairwise-additive potentials. These express how the composite atoms stretch, vibrate, and rotate about the bonds in response to intramolecular and intermolecular forces. The resultant potential energy E, as described in Chapters 8 and 9, is still highly approximate for biomolecules and undergoes continuous improvements. Uncertainties are well recognized in the representation of solvent [845], polarization effects [75], and electrostatic interactions [593, 592], and in the functional form of the local potentials (i.e., lack of anharmonic [846] and cross terms [847]).

Statics Vs. Dynamics

Minimization of this approximate energy function yields information on favorable regions in configuration space (this approach is termed molecular mechanics or statics). The numerical integration of the differential equations of motion reveals the intrinsic motions of the system under the influence of the associated force field. Thus, in principle, MD simulations can combine both the spatial and temporal aspects of conformational sampling. They are thus used in many cases as conformational search tools — to bypass the multiple-minimum problem — and as vehicles to refine low-resolution X-ray or NMR (nuclear magnetic resonance) data (e.g., [50]). However, for this purpose, special strategies, such as energy modifications or high-temperature settings, are required if substantial movements are desired. This is because MD simulations are severely limited by the very small timesteps that must be used relative to the timescales of major biological interest.

Range of Timescales

Indeed, the motion of biomolecules involves an *extraordinary* range of timescales (see Table 12.3). In general, the higher frequencies have smaller associated amplitude displacements. For example, while bond vibrations have characteristic amplitudes of a tenth of an Ångstrom, global deformations can be in the order of 100 Å (see Table I in [38] for classes of timescales and associated amplitudes in biomolecular motions). The energies associated with these motions also span a large range.

Though the fastest, high-frequency modes have the smallest amplitudes, these motions affect other modes and thus their effect must be approximated in some way. Yet, the existence of high-frequency modes severely affects the timestep that can successfully be used in biomolecular simulations. Their timescale dictates a timestep of 1 fs or less in standard explicit schemes for acceptable resolution (a tenth or less of a period). This stepsize already implies one million steps to cover only a nanosecond and falls short, by more than *ten orders of magnitude*, of the slow and large-amplitude processes of major biological interest. Dealing with this severe timestep problem has been the focus of many research groups, and some results will be the subject of sections that follow.

Challenges

Thus, in summary, although the basic idea is simple, the art of MD simulations is challenging in practice. The practical difficulties arise from various components that enter into biomolecular simulations: setting initial conditions, implementing various simulation protocols to ensure reliability, using suitable numerical integrators, considering the sensitivity of trajectories to initial conditions and other choices, meeting the large computational requirements, and visualizing and analyzing the voluminous data that are generated.

12.3.3 Initial System Settings

A molecular dynamics trajectory consists of three essential parts: initialization, equilibration, and production. Initialization requires specifying the initial coordinates and velocities for the solute macromolecule, as well as for the solvent and ion atoms.

Structure

Even when initial coordinates are available from experiment (e.g., crystal structure), the starting vector may not correspond to a minimum in the potential energy function used, and hence minimization (further refinement) is needed to relax strained contacts. When an experimental structure is not available, a *build-up* technique may be used to construct a structure on the basis of the known building blocks, and minimization again is required.

Solvation

When water molecules and salt ions are also used, special care is needed to carve an appropriate solvation model around the biopolymer. The water coordinates are typically taken from a pure water simulation (which reproduces the experimentally determined density), with initial ion placement guided by the biopolymer charge distribution and experimental measurements. Special iterative ion-placement algorithms are available in molecular dynamics programs that place positive ions in electronegative pockets and negative ions in positively-charged cavities so as to achieve a desired total ionic concentration. The final model is constructed following appropriate superimpositioning, removal of overlapping atoms, and energy relaxation. See [272, 273], for example, for detailed descriptions of such equilibration procedures for B-DNA.

Velocity

The initial velocity vector is typically set pseudorandomly so that the total kinetic energy of the system, E_k, corresponds to the expected value at the target temperature T. According to the classical equipartition theorem, each normal mode has $(k_B\mathrm{T})/2$ energy, on average, at thermal equilibrium. Thus

$$\langle E_k \rangle = \frac{1}{2}\sum_{i=1}^{3N} m_i v_i^2 \equiv \frac{1}{2}(V^0)^T \mathbf{M}(V^0) = (N_F\, k_B \mathrm{T})/2\,, \tag{12.4}$$

where N_F is the total number of degrees of freedom in the system ($3N$ or $3N-3$; see below).

Such a velocity setting can be accomplished by assigning velocity components from a Gaussian distribution. Namely, each component v_i can be chosen from a Gaussian distribution with zero mean and a variance of $(k_B\mathrm{T})/m_i$. This implies that $\langle v_i^2 \rangle = (k_B\mathrm{T})/m_i$, where the bracket $\langle\cdot\rangle$ denotes the expected value. Utilities to generate such distributions are based on uniform pseudorandom number generators and are available in many software packages (see discussion in the Monte Carlo chapter).

According to the *Central Limit theorem* (see Chapter 11), the sum of independent random variates, each chosen from a normal distribution, has itself a normal density with a mean and variance that are simply the sum of the individual means and variances. Thus, the expected value of E_k will correspond to the target temperature:

$$\langle E_k \rangle = \frac{1}{2} \langle \sum_{i=1}^{3N} m_i v_i^2 \rangle = 3N \, (k_B \mathrm{T})/2 \, . \tag{12.5}$$

Often, the three translational degrees of freedom are removed by modifying the initial velocity vector so that the initial momentum is zero. This is done by subtracting from each component of V^0 the center of mass velocity, v_{cm} ($v_{\mathrm{cm}} = (\sum_i m_i v_i)/\sum_i m_i$, where $i = 1, \ldots, N$). This modification leaves the final temperature, measured with $N_F = 3N - 3$, unchanged.

Equilibration

Once the system is set and initial coordinates and velocities are assigned, an initial round of *equilibration* is necessary before the production phase of the simulation can begin. In this equilibration period, there is an exchange between kinetic and potential energies. Stability is evident when both terms, and the total energy, appear to converge, with fluctuations occurring about the mean value. This is the expected behavior from a *microcanonical* ensemble (fixed total energy, volume, and particle number, or NVE for short); see Chapter 11, Subsection 11.5.3 and Section 12.6.

Illustration

Figure 12.2 illustrates the convergence phase of a simulation, as obtained for a hydrated DNA dodecamer system (12389 atoms: 760 of DNA, 22 sodium, and 3869 water molecules). A heating phase of 15,000 steps (5000 steps, 3°K per every 50 steps) first brings the system to the target temperature. Note that during the equilibration phase the potential and kinetic energies fluctuate and the total energy remains nearly constant (see enlargement at bottom). Second, the leapfrog Verlet integrator (see later in chapter) with a timestep of 2 fs is applied, in combination with bond-length restraints (see below), to all water molecules. The water model consists of a periodic hexagonal-prism system of height 65 Å and side length of 26 Å. In general, the equilibration time is system and protocol dependent. A stochastic approach such as Langevin dynamics (as introduced in Subsection 9.6.3 of Chapter 9) can simplify this equilibration task because the random forces quickly lead the system to thermal equilibrium, even with a zero initial velocity vector.

12.3.4 Sensitivity to Initial Conditions and Other Computational Choices

The chaotic nature of individual MD trajectories is well appreciated. It is especially important to keep this aspect in mind when analyzing data for complex, many-body systems. Roughly speaking, chaotic behavior means that a small change in initial conditions (e.g., a fraction of an Ångstrom difference in Cartesian

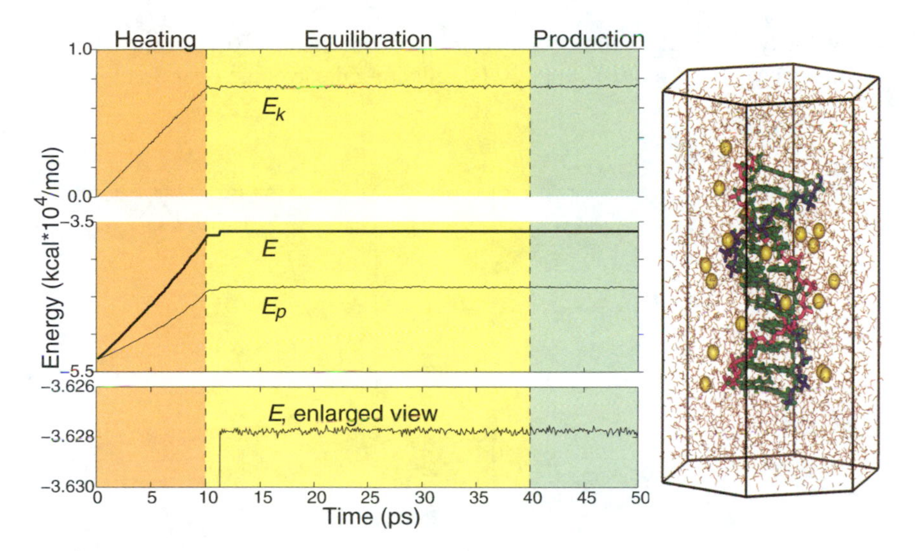

Figure 12.2. Heating and equilibration of a hydrated DNA dodecamer system of 12389 atoms in a hexagonal-prism periodic domain.

coordinates) can lead to exponentially-diverging trajectories in a relatively short time. The larger the initial difference and/or the timestep, the more rapid this *Lyapunov instability* [848].[1] Though chaos is a feature of the analytical equations themselves, the finite arithmetic used by computers and the discretization errors of the integrators used to simulate the trajectory are other factors that produce divergence.

Chaos and Saturation

Figure 12.3 illustrates this sensitivity to initial conditions. Shown are the root-mean-square (RMS) differences in atomic coordinates as a function of time t for a water tetramer system:

$$e(t) = \left[\sum_{i=1}^{3N} (x_i(t) - x_i'(t))^2 \right]^{1/2} ; \tag{12.6}$$

here, the coordinate of the perturbed trajectory is distinguished from the reference trajectory by the prime superscript.

For each curve shown in the figure, the initial coordinates of the second trajectory were perturbed by $\pm\epsilon$ from the reference trajectory ($\epsilon = 10^{-8}$, 10^{-6}, 10^{-4}).

[1]Lyapunov exponents are related to the eigenvalues of the Jacobian of the transformation associated with a dynamical system; they describe the rate of growth of an instability. For example, a system that possesses local instability has a direction along which the distance between two points in phase space at time t, $|\delta(t)|$, grows exponentially at the rate λ_0, the largest Lyapunov exponent: $|\delta(t)| = |\delta(0)| \exp(\lambda_0\, t)$.

The velocity Verlet algorithm (see below) used timesteps of 0.1 fs and 1 fs, and the difference $e(t)$ is plotted every 20 fs, reflecting the average over that interval.

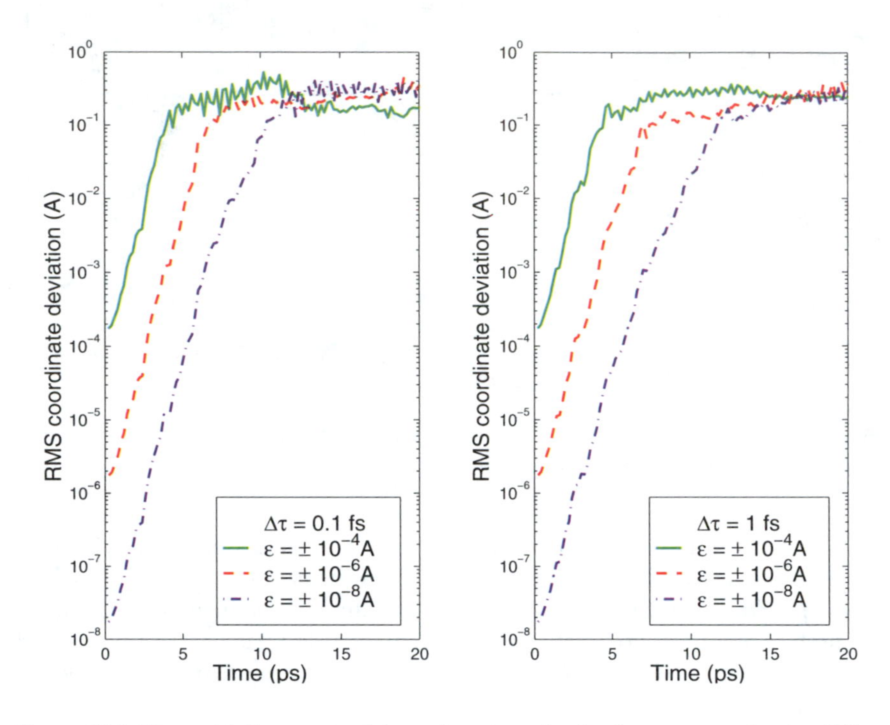

Figure 12.3. The rapid divergence of dynamic trajectories for four water molecules differing slightly in initial conditions.

Several interesting features can be gleaned from this figure. A short period of rapid exponential divergence is evident, followed by linear oscillations around a threshold value. The 'chaos' often mentioned in MD texts refers to the *initial period* of exponential growth. This is in fact *local instability* [848]. It is well known from chaos theory that compact systems, like biomolecules, whose phase space is finite, reach *saturation* for the trajectory error when the magnitude of the deviations corresponds to a system-dependent value.

In other words, the phase trajectories cannot be found beyond the distance characteristic of the phase space. As the plots show, the smaller the initial perturbation (ϵ), the longer the time required to reach saturation. The same threshold is reached for both timestep values.

This stabilization behavior realized in practice for MD simulations can also be understood by the different mathematical limit involved in the strict notion of chaos compared to that relevant for MD. Physicists define chaos rigorously for *finite times* but in the limit of *timesteps approaching zero*; in MD applications, we instead work with *finite timesteps* and simulation *times approaching infinity*.

Statistical View

For many other reasons than inherent chaos, including the approximate nature of biomolecular simulations, MD trajectories are not intended to animate the life of a biomolecule faithfully, as if by an unbiased video camera. Rather, our experimental-by-nature computer 'snapshots' aim at gaining structural insights, which can be tested, and predicting meaningful *statistical properties* of a complex system in the ultimate goal of relating structure to function. The more reliable quantities from MD simulations are thus space and time averaged. This is unlike differential-equation applications in astrophysics, where precise trajectory pathways are sought (e.g., planetary orbits). Often, more useful data are obtained by averaging over, or calculating from, several trajectories rather than from one long trajectory (e.g., [849, 850, 40]).

Let us consider the convergence of the end-to-end distance of a butane molecule, as first illustrated in [851]. Figure 12.4 shows the *time evolution* of the end-to-end-distance of the butane molecule for different choices of timesteps in the range of 0.02 to 2.0 fs. Clearly, different paths are realized even for this very small system. The associated *average end-to-end distance* values as a function of the timestep are the points corresponding to the dashed curve in Figure 12.5.

These data reveal convergence only for small timesteps, less than 0.2 fs. This value is typically *much smaller* than standard timesteps used in biomolecular simulations (0.5 fs or greater). Still, larger values can be quite reasonable depending on the specific questions being asked from a simulation.

12.3.5 Simulation Protocol

A careful simulation protocol is essential for MD simulations, to ensure not only the equilibration phases but also proper enforcement of the boundary conditions (when, for example, a biomolecule is placed in a box of water molecules), positioning of solvent and salt molecules, or computation of the nonbonded terms and associated pairlist arrays. Also important is monitoring the kinetic temperature and the energy fluctuations, for the detection of systematic drifts or abrupt changes, both of which may indicate numerical problems.

For example, severe artifacts arise when the nonbonded energy interactions are truncated abruptly. This is because the associated force terms rise violently at the truncation boundary. Improved techniques for smoothly handling the nonbonded terms are necessary [648], as described in Chapter 9. The alternative approach described also in that chapter is to forego cutoffs and compute all the nonbonded terms by fast summation techniques such as multipole or Ewald [852, 688]. Fortunately, such schemes approach linear, rather than quadratic, dependence on system size.

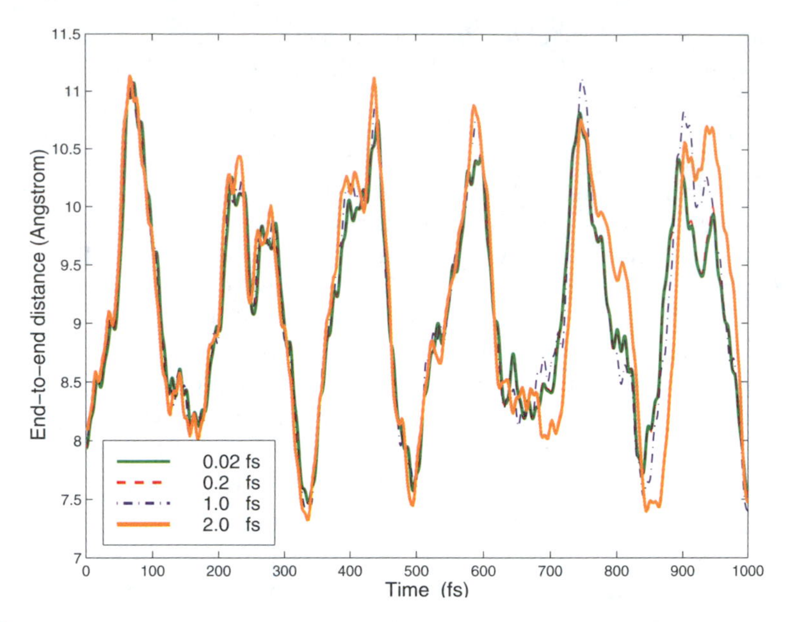

Figure 12.4. Time evolution of the end-to-end distance of butane for different timesteps.

12.3.6 High-Speed Implementations

High-speed computers are essential for performing the computationally-intensive MD simulations [853]. The dynamics of condensed systems was simulated much earlier this century, but macromolecular applications only gained momentum in the mid-to-late 1980s with the advent of high-speed computing. There are many reasons why this lag occurred, and the issue at heart is best captured by the following statement by Frauenfelder and Wolynes [854]:

> *Whatever complexity means, most people agree that biological systems have it.*

Indeed, the energy landscape is complex for biomolecules [855, 856]. The various contacts — be they hydrogen bonds, disulfide bonds, or noncovalent interactions like stacking and favorable electrostatics — are difficult to predict *a priori*. Thus, the multidimensional potential energy surface that governs biomolecular structure has many maxima, minima, and saddle points. The distributions about each favorable or unfavorable state are highly anisotropic, with the width depending on the entropy associated with that state.

Biomolecules are also asymmetric in comparison to simple systems, such as homogeneous liquid clusters, which were successfully simulated much earlier. Certainly, there are symmetries in many aspects of protein and nucleic acid structure (e.g., many proteins are dimers, and the "ideal" DNA double helix has an axis of symmetry), but in realistic environments there are many sequence-specific

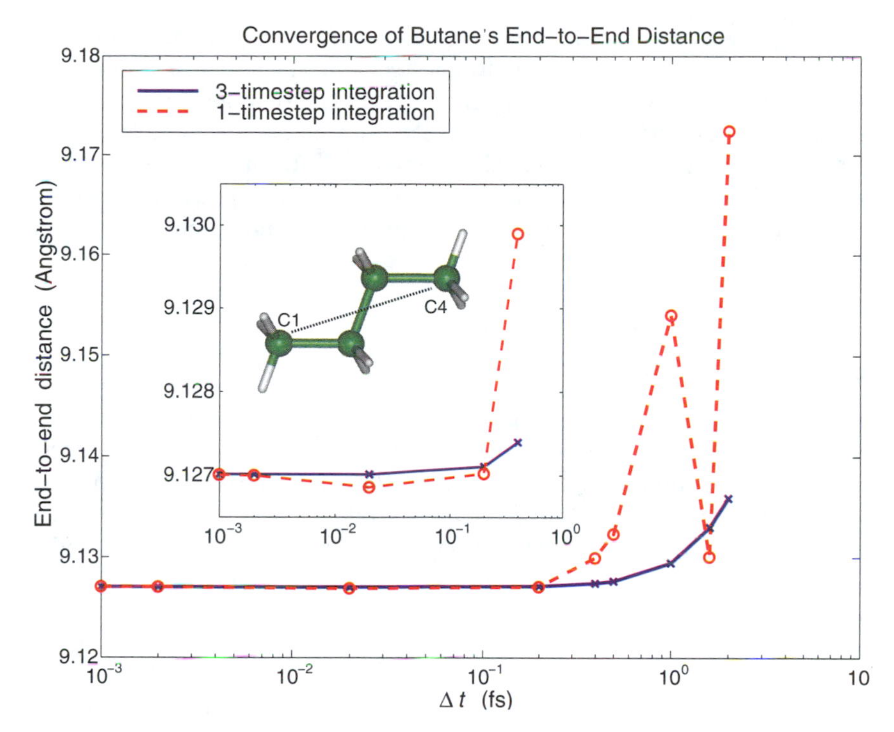

Figure 12.5. Average butane end-to-end distance for single (dashed line) and triple (solid line) timestep protocols (note logarithmic scale) as computed over 1 ps (see Figure 12.4). In the latter, the timestep used to define the data point is the outermost timestep Δt (the interval of updating the nonbonded forces), with the two smaller values used as $\Delta t/2$ and $\Delta t/4$ (for updating the dihedral-angle terms and the bond-length and angle terms, respectively).

motifs and binding interactions with other biomolecules in the environment that induce local structural variations in macromolecules. These local trends can produce profound global effects.

The motion of biomolecules is also more complex than that of small or homogeneous systems. The collective motion is a superposition of many fundamental motions that characterize the dynamics of a biomolecule: bond stretches, angle bends, torsions, and combinations of those "normal modes" (see Chapter 8 for definitions and examples). Although these components have associated frequencies of disparate magnitudes, the overall motion of a biomolecule is highly cooperative. That is, small local fluctuations can trigger a chain of events that will lead to a large global rearrangement. Energy transfer among vibrational modes may also be involved.

In addition to the overall asymmetry and complexity (of energy landscape and motion), the solvent and ions in the environment influence the structure and dynamics of biomolecules profoundly. Solvent molecules surround and interpenetrate proteins and nucleic acids and damp many characteristic motions that might be present in vacuum. Similarly, ions (sodium, calcium, magnesium, etc.) influence structure and dynamics significantly, especially of the polyelectrolytic nucleic acids [384, for example]. Thus, for physical reliability, these long-range electrostatic interactions are essential to consider in MD simulations of macromolecules. This has proven especially important for stabilizing DNA simulations and accurately describing kinetic processes involving conformational transitions of DNA [61, 276].

12.3.7 Analysis and Visualization

Careful analysis of the results and visualization techniques are also essential components of biomolecular simulations today. *With the increasing ease and accessibility of generating computer trajectories, the challenge remains of carefully analyzing the voluminous data to distill the essential findings.* While many scalar and vector functions can be computed from a long series of configurations, understanding the dynamics behavior requires a combination of sophisticated analysis and chemical/biological intuition.

Molecular graphics techniques were more of a problem in the early days of molecular mechanics than they are now. Before the surge of graphics innovations, researchers relied more heavily on mechanical models and physical sense (see related comment in this book's Preface).

These days, molecular modeling software and analysis tools have become a large industry. Many sophisticated and useful tools are available to both experimentalists and computational/theoretical chemists. The dazzling capability of computer graphics today to render and animate a large, complex three-dimensional image — often so "real" in appearance that the source may be obscured — has certainly made biological interpretation much easier, but one still has to know exactly where, and for what, to look!

12.3.8 Reliable Numerical Integration

To increase the reliability of macromolecular simulations, special care is needed to formulate efficient numerical procedures for generating dynamic trajectories.

Mathematically, there are classes of methods for conservative Hamiltonian systems termed *symplectic* that possess favorable numerical properties in theory and practice [837]. In particular, these schemes preserve volumes in phase space (as measured rigorously by the Jacobian of the transformation from one set of coordinates and momenta to the next). This preservation in turn implies certain physical invariants for the system.

Another view of symplectic integration is that the computed trajectory (with associated Hamiltonian $H(X^n, V^n)$) remains close in time to the solution of a nearby Hamiltonian $\widetilde{H}$. This proximity is rigorously defined as a trajectory whose energy is order $\mathcal{O}(\Delta t)^p$ away from the initial value of the true Hamiltonian H:

$$H \equiv \frac{1}{2}(V^0)^T \mathbf{M}(V^0) + E(X^0)\,,$$

where p is the order of the integrator, and X^n and V^n are the collective position and velocity vectors at time $n\Delta t$. This symplectic property translates to good long-time behavior in practice: small fluctuations about the initial (conserved in theory) value of H, and no systematic drift in energy, as might be realized by a nonsymplectic method.

Figure 12.6 illustrates this favorable symplectic behavior for a simple nonlinear system, a cluster of four water molecules, integrated by Verlet and by the classical fourth-order (nonsymplectic) Runge-Kutta method. A clear damping trend (i.e., decrease of energy with time) is seen by the latter, non-symplectic integrator, especially at the larger timestep (top).

While symplectic methods can follow kinetics accurately (assuming the force field is adequate), other approaches as discussed in the next chapter — Langevin and Brownian dynamics — can be effective for thermodynamic sampling. Though additional terms are added (e.g., random thermal forces), enhanced sampling made possible by larger timesteps can yield more information on the thermally-accessible configuration space.

12.3.9 Computational Complexity

Intensive Requirements

Trimming the computational time of MD simulations remains a challenge [857]. As seen from Table 1.2 of Chapter 1, the all-atom molecular models used for biomolecules are quite large (thousands of atoms). The energy and derivative computations required at each dynamic step are expensive and limit the total time that can be simulated: typical timesteps (1 fs) imply one million integration steps per nanosecond. Interestingly, Table 1.2 also shows our tendency to increase system size rather than the simulation time. This can be explained, in part, by the desire to model more complex and realistic systems and by the timespan/sampling limitation of MD. Dynamic motions over substantially longer times, as well as larger spatial scales, can only be approximated through macroscopic models of proteins and DNA (see [269] for example).

To reduce computational time of MD simulations, two basic routes can be taken and combined: reducing the work per timestep, and/or increasing the integration timestep, Δt. In the latter case, the parameter we aim to lengthen is that associated with long-range force calculations, that is, the 'outermost timestep' in multiple-timestep (MTS) methods.

Less Work Per Step

Reducing the work per step can be achieved with various protocols used in practice: employing cutoff ranges for the nonbonded interactions, updating the

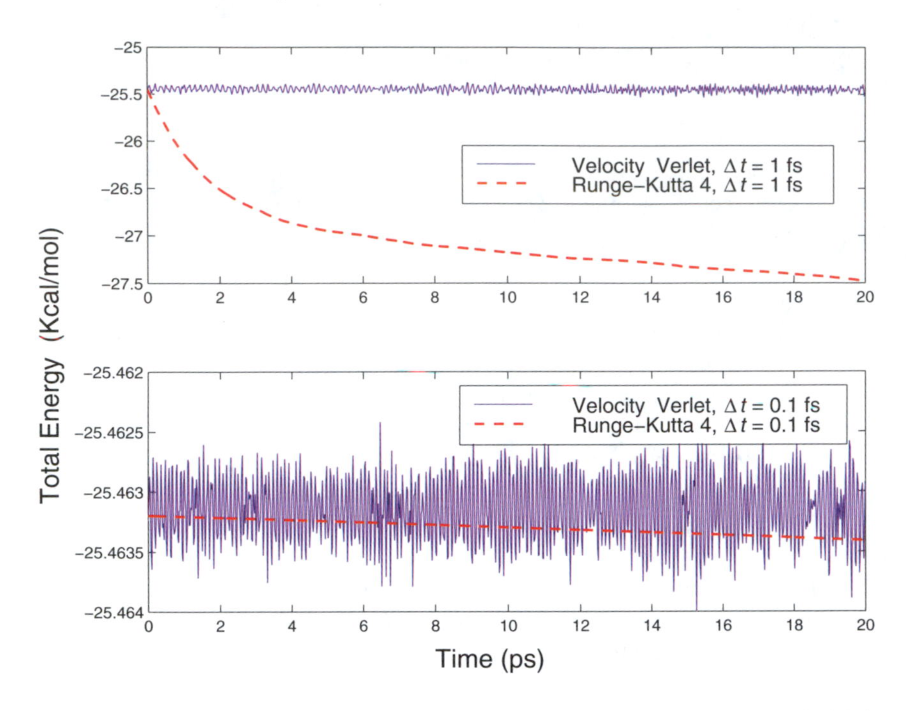

Figure 12.6. Energy evolution of a water tetramer simulated by the symplectic Verlet scheme (solid line) versus the nonsymplectic Runge-Kutta integrator (dashed line) at two timesteps Δt (0.1 and 1 fs). Note the different energy scales of the plots.

nonbonded pairlist infrequently (i.e., less often than every outer timestep), and reducing the number of degrees of freedom by constraining certain molecular regions (e.g., protein residues revealed by experiment not to be important for a certain localized transition).

Larger Timesteps: Accuracy vs. Stability

The maximum value for the outer timestep that can be successfully used is limited by both *accuracy* and *stability* considerations, as will be detailed below. Essentially, *instability* of a trajectory is easily detected when coordinates and energies grow uncontrollably. This often signals that the timestep is too large. However, unless the timestep is far too large, this instability may only be apparent from a trajectory of length of hundreds of picoseconds or longer. The guidelines for stability for a given integrator are often derived from linear analysis (i.e., theoretical analysis for a harmonic oscillator) [858, 859, 860, for example]. In practice, the timestep must typically be smaller than this limit for complex nonlinear problems to ensure stability as well as reasonable accuracy.

Accuracy limits the timestep Δt in the sense that the resolution produced must be adequate for the processes being followed and the measurements targeted. Thus, a simulation may be stable but the timestep too large to capture high-frequency fluctuations accurately.

Closely related to stability and accuracy are resonance artifacts. This topic of recent interest will be discussed separately.

Techniques for increasing the timestep range from constraining the fastest degrees of freedom via algebraic constraints to using multiple-timestep (MTS) schemes to formulating rigid-body dynamics. For recent reviews, see [861, 857], for example.

Constraining Fastest Motions

The technique of constrained dynamics has been used for a long time to enforce rigidity of bond lengths [862, 863, 51, 864]. This approach helps increase the timestep by a typical factor of two (e.g., from 1 to 2 fs), with similar gains in the computational time. Freezing bond angles, however, alters the overall motion substantially [865]. See the last section of this chapter.

Splitting Forces in MTS Schemes

In MTS schemes, we divide the force components into classes and update the associated forces at timesteps appropriate for the class. Three-class partitioning schemes are common in biomolecular simulations, though more classes can be used for added accuracy/efficiency.

For example, a simple partitioning based on potential-energy components [866] can place bond-length and bond-angle terms into the fastest class, dihedral-angle terms into the medium class, and the remaining forces into the slow-force class. Incorporating distance classes (e.g., interactions within 6 Å into the medium-force class) involves more complex bookkeeping work for the nonbonded pairlist generation but often works better. The numerical details of how these different updates are merged affect the stability of the trajectory significantly [867, 868]; see the separate section on MTS methods in the next chapter for details.

To illustrate a simple MTS scheme, consider using three timesteps for the butane system analyzed above for computing the end-to-end distance. We set the three values as $\Delta\tau$, $2\Delta\tau$, and $4\Delta\tau \equiv \Delta t$; thus a 3-timestep integrator associated with the outer timestep of $\Delta t = 2$ fs has an inner timestep of 0.5 fs and a medium timestep of 1 fs. Bond-length and angle terms are updated every $\Delta\tau$; dihedral-angle terms are updated every $2\Delta\tau$; and nonbonded interactions are recalculated every $4\Delta\tau$ for this butane model. This partitioning should thus save work for this MTS protocol $\{\Delta\tau, 2\Delta\tau, 4\Delta\tau\}$ relative to the single-timestep integrator at $\Delta\tau$.

The data shown in Figure 12.5 for the average end-to-end distance of butane compares values obtained from single-timestep methods to those obtained by Verlet-based MTS simulations. We see that the accuracy is determined by the innermost timestep. This is precisely the advantage exploited in MTS schemes: same accuracy at less computation (under the appropriate comparison) [869, 870].

For the small butane system, the MTS force splitting is not computationally advantageous because the nonbonded interactions represent a small fraction of the total force-evaluation work. However, for biomolecules, non-negligible speedups (e.g., factor of 2 or more) can be reaped with respect to single-timestep simulations performed at the inner timestep ($\Delta\tau$) used in the MTS protocol.

12.4 The Verlet Algorithm

One of the simplest and best family of integrators for biomolecular dynamics is the leapfrog/ Verlet/ Störmer group. The basic scheme was derived as a truncation of a higher-order method used by Störmer in 1907 [871] to integrate trajectories of electrons. The scheme was adapted by Verlet 60 years later [872]. A favorable numerical property of this family is exceptional stability over long times (recall the example in Figure 12.6, where the Verlet trajectory shows favorable energy conservation compared to the Runge-Kutta trajectory).

Stability far more than accuracy limits the timestep in simulations of biomolecular motion governed by approximate force fields. Programming ease is a boon for weary simulators who must wade through thousands of lines of molecular mechanics and dynamics programs. Far from mere convenience, simplicity in the integration also facilitates the implementation of variations in the basic method (e.g., constrained dynamics, extensions to various statistical ensembles) and the merging of integration schemes with other program components (e.g., fast electrostatics by the particle-mesh Ewald method).

For both these practical and theoretical reasons, it has become very difficult for alternative schemes — such as higher-order and implicit discretizations — to compete with Verlet in popularity [857]. For the same reason, the single-timestep Verlet method is still more popular in general than MTS variants (as introduced in Subsection 12.3.9 and discussed in detail in the next chapter), but this is likely to change as comprehensive and reliable MTS routines for various statistical ensembles enter into the popular biomolecular dynamics packages.

Not surprisingly, generalizations of the Verlet method have been applied to other formulations, such as constrained and stochastic dynamics. The Verlet scheme is also the basis of the common force-splitting MTS methods and for approaches using various thermodynamic ensembles.

Below we describe how the scheme is formulated and derive its leapfrog, velocity, and position variants.

12.4.1 Position and Velocity Propagation

Iterative Recipe

To derive the Verlet propagation scheme for Newtonian dynamics, we write the equation of motion, eq. (12.3), as

$$\mathbf{M}\,\ddot{X}(t) = F(X(t)) \tag{12.7}$$

where

$$F(X(t)) = -\nabla E(X(t)). \tag{12.8}$$

We also use the symbol $\widetilde{F}$ (the acceleration, or the force scaled by the inverse-mass matrix) to simplify the formulas below, where

$$\widetilde{F}(X(t)) = \mathbf{M}^{-1} F(X(t)) = -\mathbf{M}^{-1} \nabla E(X(t)). \tag{12.9}$$

In a numerical discretization, a timestep Δt is chosen and the continuous variables $X(t)$ and $V(t)$ are approximated by values at discrete time intervals $n\Delta t$ for $n = 0, 1, 2, \dots$, denoted as X^n and V^n. Below we also use the notation F^n as short hand for the force at time $n\Delta t$, namely $F(X^n)$. Similarly, $\widetilde{F}^n$ represents the numerical estimate to the acceleration at time $n\Delta t$: $\widetilde{F}(X^n)$.

An iterative formula is then chosen to define recursively the positions and velocities of the molecular system. If the formula defines the solution at timestep $n+1$ in terms of quantities determined at previous iterates, as in

$$V^{n+1} = V^n + \Delta t \widetilde{F}^n,$$

the discretization method is *explicit*. If instead the new solution is defined implicitly, in terms of known and unknown quantities, as in

$$V^{n+1} = V^n + \Delta t \widetilde{F}^{n+1},$$

the method is *implicit*. The explicit versions generally involve simple algorithms that (for propagation only) use modest memory, while implicit methods involve more complex algorithms but may enjoy enhanced stability properties.

Position Update

The original Verlet algorithm [872] describes an update for trajectory positions, as follows:

$$X^{n+1} = 2X^n - X^{n-1} + \Delta t^2 \widetilde{F}^n. \tag{12.10}$$

This is a two-step propagation scheme since positions from two prior steps (n and $n-1$) must be stored. A one-step scheme — preferable in terms of storage — can be obtained by using V^n instead of X^{n-1} as a variable on the left-hand side (see below).

The Verlet algorithm is fourth-order accurate in position. To see this, use the Taylor expansion around $X(t)$ to obtain:

$$\begin{aligned} X(t+\Delta t) \quad &+ \quad X(t-\Delta t) \\ &= \quad X(t) + \Delta t\, V(t) + \frac{\Delta t^2}{2}\widetilde{F}(X(t)) + \frac{\Delta t^3}{6}\ddot{V}(t) + \mathcal{O}(\Delta t^4) \\ &+ \quad X(t) - \Delta t\, V(t) + \frac{\Delta t^2}{2}\widetilde{F}(X(t)) - \frac{\Delta t^3}{6}\ddot{V}(t) + \mathcal{O}(\Delta t^4) \\ &= \quad 2X(t) + \Delta t^2\, \widetilde{F}(X(t)) + \mathcal{O}(\Delta t^4). \end{aligned} \tag{12.11}$$

Velocity Update

The flexibility in defining compatible velocity formulas from the Verlet position propagation formula (eq. (12.10)) has led to Verlet variants. These variants rely on a velocity formula consistent with eq. (12.10).

We derive such a formula by subtracting the Taylor expansion for $X(t - \Delta t)$ from that for $X(t + \Delta t)$, instead of adding as done above, to obtain:

$$\frac{X(t + \Delta t) - X(t - \Delta t)}{2\Delta t} = V(t) + \mathcal{O}(\Delta t^2)\,. \tag{12.12}$$

This definition can be written as the *central difference approximation*

$$V^n = (X^{n+1} - X^{n-1})/(2\Delta t)\,. \tag{12.13}$$

Though the position update formula of eq. (12.10) is fourth-order accurate, the Verlet method is overall *second-order accurate*. This is because eq. (12.10) divided by Δt^2 represents an $\mathcal{O}(\Delta t^2)$ approximation to $\ddot{X} = \widetilde{F}(X)$. Though the same trajectory positions are followed in theory, differences are realized in practice among Verlet variants due to finite computer arithmetic. Local accuracy together with stability considerations affect global measurements, like the kinetic temperature (as computed from a simulation by the relation of eq. (12.4)). This is because such global properties must be interpreted statistically.

12.4.2 Leapfrog, Velocity Verlet, and Position Verlet

Three popular variants of Verlet are the *leapfrog* scheme [873], *velocity Verlet* [874], and *position Verlet* [875, 876, 877]. All originate from the same leapfrog/Störmer/Verlet method [871, 872] and have favorable numerical properties like symplecticness (e.g., [878]) and time-reversibility [875].

The leapfrog scheme is so called because the velocity is defined at half steps ($V^{n+1/2}$) while positions are defined at whole steps (X^n) of Δt. The propagation formula of leapfrog thus transforms one point in phase space to the next as $\{V^{n-1/2}, X^n\} \Longrightarrow \{V^{n+1/2}, X^{n+1}\}$. The velocity Verlet scheme transforms instead $\{X^n, V^n\} \Longrightarrow \{X^{n+1}, V^{n+1}\}$.

We first present these three methods using a symmetric *propagation triplet*. Both leapfrog and velocity Verlet can be written as:

$$\begin{aligned} V^{n+1/2} &= V^n + \frac{\Delta t}{2}\,\widetilde{F}^n \\ X^{n+1} &= X^n + \Delta t\, V^{n+1/2} \\ V^{n+1} &= V^{n+1/2} + \frac{\Delta t}{2}\,\widetilde{F}^{n+1}\,. \end{aligned} \tag{12.14}$$

The position Verlet scheme can be written as:

$$X^{n+1/2} = X^n + \frac{\Delta t}{2}\,V^n$$

$$\begin{aligned} V^{n+1} &= V^n + \Delta t\, \widetilde{F}^{n+1/2} && (12.15)\\ X^{n+1} &= X^{n+1/2} + \frac{\Delta t}{2}\, V^{n+1}\,. \end{aligned}$$

Note that the force is evaluated at the endpoints in leapfrog and velocity Verlet but at the midpoint of position Verlet.

Leapfrog

The leapfrog Verlet method as described above can also be written as

$$V^{n+1/2} = V^{n-1/2} + \Delta t\, \widetilde{F}^n\,, \qquad (12.16)$$

$$X^{n+1} = X^n + \Delta t\, V^{n+1/2}\,. \qquad (12.17)$$

It can be derived from the position propagation formula of eq. (12.10) in combination with the velocity formula (12.13) approximated at half steps:

$$V^{n-1/2} = (X^n - X^{n-1})/\Delta t\,. \qquad (12.18)$$

It is transparent to see the equivalence of the position update formula of leapfrog (eq. (12.17)) to that used in the triplet scheme (eq. (12.14)). To see that the velocity updates are equivalent as well, replace the $n+1$ superscript by n in the third equation of the triplet (12.14) to get:

$$V^n = V^{n-1/2} + \frac{\Delta t}{2}\, \widetilde{F}^n\,, \qquad (12.19)$$

and add this equation to the first equation (for $V^{n+1/2}$) of the triplet (12.14). The result $V^{n+1/2} = V^{n-1/2} + \Delta t\, \widetilde{F}^n$ is the velocity formula of leapfrog (eq. (12.16)).

Velocity Verlet

Besides the triplet equation (12.14), the velocity Verlet scheme can also be written as:

$$X^{n+1} = X^n + \Delta t\, V^n + \frac{\Delta t^2}{2} \widetilde{F}^n\,, \qquad (12.20)$$

$$V^{n+1} = V^n + \frac{\Delta t}{2}\, (\widetilde{F}^n + \widetilde{F}^{n+1})\,. \qquad (12.21)$$

The first equation can be derived by using eq. (12.13) to express X^{n-1} as $(X^{n+1} - 2\Delta t\, V^n)$ and plugging this substitution for X^{n-1} in eq. (12.10). The second equation of velocity Verlet is obtained by applying the velocity equation (12.13) to express V^{n+1} in terms of positions and then using eq. (12.20) to write X^{n+2} in terms of X^{n+1}, V^{n+1}, and $\widetilde{F}^{n+1}$, and X^n in terms of X^{n+1}, V^n, and $\widetilde{F}^n$. That is, we write V^{n+1} as

$$V^{n+1} = \frac{X^{n+2} - X^n}{2\Delta t}$$

$$= \left[\frac{X^{n+1} + \Delta t\, V^{n+1} + \frac{\Delta t^2}{2} \widetilde{F}^{n+1}}{2\Delta t}\right] - \left[\frac{X^{n+1} - \Delta t\, V^{n} - \frac{\Delta t^2}{2} \widetilde{F}^{n}}{2\Delta t}\right]$$

$$= \frac{V^{n+1} + V^n}{2} + \frac{\Delta t}{4} (\widetilde{F}^n + \widetilde{F}^{n+1}). \tag{12.22}$$

Collecting the two V^{n+1} terms yields

$$\frac{V^{n+1}}{2} = \frac{V^n}{2} + \frac{\Delta t}{4} (\widetilde{F}^n + \widetilde{F}^{n+1}), \tag{12.23}$$

leading immediately to the position update formula of eq. (12.21).

The equivalence of the triplet version (eq. (12.14)) of velocity Verlet to the form in eqs. (12.20), (12.21) is straightforward.

Position Verlet

Besides the triplet (12.15), we can express position Verlet as:

$$X^{n+1} = X^n + \Delta t\, V^n + \frac{\Delta t^2}{2} \widetilde{F}^{n+1/2}, \tag{12.24}$$

$$V^{n+1} = V^n + \Delta t\, \widetilde{F}^{n+1/2}. \tag{12.25}$$

MTS Preference

The velocity Verlet scheme has been customarily used in MTS schemes because the forces are evaluated at the ends of the interval and this is more amenable to force splitting. Moreover, it only requires one constrained dynamics application per inner timestep cycle (when position is updated) when constrained dynamics techniques are used; two such constrained-dynamics iterations are required for position Verlet. However, position Verlet — first suggested to be preferable in large-timestep methods [876] — was recently shown to have advantages in connection with moderate to large timesteps [667, 877].

12.5 Constrained Dynamics

In constrained dynamics, the equations of motion are augmented by algebraic constraints to freeze the highest-frequency interactions. To illustrate, consider a typical modeling of bond-length potentials as the harmonic form

$$E_{\text{bond}} = \frac{S_h}{2} (r_{ij} - \overline{r_{ij}})^2,$$

where r_{ij} is an interatomic distance with equilibrium value $\overline{r_{ij}}$, and S_h is a force constant. The constraint implemented in the context of MD simulations is thus

$$g_k = r_{ij}^2 - \overline{r_{ij}}^2 = 0\,.$$

Using the formalism of Lagrange multipliers, we have then in place of eq. (12.3) the following system to solve:

$$\begin{aligned} \mathbf{M}\dot{V}(t) &= -\nabla E(X(t)) - g'(X(t))^T \Lambda \,, \\ \dot{X}(t) &= V(t)\,, \\ g(X(t)) &= 0\,. \end{aligned} \tag{12.26}$$

Here $g(X(t))$ is a vector with entries g_i containing the individual constraints, and Λ is the vector of Lagrange multipliers.

SHAKE

In 1977, the SHAKE algorithm was introduced on the basis of the leapfrog Verlet scheme of eqs. (12.16, 12.17) [51]:

$$\begin{aligned} V^{n+1/2} &= V^{n-1/2} - \Delta t \nabla E(X^n) - \Delta t g'(X^n)^T \Lambda^n \,, \\ X^{n+1} &= X^n + \Delta t V^{n+1/2}\,, \\ g(X^{n+1}) &= 0\,. \end{aligned} \tag{12.27}$$

Like the Verlet method, the forces $-\nabla E$ must be computed only once per step. However, the m constraints require at each timestep, in addition, the solution of a nonlinear system of m equations in the m unknowns $\{\lambda_i\}$. Thus, the SHAKE method is *semi*-explicit.

RATTLE

A variant termed RATTLE was later introduced [879]. Similar techniques have been proposed for constraining other internal degrees of freedom [880], and direct methods have been developed for the special case of rigid water molecules [864]. Leimkuhler and Skeel [881] have shown that the RATTLE method is symplectic and that SHAKE, while not technically symplectic, yields solutions identical to those of RATTLE for the positions and only slightly perturbed velocities.

Computational Advantage

With the fastest vibrations removed from the model, the integration timestep can be lengthened, resulting in a typical force calculation per time ratio of 0.5 per fs. This computational advantage must be balanced against the additional work per step required to solve the nonlinear equations of constraints. Each proposed constrained model is thus accompanied by a practical numerical algorithm.

For the time discretization of eq. (12.27), an iterative scheme for solution of the nonlinear equations was presented in [51], where the individual constraints are imposed sequentially to reset the coordinates that result from an unconstrained Verlet step. This iterative scheme is widely used due to its simplicity and frugal consumption of computer memory. However, the SHAKE iteration can converge very slowly, or even fail to converge, for complex bond geometries [882]. Improved convergence was later reported by Barth *et al.* [882], who proposed enhancements based on the relationship between the SHAKE iteration process and the iterative Gauss-Seidel/Successive-Over-Relaxation techniques for solving nonlinear systems. This was also independently developed by Xie and Scott, as reported later by Xie *et al.* [883].

Limitations

The computational advantage of constrained models is clear, but the verity of the constrained model for reproducing the dynamics of unconstrained systems is a separate issue. Van Gunsteren and Karplus [865] showed through simulations of the protein BPTI in vacuum that the use of fixed bond lengths does not significantly alter the dynamical properties of the system, whereas fixing bond angles does. Similar conclusions were reported by Toxvaerd [884] for decane molecules.

Though it is possible that the former study was also influenced by poor convergence of SHAKE for bond-angle constraints (which can lead to overdetermination and singularities in the constraints equations), the intricate vibrational coupling in biomolecules argues against the general use of angle constraints. As shown in [885], from the point of view of increasing the timestep in biomolecular simulations, only constraints of light-atom angle terms in addition to all bond lengths might be beneficial. Furthermore, it is common practice not to constrain motion beyond that associated with light-atom bonds because the overall motion can be disturbed due to vibrational coupling.

12.6 Various MD Ensembles

12.6.1 Need for Other Ensembles

The algorithmic framework discussed thus far is appropriate for the microcanonical (constant NVE) ensemble (see the equilibration illustration in Figure 12.2, for example), where the total energy E is a constant of the motion. This assumes that the time averages are equivalent to the ensemble averages.

The study of molecular properties as a function of temperature and pressure — rather than volume and energy — is of general importance. Thus, microcanonical ensembles are inappropriate for simulating certain systems which require constant pressure and/or temperature conditions and allow the energy and volume to fluctuate. This is the case for a micelle or lipid bilayer complex (proteins and lipids in an ionic solution) under constant pressure (isotropic pressure or the pressure tensor), and for crystalline solids under constant stress (i.e., constant pressure tensor). For such systems, other ensembles may be more appropriate, such as *canonical* (constant temperature and volume, NVT) *isothermal-isobaric* (constant temperature and pressure, NPT), or constant pressure and enthalpy (NPH). Special techniques have been developed for these ensembles, and most rely on the Verlet framework just described, including the MTS variants discussed in the next chapter [886].

For pioneering works, see the articles by Andersen [887] and the extension by Parrinello and Rahman [888] for NPH, and that of Nosé [889] and Hoover [890] for NVT and NPT ensembles (started in 1984). For details on algorithms for these extended ensembles, see the texts [30, 31], a review [891], and the recent literature.

In this section we only give a flavor of some of these methods. Methods for handling these thermodynamic constraints or generating the various ensembles can be grouped into simple *constrained formulations*, including approaches that involve stochastic-motivated models, and more sophisticated techniques that involve additional degrees of freedom (*extended system methods*). The former group, while easy to implement, does not generally generate the desired ensemble in rigorous terms.

12.6.2 Simple Algorithms

Constraint-based methods can involve Lagrange-multiplier coupling, or simple scaling of phase-space variables. For example, the velocity vector might be simply scaled at each timestep to fix the desired kinetic temperature T at the target value T_0 [30]. This can be achieved by the scaling

$$V^{\text{new}} \leftarrow c_t \, V^{\text{old}} \,, \tag{12.28}$$

where

$$c_t = \sqrt{T_0/T} \,. \tag{12.29}$$

This drastic approach, however, implies rapid energy transfer to, from, and among the various degrees of freedom in the system. In particular, it can be shown that velocity rescaling leads to an artifactual pumping of energy into low-frequency modes [892]; occasional resetting of velocities can mitigate this problem.

Weak Coupling Thermostat for Constant T

Similar algorithms that are motivated by stochastic approaches are also easy to implement but can steer the system toward the desired constant (e.g., target temperature or pressure) more gently. This can be accomplished by mimicking a diffusive process governed by a fictitious frictional coefficient γ, whose value controls the relaxation rate of this coupling [893].

For example, consider the weak coupling thermostat proposed by Berendsen and coworkers [893], widely used for constant-temperature MD. In this approach, the equations of motion (eq. (12.3)) are modified as:

$$\begin{aligned}
\dot{X}(t) &= V(t) \,, \\
\mathbf{M}\dot{V}(t) &= -\nabla E(X(t)) - \gamma_t \, \mathbf{M}V(t) \,, \\
\gamma_t &= \frac{1}{2\tau}\left(1 - \frac{T_0}{T}\right) ,
\end{aligned} \tag{12.30}$$

where γ_t has units of inverse time, T is the instantaneous kinetic temperature, and τ is the time constant of the coupling to the heat bath. (Compare to the Langevin formalism described in the next chapter, Section 13.4).

This augmentation effectively scales the velocity vector via eq. (12.28) by the factor:

$$c_t = \sqrt{1 - \frac{\Delta t}{\tau}\left(1 - \frac{\mathrm{T}_0}{\mathrm{T}}\right)}, \tag{12.31}$$

where Δt is the (Verlet) timestep. This scaling produces a least-squares local disturbance to the velocity vector, so as to satisfy the global temperature constraint.

The advantage of this weak bath coupling approach is the introduction of added thermostat parameters (not dynamic variables, as below). The factor τ controls the characteristic decay time — and hence the strength — of the coupling to the heat bath. When τ is large, γ_t is small, and the scaling factor c_t approaches unity (the microcanonical ensemble is approached); when τ is small, γ_t is large — that is, the coupling to the heat bath is strong — and the energy exchange between the molecular system and the thermal reservoir is significant.

In particular, when $\tau = \Delta t$, $c_t = \sqrt{\mathrm{T}_0/\mathrm{T}}$, producing the simplest scaling approach for constant-T simulations, but also the most artificial, as mentioned above. The advantage of the flexible coupling τ parameter (or γ_t) is the control of the *rate* at which the target temperature is reached.

Still, this approach, while convenient, does not generate the true canonical ensemble; it can also lead to artifacts as simple velocity scaling [892]. See below for a more rigorous approach that reproduces the desired thermal ensemble properly.

Weak Coupling Barostat for Constant P

In the same spirit, for the purpose of constant pressure simulations for isotropic systems modeled in a finite volume (generally via periodic boundary conditions), Berendsen and coworkers suggest a simple scaling to the Cartesian positions,

$$X^{\text{new}} \leftarrow c_p\, X^{\text{old}}, \tag{12.32}$$

as well as a volume scaling (see below), where the scale factor is:

$$c_p = \left[1 - \frac{\beta \Delta t}{\tau}\,(P_0 - P)\right]^{1/3}. \tag{12.33}$$

Here β is the *isothermal compressibility*, P is the *instantaneous (or internal) pressure*, P_0 is the target pressure (or external pressure P_{ex}), and τ is the pressure coupling time (see Box 12.2). The goal of this formulation is to allow the volume of the macromolecular system (ν_l) to fluctuate as the instantaneous (internal) pressure (P) approaches the applied (external) pressure, P_0.

This procedure is applied to the coordinates of particles under periodic boundary conditions, and to the box length L for an isotropic system in a cubic box

($L^{\text{new}} \leftarrow c_p\, L^{\text{old}}$). From the point of view of the guiding dynamic equations, this coordinate and volume scaling protocol effectively solves the augmented equation of motion for the time derivative of position ($\dot{X}(t) = V(t)$), where the additional term is proportional to X as follows:

$$\begin{aligned} \dot{X}(t) &= V(t) - \widetilde{\beta} X(t)\,, \\ \widetilde{\beta} &= \frac{\beta\,(P_0 - P(t))}{3\tau}\,. \end{aligned} \tag{12.34}$$

This equation can also be written as a pair of differential equations describing the evolution of the position (X) and volume (ν_l) by using eq. (12.36) in Box 12.2 to produce the following equations (we suppress the time dependency for clarity):

$$\begin{aligned} \dot{X} &= V + \frac{1}{3}\frac{\dot{\nu}_l}{\nu_l}\,, \\ \dot{\nu}_l &= \frac{\beta\,(P - P_0)}{\tau}\,\nu_l\,. \end{aligned} \tag{12.35}$$

Thus, this approach to constant pressure simulations allows the volume of the system to fluctuate as the pressure is held constant by changing the cell size uniformly but not its shape, as appropriate for an isotropic system. For an anisotropic system, instead of the scalar pressure variable, the 3×3 pressure matrix (see Box 12.2) is relevant [29].

As for the weak coupling thermostat method, the above approach does not generally produce trajectories from any known ensemble. A rigorous approach using an extended system method which includes dynamic thermostat and barostat variables is outlined below. See also [894] for an improved constant-pressure simulation protocol using a Langevin piston that is straightforward to implement. The Langevin piston method attempts to eliminate the nonphysical vibrations of the volume associated with the approach above and to generate in theory the correct NPT ensemble.

Box 12.2: Pressure Variables Definitions

The *isothermal compressibility* describes the change in volume (ν_l) of a substance under external pressure P via:

$$\frac{dP}{dt} = \frac{-1}{\beta\,\nu_l}\left(\frac{d\nu_l}{dt}\right) \tag{12.36}$$

or

$$\beta\dot{P} = -\dot{\nu}_l/\nu_l\,. \tag{12.37}$$

The *instantaneous pressure* P is calculated from the thermodynamic relation between the pressure, temperature, volume, and internal virial (vir) of a system by [30]:

$$P = \frac{2}{3\,\nu_l}(E_k - vir)\,. \tag{12.38}$$

The internal virial is proportional to the inner product of the each atom's position vector ($\mathbf{r}_i$) with the corresponding force component acting on atom i due to all particles (F_i):

$$vir = -\sum_i(\mathbf{r}_i^T\,F_i)\,. \tag{12.39}$$

(Intramolecular forces are thus considered in addition to intermolecular forces). Note that in the expression for the pressure (eq. (12.38)) we have used the relation between the temperature T and the kinetic energy E_k ($\langle E_k\rangle = \frac{N_F}{2}k_B\mathrm{T}$), where the number of degrees of freedom, N_F, is three times the number of atoms in the system.

For an anisotropic system, the pressure becomes a 3×3 matrix called the *pressure tensor* **P** [29]. This tensor is defined as:

$$\mathbf{P} = \frac{1}{\nu_l}\left[\sum_i(m_i\mathbf{v}_i\mathbf{v}_i^T) + \sum_i(\mathbf{r}_iF_i^T)\right], \tag{12.40}$$

where m_i and $\mathbf{v}_i$ are the scalar mass and 3-component velocity vector corresponding to atom i, respectively; the result of an outer product between a 3×1 vector and its transpose is a 3×3 matrix.

12.6.3 Extended System Methods

Extended-system methods introduce additional degrees of freedom to represent the environment of the macromolecular system (e.g., pressure piston, thermostat, etc.). These degrees of freedom have positions, momenta, and/or other associated variables. Examples include an external volume variable ν_l associated with a piston of given mass, whose potential energy is expressed in terms of the desired pressure [895], or effective thermodynamic variables that are functions of the positions and momenta of the added dynamic variables.

The equations of motion for the extended system are expressed as augmented versions of the standard equations of motions, to represent the evolution of both the internal and external variables in the desired ensemble. The solutions for the time evolution of the extended system by standard numerical methods (for the *microcanonical* ensemble) then produce the phase-space variables appropriate for the desired ensemble.

Canonical Ensemble

For example, to produce the true canonical (NVT) ensemble — which the simple scaling and thermostat approaches above do not accomplish — the method termed Nosé-Hoover [895, 890] adds a fictitious degree of freedom to the physical system with 'coordinate' parameter x_t (effectively a scaling parameter [31]), mass m_t, and thermodynamic friction coefficient ζ_t. (This friction coefficient is related to x_t and the corresponding momentum $\dot{x}_t$).

After appropriate scaling of variables (see [31], for example), the equations of motion for the real and artificial system combine to yield:

$$\begin{aligned} \dot{X}(t) &= V(t)\,, \\ \mathbf{M}\dot{V}(t) &= -\nabla E(X(t)) - \zeta_t\,\mathbf{M}V(t)\,, \\ m_t\,\dot{\zeta}_t(t) &= 2V^T\mathbf{M}V - g\,k_B\mathrm{T}_0\,, \end{aligned} \tag{12.41}$$

where g is the number of degrees of freedom in the system. The conserved quantity under these augmented equations of motion has two energy terms in addition to the internal kinetic and potential energies:

$$\widehat{H}^{\mathrm{NVT}} = \frac{1}{2}(V^T\mathbf{M}V) + E(X) + \frac{1}{2}(m_t\,\zeta_t^2) + g\,k_B\mathrm{T}_0\,x_t\,. \tag{12.42}$$

The choice of the fictitious mass m_t should ensure that the thermalization process is efficient. Too small a value implies large harmonic motion and rapid thermal fluctuations for the extended degree of freedom, and this in turn restricts the timestep too severely. Too large a value, on the other hand, makes the thermalization process slow. In general, m_t is chosen to be proportional to $g\,k_B\mathrm{T}$, and the integration of the above system is performed fairly accurately so that the energy is well conserved. See the original literature and [31, 891] for algorithmic details.

Isothermal-Isobaric Ensemble

Controlling the pressure in MD simulations is more involved. Instead of fixing the volume, as in microcanonical simulations, the volume of the system is considered as a dynamic variable, and it is allowed to fluctuate while the pressure is held constant. An approach analogous the Nosé-Hoover NVT form above has been described by Andersen [887] and Hoover [890] for the isothermal-isobaric, or NPT ensembles.

Essentially, in addition to the effective coordinate, mass, and friction set (x_t, m_t, ζ_t) associated with the fictitious thermostat variable, a set (x_p, m_p, ζ_p) is introduced and associated with a virtual pressure piston ("barostat").

The effective equations of motion for a 3-dimensional system become:

$$
\begin{aligned}
\dot{X}(t) &= V(t) + \zeta_p X\,, \\
\mathbf{M}\dot{V}(t) &= F(X(t)) - \mathbf{M}\dot{V}(t)\left[\left(1+\frac{3}{g}\right)\zeta_p + \zeta_t\right], \\
\dot{\nu_l} &= 3\,\nu_l\,\zeta_p\,, \\
m_p\,\dot{\zeta_p}(t) &= 3\,\nu_l\,(P_{\rm in} - P_0) + \frac{3}{g}\,(2V^T\mathbf{M}V) - m_p\,\zeta_t\,\zeta_p\,, \\
m_t\,\dot{\zeta_t}(t) &= 2V^T\mathbf{M}V + \frac{\zeta_p^2}{m_p} - (g+1)\,k_B\mathrm{T}_0\,,
\end{aligned}
\tag{12.43}
$$

where g is the number of degrees of freedom in the system, P_0 is the external applied pressure, and $P_{\rm in}$ is the internal pressure, defined as:

$$
P_{\rm in} = \frac{2}{3\,\nu_l}\left[E_k - vir - \left(\frac{3\,\nu_l}{2}\right)\frac{\partial E(X,\nu_l)}{\partial \nu_l}\right]; \tag{12.44}
$$

the virial vir is defined in equation (12.39). The conserved quantity under these augmented equations of motion is:

$$
\widehat{H}^{\rm NPT} = \frac{1}{2}(V^T\mathbf{M}V) + E(X,\nu_l) + \frac{1}{2}(m_t\,\zeta_t^2 + m_p\,\zeta_p^2) + (g+1)\,k_B\mathrm{T}_0\,x_t + P_0\,\nu_l. \tag{12.45}
$$

By this approach, the volume of the system fluctuates under the applied thermostat and barostats so that the system is driven to the steady state at which the average internal pressure P is equal to the external applied force P_0.

Algorithmic extensions to anisotropic systems, to allow changes in cell shape in addition to size, have also been developed using the pressure tensor and matrix analogs of the above approaches [893, 891, for example].

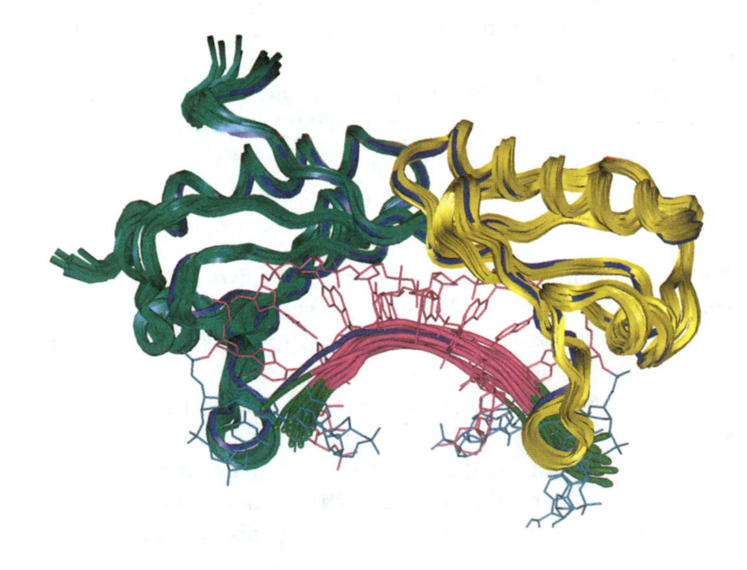

13

Molecular Dynamics: Further Topics

Chapter 13 Notation

SYMBOL	DEFINITION
Matrices	
$\mathbf{D}$	diffusion tensor (with sub-block matrices $\overline{\mathbf{D}}_{ij}$)
$\widetilde{\mathbf{H}}$	local Hessian approximation
$\mathbf{I}$	identity matrix
$\mathbf{J}$	constant matrix used in defining symplectic transformation
$\mathbf{M}$	mass matrix
$\mathbf{S}$	phase space transformation ($\mathbf{S} = \mathbf{DQD}^{-1}$; also Cholesky factor of $\mathbf{D}$ in BD)
$\mathbf{T}$	hydrodynamic tensor
$\mathbf{Z}$	friction tensor (related to $\mathbf{D}$ via $k_B T\mathbf{D}^{-1}$)
$\mathbf{\Psi_J}$	Jacobian matrix of $\mathbf{J}$
Vectors	
F	force
$\widetilde{F}$	acceleration ($\ddot{X}$), or $\mathbf{M}^{-1}\, F(X(t))$
$F_{\text{fast}}, F_{\text{med}}, F_{\text{slow}}$	forces of 'fast', 'medium' and 'slow' components
P, $\dot{P}$	momentum and its first time derivative
R, R_B	random forces
V	velocity, components $\{v_i\}$ (also $\dot{X}$)
X, $\dot{X}$, $\ddot{X}$	position (components $\{x_i\}$), and its first and second time derivatives
X_r	reference position for MTS extrapolation
$Y, \dot{Y}$	vector and its first time derivative
∇E	energy gradient

Chapter 13 Notation Table (continued)

SYMBOL	DEFINITION
Scalars & Functions	
a	hydrodynamic bead radius
k_1, k_2	integers (used in MTS to relate timesteps; $r = k_1 k_2$)
n, m	integers (defining resonance condition); also m = mass
D_t	translational diffusion constant
E_k	kinetic energy
H	Hamiltonian
$\mathcal{L}$	Liouville operator
N	number of atoms
N_b	number of beads (Brownian dynamics)
T	temperature
T_p	period (of characteristic frequency)
γ	Langevin damping constant
ζ	frictional coefficient ($m\gamma$ for example)
η	solvent viscosity
θ	phase space rotation defined by integrator
θ_{eff}	effective phase space rotation
ω	natural frequency ($\epsilon = \omega\Delta t$)
ω_{eff}	effective (scheme-dependent) frequency
$\Delta\tau$, Δt_m, Δt	timesteps (inner, medium, and outer)
$\Phi(X)$	'dynamics function' for implicit integration

Only by taking an infinitesimally small unit for observation (the differential of history ...) and attaining to the art of integrating them (that is, finding the sum of these infinitesimals) can we hope to arrive at the laws of history.

Lev Tolstoy, in *War and Peace* (1828–1910).

13.1 Introduction

In this chapter we survey more advanced aspects of integration approaches for biomolecular dynamics that are suitable for students interested in the mathematical issues of numerical schemes. We begin in Section 13.2 by introducing symplectic integrators in terms of an effective rotation in phase space and illustrate basic concepts by simple harmonic analysis.

The multiple-timestep (MTS), or force splitting, methods are presented next (Section 13.3), and the extreme variants of splitting by extrapolation versus splitting by impulses are contrasted with regard to resonance artifacts. Understanding favorable and unfavorable features of MTS schemes with respect to stability, accuracy, and resonance artifacts has led to important developments in recent years of efficient, long-timestep methods for biomolecular dynamics, as well as to new frameworks for method design and interpretation.

In this connection, we introduce in Sections 13.4 and 13.5, in turn, the stochastic dynamics approaches of Langevin and Brownian dynamics; both are in fact closely related, and the separation of terms may be arbitrary. These stochastic formulations may be viewed as constructs that enhance sampling, or that, through the introduction of random forces, make possible large timesteps by masking mild instabilities resulting from Newtonian integration.

Stochastic approaches also constitute approximate models for following large-scale motions in systems where random fluctuations (e.g., introduced by the bulk solvent) are at least as important as the systematic forces. This is the case, for example, in long DNA polymers that move with agility in solution, sampling many equilibrium configurations, rather than remaining near a single state. Studies of DNA supercoiling and of diffusion-controlled ligand gating events in enzyme catalysis, for example, have relied on such Brownian dynamics approaches.

We also mention implicit integration schemes in Section 13.6. This class of generally-more-expensive schemes has been explored as a possible way to increase the timestep in biomolecular dynamics simulations. We conclude in Section 13.7 with future perspectives on MD algorithm developments and with a description of some promising integration alternatives.

The reader should have read Chapter 12, where basic aspects of molecular dynamics and associated notation have been presented.

13.2 Symplectic Integrators

As introduced in the last chapter, symplectic or canonical integrators *preserve* special properties associated with the Hamiltonian system of differential equations. These properties include *volume elements in phase space* and the Hamiltonian value (*energy*). In practice, the total energy is not preserved exactly, but the energy error remains constant over long times. This is different from nonsymplectic methods, which typically display a systematic energy drift in time (usually damping). See Figure 12.6 for such an example of integration by the symplectic Verlet versus the nonsymplectic Runge-Kutta method.

Such favorable properties can be explained by *backward error analysis*,[1] which shows that the numerical trajectory of a symplectic integrator is the *exact* solution of a nearby Hamiltonian. The proximity measure depends on the timestep of the integrator, Δt, and on the scheme's order of accuracy, p. That is, it can

[1]See [896, Section 2.4.1] or [897, Section 2.7], for example. Roughly speaking, backward error analysis transforms the problem of estimating computational errors (in our case due to the finite-difference approximation) back to the problem of estimating the effect of changing the input data slightly.

be shown that the Hamiltonian corresponding to the numerical trajectory of a symplectic method remains order $\mathcal{O}(\Delta t)^p$ away from the initial value of the true Hamiltonian.

A comprehensive treatment of symplectic integration can be found in [837]. Here only a few key concepts will be introduced, many of which are relevant to future sections.

13.2.1 Symplectic Transformation

A Hamiltonian system is generally described by the motion of the collective position and momenta vectors $X, P \in \mathcal{R}^{3N}$ in time (here $P = \mathbf{M}V$ where $\mathbf{M}$ is the diagonal mass matrix; see last chapter). The Hamiltonian $H(X, P)$ is given as the sum of the kinetic and potential energy of the system:

$$H(X, P) = \frac{1}{2}(P^T \mathbf{M}^{-1} P) + E(X)\,. \tag{13.1}$$

The evolution of this system is governed by the equations:

$$\dot{P} = -\partial H/\partial X\,, \qquad \dot{X} = \partial H/\partial P\,, \tag{13.2}$$

or, equivalently, in vector form as:

$$\begin{pmatrix} \dot{X} \\ \dot{P} \end{pmatrix} = \begin{bmatrix} 0 & \mathbf{I} \\ -\mathbf{I} & 0 \end{bmatrix} \begin{pmatrix} \partial H/\partial X \\ \partial H/\partial P \end{pmatrix} \equiv \mathbf{J} \cdot \nabla H\,, \tag{13.3}$$

where $\mathbf{I}$ is the $3N \times 3N$ identity matrix.

An integrator defines a mapping Ψ that transforms coordinates and momenta at time t, $\{X^n, P^n\}$, to coordinates and momenta at time $t + \Delta t$, $\{X^{n+1}, P^{n+1}\}$. This transformation is *symplectic* if and only if its Jacobian matrix, $\mathbf{\Psi_J}$, satisfies:

$$\mathbf{\Psi_J}^T \, \mathbf{J} \, \mathbf{\Psi_J} = \mathbf{J}\,, \tag{13.4}$$

where $\mathbf{\Psi_J}$ is the matrix

$$\mathbf{\Psi_J} \equiv \begin{bmatrix} \partial X^{n+1}/\partial X^n & \partial X^{n+1}/\partial P^n \\ \partial P^{n+1}/\partial X^n & \partial P^{n+1}/\partial P^n \end{bmatrix}. \tag{13.5}$$

It can be shown that a *composition* of symplectic transformations is also symplectic and that the inverse of a symplectic mapping is symplectic. These properties are often used in practice to prove that an integrator is symplectic.

For a proof that the Verlet scheme is symplectic, see [898]. Below we only illustrate how the Verlet transformation can be interpreted for a linear system as a rotation in phase space. Analysis of a harmonic oscillator is often a first step in analyzing behavior of a numerical scheme and already sheds considerable insight [858, 860, 899, 859, 878, 900, for example].

13.2.2 Harmonic Oscillator Example

Consider the Verlet scheme of eq. (12.15) for linear forces $F(X) = -\omega^2 X$, where ω is the natural frequency of the oscillator. The transformation $\mathbf{S}$ can be used to relate one phase point to the next [878]. That is,

$$\begin{pmatrix} \omega X^{n+1} \\ V^{n+1} \end{pmatrix} = \mathbf{S} \begin{pmatrix} \omega X^{n} \\ V^{n} \end{pmatrix}, \tag{13.6}$$

where $\mathbf{S}$ is defined as

$$\mathbf{S} = \begin{bmatrix} 1-\epsilon^2/2 & \epsilon\,(1-\epsilon^2/4) \\ -\epsilon & 1-\epsilon^2/2 \end{bmatrix}, \tag{13.7}$$

with the parameter ϵ as the frequency times the timestep:

$$\epsilon = \omega \Delta t\,. \tag{13.8}$$

It is simple to show[2] that the product of the two eigenvalues is equal to 1 ($\lambda_1\,\lambda_2 = 1$) and that the absolute value of their sum is less than or equal to 2 ($|\lambda_1+\lambda_2| \leq 2$). Thus, powers of $\mathbf{S}$ (which are matrices) are bounded if and only if

$$\epsilon^2 < 4\,, \tag{13.9}$$

or equivalently

$$\Delta t < 2/\omega\,. \tag{13.10}$$

13.2.3 Linear Stability

The above restriction on the timestep size is the *stability condition* for Verlet. Thus, if the period of the oscillator is $T_p = 2\pi/\omega$, this *linear stability* condition becomes

$$\Delta t < T_p/\pi\,. \tag{13.11}$$

Table 13.1 gives corresponding linear stability limits on the timestep for the high-frequency end of biomolecular vibrational modes. Clearly, in general, we have:

T_p (stretch) $< T_p$ (bend), and

T_p (light atoms) $< T_p$ (heavy atoms).

We also note that the light-atom bends (e.g., H–O–H) and the heavy-atom bond stretches (e.g., C=C) have very similar periods.

In general, the table emphasizes the *lack of clear gaps in timescales* among these modes. This explains why constraining some high-frequency motions in a dynamic formulation typically affects other vibrational modes and why the computational benefit (larger permitted timestep) is relatively modest.

For example, if we constrain all bonds for a water model, thereby eliminating the motion and stability limit corresponding to the O–H stretch (period $\sim$ 10 fs),

---

[2]Hint: For a 2×2 matrix, the matrix trace (sum of diagonals) is the sum of the eigenvalues, here $2 - \epsilon^2$, and the matrix determinant is their product, here 1.

the next relevant period that limits the timestep is the H–O–H bend (period ~ 21 fs), followed by water libration (~ 42 fs). Thus, the effective timestep can at most be doubled by constrained dynamics.

Table 13.1. The timestep limit for Verlet based on a harmonic oscillator analysis.

Vibrational Mode[a]	Wave number $(1/\lambda)$ [cm^{-1}]	Period T_p (λ/c) [fs][b]	T_p/π [fs]
O–H, N–H stretch	3200-3600	9.8	3.1
C–H stretch	3000	11.1	3.5
O–C–O asymm. stretch	2400	13.9	4.5
C≡C, C≡N stretch	2100	15.9	5.1
C=O (carbonyl) stretch	1700	19.6	6.2
C=C stretch			
H–O–H bend	1600	20.8	6.4
C–N–H, H–N–H bend	1500	22.2	7.1
C=C (aromatic) stretch			
C–N stretch (amines)	1250	26.2	8.4
Water Libration (rocking)	800	41.7	13
O–C–O bending	700	47.6	15
C=C–H bending (alkenes)			
C=C–H bending (aromatic)			

[a] All values are approximate.
[b] The value of the speed of light is taken as $c = 3.00 \times 10^{10}$ cm s^{-1}.

13.2.4 *Timestep-Dependent Rotation in Phase Space*

Under the linear stability assumption, the matrix $\mathbf{S}$ of eq. (13.7) has the two eigenvalues $\exp(\pm i\theta)$ ($i = \sqrt{-1}$), where

$$\theta = 2\sin^{-1}(\omega\,\Delta t/2) \tag{13.12}$$

$$= \omega\Delta t + \frac{1}{24}(\omega\,\Delta t)^3 + \mathcal{O}(\omega\,\Delta t)^5\,. \tag{13.13}$$

Thus, the angle θ depends on the timestep (and on ω). To see that this transformation defines a rotation in phase space, we decompose the phase-space transforming matrix $\mathbf{S}$ as

$$\mathbf{S} = \mathbf{D}\mathbf{Q}\mathbf{D}^{-1}\,. \tag{13.14}$$

In this definition, the matrix

$$\mathbf{Q} = \begin{bmatrix} \cos\theta & \sin\theta \\ -\sin\theta & \cos\theta \end{bmatrix} \tag{13.15}$$

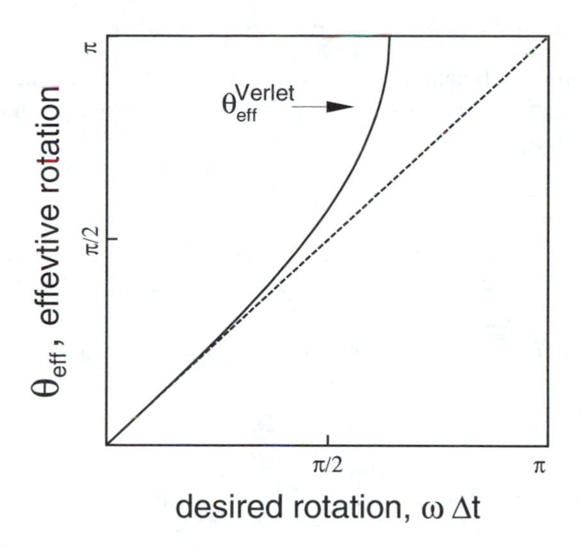

Figure 13.1. The effective rotation $\theta_{\text{eff}}^{\text{verlet}}$ (in radians), eq. (13.19), plotted against the desired rotation, $\omega\Delta t$, for the Verlet scheme.

defines a rotation of $-\theta$ radians in phase space, and **D** is the diagonal matrix

$$\mathbf{D} = \begin{bmatrix} 1 & 0 \\ 0 & 1 + \tan^2 \frac{\theta}{2} \end{bmatrix} . \tag{13.16}$$

Behavior of the integrator in time can be interpreted through analysis of the powers of **S** given by

$$\mathbf{S}^n = \mathbf{D}\mathbf{Q}^n\mathbf{D}^{-1} ,$$

where

$$\mathbf{Q} = \begin{bmatrix} \cos n\theta & \sin n\theta \\ -\sin n\theta & \cos n\theta \end{bmatrix} . \tag{13.17}$$

The timestep-dependent behavior of the transformation **S** can be interpreted as follows. Equations (13.12) and (13.13) show that the integrator is using θ as an approximation to the exact rotation $\omega\Delta t$. The smaller the timestep, the closer the approximation. This is shown in Figure 13.1, where rapid divergence from the target value (dashed diagonal line) is evident as Δt is increased.

The *effective rotation* θ_{eff} can thus be defined as:

$$\theta_{\text{eff}} = \omega_{\text{eff}}\, \Delta t . \tag{13.18}$$

For the Verlet method, the effective rotation is given by equation (13.12):

$$\theta_{\text{eff}}^{\text{verlet}} = 2 \sin^{-1}(\omega\Delta t/2) . \tag{13.19}$$

13.2.5 Resonance Condition for Periodic Motion

For periodic motion with natural frequency ω, nonphysical resonance — an artifact of the symplectic integrator — can occur when ω is related by integers n and m to the forcing frequency $(2\pi/\Delta t)$:

$$\frac{n}{m}\,\omega = \frac{2\pi}{\Delta t}\,. \tag{13.20}$$

Here n is the resonance order, and the integers n and m are relatively prime. Now recall that above we have shown that the Verlet method has the timestep-dependent frequency ω_{eff} given by $\theta_{\text{eff}}^{\text{verlet}}/\Delta t$ where $\theta_{\text{eff}}^{\text{verlet}}$ is defined in eq. (13.19); this frequency thus depends on the timestep in a nonlinear way. Given this frequency, the integrator-dependent resonance condition becomes

$$\frac{n}{m}\,\omega_{\text{eff}} = \frac{2\pi}{\Delta t}\,. \tag{13.21}$$

A "resonance of order $n:m$" means that n phase space points are sampled in m revolutions:

$$n\theta_{\text{eff}} = n\Delta t\,\omega_{\text{eff}} = 2\pi m\,.$$

This special, finite-coverage of phase space can lead to incorrect, limited sampling of configuration space. See Figure 13.2 for an illustration.

As was shown in [899], equation (13.19) can be used to formulate *a condition for a resonant timestep* for the harmonic oscillator system. That is, using

$$\omega_{\text{eff}}^{\text{verlet}} = \frac{2\sin(\omega\Delta t/2)}{\Delta t} \tag{13.22}$$

with the resonance condition of eq. (13.21), we have

$$\frac{\omega\Delta t}{2} = \sin\left(\frac{m\pi}{n}\right)\,. \tag{13.23}$$

Equivalently,

$$\Delta t_{n:m}^{\text{verlet}} = \frac{2}{\omega}\sin\left(\frac{m\pi}{n}\right) = \frac{T_p}{\pi}\sin\left(\frac{m\pi}{n}\right)\,, \tag{13.24}$$

where T_p is the natural period of the oscillator.

Table 13.2 lists the values for low-order resonances (which are the most severe) for $m = 1$ based on a period $T_p = 10$ fs (or $\omega = 0.63$ fs^{-1}). This period corresponds to the fastest period in biomolecular simulations, such as an O–H bond stretch (see Table 13.1). For $n = 2$, we have *linear stability*.

Clearly, since the limiting timesteps $\Delta t_{n:1}$ for resonance orders $n > 2$ are *smaller* than the linear stability limit $(\Delta t_{2:1})$, resonance limits the timestep to values *lower* than classical stability. Since the third-order resonance leads to instability and the fourth-order resonance often leads to instability, in practice we require $\Delta t < \Delta t_{4:1}$. This implies for Verlet a stricter restriction than eq. (13.10), which comes from the second-order resonance (or linear stability), to

$$\Delta t < \sqrt{2}/\omega\,, \tag{13.25}$$

which corresponds to the fourth-order resonance ($\Delta t_{2:1}$ and $\Delta t_{4:1}$ are listed in Table 13.2).

13.2.6 *Resonance Artifacts*

For nonlinear systems, the effective forcing frequency for Verlet is not known, but for relatively small energy values, it is reasonable to approximate this effective frequency by expressions known for harmonic oscillators, as derived above. Such approximations can be appropriate, as shown in [899] and Figure 13.2.

This figure from [899] shows the resonance artifacts of the Verlet scheme in terms of peculiar phase-space diagrams (position versus momentum plot) obtained for a Morse oscillator[3] integrated at $\Delta t = 2$ fs for increasing initial energies (or corresponding temperatures). Around this timestep, resonances of order 6 and higher are captured (see the reference resonant timesteps based on linear analysis in Table 13.2). The 6:1 resonance shows the 6 islands (each a closed orbit; see inset) covered in one revolution. As the initial energy increases, so does the resonance order: see the 13:2 and 20:3 resonances near the box periphery. Analysis of stability and resonance for MTS schemes can be similarly performed by extending the model to more than one frequency [868, 609].

Section 13.6 presents a resonance analysis for the implicit-midpoint (IM) scheme, and Box 13.5 derives IM's resonance condition (based on its phase-space rotation $\omega_{\text{eff}}^{\text{IM}}$). Figure 13.16 illustrates resonance for a nonlinear system for both Verlet and IM.

Table 13.2. Stability limit and resonant timestep formulas for the Verlet scheme, with numerical values for a vibrational mode with period $T_p = 10$ fs.

n, res. order[a]	$\Delta t_{n:1}(\omega)$	$\Delta t_{n:1}(T_p)$	$\Delta t_{n:1}$ for MD[b]
2	$2/\omega$	$0.318\,T_p$ $[= T_p/\pi]$	3.2 fs
3	$\sqrt{3}/\omega$	$0.276\,T_p$	2.8 fs
4	$\sqrt{2}/\omega$	$0.225\,T_p$	2.3 fs
5	$1.176/\omega$ $[= 2\sin(\pi/5)/\omega]$	$0.188\,T_p$	1.9 fs
6	$1/\omega$	$0.159\,T_p$ $[= T_p/2\pi]$	1.6 fs

[a] For $n = 2$, we have the linear stability condition.
[b] To compare resonance in Verlet to implicit-midpoint integration (see end of chapter), consider the analogous resonant timesteps $\Delta t_{n:1}$ for $n = 3, 4, 5, 6$ of 5.5, 3.2, 2.3, and 1.8 fs, respectively, for the implicit-midpoint scheme.

[3] The Morse bond potential and its relation to the harmonic potential is discussed in the Bond Length Potentials section of Chapter 8 (see eq. (8.7)).

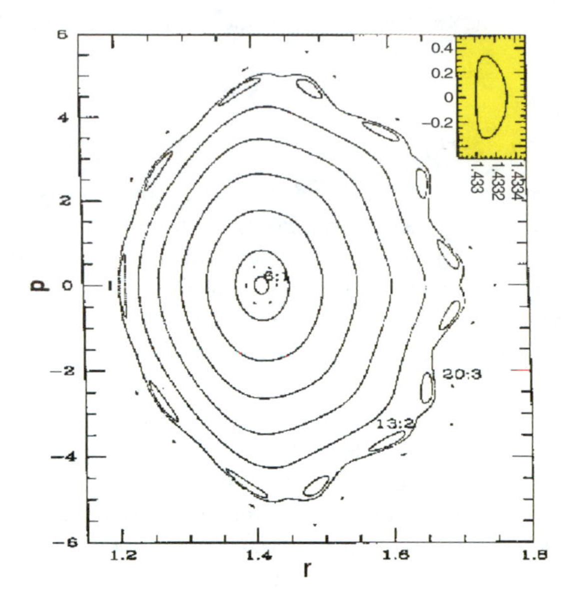

Figure 13.2. Morse oscillator phase diagrams obtained from the leapfrog Verlet integrator with $\Delta t = 2$ fs at increasing initial energies. Resonances of order 6 and higher can be noted. One of the inner-most islands (near the 6:1 label) is enlarged in the inset to show that each corresponds to a closed orbit. See [899] for details.

13.3 Multiple-Timestep (MTS) Methods

13.3.1 Basic Idea

MTS schemes were introduced in the 1970s [901, 902] in an effort to reduce the computational cost of molecular simulations. Savings can be realized if the forces due to distant interactions are held constant over longer intervals than the short-range forces. Standard integration procedures can then be modified by evaluating the long-range forces less often than the short-range terms. Between updates, the slow forces can be incorporated into the discretization as a piecewise constant function (via *extrapolation*) or as a sum of delta functions (via *impulses*). See Figure 13.4 for a schematic illustration of these two approaches.

To illustrate, in the discussion below we consider below a force splitting scheme that involves three components for updating the fast, medium, and slow forces:

$$F_{\text{fast}},\ F_{\text{med}},\ \text{and}\ F_{\text{slow}},$$

with corresponding timesteps

$$\Delta\tau \leq \Delta t_m \leq \Delta t.$$

The timesteps are related via the ratios

$$\begin{aligned} k_1 &= \Delta t_m / \Delta\tau\,, \\ k_2 &= \Delta t / \Delta t_m\,, \end{aligned} \tag{13.26}$$

Extrapolative MTS based on Position Verlet	**Impulse MTS based on Velocity Verlet**
$X_r^0 \equiv X + \frac{\Delta t_m}{2} V$ $\widetilde{F}_{\text{slow}} \equiv -\mathbf{M}^{-1}\nabla E_{\text{slow}}(X_r)$ **For** $j = 1$ to k_2 $X_r \equiv X_r^j \leftarrow X + \frac{\Delta t_m}{2} V$ $\widetilde{F}_{\text{med}} \equiv -\mathbf{M}^{-1}\nabla E_{\text{med}}(X_r)$ $\widetilde{F} \leftarrow \widetilde{F}_{\text{med}} + \widetilde{F}_{\text{slow}}$ **For** $i = 1$ to k_1 $X \leftarrow X + \frac{\Delta\tau}{2} V$ $V \leftarrow V + \Delta\tau\,(\widetilde{F} + \widetilde{F}_{\text{fast}})$ $X \leftarrow X + \frac{\Delta\tau}{2} V$ **End** **End**	$\widetilde{F}_{\text{slow}} \equiv -\mathbf{M}^{-1}\nabla E_{\text{slow}}(X)$ $V \leftarrow V + \frac{\Delta t}{2} \widetilde{F}_{\text{slow}}$ **For** $j = 0$ to $k_2 - 1$ $\widetilde{F}_{\text{med}} \equiv -\mathbf{M}^{-1}\nabla E_{\text{med}}(X)$ $V \leftarrow V + \frac{\Delta t_m}{2} \widetilde{F}_{\text{med}}$ **For** $i = 0$ to $k_1 - 1$ $V \leftarrow V + \frac{\Delta\tau}{2} \widetilde{F}_{\text{fast}}$ $X \leftarrow X + \Delta\tau\, V$ $V \leftarrow V + \frac{\Delta\tau}{2} \widetilde{F}_{\text{fast}}$ **End** $V \leftarrow V + \frac{\Delta t_m}{2} \widetilde{F}_{\text{med}}$ **End** $V \leftarrow V + \frac{\Delta t}{2} \widetilde{F}_{\text{slow}}$

Figure 13.3. Algorithmic sketches of molecular dynamics integration by two force-splitting variants: *extrapolation* (left) versus *impulses* (right), based on position Verlet (extrapolation) and velocity Verlet (impulses).

$$r \;=\; k_1\, k_2 = \Delta t / \Delta\tau\,,$$

where k_1 and k_2 are integers. Note that when $k_1 = k_2 = 1$ we have a single-timestep (STS) method.

13.3.2 Extrapolation

A simple, Verlet-based, extrapolative force-splitting approach is best formulated on the basis of *position Verlet* (eq. (12.15)) since the fast and medium forces are evaluated at the *middle* of the corresponding interval rather than at the beginning (and end). In programming style, where new iterates overwrite the old (i.e., $X \leftarrow X + \cdots$), this extrapolative force-splitting scheme can be written as the doubly-nested loop for covering each Δt sweep shown in Figure 13.3, left-hand side.

Note that in this loop the *slow force* is evaluated *once* (at the point X_r^0), the *medium force* is evaluated k_2 *times* (for each X_r^j), and the *fast force* is evaluated $k_1 k_2$ (or r) *times* at a corresponding midpoint. If the *slow force* calculations take the majority of the CPU time, the MTS approach will result in significant computational savings.

In practical implementations, the force distance classes are best treated by a smooth force-switching approach; see Chapter 9, Spherical-Cutoff section.

Simple extrapolation formulations, as above, were first tried for molecular dynamics [901, 903, 902, 852, 904]. However, these variants exhibited systematic energy drifts, a result of their nonsymplecticness.

13.3.3 Impulses

Work continued in the 1980s on MTS methods in a variety of contexts [905, 906], leading to the introduction in 1991/1992 (by Schulten and coworkers [852] and independently by Berne and coworkers [875]) to an MTS method which is symplectic and time-reversible. This similar MTS variant, based on impulses rather than extrapolation, was termed Verlet-I by the former group [852] and r-Respa [875] by the latter.

The reversible Respa method was derived from a general Trotter factorization associated with the Liouville operator $\mathcal{L}$. Liouville operators are fundamental tools in statistical mechanics for the description of the canonical equations of motion of Hamiltonian systems. The Liouville operator can be decomposed into parts corresponding to different components of the energy using the reversible Trotter factorization. Here, we write the Liouville operator $\mathcal{L}$ as a sum of three operators that characterize the scales of motions associated with different potential components:

$$\mathcal{L} = \mathcal{L}_{\text{fast}} + \mathcal{L}_{\text{med}} + \mathcal{L}_{\text{slow}} \,.$$

We then use a symmetric factorization of the components to arrive at (recall that $k_1 \Delta\tau = \Delta t_m$ and $k_2 \Delta t_m = \Delta t$):

$$\begin{aligned}
\exp\left[i\Delta t \mathcal{L}\right] &= \exp\left[i\Delta t(\mathcal{L}_{\text{fast}} + \mathcal{L}_{\text{med}} + \mathcal{L}_{\text{slow}})\right] \\
&= \exp\left[i\left(\frac{\Delta t}{2}\right)\mathcal{L}_{\text{slow}}\right] \times \\
&\left(\exp\left[i\left(\frac{\Delta t_m}{2}\right)\mathcal{L}_{\text{med}}\right] \left(\exp\left[i\Delta\tau\mathcal{L}_{\text{fast}}\right]\right)^{k_1} \exp\left[i\left(\frac{\Delta t_m}{2}\right)\mathcal{L}_{\text{med}}\right]\right)^{k_2} \\
&\times \exp\left[i\left(\frac{\Delta t}{2}\right)\mathcal{L}_{\text{slow}}\right] + \mathcal{O}(\Delta t^3).
\end{aligned}$$

This factorization effectively shows that the propagation of the solution can be approximated by a combination of terms corresponding to several force components, each of which is resolved on a suitable timescale (e.g., $\Delta\tau/\Delta t_m/\Delta t$ for fast/medium/slow components). The middle term, corresponding to the fast components of the motion, is discretized with the Verlet method at a small timestep ($\Delta\tau$).

A sweep over one Δt interval by an impulse-MTS Verlet approach, based on the leapfrog or velocity Verlet triplet (eq. 12.14), can be written as the doubly-nested iteration process shown in Figure 13.3, right-hand side.

13.3.4 Vulnerability of Impulse Splitting to Resonance Artifacts

Note that the application of the slow force results in an *impulse*. The velocities are modified only outside of the inner loop (i.e., at the onset and at the end of a sweep covering Δt) by a term proportional to $r\,\Delta\tau$, thus r times larger than the changes made to X and V in the inner loop. This is shown schematically in Figure 13.4 for a dual-timestep method with $r = 4$ (see tall spikes at bottom).

Thus, as the time interval between slow-force updates increases, the size of these "impulses" grows. This causes a *resonance artifact* when the impulse frequency, or the MTS outer timestep Δt, occurs near a natural frequency of the system. This artifact results because the long-range forces add energy at pulses that correspond to the natural frequency of the system. Of course, the true physical system experiences continuous variation of the long-range forces. Next we analyze resonance artifacts of MTS methods in more detail for a harmonic model.

13.3.5 Resonance Artifacts in MTS

Analysis of a simple linear system is instructive to illustrate resonance in MTS schemes [907, 868]. Recent works [868, 609, 908] have shown that impulse methods are *generally stable* except at integer multiples of half the period of the fastest motion, with the severity of the *instability worsening* with the timestep. Extrapolation methods are *generally unstable* for the Newtonian model problem, but the *instability is bounded* for increasing timesteps. Similar results hold for stochastic extensions of MTS [868].

Simple Example

Figure 13.5 from [609] illustrates this behavior as analyzed on a one-dimensional linear oscillator obeying the equations

$$\dot{X} = V, \qquad \dot{V} = -(\lambda_1 + \lambda_2)X\,,$$

where X and V denote the scalar position and velocity, respectively, and a unit mass for the particle is assumed. The scalars $\lambda_1 > \lambda_2$ represent two motion components differing in timescales. The characteristic angular frequencies associated with these components and respective periods are

$$\omega_i = \sqrt{\lambda_i}\,, \qquad T_{p_i} = 2\pi/\omega_i\,, \qquad i = 1, 2\,.$$

For the analysis of Figure 13.5, we set $\lambda_1 = \pi^2$ and $\lambda_2 = (\pi/5)^2$ to produce $T_{p_2} = 10 = 5\,T_{p_1}$ time units. The slow force component is defined as $-\lambda_1 X$ and

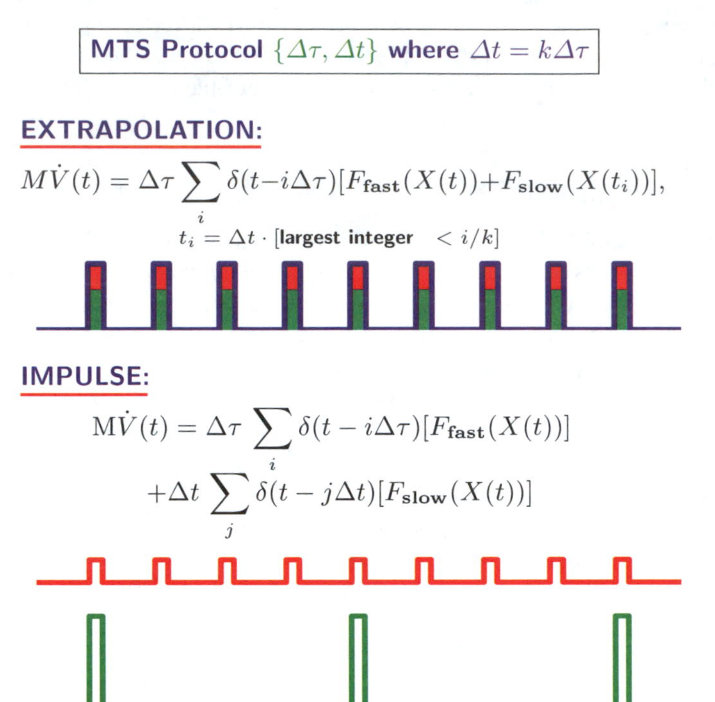

Figure 13.4. Schematic illustration of extrapolative vs. impulse force splitting integration for a dual-timestep protocol with inner timestep $\Delta\tau$ and outer timestep $\Delta t = k\Delta\tau$ ($k = 4$ used). In extrapolative splitting, a slow-force contribution (green) is made each time the fast force (red) is evaluated (i.e., every $\Delta\tau$ interval). In impulse splitting, in contrast, contributions of the slow forces (tall spikes) are considered only at the time of their evaluation (e.g., every four fast-force evaluations).

is updated at timesteps Δt that are k times larger than those ($\Delta\tau$) used for the fast components, $-\lambda_2 X$.

In the linear analysis of [609], eigenvalue magnitudes derived from the propagation matrices associated with the impulse and extrapolation force splitting schemes are plotted against the outer timestep. A scheme is *unstable* if the eigenvalue magnitude exceeds unity. Figure 13.5 shows results for both Newtonian (left) and Langevin (right) dynamics; the latter, stochastic dynamics approach is described in more detail below. The inner timestep was set to $\Delta\tau = 0.001$ and the outer timestep to $k\Delta\tau$ (compare to $T_{p_1} = 2$ and $T_{p_2} = 10$).

We see that the impulse method (dashed line) is unstable at integral multiples (m) of half the fast period ($m\,T_{p_1}/2$, where $T_{p_1}/2 = 1$ here). Furthermore, the severity of these corruptions increases with k, with the amplitudes of the resonant spikes increasing linearly with Δt and becoming wider.

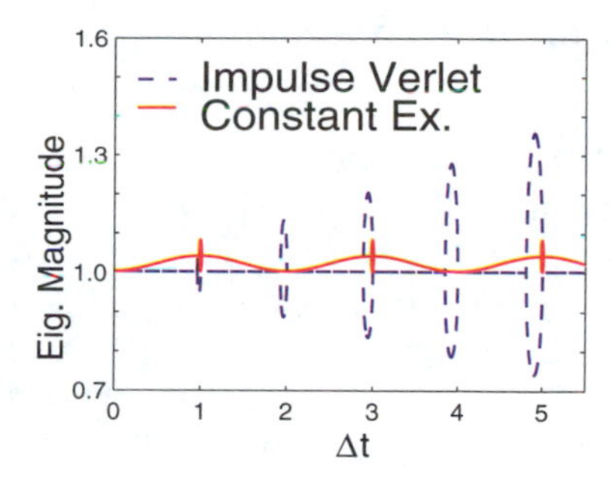

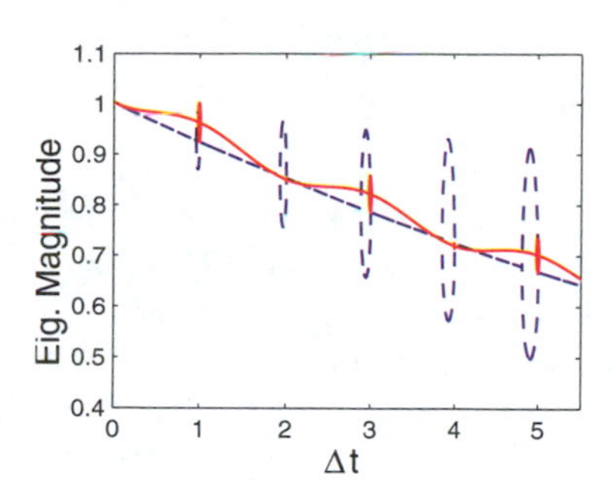

Figure 13.5. Propagation-matrix eigenvalue magnitudes for Newtonian (left) and Langevin (right, $\gamma = 0.162$) *impulse* and *extrapolative* force-splitting methods as a function of the outer timestep as calculated for a one-dimensional linear test system governed by $\ddot{X} = -(\lambda_1 + \lambda_2)X$ with $\lambda_1 = \pi^2$, $\lambda_2 = (\pi/5)^2$, where X denotes the scalar position of the particle with unit mass. These settings produce associated characteristic periods for the fast and slow components of $T_{p_1} = 2\pi/\sqrt{\lambda_1} = 2$ and $T_{p_2} = 2\pi/\sqrt{\lambda_2} = 10$ time units. The inner timestep is $\Delta\tau = 0.001$ [609].

While the use of impulses clearly leads to serious artifacts at certain timesteps, for other timesteps the impulse method is stable for this one-dimensional linear model. It might seem that avoiding the resonant timesteps is possible in practice. Unfortunately, experiments on a large nonlinear system show that once a resonant timestep is reached, good numerical behavior cannot be restored by increasing the timestep [900]. See Figure 13.7 for an illustration of resonance on a nonlinear system.

For the extrapolation method, in contrast, we note in Figure 13.5 (solid curves) deviations from unit eigenvalues except around isolated outer timesteps. Since both eigenvalues exceed one in magnitude, the method is not volume-preserving or symplectic, explaining the energy drift observed in practice for force splitting by extrapolation. Still, the resonant spikes for constant extrapolation have magnitudes that are *independent of the outer timestep*, of values approximately $1 + \lambda_2/\lambda_1$ ($\lambda_2/\lambda_1 = 0.04$ here). Therefore, the larger the separation of frequencies, the more benign the numerical artifacts in practice.

Extensions to Stochasticity and Nonlinearity

These resonance patterns generalize to stochastic dynamics, where friction and random terms are added (see next section), as seen in the same figure (the eigenvalue magnitudes decrease due to frictional damping). The value of γ used here ensures that the first spike for the impulse method has magnitude less than unity [609]. Behavior is more complicated for a linear three-dimensional test problem [609]. Still, applications to a nonlinear potential function for a protein, as shown in Figure 13.7 (discussed below), exhibit similar overall patterns for impulse versus extrapolative force splitting treatments.

13.3.6 Limitations of Resonance Artifacts on Speedup; Possible Cures

The original work of Schulten and coworkers [852] expressed reservations regarding force impulses. A subsequent study by Biesiadecki and Skeel [867] discussed resonance as well as the systematic drift of extrapolative treatment in simple oscillator systems, though it conjectured that resonance artifacts were not likely to be so clear in nonlinear systems. This, unfortunately, turned out not to be true. Other articles noted a rapid energy growth [870, 866] when the interval between slow-force updates approaches and exceeds 5 fs, half the period of the fastest oscillations in biomolecules

The presence of these resonances in impulse splitting limits the outermost timestep, Δt (or the interval of slow-force update), in MTS schemes. In turn, the overall speedup over a standard Verlet trajectory is capped. The achievable speedup factor has been reported to be around 4–6 [875, 870, 866] for biomolecules. However, this number depends critically on the reference STS (single-timestep) method to which MTS performance is compared. Since the inner timestep in MTS schemes is often small (e.g., 0.5 or even 0.25 fs), such speedup factors around 5 are obtained with respect to STS trajectories at 0.5 fs (10 with respect to 0.25 fs!). While accuracy is comparable when this small timestep is used, typically STS methods use larger timesteps, such as 1 or 2 fs (often with SHAKE). This means that the actual computational benefit from the use of MTS schemes in terms of CPU time per nanosecond, for example, is much less. Still, a careful MTS implementation with inner timestep $\Delta\tau = 0.5$ fs can yield better accuracy than a STS method using 1 or 2 fs.

Given this resonance limitation on speedup, it is thus of great interest to revise these methods to yield larger speedups.

The mollified impulse method of Skeel and coworkers [907, 909, 910] has extended the outer timestep by roughly a factor of 1.5 (e.g., to 8 fs). With additional Langevin coupling, following the LN approach [876], the Langevin mollified method (termed LM) can compensate for the inaccuracies due to the use of impulses to approximate a slowly varying force. This is accomplished by substituting $\mathcal{A}(X)^T F_{\text{slow}}(\mathcal{A}(X))$ for the slow force term where $\mathcal{A}(X)$ is a time-averaging function.

Another approach altogether is to use extrapolation in the context of a stochastic formulation, as in the LN method [885, 876, 868] (see next section). This combination avoids the systematic energy drift, alleviates severe resonance, and allows much longer outer timesteps; see also discussion of masking resonances via the introduction of stochasticity in an editorial overview [911] and review [912].

Though the Newtonian description is naturally altered, the stochastic formulation may be useful for enhanced sampling. The contribution of the stochastic terms can also be made as small as possible, just to ensure stability [913, 877]. For example, unlike the predictions in the above review [912], a smaller stochastic contribution (damping constant of 5 to 10 ps^{-1}) has been used in the LN scheme without reducing the outer timestep, and hence without compromising the speedup [334].

13.4 Langevin Dynamics

13.4.1 *Many Uses*

A stochastic alternative to Newtonian dynamics, namely Langevin dynamics, has been used in a variety of biomolecular simulation contexts for various numerical and physical reasons. The Langevin model has been employed to eliminate explicit representation of water molecules [715], treat droplet surface effects [728, 914], represent hydration shell models in large systems [915, 916, 917], enhance sampling [731, 725, 918, 500, 919, 912, 40], and counteract numerical damping while masking mild instabilities of certain long-timestep approaches [920, 921, 922, 923, 876]. See Pastor's comprehensive review on the use of the Langevin equation [715]. The Langevin equation is also discussed in Subsection 9.6.3 of Chapter 9 in connection with continuum solvation representations.

13.4.2 *Phenomenological Heat Bath*

The Langevin model is phenomenological [29] — adding friction and random forces to the systematic forces — but with the physical motivation to represent a simple heat bath for the macromolecule by accounting for molecular collisions. The continuous form of the simplest Langevin equation is given by:

$$\mathbf{M}\ddot{X}(t) = -\nabla E(X(t)) - \gamma \mathbf{M}\dot{X}(t) + R(t), \qquad (13.27)$$

where γ is the collision parameter (in reciprocal units of time), also known as the damping constant. The random-force vector R is a stationary Gaussian process with statistical properties given by:

$$\langle R(t) \rangle \;=\; 0, \qquad \langle R(t) R(t')^T \rangle \;=\; 2\gamma k_B \mathrm{T} \mathbf{M}\, \delta(t-t'), \qquad (13.28)$$

where k_B is the Boltzmann constant, T is the target temperature, and δ is the usual Dirac symbol.

13.4.3 *The Effect of γ*

The magnitude of γ determines the relative strength of the *inertial* forces with respect to the random (external) forces. Thus, as γ increases, we span the inertial to the *diffusive* (Brownian) regime. The Brownian range is used (with suitable algorithms) to explore configuration spaces of floppy systems efficiently. See [723, 512], for instance, for applications to the large-scale opening/closing lid motion of the enzyme triosephosphate isomerase (TIM), and to the juxtaposition of linearly-distant segments in long DNA systems, respectively.

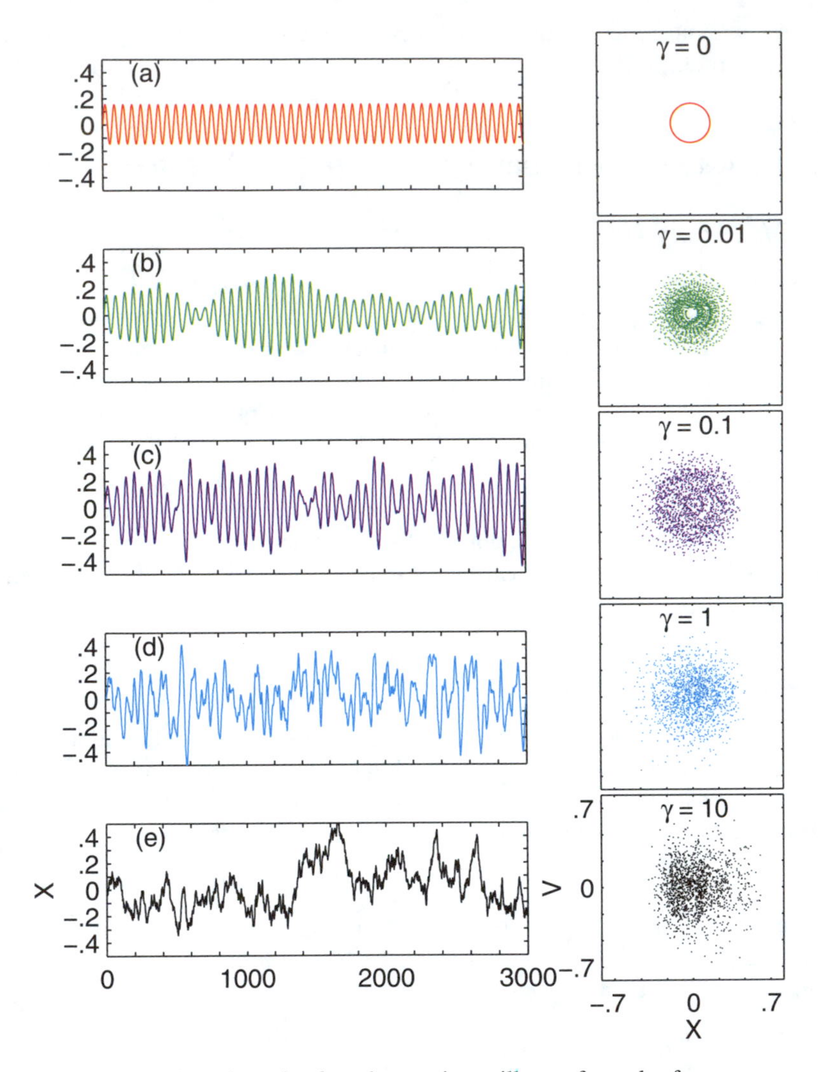

Figure 13.6. Langevin trajectories for a harmonic oscillator of angular frequency $\omega = 1$ and unit mass simulated by a Verlet extension to Langevin dynamics at a timestep of 0.1 (about 1/60 the period) for various γ. Shown for each γ are plots for position versus time (left panels) and phase-space diagrams (right squares). Note that points in the phase-space diagrams are not connected and only appear connected for $\gamma = 0$ because of the closed orbit.

Figure 13.6 illustrates the effects of increasing γ on the trajectories and phase diagrams of a harmonic oscillator. The systematic harmonic motion and the closed, circular trajectories characteristic at zero viscosity change as the relative contribution of the random to systematic forces increases.

Since the stochastic Langevin forces mimic collisions between solvent molecules and the biomolecule (the solute), we see that the characteristic vibrational frequencies of a molecule in vacuum are damped. In particular, the low-frequency vibrational modes are overdamped, and various correlation functions are smoothed (see Case [924] for a review and further references). The magnitude of such disturbances with respect to Newtonian behavior depends on γ [876].

A physical value for γ for each particle can be chosen according to Stokes' law for a hydrodynamic particle of radius a:

$$\gamma = 6\pi\eta a/m\,, \tag{13.29}$$

where m is the particle's mass (not to be confused with the integer m used to define resonance), and η is the solvent viscosity. For example, $\gamma = 50$ ps^{-1} is a typical collision frequency for protein atoms exposed to solvent having a viscosity of 1 cp at room temperature [858]; this value is also in the range of that estimated for water ($\gamma = 54.9$ ps^{-1}).

It is also possible to choose an appropriate value for γ for a system modeled by the simple Langevin equation so as to reproduce observed experimental translation diffusion constants, D_t. Namely, in the *diffusive limit*, D_t is related to γ by

$$D_t = k_B \mathrm{T} / \sum m\gamma\,.$$

See [715, 732] for example, for the use of these relations in studies of supercoiled DNA.

13.4.4 Generalized Verlet for Langevin Dynamics

The Verlet algorithm can easily be generalized to include the friction and stochastic terms above. A common discretization is that described by Brooks, Brünger and Karplus, known as BBK [728, 858]:

Generalized Verlet Algorithm for Langevin Dynamics

$$\begin{aligned} V^{n+1/2} &= V^n + \mathbf{M}^{-1}\frac{\Delta t}{2}\left[-\nabla E(X^n) - \gamma \mathbf{M} V^n + R^n\right] \\ X^{n+1} &= X^n + \Delta t V^{n+1/2} \\ V^{n+1} &= V^{n+1/2} + \mathbf{M}^{-1}\frac{\Delta t}{2}\left[-\nabla E(X^{n+1}) - \gamma \mathbf{M} V^{n+1} + R^{n+1}\right]. \end{aligned} \tag{13.30}$$

This Langevin scheme reduces to velocity Verlet (triplet eq. (12.14)) when γ and hence R^n are zero.

Note that the third equation above defines V^{n+1} implicitly; the linear dependency, however, allows solution for V^{n+1} in closed form (i.e., explicitly). The

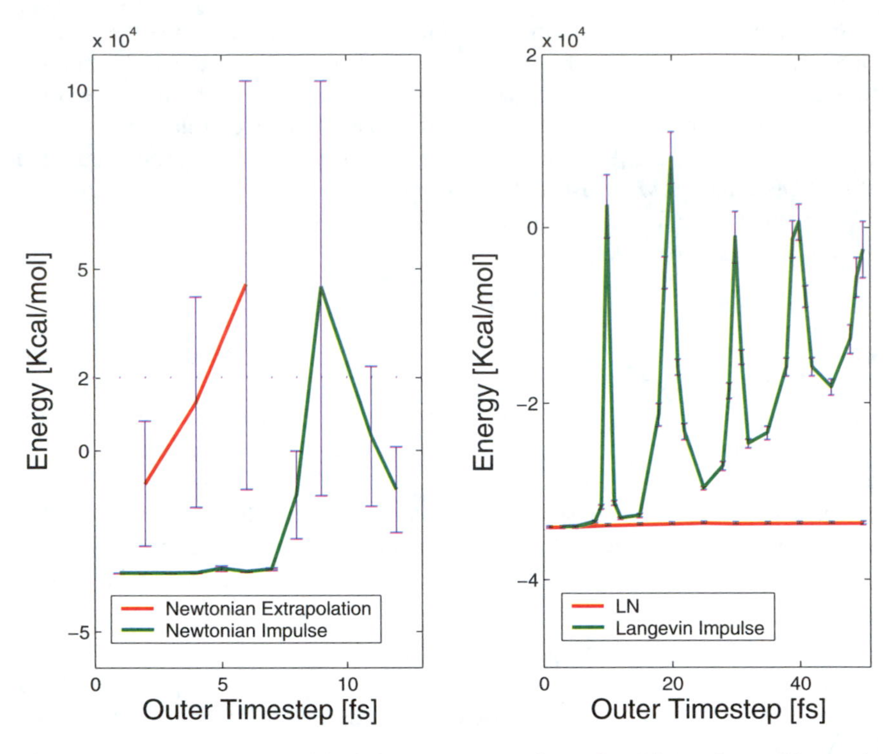

Figure 13.7. Energy means and deviations as computed over 5-ps Newtonian and Langevin impulse and extrapolative force splitting schemes for the protein BPTI with the damping constant $\gamma = 20\ \text{ps}^{-1}$, as functions of the outer timestep for a fixed inner timestep of 0.5 fs and medium timestep of 1 fs [609].

superscript used for R has little significance, as the random force is chosen independent at each step.

When the Dirac delta function of eq. (13.28) is discretized, $\delta(t-t')$ is replaced by $\delta_{nm}/\Delta t$.

The BBK method is only appropriate for the small-γ regime. A more versatile algorithm was derived in [729]. See also [730] for a related Langevin discretization.

13.4.5 The LN Method

The idea of combining force splitting via extrapolation with Langevin dynamics can alleviate severe resonance effects — as discussed in the MTS section (and shown in Figure 13.5 for a simple harmonic model) — and allow larger outer timesteps to be used than impulse-based MTS schemes [876, 868, 885].

The LN algorithm based on this combination (see background in Box 13.1) is sketched in Figure 13.8 in the same notation used for the MTS schemes. The

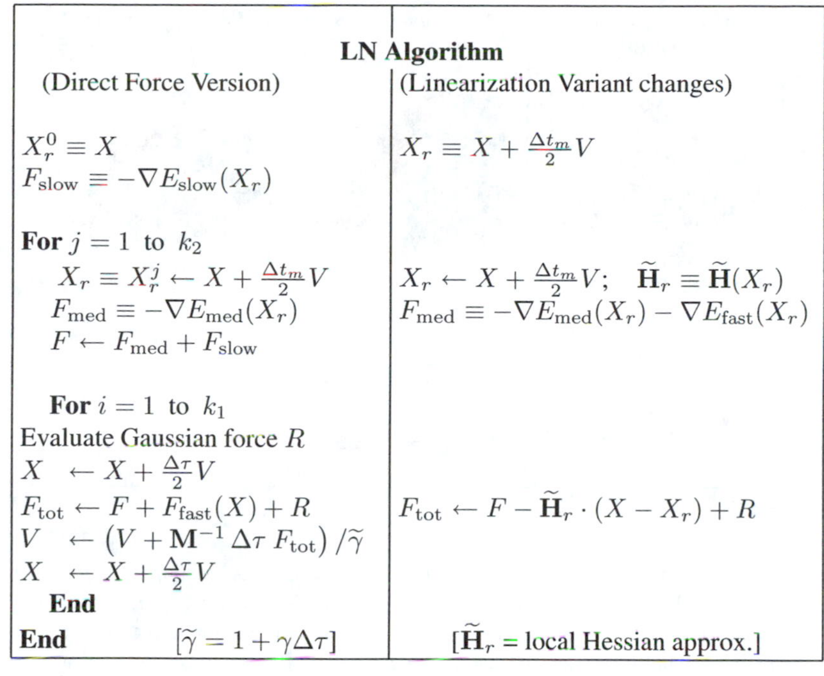

Figure 13.8. Algorithmic sketch of the LN scheme [876] for Langevin dynamics by extrapolative force-splitting based on position Verlet. The version on the left uses direct fast-force evaluations, while the reference version on the right (only alternative statements shown) uses linearization to approximate the fast forces over a Δt_m interval. The latter requires a local Hessian formulation, $\widetilde{\mathbf{H}}_r = \widetilde{\mathbf{H}}(X_r)$, at point X_r every time the medium forces are evaluated. See Box 13.1 for background details.

direct-force algorithms on the left forms the basis for the algorithms implemented in the CHARMM [876] and AMBER [667, 650, 671] programs.

Note that LN is based on position Verlet rather than velocity Verlet. If a constrained dynamics formulation is used (e.g., SHAKE), this splitting version requires on average two SHAKE iterations per inner loop rather than the one required by velocity Verlet. However, we have found stability advantages in practice for the position Verlet scheme over velocity Verlet in the unusual limit of moderate to large timesteps and large timescale separations [877].

Resonance Alleviation

Applications to the solvated model of the protein BPTI in Figure 13.7 show how the Langevin extrapolative treatment alleviates severe resonances present in impulse treatments for both Newtonian and Langevin dynamics. Newtonian extrapolation yields large energy increases as the timestep increases. Both Newtonian and Langevin impulse splitting exhibit resonant spikes: beyond half the

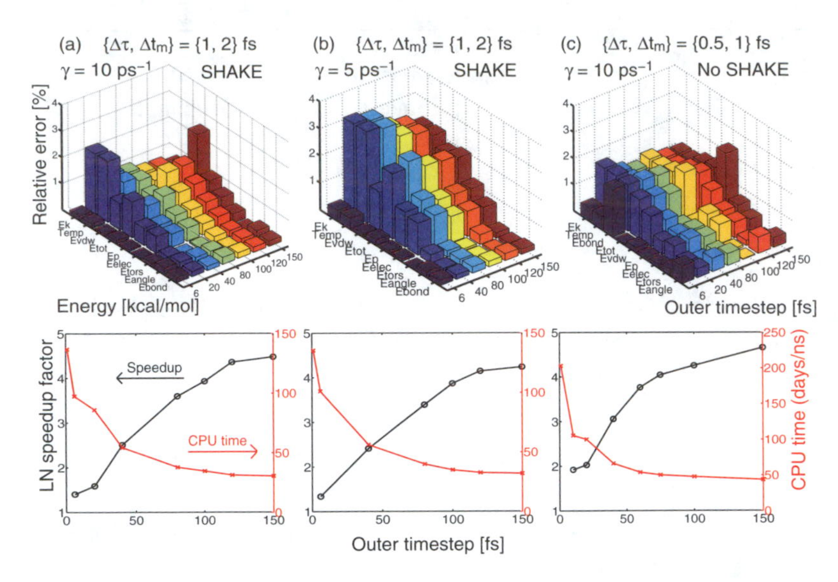

Figure 13.9. 'Manhattan plots' for a solvated polymerase/DNA system showing the LN relative errors in different energy components (bond, bond angle, torsion, electrostatic, potential, total, van der Waals) and the kinetic temperature as a function of the outer timestep Δt compared to single-timestep Langevin trajectories at the same $\Delta\tau$, as well as corresponding speedups (scale at left) and CPU times per nanosecond (scale at right). 'SHAKE' refers to constraining all bonds involving hydrogen. The data are based on simulations of length 12 ps started from the intermediate polymerase/DNA complex [925, 656], pictured in Figure 13.10. See also Box 13.1.

fastest period for Newtonian impulse, and beyond the fastest period for Langevin impulse splitting. Unfortunately, once such resonance is encountered, good behavior cannot be restored after the first resonant timestep. The LN combination, in contrast, does not exhibit resonant spikes for this value of damping constant, 20 ps^{-1}. It stabilizes the energy growth seen for Newtonian extrapolation. See also [656] for detailed performance evaluations for a large protein/DNA complex.

The exchange of Hamiltonian dynamics for stochastic dynamics can guarantee better numerical behavior, but the resulting dynamics are altered. Stochastic dynamics are, however, suitable for many thermodynamic and sampling questions. Small-timestep dynamics simulations can always be performed in tandem once an interesting region of conformation space is identified.

Skeel and coworkers have also recently adopted the stochastic coupling [909, 910] in the context of the mollified impulse method [67] to extend the timestep beyond half the fast period. Because their γ is yet smaller, the agreement with Hamiltonian dynamics is better, but the largest possible outer timesteps and hence speedup factors are reduced.

Box 13.1: LN Background and Assessment

Background. LN arose fortuitously [885] upon analysis of the range of harmonic validity of the Langevin/Normal-mode method LIN [922, 923]. Essentially, in LIN the equations of motion are linearized using an approximate Hessian and solved for the harmonic component of the motion; an implicit integration step with a large timestep then resolves the residual motion (see subsection on implicit methods). Approaches based on linearization of the equations of motion have been attempted for MD [926, 927, 928, 929], but computational issues ruled out general macromolecular applications. Indeed, the LIN method is stable over large timesteps such as 15 fs but the speedup is modest due to the cost of the minimization subproblem involved in the implicit discretization. The discarding of LIN's implicit-discretization phase — while reducing the frequency of the linearization — in combination with a force splitting strategy forms the basis of the LN approach [876].

Performance: Energetics and Speedup. Performance of MTS schemes can be analyzed by 'Manhattan plots', as shown in Figure 13.9 for a large polymerase/DNA system of 41,973 atoms (illustrated in Figure 13.10) [656]; that is, differences of mean energy components are reported as a function of the outer timestep Δt relative to STS Langevin simulations. For three LN protocols — using different combinations of $\Delta\tau$, Δt_m, γ, and bond constraints (SHAKE on or off) — these plots, along with corresponding CPU times and speedup, show that the first protocol has the optimal combination of low relative error in all energy components (below 3%) and low CPU time per physical time unit. The computational speedup factor is 4 or more in all cases.

Performance: Dynamics. The assignment of the Langevin parameter γ in the LN scheme ensures numerical stability on one hand and minimizes the perturbations to Hamiltonian dynamics on the other; we have used $\gamma = 10$ ps^{-1} or smaller in biomolecular simulations. To assess the effect of γ of dynamic properties, the protocol-sensitive spectral density functions computed from various trajectories can be analyzed (see Figure 13.11 caption). The densities for solvated BPTI in Figure 13.11 show how the characteristic frequencies can be more closely approximated as γ is decreased; the densities for the large polymerase system in Figure 13.12 show, in addition, the good agreement between the STS Langevin and LN-computed frequencies for the same γ. This emphasizes the success of MTS integrators as long as the inner timestep is small. (Recall another illustration of this point for butane in Figure 12.5, where the average butane end-to-end distance is shown for STS versus MTS protocols).

Detailed comparisons of the evolution of various geometric variables (Figure 13.13) reflect the agreement between LN and the reference Langevin simulation as well [656]. As expected, individual trajectories diverge, but the angular fluctuations are all in reasonable ranges. The flexibility of the DNA backbone angles is expected at the base pair near the kink induced by the polymerase [656].

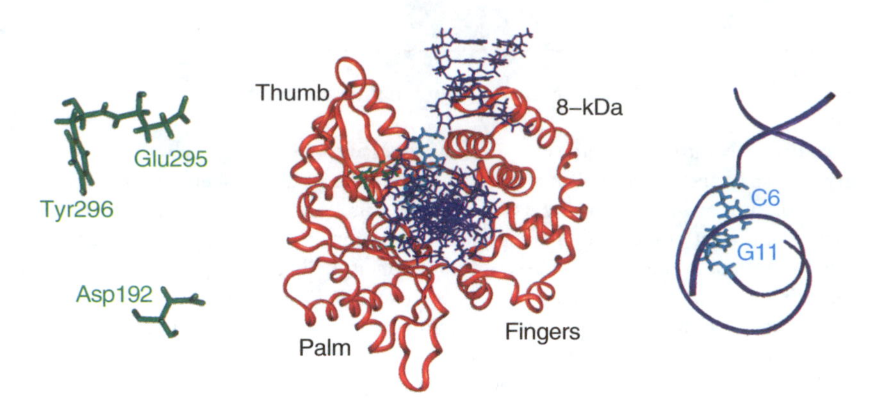

Figure 13.10. Solvated polymerase/DNA system used for evaluating LN [925, 656].

Testing and Application

Results have shown that good parameter choices for a 3-class LN scheme are $\Delta\tau$ = 0.5 fs, Δt_m = 1 fs, and Δt up to 150 fs. If constrained dynamics for the light atom bonds are used, the inner timestep can be increased to 1 fs and the medium timestep to around 2 fs. Various biomolecular applications have shown good agreement of LN trajectories to small-timestep Langevin simulations and significant speedup factors [876, 334, 656]. See Box 13.1 for performance assessment of energetics, dynamics, and speedup on a biomolecule.

Recent work has tailored elements of the LN integrator to particle-mesh Ewald protocols [667, 877, 650, 908]. Challenges remain regarding the most effective MTS splitting procedures for Ewald formulations, given the fast terms present in the Ewald reciprocal component [669, 857, 930, 667, 670, 650, 671] and the numerical artifact stemming from subtraction of the term accounting for excluded-nonbonded atoms pairs [668, 650, 671].

13.5 Brownian Dynamics (BD)

13.5.1 Brownian Motion

The mathematical theory of Brownian motion is rich and subtle, involving high-level physics and mathematics. Important contributors to the theory include Einstein (who explained Brownian motion in 1905) and Planck. Here we present the BD framework as an extension to the Langevin model presented above and focus on numerical algorithms for Brownian dynamics. See statistical mechanics texts, such as [29], for more comprehensive presentations.

The term *Brownian* is credited to the botanist Robert Brown, who in 1827 observed that fine particles — like pollen grains, dust, and soot — immersed in a fluid undergo a continuous irregular motion due to collisions of the particles with

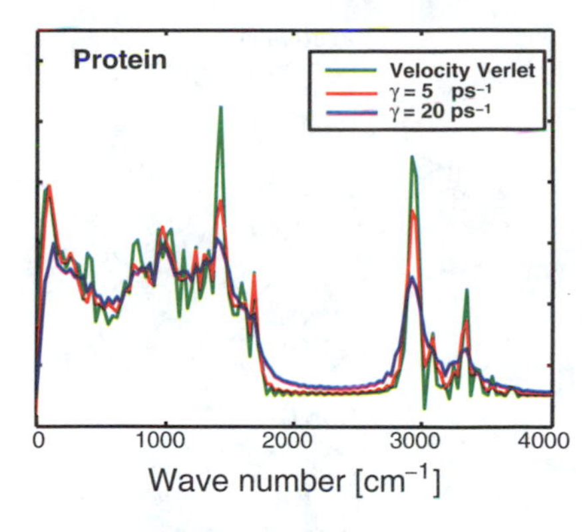

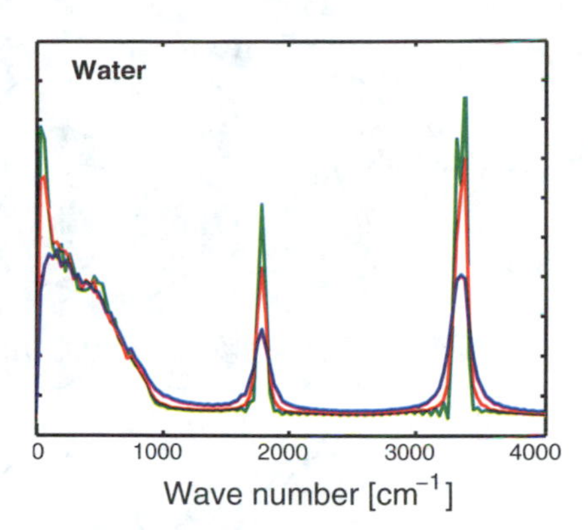

Figure 13.11. Spectral density functions calculated by Verlet and LN over 5 ps runs for a solvated system of the protein BPTI (protein and water frequencies shown separately) at three γ values: 0 (by the velocity Verlet scheme), 5 and 20 ps^{-1} by the LN scheme [609]. The functions are computed by Fourier transforming the velocity autocorrelation time series for each atom in the system to obtain a power spectrum for each atom. These spectra are averaged over the water and biomolecule atoms separately for global characterization of the motion. See [609] for the detailed protocol. Note that the characteristic frequencies obtained by this procedure reflect the force field constants rather than physical frequencies *per se*.

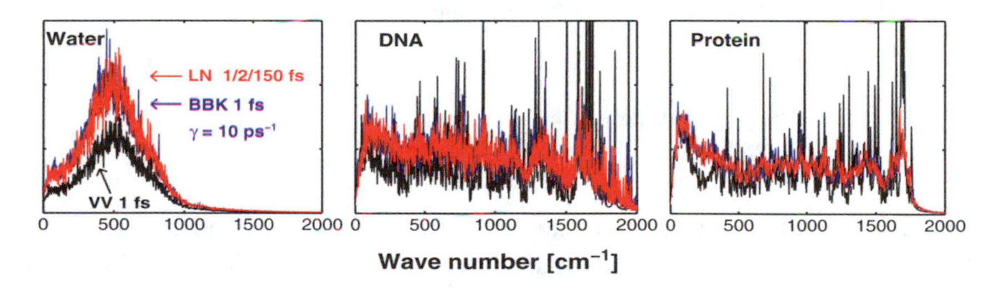

Figure 13.12. Spectral density functions calculated by the STS Newtonian (velocity Verlet) and STS Langevin (BBK, $\gamma = 10$ ps^{-1}) schemes versus the MTS LN scheme at $\gamma = 10$ ps^{-1} (triple timestep protocol 1/2/150 fs) for a solvated polymerase/DNA system simulated over 12 ps and sampled every 2 fs. Spectral densities for the protein, DNA, and water atoms are shown separately. See also caption to Figure 13.11.

the solvent molecules. Dutch physician Jan Ingenhausz was actually the first, in 1785, to report this motion for powdered charcoal on an alcohol surface. The effective force on such particles originates from friction, as governed by Stokes' law, and a fluctuating random force.

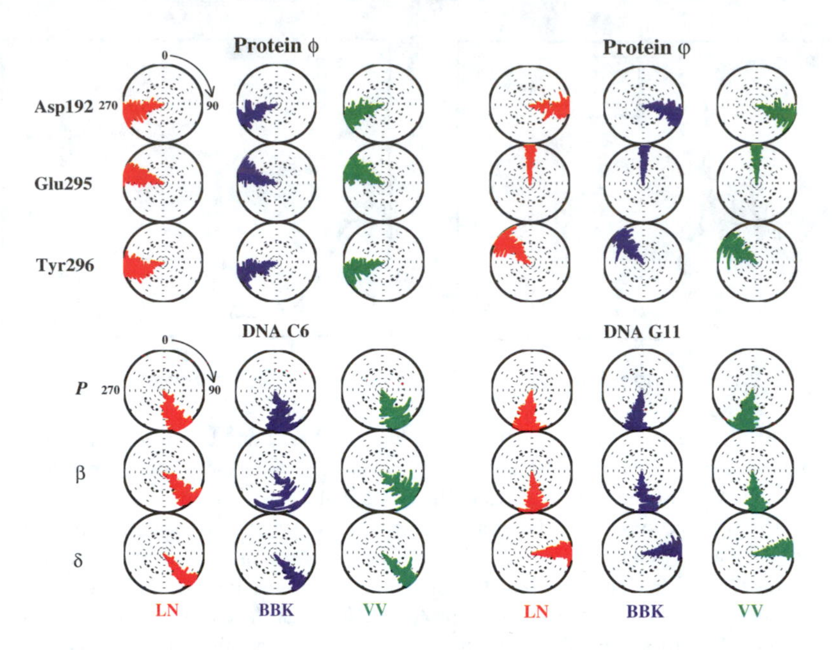

Figure 13.13. Comparisons of the evolution of representative dihedral angles (from protein ϕ and ψ and DNA sugar and backbone near one CG base pair) over 12 ps for the STS velocity Verlet, STS Langevin (BBK), and the MTS LN scheme for a solvated polymerase/DNA system. The LN protocol used is $\Delta\tau = 1$ fs, $\Delta t_m = 2$ fs, and $\Delta t = 150$ fs, with SHAKE. The polar vector coordinates correspond to time and angle. See Figure 13.10 for location of residues.

13.5.2 Brownian Framework

Generalized Friction

Generalized frictional interactions among the particles can be incorporated into the Langevin equation introduced in Section 13.4 using a friction tensor **Z**. This matrix replaces the single parameter γ, so as to describe the action of the solvent by:

$$\mathbf{M}\ddot{X}(t) = -\nabla E(X(t)) - \mathbf{Z}\dot{X}(t) + R(t), \tag{13.31}$$

where the mean and covariance of the random force R are given by:

$$\langle R(t) \rangle \quad = \quad 0, \qquad \langle R(t)R(t')^T \rangle \; = \; 2k_B \mathrm{T}\mathbf{Z}\,\delta(t-t')\,. \tag{13.32}$$

The above relation is based on the fluctuation/dissipation theorem, a fundamental result showing how friction is related to fluctuations of the random force, assuming the Brownian particle is randomly moving about thermal equilibrium. This description ensures that the ensembles of the trajectories generated from eq. (13.31) are governed by the Fokker-Planck equation, a partial-differential equation describing the density function in phase space of a particle undergoing diffusive motion.

The random force R is white noise and has no natural timescale. Thus, the inertial relaxation times given by the inverses of the eigenvalues of the matrix $\mathbf{M}^{-1}\,\mathbf{Z}$ define the characteristic timescale of the thermal motion in eq. (13.31).

Neglect of Inertia

When the inertial relaxation times are short compared to the timescale of interest, it is often possible to ignore inertia in the governing equation, that is, discard the momentum variables, assuming $\mathbf{M}\ddot{X}(t) = 0$. From eq. (13.31), we have:

$$\dot{X}(t) = -\mathbf{Z}^{-1}\,\nabla E(X(t)) + \mathbf{Z}^{-1}\,R(t)\,, \tag{13.33}$$

and this can be written in the *Brownian dynamics* form:

$$\dot{X}(t) = -\frac{\mathbf{D}}{k_B \mathrm{T}}\,\nabla E(X(t)) + R_B(t)\,; \tag{13.34}$$

here $\mathbf{D}$ is the diffusion tensor

$$\mathbf{D} = k_B \mathrm{T} \mathbf{Z}^{-1}\,, \tag{13.35}$$

and the mean and covariance of the random force R_B depend on $\mathbf{D}$ as:

$$\langle R_B(t)\rangle \;=\; 0, \qquad \langle R_B(t) R_B(t')^T\rangle \;=\; 2\mathbf{D}\;\delta(t-t')\,. \tag{13.36}$$

Thus, solvent effects are sufficiently large to make inertial forces negligible, and the motion is overall *Brownian* and random in character. This description is effective for very large, dense systems whose conformations in solution are continuously and significantly altered by the fluid flow in their environment.

Transport Properties

Brownian theory allows us to determine average behavior, such as transport properties, for systems governed by such diffusional motion. For example, a molecule modeled as a free Brownian particle that moves a net distance x over a time interval Δt has an expected mean square distance $\langle x^2 \rangle$ (analogous to the mean square end-to-end distance in a polymer chain) proportional to Δt: $\langle x^2 \rangle \propto \Delta t$. Einstein showed that the proportionality constant is $2D$, where D is the diffusion coefficient:

$$\langle x^2 \rangle = 2D\,\Delta t\,.$$

Algorithms

An appropriate simulation algorithm for following Brownian motion can be defined by assuming that the timescales for momentum and position relaxation are well separated, with the former occurring much faster. BD algorithms then prescribe recursion recipes that displace each current position by a random force — similar in flavor to the Langevin random force — and additional terms that depend on the diffusion tensor.

It is not always possible to neglect inertial contributions. For example, it was shown that inertial contributions can affect long-time processes in long DNA due to mode coupling [722]. An "inertial BD" algorithm termed IBD has been developed, tested on simple systems [734], and applied to bead models of long DNA [722]. IBD more accurately approximates long-time kinetic processes that occur within equilibrium ensembles, as long as the timestep is not too small.

In practice, the IBD scheme adds a mass-dependent correction term to the usual BD propagation scheme (see below). Though IBD has an additional computational cost of a factor of two over BD, the computational complexity is the same as for the usual BD scheme based on the Cholesky factorization (consult a numerical methods textbook like [896] for the Cholesky factorization, and see Box 13.4 and below for IBD details).

13.5.3 General Propagation Framework

In general, larger timesteps are used for Brownian dynamics simulations than for molecular and Langevin dynamics simulations. As a first example, consider a free particle in one dimension whose diffusion constant D is (by definition) the mean square displacement divided by $2t$:

$$2tD \approx \langle |x(t) - x(0)|^2 \rangle$$

over sufficiently long times. This diffusional motion can be simulated by the simple scheme:

$$x^{n+1} = x^n + R^n \,, \tag{13.37}$$

where the random force R is related to D by:

$$\langle R^n \rangle = 0, \qquad \langle (R^n)^2 \rangle = 2D\Delta t \,. \tag{13.38}$$

This propagation scheme reproduces D over long times t; see homework assignment 12.

Ermak/McCammon

Ermak and McCammon [733] derived a basic BD propagation scheme from the generalized Langevin equation in the high-friction limit, where it is assumed that momentum relaxation occurs much faster than position relaxation. For a three-dimensional particle diffusing relative to other particles and subject to a force $F(X)$, the derived BD scheme becomes:

$$X^{n+1} = X^n + \frac{\Delta t}{k_B \mathrm{T}} \mathbf{D}^n F(X^n) + R^n \,, \tag{13.39}$$

$$\langle R^n \rangle = 0, \qquad \langle R^n (R^m)^T \rangle = 2\mathbf{D}^n \Delta t \, \delta_{nm} \,. \tag{13.40}$$

For reference, the IBD algorithm developed in [734] has the following mass-dependent correction term:

$$X^{n+1} = X^n + \frac{\Delta t}{k_B\,\mathrm{T}}\,\mathbf{D}^n\,F(X^n) + \frac{1}{(k_B\,\mathrm{T})^2}\,\mathbf{D}^n\mathbf{M}\mathbf{D}^n\,[F(X^{n-1}) - F(X^n)] + R^n\,, \tag{13.41}$$

with the same random-force properties for R as above (eq. (13.40)).

13.5.4 Hydrodynamic Interactions

Various approaches have been developed to define the diffusion tensor $\mathbf{D}$ in these equations. Recall from Subsection 13.2.5 that the simple Langevin formulation offers only a simple isotropic description of viscous effects: frictional effects are taken to be isotropic, so γ is a scalar. Furthermore, the random force R acts independently on each particle, thereby ignoring changes in force due to the solvent-mediated dynamic interparticle interactions.

Tensor $\mathbf{T}$

Account of these modified particle interactions due to solvent structure requires formulation of a configuration and momenta-dependent *hydrodynamic tensor* $\mathbf{T}$ to express the instantaneous effective solute force. This is because each atom's force changes the solvent flow, and this in turn affects forces on other atoms through the frictional forces affecting them. The tensor $\mathbf{T}$ is related to $\mathbf{D}$ by the $k_B\mathrm{T}$ factor (eq. (13.43) below).

Various expressions for hydrodynamic tensors are derived from hydrodynamic theories (due to Kirkwood and Riseman, Oseen, Burgers, and others) so as to describe the effective velocity perturbation term, ΔV^n, as the product of the tensor and force, $\mathbf{T}(X^n)\,F(X^n)$. Typically, the dependence of the tensor on momenta is ignored, and only configuration-dependent effects are considered. The governing BD equations that include hydrodynamic interactions then become:

$$\dot{X}(t) = \mathbf{T}(X)\,F(X(t)) + k_B\,\mathrm{T}\,\nabla_X \cdot \mathbf{T}(X) + R_B(t)\,, \tag{13.42}$$

where the gradient term $\nabla_X \cdot \mathbf{T}(X)$ is the vector with components i given by $\sum_j \partial \mathbf{T}_{ij}(r_{ij})/\partial X_j$.

The Oseen and Rotne-Prager tensors expressed below have the favorable property that the derivative of the tensor with respect to the displacement vector is zero, a term omitted in the BD propagation scheme of eq. (13.39) [733]. These tensors have been derived for polymer systems modeled as hydrodynamic (spherical) beads of radius a immersed in a fluid of viscosity η [719]. For a polymer system of N beads, the diffusion tensor $\mathbf{D}$ is then a configuration-dependent, symmetric $3N \times 3N$ matrix in which each 3×3 subblock $\overline{\mathbf{D}}_{ij}$ is defined as:

$$\mathbf{D} = \begin{pmatrix} \overline{\mathbf{D}}_{11} & \overline{\mathbf{D}}_{12} & . & . & . & \overline{\mathbf{D}}_{1N} \\ \overline{\mathbf{D}}_{21} & \overline{\mathbf{D}}_{22} & . & . & . & \overline{\mathbf{D}}_{2N} \\ . & . & . & . & . & \\ . & . & . & . & . & \\ \overline{\mathbf{D}}_{N1} & \overline{\mathbf{D}}_{N2} & . & . & . & \overline{\mathbf{D}}_{NN} \end{pmatrix} ; \qquad \overline{\mathbf{D}}_{ij} = (k_B \mathrm{T})\, \mathbf{T}_{ij} \,. \qquad (13.43)$$

Box 13.2 outlines the principles on the basis of which the tensor approximations are derived.

The frictional coefficient ζ introduced in the derivation of Box 13.3 can be related to the Langevin parameter γ by: $\zeta = \gamma m$ if all beads have the same mass m. In fact, we can approximate roughly the reduction of the effective total friction for a polymer due to hydrodynamic effects using the Kirkwood-Riseman (KR) approximation for bead hydrodynamics based on the Oseen tensor [715]. See Box 13.3 for this approximation.

Box 13.2: Derivation of Tensor Expressions

For macromolecules in solution, the velocity of the solvent at a polymer segment differs from the velocity of the bulk solvent since the polymer alters the local flow. The solvent motion can thus be considered as an applied force. Let u_i be the velocity of a polymer particle (e.g., bead), v_i be the velocity of the solvent at that particle, and v^0 be the velocity of the bulk solvent. Then $v_i - u_i$ represents the net velocity, proportional to the effective force on the solvent at particle i:

$$F_i = -\zeta\,(v_i - u_i)\,, \qquad (13.44)$$

where ζ is the translational friction coefficient of a particle. For example, $\zeta = 6\pi\eta a$ for a bead of radius a and solvent viscosity η; assume for simplicity that ζ is the same for all beads. The velocity at the ith bead can be written as v^0 plus some perturbation due to all other segments:

$$v_i = v^0 + \sum_{j \neq i} \mathbf{T}_{ij} F_j \,. \qquad (13.45)$$

By inserting eq. (13.45) into eq. (13.44), we have:

$$F_i = -\zeta \left(v^0 + \sum_{j \neq i} \mathbf{T}_{ij} F_j - u_i \right) = -\zeta\,(v^0 - u_i) - \zeta \sum_{j \neq i} \mathbf{T}_{ij} F_j \,. \qquad (13.46)$$

It follows that

$$F_i + \zeta \sum_{j \neq i} \mathbf{T}_{ij} F_j = -\zeta\,(u_i - v^0)\,. \qquad (13.47)$$

This equation is the basis for deriving the hydrodynamic tensor $\mathbf{T}$, an equation that must be solved to find the net force for a given system.

Oseen Tensor

One of the simplest hydrodynamic tensors is the Oseen tensor, developed by Oseen and Burgers (circa 1930) for the hydrodynamic interaction between a point of friction and a fluid, defined as:

$$\mathbf{T}_{ij} = \begin{cases} \frac{1}{6\pi\eta a}\,\mathbf{I}_{3\times 3} & i = j \\ \frac{1}{8\pi\eta r_{ij}}\left[\mathbf{I}_{3\times 3} + \frac{\mathbf{r}_{ij}\,\mathbf{r}_{ij}^T}{r_{ij}^2}\right] & i \neq j \end{cases} . \tag{13.48}$$

Here $\mathbf{r}_{ij}$ denotes an interbead distance vector, and r_{ij} is the corresponding scalar distance. The expression for $i \neq j$ is the first term in an expansion corresponding to the pair diffusion for an incompressible fluid in inverse powers of r_{ij}.

Rotne-Prager Tensor

Another term in the expansion produces the Rotne-Prager hydrodynamic tensor $\mathbf{T}_{ij}$. For nonoverlapping beads ($r_{ij} > 2a$), this matrix is defined as [931]:

$$\mathbf{T}_{ij} = \begin{cases} \frac{1}{6\pi\eta a}\,\mathbf{I}_{3\times 3} & i = j \\ \frac{1}{8\pi\eta r_{ij}}\left[\left(\mathbf{I}_{3\times 3} + \frac{\mathbf{r}_{ij}\,\mathbf{r}_{ij}^T}{r_{ij}^2}\right) + \frac{2a^2}{r_{ij}^2}\left(\frac{\mathbf{I}_{3\times 3}}{3} - \frac{\mathbf{r}_{ij}\,\mathbf{r}_{ij}^T}{r_{ij}^2}\right)\right] & i \neq j \end{cases} \tag{13.49}$$

Note that both tensors are a generalization of the scalar Langevin friction term expressed in terms of the γ set according to Stokes' law.

Box 13.3: Approximation of The Effective BD Friction Coefficient

In the absence of hydrodynamic interactions (the "free-draining" limit), effective friction constant f_T is the sum of the friction constants of the N_b individual beads, i.e., $f_T = N_b\zeta = N_b\,(\gamma m)$. The Kirkwood-Riseman (KR) equation for f_T, which approximately incorporates hydrodynamic interactions, can be written as:

$$f_T = \frac{N_b\zeta}{1 + \frac{\zeta}{6\pi\eta N_b}\sum_{i\neq j}\langle\frac{1}{r_{ij}}\rangle}, \tag{13.50}$$

where $\langle 1/r_{ij}\rangle$ is the mean inverse distance between beads i and j averaged over an ensemble of configurations. Thus, the KR approximation is based on a preaveraged Oseen configuration tensor [715]. If we use Stokes' law to set ζ as $6\pi\eta a$, we obtain:

$$f_T = \frac{N_b\zeta}{1 + \frac{a}{N_b}\sum_{i\neq j}\langle\frac{1}{r_{ij}}\rangle} . \tag{13.51}$$

Thus, the denominator of this expression reflects the reduction by hydrodynamics of the effective friction from the reference value of $N_b\,\zeta$. This approximation has been used to estimate hydrodynamic effects for long DNA [732].

13.5.5 BD Propagation Scheme: Cholesky vs. Chebyshev Approximation

Once the matrix $\mathbf{D}$ is formulated at each step n of the BD algorithm from the hydrodynamic tensor (eq. (13.43)) the random force R must be set to satisfy eq. (13.40). *This is actually the computationally intensive part of the BD propagation scheme when hydrodynamics interactions are considered.*

The traditional way, based on a Cholesky decomposition of $\mathbf{D}$ (see [896] and Box 13.4) increases in computational time as the cube of the system size, since a Cholesky factorization of $\mathbf{D}$ is required at every step. The alternative approach based on Chebyshev polynomials proposed by Marshall Fixman [735] only increases in complexity roughly with the square of the number of variables. See Box 13.4 for details.

Essentially, both methods for determining the random force R first compute a $3N$-vector Z from a Gaussian distribution so that

$$\langle Z_i \rangle = 0, \qquad \langle Z_i \, Z_j \rangle = 2\Delta t \delta_{ij} \, ;$$

(the indices run from 1 to $3N$). The second step is different. In the Cholesky-based approach, R is computed as $R^n = \mathbf{S}^n Z$, where $\mathbf{S}$ is the Cholesky factor of $\mathbf{D}$. In the Chebyshev procedure, we compute the random force vector instead as $R^n = \widetilde{\mathbf{S}}^n Z$, where $\widetilde{\mathbf{S}}$ is the square root matrix of $\mathbf{D}$, and the product is computed as a series of Chebyshev polynomials (see Box 13.4).

A recent application of the Chebyshev approach for computing the Brownian random force in simulations of long DNA demonstrates computational savings for large systems [269]. Figure 13.14 compares the percentage CPU work required for the systematic versus hydrodynamic forces in BD simulations of DNA modeled as macroscopic hydrodynamic beads using the Cholesky versus Chebyshev approaches. The figure also shows the total CPU time required to simulate 10 ms in both cases.

We see that for the largest, 12,000 base-pair DNA system studied, the Chebyshev approach is twice as fast. The overall speedup is not more dramatic since the system size (in terms of beads) is relatively small.

Perhaps more significantly, the Chebyshev alternative to the Cholesky factorization also opens the door to other BD protocols (such as the recent inertial BD idea [734, 722]) and is crucial to BD studies of finer macroscopic models of DNA, such as residue-based rather than bead-based.

Finally, note that once the BD computational bottleneck is reduced to electrostatics and hydrodynamics (roughly $\mathcal{O}(N^2)$ for both), fast electrostatic methods, such as described in Chapter 9, will help accelerate such BD computations further.

Box 13.4: BD Implementation: Cholesky vs. Chebyshev Approach

In the Cholesky approach to computing the random force to satisfy the properties of eq. (13.40), the Cholesky decomposition of the diffusion tensor **D** is determined:

$$\mathbf{D} = \mathbf{S}\mathbf{S}^T ,$$

where $\mathbf{S}$ is a lower triangular matrix. The desired vector R^n is then computed from the following matrix/vector product:

$$R^n = \mathbf{S}^n Z . \tag{13.52}$$

It can be easily shown that this R^n satisfies $\langle R^n (R^m)^T \rangle = 2\mathbf{D}^n \Delta t \, \delta_{nm}$, as desired.

Note that the Cholesky decomposition of $\mathbf{D}$ can be written in closed form as the following procedure for determining the elements of $\mathbf{S}$, s_{ij}, row by row (i.e., $s_{11}, s_{21}, s_{22}, s_{31}, \ldots s_{3N\,3N}$):

$$s_{ij} = \begin{array}{ll} \left(D_{ii} - \sum_{k=1}^{i-1} s_{ik}^2\right)^{1/2} & \\ \left(D_{ij} - \sum_{k=1}^{j-1} s_{ik}\, s_{jk}\right) / s_{jj} & i > j \end{array} . \tag{13.53}$$

An advantage of the Cholesky approach is that the factors can be reused. Thus, it is possible to reduce the overall cost of the BD simulation by less-frequent updating of the hydrodynamic tensor; parallelization of some of the numerical linear algebra tasks can also further accelerate the total computational time [512, 499].

The Chebyshev alternative was proposed over a decade ago by Marshall Fixman [735] and recently applied [269, 932, 933]. Instead of a Cholesky decomposition $\mathbf{D} = \mathbf{S}\mathbf{S}^T$, Fixman suggests to calculate R from the relation

$$R^n = \widetilde{\mathbf{S}}^n Z, \tag{13.54}$$

rather than eq. (13.52) where $\widetilde{\mathbf{S}}$ is the *square root matrix* of $\mathbf{D}$:

$$\mathbf{D} = \widetilde{\mathbf{S}}^2 .$$

This idea is based on expanding the matrix/vector product $\widetilde{\mathbf{S}}Z$ as a series of Chebyshev polynomials which approximate the function $\sqrt{x}$ on some given interval believed to contain eigenvalues. This calculation requires about $\mathcal{O}(N^2)$ operations, thus reflecting substantial savings with respect to the Cholesky approach.

The details of computing the expansion of R in this Chebyshev approximation were recently given in the appendix of [269] in an application of BD to supercoiled DNA. See also [932, 933] for other applications. The Chebyshev implementation requires determining bounds on the maximum and minimum eigenvalues of $\mathbf{D}^n$ and then computing an expansion of desired order (according to some error criterion) of R in terms of polynomials, with coefficients determined for the square-root function.

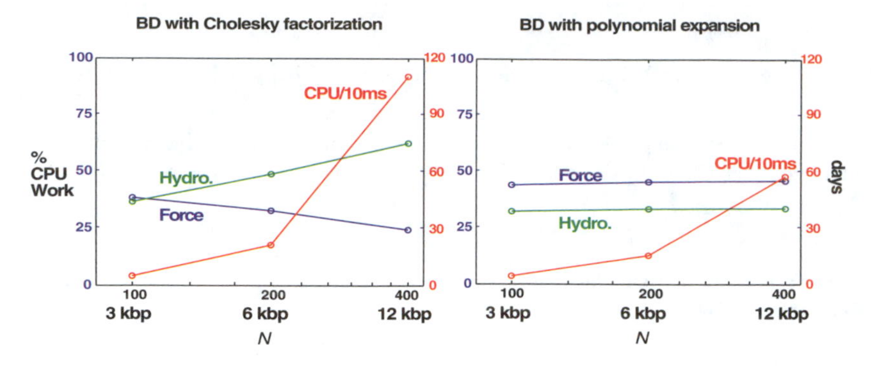

Figure 13.14. Computational complexity for BD schemes with hydrodynamics as obtained for a bead model of DNA using Cholesky and Chebyshev approaches for computing the Brownian random force. The fraction of CPU associated with the hydrodynamics and systematic force calculations is shown in each case (% scales at left) as a function of system size, with corresponding total number of days required to compute a 10 ms trajectory (shown with the scale at right). Computations are reported on an SGI Origin 2000 machine with 300 MHz R12000 processor [269].

13.6 Implicit Integration

A reasonable approach for allowing long timesteps is the use of *implicit* integration schemes [934]. These methods are designed specifically for problems with disparate timescales where explicit methods do not usually perform well, such as chemical reactions [935]. The integration formulas of implicit methods are designed to increase the range of stability for the difference equation.

Typically, implicit methods work well when the fast components are rapidly decaying and largely decoupled from the slower components. This is not the case for biomolecules, where vibrational modes are intricately coupled and the fast motions are oscillatory.

13.6.1 *Implicit vs. Explicit Euler*

To illustrate the use of implicit methods, we consider a simple example for the solution of the general differential equation

$$\dot{Y}(t) = \mathcal{F}[Y(t)]\,, \tag{13.55}$$

where $Y = (X, V)$ is a vector and $\mathcal{F}$ is a vector function. The Newtonian system of equations (12.7, 12.8) can be written in this form with the composite vector $Y = (X, \dot{X})$ and the function

$$\mathcal{F}[Y(t)] = [V(t), -\mathbf{M}^{-1}\nabla E\,(X(t))]\,.$$

The *implicit-Euler* (IE) scheme discretizes eq. (13.55) as

$$(Y^{n+1} - Y^n)/\Delta t = \mathcal{F}(Y^{n+1})\,, \tag{13.56}$$

while the *explicit Euler* (EE) analog has the different right-hand-side:

$$(Y^{n+1} - Y^n)/\Delta t = \mathcal{F}(Y^n)\,. \tag{13.57}$$

In the former case, the solution Y^{n+1} is derived implicitly, since $\mathcal{F}(Y^{n+1})$ is not known and must be solved by some procedure. In the latter, the solution can be explicitly formulated in terms of quantities known from previous steps.

Though the explicit approach is simpler to solve, it imposes a severe restriction on the timestep. Implicit schemes can yield much better stability behavior for 'stiff' problems [935]. This can be seen in Figure 13.15, where we solve the one-dimensional problem $y' = -ay, a > 0$, whose exact solution is $y = \exp(-at)$ by the implicit and explicit Euler methods.

It can be shown that the former gives the recursion relation

$$y^n = y^0/(1 + a\Delta t)^n\,,$$

while the latter gives

$$y^n = y^0(1 - a\Delta t)^n\,.$$

The IE scheme is always stable since $a\Delta t > 0$, but EE requires that $a\Delta t < 2$.

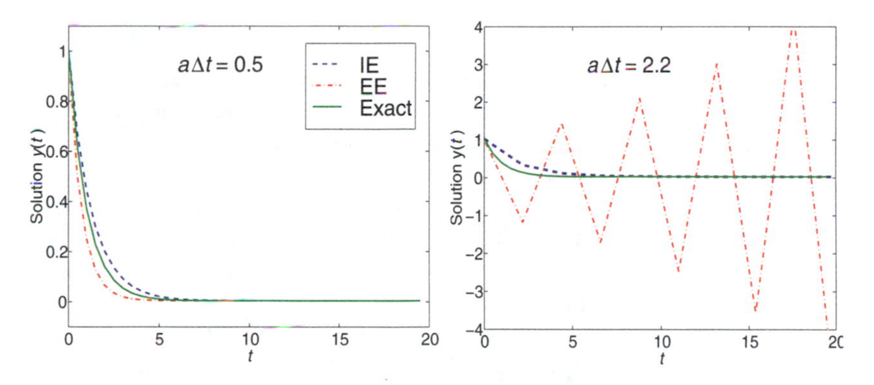

Figure 13.15. Implicit (IE) and explicit (EE) Euler solutions to the one-dimensional differential equation $y' = -ay$ with $a > 0$ are shown. They indicate stability of IE at small and large timesteps but growing instability of EE with time if the timestep is not sufficiently small. Here we use $y^0 = 1$ and two ratios for $a\Delta t$ (0.5, left, and 2.2, right). The corresponding IE and EE solutions (see text) yield, respectively, $y^n = 1/(1.5)^n$ and $y^n = (0.5)^n$ for the smaller timestep, and $y^n = 1/(3.2)^n$ and $y^n = (-1.2)^n$ for the larger timestep.

The experience with implicit methods in the context of biomolecular dynamics has been limited and rather disappointing (e.g., [936, 859]). The disappointment stems from three basic reasons: intrinsic damping, large computational demands, and resonance artifacts.

13.6.2 Intrinsic Damping

Despite the high stability of the IE discretization, the IE scheme is nonconservative and exhibits intrinsic damping. For system (13.27), the IE discretization is:

$$\begin{aligned} \mathbf{M}(V^{n+1} - V^n)/\Delta t &= F(X^{n+1}) - \gamma \mathbf{M} V^{n+1} + R^{n+1}, \\ (X^{n+1} - X^n)/\Delta t &= V^{n+1}. \end{aligned} \tag{13.58}$$

This scheme was introduced to molecular dynamics with the Langevin restoring force over a decade ago [920, 921], but its γ and timestep-dependent damping [920, 937] ruled out general applications to biomolecules. This was concluded from the tendency of the scheme to preserve 'local' structure (as measured by ice-like features of liquid water [228], the much slower decay of autocorrelation functions [938, 725], or lower effective temperature [725]).

Thus, at large timesteps, the IE method was numerically stable but behaved like an energy minimizer. Still, the IE scheme was incorporated as an ingredient in dynamic simulations of macroscopic separable models in which high-frequency modes are not relevant [482, 939, 350, 486], and in enhanced sampling schemes with additional mechanisms that counteract numerical damping [725, 918, 726].

13.6.3 Computational Time

In addition to the general damping problem, there is often no net computational advantage in implicit schemes. Each timestep, albeit longer, is much more expensive than an explicit timestep because a subproblem must be solved at each step. This subproblem arises because the solution for X^{n+1} and V^{n+1} cannot be solved in closed form (i.e., in terms of previous iterates) as in explicit schemes. Note that the $(n+1)$ iterates of V or X appear on both the right and left-hand sides of eq. (13.58). Solving for X^{n+1} requires solution of a nonlinear system.

It has been shown that this solution can be formulated as a nonlinear optimization problem and solved using efficient Newton schemes [920, 859], such as the truncated-Newton method introduced in Chapter 10. As fast as the minimization algorithm might be, the added cost of the minimization makes implicit schemes competitive with explicit, small-timestep integrators only at very large timesteps, a regime where reliability is questionable given the damping and resonance effects.

13.6.4 Resonance Artifacts

Resonance, the third problem associated with implicit methods, is relevant to implicit schemes since large timesteps may yield stable methods [899].

A detailed examination of resonance artifacts with the implicit midpoint (IM) was described in [899]. IM differs from IE in that it is symmetric and symplectic. It is also special in the sense that the transformation matrix for the model linear

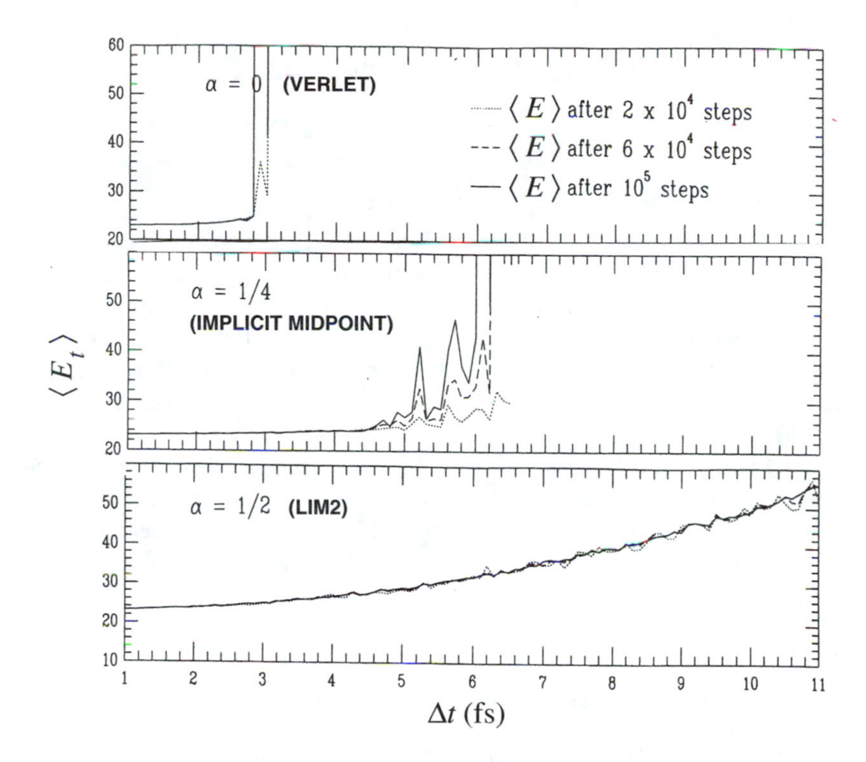

Figure 13.16. Mean total energy for a blocked alanine model as a function of the timestep, as obtained by Verlet, implicit-midpoint, and the symplectic implicit scheme LIM2 [859]; see [900].

problem is unitary, partitioning kinetic and potential-energy components identically. For these reasons, several researchers have explored the application of IM to long-timestep dynamics [940, 941, 942, 943].

IM Analysis

The IM discretization applied to system (13.27) is:

$$\begin{aligned} \mathbf{M}\,(V^{n+1}-V^n)/\Delta t &= \frac{-\nabla E}{2}\,(X^n+X^{n+1}) - \frac{\gamma \mathbf{M}}{2}\,(V^n+V^{n+1}) \,+\, R^n\,, \\ (X^{n+1}-X^n)/\Delta t &= (V^n+V^{n+1})/2\,. \end{aligned} \tag{13.59}$$

This nonlinear system can be solved, following [920], by obtaining X^{n+1} as a minimum of the "dynamics function" $\Phi(X)$:

$$\Phi(X) = \frac{\widetilde{\gamma}}{2}\,(X-X_0^n)^{\mathrm{T}}\,\mathbf{M}\,(X-X_0^n) + (\Delta t)^2\,E\left(\frac{X+X^n}{2}\right), \tag{13.60}$$

where

$$X_0^n = X^n + (\Delta t\, V^n)/\widetilde{\gamma} + \mathbf{M}^{-1}(\Delta t)^2 R^n/(2\widetilde{\gamma}),$$

and

$$\widetilde{\gamma} = 1 + (\gamma \Delta t)/2\,.$$

Hence, for IM applied to Newtonian dynamics $\widetilde{\gamma} = 1$, and the R^n term in X_0^n is absent. Following minimization of the IM dynamics function to obtain X^{n+1}, the new velocity, V^{n+1}, is obtained from the second equation of system (13.59).

An examination of the application of IM to MD showed good numerical properties (e.g., energy conservation and stability) for moderate timesteps, larger than Verlet [899, 859]. However integrator-induced *resonance* artifacts were found to be severe as the timestep approaches half the characteristic period [899] (see Box 13.5).

Figure 13.16 shows the mean total energy for Verlet and IM for a blocked alanine model [900] (misnamed 'dipeptide' in some works). Note that Verlet becomes unstable around 2.8 fs (recall Table 13.2 for reference linear-analysis values), and IM exhibits resonance peaks around 5.2 and 5.7 fs and instability around 6 fs, all near half the period of the fast motion.

Box 13.5: Analysis of Implicit-Midpoint (IM) Resonance

The linear analysis of [899] has shown that the *effective angular rotation* θ_{eff} of IM is:

$$\theta_{\text{eff}}^{\text{IM}} = \frac{2}{\Delta t}\tan^{-1}(\omega \Delta t/2)\,. \tag{13.61}$$

Recall that the corresponding value for the Verlet scheme (eq. (13.19)) is:

$$\theta_{\text{eff}}^{\text{verlet}} = 2\sin^{-1}(\omega \Delta t/2)\,.$$

As derived above, the resonant timestep of order $n{:}m$ can be obtained for the model linear problem by using the function for ω_{eff} in the expression

$$n \Delta t\, \omega_{\text{eff}} = 2\pi m\,,$$

since those timesteps correspond to a sampling of n phase-space points in m revolutions.

This substitution yields for the IM scheme

$$\Delta t_{n:m}^{\text{IM}} = \frac{2}{\omega}\tan\left(\frac{\pi m}{n}\right) = \frac{T_p}{\pi}\tan\left(\frac{\pi m}{n}\right). \tag{13.62}$$

For $n = 3, 4$ and $m = 1$, we have, for example: $\Delta t_{3:1} = 2\sqrt{3}/\omega$ and $\Delta t_{4:1} = 2/\omega$ for IM. These resonance times are larger than corresponding values for Verlet: $\sqrt{3}/\omega$ and $\sqrt{2}/\omega$, respectively. Hence IM confers better stability than Verlet.

If we consider a fast period of 10 fs (see Table 13.2), the IM resonant timesteps for orders $n = 3, 4, 5, 6$ are 5.5, 3.2, 2.3, and 1.8 fs, respectively, compared to 2.8, 2.3, 1.9, and 1.6 fs for Verlet. This analysis explains the results shown in Figure 13.16.

A Family of Symplectic Implicit Schemes

The IE and IM methods described above turn out to be quite special in that IE's damping is extreme and IM's resonance patterns are quite severe relative to related symplectic methods. However, success was not much greater with a symplectic implicit Runge-Kutta integrator examined by Janežič and coworkers [936].

A family of implicit symplectic methods including IM and Verlet (as a special case) was also explored in subsequent works by Skeel, Schlick, and coworkers [907, 900]. The family of methods can be parameterized via α (see Box 13.6). The values $\alpha = 0$, $\alpha = 1/2$, and $1/4$ correspond to the Verlet, 'LIM2', and IM methods, respectively. The LIM2 method appeared attractive based on the effective rotation analysis (see Box 13.6), since it should exhibit no erratic resonance patterns; this was indeed observed in practice (Figure 13.16.) Unfortunately, this resonance removal comes at the price of large increases in energy with the timestep, as also seen in Figure 13.16.

These scheme-dependent resonances were concluded by the analysis in [900] on the basis of the effect of the parameter α on the numerical frequency of the integrator for the system simulated. Specifically, the maximum possible phase angle change per timestep *decreases* as the parameter α *increases*. Hence, the angle change can be limited by selecting a suitable α.

The choice $\alpha \geq \frac{1}{2}$, as in the LIM2 method, restricts the phase angle change to less than one quarter of a period and thus is expected to eliminate notable disturbances due to fourth-order resonance.

The requirement that $\alpha \geq \frac{1}{3}$ guarantees that the phase angle change per timestep is less than one third of a period and therefore should also avoid third-order resonance for the model problem.

This was verified in an application to a representative nonlinear system, a blocked alanine model [900] in Figure 13.16. This figure contrasts the early (i.e., with small timestep) instability of Verlet with the resonance of IM around one half the fastest period, and the absence of resonances in LIM2 ($\alpha = 1/2$).

Unfortunately, the increases of the average energy in LIM2 with the timestep are unacceptable: they reflect values that are approximately 30% and 100% larger than the small-timestep value, for $\Delta t = 5$ and 9 fs, respectively, for this system. Part of this behavior is also due to an error constant for LIM2 that is greater than that of leapfrog/Verlet.

Perspective

In sum, it is difficult to expect reasonable resolution by implicit methods in the large-timestep regime given the stability limit and resonance problems mentioned above. Perhaps semi-implicit [944] or cheaper implementations of implicit schemes [945] will better handle this problem: it might be possible to treat the local terms implicitly and the nonlocal terms explicitly. Exploitation of parallel machine architecture has potential for further speedup but, if experience to date on parallelization of linear algebra codes can be considered representative, parallel computers tend to favor explicit methods.

Box 13.6: Implicit Symplectic Integrator Family

The family of implicit symplectic methods parameterized by α can be formulated as follows [907, 900]:

$$\begin{aligned} X^{n+1/2} &= X^n + \frac{\Delta t}{2} V^n \\ V^{n+1/2} &= V^n + \frac{\Delta t}{2} \mathbf{M}^{-1} F^{n+1/2} \\ V^{n+1} &= V^{n+1/2} + \frac{\Delta t}{2} \mathbf{M}^{-1} F^{n+1/2} \\ X^{n+1} &= X^{n+1/2} + \frac{\Delta t}{2} V^{n+1}, \end{aligned} \tag{13.63}$$

where

$$F^{n+1/2} = F(X^{n+1/2} + \alpha \Delta t^2 \mathbf{M}^{-1} F^{n+1/2}). \tag{13.64}$$

The values $\alpha = 0$ and $1/4$ correspond to the Verlet and IM methods, respectively, and $\alpha = 1/2$ corresponds to the LIM2 method introduced in the text.

The above implicit system can be solved by minimizing the dynamics function Φ to obtain X and then evaluating F at this minimum point. Here Φ is defined as

$$\Phi(X) = \frac{1}{\gamma \Delta t^2} (X - X_0^n)^T \mathbf{M} (X - X_0^n) + \alpha E(X), \tag{13.65}$$

with

$$X_0^n = X^n + \frac{\Delta t}{2} V^n.$$

The effective rotation angle for this α-family is [900]:

$$\begin{aligned} \theta_{\text{eff}}^{\alpha \text{ family}} &= 2 \sin^{-1} \left(\frac{\omega \Delta t}{2} \sqrt{\phi} \right) \\ &= \omega \Delta t + \left(\frac{1}{24} - \frac{\alpha}{2} \right) (\omega \Delta t)^3 + \mathcal{O}((\omega \Delta t)^5) \end{aligned} \tag{13.66}$$

where

$$\phi = 1 \,/\, [1 + \alpha (\omega \Delta t)^2]. \tag{13.67}$$

Thus the maximal effective rotation can be controlled by the choice of α (see text).

13.7 Future Outlook

13.7.1 Integration Ingenuity

Many Approaches

In this chapter we have discussed various numerical integration techniques for Newtonian and Langevin/Brownian dynamics and focused on variations that can increase the timestep and lead to overall computational gains. For very detailed dynamic pathways, the standard Newtonian approach is necessary, but for certain systems (like long DNA supercoils or diffusion-controlled mechanisms in enzyme catalysis), stochastic models are appropriate and computationally advantageous.

The approaches discussed here to ameliorate the severe timestep restriction in biomolecular dynamics involve force splitting or multiple timesteps, various harmonic approximations, implicit integration, and other separating frameworks. Each such avenue has encountered different problems along the way, but these obstacles serve to increase our understanding of the intriguing numerical, computational, and accuracy issues involved in propagating the motion of complex nonlinear systems.

Resonance

The heightened appreciation of resonance problems [867, 899], in particular, contrasts with the more systematic error associated with numerical stability that grows monotonically with the discretization size. Ironically, resonance artifacts are worse in the modern impulse multiple-timestep methods, formulated to be symplectic and reversible; the earlier extrapolative variants were abandoned due to energy drifts. Stochasticity and slow-force averaging, as in the LM method [909, 910], or stochasticity and extrapolation, as in the LN method [876], can dampen and/or remove resonance artifacts and allow larger timesteps, but this requires added forces to standard Newtonian dynamics.

Ultimately, the compromise between the realized speedup and the accuracy obtained for the governing dynamic model should depend on the applications for which the dynamic simulations are used (e.g., configurational sampling versus detailed dynamics).

13.7.2 Current Challenges

PME Protocols

A problem that remains unsolved in part in connection with MTS methods is their optimal integration with Ewald and particle-mesh Ewald protocols. The problem is due to the presence of fast terms in the reciprocal Ewald component; this limits the outer MTS timestep and hence the speedup. For some discussion and approaches, see [669, 668, 930, 667, 670, 650, 671]. Improving such algorithms will allow us to use larger outer timesteps and hence simulate systems over longer times. In 2000, Daggett [74] suggested the cutoff criterion of 5 ns beyond which trajectories of peptides and proteins can be considered "long"; at the very least, this threshold likely guarantees equilibration of the initial systems.

Technology's Role

Not to be downplayed as a factor in the increasing of simulation scope is technology improvement. The steady increase in computer power, the declining cost of fast and highly-modular processors, and the rise of efficient parallelization of MD codes surely are helping to bridge the timescale gap between simulation range and experimental durations. The triumphant report of Duan and Kollman's 1 μs trajectory approaching the folding state of a 36-residue villin headpiece, from a disordered configuration (4 dedicated months of 256 processors on the Cray T3D/E), is a case in point [69]. However, as discussed in Chapter 1 (see Table 1.2 and Figures 1.1 and 1.2 of Chapter 1) and suggested in [74], simulation times are unfortunately not increasing commensurately with the rises in available computer power. Designing reliable, efficient, and general software for large-scale MD applications on multiprocessors remains a challenge.

Enhancing Sampling

As suggested in Chapter 1, the slow rise in simulation times may be partially explained by our heightened appreciation of the need to further develop force fields as well as to sample biomolecular configuration space more globally. To address the sampling, multiple trajectories can be effective for improving statistics [368], especially when applied to small systems, as in folding studies of peptides [39] or protein segments [40].

Besides stochastic models, many other routes can be followed to generate long-time views of biomolecules and address various conformational sampling questions (e.g., Figure 12.1).

New Monte Carlo techniques that survey conformational space efficiently are also key players in this sampling quest [164] (see also Monte Carlo chapter for MC/MD hybrids), as are alternative coordinate frameworks designed to extract collective motions for biomolecules that are essential to function [946, 947, 948].

If altering the model — through biasing forces or guiding restraints — is considered fair game, much room for ingenuity remains.

For example, diffusion-limited processes associated with high-energy or entropy barrier can be 'accelerated' through biasing forces in Brownian dynamics simulations, with rate calculations adjusted by associating lower weights with movements along high biases and vice versa [949]. Biomolecular systems can be 'steered' [950, 951] or 'guided' [952, 953] — subjected to time-dependent external forces along certain degrees of freedom or along a local free-energy gradient — to probe molecular details of certain experiments, such as atomic force microscopy and optical tweezer manipulations (see citations in [950, 951]), or to study folding/unfolding events [954].

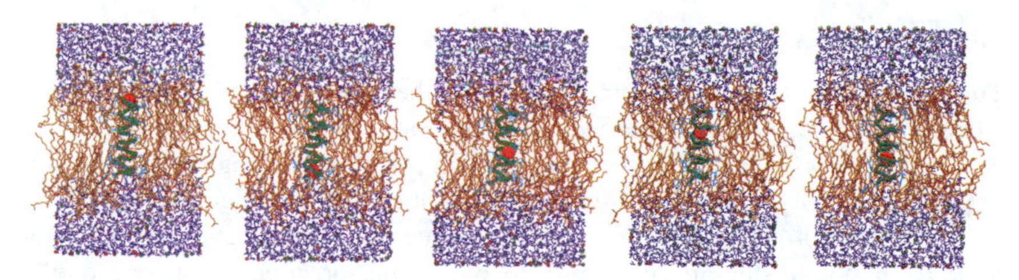

Figure 13.17. Stochastic-path approach snapshots describing the permeation of a sodium ion (huge red sphere) through the gramicidin channel embedded in a DMPC membrane (K. Siva and R. Elber, unpublished).

Targeted MD

If generating pathways between known structures is the goal (e.g., a closed form and an open form of a polymerase enzyme [955], or unfolded and folded state of a peptide [835, 836]), the dynamic pathway may be 'targeted' by use of restraints, which monitor the distance to the reference structure [835, 956]. Despite the fictitious trajectory that results from such 'targeted MD', insights into disallowable configurational states can be obtained (e.g., as illustrated in Figure 12.1), as well as conclusions regarding common pathway themes (e.g., [656]), though individual trajectories diverge from one another.

Many other methods for simulating transitions between defined conformations have been developed (e.g., [957, 958]). Umbrella sampling methods can also be very effective in combination with such biased trajectories (e.g., [959]).

Stochastic Path Approach

Similar in spirit to the TMD approach is the successful pathway generation approach of Elber and coworkers [960, 836, 961] developed over more than a decade (see Table 12.1 and the illustration in Figure 13.17). The technique combines in a new way the notions of stochastic difference equations — as an approximate approach to generate long-time trajectories — with path integral formulations — for solving boundary-value, rather than initial-value, problems. In practice, global optimization is applied to the entire trajectory using energy functionals defined by the boundary-value problem solved. Though high-speed parallel computers are required to optimize the entire trajectory for large systems [961] and the trajectories cannot be attributed with a known statistical ensemble, this promising approach succeeds in filtering high-frequency modes, even with very large effective time steps, and in approximating dynamic pathways of biomolecules [836].

Noteworthy is also the transition path sampling approach of Chandler and coworkers, which aims to capture rare events without prior knowledge of the mechanism, reaction coordinate, and transition states involved, using statistical sampling of trajectory space [827].

Tip of the Iceberg

These long-time and greater-sampling approaches represent only the tip of the iceberg of those possible. Undoubtedly, motivated by fundamental biological problems like protein folding, the zealous computational biologists and the scientists they enlist in these quests (like IBM's Blue-Gene team and the thousands of Folding@home contributors!) will continue to combine algorithmic ingenuity with state-of-the-art computing to overcome, or circumvent, one of the fundamental challenges in molecular dynamics simulations.

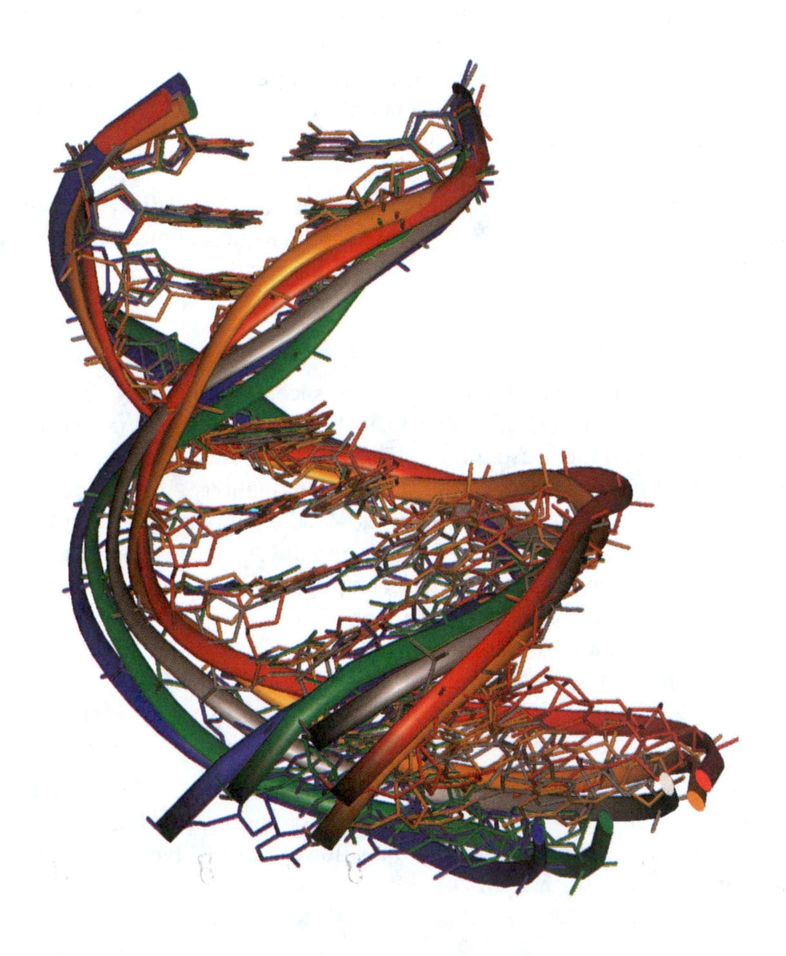

14

Similarity and Diversity in Chemical Design

Chapter 14 Notation

SYMBOL	DEFINITION
Matrices	
A	Dataset matrix $(n \times m)$
A_k	Rank k approximation to A
C	covariance matrix $(m \times m)$, elements $c_{jj'}$
P_k	projection matrix
U	SVD factor of A $(n \times n)$, contains left singular values
V	SVD factor of A $(m \times m)$, contains right singular values; also eigenvector matrix of C
V_k	low-rank approximation to eigenvector matrix $(m \times k)$
Σ	SVD factor $(n \times m)$, contains singular values
Σ_k	low-rank approximation to Σ
Vectors	
u_j	left singular value
v_j	right singular value
Xi	vector of compound i (components $Xi_1, Xi_2, \cdots Xi_m$)
$\hat{X}i$	scaled version of X_i
Yi	projection of X_i; also principal component of C
Scalars & Functions	
d_{ij}	intercompound distance ij in the projected representation
f, E	target optimization functions
l_{ij}	lower bounds on intercompound distance ij
u_{ij}	upper bounds on intercompound distance ij
m	number of dataset descriptors
n	number of dataset components
N	number of variables

Chapter 14 Notation Table (continued)

SYMBOL	DEFINITION
T_d	total number of distance segments satisfying a given deviation from target
α, β	scaling factors
δ	Euclidean distance (with upper/lower bounds u, l)
λ	eigenvalues
μ	mean value
ω	weights used in target optimization function
σ	singular values

Every sentence I utter must be understood not as an affirmation but as a question.

Niels Bohr (1885–1962).

14.1 Introduction to Drug Design

Following an introduction to drug discovery research, this chapter presents some mathematical formulations and approaches to a few problems involved in chemical database analysis that might interest mathematical/physical scientists. This material, though preliminary, is included since chemical design (and the closely related field of combinatorial chemistry) is a rapidly growing field of wide interest, for example in medicinal chemistry; see subsection 14.1.5 for an overview of this chapter and its limitations.

For a historical perspective of drug discovery, see [962, 963], for example, and for specialized treatments in drug design modeling consult the texts by Leach [964] and Cohen [965]. The review in [143] provides a useful industrial perspective of the science of lead-drug generation, with references to available software and web resources.

14.1.1 Chemical Libraries

The field of combinatorial chemistry was recognized by *Science* in 1997 as one of nine "discoveries that transform our ideas about the natural world and also offer potential benefits to society". Indeed, the systematic assembly of chemical building blocks to form potential biologically-active compounds and their rapid testing for bioactivity has experienced a rapid growth in both experimental and theoretical approaches (e.g., [966, 967, 968]); see the editorial overview on combinatorial chemistry [969] and the associated group of articles. Two combinatorial chemistry journals were launched in 1997, with new journals since then, and a Gordon Research conference on Combinatorial Chemistry was created. The number of new-drug candidates reaching the clinical-trial stage is greater than ever. Indeed, Stu Borman writes in the March 8, 1999 issue of *Chemical & Engineering News*: "Recent advances in solid-phase synthesis, informatics, and high-throughput screening suggest combinatorial chemistry is coming of age" [970].

Accelerated (automated and parallel) synthesis techniques combined with screening by molecular modeling and database analysis are the tools of combinatorial chemists. These tools can be applied to propose candidate molecules that resemble antibiotics, to find novel catalysts for certain reactions, to design inhibitors for the HIV protease, or to construct molecular sieves for the chemical industries based on zeolites. Thus, combinatorial technology is used to develop not only new drugs but also new materials, such as for electronic devices. Indeed, as electronic instruments become smaller, thin insulating materials for integrated circuit technology are needed. For example, the design of a new thin-film insulator at Bell Labs of Lucent Technologies [971] combined an optimal mixture of the metals zirconium (Zr), tin (Sn), and titanium (Ti) with oxygen.

As such experimental synthesis techniques are becoming cheaper and faster, huge chemical databases are becoming available for computer-aided [963] and structure-based [972] drug design; the development of reliable computational tools for the study of these database compounds is thus becoming more important than ever. The term *cheminformatics* (*chemical informatics*, also called *chemoinformatics*), has been coined to describe this emerging discipline that aims at transforming such data into information, and that information into knowledge useful for faster identification and optimization of lead drugs [143].

14.1.2 Early Drug Development Work

Before the 1970s, proposals for new drug candidates came mostly from laboratory syntheses or extractions from Nature. A notable example of the latter is Carl Djerassi's use of locally grown yams near his laboratory in Mexico City to synthesize cortisone; a year later, this led to his creation of the first steroid effective as a birth control pill [973]. Synthetic technology has certainly risen, but natural products have been and remain vital as pharmaceuticals (see [974] and Box 14.1 for a historical perspective).

A pioneer in the systematic development of therapeutic substances is James W. Black, who won the Nobel Prize in Physiology or Medicine in 1988 for his research on drugs beginning in 1964, including histamine H_2-receptor antagonists. Black's team at Smith Kline & French in England synthesized and tested systematically compounds to block histamine, a natural component produced in the stomach that stimulates secretion of gastric juices. Their work led to development of a classic 'rationally-designed' drug in 1972 known as Tagamet (cimetidine). This drug effectively inhibits gastric-acid production and has revolutionized the treatment of peptic ulcers.

Later, the term *rational drug design* was introduced as our understanding of biochemical processes increased, as computer technology improved, and as the field of molecular modeling gained wider acceptance. 'Rational drug design' refers to the systematic study of correlations between compound composition and its bioactive properties.

Box 14.1: Natural Pharmaceuticals

Though burdened by political, environmental, and economic issues, pharmaceutical industries have long explored unusual venues for disease remedies, many in remote parts of the world and involving indigenous cures. Micro-organisms and fungi, in particular, are globally available and can be reproduced readily. For example, among the world's 25 top-selling drugs in 1997, seven were derived from natural sources. Some notable examples of products derived from Nature are listed below.

- A fungus found on a Japanese golf course is being used by Merck to make the cholesterol lowering drug mevacor, one of the 25 top-sellers of 1997.
- A fungus found on a Norwegian mountain is the basis for another 1997 top-seller, the transplant drug cyclosporin, made by Novartis.
- A fungus from a Pacific yew tree is also the source of the anticancer agent paclitaxel (taxol).
- The rosy periwinkle of Madagascar is the source of Eli Lilly's two cancer drugs vincristine and vinblastine, which have helped fight testicular cancer and childhood leukemia since the 1960s.
- A microbe discovered in a Yellowstone hot spring is the source of a heat-resistant enzyme now key in DNA amplification processes.
- Ocean salmon is a source for osteoporosis drugs (calcimar and miacalcin), and coral extracts are used for bone replacement.
- The versatile polymer chitosan, extracted from crab and shrimp shells, is a well known fat-binding weight-loss aid, in addition to its usage in paper additives, pool cleaners, cosmetics, and hair gels.
- Frog-skin secretions serve as models for development of painkillers with fewer side effects than morphine. This chemical secret, long exploited by Amazon rain forest tribesmen, is now being pursued with frogs from Ecuador by Abbott Labs.
- Marine organisms from the Philippines are being investigated as sources of chemicals toxic to cancer cells.
- The venomous lizard termed Gila monster inhabiting Phoenix, Arizona, may provide a powerful peptide, exenden, for treating diabetes, because it stimulates insulin secretion and aids digestion in lizards that gorge thrice-yearly.
- A compound isolated from a flowering plant in a Malaysian rainforest, calanolide A, is a promising drug candidate for AIDS therapy, in the class of non-nucleoside reverse transcriptase inhibitors.

- A protein from a West African berry was identified by University of Wisconsin scientists as 2000 times sweeter than sugar; sweeteners are being developed from this source to make possible sweeter food products by gene insertion.
- A natural marine product (ecteinascidin 743) derived from the Caribbean sea squirt *Ecteinascidia turbinata* was found to be an active inhibitor of cell proliferation in the late 1960s, but only recently purified, synthesized, and tested in clinical trials against certain cancers.
- A Caribbean marine fungus extract (developed as halimide) shows early promise against cancer, including some breast cancers resistant to other drugs.

One of the most challenging aspects of using natural products as pharmaceutical agents is a sourcing problem, namely extracting and purifying adequate supplies of the target chemicals. For example, biochemical variations within specifies combined with international laws restricting collection (e.g., of frogs from Ecuador whose skins contain an alkaloid compound with powerful painkilling effects) limit available natural sources. In the case of the frog skin chemical, this sourcing problem prompted the synthetic design of a new type of analgesic that is potentially nonaddictive [974].

14.1.3 Molecular Modeling in Rational Drug Design

Since the 1980s, further improvements in modeling methodology, computer technology, as well as X-ray crystallography and NMR spectroscopy for biomolecules, have increased the participation of molecular modeling in this lucrative field. Molecular modeling is playing a more significant role in drug development [975, 976] as more disease targets are being identified and solved at atomic resolution (e.g., HIV-1 protease, HIV integrase, adenovirus receptor), as our understanding of the molecular and cellular aspects of disease is enhanced (e.g., regarding pain signaling mechanisms, or the immune invasion mechanism of the HIV virus), and as viral genomes are sequenced [118]. Indeed, in analogy to genomics and proteomics — which broadly define the enterprises of identifying and classifying the genes and the proteins in the genome — the discipline of *chemogenomics* [977] has been associated with the delineation of drugs for all possible drug targets.

As described in the first chapter, examples of drugs made famous by molecular modeling include HIV-protease inhibitors (AIDS treatments), thrombin inhibitors (for blood coagulation and clotting diseases), neuropeptide inhibitors (for blocking the pain signals resulting from migraines), and PDE-5 inhibitors (for treating impotence by blocking a chemical reaction which controls muscle relaxation and resulting blood flow rate). See Figure 14.1 for illustrations of popular drugs for migraine, HIV/AIDS, and blood-flow related diseases.

Such computer modeling and analysis — rather than using trial and error and exhaustive database studies — was thought to lead to dramatic progress in the design of drugs. However, some believe that the field of rational drug design has not lived up to its expectations.

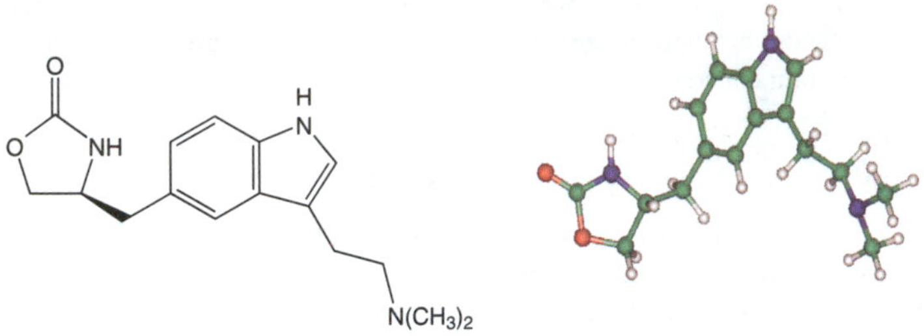

Zomig: $C_{16}H_{21}N_3O_2$

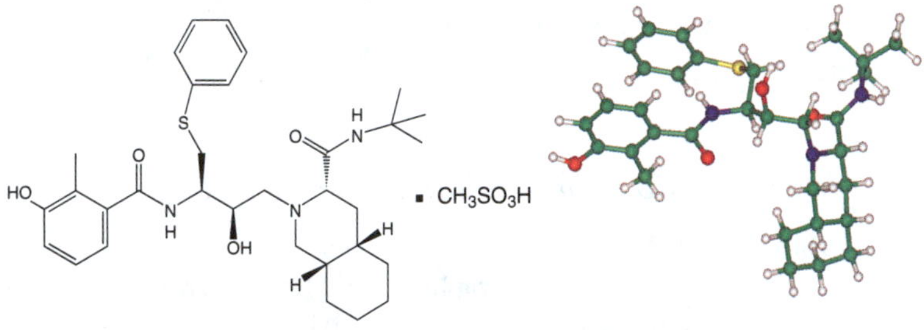

Viracept: $C_{33}H_{49}O_7N_3S_2$

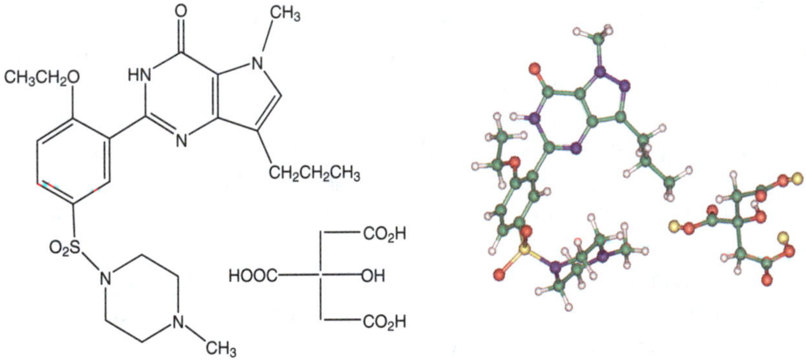

Viagra: $C_{28}H_{37}O_{11}N_6S$

Figure 14.1. Popular drug examples. Top: *zolmitriptan* (zomig) for migraines, a 5-HT_1 receptor agonist that enhances the action of serotonin. Middle: *nelfinavir mesylate* (viracept), a protease inhibitor for AIDS treatment. Bottom: *sildenfel citrate* (viagra) for penile dysfunctions, a temporary inhibitor of phosphodiesterase-5, which regulates associated muscle relaxation and blood flow by converting cyclic guanosine monophosphate to guanosine monophosphate. See other household examples in Figure 14.3.

One reason for the restrained success is the limited reliability of modeling molecular interactions between drugs and target molecules; such interactions must be described very accurately energetically to be useful in predictions.

Another reason for the limited success of drug modeling is that the design of compounds with the correct binding properties (e.g., dissociation constants in the micromolar range and higher) is only a first step in the complex process of drug design; many other considerations and long-term studies are needed to determine the drug's bioactivity and its effects on the human body [978]. For example, a compound may bind well to the intended target but be inactive biologically if the reaction that the drug targets is influenced by other components (see Box 14.2 for an example). Even when a drug binds well to an appropriate target, an optimal therapeutic agent must be delivered precisely to its target [979], screened for undesirable drug/drug interactions [980], lack toxicity and carcinogenicity (likewise for its metabolites), be stable, and have a long shelf life.

14.1.4 The Competition: Automated Technology

Even accepting those limitations of computer-based approaches, rational drug design has avid competition from automated technology: new synthesis techniques, such as robotic systems that can run hundreds of concurrent synthetic reactions, have emerged, thereby enhancing synthesis productivity enormously. With "high-throughput screening", these candidates can be screened rapidly to analyze binding affinities, determine transport properties, and assess conformational flexibility.

Many believe that such a production *en masse* is the key to establishing diverse databases of drug candidates. Thus, at this time, it might be viewed that *drug design need not be 'rational' if it can be exhaustive*. Still, others advocate a more focused design approach.

Another convincing argument for the focused design approach is that the amount of synthesized compounds is so vast (and rapidly generated) that computers will be essential to sort through the huge databases for compound management and applications. Such applications involve clustering analysis and *similarity* and *diversity sampling* (see below), preliminary steps in generating drug candidates or optimizing bioactive compounds.

This information explosion explains the resurrection of computer-aided drug design and its enhancement in scope under the new title **combinatorial chemistry**, affectionately endorsed as 'the darling of chemistry' [976]. The latest developments in this rapidly growing field can be gleaned from the Diversity Information pages: www.5z.com/divinfo/.

14.1.5 Chapter Overview

In this chapter, a brief introduction into a limited number of the mathematical questions involved in this discipline of chemical library design is presented, namely *similarity and diversity sampling*. Some ideas on cluster analysis and database searching are also described. This chapter is only intended to whet the appetite for chemical design and to invite mathematical scientists to work on related problems.

See extended treatments in the texts by Leach [964] and Cohen [965] and in reviews in the field, such as [143, 981, 982, 963, 983, 984, 985].

Because medicinal chemistry applications are an important subfield of chemical design, this last chapter also provides some perspectives on current developments in drug design, as well as mentioning emerging areas such as personalized medicine and biochips.

14.2 Problems in Chemical Libraries

Chemical libraries consist of compounds (known chemical formulas) with potential and/or demonstrated therapeutic activities. Most libraries are proprietary, residing in pharmaceutical houses, but public sources also exist.[1] Both target-independent and target-specific libraries exist. The name 'combinatorial libraries' stems from the important combinatorial problems associated with the experimental design of compounds in chemical libraries, as well as computational searches for potential leads using concepts of **similarity** and **diversity** as introduced below.

14.2.1 Database Analysis

In broad terms, two general problem categories can be defined in chemical library analysis and design:

Database systematics: analysis and compound grouping, compound classification, elimination of redundancy in compound representation (dimensionality reduction), data visualization, etc., and

Database applications: efficient formulation of quantitative links between compound properties and biological activity for compound selection and design optimization experiments.

---

[1] Some examples of public databases are:

Derwent World Drug Index (www.derwent.com), with information on the chemical structure and activity data of marketed and development drugs;

Comprehensive Medicinal Chemistry (CMC) (www.mdli.com/cgi/dynamic/product.html?uid=$uid&key=$key&id=21), providing 3D models and various biochemical properties of drugs;

MDL Drug Data Report (MDDR) (www.mdli.com/cgi/dynamic/product.html?uid=$uid&key=$key&id=20), with various information on known drugs and those under development, including descriptions of therapeutic action and cross-reference search utilities; and

NCI's Drug database (cactus.cit.nih.gov), a large public chemical structure database, with sophisticated browser, search capabilities, and interfaces to other databases.

A text reference for known drugs is the Physicians Desk Reference guide.

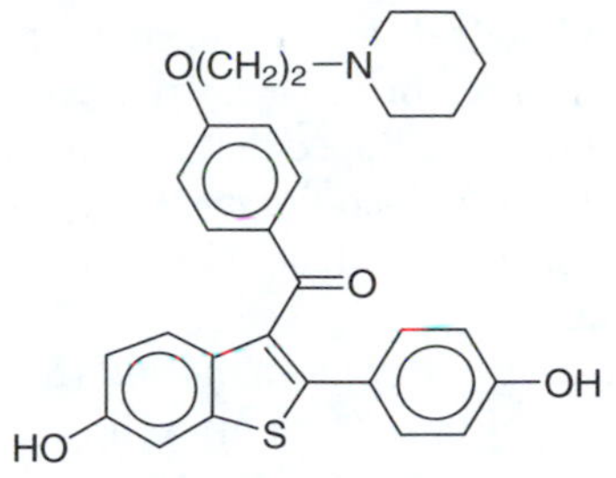

Raloxifene

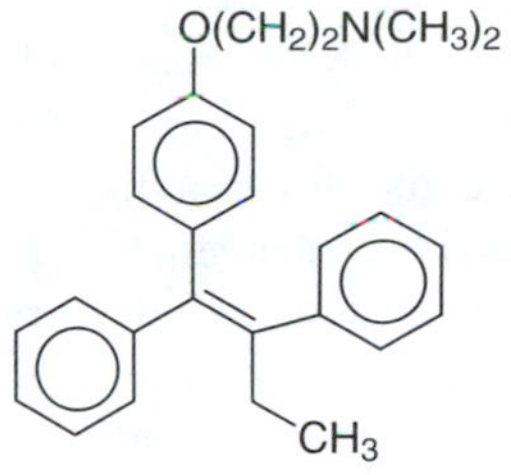

Tamoxifen

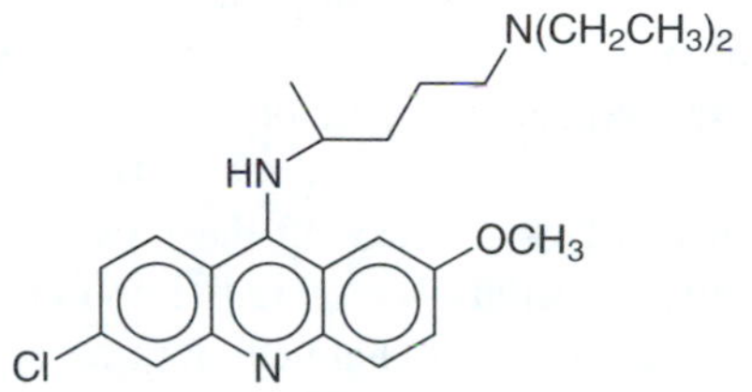

Quinacrine

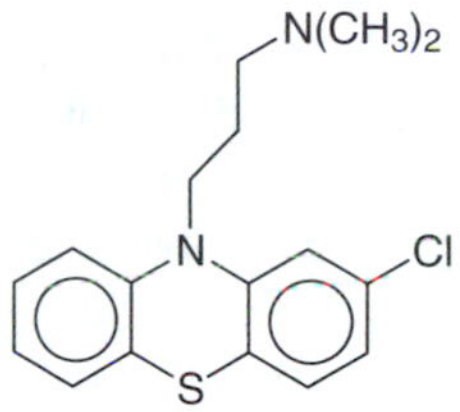

Chlorpromazine

Figure 14.2. Related pairs of drugs: the antiestrogens raloxifene and tamoxifen, and the tricyclic compounds with aliphatic side-chains at the middle ring quinacrine and chlorpromazine.

Both of these general database problems involved in chemical libraries are associated with several mathematical disciplines. Those disciplines include multivariate statistical analysis and numerical linear algebra, multivariate nonlinear optimization (for continuous formulations), combinatorial optimization (for discrete formulations), distance geometry techniques, and configurational sampling.

14.2.2 Similarity and Diversity Sampling

Two specific problems, described formally in the next section after the introduction of chemical descriptors, are the similarity and diversity problems.

The *similarity* problem in drug design involves finding molecules that are 'similar' in physical, chemical, and/or biological characteristics to a known target compound. Deducing compound similarity is important, for example, when one drug is known and others are sought with similar physiochemical and biological properties, and perhaps with reduced side effects.

One example is the target bone-building drug *raloxifene*, whose chemical structure is somewhat related to the breast cancer drug *tamoxifen* (see Figure 14.2) (e.g., [986]). Both are members of the family of *selective estrogen receptor modulators* (SERMs) that bind to estrogen receptors in the breast cancer cells and exert a profound influence on cell replication. It is hoped that raloxifene will be as effective for treating breast tumors but will reduce the increased risk of endometrial cancer noted for tamoxifen. Perhaps raloxifene will also not lose its effectiveness after five years like tamoxifen.

Another example of a related pair of drugs is *chlorpromazine* (for treating schizophrenia) and *quinacrine* (antimalarial drug). These tricyclic compounds with aliphatic side chains at the middle ring group (see Figure 14.2) were suggested as candidates for treating Creutzfeldt-Jakob and other prion diseases [987].

Because similarity in structure might serve as a first criterion for similarity in activity/function, similarity searching can be performed using 3D structural and energetic searches (e.g., induced fit or 'docking' [972, 988]) or using the concept of molecular descriptors introduced in the next section, possibly in combination with other discriminatory criteria.

The *diversity* problem in drug design involves delineating the most diverse subset of compounds within a given library. Diversity sampling is important for practical reasons. The smaller, representative subsets of chemical libraries (in the sense of being most 'diverse') might be searched first for lead compounds, thereby reducing the search time; representative databases might also be used to prioritize the choice of compounds to be purchased and/or synthesized, similarly resulting in an accelerated discovery process, not to speak of economic savings.

Box 14.2: Treatments for Chronic Pain

Amazing breakthroughs have been achieved recently in the treatment of chronic pain. Such advances were made possible by an increased understanding of the distinct cellular mechanisms that cause pain due to different triggers, from sprained backs to arthritis to cancer.

What is Pain? Pain signals start when nerve fibers known as nociceptors, found throughout the human body, react to some disturbances in nearby tissues. The nerve fibers send chemical pain messengers that collect in the dorsal horn of the spinal cord. Their release depends on the opening of certain pain gates. Only when these messengers are released into the brain is pain felt in the body.

Natural Ammunition. Fortunately, the body has a battery of natural painkillers that can close those pain gates or send signals from the brain. These compensatory agents include endorphins, adrenaline, and serotonin (a peptide similar to opium). Many painkillers enhance or mimic the action of these natural aids (e.g., opium-bases drugs such as *morphine*, *codeine*, and *methadone*). However, these opiates have many undesirable side effects.

Painkiller Targets. To address the problem of pain, new treatments are targeting *specific* opiate receptors. For example, Actiq, developed by Anesta Corp. for intense cancer pain, is a lozenge placed in the cheek that is absorbed quickly into the bloodstream, avoiding the gut. Other pain relievers include a class of drugs known as COX-2 inhibitors, like Monsanto's *celebrex* (celecoxib) and Merck's *vioxx*, which relieve aches and inflammation with fewer stomach-damaging effects. They do so by targeting only one (COX-2) of two enzymes called cyclo-oxegenases (COX), which are believed to cause inflammation and thereby trigger pain.

While regular non-steroidal anti-inflammatory drugs (NSAIDs, like *aspirin*, *ibuprofen*, and *naproxen*) and others available by prescription attack both COX-1 and COX-2, COX-1 is also known to protect the stomach lining; this explains the stomach pain that many people experience with NSAIDs and the pain relief without the side effects that COX-2 inhibitors can offer.

Modern pain treatment also involves compounds that stop pain signals before the brain gets the message, either by intercepting the signals in the spinal cord or by blocking their route to the spine. Evidence is emerging that a powerful chemical called 'substance P' can be used as an agent to deliver pain blockers to receptors found throughout the body; an experimental drug based on this idea (marketed by Pfizer) has proven effective at easing tooth pain.

14.2.3 Bioactivity Relationships

Besides database systematics, such as similarity and diversity sampling, the establishment of clear links between compound properties and bioactivity is, of course, the heart of drug design. In many respects, this association is not unlike the protein prediction problem in which we seek some target energy function that upon global minimization will produce the biologically relevant, or native, structure of a protein.

In our context, formulating that 'function' to relate sequence and structure while not ignoring the environment might even be more difficult, since we are studying small molecules for which the evolutionary relationships are not clear as they might be for proteins. Further, the bioactive properties of a drug depend on much more than its chemical composition, three-dimensional (3D) structure, and energetic properties. A complex orchestration of cellular machinery is often involved in a particular human ailment or symptom, and this network must be understood to alleviate the condition safely and successfully.

A successful drug has usually passed many rounds of chemical modifications that enhanced its potency, optimized its selectivity, and reduced its toxicity. An example involves obesity treatments by the hormone *leptin*. Limited clinical studies have shown that leptin injections do not lead to clear trends of weight loss in people, despite demonstrating dramatic slimming of mice. Though not a quick panacea in humans, leptin has nonetheless opened the door to pharmacological manipulations of body weight, a dream with many medical — not to speak of monetary — benefits. Therapeutic manipulations will require an understanding of the complex mechanism associated with leptin regulation of our appetite, such as its signaling the brain on the status of body fat.

Box 14.2 contains another illustration of the need to understand such complex networks in connection with drug development for chronic pain. These examples clearly show that *lead generation*, the first step in drug development, is followed by *lead optimization*, the challenging, slower phase.

In fact, this complexity of the molecular machinery that underlies disease has given rise to the subdisciplines of *molecular medicine* and *personalized medicine* (see Box 14.3), where DNA technology plays an important role. Specifically, DNA chips — small glass wafers like computer chips studded with bits of DNA instead of transistors — can analyze the activities of thousands of genes at a time, helping to predict disease susceptibility in individuals, classify certain cancers, and to design treatments [128].

For example, DNA chips can study expression patterns in the tumor suppressor gene *p53* (the gene with the single most common mutations in human cancers), and such patterns can be useful for understanding and predicting response to chemotherapy and other drugs. DNA microarrays have also been used to identify genes that selectively stimulate metastasis (the spread of tumor cells from the original growth to other sites) in melanoma cells.

Besides developments on more personalized medicine, which will also be enhanced by a better understanding of the human body and its ailments, new advances in drug delivery systems may be important for improving the rate and period of drug delivery in general [989].

Box 14.3: Molecular and Personalized Medicine

Pauling's Groundwork. Molecular medicine seeks to enhance our therapeutic solutions by understanding the molecular basis of disease. Linus Pauling lay the groundwork for this field in his seminal 1949 paper [990] which demonstrated that the hemoglobin from sickle cell anemia sufferers has a different electric charge than that from healthy people. This difference was later explained by Vernon Ingram as arising from a single amino acid difference [991]. Those pioneering works relied on electrophoretic mobility measurements and fingerprinting techniques (electrophoresis combined with paper chromatography) for peptides.

Disease Simulations. A modern incarnation of molecular medicine involves conducting virtual experiments by computer simulation with the goal of developing new hypotheses regarding disease mechanisms and prevention. For example, scientists at Entelos Inc. (Menlo Park, California) are simulating cell inflammation caused by asthma to try to learn how blocking certain inflammation factors might affect cellular receptors and then to identify targets for steroid inhalers.

From SNPs to Tailored Drugs. Another trend of personalized medicine refers to tailoring drugs to individual genetic makeup. User-specific drugs have great potential to be more potent and to eliminate ill side effects experienced by some individuals.

The tailoring is based on identifying the small variations in people's DNA where a single nucleotide differs from the standard sequence. These mutations, or individual variations in genome sequence that occur once every couple of hundred of base pairs, are called single-nucleotide polymorphisms known as SNPs (pronounced "snips"). The presence of SNPs can be signaled visually using DNA chips or biochips, instruments of fancy of the biotechnology industry (See [128] and Box 1.5 of Chapter 1). Given their enormous potential in genomics and drug development, companies like Affymetrix, Agilent, Motorola, Texas Instruments, and IBM. all have biochips under development. The approach of personal genomics or customized therapy as a whole has been coined *pharmacogenomics*, literally *pharmacogenetics* (the science of studying variations in drug responses among individuals) on a genome-wide scale (e.g., [992, 993]).

The field gained momentum in April 1999 when eleven pharmaceutical and technology companies and the Wellcome Trust announced a genome *mapping* consortium for SNPs. The consortium's goal is to construct a fine-grained map of order 300,000 SNPs to permit searching for SNP patterns that correlate with particular drug responses.

Examples of drugs fine-tuned to people by this approach include a breast-cancer drug *herceptin* (Genetech Inc.) that works on patients with elevated levels of the protein *her-2*, and a drug for chronic asthma, *zyflo* (Abbott Labs), safe for most people but can be fatal to 4% of the patients. Directed drugs are also being developed to treat or diagnose diabetics, neurological diseases like Alzheimer's, prostate cancer, and ailments requiring antibiotics. Though there are many hurdles to this new field, not to mention possible financial drawbacks, it is hoped that some benefits could be realized in the not-too-distant future.

14.3 General Problem Definitions

14.3.1 The Dataset

Our given dataset of size n contains information on compounds with potential *biological activity* (drugs, herbicides, pesticides, etc.). A schematic illustration is presented in Figure 14.3. The value of n is large, say one million or more. Because of the enormous dataset size, the problems described below are simple to solve in principle but extremely challenging in practice because of the large associated computational times. Any systematic schemes to reduce this computing time can thus be valuable.

14.3.2 The Compound Descriptors

Each compound in the database is characterized by a vector (the *descriptor*). The vector can have real or binary elements. There are many ways to formulate these descriptors so as to reduce the database search time and maximize success in generation of lead compounds.

Chemical Library ($n >> m$)

Compound ($i = 1,\ldots, n$)	Vectorial Descriptors ($k = 1,\ldots, m$)				Biological Targets ($j = 1,\ldots, m_B$)			
	Xi_1	Xi_2	$\cdots$	Xi_m	Bi_1	Bi_2	$\cdots$	Bi_{m_B}
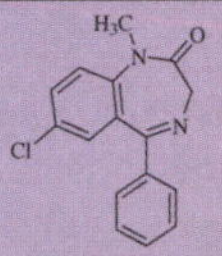 1 Valium	0.873	0.763	...	0.531	0	1	...	0
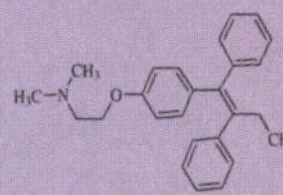 2 Tamoxifen	0.912	0.131	...	0.834	0	0	...	1
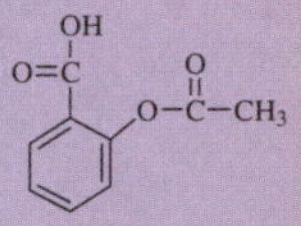 3 Aspirin	0.763	0.214	...	0.533	0	0	...	0
⋮	⋮	⋮	⋮	⋮	⋮	⋮	⋮	⋮
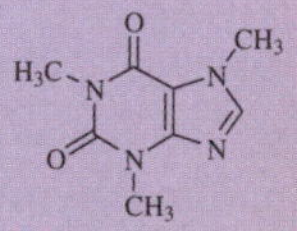 i Caffeine	0.925	0.237	...	0.742	1	0	...	1
⋮	⋮	⋮	⋮	⋮	⋮	⋮	⋮	⋮
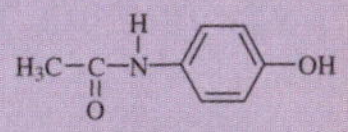 n Acetaminophen	0.347	0.279	...	0.846	1	1	...	0

Figure 14.3. A chemical library can be represented by n compounds i (known or potential drugs), each associated with m characteristic descriptors ($\{Xi_k\}$) and activities $\{Bi_j\}$ with respect to m_B biological targets (known or potential).

Conventionally, each compound i is described by a list of **chemical descriptors**, which may reflect *molecular composition*, such as atom number, atom connectivity, or number of functional groups (like aromatic or heterocyclic rings, tertiary aliphatic amines, alcohols, and carboxamides), *molecular geometry*, such as number of rotatable bonds, *electrostatic properties*, such as charge distribution, and various *physiochemical measurements* that are important for bioactivity.

These descriptors are currently available from many commercial packages like MolConn-X and MolConn-Z [994, 995]. Descriptors fall into many classes. Examples include:

2D descriptors — also called molecular connectivity or topological indices — reflecting molecular connectivity and other topological invariants;

binary descriptors — simpler encoded representations indicating the presence or absence of a property, such as whether or not the compound contains at least three nitrogen atoms, doubly-bonded nitrogens, or alcohol functional groups;

3D descriptors — reflecting geometric structural factors like van der Waals volume and surface area; and

electronic descriptors — characterizing the ionization potential, partial atomic charges, or electron densities.

See also [985] for further examples.

Binary descriptors allow rapid database analysis using Boolean algebra operations. The popular MolConn-X and MolConn-Z programs [994, 995], for example, generate topological descriptors based on molecular connectivity indices (e.g., number of atoms, number of rings, molecular branching paths, atoms types, bond types, etc.). Such descriptors have been found to be a convenient and reasonably successful approximation to quantify molecular structure and relate structure to biological activity (see review in [996]). These descriptors can be used to characterize compounds in conjunction with other selectivity criteria based on activity data for a training set (e.g., [997, 998]). The search for the most appropriate descriptors is an ongoing enterprise, not unlike force-field development for macromolecules.

The number of these descriptors, m, is roughly on the order of 1000, thus much smaller than n (the number of compounds) but too large to permit standard systematic comparisons for the problems that arise.

Let us define the vector Xi associated with compound i to be the row m-vector

$$\{Xi_1, Xi_2, \ldots, Xi_m\}\,.$$

Our dataset $\mathcal{S}$ can then be described as the collection of n vectors

$$\mathcal{S} = \{X1, X2, X3, \ldots, Xn\}\,,$$

or expressed as a rectangular matrix $A_{n \times m}$ by listing, in rows, the m chemical descriptors of the n database compounds:

$$A = \begin{pmatrix} X1_1 & X1_2 & \cdots & \cdots & \cdots & X1_m \\ X2_1 & X2_2 & \cdots & \cdots & \cdots & X2_m \\ \vdots & & & \cdots & & \\ \vdots & & & \cdots & & \\ \vdots & & & \cdots & & \\ \vdots & & & \cdots & & \\ \vdots & & & \cdots & & \\ Xn_1 & Xn_2 & \cdots & \cdots & \cdots & Xn_m \end{pmatrix}. \qquad (14.1)$$

In practice, this rectangular $n \times m$ matrix has $n \gg m$ (i.e., the matrix is long and narrow), where n is on the order of millions and m is several hundreds.

The compound descriptors are generally *highly redundant*. Yet, it is far from trivial how to select the "principal descriptors". Thus, various statistical techniques (principal component analysis, classic multivariate regression; see below) have been used to assess the degree of correlation among variables so as to eliminate highly-correlated descriptors and reduce the dimension of the problems involved.

14.3.3 Characterizing Biological Activity

Another aspect of each compound in such databases is its *biological activity*. Pharmaceutical scientists might describe this property by associating a simple *affirmative* or *negative* score with each compound to indicate various areas of activity (e.g., with respect to various ailments or targets, which may include categories like headache, diabetes, protease inhibitors, etc.).

Drugs may enhance/activate (e.g., *agonists*) or inhibit (e.g., *antagonists, inhibitors*) certain biochemical processes. This bioactivity aspect of database problems is far less quantitative than the simple chemical descriptors. Of course, it also requires synthesis and biological testing for activity determination. Studies of several drug databases have suggested that active compounds can be associated with certain ranges of physiochemical properties like molecular weight and occurrence of functional groups [999].

For the purpose of the problems outlined here, it suffices to think of such an additional set of descriptors associated with each compound. For example, a matrix $B_{n \times m_B}$ may complement the $n \times m$ database matrix A; see Figure 14.3. Each row i of B may correspond to measures of activity of compound i with respect to specific targets (e.g., binary variables for active/nonactive target response).

The ultimate goal in drug design is to find a compound that yields the desired pharmacological effect. This quest has led to the broad area termed SAR, an acronym for Structure/Activity Relationship [964]. This discipline applies various statistical, modeling, or optimization techniques to relate compound properties to associated pharmacological activity. A simple linear model, for example, might attempt to solve for variables in the form of a matrix $X_{m \times m_B}$, satisfying

$$AX = B\,. \tag{14.2}$$

Explained more intuitively, SAR formulations attempt to relate the given compound descriptors to experimentally-determined bioactivity markers. While earlier models for 'quantitative SAR' (QSAR) involved simple linear formulations for fitting properties and various statistical techniques (e.g., multivariate regression, principal component analysis), nonlinear optimization techniques combined with other visual and computational techniques are more common today [1000]. The problem remains very challenging, with rigorous frameworks continuously being sought.

14.3.4 The Target Function

To compare compounds in the database to each other and to new targets, a quantitative assessment can be based on common structural features. Whether characterized by topological (chemical-formula based) or 3D features, this assessment can be broadly based on the vectorial chemical descriptors provided by various computer packages. A target function f is defined, typically based on the *Euclidean distance* function between vector pairs, δ, where

$$f(Xi, Xj) = \delta_{ij} \equiv \|Xi - Xj\| = \sqrt{\sum_{k=1}^{m} (Xi_k - Xj_k)^2}\,. \tag{14.3}$$

Thus, to measure the similarity or diversity for each pair of compounds Xi and Xj, the function $f(Xi, Xj)$ is often set to the simple distance function δ_{ij}. Other functions of distance are also appropriate depending upon the objectives of the optimization task.

14.3.5 Scaling Descriptors

Scaling the descriptor components is important for proper assessment of the score function [984]. This is because the individual chemical descriptors can vary drastically in their magnitudes as well as the variance within the dataset. Subsequently, a few large descriptors can overwhelm the similarity or diversity measures. For example, actual descriptor components of a database compound may look like the following:

```
11.0000  0.6433 4.5000  0.0833 150.2200  8.4831  0.0159 -1.0000 113.2239 ..
 1.000   0.2917 0.5000  0.0000  40.0000  7.2566  0.0801  1.0000 782.7121 ..
-8.0000  0.2081 0.5000  0.0186  80.0000  0.0000  0.0017  1.0000  62.2016 ..
 2.0000  0.0000 2.5000 -0.9010   0.0000  1.3867  0.2500  1.0000 120.0030 ..
 0.0000  0.0000 3.0000  0.0326   0.0000 -4.3984  0.1759  1.0000  11.2189 ..
80.0000 -0.0442 6.0000  0.7002 210.0000 -1.9784  0.0026 -1.0000 370.3473 ..
```

```
-5.0000 -0.1491 0.0000  0.0000  10.0000  9.0909  0.1641  1.0000  98.2782 ..
-1.0000  0.5427 4.5000  0.8963  35.0000  2.0061  0.0720  1.0000 119.8090 ..
17.0000 -0.3209 0.5000  0.0803   0.0000  9.4765  0.0000 -1.0000  11.7011 ..
19.0000  0.2690 1.0000 -0.3420  90.0000  0.0000  0.0000 -1.0000 201.0180 ..
 0.0000  0.0000 0.0000  0.2000  40.0000  9.1702  0.0429 -1.0000  23.2423 ..
 4.0000  0.3061 0.5000  0.6670  10.0000  2.3820  0.0023  1.0000   0.0000 ..
 4.0000  0.7702 1.5000  0.1870   0.0000  0.0000  0.7290  1.0000   0.0000 ..
 1.0000 -0.1134 1.5000  0.3356  40.0000  0.0000  0.7782 -1.0000 314.6658 ..
 0.0000  0.0000 0.0000  0.7842   0.0000 -6.1659  0.0000  1.0000  85.2285 ..
 3.0000  0.0000 0.0000  0.2382  75.0000  4.2276  0.1260  1.0000   7.2854 ..
15.0000  0.3479 4.0000  0.0034   0.0000  0.5152  0.3018  1.0000 280.8721 ..
 7.0000  0.6945 3.5000  0.4552   0.0000  3.5315  0.3065 -1.0000   0.0000 ..
   .        .      .       .        .       .       .       .        .    ..
   .        .      .       .        .       .       .       .        .    ..
```

Clearly, the ranges of individual descriptors vary (e.g., 0 to 1 versus 0 to 1000). Thus, given no chemical/physical guidance, it is customary to scale the vector entries before analysis. In practice, however, it is very difficult to determine the appropriate scaling and displacement factors for the specific application problem [984]. A general scaling of each Xi_k to produce $\hat{X}i_k$ can be defined using two real numbers α_k and β_k, for $k = 1, 2, \ldots, m$, termed the *scaling* and *displacement* factors, respectively, where $\alpha_k > 0$. Namely, for $k = 1, 2, \ldots, m$, we define the scaled components as

$$\hat{X}i_k = \alpha_k \,(Xi_k - \beta_k), \qquad 1 \le i \le n\,. \tag{14.4}$$

The following two scaling procedures are often used. The first makes each column in the range $[0, 1]$: each column of the matrix A is modified using eq. (14.4) by setting the factors as

$$\begin{aligned} \beta_k &= \min_{1 \le i \le n} Xi_k\,, \\ \alpha_k &= 1/\left(\max_{1 \le i \le n} Xi_k - \beta_k \right). \end{aligned} \tag{14.5}$$

This scaling procedure is also termed “standardization of descriptors”.

The second scaling produces a new matrix A where each column has a mean of zero and a standard deviation of one. It does so by setting the factors (for $k = 1, 2, \ldots, m$) as

$$\begin{aligned} \beta_k &= \frac{1}{n} \sum_{i=1}^{n} Xi_k\,, \\ \alpha_k &= 1/\sqrt{\frac{1}{n} \sum_{i=1}^{n} (Xi_k - \beta_k)^2}\,. \end{aligned} \tag{14.6}$$

Both scaling procedures defined by eqs. (14.5) and (14.6) are based on the assumption that no one descriptor dominates the overall distance measures.

14.3.6 The Similarity and Diversity Problems

The Euclidean distance function $f(Xi, Xj) = \delta_{ij}$ based on the chemical descriptors can be used in performing similarity searches among the database compounds and between these compounds and a particular target. This involves optimization of the distance function over $i = 1, \ldots, n$, for a fixed j:

$$\text{Minimize }_{Xi \in \mathcal{S}} \{f(\delta_{ij})\}. \tag{14.7}$$

More difficult and computationally-demanding is the diversity problem. Namely, we seek to reduce the database of the n compounds by selecting a "representative subset" of the compounds contained in $\mathcal{S}$, that is one that is "the most diverse" in terms of potential chemical activity. We can formulate the diversity problem as follows:

$$\text{Maximize} \sum_{Xi, Xj \in \mathcal{S}_0} \{f(\delta_{ij})\} \tag{14.8}$$

for a given subset $\mathcal{S}_0$ of size n_0.

The molecular diversity problem naturally arises since pharmaceutical companies must scan huge databases each time they search for a specific pharmacological activity. Thus reducing the set of n compounds to n_0 representative elements of the set $\mathcal{S}_0$ is likely to accelerate such searches. 'Combinatorial library design' corresponds to this attempt to choose the best set of substituents for combinatorial synthetic schemes so as to maximize the likelihood of identifying lead compounds.

The molecular diversity problem involves maximizing the volume spanned by the elements of $\mathcal{S}_0$ as well as the separation between those elements. Geometrically, we seek a well separated, uniform-like distribution of points in the high-dimensional compound space in which each chemical cluster has a 'representative'.

A simple, heuristic formulation of this problem might be based on the similarity problem above: successively minimize $f(\delta_{ij})$ over all i, for a fixed (target) j, so as to eliminate a subset $\{Xi\}$ of compounds that are similar to Xj. This approach thus identifies groupings that *maximize intracluster similarity* as well as *maximize intercluster diversity*.

The *combinatorial optimization* problem, an example of a very difficult computational task, has *non-polynomial computational complexity* ('NP-complete') (see footnote in Chapter 10, Section 10.2). This is because an exhaustive calculation of the above distance-sum function over a *fixed set* $\mathcal{S}_0$ of n_0 elements requires a total of $\mathcal{O}(n_0^2 m)$ operations. However, there are many possible subsets of $\mathcal{S}$ of size n_0, namely $C_n^{n_0}$ of them, where

$$\begin{aligned} C_n^{n_0} &= \frac{n!}{n_0!\,(n - n_0!)} \\ &= \frac{n(n-1)(n-2)\cdots(n-n_0+1)}{n_0!}. \end{aligned} \tag{14.9}$$

As a simple example, for $n = 4$, we have $C_4^1 = 4/1 = 4$ subsets of one element; $C_4^2 = (4 \times 3)/2 = 6$ different subsets of two elements, $C_4^3 = (4 \times 3 \times 2)/(3!) = 4$

subsets of three elements, and $C_4^4 = (4 \times 3 \times 2)/(4!) =$ one subset of four elements.

Typically, these combinatorial optimization problems are solved by stochastic and heuristic approaches. These include genetic algorithms, simulated annealing, and tabu-search variants. (See Agrafiotis [1001], for example, for a review).

As in other applications, the efficiency of simulated annealing depends strongly on the choice of cooling schedule and other parameters. Several potentially valuable annealing algorithms such as deterministic annealing, multiscale annealing, and adaptive simulated annealing, as well as other variants, have been extensively studied.

Various formulations of the diversity problem have been used in practice. Examples include the maximin function — to maximize the minimum intermolecular similarity:

$$\text{Maximize}_{i,\ Xi \in \mathcal{S}_0} \left\{ \min_{\substack{j \neq i \\ Xj \in \mathcal{S}_0}} (\delta_{ij}) \right\} \tag{14.10}$$

or its variant — maximizing the sum of these distances:

$$\text{Maximize}_{Xi, Xj \in \mathcal{S}_0} \sum_i \{ \min_{j \neq i} (\delta_{ij}) \} \,. \tag{14.11}$$

The maximization problem above can be formulated as a minimization problem by standard techniques if $f(x)$ is normalized so it is monotonic with range $[0, 1]$, since we can often write

$$\max[f(x)] \Leftrightarrow \min[-f(x)] \quad \text{or} \quad \min[1 - f(x)] \,.$$

In special cases, combinatorial optimization problems can be formulated as integer programming and mixed-integer programming problems. In this approach, linear programming techniques such as interior methods can be applied to the solution of combinatorial optimization problems, leading to branch and bound algorithms, cutting plane algorithms, and dynamic programming algorithms. Parallel implementation of combinatorial optimization algorithms is also important in practice to improve the performance.

Other important research areas in combinatorial optimization include the study of various algebraic structures (such as matroids and greedoids) within which some combinatorial optimization problems can more easily be solved [1002].

Currently, practical algorithms for addressing the diversity problem in drug design are relatively simple heuristic schemes that have computational complexity of at most $\mathcal{O}(n^2)$, already a huge number for large n.

14.4 Data Compression and Cluster Analysis

Dimensionality reduction and data visualization are important aids in handling the similarity and diversity problems outlined above. Principal component analysis (PCA) is a classic technique for data compression (or dimensionality reduction).

It has already shown to be useful in analyzing microarray data (e.g., [131]), as discussed in Chapter 1. The singular value decomposition (SVD) is another closely related approach. Data visualization for cluster analysis requires dimensionality reduction in the form of a projection from a high-dimensional space to 2D or 3D so that the dataset can be easily visualized. Cluster analysis is heuristic in nature.

In this section we outline the PCA and SVD approaches for dimensionality reduction in turn, continue with the distance refinement that can follow such analyses, and illustrate projection and clustering results with some examples.

14.4.1 Data Compression Based on Principal Component Analysis (PCA)

PCA transforms the input system (our database matrix A) into a smaller matrix described by a few uncorrelated variables called the **principal components** (PCs). These PCs are related to the eigenvectors of the covariance matrix defined by the component variables. The basic idea is to choose the orthogonal components so that the original data variance is well approximated. That is, the relations of similarity/dissimilarity among the compounds can be well approximated in the reduced description. This is done by performing eigenvalue analysis on the covariance matrix that describes the statistical relations among the descriptor variables.

Covariance Matrix and PCs

Let a_{ij} be an element of our $n \times m$ database matrix A. The covariance matrix $C_{m \times m}$ is formed by elements $c_{jj'}$ where each entry is obtained from the sum

$$c_{jj'} = \frac{1}{n-1} \sum_{i=1}^{n} (a_{ij} - \mu_j)\,(a_{ij'} - \mu_{j'})\,. \tag{14.12}$$

Here μ_j is the mean of the column associated with descriptor j:

$$\mu_j = \frac{1}{n} \sum_{i=1}^{n} a_{ij}\,. \tag{14.13}$$

C is a symmetric semi-definite matrix and thus has the spectral decomposition

$$C = V \Sigma V^T\,, \tag{14.14}$$

where the superscript T denotes the matrix transpose, and the matrix V $(m \times m)$ is the orthogonal eigenvector matrix satisfying $VV^T = I_{m \times m}$ with m component vectors $\{v_i\}$. The diagonal matrix Σ of dimension m contains the m ordered eigenvalues

$$\lambda_1 \geq \lambda_2 \geq \cdots \geq \lambda_m \geq 0\,.$$

We then define the m PCs Yj for $j = 1, 2, \cdots, m$ as the product of the original matrix A and the eigenvectors v_j:

$$Yj = Av_j\,, \qquad j = 1, 2, \cdots, m\,. \tag{14.15}$$

We also define the $m \times m$ matrix Y corresponding to eq. (14.15), related to V, as the matrix that holds the m PCs $Y1, Y2, \cdots, Ym$; this allows us to write eq. (14.15) in the matrix form $Y = AV$. Since $VV^T = I$, we then obtain an expression for the dataset matrix A in terms of the PCs:

$$A = YV^T\,. \tag{14.16}$$

Dimensionality Reduction

The problem dimensionality can be reduced based on eq. (14.16). First note that eq. (14.16) can be written as:

$$A = \sum_{j=1}^{m} Yj \cdot v_j^T\,. \tag{14.17}$$

Second, note that Xi, the vector of compound i, is the transpose of the ith row vector of A:

$$Xi = A^T\, e_i\,, \tag{14.18}$$

where e_i is an $n \times 1$ unit vector with 1 in the ith component and 0 elsewhere. Thus, compound Xi is expressed as the linear combination of the orthonormal set of eigenvectors $\{v_j\}$ of the covariance matrix C derived from A:

$$Xi = \sum_{j=1}^{m} (Yj_i)\, v_j\,, \quad i = 1, 2, \cdots, n\,, \tag{14.19}$$

where Yj_i is the ith component of the column vector Yj.

Based on eq. (14.19), the problem dimensionality m can be reduced by constructing a k-dimensional approximation to Xi, Xi^k, in terms of the first k PCs:

$$Xi^k = \sum_{j=1}^{k} (Yj_i)\, v_j\,, \quad i = 1, 2, \cdots, n\,. \tag{14.20}$$

The index k of the approximation can be chosen according a criterion involving the threshold variance γ, where

$$\left(\sum_{i=1}^{k} \lambda_i \right) \Big/ \left(\sum_{i=1}^{m} \lambda_i \right) \geq \gamma\,. \tag{14.21}$$

The eigenvalues of C represent the variances of the PCs. Thus, the measure $\gamma = 1$ for $k = m$ reflects a 100% variance representation. In practice, good approximations to the overall variance (e.g., $\gamma > 0.7$) can be obtained for $k \ll m$ for large databases.

For such a suitably chosen k, the smaller database represented by components $\{Xi^k\}$ for $i = 1, 2, \cdots, n$ approximates the variance of the original database A reasonably, making it valuable for cluster analysis.

As we show below, the singular value decomposition can be used to compute the factorization of the covariance matrix C when the 'natural scaling' of eq. (14.6) is used.

14.4.2 Data Compression Based on the Singular Value Decomposition (SVD)

SVD is a procedure for data compression used in many practical applications like image processing and cryptanalysis (code deciphering) [1003, for example]. Essentially, it is a factorization for rectangular matrices that is a generalization of the eigenvalue decomposition for square matrices. Image processing techniques are common tools for managing large datasets, such as digital encyclopedias, or images transmitted to earth from space shuttles on limited-speed modems.

SVD defines two appropriate *orthogonal coordinate systems* for the domain and range of the mapping defined by a rectangular $n \times m$ matrix A. This matrix maps a vector $x \in \mathcal{R}^n$ to a vector $y = Ax \in \mathcal{R}^m$. The SVD determines the orthonormal coordinate system of $\mathcal{R}^n$ (the columns of an $n \times n$ matrix U) and the orthonormal coordinate system of $\mathcal{R}^m$ (the columns of an $m \times m$ matrix V) so that A is diagonal.

The SVD is used routinely for storing computer-generated images. If, a photograph is stored as a matrix where each entry corresponds to a pixel in the photo, fine resolution requires storage of a huge matrix. The SVD can factor this matrix and determine its *best rank-k approximation*. This approximation is computed not as an explicit matrix but rather as a sum of k outer products, each term of which requires the storage of two vectors, one of dimension of n and another of dimension m ($m+n$ storage for the pair). Hence, the total storage required for the image is reduced from nm to $(m + n)k$.

The SVD also provides the *rank of* A (the number of independent columns), thus specifying how the data may be stored more compactly via the best rank-k approximation. This reformulation can reduce the computational work required for evaluation of the distance function used for similarity or diversity sampling.

SVD Factorization

The SVD decomposes the real matrix A as:

$$A = U \Sigma V^T, \tag{14.22}$$

where the matrices U ($n \times n$) and V ($m \times m$) are orthogonal, i.e., $UU^T = I_{n \times n}$ and $VV^T = I_{m \times m}$. The matrix Σ ($n \times m$) contains at most m nonzero entries

$(\sigma_i, i = 1, \cdots, m)$, known as the *singular values*, in the first m diagonal elements:

$$\Sigma = \begin{pmatrix} \sigma_1 & 0 & 0 & 0 & \cdots & \cdots & \cdots & 0 \\ 0 & \sigma_2 & 0 & 0 & \cdots & \cdots & \cdots & 0 \\ 0 & 0 & \cdots & \sigma_r & \cdots & \cdots & \cdots & 0 \\ \vdots & & & & & & & \\ 0 & 0 & 0 & 0 & \cdots & \cdots & \cdots & \sigma_m \\ 0 & 0 & 0 & 0 & 0 & 0 & 0 & 0 \\ \vdots & & & & & & & \\ 0 & 0 & 0 & 0 & \cdots & \cdots & \cdots & 0 \end{pmatrix} \tag{14.23}$$

where

$$\sigma_1 \geq \sigma_2 \geq \ldots \geq \sigma_r \ldots \geq \sigma_m \geq 0\,.$$

The columns of U, namely $u_1, \ldots, u_n$, are the *left singular vectors*; the columns of V, namely $v_1, \ldots, v_m$, are the *right singular vectors*. In addition, r = rank of A = number of nonzero singular values. Thus if $r \ll m$, a rank-r approximation of A is natural. Otherwise, we can set k to be smaller than r by neglecting the singular values beyond a certain threshold.

Low-Rank Approximation

The rank-k approximation to A can be obtained by noting that A can be written as the sum of rank-1 matrices:

$$A = \sum_{j=1}^{r} \sigma_j \, u_j \, v_j^T\,. \tag{14.24}$$

The rank-k approximation, A_k, is simply formed by extending the summation in eq. (14.24) from 1 to k instead of 1 to r. In practice, this means storing k left singular vectors and k right singular vectors. This matrix A_k can also be written as

$$A_k = \sum_{j=1}^{k} \sigma_j \, u_j \, v_j^T = U \Sigma_k V^T \tag{14.25}$$

where

$$\Sigma_k = \text{diag}\,(\sigma_1, \ldots, \sigma_k, 0, \ldots, 0)\,.$$

This matrix is closest to A in the sense that

$$\|A - A_k\| = \sigma_{k+1}$$

for the standard Euclidean norm.

Recall that we can express each Xi as:

$$\text{Row } i \text{ of } (A) = (A^T e_i)^T\,,$$

where e_i is an $n \times 1$ unit vector with 1 in the ith component and 0 elsewhere. Using the decomposition of eq. (14.24), we have:

$$A^T e_i = \sum_{j=1}^{r} \sigma_j \, v_j \, u_j^T \, e_i = \sum_{j=1}^{r} (\sigma_j \, u_{j_i}) \, v_j \, .$$

The SVD transforms this row vector to $[(A_k)^T \, e_i]^T$, where:

$$(A_k)^T \, e_i = \sum_{j=1}^{k} (\sigma_j \, u_{j_i}) \, v_j \, . \tag{14.26}$$

Projection

This transformation can be used to project a vector onto the first k principal components. That is, the projection matrix $P_k = \sum_{j=1}^{k} [v_j v_j^T]$ maps a vector from m to k dimensions. For example, for $k = 2$, we have:

$$\begin{aligned} P_2 A^T \, e_i \quad &= \quad \sum_{j=1}^{r} (v_1 \, v_1^T + v_2 \, v_2^T)(\sigma_j \, u_{j_i}) \, v_j \\ &= \quad (\sigma_1 \, u_{1_i}) \, v_1 \; + \; (\sigma_2 \, u_{2_i}) \, v_2 \, . \end{aligned} \tag{14.27}$$

Thus, this projection maps the m-dimensional row vector Xi onto the two-dimensional (2D) vector Yi with components $\sigma_1 u_{1_i}$ and $\sigma_2 u_{2_i}$. This mapping generalizes to a projection onto the k-dimensional space where $k \ll m$:

$$Yi^k = (\sigma_1 \, u_{1_i} \, , \sigma_2 \, u_{2_i} \, , \cdots , \sigma_k \, u_{k_i}) \, . \tag{14.28}$$

14.4.3 Relation Between PCA and SVD

It can be shown that the eigenvectors $\{v_i\}$ of the covariance matrix (eq. (14.14)) coincide with the right eigenvectors $\{v_i\}$ defined above when the second scaling (eq. (14.6)) is applied to the database matrix. Recall that this scaling makes all columns have zero means and a variance of unity.

Moreover, the left SVD vectors $\{u_i\}$ can be related to the singular values $\{\sigma_i\}$ and PC vectors $\{Yi\}$ of eq. (14.15) by

$$u_i = A \, v_i / \sigma_i = Yi / \sigma_i \, . \tag{14.29}$$

Therefore, we can use the SVD factorization as defined above (eq. (14.22)) to compute the PCs $\{Yi\}$ of the covariance matrix C. The SVD approach is more efficient since formulation of the covariance matrix is not required.

The algorithm ARPACK [1004] can compute the first k PCs, saving significant storage. It requires an order $\mathcal{O}(nk)$ memory and $\mathcal{O}(nm^2)$ floating point operations.

14.4.4 Data Analysis via PCA or SVD and Distance Refinement

The SVD or the PCA projection is a first step in database visualization. The second step refines this projection so that the original Euclidean distances $\{\delta_{ij}\}$ in the m-dimensional space are closely related to the corresponding distances $\{d_{ij}\}$ in the reduced, k-D space. Here,

$$\delta_{ij} \equiv ||Xi - Xj||$$

and

$$d_{ij} \equiv ||Yi - Yj||$$

for all i, j, where the vectors $\{Y_i\}$ are the k-D vectors produced by SVD defined by eq. (14.28).

Projection Refinement

This distance refinement is a common task in distance geometry refinement of NMR models. In the NMR context, a set of interatomic distances is given and the objective is to find the 3D coordinate vector (the molecular structure) that best fits the data. Since such a problem is typically overdetermined — there are $\mathcal{O}(n^2)$ distances but only $\mathcal{O}(n)$ Cartesian coordinates for a system of n atoms — an optimal *approximate solution* is sought.

For example, optimization work on evolutionary trees [1005] solved an identical mathematical problem in an unusual context that is closely related to the molecular similarity problem here. Specifically, the experimental distance-data in evolutionary studies reflect complex factors rather than simple spatial distances (e.g., interspecies data arise from immunological studies which compare the genetic material among taxa and assign similarity scores). Finding a 3D evolutionary tree by the distance-geometry approach, rather than the conventional 2D tree which conveys evolutionary linkages, helps identify subgroup similarities.

Distance Geometry

The distance-geometry problem in our evolutionary context can be formulated as follows. We are given a set of pairwise distances with associated lower and upper bounds:

$$\{l_{ij} \leq \delta_{ij} \leq u_{ij}\}, \quad \text{for} \quad i, j = 1, 2, \ldots, n,$$

where each δ_{ij} is a target interspecies distance with associated lower and upper bounds l_{ij} and u_{ij}, respectively, and n is the number of species. Our goal is to compute a 3D "tree" for those species based on the measured distance/similarity data.

This distance geometry problem can be reduced to finding a coordinate vector Y that minimizes the objective function

$$E(Y) = \sum_{i<j} \omega_{ij} \left(d_{ij}^2(Y) - \delta_{ij}^2\right)^2 , \tag{14.30}$$

where $d_{ij}(Y)$ is Euclidean distance between points i and j in the vector Y, and the $\{\omega_{ij}\}$ are appropriately-chosen weights.

In the combinatorial chemistry context, we use the same function $E(Y)$ where Y is the vector of $2n$ components, listing the 2D projections of each compound in turn. Details of this data clustering approach are described in [1006, 1007]. Minimization can be performed so that the high-dimensional distance relationships are approximated.

Besides the value of the objective function (eq. (14.30)), a useful measure of the distance approximation in the low-dimensional space is the percentage of intercompound distances $\{i, j\}$ (out of $n(n-1)/2$) that are within a certain threshold of the original distances. We first define the deviations from the targets by a percentage η so that

$$\begin{aligned} |d(Yi, Yj) - \delta_{ij}| \leq \eta\, \delta_{ij} \qquad & \text{when} \quad \delta_{ij} > d_{\min}\,, \\ d(Yi, Yj) \leq \tilde{\epsilon} \qquad & \text{when} \quad \delta_{ij} \leq d_{\min}\,, \end{aligned} \tag{14.31}$$

where η, $\tilde{\epsilon}$, and $d_{\min}$ are given small positive numbers less than one. For example, $\eta = 0.1$ specifies a 10% accuracy; the other values may be set to small positive numbers such as $d_{\min} = 10^{-12}$ and $\tilde{\epsilon} = 10^{-8}$. The second case above (very small original distance) may occur when two compounds in the datasets are highly similar.

With this definition, the total number T_d of the distance segments $d(Yi, Yj)$ satisfying eq. (14.31) can be used to assess the degree of distance preservation of our mapping. We define the percentage ρ of the distance segments satisfying eq. (14.31) as

$$\rho = \frac{T_d}{n(n-1)/2} \times 100\,. \tag{14.32}$$

The greater the ρ value (the maximum is 100), the better the mapping and the more information that can be inferred from the projected views of the database compounds.

This minimization procedure (projection refinement) is quite difficult for scaled datasets. Experiments with several chemical datasets of size 58 to 27255 compounds show that the percentage of distances satisfying a threshold deviation ρ of 10% (eq. (14.31)) is in the range of 40% [1006, 1007]. Nonetheless, these low values can be made close to 100% with projections onto 10-dimensional space. This is illustrated in Figure 14.4, which shows the percentage of distances satisfying eq. (14.31) for $\eta = 0.1$ as a function of the projection dimension for a database ARTF.

A similar improvement can be achieved with larger tolerances η (e.g., distances that are within 25% of the original values rather than 10%) [1006, 1007].

14.4.5 *Projection, Refinement, and Clustering Example*

As an illustration, consider the model database ARTF of 402 compounds and $m = 312$ descriptors containing eight chemical subgroups. We have analyzed

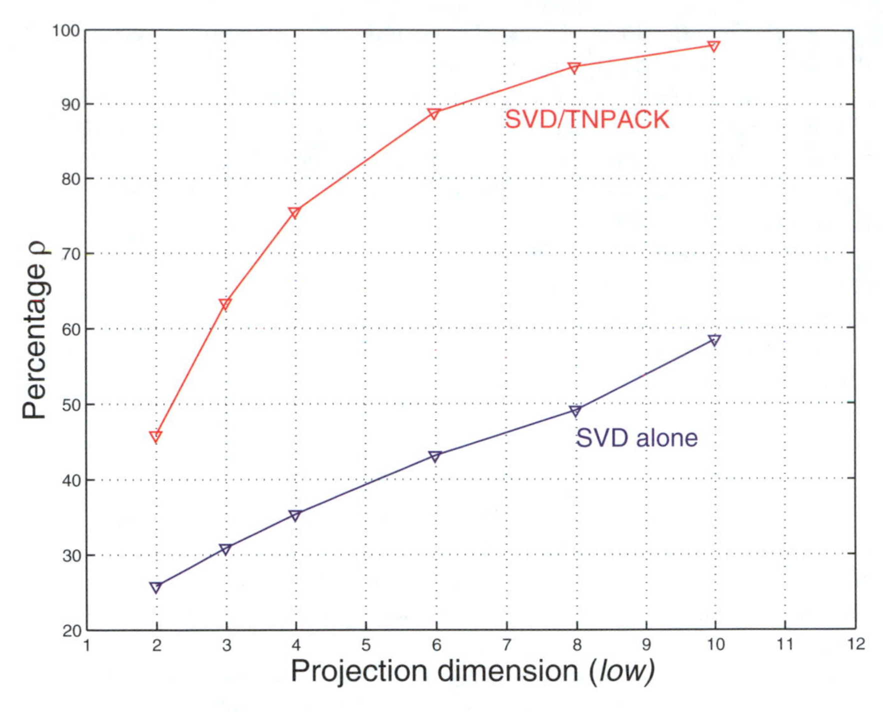

Figure 14.4. Performance of the SVD and SVD/minimization protocols for the ARTF chemical database in terms of the percentage of distances satisfying eq. (14.31) for $\eta = 0.1$ (reflecting 10% distance deviations) as a function of the projection dimension [1006, 1007].

this database by performing 2D and 3D projections based on the SVD factorization followed by minimization refinement by TNPACK [772, 773, 793] for performance assessment in terms of accuracy as well as visual analysis of the compound interrelationships.

From Figure 14.4 we note that the refinement stage that follows the SVD projection is important for increasing the accuracy in every dimension. Namely, the accuracy is increased by 25–40% in this example.

The 2D and 3D projection patterns obtained for ARTF in Figure 14.5 show the utility of such a projection approach. The resemblance between the 2D and 3D views is evident, and the various 3D views offer different perspectives of the intercompound relationships.

We note that clusters corresponding to individual pharmacological subsets appear very close to each other, though partial overlap of clusters is evident. The *ecdysteroid* group forms a diverse but separate set of points. The *estrogen* class is also clustered and somewhat separate from the others. The strong overlap of the three clusters corresponding to *D1 agonists*, *D1 antagonists*, and *H1 receptor ligands* is reasonable given the relative chemical similarity of these compounds: all act at receptors of the same pharmacological class (i.e., G-protein coupled re-

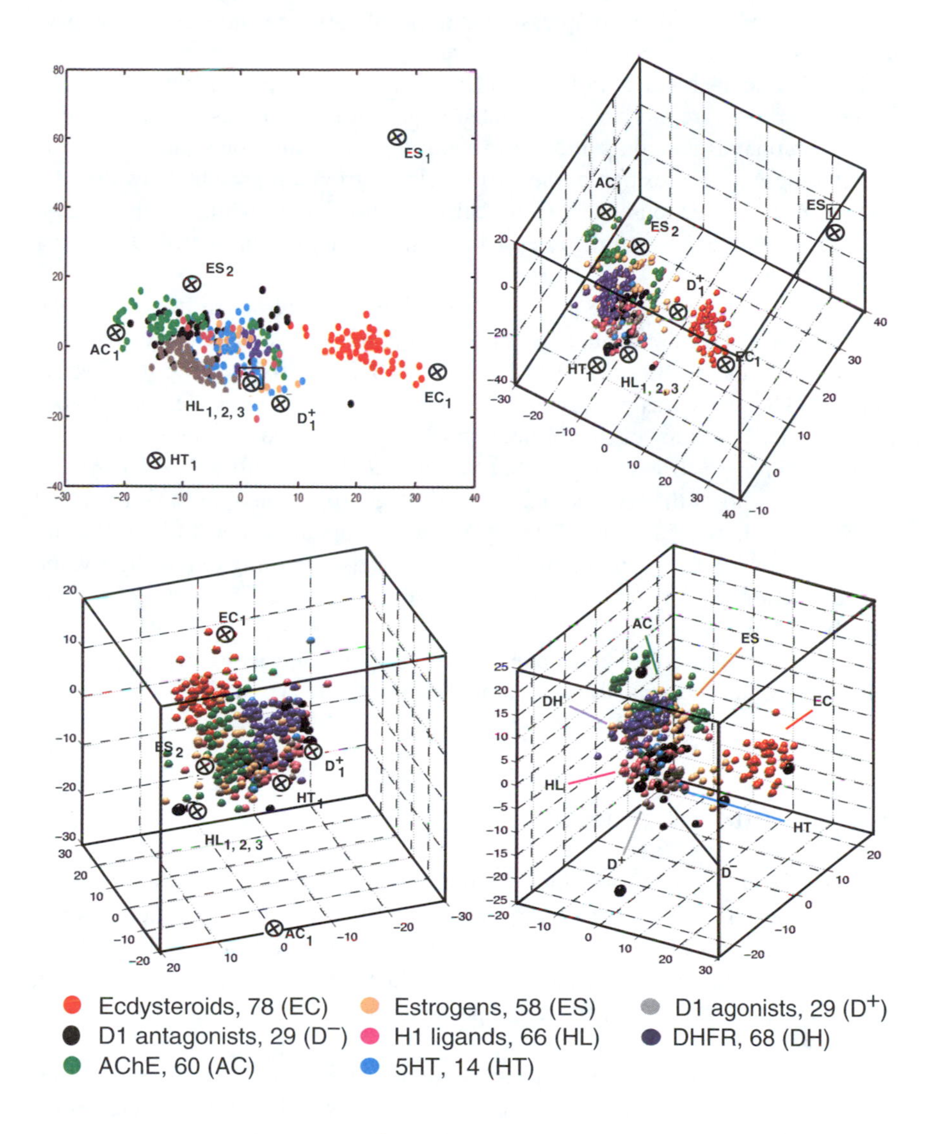

Figure 14.5. Two and three-dimensional projections of the chemical database ARTF of 402 compounds composed of the eight chemical subgroups ecdysteroids (EC), estrogens (ES), D1 agonists (D^+), D1 antagonists (D^-), H1 ligands (HL), DHFR inhibitors (DH), AchE inhibitors (AC), and 5HT ligands (HT) using the projection/refinement SVD/TNPACK approach [1006, 1007]. Three views are shown for the 3D projection. The accuracy of the 2D projection is about 46% and that of the 3D is 63% (with $\eta = 0.1$); see eq. (14.31). The 2D projection was obtained by refining the 3D projection. The nine chemical structures labeled in the projections are drawn in Figure 14.6.

ceptors). Thus, such data compression and visualization techniques can be used as a quick analysis tool of the database structure.

The chemical structures in Figure 14.6 reveal that compounds that are nearer in the projection are more closely related than those that are distant; this is seen when compounds are compared both within the same subgroup and within different subgroup. For example, the two labeled estrogen representatives that are distant in the projection appear chemically quite different, while the three clustered H1 ligands appear similar to each other and perhaps to the nearby D1 agonist representative.

An example of a database projection in 2D by the alternative PCA approach followed by distance refinement is shown in Figures 14.7 and 14.8 for 832 compounds from the MDL Drug Data Report (MDDR) database using topological indices. (This work was performed in collaboration with Merck Research Laboratories). The accuracy of this projection (the percentage of distances satisfying eq. (14.31) for $\eta = 0.1$) is only 0.2% after PCA and 24.8% after PCA/TNPACK. Figure 14.7 shows that compounds close in the projection appear similar, and Figure 14.8 shows that more distantly related compounds tend to be different. Without knowing the grouping of these compounds according to bioactivity, the clusters identified in Figure 14.8 suggest a 'diversity subset' consisting of a few members from each cluster.

The approach described here appears promising, but further work is required to make the technique viable for very large databases.

14.5 Future Perspectives

Similarity and diversity sampling of combinatorial chemistry libraries is a field in its infancy. The choice of descriptors as well as metrics used to define similarity and diversity are empirical and perhaps application dependent. Thus, many challenges remain for future developments in the field, and the added involvement of mathematical scientists and new approaches borrowed from allied disciplines might be fruitful.

Developments are needed for formulation of descriptor sets, rigorous mathematical frameworks for their analysis, and efficient algorithms for very large-scale problems based on statistics, cluster analysis, and optimization. The algorithmic challenge of manipulating large datasets might also explain the tendency toward smaller and focused libraries [1008]; still, as argued in [1009], this assumed defeat is premature!

The central assumption of structure/activity relationships of course remains a challenge to validate, develop, and further apply.

More broadly, structure-based drug design is likely to increase in importance as many more protein targets are identified and synthesized [975], and as modeling programs improve in their ability to predict binding affinities of certain ligands (e.g., peptide-like) that share chemical groups with macromolecules, the focus

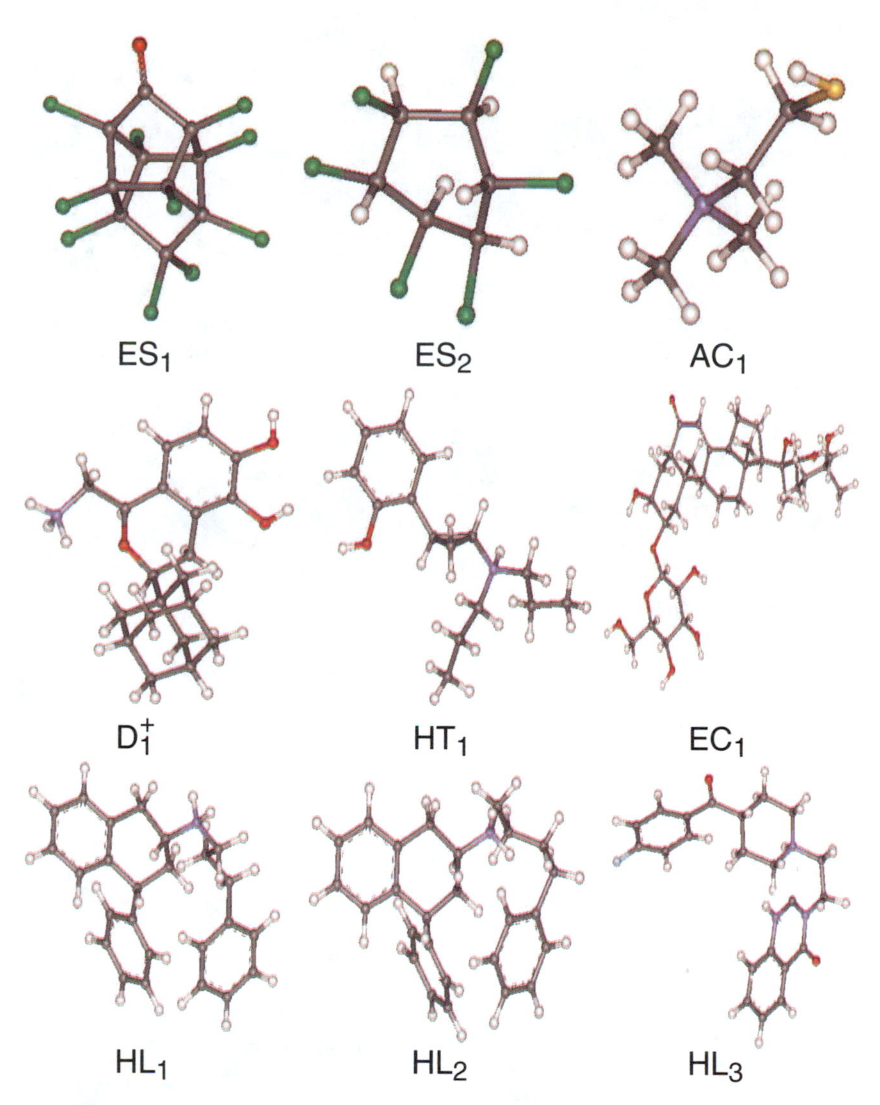

Figure 14.6. Selected chemical structures from the ARTF projection shown in Figure 14.5 reveal similarity of nearby structures and dissimilarity of distant compounds.

of many biomodeling packages. The difficulty in determining membrane protein structures continues to be a limitation since membrane receptors are important pharmacological targets.

While perhaps not the dominant technique, it is clear that structure-based drug design will be an important component of drug modification and optimization after available leads have been generated. The search for the needle in the haystack

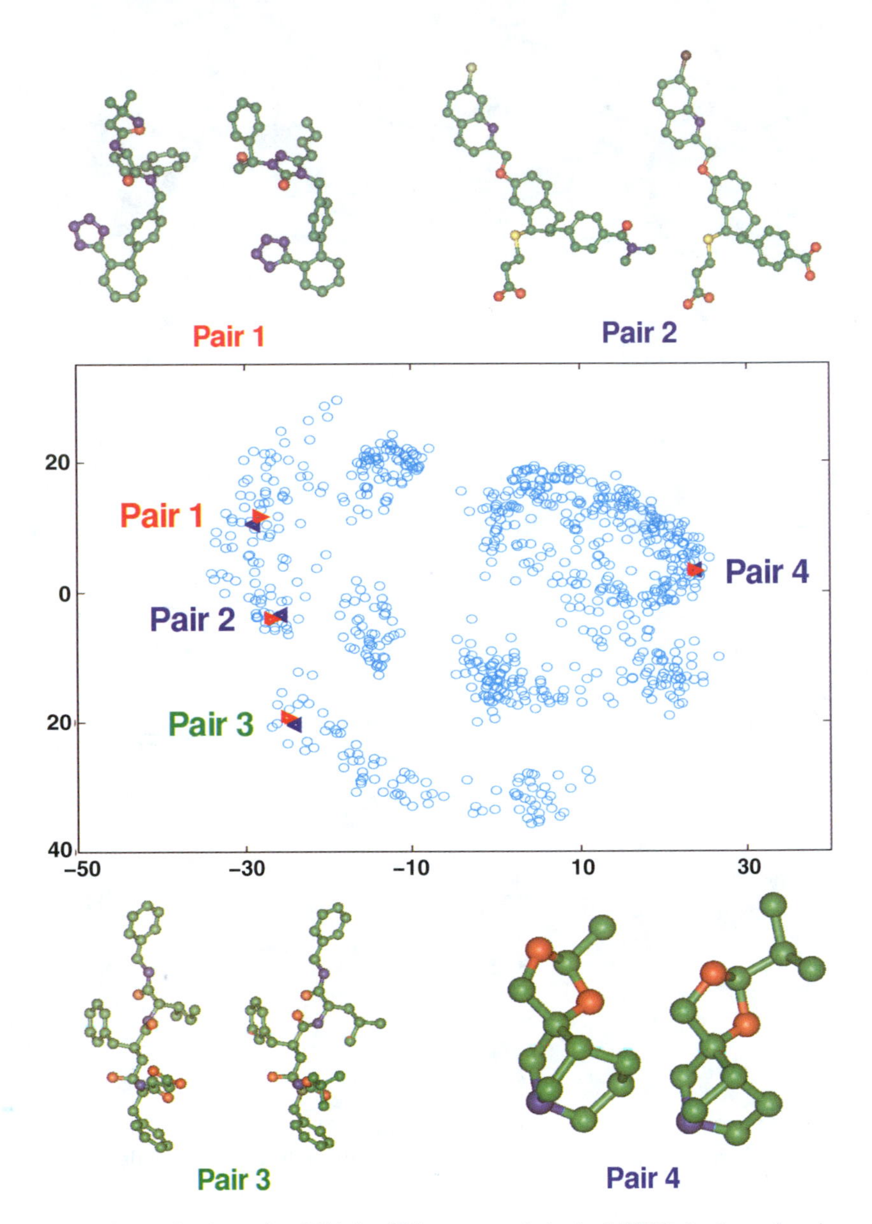

Figure 14.7. 2D projection using PCA for 832 compounds in the MDDR database showing the similarity of four compound pairs that are near in the projection.

(i.e., a successful drug) will likely be guided by the steady light generated by computer modeling.

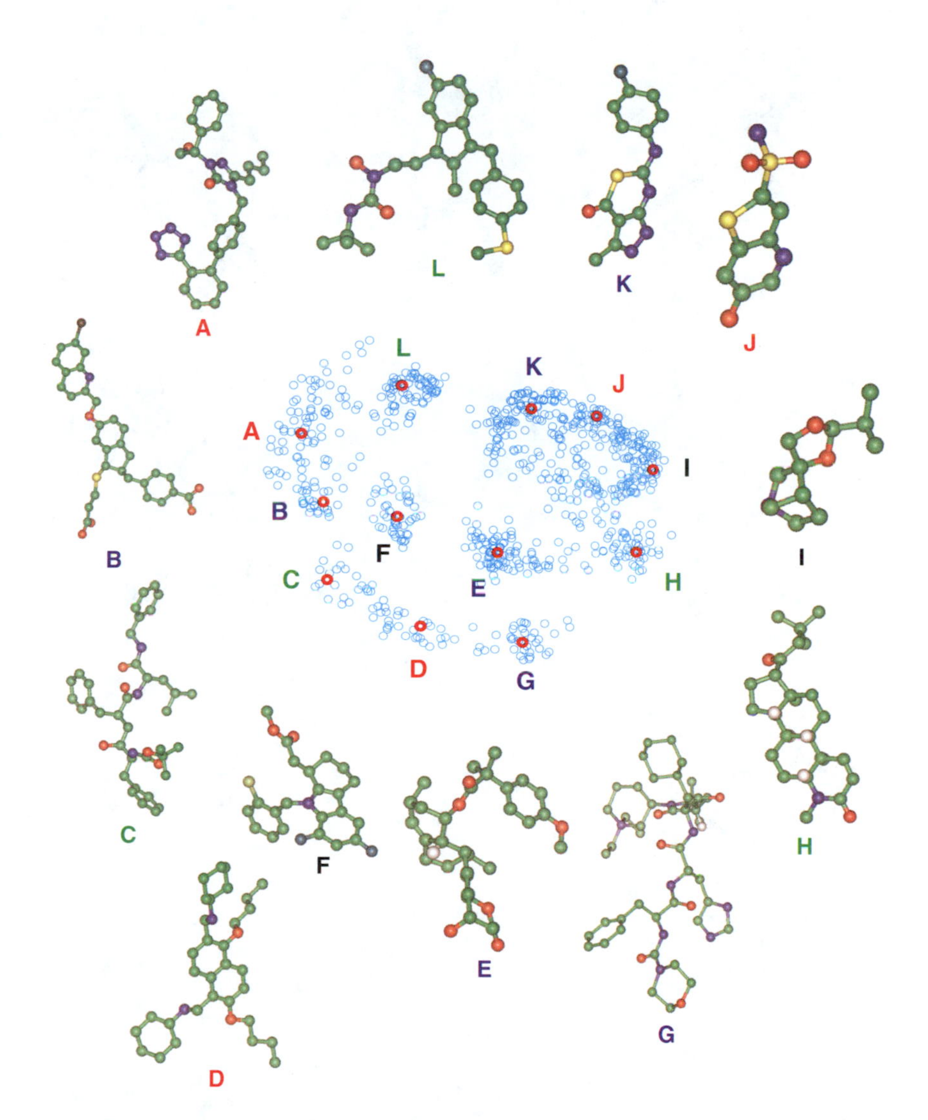

Figure 14.8. 2D projection using PCA for 832 compounds in the MDDR database showing the diversity of compounds that represent different clusters in the projection (distinguished by letters). A representative subset may thus consist of one or only a few members from each cluster.

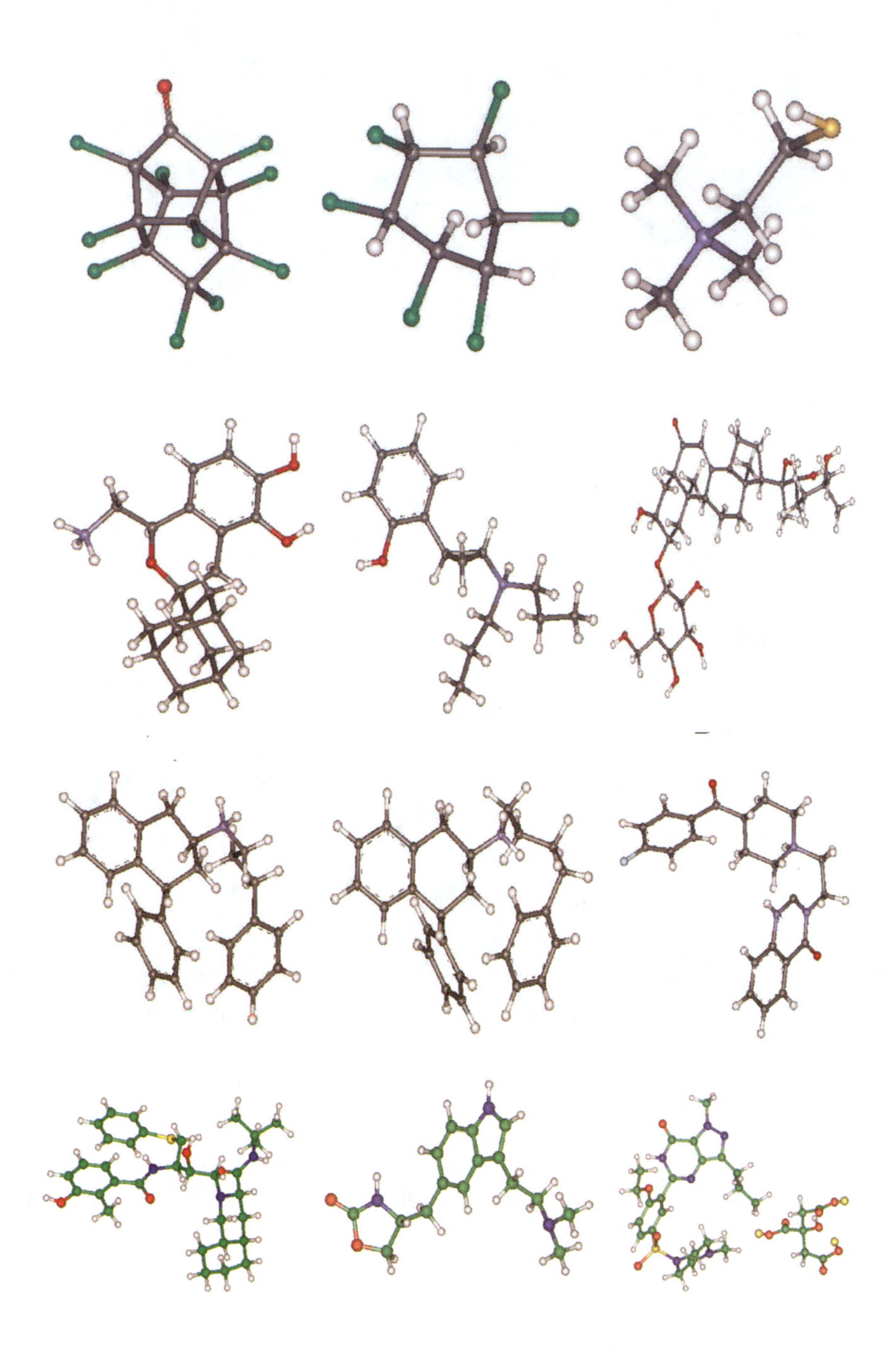

Epilogue

The need for accuracy must be weighed against the need for finality.

Comments by the Justices of the U.S. Supreme Court regarding bringing to a close the uncertain outcome, one month after the elections, of the U.S. Presidential elections in 2000 between George W. Bush and Albert Gore (2000).

Appendix A

Molecular Modeling Sample Syllabus

Please Note: Article numbers in the Homework column refer to items in the course reading list (Appendix B).

Class	**Subject**	Homework
1	Course and Field Overview: • What is molecular modeling and how has it evolved? • What are the practical applications and important questions? (**Preface** and **Chapter 1**)	**1**: Introduction to sequence and structural databases and to the early molecular modeling literature. *Read papers 1,2,3,20,33.*
2	• Continuation of the overview on biomolecular modeling and simulation, from drug design to new materials. • Discussion of the 1959 paper of Alder & Wainwright and 1971 work of Rahman & Stillinger: difficulties then and now. • Introduction to interesting biomolecular modeling problems: protein folding, protein misfolding, nucleic acid/protein interactions, and RNA folding. (**Chapter 2**)	**2**: Retrieval of structural information from the Protein Data Bank (PDB), and the display, manipulation, and analysis of three-dimensional biomolecular structures with the Insight II molecular graphics package. Explore kinemage tutorials. *Read papers 5,8,30,31,34.*
3	• Minitutorial on protein structure: amino acid repertoire, primary to quaternary structure, protein structure classification. • Kinemage tutorial demonstration: folding motifs and major protein classes. (**Chapters 3 & 4**)	**3**: Construction and analysis of the pentapeptide Met-enkephalin with the Insight II program. *Read papers 2,4,6.*
4	• Discuss homework assignments 1 and 2. • Minitutorial on nucleic acid structure: building blocks, backbone conformational flexibility, helical parameters, and DNA supercoiling. (**Chapters 5 & 6**)	**4**: Generation and analysis of Ramachandran plots for proteins and introduction to the NDB. *Read papers 23, 29.*
5	Guest Lecturer: *The Nucleic Acid Database and the 'New Protein Databank' (RCSB)*, Prof. Helen Berman (Rutgers University, Department of Chemistry), Director of NDB and RCSB.	**5**: Analysis of Protein/DNA Complexes with Insight and NDB. *Read papers 7, 21.*
6	• Discuss homework assignment 4. • Computational and theoretical approaches to structure prediction (from the quantum-mechanical to the molecular mechanical description). (**Chapters 7 & 8**)	**6** (**MIDTERM**): Sequence/Structure/Function Relationships in Proteins, A Contest! *Read papers 22, 35.*

Table A.1: (continued)

Week	**Subject**	Homework
7	Guest Lecturer: *Protein Structure Modeling*, Dr. Andrej Sali (Rockefeller University), expert in protein modeling.	**7**: Molecular mechanics force fields: approximations, variations, and the assessment of results with respect to experiment and other simulations (*papers 10,11,15,16*). *Read papers 13,14,17,24.*
8	Amer. Chem. Soc. 1990 videotapes: *Molecular Modeling in Biological Systems*: 1 – Peter Kollman, "Methods in Molecular Modeling", 4 – Panel Discussion.	
9	Guest Lecturer: *Ab Initio Calculation of Protein Structure by Global Optimization of Potential Energy*, Prof. Harold Scheraga (Cornell University, Department of Chemistry), pioneer of protein force fields and computation of protein structure.	**8**: A bit of programming: nonbonded versus bonded energy computations. *Read paper 36.*
10	MIDTERM class presentations	
11	• Continue MIDTERM presentations. • Force Field Debate!	**9 (TERM PROJECT)**: The Successes (Failures?) of Molecular Modeling. *Read papers 9,18,19.*
12	• Molecular mechanics force fields — origin, variations, and parameterization. • Special topics — molecular topology: bookkeeping and data structures, potential energy differentiation. • Special issues in nonbonded energy computations — spherical cutoffs, fast electrostatics by the multipole method, periodic boundary conditions, and the Ewald summation. (**Chapter 9**)	**10**: Experiments in Geometry Structure Optimization: Minimization of Biphenyl with Insight II/Discover. *Read paper 12.*
13	• Go over Assignment 8, including general discussion of programming and timing strategies. • Optimization techniques for multivariate functions in computational chemistry. (**Chapter 10**)	**11**: A global optimization contest for a pentapeptide! *Read papers 25, 26.*
14	Monte Carlo Simulations. (**Chapter 11**)	**12**: An exercise in Monte Carlo. **13** (Optional): Advanced exercises in minimization and MC. *Read papers 27,28,32.*
15	Molecular dynamics simulations — theory and practice. (**Chapters 12 & 13**)	**14** (Optional): Advanced exercises in molecular dynamics. **15** (Optional): Scaling of protein conformations and folding simulations.

Appendix B

Article Reading List

Before 1970

1. B. J. Alder and T. E. Wainwright, "Studies in Molecular Dynamics. I. General Method", *J. Chem. Phys.* **31**, 459–466 (1959).

2. G. Némethy and H. A. Scheraga, "Theoretical Determination of Sterically Allowed Conformations of a Polypeptide Chain by a Computer Method", *Biopolymers* **3**, 155–184 (1965).

1970s

3. A. Rahman and F. H. Stillinger, "Molecular Dynamics Study of Liquid Water", *J. Chem. Phys.* **55**, 3336–3359 (1971).

4. P. Y. Chou and G. D. Fasman, "Prediction of Protein Conformation", *Biochemistry* **13**, 222–245 (1974).

5. M. Levitt and A. Warshel, "Computer Simulation of Protein Folding", *Nature* **253**, 694–698 (1975).

6. M. Levitt and C. Chothia, "Structural Patterns in Globular Proteins", *Nature* **261**, 552–558 (1976).

1980s

7. S. Lifson, "Potential Energy Functions for Structural Molecular Biology", in *Methods in Structural Molecular Biology*, pp. 359–385, D. B. Davies, W. Saenger, and S. S. Danyluk, Eds., Plenum Press, London (1981).

8. M. Karplus and J. A. McCammon, "The Dynamics of Proteins", *Sci. Amer.* **254**, 42–51 (1986).

9. M. S. Friedrichs and P. G. Wolynes, "Toward Protein Tertiary Structure Recognition by Means of Associative Memory Hamiltonians", *Science* **246**, 371–373 (1989).

10. I. K. Roterman, M. H. Lambert, K. D. Gibson, and H. A. Scheraga, "Comparison of the CHARMM, AMBER and ECEPP Potentials for Peptides. I. Conformational Predictions for the Tandemly Repeated Peptide (Asn-Ala-Asn-Pro)$_9$", *J. Biomol. Struct. Dyn.* **7**, 391–419 (1989a).

11. I. K. Roterman, M. H. Lambert, K. D. Gibson, and H. A. Scheraga, "Comparison of the CHARMM, AMBER and ECEPP Potentials for Peptides. II. ϕ–ψ Maps

for N-Methyl Amide: Comparisons, Contrasts and Simple Experimental Tests", *J. Biomol. Struct. Dyn.* **7**, 421–453 (1989b).

1990–1992

12. M. Karplus and G. A. Petsko, "Molecular Dynamics Simulations in Biology", *Nature* **347**, 631–639 (1990).

13. J. Skolnick and A. Kolinski, "Simulations of the Folding of a Globular Protein", *Science* **250**, 1121–1125 (1990).

14. F. M. Richards, "The Protein Folding Problem" *Sci. Amer.* **264**, 54–63 (1991).

15. P. A. Kollman and K. A. Dill, "Decisions in Force Field Development: An Alternative to Those Described by Roterman *et al.*", *J. Biomol. Struct. Dyn.* **8**, 1103–1107 (1991).

16. K. B. Gibson and H. A. Scheraga", "Decisions in Force Field Development: Reply to Kollman and Dill", *J. Biomol. Struct. Dyn.* **8**, 1109–1111 (1991).

17. H. A. Scheraga, "Predicting Three-Dimensional Structures of Oligopeptides", in Reviews in Computational Chemistry, K. B. Lipkowitz and D. B. Boyd, Editors, Vol. 3, pp. 73–142, VCH Publishers, New York (1992).

18. T. Schlick, "Optimization Methods in Computational Chemistry", in Reviews in Computational Chemistry, K. B. Lipkowitz and D. B. Boyd, Editors, Vol. 3, pp. 1–71, VCH Publishers, New York (1992). See also T. Schlick, "Geometry Optimization", in the Encyclopedia of Computational Chemistry, P. von Ragué Schleyer (Editor-in-Chief) and N. L. Allinger and T. Clark and J. Gasteiger and P. A. Kollman and Schaefer, III, H. F., Editors, Vol. 3, pp. 1136–1157, John Wiley & Sons, West Sussex, England (1998).

1993–1995

19. R. A. Abagyan and M. M. Totrov, "Biased Probability Monte Carlo Conformational Searches and Electrostatic Calculations for Peptides and Proteins", *J. Mol. Biol.* **235**, 983–1002 (1994).

20. J. A. Board, Jr., L. V. Kalé, K. Schulten, R. D. Skeel, and T. Schlick, "Modeling Biomolecules: Larger Scales, Longer Durations", *IEEE Comp. Sci. Eng.* **1**, 19–30 (Winter 1994).

21. K. B. Lipkowitz, "Abuses of Molecular Mechanics. Pitfalls to Avoid", *J. Chem. Educ.* **72**, 1070–1075 (1995).

22. B. Honig and A. Nicholls, "Classical Electrostatics in Biology and Chemistry", *Science* **268**, 1144–1149 (1995).

1996–1998

23. B. Cipra, "Computer Science Discovers DNA", in *What's Happening in the Mathematical Sciences*, pp. 26–37 (P. Zorn, Ed.), American Mathematical Society, Colonial Printing, Cranston, RI (1996).

24. A. Neumaier, "Molecular Modeling of Proteins and Mathematical Prediction of Protein Structure", *SIAM Review* **39**, 407–460 (1997).

25. K. A. Dill and H. S. Chan, "From Levinthal to Pathways to Funnels", *Nature Struc. Biol.* **4**, 10–19 (1997).

26. T. Lazaridis and M. Karplus, " 'New View' of Protein Folding Reconciled with the Old Through Multiple Unfolding Simulations", *Science* **278**, 1928–1931 (1997).

27. T. Schlick, E. Barth, and M. Mandziuk, "Biomolecular Dynamics at Long Timesteps: Bridging the Timescale Gap Between Simulation and Experimentation", *Ann. Rev. Biophys. Biomol. Struc.* **26**, 179–220 (1997).

28. E. Barth and T. Schlick, "Overcoming Stability Limitations in Biomolecular Dynamics: I. Combining Force Splitting via Extrapolation with Langevin Dynamics in LN", *J. Chem. Phys.* **109**, 1617–1632 (1998).

29. M. Gerstein and M. Levitt, "Simulating Water and the Molecules of Life", *Sci. Amer.* **279**, 101–105 (1998).

30. Y. Duan and P. A. Kollman, "Pathways to a Protein Folding Intermediate Observed in a 1-Microsecond Simulation in Aqueous Solution", *Science* **282**, 740–744 (1998).

31. H. J. C. Berendsen, "A Glimpse of the Holy Grail", *Science* **282**, 642–643 (1998).

32. L. S. D. Caves, J. D. Evanseck, and M. Karplus, "Locally Accessible Conformations of Proteins: Multiple Molecular Dynamics Simulations of Crambin", *Prot. Sci.* **7**, 649–666 (1998).

33. W. F. van Gunsteren and A. E. Mark, "Validation of Molecular Dynamics Simulation", *J. Chem. Phys.* **108**, 6109–6116 (1998).

34. X. Daura, B. Juan, D. Seebach, W. F. Van Gunsteren, and A. Mark, "Reversible Peptide Folding in Solution by Molecular Dynamics Simulation", *J. Mol. Biol.* **280**, 925–932 (1998).

1999–

35. D. Baker and A. Sali, "Protein Structure Prediction and Structural Genomics", *Science* **294**, 93–96 (2001).

36. R. M. Karp, "Mathematical Challenges from Genomics and Molecular Biology", *Notices Amer. Math. Soc.* **49**, 544–553 (2002).

Appendix C
Supplementary Course Texts

1. M. P. Allen and D. J. Tildesley. *Computer Simulation of Liquids.* Oxford University Press, New York, NY, 1987.

   [Good advanced reference book for molecular simulations.]

2. A. D. Bates and A. Maxwell. *DNA Topology*, In Focus series, Oxford University Press, New York, NY, 1993.

   [Beautiful paperback on the higher organizational forms of DNA.]

3. V. A. Bloomfield, D. M. Crothers, and I. Tinoco, Jr. *Nucleic Acids: Structures, Properties, and Functions.* University Science Press, New York, NY, 2000.

   [Comprehensive account of nucleic acids at an advanced level, with emphasis on biological function and experimental techniques. The first part describes nucleic acid properties on the atomic and molecular levels as deduced by various experimental techniques. The second part presents macromolecular features of nucleic acids in solution (e.g., dynamics behavior, DNA supercoiling). The third part covers noncovalent interactions of nucleic acids and other molecules (ions, drugs, proteins) and higher-order structures due to cellular packing.]

4. C. Branden and J. Tooze. *Introduction to Protein Structure*, second edition, Garland Publishing Inc., New York, NY, 1999.
   (www.proteinstructure.com/).

   [A modern and nicely illustrated protein structure textbook dealing with basic structural principles (part 1) and other topics (part 2, broadly grouped under the heading of relationships among structure, function, and engineering). Part 2 includes chapters on transcription regulation, signal transduction, immune regulation, membrane and fibrous proteins, and virus structures.]

5. P. Bratley, B. L. Fox, and L. E. Schrage. *A Guide to Simulation*. Springer-Verlag, New York, NY, 1987.

[Good introduction to Monte Carlo simulations.]

6. C. L. Brooks, III, M. Karplus, and B. M. Pettitt. *Proteins: A Theoretical Perspective of Dynamics, Structure, and Thermodynamics*. Wiley Interscience, New York, NY, 1988.

[Nice collection on protein simulations.]

7. U. Burkert and N. L. Allinger. *Molecular Mechanics*. American Chemical Society, Washington D.C., 1982.

[Excellent introduction to molecular mechanics.]

8. C. R. Cantor and P.R. Schimmel. *Biophysical Chemistry*. Vol. 1,2,3. W.H. Freeman and Company, San Francisco, CA, 1980.

[Excellent text and reference book on many aspects of biophysical chemistry.]

9. N. R. Cohen, Editor. *Guidebook on Molecular Modeling in Drug Design*. Academic Press, San Diego, CA, 1996.

[Modern reference for molecular modeling as applied to drug design problems, containing contributed chapters by industrial and academic scientists, on problem formulation (database analysis, docking), modeling tools, and medicinal chemistry applications.]

10. P. Deuflhard, J. Hermans, B. Leimkuhler, A. E. Mark, S. Reich,. R. D. Skeel, Editors. *Computational Molecular Dynamics: Challenges, Methods, Ideas – Proceedings of the 2nd International Symposium on Algorithms for Macromolecular Modelling, Berlin, May 21-24, 1997*, Lecture Notes in Computational Science and Engineering, Vol. 4 (Series Editors M. Griebel, D. E. Keyes, R. M. Nieminen, D. Roose, and T. Schlick), Springer-Verlag, Berlin, 1999.

[Collection of articles from invited presentations to the May 1997 Berlin workshop on methods for macromolecular modeling. The book contains sections on the following topics: conformational dynamics, thermodynamic modeling, enhanced time-stepping algorithms, quantum-classical simulations, and parallel force field evaluation.]

11. T. E. Creighton, Editor. *Protein Folding*. W.H. Freeman & Company, New York, NY, 1992.

[Nice collection with general topics regarding proteins covered.]

12. D. Eisenberg and D. Crothers. *Physical Chemistry with Applications to the Life Sciences*. Benjamin Cummings, Menlo Park, CA, 1979.

[Wonderful physical chemistry textbook, with useful biomolecular information.]

13. A. Fersht. *Structure and Mechanism in Protein Science: A Guide to Enzyme Catalysis and Protein Folding*. W. H. Freeman and Company, New York, NY, 1999.

[A comprehensive perspective of both enzyme catalysis and protein folding by a pioneer researcher, an updated version of the author's 1995 text on

Enzyme Structure and Mechanism; the text reviews protein structure, emphasizing general principles, as well as mentioning recent advances and insights from theoretical approaches.]

14. D. Frenkel and B. Smit. *Understanding Molecular Simulations. From Algorithms to Applications*. second edition, Academic Press, San Diego, CA, 2001.

    [Excellent introduction to computer simulation of molecular systems, containing a nice mix of mathematical details and more informal, descriptive text. The focus is on Monte Carlo and molecular dynamics methodologies, including simple algorithms and numerical illustrations. Some important recent methodological advances are also included.]

15. L. M. Gierasch and J. King, Editors. *Protein Folding, Deciphering the Second Half of the Genetic Code.* AAAS, Washington D.C., 1990.

    [Interesting and beautifully illustrated collection of articles; best bet: Jane Richardson's origami analogues of protein folding motifs!]

16. H. Gould and J. Tobochnik. *An Introduction to Computer Simulation Methods: Applications to Physical Systems. Parts 1 and 2.* Addison-Wesley, Reading, MA, 1988.

    [A good introduction to computer simulations, with a focus on classical mechanics in Part 1 and statistical physics in Part 2. The material is made highly accessible to undergraduates by the inclusion of many simple numerical examples, useful illustrations, and programming segments.]

17. A. Y. Grosberg and A. R. Khokhlov. *Giant Molecules. Here, There, and Everywhere* . . .. Academic Press, San Diego, CA, 1997.

    [A lively introduction to polymer physics, with nice illustrations and enticing color plates, aptly fitting a beautiful subject. In the format of a coffee-table book, the authors cover important subjects like the wide range of polymeric subjects, ideal chain models and their properties, Brownian motion, biological polymers, and polymer dynamics. An accompanying CD-ROM animates polymer motion, including reptation and coil collapse.]

18. J. M. Haile. *Molecular Dynamics Simulations: Elementary Methods.* Wiley, New York, NY, 1992.

    [Elementary text on molecular dynamics.]

19. M. Kalos and P. A. Whitlock. *Monte Carlo Methods*. John Wiley & Sons, New York, NY, 1986.

    [Good introduction to Monte Carlo techniques.]

20. A. R. Leach. *Molecular Modelling. Principles and Applications*. Pearson Education Limited, Harlow, England, 2001.

    [Broad introduction to many aspects of molecular modeling and computational chemistry techniques, covering basic concepts, quantum and molecular mechanics models, techniques for energy minimization, molecular dynamics, Monte Carlo sampling, protein structure prediction, free energies, solvation, and drug design applications.]

21. K. B. Lipkowitz and D. B. Boyd, Editors *Reviews in Computational Chemistry.* VCH Publishers, New York, NY, 1990 –

[Nice series of books, with volumes appearing annually with comprehensive reviews and tutorials on many aspects of computational chemistry.]

22. J. A. McCammon and S. C. Harvey. *Dynamics of Proteins and Nucleic Acids.* Cambridge University Press, Cambridge, MA, 1987.

[First book on biomolecular dynamics simulations. Excellent overview of field.]

23. D. A. McQuarrie. *Statistical Mechanics.* Harper Collins Publishers, New York, NY, 1976.

[Good reference text.]

24. National Research Council report. Mathematical Challenges from Theoretical / Computational Chemistry, National Academy Press, Washington D.C., 1995. (www.nap.edu/readingroom/books/mctcc/).

[Panel report on the opportunities for collaboration, past achievements, and future possibilities between mathematical and chemical scientists.]

25. P. A. Pevzner. *Computational Molecular Biology. An Algorithmic Approach.* MIT Press, Cambridge, MA, 2000.

[Modern text in computational molecular biology mainly addressed to computer scientists and mathematicians interested in discrete algorithms but accessible to biologists with mathematical grounding. Topics covered include gene hunting, sequencing and mapping, DNA microarray analysis, sequence comparison and alignment, genome evolution, and proteomics.]

26. P. von Ragué Schleyer, Editor-in-Chief, and N. L. Allinger, T. Clark, J. Gasteiger, P. A. Kollman, H. F. Schaefer, III, and P. R. Schreiner, Editors. *Encyclopedia of Computational Chemistry* John Wiley & Sons, West Sussex, England, 1998.

[Comprehensive reference series (five large volumes) written by experts in the field.]

27. D. C. Rapaport. *The Art of Molecular Dynamics Simulation.* Cambridge University Press, Cambridge, England, 1995.

[Elementary text on molecular dynamics focusing on software details.]

28. W. Saenger. *Principles of Nucleic Acid Structure.* Springer Advanced Texts in Chemistry, Springer-Verlag, New York, NY, 1984.

[Wonderful guide to the richness of DNA structure, with an amazing breadth of topics.]

29. T. Schlick and H. H. Gan, Editors. *Computational Methods for Macromolecules: Challenges, Methods, and Applications – Proceedings of the 3rd International Workshop on Algorithms for Macromolecular Modelling, New York, October 12–14, 2000*, Lecture Notes in Computational Science and Engineering, Vol. 24 (Series Editors M. Griebel, D. E. Keyes, R. M. Nieminen, D. Roose, and T. Schlick), Springer-Verlag, Berlin, 2002.

[Collection of articles from presentations at the 2000 international workshop on macromolecular modeling, covering biomolecular simulation methods and applications. The contributions are grouped under the following subjects: field perspective (preface), biomolecular dynamics applications, molecular dynamics methods, Monte Carlo methods, other conformational sampling approaches, free energy methods, long-range interactions and fast electrostatics, and statistical approaches to protein structures.]

30. R. R. Sinden. *DNA Structure and Function.* Academic Press, San Diego, CA, 1994.

    [Nice modern textbook on DNA structure.]

31. G. E. Schulz and R. H. Schirmer. *Principles of Protein Structure.* Springer Advanced Texts in Chemistry, Springer-Verlag, New York, NY, 1990.

    [Nice advanced text on the rapidly-changing field of protein folding.]

32. L. Stryer, *Biochemistry.* W.H. Freeman, New York, NY, latest edition (fifth in December 2001).

    [Wonderful biochemistry textbook, up to date.]

33. W. Van Gunsteren and P. Weiner, Editors (1989) and W. Van Gunsteren, P. Weiner, and A.T. Wilkinson, Editors (1993, 1996): *Computer Simulation of Biomolecular Systems: Theoretical and Experimental Applications.* Vol. 1,2,3. ESCOM, Leiden, The Netherlands, 1989, 1993, 1996.

    [Good series on biomolecular simulations, covering both algorithms and applications.]

Appendix D
Homework Assignments

Please Note: (1) Files mentioned throughout the homeworks can be obtained from the course site, through our group's home page monod.biomath.nyu.edu *or by contacting the author at* schlick@nyu.edu. (2) Insight modeling commands may have changed since the time of this writing and may need updating by instructor and/or students.

Assignment 1: Sequence and Structural Databases, Molecular Modeling Perspective

1. **Molecular Modeling Resources**. Search the web for resources in *molecular modeling*. Look, in particular, for tutorials and instructional material. A good place to start is the NIH site: cmm.info.nih.gov/modeling/, which also provides links to many "Related Web Sites". (Make use of bookmark-type browser utilities to keep useful web sites handy for future use).
   Submit information from two of the most valuable sites you discover (printout) along with a description of how you found the material and what you found most useful.

2. **Sequence and Structure Information Databases**. Search the web for protein (amino-acid) sequence and structure databases. Examples of sequence databases are: PIR, Swiss-Prot, GenPept, and NRPR.[1]

   (a) Plot the amount of available *sequence* database as a function of year, going back as far as possible, to the 1970s. Plot the information on both a regular scale and on a logarithm scale.

   (b) Similarly, plot the amount of *structural* information available as a function of year, on both a regular and a logarithm scale.

   (c) Plot the *sequence and structure* information on the *same plot* in both standard and logarithm views. What can you say about the rate of growth of sequence and structural information? Discuss these finding in relation to the Human Genome Project.

3. **Early Molecular Modeling Literature and Current Progress**. Read two articles dealing with early molecular modeling work:

   - B. J. Alder and T. E. Wainwright, "Studies in Molecular Dynamics. I. General Method", *J. Chem. Phys.* **31**, 459–466 (1959).
   - G. Némethy and H. A. Scheraga, "Theoretical Determination of Sterically Allowed Conformations of a Polypeptide Chain by a Computer Method", *Biopolymers* **3**, 155–184 (1965).

   Do not worry about not understanding the technical details for now.
   Then also read later articles describing current progress in the field and another discussing issues in validating simulation results:

---

[1] PIR was established in 1984 by the National Biomedical Research Foundation (NBRF) and is a good starting point for protein database searching. PIR is somewhat more comprehensive than SwissProt but smaller and better annotated than GenPept (which also includes many hypothetical sequences of unknown function. Since 1999, the NBRF has added a new section to PIR called PATCHX which contains a non-redundant set of all other protein sequences not included in PIR (from other databases), with subsequences removed. Thus, PIR supplemented by PATCHX provides a comprehensive collection of protein sequence data in the public domain. Any search through PIR will automatically include PATCHX. The SwissProt database is useful for searches limited to well annotated sequences, and GenPept is useful for searching all possible sequences, including those that have unknown functions.

The best known structural databases are the Protein Data Bank (PDB) and the Nucleic Acid Database (NDB).

The PDB, managed from 1971 through June 1999 by the Brookhaven National Laboratory, is now operated by the Research Collaboratory for Structural Bioinformatics (RCSB) (www.rcsb.org/), a consortium among Rutgers University, the University of California at San Diego, and the National Institute of Standards and Technology. The RCSB has introduced new features, such as a web-based tool for data deposition, fast data processing systems, and new search engines (text-based and data-based), both with extensive reporting capabilities.

The NDB, pioneered in 1992 by the Rutgers RCSB leader Helen Berman, similarly assembles and distributes structural information about nucleic acids (ndbserver.rutgers.edu/). NDB contains an atlas, an archive, and a sophisticated search engine to access the data.

- J. A. Board, Jr., L. V. Kalé, K. Schulten, R. D. Skeel, and T. Schlick, "Modeling Biomolecules: Larger Scales, Longer Durations", *IEEE Comp. Sci. Eng.* **1**, 19–30 (1994).
- W. F. van Gunsteren and A. E. Mark, "Validation of Molecular Dynamics Simulation", *J. Chem. Phys.* **108**, 6109–6116 (1998).

First describe (in about two pages) the difficulties that Alder and Wainwright enumerate in 1959 regarding molecular dynamics simulations. Then discuss the issues that are still serious limiting factors today, as well as others, such as those noted by van Gunsteren and Mark. Have any of the original limitations been resolved or are likely to be resolved in the near future?

Assignment 2: Introduction to the Insight II Modeling Package and the PDB File Structure and Retrieval

1. **Introduction to Insight II**. If you are not familiar with Insight II you might start by reading the short manual and the tutorial from the web site of the home page of Molecular Simulations (MSI) (www.accelrys.com/insight/index.html). The site URLs may change.
   The manual also contains a short list of basic UNIX commands and a description of simple text editors. Some of the displays on our computers are slightly different from those described in the tutorial but the differences are not critical. For example, in our version the help window automatically follows the tasks invoked from the pulldown menus, and the dial boxes are located on the left, rather than the right, side of the screen. For more information, refer to the Insight II User Guide.

   Note: after opening Insight II, ignore the message about detection of the unlicensed mode. Our site does not have license for the Sketch module. Repeated warning messages might occur during the "build in" Insight II Pilot *tutorial.*

2. **Running Insight II**. Before starting Insight II, you must define the set of environmental variables by running two commands:
   `source /local/msi/cshrc`
   Alternately, you can insert these lines into your `.cshrc` file and they will be executed by the system automatically every time you log on. In that case, you will only need to specify the command
   `source .cshrc`
   after editing your `.cshrc` file the first time. When this insertion is complete, you can run Insight II by specifying the command
   `insightII`
   at the UNIX prompt. Remember, UNIX commands are case sensitive.

   Note: NYU staff may have already inserted these commands in your `.cshrc` *file.*

3. PDB **Structures**. Check the PDB web site for information about the type of stored 3D structures (i.e., proteins, DNA, RNA, DNA/protein complexes, etc.) and the amount in each group. Report your findings.

4. **Retrieval of** PDB **Files**. Using the web PDB browser, find coordinate files for the crystal forms II and III of Bovine Pancreatic Trypsin Inhibitor (BPTI) among the many BPTI entries. From the PDB Home page, go to **Searching and Browsing PDB** and then choose **PDB's web Browser**. You can search by specifying the abbreviation "BPTI" in the Compound window. When the search is completed, records containing BPTI (including its mutated forms) will be displayed at the bottom of the page. Note their ID codes (a number followed by three letters). The two middle letters in this code constitute the name of the subdirectory where the file of

interest resides. For example, the subdirectory name for ID code `1abb` is `ab` and the file name `pdb1abb.ent` . Ftp to `ftp.rcsb.org` and login as `anonymous` (the password instructions will be on the screen). Change directory to `pub/pdb/data/structures/divided/pdb/ab` and get the desired file with the command
`get pdb1abb.ent.Z`

5. **Format of PDB Files**. Read the text in the top of both PDB files and describe the differences in the number of recorded residues, structure resolution, number of solvent molecules, experimental conditions, etc.
   Attach to the assignment sheet a printout of a few lines, starting with the word ATOM, from a PDB file; mark with arrows and describe the content of each specific format field. (The PDB browser contains information about the format; see also the original paper on PDB files: *J. Mol. Biol.* **112**, 535–542, 1977).

6. **Displaying a Protein in Insight II**. Retrieve from PDB the file of mutated form of BPTI (ID = `7pti`), and start Insight II. From the top menu bar select [Molecule][2] and then choose Get. Press the PDB button in Get File Type and the User button specify the directory. Select the file of the `7pti` structure. (Do not press Heteroatom button). Execute. The structure of BPTI should now be on the screen. Use [Object] / DepthCue and then [Transform] / Clip for viewing the protein.
   Change the display ([Molecule] / Display) to Backbone only to speed up the response time. Label the mutated residues ([Molecule] / Label).
   Repeat the operations described above to display the structure of BPTI's crystal form II (keep both structures on the screen).
   Now, overlay both structures by selecting Overlay from [Transform]. In few paragraphs describe the structural differences between both forms of BPTI.

7. **Ramachandran Plots**. To create a file listing with all the dihedral angles ϕ and ψ for a protein, you can use [Protein] / List from the **Biopolymer** module. Select Dihedrals and the protein (any of the structures). Press List_to_file button, specify the file name, and execute. From the recorded data create a scatter plot (phase diagram), so that each point corresponds to one (ϕ, ψ) value.

[2] The following notation will be used throughout the homework assignments:

- [Pulldown] corresponds to a menu bar pulldown.
- Command corresponds to a command from the pulldown menu.
- Option corresponds to an option in the dialog box of a given command.

Summary of Items to Hand in:

(a) Data with the amount of PDB 3D structures for each category of biomolecules.
(b) Description of the differences in the informational part of the PDB files for form II and form III of BPTI.
(c) Explanation of the PDB format for storing atomic coordinates.
(d) Description of the structural differences between BPTI (form II) and the mutated form (7pti).
(e) The table with dihedral angles ϕ and ψ listed for BPTI.
(f) Scatter plot with points (ψ, ϕ) for each residue of BPTI.

Background Reading from Coursepack (Appendix B)

- M. Levitt and A. Warshel, "Computer Simulation of Protein Folding", *Nature* **253**, 694–698 (1975).
- M. Karplus and J. A. McCammon, "The Dynamics of Proteins", *Sci. Amer.* **254**, 42–51 (1986).
- Y. Duan and P. A. Kollman, "Pathways to a Protein Folding Intermediate Observed in a 1-Microsecond Simulation in Aqueous Solution", *Science* **282**, 740–744 (1998).
- H. J. C. Berendsen, "A Glimpse of the Holy Grail", *Science* **282**, 642–643 (1998).
- X. Daura, B. Juan, D. Seebach, W. F. Van Gunsteren, and A. Mark, "Reversible Peptide Folding in Solution by Molecular Dynamics Simulation", *J. Mol. Biol.* **280**, 925–932 (1998).

Assignment 3: Construction and Analysis of the Pentapeptide Met-enkephalin with the Insight II Program

1. **Building a Pentapeptide**. To build the molecule met-enkephalin, whose amino acid sequence is **Tyr–Gly–Gly–Phe–Met**, invoke the **Biopolymer** module. From Residue select Append.
   Specify the molecule's name and choose Extended to specify the structural motif of the backbone. (At this stage, do not worry whether the structure is correct). Select the first residue, **Tyr**. Then add each of the remaining four residues in turn.
   You can *center* the molecule on your screen by clicking on it with the center mouse button and dragging it to the desired position. You can also *rotate* the molecule by pressing the right mouse button and dragging the molecule. By pressing both (center and right) mouse buttons and dragging the molecule you can change its position along the z-axis, perpendicular to the screen. You can rotate the molecule around the z-axis by dragging the mouse while both left and right mouse buttons are pressed. To change the representation style of the molecule, select Molecule / Render and choose any of the options. Note how the speed of executing commands (e.g., translation, rotation) is affected by the representation display.
   Now you must amend the ends of the structure you built. Switch to Protein and choose Cap. Change both the N-terminus and C-terminus moieties to the zwitterionic form (NH_3^+ and COO^-) to get a proper oligopeptide.

2. **Measurements of Met-enkephalin's Structural Parameters**. Generate a table listing all the dihedral angles in your met-enkephalin molecule by using Protein / List from the **Biopolymer** module. Select the appropriate command (Dihedrals) and direct the data to a file (List_to_file button on). This file can be viewed and edited later.
   To measure individual bond lengths or distances, bond angles, dihedral angles, etc., use Measure. Select the atoms for the measurement by clicking on them with the left mouse button.

3. **Rotameric Structures**. By a *rotamer*, Insight II refers to a different conformational arrangement of a side chain of a given amino acid. The rotameric structures identified in Insight II are correlated with the ϕ and ψ angles for a given residue (i.e., are sterically compatible).
   Select Manual_Rotamer from Residue of the **Biopolymer** module. Press the Evaluate_Energy button. Energy for a given rotameric structure will be printed in the information window at the bottom of the screen (you can scroll by pressing on arrows to its right). Keep the Nonbond Cutoff value — an option which is displayed on the screen once Evaluate_Energy is chosen — at the default value of 8.0 Å.
   Select a residue by clicking on it and sweep through all the rotamers

of that residue (while holding other rotamer conformations fixed). Next will trigger the execution of the command. In a table, report the rotamer structures for each residue by specifying the dihedral angles χ_1, χ_2, χ_3, ... (Protein / List), along with associated energies.
Now assemble the pentapeptide structure with the lowest rotamer energy and save it. Thus far we have ignored a global optimization of the structure and have only built it up from low-energy conformations of its components. *We will return to optimization later in the course, after studying minimization techniques.*

4. **Main-chain Structure**. Choose Protein / Secondary from the **Biopolymer** module. Change the main-chain configuration by selecting different motifs (Alpha_R_Helix, Alpha_L_Helix, 3–10_Helix, etc.). For each motif:

   (a) prepare a table with the dihedral angles ϕ and ψ for each of the residues (Protein / List),
   (b) list all the hydrogen bonds present in the structure. Measure / HBonds and Molecule / Label will be helpful here.

5. **Torsional Rotation**. Bring back your Met-enkephalin's backbone structure in the extended form. You can use the structure saved in part 3 of this assignment (File / Restore_Folder). Change the ψ dihedral angle on the second residue, **Gly**, to $60°$ and the ϕ dihedral angle on the forth residue, **Phe**, to $-60°$.
   Torsional motion around a chosen dihedral angle can be performed with Transform / Torsion command or by pressing the Torsion button on the left of the Insight II screen.
   First, click with the left mouse button on the bond which constitutes the axis of rotation, then press the Torsion button. A little cone, defining the direction of the torsion, will pop up on the screen at one of the bond ends. Now you can change the torsion angle by horizontally dragging the mouse with the center button pressed in. To exit the torsion mode press the Torsion button again (the cone will disappear).
   Calculate the distance between the N-terminus (N atom) and C-terminus (C atom) atoms. Use the Measure / Distance command. Keep the Monitor button on, and select Monitor Mode/Add. Atom 1 and Atom 2 can be selected by clicking on them with the left mouse button.
   Print a picture of your molecule (keep the distance between the N-terminus and C-terminus atoms on the screen). To save ink, the background color should be changed to white every time you print a color or black/white picture. To do so, go to Session / Environment, press the Background button and change the color to white. Now go to File and choose Export_Plot.

Select postscript, Gray_Scale and optionally Ball_and_Stick. Save the file as postscript by using Save_Device_File (the file will have the ".ps" extension). This file can be printed on any postscript printer. Hand in your printout as part of the homework.

Summary of Items to Hand In:

(a) The table with the dihedral angles for met-enkephalin.
(b) The table with the dihedral angles and energies for the rotameric structures of met-enkephalin.
(c) The table with the $\{\phi, \psi\}$ dihedral angles for each different backbone motif of met-enkephalin.
(d) The listing of the hydrogen bonds for each of the backbone motif for met-enkephalin.
(e) Printout of the met-enkephalin structure with the end-to-end link marked.

Background Reading from Coursepack

- G. Némethy and H. A. Scheraga, "Theoretical Determination of Sterically Allowed Conformations of a Polypeptide Chain by a Computer Method", *Biopolymers* **3**, 155–184 (1965).
- P. Y. Chou and G. D. Fasman, "Prediction of Protein Conformation", *Biochemistry* **13**, 222–245 (1974).
- M. Levitt and C. Chothia, "Structural Patterns in Globular Proteins", *Nature* **261**, 552–558 (1976).

Assignment 4: Creating Ramachandran Plots from Known Protein Structures and the NDB

1. **Ramachandran Plots**. Our goal is to generate Ramachandran plots for a particular amino acid residue or a group of residues. We have assembled a database of 50 proteins based on the article: M.A. Williams, J.M. Goodfellow, and J.M. Thornton, "Buried Water and Internal Cavities", *Protein Science* **3**, 1224 (1994). These files can be found in the PDB directory of Insight II prepared for our course.
   Each of you must generate two Ramachandran plots. Check the chart below for your particular assignment (on the basis of your last name).

First letter of your last name	Subgroup 1	Subgroup 2
A–N	**Ala, Val, Leu, Ile**	**Gly**
O–R	**Asp, Asn, Glu, Gln**	**Pro**
S–V	**Lys, Arg, His**	**Ser, Thr**
W–Z	**Trp, Tyr, Phe**	**Cys, Met**

   Each plot should have the data points for the (ϕ, ψ) dihedral angles, corresponding to the assigned group of residues from all the proteins in the database. To find the values of the ϕ and ψ dihedral angles in a protein you can use Protein / List command from the **Biopolymer** module. Record these angles to a file. You can use the Fortran code posted on the website (aa_select.f) for searching the Protein / List output files (called PDA files) for the dihedral angles of selected residues. Alternatively, you can write a suitable program in a different language.
   If you use the code from the website, you will need to edit it to replace all occurrences of the names `ALA, VAL`, etc. by the abbreviations of the residues you are searching for. These abbreviations must be capitalized. Compile the code and execute it with each PDA file as input. The output file should contain only two numbers per line, the ϕ and ψ angles for the specified residues. Check the numbers for correctness by comparing a few lines from this output with the numbers in the PDA file.

   Note: For plotting, you must use Insight II *so that all plots are uniform in size. We will overlay them in class! Follow the instructions below.*

   Prepare a file with your data points collected from all the proteins for each of the assigned groups of residues. These files should have the extension `tbl (filename.tbl)`. In addition, you must format these files for Insight II by inserting the 12 lines as indicated:

```
#
TITLE: Phi (deg)
MEASUREMENT TYPE: Ang
UNITS OF MEASUREMENT: Deg
FUNCTION: dihedral
#
TITLE: Psi (deg)
MEASUREMENT TYPE: Ang
UNITS OF MEASUREMENT: Deg
FUNCTION: dihedral
#
#
−34.5     144.9               This is the first line of your numeric data.
⋮
```

Note that at the top of the file there is space for your own comments and you can use as many lines as you need. Insight II will start reading the data from the line without the character # on the first column. That first line should be as indicated above.

If you have done everything correctly, you are ready for plotting. Press the Graph button on the left of the screen and select Get. Specify the data file name, dihedral1 as X_Function and dihedral2 as Y_Function. Keep the New_Graph on and execute. The zigzag appearance of the plot now requires fixing. Move the graph box near the bottom-left corner of the screen. First, connect to the object (your plot) with the command Transform / Connect. Second, move the plot by clicking on it and dragging to the desired position. The scatter plot will be produced in the next step. Select Point (only!) from the Graph / Modify_Display dialog box. Choose Star as the Point Symbol, scale it ten times and execute. Use Graph / Threshold to change the minimum value to −180.0 and the maximum value to 180.0 for both, X Graph Axis and Y Graph Axis. Scale each axis 4 times with Graph / Scale_Axis command. Change the color of any white elements in your plot with Graph / Color command. Change the background color to white. Print your plot (see instructions below) and repeat the procedure for the second data file.

Prepare a transparency for each of the two plots and bring to class. (Hand in the printed version of the plots with your homework.) Also e-mail the $\{\phi, \psi\}$ files you generated, sending each file separately and specifying your name and the group of residues in the `Subject` line.

2. **The NDB**. The next part of the homework will acquaint you to the Nucleic Acid Database, NDB, on which we will have a guest lecturer. The web address for NDB is: ndbserver.rutgers.edu/

   First, explore the database to discover what is available, and look through the newsletter archives for current update information. Then, describe the structures available in the database and the numbers in each class (e.g., B-DNA, ribozymes). Explore the different features of NDB. There are many exciting structures and utilities.

3. **Sugar Conformations for Canonical A, B, and Z-DNA**. To appreciate the different features of canonical A, B, and Z-DNA forms, choose through the NDB entries one unmodified form in each DNA class with the largest number of residues possible. You can view the structure within NDB, or by porting the PDB file to Insight II and using the Molecule / Get from the **Viewer** module (though the latter may be more difficult).
   Learn how to use the Report facility in the NDB Query Interface to generate a list of all the sugar pseudorotation angles (P) for each deoxyribose in each of the three structures you have chosen for your study. (Remember that there are two sugars per base pair). Print a table for each structure.
   Now, on one figure for all the three structures, plot P versus the residue number. Exclude the two terminal base pairs of each structure for plotting purposes. You may need to connect the points corresponding to each structure for clarity.
   Label the pattern clearly to indicate the A, B, and Z-DNA data. Hand in this plot, with a description of the patterns you had identified for the sugar conformation for the three canonical DNA forms, indicating the specific structures you have chosen.

Summary of Items to Hand In:

(a) Ramachandran plots for the two subgroups of amino acid residues assigned to you.
(b) Data on the structure class types and the amount of structures in each class available in NDB.
(c) The figure with P versus residue number for the A, B, and Z-DNA forms, with complete discussion.

Background Reading from Coursepack

- B. Cipra, "Computer Science Discovers DNA", in *What's Happening in the Mathematical Sciences*, pp. 26–37 (P. Zorn, Ed.), American Mathematical Society, Colonial Printing, Cranston, RI (1996).
- M. Gerstein and M. Levitt, "Simulating Water and the Molecules of Life", *Sci. Amer.* **279**, 101–105 (1998).

Printing Instructions

To save ink, change the background color from black to white at each printing of a color or black/white image. Go to Session / Environment, press the Background button and change the color to white.

Proceed to File, choose Export_Plot, postscript, Gray_Scale and optionally Ball_and_Stick. Use Save_Device_File to save a postscript file (the file will have the ".ps" extension). This file can be printed on any postscript printer you can access. Hand in your printout as part of the homework.

Fortran program for selecting Ala, Ile, Val, and Leu data
(see website)

Assignment 5: Analysis of Protein/DNA Complexes with Insight and NDB: Canonical vs. Protein/Bound DNA, and DNA/Protein Interactions

In this assignment, we will study three nucleic acids structures, two of which have been crystallized with regulatory proteins: a nucleic acid dodecamer, a DNA oligomer bound to a prokaryotic protein in the helix-turn-helix (HTH) motif, and a DNA oligomer (known as the TATA-box sequence) bound to a eukaryotic transcription factor.

NDB ID code	Structure
• **BDL078**	DNA dodecamer
• **PDR010**	DNA (20 bp) bound to bacteriophage λ cI repressor
• **PDT034**	DNA (16 bp) bound to human TATA-Box Binding protein

1. **Structure Downloading**. Download each structure from the NDB, already explored in Assignment 4 (ndbserver.rutgers.edu:80/). Use the Archives, Atlas or Search entry points to the NDB; all are useful. In particular, the Atlas allows you to quickly view the structures. The Search entry point will be utilized later in this assignment.
   Load each structure into Insight II. Separate the two complexes into DNA and Protein objects using the Modify / Unmerge command from either the **Biopolymer** or **Builder** modules. Both the DNA and Protein objects will be used later in this assignment. Use caution with the wildcard character * in selecting multiple atoms/residues/nucleotides/ proteins in Insight II. A quick test which you can used to test which object you've created with the Modify / Unmerge is to blank or blink the object with the Object / Blank or Object / Blink commands.
   In order to create the B-DNA models for specified nucleic acid sequences (see below) and manipulate the downloaded structures (as in the Unmerge command), you will need to understand certain general parameters related to the structure as used in both the PDB file and Insight. These parameters refer to the DNA sequence, strand and residue labels, and so on. Explore the text in the downloaded files as well as in the structural displays. Information about relevant formatted lines in the coordinate files (such as SEQRES and ATOM) is available from the PDB web site (see Assignment 2).

2. **Canonical vs. Protein/Bound DNA Analysis**. Create an idealized B-DNA structure using the Nucleotide / Append command in the **Biopolymer** module corresponding to each of the three nucleic acid sequences in the above structures. The Nucleotide / Append command both creates new nucleic acid molecules and appends to existing molecules; when it is creat-

ing a new molecule, the Append Point is None.[3] You will need to use the Nucleic_Acid / Cap pulldown menu to replace the phosphate group with a hydroxyl group at the 5′ ends of each strand.

Nucleotide / Append will create your B-DNA model, defining each strand as a separate object. Your downloaded DNAs, however, are each composed of a single object in Insight II. To properly superimpose the two structures (the task in the next part), you will need to Modify / Merge the two strands as one object.

Superimpose each B-DNA model upon its respective crystal structure from the NDB using the Transform / Superimpose pulldown menu. If your DNAs are each composed of a single object in Insight II, you will need to Modify / Merge the two strands as one object to properly superimpose the two structures. Use the Heavy option to avoid superpositioning of hydrogen atoms. After selecting the B-DNA model and the crystal structure in the selection boxes, you will need to click the End Definition box to Execute the Superimpose command. The root-mean-square deviation (RMSD) value will be printed at the bottom of your screen.

(a) Record the RMSD values relative to idealized B-DNA for each superimposed model/structure by repeating this procedure for each crystal structure.

(b) Use the Search entry point to the NDB to extract the following parameters for each base pair of the three structures: P (pseudorotation sugar pucker); the dihedral angles χ, α, β, γ, δ, ϵ, ζ; and the helical parameters twist, tilt, roll, and propeller twist: Ω, τ, ρ, and ω, respectively.

For each conformational variable (excluding those from the end residues), calculate the average (μ) and standard deviation (σ) from the data per structure. Prepare a table in the following form:

Now discuss your results in terms of the differences noted between the protein-bound DNA and canonical B-DNA (which the BDL078 structure represents):

(c) Which structure is most deformed from B-DNA? Which parameters display the largest changes from B-DNA (consider both μ and σ)? Based on these parameters, what is similar in the way the two complexes deform their recognition sites away from B-DNA (look for

[3]The **PDR010** structure has an overhanging base at each end, that is a base without a Watson-Crick partner on the other strand. (This procedure promotes crystal formation.) The recommended procedure for creating the overhanging base is to use Nucleotide / Append to create a 21-base-pair duplex of the correct sequence and then use Nucleotide / Delete command to delete one base from each strand prior to using Nucleic_Acid / Cap.

	BDL078 μ σ	PDR010 μ σ	PDT034 μ σ
P			
χ			
α			
β			
γ			
δ			
ϵ			
ζ			
Ω			
τ			
ρ			
ω			
RMSD			

similarities in columns 3 and 4 above which are different from column 2)?

(d) Are any of the changes observed localized to particular regions in the DNA? (Consider properties with μ values similar to B-DNA but large associated σ values). Plot one of these parameters as a function of position (base pair) along the DNA.

(e) Generate a *side-by-side* picture of the three DNA structures. A recommended utility for this is the File / Export_Plot facility.

3. **Analysis of Interface Between Proteins and DNA**. Next, we will examine some of the interactions formed at the interface between the regulatory proteins and their DNA binding sites. Load the PDB files of each DNA/protein complex in turn and unmerge the DNA part (but leave the DNA and protein together in space).

The main tool used here is the Subset / Interface pulldown in the central **Viewer** module. This menu allows subsets to be defined in one molecule that satisfy a certain spatial relationship with respect to the other molecule. For example, we would like to use this menu to define subsets of atoms in the protein that are near functional groups of the DNA. A contour level of 3.5 Å is useful in this menu for defining interactions between non-hydrogen atoms, since it roughly corresponds to distances for strong interactions.

Open the Subset / Interface pulldown menu. You can define the Subset Name as you please. You can define the Center of Subset to be a specific functional group in DNA. For example, DNA:T:C5M refers to atom C5M of thymine's methyl group in the DNA. Define the Search_Domain to be the protein. The Radius of Subset should be set to the value 3.5.

For your reference, these are names of some DNA atoms:

- Phosphate groups: Atoms P, O1P, O2P
- Thymine methyl groups: Atom C5M
- Adenine amino groups: Atom N6
- Pyrimidine carbonyls: Atom O2
- Purine amines: Atom N3

(a) Save listings of these subsets into output files. Subset / List is recommended for this task.

> [Do not be alarmed if you get the error message "Invalid Comparison Object"; it simply means that the comparison could not be performed since no member of the set fulfilled the criteria. If an attempt is made to analyze all atoms B that are 3.5 Å from protein A, but all atoms of protein A are more than 3.51 Å from B, this error will occur.]

Combine the analyses of the two complexes (if you can) and construct *histograms* of the residue types in each subset. That is, from the listings of all residues within 3.5 Å of the atom groups above, count the number of times each residue appears (e.g., Methionine appears 60 times, Glutamate 3) and generate histogram plots as illustrated below; use the one-letter mnemonic for the amino acids. (You may also want to count the frequencies of residues grouped by type, like polar, hydrophobic, charged, etc.).

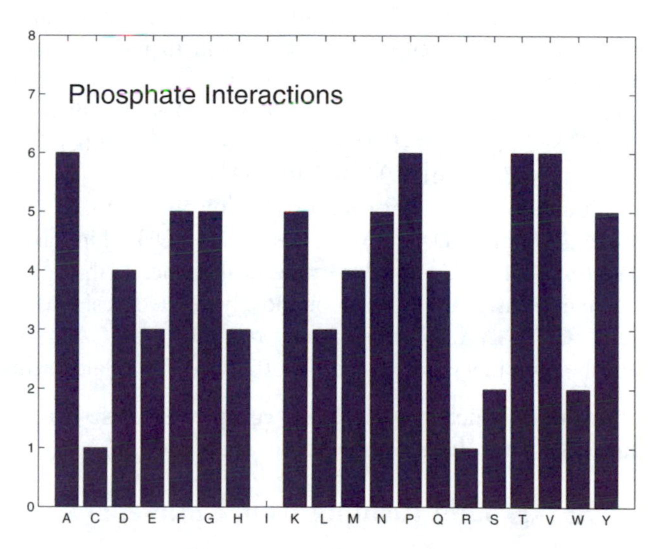

Figure D.1. Sample histogram for protein/DNA interaction analysis

(b) Do you observe common patterns in the two complexes? Are certain amino acids likely to be found interacting with a particular functional group? What types of interactions are being formed between these nucleic acid functional groups and the regulatory protein (e.g. attribute to each of the nucleic acid groups above a type of interaction such as hydrophobic, hydrogen bonding, electrostatic, intercalation/insertion motif, etc.)?

(c) Is there anything unusual about the subsets formed between the proteins and the O2 carbonyl/N3 amine atoms?

(d) Is there any relation between the interactions observed in these subsets and the deviations from canonical B-DNA structure observed above (i.e. how do the interactions you observe explain any of the parameter variances you diagnosed)?

Note: A trick to identify atoms/residues is via Molecule / Color *for assigning a color to an atom/residue. Other labeling tools such as* Molecule / Render *and* Molecule / Label *can similarly be used.*

(e) **Bonus Question:**[4] **BDL078 Homologues.** We have used the BDL078 structure as an example of B-DNA in the analysis above. However, sequence-dependent variations in local structure are also important. Therefore, a more sensitive analysis of free versus protein/bound DNA employs the same nucleotide sequence, both with and without bound proteins. There are few cases, however, in which high-resolution DNA structures are available in both the free and protein-bound states. Such analyses are illuminating and show how intrinsic DNA preferences are amplified in the DNA/protein complexes. See recent reports regarding the complex between DNA and the bovine papillomavirus E2 protein in D.M. Crothers (*Proc. Natl. Acad. Sci.* **95**:15163–15165, 1998) and H. Rozenberg *et al.* (*Proc. Natl. Acad. Sci.* **95**:15194–15199, 1998).

Such analyses have not been done with our BDL078 sequence, but there are protein-DNA complexes in the NDB which are closely related to BDL078. This close relationship means that: (*i*) the related sequence has many similar or closely related residues to BDL078 (e.g., GG**G**AA**A**T**T**T is closely related to GG**C**AT**A**A**C**TT), and (*ii)* the protein would bind to BDL078 and this related sequence.

Determine which protein/DNA complexes these are, and briefly describe what these complexed proteins do.

Background Reading from Coursepack

- K. B. Lipkowitz, "Abuses of Molecular Mechanics. Pitfalls to Avoid", *J. Chem. Educ.* **72**, 1070–1075 (1995).

[4]The correct solution will allow you to drop lowest homework grade in any assignment.

- S. Lifson, "Potential Energy Functions for Structural Molecular Biology", in *Methods in Structural Molecular Biology*, pp. 359–385, D. B. Davies, W. Saenger, and S. S. Danyluk, Eds., Plenum Press, London (1981).

Assignment 6: MIDTERM: Homology Contest! Exploring Sequence/Structure/Function Relationships (& Related Tools/Databases like SCOP, IMAGE, BLAST, NDB, PDB)

With the rapidly growing information on genomic sequences, *comparative modeling* — structure prediction based on sequence similarity — is becoming increasingly valuable. Indeed, structural and functional genomics, the three-dimensional (3D) structure and functional analysis of genomic products, are rising disciplines in bioinformatics. It has been reported, for example, that a sequence homology of larger than 40% usually implies more than 90% 3D-structure overlap (see below for precise definitions of similarity). Thus, with the growing amount of genomic information, we may eventually be able to predict reliably 3D structures of proteins. Since structural similarity is often preserved more strongly than sequence through evolution, reliable homology-based predictions might provide crucial functional properties of new gene products in the near future.

Through this assignment, you will gain some experience in quantifying and analyzing sequence and 3D structure similarity for proteins. You will also explore sequence and structure databases in search of interesting examples, and learn how to use important computational and database resources. You will have to be resourceful in looking for suitable programs for alignment and structure analysis besides those below; no simple recipes will be given here.

This assignment can be done by teams of two students; choose a partner with complementary skills. You will have to present your results to the class.

The 5 Tasks

Find and demonstrate the following four relationships for proteins:

1. [**EASY**] Two proteins with very *high sequence similarity* (but less than 95%) and very *high structural similarity*. Excluded from consideration are trivial examples, such as involving multiple PDB entries for the same protein.

2. [**EASY**] Two proteins with very *high sequence similarity* (but less than 95%) and *very high structural similarity* but markedly *different biological/functional* properties.

3. [**MODERATE**] Two proteins with *low sequence similarity* but *high structural similarity*. Also comment on the *functional* properties of the pair.

4. [**HARD**] Two proteins with very *high sequence similarity* but very *low structural similarity*. Also comment on the *functional* properties in your example.

For problems 3 and 4 above, the class contest will be won by the students that find the most extreme examples (i.e., the maximal sequence similarity / minimal structural similarity, minimal sequence similarity / maximal structural similarity).

5. [**EASY WARMUP**] Search and identify all the determined structures in the PDB/NDB that contain the nucleic acid sequence TATAAAAG. Discuss these structures and their significance.

For each task, generate color molecular views, report the analyses in detail, and include a description of how you found the example. Also discuss your similarity/dissimilarity criteria (see below), and prepare a class presentation on your results.

Ground Rules

1. Homology, or sequence similarity, will be defined by the percentage of sequence identity.

2. 3D-structure similarity will be defined in two ways:

   (a) the percentage of C^{α} atoms of the proteins that "overlap", i.e., are within 3.5 Å of each other in a rigid-body alignment of the protein;
   (b) the root-mean-square-deviation (RMSD) between C^{α} atoms of the proteins in a rigid-body alignment of the protein. (Recall your experience with RMSD measurements in the previous assignment).

You should first experiment with overlapping several protein structures to determine what RMSD values and/or percentages of C^{α} overlap indicate random similarity. *Discuss this in your submission.*

Tools of the Trade

1. **Sequence and Structure Databases**. You have already navigated through the structural PDB and NDB databases and various sequence databases. Continue to work with these and the RCSB facilities.

2. **SCOP**. This site for the *Structural Classification of Proteins* (scop.mrc-lmb.cam.ac.uk/scop/ categorizes proteins according to the levels (top-to-bottom) of: class, fold, superfamily, family, domain, and reference PDB structure.

3. **Insight II**. Continue to use Insight II for structure display and analysis.

4. **NCBI Tools like BLAST and Its Cousins**. BLAST is a library of heuristic similarity search programs (Basic Local Alignment Search Tools) that explore relationships involving protein and nucleic-acid sequences and 3D structures. This library contains blastp, blastn, blastx, tblastn, tblastx,

and others, developed at the National Center for Biotechnology Information at the National Library of Medicine of the National Institutes of Health. Get started at their web site www.ncbi.nlm.nih.gov/BLAST/. This page leads to the BLAST suites as well as contains usage information. See, for example, Overview, Manual, BLAST FAQs, References.

BLAST, one of the most popular tools among molecular biology researchers, has evolved rapidly since its inauguration in 1990. BLAST searches a database in two stages, finding small sequence lengths that match the target exactly and then attempting to extend the length of the match from this subset of sequences in the database. Not only are the alignment algorithms improving continuously (e.g., allowing alignments of DNA or protein sequences with insertions or deletions in Gapped BLAST; forming families of aligned sequences and quick profiles of them in Position-Specific Iterated (PSI)-BLAST; or incorporating biological-function hypotheses into sequence queries to restrict the analysis to subset of protein sequences as in Pattern-Hit Initiated (PHI)-BLAST), but performance has been greatly accelerated. Algorithmic features include dynamic programming tools, hidden Markov models, and various optimization strategies.

To align two protein or nucleotide sequences, go to the link of BLAST 2 sequences (www.ncbi.nlm.nih.gov/gorf/bl2.html) and set up the computation according to the instructions. Take care to choose the options of the computation with care, and explore different options. The server will send the results to the web browser being used.

Some available programs are:

blastp: compares an amino acid query sequence against a protein sequence database.

blastn: compares a nucleotide query sequence against a nucleotide sequence database.

blastx: compares the six-frame conceptual translation products of a nucleotide query sequence (both strands) against a protein sequence database.

tblastn: compares a protein query sequence against a nucleotide sequence database dynamically translated in all six reading frames (both strands).

tblastx: compares the six-frame translations of a nucleotide query sequence against the six-frame translations of a nucleotide sequence database.

See www.ncbi.nlm.nih.gov/BLAST/newblast.html#introduction for further information.

Other similarity programs are available (such as MEME *and* MAST *from SDSC); use anything appropriate for the task.*

5. **An Image Library**. The *Image Library of Biological Macromolecules* organized by the Institute for Molecular Biotechnology in Jena, Germany (www.imb-jena.de/IMAGE.html) offers a colorful library of biomolecular images corresponding to structures available in databases like the NDB and PDB. Besides detailed colorful illustrations of the structure in a variety

of styles, relevant structural information and publication links are available. Basic tutorials on structural biology are under preparation at this site.

HINTS for the Assignment

1. Scan the literature for related papers on comparative or homology modeling but do not repeat known examples. You **CAN** be original.

2. Large changes in 3D structure despite high sequence similarity can result from the following situations:

   - mutations in critical regions of the proteins such as active sites
   - mutations in ligand binding sites (as in immunoglobulins)
   - mutations in regions that connect two secondary-structural elements (as in helix-loop-helix motifs)
   - structure determination of the same system at different environmental conditions (e.g., different solvent, different crystal packing forms for mutant proteins)
   - proteins containing the same subunits but a different number of subunits, with a structure/fold/topology that depends critically on that number.

   Search PDB and SCOP for examples in this spirit.

3. Look for groups of proteins in the same family, or for proteins sharing the same fold in the SCOP site. The structural classification information should generate ideas.

4. General structure alignment via Insight is not very sophisticated and may be entirely unsuitable for sequences of disparate lengths and for structures with two similar subdomains adopting a different relative orientation. Search for suitable programs for these cases (e.g., from the RCSB, www.rcsb.org and from SDSC) and also write/use your own programs to perform certain analyses, such as structure similarity measurements upon alignment (e.g., criterion 2a under **Ground Rules**).

Background Reading

- D. Baker and A. Sali, "Protein Structure Prediction and Structural Genomics" *Science* **294**, 93–96 (2001). [**From Coursepack**].
- B. Honig and A. Nicholls, "Classical Electrostatics in Biology and Chemistry", *Science* **268**, 1144–1149 (1995) [**From Coursepack**].
- D. Case, "NMR Refinement", in P. von Ragué Schleyer (Editor-in Chief), N. L. Allinger, T. Clark, J. Gasteiger, P. A. Kollman, and H. F. Schaefer, III, editors, *Encyclopedia of Computational Chemistry*, volume 3, pages 1866–1876. John Wiley & Sons, West Sussex, England, 1998.

Assignment 7: Molecular Mechanics Force Fields: Approximations, Variations, and the Assessment of Results with respect to Experiment and other Simulations

1. **Reading**. This assignment deals with the series of four articles below, which raise both general and specific problems in biomolecular simulations. At issue is the validation of conformational predictions by various molecular mechanics force fields. You may also wish to refer to the Lipkowitz article from Assignment 5 (on the pitfalls of molecular mechanics) and the van Gunsteren and Mark article from Assignment 1 (on validating molecular dynamics simulations). Begin by reading these papers (included in the Coursepack, see Appendix B) and thinking about the modeling issues as you read them.

   - I. K. Roterman, M. H. Lambert, K. D. Gibson, and H. A. Scheraga, "Comparison of the CHARMM, AMBER and ECEPP Potentials for Peptides. I. Conformational Predictions for the Tandemly Repeated Peptide (Asn-Ala-Asn-Pro)$_9$", *J. Biomol. Struct. Dyn.* **7**, 391–419 (1989a).
   - I. K. Roterman, M. H. Lambert, K. D. Gibson, and H. A. Scheraga, "Comparison of the CHARMM, AMBER and ECEPP Potentials for Peptides. II. ϕ–ψ Maps for N-Methyl Amide: Comparisons, Contrasts and Simple Experimental Tests", *J. Biomol. Struct. Dyn.* **7**, 421–453 (1989b).
   - P. A. Kollman and K. A. Dill, "Decisions in Force Field Development: An Alternative to Those Described by Roterman *et al.*", *J. Biomol. Struct. Dyn.* **8**, 1103–1107 (1991).
   - K. B. Gibson and H. A. Scheraga", "Decisions in Force Field Development: Reply to Kollman and Dill", *J. Biomol. Struct. Dyn.* **8**, 1109–1111 (1991).

2. **Preparation for Class Discussion.** You will be divided into three groups (assignments will be given in class): (1) the moderators, (2) the ECEPP group, and (3) the AMBER and CHARMM group. Each group will have to prepare material, as described below, for class presentation and discussion. *All materials should be prepared on overhead projector slides.* You should meet with your group members in advance to plan your presentation and debate strategies.

The *moderators* will be in charge of presenting in detail the *facts*: what studies were performed, what questions were asked, and what analyses were made. You should be prepared to answer any background questions (e.g., definitions of polymer quantities analyzed).

The *ECEPP* group will endorse the point of view taken by Roterman, Gibson, Scheraga, and co-workers. Besides understanding your position well, you will need to bring to the debate *concrete examples from the literature* to support your position. Be creative and try to find interesting examples.

The *AMBER* folks and *CHARMMers* will endorse the approach taken in these two molecular packages and, in particular, the point of view taken by Kollman and Dill in their reply to Roterman *et al.*. As above, besides understanding well your molecular mechanics packages and position taken in the reply, you will need to bring to the debate *concrete examples from the literature* to support your position. Be creative in your supporting materials and strategies.

3. **Useful Recommendations**. Summarize in brief the useful recommendations and comments that emerged from all the above articles, as well as additional ones, for practitioners of molecular modeling. That is, propose *concrete procedures* that biomolecular simulators can use to gain as much confidence as possible in their conclusions and predictions.

   Remember, uncertainties and approximations in numerical modeling and simulations will always exist! The field of modeling biomolecules on modern computers involves as much art as science. But despite their obvious limitations, modeling methodologies are improving continuously. The goal of every practitioner should be to realize the highest possible accuracy as is compatible with the model and methods utilized. Like any calculation, 'error bars' in the broad sense should be attributed to the results and conclusions claimed.

4. **Points to Keep in Mind**. Throughout this assignment, think about the following important issues in molecular modeling:
   - Accuracy versus approximation
   - Theory versus experiment
   - Dependence of simulation results on the protocols used
     - starting configuration
     - model assumptions
     - force field
     - algorithms (minimization, adiabatic mapping, etc.)
   - Assessment of Results:
     - How can you distinguish between bona fide physical *trends* and numerical *artifacts*?
     - How can you decide whether the model is wrong (energy, assumptions, etc.) or the method is inappropriate?
     - What are appropriate comparisons with experimental results?

Summary of Items to Hand in:

(a) Brief description of the issues raised in the four articles regarding molecular mechanics predictions.
(b) Your work in preparation of the class debate.
(c) Proposals of procedures to be used to attain the maximum possible confidence from biomolecular simulations.

Have Fun!

Background Reading from Coursepack

- J. Skolnick and A. Kolinski, "Simulations of the Folding of a Globular Protein", *Science* **250**, 1121–1125 (1990).
- F. M. Richards, "The Protein Folding Problem", *Sci. Amer.* **264**, 54–63 (1991).
- H. A. Scheraga, "Predicting Three-Dimensional Structures of Oligopeptides", in Reviews in Computational Chemistry, K. B. Lipkowitz and D. B. Boyd, Editors, Vol. 3, pp. 73–142, VCH Publishers, New York (1992).
- A. Neumaier, "Molecular Modeling of Proteins and Mathematical Prediction of Protein Structure", *SIAM Review* **39**, 407–460 (1997).

Background Reading for Scheraga's Lecture

- J. Pillardy, Y. A. Arnautova, C. Czaplewski, K. D. Gibson, and H. A. Scheraga, "Conformation-Family Monte Carlo: A New Method for Crystal Structure Prediction", *Proc. Natl. Acad. Sci., USA* **98**, 12351–12356 (2001).
- J. Pillardy, C. Czaplewski, A. Liwo, W. J. Wedemeyer, J. Lee, D. R. Ripoll, P. Arlukowicz, S. Oldziej, Y. A. Arnautova and H. A. Scheraga, "Development of Physics-Based Energy Functions that Predict Medium-Resolution Structures for Protein of the α, β, and α/β Structural Classes", *J. Phys. Chem. B* **105**, 7299–7311 (2001).
- J. Lee, D. R. Ripoll, C. Czaplewski, J. Pillardy, W. J. Wedemeyer and H. A. Scheraga, "Optimization of Parameters in Macromolecular Potential Energy Functions by Conformational Space Annealing", *J. Phys. Chem. B* **105**, 7291–7298 (2001).
- A. Liwo, C. Czaplewski, J. Pillardy and H. A. Scheraga, "Cumulant-Based Expressions for the Multibody Terms for the Correlation Between Local and Electrostatic Interactions in the United-Residue Force Field", *J. Chem. Phys.* **115**, 2323–2347 (2001).

- J. Pillardy, C. Czaplewski, A. Liwo, J. Lee, D. R. Ripoll, R. Kazmierkiewicz, S. Oldziej, W. J. Wedemeyer, K. D. Gibson, Y. A. Arnautova, J. Saunders, Y.-J. Ye and H. A. Scheraga, "Recent Improvements in Prediction of Protein Structure by Global Optimization of a Potential Energy Function", *Proc. Natl. Acad. Sci., USA* **98**, 2329–2333 (2001).
- J. Pillardy, C. Czaplewski, W. J. Wedemeyer and H. A. Scheraga, "Conformation-Family Monte Carlo (CFMC): An Efficient Computational Method for Identifying the Low-Energy States of a Macromolecule", *Helv. Chim. Acta* **83**, 2214–2230 (2000).
- J. Lee, J. Pillardy, C. Czaplewski, Y. Arnautova, D. R. Ripoll, A. Liwo, K. D. Gibson, R. J. Wawak, and H. A. Scheraga, "Efficient Parallel Algorithms in Global Optimization of Potential Energy Functions", *Comput. Physics Commun.* **128**, 399–411 (2000).
- J. Lee, A. Liwo, D. R. Ripoll, J. Pillardy, J. A. Saunders, K. D. Gibson and H. A. Scheraga, "Hierarchical Energy-Based Approach to Protein-Structure Prediction: Blind-Test Evaluation with CASP3 Targets", *Intl. J. Quantum Chem.* **71**, 90–117 (2000).
- J. Pillardy, R. J. Wawak, Y. A. Arnautova, C. Czaplewski, and H. A. Scheraga, "Crystal Structure Prediction by Global Optimization as a Tool for Evaluating Potentials: Role of the Dipole Moment Correction Term in Successful Predictions", *J. Am. Chem. Soc.* **122**, 907–921 (2000).
- H. A. Scheraga, J. Lee, J. Pillardy, Y.-J. Ye, A. Liwo, and D. R. Ripoll, "Surmounting the Multiple-Minima Problem in Protein Folding", *J. Global Optimization* **15**, 235–260 (1999).
- J. Lee, A. Liwo and H. A. Scheraga, "Energy-Based *de novo* Protein Folding by Conformational Space Annealing and an Off-lattice United-Residue Force Field: Application to the 10-55 Fragment of Staphylococcal Protein A and to apo calbindin D9K", *Proc. Natl. Acad. Sci., USA* **96**, 2025–2030 (1999).
- J. Lee, H. A. Scheraga and S. Rackovsky, "Conformational Analysis of The 20-Residue Membrane-Bound Portion of Melittin by Conformational Space Annealing", *Biopolymers* **46**, 103–115 (1998).
- R. J. Wawak, J. Pillardy, A. Liwo, K.D. Gibson and H. A. Scheraga, "Diffusion Equation and Distance Scaling Methods of Global Optimization: Applications to Crystal Structure Prediction", *J. Phys. Chem.* **102**, 2904–2918 (1998).
- A. Liwo, R. Kazmierkiewicz, C. Czaplewski, M. Groth, S. Oldziej, R. J. Wawak, S. Rackovsky, M. R. Pincus, and H. A. Scheraga, "United-Residue Force Field for Off-Lattice Protein-Structure Simulations; III. Origin of Backbone Hydrogen-Bonding Cooperativity in United-Residue Potentials", *J. Comput. Chem.* **19**, 259–276 (1998).

Assignment 8: A Bit of Programming: Nonbonded Versus Bonded Energy Computations

This is a small programming assignment. At least one assignment in this modeling course should give you such first-hand experience! If you are a novice in programming, NYU staff, the TA, and the course assistant can help you, so set up an appointment with them early. You will have *two weeks* for this assignment.

1. **Programming Nonbonded Energy Computations.**
We will begin with the nonbonded energy computations since they are most straightforward (but most expensive!)

Write a simple program to compute the nonbonded energy of a system of 1000 atoms. The nonbonded energy, Lennard Jones and Coulomb terms, should have the form:

$$E_{NONB} = \sum_{i<j} \left[\frac{-A_{ij}}{R_{ij}^3} + \frac{B_{ij}}{R_{ij}^6} + \frac{q_i q_j}{\sqrt{R_{ij}}} \right]. \quad \text{(D.1)}$$

Here R_{ij} is an interatomic distance *squared*, and A_{ij}, B_{ij}, q_i, and q_j are the familiar energy parameters.
For an atom $\mathbf{x}_k$ of Cartesian components $\{x_{k1}, x_{k2}, x_{k3}\}$,

$$R_{ij} = (x_{j1} - x_{i1})^2 + (x_{j2} - x_{i2})^2 + (x_{j3} - x_{i3})^2 , \quad \text{(D.2)}$$

and the interatomic distance r_{ij} is

$$r_{ij} = \sqrt{R_{ij}}.$$

Set up your program to read in arbitrary atomic coordinate data — a file will be sent to you electronically in case you want to use it[5] — and *repeat the nonbonded energy calculation* 10,000 times for all $\{i, j\}$ pairs with $i < j$ for $j = 1, \ldots 1000$. The R_{ij} calculations can be placed in some inline function.
Perform the calculations for the nonbonded energy evaluations for both single and double precision, and record the total CPU time in each case.
Also report how much CPU time and CPU percentage the square-root ($\sqrt{\ }$) operation consumes. There are special timing functions that describe the distribution of CPU time among the various program parts.
Describe the machine you are using, the precision, and attach the subroutine and program output, along with the above results.

[5]The file can be obtained through the link to the course web site or directly from the author.

2. **Programming the Bonded Energy Computations.**
Next we will write three additional subprograms to compute the bond energy, bond-angle energy, and dihedral-angle energy of a molecular system. For now, we will assume there are 1000 bonds, 1000 bond angles, and 1000 dihedral angles. We will again perform 10,000 energy evaluations of each energy term, with each sweep here involving 1000 internal variables. This large number is necessary to get reliable timing values for the bonded interactions.

You can choose in each case any representative potential form. For example, you may use:

$$E_{BOND} = \sum_{i,j \in S_B} S_{ij} \left(r_{ij} - \bar{r}_{ij} \right)^2, \tag{D.3}$$

$$E_{BANG} = \sum_{i,j,k \in S_{BA}} K_{ijk} (\cos\theta_{ijk} - \cos\bar{\theta}_{ijk})^2, \tag{D.4}$$

$$E_{TOR} = \sum_{ijk\ell \in S_{DA}} \left(\frac{V3_{ijk\ell}}{2} \left[1 + \cos(3\tau_{ijk\ell}) \right] \right), \tag{D.5}$$

where the sets S_B, S_{BA}, and S_{DA} contain all bonds, bond angles, and dihedral angles, respectively. Here θ and τ denote a bond angle and dihedral angle, respectively, of a given triplet or quadruplet of atoms. The values with overhead bar symbols indicate reference values.

Since we are interested only in timing for now, you can use any pairs, triplets, or quadruplets in your sample energy routines — even the same sequence — repeatedly, as long as the total number of interactions used to obtain each energy term is 1000.

Some program segments which you may find helpful are posted on the website. The derivative components are present in the angle routines, *but you do not need them* for this assignment. You can find details of the cosba and cosda subroutines in an article.[6]

For your convenience, an addendum to this assignment also summarizes the basic geometric relations involved in defining internal variables.

Report the CPU time required for each routine in a table, including absolute time as well as percentage of the total time of *bonded* energy components. Again, attach your programs and output to the report of the results.

[6] "A Recipe for Evaluating and Differentiating $cos\phi$ Expressions", *J. Comp. Chem.* **10**, 951–956, 1989.

3. **Setting up A Polymer Model**. For obtaining realistic CPU estimates, we will now consider a simple n-alkane chain with the chemical formula CH_3–$(CH_2)_m$–CH_3, where m is an integer. For large m, this is polyethylene. For $m = 2$, for example, we have butane, chemical formula C_4H_{10}. To have about 1000 atoms, we will use $m = 330$ for our model calculations.

   Determine the number of bonds, bond angles, dihedral angles, and unique interatomic distances (atom pairs) that polyethylene has as a function of m. Consider all the distinct possibilities for the bonds and angles. Report these expressions.

   Then report how many bonds, bond angles, dihedral angles, and unique atom pairs the polymer has for the case $m = 330$.

4. **Bonded Versus Nonbonded Energy Computations**.
   Now we will combine the timing above to estimate the CPU time spent in bonded versus nonbonded energy computations for 10,000 iterations (of energy evaluations) for our polymer of 998 atoms.
   Scale the timing you obtained above (10,000 iterations for 1000 atoms for the nonbonded terms, and 10,000 iterations for 1000 bonds, bond angles, and dihedral angles) so that they correspond to the numbers relevant for our polymer with $m = 330$, as determined in item 3 above.

   Collect the data in one table which reports the CPU time and percentage required for each of the four subroutines.

   What can you conclude? What can you suggest to speed up the nonbonded computations, especially if derivatives are also required?

5. **Extra Credit!**
   For extra credit (the grade on this will replace your lowest homework grade), write the four subroutines above specifically for polyethylene. This means that you should use realistic coordinates, as well as correct data structures so that you consider all relevant bonds and angles for this polymer. Similarly, for energy parameters, associate values according to atom, bond, and angle types (e.g., C–C and C–H bonds, C–C–C, H–C–H, and H–C–C bond angles, and rotations about C–C bonds). You can use any resources on Insight to help you.

   Hand in all programs and results as requested above.

Addendum to Assignment 8:
Definitions of Internal Variables in Molecules

A bond angle θ_{ijk} formed by a bonded triplet of atoms i–j–k is expressed as an inner product:

$$\cos\theta_{ijk} = \frac{(\mathbf{x}_k - \mathbf{x}_j) \bullet (\mathbf{x}_i - \mathbf{x}_j)}{r_{jk}\; r_{ji}}\,, \tag{D.6}$$

or

$$\cos\theta_{ijk} = (\mathbf{r}_{jk} \bullet \mathbf{r}_{ji})/r_{jk}\; r_{ji}\,,$$

where the distance *vector* from atom j to i is given by

$$\mathbf{r}_{ji} \;=\; \mathbf{x}_i - \mathbf{x}_j \;=\; [x_{i1} - x_{j1},\;\; x_{i2} - x_{j2},\;\; x_{i3} - x_{j3}]^{\mathrm{T}}. \tag{D.7}$$

A dihedral angle $\tau_{ijk\ell}$, defining the rotation of bond i–j about bond j–k with respect to k–l, is expressed as

$$\cos\tau_{ijk\ell} = \mathbf{n}_{ab} \bullet \mathbf{n}_{bc}. \tag{D.8}$$

The vectors $\mathbf{n}_{ab}$ and $\mathbf{n}_{bc}$ denote unit normals to planes spanned by vectors $\{\mathbf{a}\;,\mathbf{b}\}$ and $\{\mathbf{b}\;,\mathbf{c}\}$, respectively, where $\mathbf{a} = \mathbf{r}_{ij}$, $\mathbf{b} = \mathbf{r}_{jk}$, and $\mathbf{c} = \mathbf{r}_{k\ell}$. Denoting θ_{ab} and θ_{bc} as angles θ_{ijk} and $\theta_{jk\ell}$, respectively, we write:

$$\cos\tau_{ijk\ell} = \frac{\mathbf{a}\times\mathbf{b}}{\|\mathbf{a}\|\;\|\mathbf{b}\|\sin\theta_{ab}} \bullet \frac{\mathbf{b}\times\mathbf{c}}{\|\mathbf{b}\|\;\|\mathbf{c}\|\sin\theta_{bc}}\,. \tag{D.9}$$

The sign of $\tau_{ijk\ell}$ is determined by the sign of the triple scalar product $\mathbf{a} \bullet (\mathbf{b}\times\mathbf{c})$.

To simplify potential energy equations (and differentiation when needed) it is convenient to work with inner product expressions and use Lagrange's identity $(\mathbf{a}\times\mathbf{b}) \bullet (\mathbf{c}\times\mathbf{d}) = (\mathbf{a}\bullet\mathbf{c})\;(\mathbf{b}\bullet\mathbf{d}) - (\mathbf{b}\bullet\mathbf{c})(\mathbf{a}\bullet\mathbf{d})$. This produces the alternative expression:

$$\cos\tau_{ijk\ell} = \frac{(\mathbf{a}\times\mathbf{b}) \bullet (\mathbf{b}\times\mathbf{c})}{[\,(\mathbf{a}\times\mathbf{b}) \bullet (\mathbf{a}\times\mathbf{b})\;\;(\mathbf{b}\times\mathbf{c}) \bullet (\mathbf{b}\times\mathbf{c})\,]^{\,1/2}}$$

$$= \frac{(\mathbf{a}\bullet\mathbf{b})(\mathbf{b}\bullet\mathbf{c}) - (\mathbf{a}\bullet\mathbf{c})(\mathbf{b}\bullet\mathbf{b})}{\left\{\,[\,(\mathbf{a}\bullet\mathbf{a})(\mathbf{b}\bullet\mathbf{b}) - (\mathbf{a}\bullet\mathbf{b})^2\,]\;[\,(\mathbf{b}\bullet\mathbf{b})(\mathbf{c}\bullet\mathbf{c}) - (\mathbf{b}\bullet\mathbf{c})^2\,]\,\right\}^{\,1/2}}\,. \tag{D.10}$$

According to this convention, $\tau = 0^{\mathrm{o}}$ defines a *cis* coplanar orientation for atoms i–j–k–l, $\tau = 180^{\mathrm{o}}$ defines a *trans* coplanar orientation, and a positive sign corresponds to a clockwise rotation of the far bond with respect to the near bond (when viewed along the j–k bond).

(See code segments on website)

Coordinate file (available electronically) for the 1000-atom molecule $CH_2OH{-}(CH_2)_{330}{-}CH_2OH$:

Atom	X	Y	Z	ID
1	4.988000	2.012136	-7.818089	O
2	5.915912	2.025234	-7.572307	H
3	4.596542	0.674197	-8.131964	C
4	5.198226	0.305558	-8.962735	H
5	4.751310	0.037135	-7.261160	H
6	3.108016	0.653187	-8.526237	C
7	2.517394	1.015322	-7.710808	H
8	2.956134	1.278341	-9.381230	H
9	2.686321	-0.787809	-8.863891	C
10	2.838205	-1.412964	-8.008898	H
..	....	....	....	.

Etc.

Assignment 9: TERM PROJECT

The Successes (Failures?) of Molecular Modeling

The year is 2006. You have graduated and moved on with your life. Due to your outstanding academic record at NYU, you have landed a high-profile job as a staff research scientist for PBS (Public Broadcasting Service) in the nation's capital.

You are now assigned to prepare for an internationally televised scientific program entitled *Biocomputing in the Third Millennium*. In this program, a team of scientific experts will respond to live questions transmitted by comphones from the general public. Since these scientists are busy traveling, consulting, reviewing papers, writing grants, researching, and teaching, your group is in charge of preparing all background information for the panelists.

Specifically, you are told to prepare for the following questions:

> *Can the panel describe some concrete examples where computational tools have significantly enhanced our understanding of molecular systems — from small organic systems to macromolecules — by offering new insights, interpretations, and predictions, of practical and scientific importance, that were impossible to obtain by experimental techniques?*
>
> *What modeling/simulation tools were used in each case, and what can be credited to each success (computing power, algorithms, intuition, right time, sheer luck, persistence, etc.)?*

You are promised by your boss a hefty bonus for each complete and satisfactory item provided. However, the minimal requirement (for obtaining a B-level mark on your monthly evaluation form, given that you produce truly outstanding examples) is detailing FOUR "SENSATIONAL" EXAMPLES.

Each example must be clearly described and entered under the following subheadings: Problem, Methodology, Success, Significance, References. The second item, Methodology, requires the most comprehensive coverage, followed by Significance. You are asked to attach to your meticulous writeup any visual aids (charts, figures, sketches) that will enhance the presentation, both to a general (nonspecialist) audience and to a highly informed scientist. Creativity is highly desired. Try also to analyze the findings in a larger context.

Back at your ergonomic desk, with your feet up and glancing at regal Washington monuments against a glorious background of blossoming cherry trees with occasional views of ambitious runners and politicians, you recall a molecular modeling course you took in the good ol' days at NYU. Memories come back of many homework assignments inflicted upon you weekly by your professor — dealing with web resources, sequence and structural databases, Insight, sequence/structure contests, force fields, tedious programming, difficult minimization, and Monte Carlo simulations. You find fragments of lecture notes and transparency copies inside an old purple gym bag and begin to follow up on, and explore, some of those key words, resources, authors, and topics. You also begin to wonder if there are any interesting and instructive examples of *failures in molecular modeling* and decide to pursue those for an extra bonus. (Maybe the boss will let *you* design the next scientific program?)

Your deadline in early May is rapidly approaching and you begin to work early and diligently. The promise that the best examples provided by the crew will be published, if appropriate, in an article provides further motivation for the assignment. You also decide to contact your professor if she is still at NYU when you get stuck or have questions.

You find the project more interesting now, and vow to become *famous* (and maybe even *rich*)!

Background Reading from Coursepack

- M. S. Friedrichs and P. G. Wolynes, "Toward Protein Tertiary Structure Recognition by Means of Associative Memory Hamiltonians", *Science* **246**, 371–373 (1989).
- T. Schlick, "Optimization Methods in Computational Chemistry", in Reviews in Computational Chemistry, K. B. Lipkowitz and D. B. Boyd, Editors, Vol. 3, pp. 1–71, VCH Publishers, New York (1992).
- See also an updated version titled "Geometry Optimization" in the Encyclopedia of Computational Chemistry, P. von Ragué Schleyer (Editor-in-Chief) and N. L. Allinger, T. Clark, J. Gasteiger, P. A. Kollman, and H. F. Schaefer, III, Editors, Vol. 3, pp. 1136–1157, John Wiley & Sons, West Sussex, England (1998).
- R. A. Abagyan and M. M. Totrov, "Biased Probability Monte Carlo Conformational Searches and Electrostatic Calculations for Peptides and Proteins", *J. Mol. Biol.* **235**, 983–1002 (1994).

Assignment 10: Experiments in Molecular Geometry Optimization: Biphenyl Minimization

See Insight II *and* Discover *manuals for reference.*

1. **Brief introduction to the Discover module of** Insight II.

The Discover[7] software performs energy minimization and molecular dynamics simulations. This program constitutes a powerful modeling tool since it offers many features such as constrained and restrained minimization, calculation of vibrational frequencies, and analysis tools. Many variations of simulation conditions (e.g., constant temperature, constant pressure) are available.

We will access the Discover software from the Insight II environment. The **Discover** module of Insight II is a convenient interface to the Discover program. This module builds Discover input files from information provided through graphical interfaces, and it allows users to run Discover jobs interactively. Though more advanced users may prefer to use the independent version of Discover, the Insight II environment is more appropriate for a novice.

Before using Discover, make sure that Insight II contains all of the necessary information to define the topology, coordinates, and force field parameters. These include, for example, atom types and partial charges (see lecture notes for structure definitions).

If you succeed in displaying the molecule correctly on the screen, the topological and coordinate information is most likely in order. However, selecting the appropriate force field and assigning atom types and parameters is a separate task.

(a) To select the force field, use Forcefield / Select.
(b) To assign atom types, use the Fix option for Potential Action in Forcefield / Potentials. Alternatively, first assign atom types with Atom / Potential in the **Biopolymer** module, and then use the Accept option for Potential Action in Forcefield / Potentials.
(c) To assign charges, use the Fix option for both Partial Chg Action and Formal Chg Action under Forcefield / Potentials.

Note that after each change in the force field you must assign atom types and charges anew.

[7]Note that the name *Discover* has two separate meanings. The first, Discover, stands for the software package with minimization and molecular dynamics routines. The second, typed in bold (**Discover**), refers to the module available in Insight II.

To check if the assigned atom types and partial charges are correct, you can select Potential or Partial_charge in Molecule / Label to label each atom. Once you specify the information about the structure and parameters, you are ready to move to the **Discover** module. (We will not use **Discover_3** in this course).

The Constraint pulldown menu contains various atom-constraining and restraining procedures that you can select. In Parameters, you select the simulation type for Discover (Minimize, Dynamics, etc.), as well as the choice for cutoff parameters for nonbonded interactions, periodic boundary conditions (Variables), and dielectric constant (Set). Take time to familiarize yourself with the first three pulldown menus Constraint, Parameters, and Run, with Insight_help active, to learn about the various commands they contain.

To start a simulation, go to Run / Run, select desired options, choose the object for calculations, and execute.
Each Discover run is assigned a number in the order of the execution start time. The files created during the execution are identified by the calculation object (molecular system name) and the job integer (appended to the name). The file extension specifies the file type. Examples are listed below.

Discover Input Files:

- Commands (`.inp`)
- Cartesian Coordinates (`.car`)
- Molecular Data (`.mdf`)
- Force field Parameters (`.frc`)
- Restraints (`.rstrnt`)

Discover Output Files:

- Standard Output (`.out`)
- Cartesian Coordinates (final structure) (`.cor`)
- Cartesian Coordinate Archive (multiple frames) (`.arc`)
- Automatic Potential Parameter Assignment (`.prm`)
- Discover Dynamics Restart Information (`.rst`)

You can specify the files to save with Run / Files. By default, all are saved.

2. **Setting up biphenyl minimization**.

We will begin to learn about potential energy minimization for a simple yet interesting system, biphenyl (see Fig. D.2).

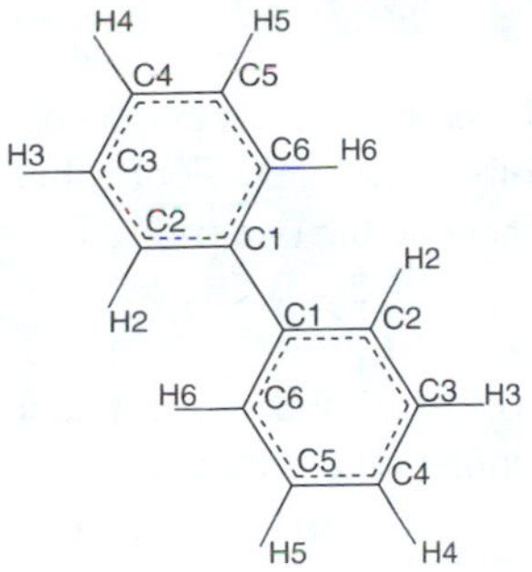

Figure D.2. Biphenyl

You will receive electronically two files containing the coordinates of biphenyl[8] The first, `biphenyl.car`, includes the structure with coplanar phenyl rings. This configuration was created with the **Builder** module by connecting two benzene rings.
The second file, `biphenyl_distorted.car`, contains the structure with each of the phenyl rings distorted from planarity.
Before displaying these structures, check that the AMBER force field is chosen. This will save work in assigning AMBER force field parameters. It will also permit you to proceed to Discover directly. (Note: For other force fields, you would have to assign parameters through Forcefield / Potentials). To open a coordinate file and display a structure, use Molecule / Get, specify Archive as the File Type, and select the desired file.

3. **Generation of energy profiles by restrained minimization**.

   A potential energy profile along some molecular coordinate, X (such as the rotamer dihedral angle χ_1), describes the dependence of the energy, minimized with respect to the remaining coordinates, on X. The simplest way to generate such a profile is to use minimization with *restraints*. Restraining a coordinate X to a specified value X^0 can be accomplished by adding harmonic penalty term,

$$E_{\text{rstr}} = K\,(X - X^0)^2,$$

   to the potential energy. After minimization, X should not deviate significantly from X^0 when the force constant K is large.[9] For a complete profile, minimum energy values must be calculated for a series of values $\{X_1^0, X_2^0,$

[8]Files can be obtained through the link to the course web site or directly from the author.

[9]*Constrained*, as opposed to restrained, minimization entails a more complex procedure to guarantee that $X = X^0$.

$X_3^0, \ldots\}$ in the range of X.

For biphenyl, we will analyze the dependence of energy on the torsion angle between the planes of phenyl rings. Four dihedral angles are defined about the C1–C1 bond connecting the two rings. They are specified by the following atom quadruplets {1B:C2, 1B:C1, 1:C1, 1:C6}, {1B:C6, 1B:C1, 1:C1, 1:C2}, {1B:C2, 1B:C1, 1:C1, 1:C2}, and {1B:C6, 1B:C1, 1:C1, 1:C6}. Restraining only one of them will result in a nonplanar geometry of phenyl rings (since the remaining dihedral angles will tend to assume values associated with a lower energy). To ensure that the phenyl-ring planes are not distorted, it is necessary to restrain a *pair* of dihedral angles to the same value. You can choose the first two or the last two atom quadruplets from the list above.

The plot for the full range of the angle, $[-180.0°, 180.0°]$, can be created by first computing energy minima for a sequence of values in the range $[0.0°, 90.0°]$ (e.g., $0.0°, 10.0°, 20.0°, \ldots, 90.0°$), and then using symmetry operations.

Start with the coplanar structure (`biphenyl.car`). Make sure that potential parameters are properly assigned.

Then select Constraint / TorsionForce. You can now proceed in different ways to calculate the energy values for the profile. For instance, you can make 10 separate minimization runs, each time specifying both restraints (Intervals set to 1). Alternatively, you can execute one run specifying the range of values for both restraints (Intervals set to 9, Starting_Angle set to 0.0, and Angle_Size set to 90.0). In the latter case, you must extract the appropriate energy values from the output file. For two restraints, defined at ten points each, 100 energy values (corresponding to all restraints) will be listed as output. Extract only those values for which the restraint targets on both angles are identical.

Use Force Constant set to the range of 2000–5000.

Switch to Parameters / Minimize and select Conjugate gradient algorithm with Gradient tolerance set to 0.001.

Note that Parameters / Set and Parameters / Variables are left at their default values. Now proceed to Run.

Another possibility is to use Run / Files to limit the number of output files. Before executing the Run / Run command check the restraints and selected minimization options using the List option.

After minimization, the dihedral angle might deviate somewhat from the value specified in the restraint. Save the final structure (both dihedral angles are around 90°) to `biphenyl.psv` using File / Save_Folder.

You can view these structures (frames) with Trajectory / Get and Trajectory / Conformation from the **Analysis** module and determine the torsion angle value.

Plotting the profile should be done only after completing the next section of the assignment.

4. **Unrestrained minimization for biphenyl**.

   In addition to the restrained minimization calculations, perform unrestrained minimization to find the "global" energy minimum, $E_{\min}$, for biphenyl.

   Now express the profile energy E from the previous section relative to the $E_{\min}$ (i.e., $E - E_{\min}$), and plot against the dihedral angle for the full range $[-180.0°, 180.0°]$.

   Note that $E_{\min}$ may be larger than some E values. Why is that?

5. **Comparison of different force fields**.

   Repeat the energy profile calculations with the cff91 force field. Plot the results obtained with the AMBER and cff91 force fields on one plot and discuss your findings.

6. **Dependence on initial conditions**.

   Perform unrestrained minimization of biphenyl starting with the structure specified in the `biphenyl.psv` file from Part 3. Use the cff91 or AMBER force field and any minimization algorithm you wish, but use the Derivative tolerance of 0.001.

   Describe the minimization algorithm briefly and discuss your results.

7. **Assessment of the performance of various minimization algorithms in** Insight II.

   For each minimization algorithm offered in Discover record the CPU time required for convergence of the energy gradient to the target values of 10.0, 0.1, 0.001, and 0.00001 kcal/Å. Use the AMBER force field.

   For each algorithm, begin with the structure contained in the file `biphenyl_distorted.car`. Select the desired Algorithm from Parameters / Minimize; set Iterations to 5000; specify the first target value of the derivative; execute; and proceed to execute the Run/Run command. After this job is completed, change the derivative tolerance to the next target value and repeat minimization. Extract the computational times and values of minima from each output file. Do not increase the number of Iterations above 5000. If the specified convergence is not reached with this threshold, note that in your report.

Repeat this procedure for each of the remaining algorithms. (Remember to start with the structure from `biphenyl_distorted.car` file.) Construct a table comparing the performance of minimization algorithms in the different regions of derivative tolerance (report the timing and energy minimum values).

On the basis of these results, and the information you have learned in class, suggest a simulation schedule to achieve an optimal minimization of a large molecule. Note that for our small system the gradient norm associated with the initial configuration of biphenyl is not extremely large.

Background Reading from Coursepack

- M. Karplus and G. A. Petsko, "Molecular Dynamics Simulations in Biology", *Nature* **347**, 631–639 (1990).

Assignment 11: A Global Optimization Contest!

Our goal is to compute the lowest energy structure for the pentapeptide met-enkephalin, whose sequence is **Tyr–Gly–Gly–Phe–Met**. Many local minima exist for this molecule, so it is a challenge to reach the global minimum. *The student who finds the structure of the lowest energy will receive a prize from the instructor.*

The rules of this contest are:

1. use a molecule with *charged* COO^- and NH_3^+ ends
2. use the AMBER force field
3. use the distance dependent dielectric constant (**Discover** module, Parameters / Set command, Dist_Dependent button on)
4. use $1/2$ as the scale factor for 1–4 nonbonded interactions (i.e., Parameters / Scale_Terms command, p1_4 button on, and specify 0.5)

You can use *any* technique mentioned in this course (energy minimization, molecular dynamics, Monte Carlo sampling), as well as any other resources (e.g., web and literature), to find the global minimum of the pentapeptide.

Be Creative.

Hand in a detailed report describing how you reached the minimum for met-enkephalin and any particular difficulties, or interesting observations, you encountered along the way. Attach the Cartesian coordinate file and the energy value reached.

Also submit a three-dimensional picture of the configuration of lowest energy along with a table specifying all associated bond lengths and bond angle values, and the $\{\phi, \psi\}$ and χ dihedral-angle values per residue.

To qualify for consideration of the prize, send electronically the coordinate file with the minimized structure to the instructor and TA.

Good Luck!

Background Reading from Coursepack

- K. A. Dill and H. S. Chan, "From Levinthal to Pathways to Funnels", *Nature Struc. Biol.* **4**, 10–19 (1997).
- T. Lazaridis and M. Karplus, " 'New View' of Protein Folding Reconciled with the Old Through Multiple Unfolding Simulations", *Science* **278**, 1928–1931 (1997).

Assignment 12: Monte Carlo Simulations

1. **Random Number Generators**. Investigate the types of random number generators available on: (a) your local computing environment and (b) a mathematical package that you frequently use. How good are they? Is either one adequate for long molecular dynamics runs? Suggest how to improve them and test your ideas.

   To understand some of the defects in linear congruential random number generators, consider the sequence defined by the formula $y_{i+1} = (a\, y_i + c) \bmod M$, with $a = 65539$, $M = 2^{31}$, and $c = 0$. (This defines the infamous random number generator known as RANDU developed by IBM in the 1960s, which subsequent research showed to be seriously flawed). A relatively small number of numbers in the sequence (e.g., 2500) can already reveal a structure in three dimensions when triplets of consecutive random numbers are plotted on the unit cube. Specifically, plot consecutive pairs and triplets of numbers in two and three-dimensional plots, respectively, for an increasing number of generated random numbers in the sequence, e.g., 2500, 50,000, and 1 million. (Hint: Figure D.3 shows results from 2500 numbers in the sequence).

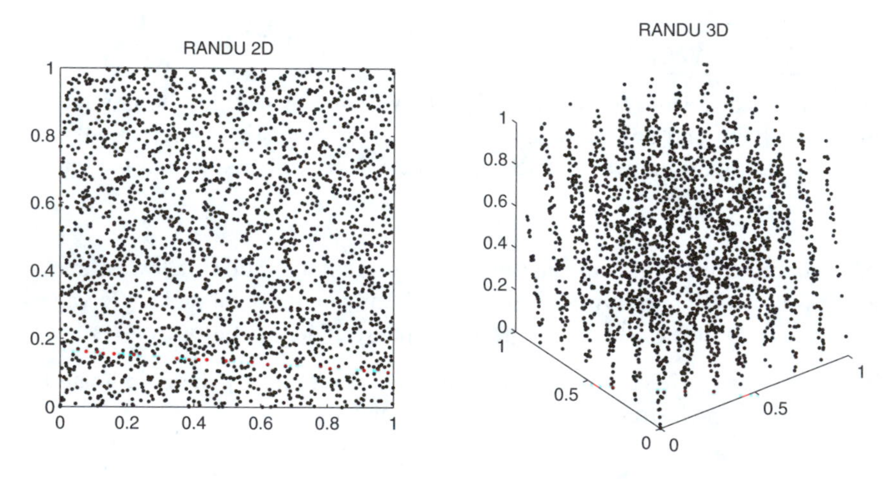

Figure D.3. Plots generated from pairs and triplets of consecutive points in the linear congruential generator known as RANDU defined by $a = 65539$, $M = 2^{31}$, and $c = 0$ when 2500 total points in the sequence are generated.

2. **MC Means**. Propose and implement a Monte Carlo procedure to calculate π based on integration. How many MC data points are needed to yield an answer correct up to 5 decimal places? What is the computational time

involved? Show a table of your results displaying the number of MC steps, the associated π estimate, and the calculated error.

3. **Gaussian Variates**. You are stranded in an airport with your faithful laptop with one hour to spare until the deadline for emailing your homework assignment to your instructor. The assignment (next item) relies on a *Gaussian random number generator*, but you have forgotten the appropriate formulas involved in the commonly used Box/Muller/Marsaglia transformation approach. Fortunately, however, you remember the powerful Central Limit Theorem in basic probability and decide to form a random Gaussian variate by sampling N uniform random variates $\{x_i\}$ on the unit interval as

$$\bar{y} = \sum_{i=1}^{N} x_i \, .$$

You quickly program the expression:

$$y = \sqrt{\frac{1}{\sigma^2(\bar{y})}} \sum_{i=1}^{N} [x_i - \mu(\bar{y})]$$

where above σ^2 is the standard deviation of $\bar{y} = N\sigma^2(x)$ and the mean $\mu(\bar{y}) = N\mu(x)$. [Recall that the uniform distribution has a mean of 1/2 and variance of 1/12].

How large should N be, you wonder. You must finish the assignment in a hurry. To have confidence in your choice, you set up some tests to determine when N is sufficiently large, and send your resulting routine, along with your testing reports, and results for several choices of N.

4. **Brownian Motion**. Now you can use the Gaussian variate generator above for propagating *Brownian motion* for a single particle governed by the biharmonic potential $U(x) = kx^4/4$. Recall that Brownian motion can be mimicked by simulating the following iterative process for the particle's position:

$$x^{n+1} = x^n + \frac{\Delta t}{m\gamma} F^n + R^n$$

where

$$\langle R^i R^j \rangle = \frac{2k_B T \Delta t}{m\gamma} \delta_{ij} , \qquad \langle R^i \rangle = 0 \, .$$

Here m is the particle's mass; γ is the collision frequency, also equal to ξ/m where ξ is the frictional constant; and F is the systematic force. You are required to test the obtained mean square atomic fluctuations against the known result due to Einstein:

$$\langle x^2 \rangle = 2 \left(\frac{k_B T}{m\gamma} \right) t = 2Dt \,,$$

where D is the diffusion constant.
The following parameters may be useful to simulate a single particle of mass $m = 4 \times 10^{-18}$ kg and radius $a = 100$ nm in water: by Stokes' law, this particle's friction coefficient is $\xi = 6\pi\eta a$ = 1.9×10^{-9} kg/s, and $D = k_B T/\xi = 2.2 \times 10^{-12}$m^2/s. You may, however, need to scale the units appropriately to make the computations reasonable.

Plot the mean square fluctuations of the particle as a function of time, compare to the expected results, and show that for $t \gg 1/\gamma = 2 \times 10^{-9}$ s the particle's motion is well described by random-walk or diffusion process.

Background Reading from Coursepack

- T. Schlick, E. Barth, and M. Mandziuk, "Biomolecular Dynamics at Long Timesteps: Bridging the Timescale Gap Between Simulation and Experimentation", *Ann. Rev. Biophys. Biomol. Struc.* **26**, 179–220 (1997).
- E. Barth and T. Schlick, "Overcoming Stability Limitations in Biomolecular Dynamics: I. Combining Force Splitting via Extrapolation with Langevin Dynamics in LN", *J. Chem. Phys.* **109**, 1617–1632 (1998).
- L. S. D. Caves, J. D. Evanseck, and M. Karplus, "Locally Accessible Conformations of Proteins: Multiple Molecular Dynamics Simulations of Crambin", *Prot. Sci.* **7**, 649–666 (1998).

Assignment 13: Advanced Exercises in Monte Carlo and Minimization Techniques

1. Study the function:

$$E(x, y) = ax^2 + by^2 + c\,(1 - \cos\gamma x) + d\,(1 - \cos\delta y)\,. \tag{D.11}$$

   Note that is has many local minima and a global minimum at $(x, y) = (0, 0)$. Minimize $E(x, y)$ with $a = 1, b = 2, c = 0.3$, $\gamma = 3\pi$, $d = 0.4$, and $\delta = 4\pi$ by the standard simulated annealing method. Use the starting point $(1, 1)$ and step perturbations $\Delta x = 0.15$, and set β in the range of 3.5 to 4.5. Limit the number of steps to ~ 150. Now implement the *variant* of the simulated annealing method where acceptance probabilities for steps with $\Delta E < 0$ are proportional to $\exp(-\beta E^g \Delta E)$, with the exponent $g = -1$. Analyze and compare the efficiency of the searches in both cases. It will be useful to plot all pairs of points (x, y) that are generated by the method and distinguish 'accepted' from 'rejected' points.

2. Devise a different variant of the basic simulated annealing minimization method that would incorporate *gradient* information to make the searches more efficient.

3. Consider the following global optimization deterministic approach based on the *diffusion equation* as first suggested by Scheraga and colleagues (L. Piela, J. Kostrowicki, and H. A. Scheraga, "The Multiple-Minima Problem in Conformational Analysis of Molecules. Deformation of the Potential Energy Hypersurface by the Diffusion Equation Method", *J. Chem. Phys.* **93**, 3339–3346 (1989)).
   The basic idea is to deform the energy surface smoothly. That is, we seek to make "shallow" wells in the potential energy landscape disappear iteratively until we reach a global minimum of the deformed function. Then we "backtrack" by successive minimization from the global minimum of the transformed surface in the hope of reaching the global minimum of the real potential energy surface. This idea can be implemented by using the heat equation where T represents the temperature distribution in space x, and t represents time:

$$\frac{\partial^2 T}{\partial x^2} = \frac{\partial T}{\partial t} \tag{D.12}$$

$$T(x, 0) = E(x)\,. \tag{D.13}$$

   Here, the boundary condition at time $t = 0$ equates the initial temperature distribution with the potential energy function $E(x)$. Under certain conditions (e.g., E is bounded), a solution exists. Physically, the application of this equation exploits the fact that the heat flow (temperature distribution) should eventually settle down.

To formulate this idea, let us for simplicity consider first a one-dimensional problem where the energy function E depends on a scalar x. Let $E^{(n)}(x)$ represent the nth derivative of E with respect to x and define the transformation operator $\mathcal{S}$ on the energy function E for $\beta > 0$ as follows:

$$\mathcal{S}[E(x)] = E(x) + \beta E^{(2)}(x) \,. \tag{D.14}$$

That is, we have:

$$\begin{aligned}
\mathcal{S}^0 E &= E \\
\mathcal{S}^1 E &= E + \beta E^{(2)} \\
\mathcal{S}^2 E &= E + 2\beta E^{(2)} + \beta^2 E^{(4)} \\
\mathcal{S}^3 E &= E + 3\beta E^{(2)} + 3\beta^2 E^{(4)} + \beta^3 E^{(8)} \\
\vdots & \qquad \vdots \\
\mathcal{S}^N E &= (1 + \beta d^2/dx^2)^N E \,.
\end{aligned}$$

Now writing $\beta = t/N$ where t is the time variable, and letting $N \to \infty$, we write:

$$\exp\left(t d^2/dx^2\right) E \equiv \exp(A(t))\, E = \left[1 + A + \frac{A^2}{2!} + \frac{A^3}{3!} + \cdots\right] . \tag{D.15}$$

Thus we can define $T(t)$ as

$$T(t) = \exp((A(t)) = \exp(t d^2/dx^2) \,. \tag{D.16}$$

In higher dimensions, let x represent the collective vector of n independent variables; we replace the differential operator above d^2/dx^2 by the *Laplacian operator*, that is

$$\Delta = \sum_{i=1}^{n} \partial^2/\partial x_i \,.$$

Using this definition, we can also write

$$T(t) = T_1(t)\, T_2(t) \cdots T_n(t)$$

where

$$T_i = \exp(t\partial^2/\partial x_i) \,.$$

This definition produces the heat equation (D.12, D.13) since

$$\begin{aligned}
\frac{\partial T(t)[E(x)]}{\partial t} &= \left[\frac{dA}{dt} + \frac{2A}{2}\frac{dA}{dt} + \frac{3A^2}{3!}\frac{dA}{dt} + \cdots\right][E] \\
&= \left[1 + A + \frac{A^2}{2} + \cdots\right]\frac{d^2}{dx^2}[E]
\end{aligned}$$

$$= \frac{\partial^2 T(t)}{\partial x^2} [E(x)] .$$

In practice, the diffusion equation method for global optimization is implemented by solving the heat equation by Fourier techniques (easy, for example, if we have dihedral potentials only) or by solving for T up to a sufficiently large time t. This solution, or approximate solution (representing $E(x,t)$ for some large t), is expected to yield a deformed surface with one (global) minimum. With a local minimization algorithm, we compute the global minimum x^* of the deformed surface, and then begin an iterative deformation/minimization procedure from x^* and $E(x,t)$ so that at each step we deform backwards the potential energy surface and obtain its associated global minimum ($E(x,t) \longrightarrow E(x,t-\Delta t)$ and x^* to x^1, $E(x,t-\Delta t) \longrightarrow E(x,t-2\Delta t)$ and x^1 to x^2, $\cdots$ $E(x,0) \longrightarrow x^k$). Of course, depending on how the backtracking is performed, different final solutions can be obtained.

(a) To experiment with this interesting diffusion-equation approach for global minimization, derive a general form for the deformation operator $T(t) = \exp(td^2/dx^2)$ on the following special functions $E(x)$: (*i*) polynomial functions of degree n, and (*ii*) trigonometric functions $sin\omega x$ and $cos\omega x$, where ω is a real-valued number (frequency). What is the significance of your result for (*ii*)?

(b) Apply the deformation operator $T(t) = \exp(td^2/dx^2)$ to the quadratic function

$$E(x) = x^4 + ax^3 + bx^2 , \tag{D.17}$$

with $a = 3$ and $b = 1$. Evaluate and plot your resulting $T(t)E(x)$ function at $t = 0, \Delta t, 2\Delta t, ...$, for small time increments Δt until the global minimum is obtained.

(c) Apply the deformation operator $T(t)$ for the two-variable function in eq. (D.11). Examine behavior of the deformation as $t \to \infty$ as a function of the constants a and b. Under what conditions will a unique minimum be obtained as $t \to \infty$?

4. Use Newton minimization to find the minimum of the two-variable function in equation (D.11) and the one-variable function in equation (D.17). It is sufficient for the line search to use simple bisection: $\lambda = 1, 0.5$, etc., or some other simple backtracking strategy. For the quartic function, experiment with various starting points.

Assignment 14: Advanced Exercises in Molecular Dynamics

1. Calculate the 'natural' time unit for molecular dynamics simulations of biomolecules from the relation: energy = mass $*$ (length/time)^2, to obtain the time unit τ corresponding to the following units:

length	(l):	1 Å = 10^{-10}m
mass	(m):	1 amu = 1 g/mol
energy	(v):	1 kcal/mol = 4.184 kJ/mol.

   Estimate the "quantum mechanical cutoff frequency", $\omega_c = k\mathrm{T}/\hbar$ at room temperature ($\sim$ 300°K).

2. Derive the amplitude decay rate of $\gamma/2$ for an underdamped harmonic oscillator due to *friction* by solving the equations of motion:

$$m\frac{d^2x}{dt} = -kx - m\gamma\frac{dx}{dt} \tag{D.18}$$

   and examining time behavior of the solution.

3. Derive the amplitude decay rate of $\omega^2(\Delta t)/2$ *intrinsic* to the *implicit-Euler* scheme by solving the discretized form of eq. (D.18).

4. Compare your answer in problem 2 above with behavior of the *explicit-Euler* solution of eq. (D.18).

5. Compare molecular and Langevin dynamics simulations of two water molecules by the Verlet discretization of the equation of motion and its Langevin analog. Use the "SPC" *intermolecular* potential, given by:

$$E = \sum_{\substack{(i,j)\equiv(\mathrm{O,O})\ \text{pairs} \\ i<j}} \left[\frac{-A}{r_{ij}^6} + \frac{B}{r_{ij}^{12}}\right] + \sum_{\substack{(k,\ell)\equiv\text{intermolecular} \\ (\mathrm{O,O}),(\mathrm{O,H}),(\mathrm{H,H})\ \text{pairs} \\ k<\ell}} \left[\frac{Q_k Q_\ell}{r_{k\ell}}\right]$$

   where

$$\begin{aligned} A &= 626\,(\mathrm{kcal}\ \text{Å}^6)/\mathrm{mol} \\ B &= 629 \times 10^3 (\mathrm{kcal}\ \text{Å}^{12})/\mathrm{mol} \\ Q_\mathrm{H} &= 0.41\,e \\ Q_\mathrm{O} &= -0.82\,e\,. \end{aligned}$$

   A numerical factor of 332 is needed in the electrostatic potential to obtain energies in kcal/mol with the coefficients above. For simplicity, assume that *intramolecular* geometries are rigid: r_{OH} = 1 Å, $\cos\theta_{\mathrm{HOH}} = -1/3$. (You can use harmonic soft constraints). Begin by first minimizing the energy of the water dimer and examining the hydrogen bond geometry (hydrogen-bond distance and angle θ between the hydrogen-bond vector and bisector of the acceptor molecule). Then study numerical behavior of the two models/schemes as a function of Δt, and examine the hydrogen bond geometry. Experiment with Δt = 1, 2, 5, and 10 fs and use γ values in the range of 1 to 50 ps^{-1}. If you are more ambitious, continue to study energy-minimized structures of water clusters of larger sizes and their dynamics. Analyze the hydrogen bonding networks of these clusters.

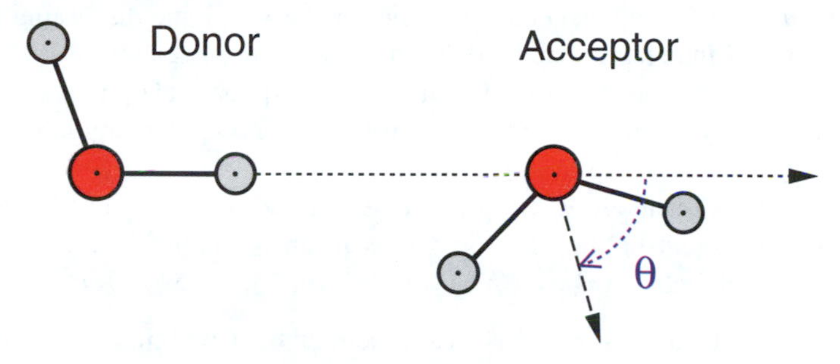

Figure D.4. Hydrogen bond geometry: the angle θ is defined between the hydrogen-bond vector and the bisector of the acceptor molecule.

Some Useful Constants and Conversion Factors

Avogadro's Number	$N_A = 6.0221 \times 10^{23}$ mol^{-1}
Planck's Constant	$h = 6.6261 \times 10^{-34}$ Jsec
	$\hbar = h/2\pi = 1.055 \times 10^{-34}$ Jsec
Boltzmann's Constant	$k_B = 1.38066 \times 10^{-23}$ JK^{-1}
Gas Constant	$R = k_B N_A = 8.3145$ JK^{-1} mol^{-1}
Atomic Mass Unit, amu	$(1/N_A) = 1$ g/mol = 1.6605 $\times 10^{-27}$ kg
	$\pi = 3.14159$
	1 kcal = 4.184 kJ

Assignment 15: BONUS PROJECT

The Scaling of Protein-Conformer Number with Size and Solvability of the Protein Folding Problem

The phone rings one morning as you sip through your Sunluck vanilla latté grande and check your emails in your office at the University of Seabeetle. The editor of the journal *Proteomics Today* is on the line.

Given your expertise in biomolecular modeling, she asks your help in writing a brief *Folding In Silico* Perspectives article for the next issue discussing the following series of interesting papers debating the nature of scaling of the number of protein conformers as function of chain length (exponential or nonexponential) and the solvability of the protein folding problem by computer simulation:

- W. F. van Gunsteren, R. Bürgi, C. Peter, and X. Daura, "The Key to Solving the Protein-Folding Problem Lies in an Accurate Description of the Denatured State", *Angew. Chem. Int. Ed.* **40**, 352–355 (2001).
- A. R. Dinner and M. Karplus, "Comment on the Communication 'The Key to Solving the Protein-Folding Problem Lies in an Accurate Description of the Denatured State' by van Gunsteren et al.", *Angew. Chem. Int. Ed.* **40**, 4615–4616 (2001).
- W. F. van Gunsteren, R. Bürgi, C. Peter, and X. Daura, "Reply", *Angew. Chem. Int. Ed.* **40**, 4616–4618 (2001).

Since many of *Proteomics Today* readers and authors perform computer simulations of proteins, both macroscopic and all-atom based, the editor wants you also to mention the strengths and weaknesses of these different approaches.

Though already overloaded with preparing final examinations for your classes, writing grant proposals, supervising your students and postdocs, and reviewing several articles for journals (long-overdue), you agree to take this challenging and potentially rewarding assignment. You immerse yourself in the papers, send your graduate students to collect background articles and information, order through your cellphone Sunluck's caramel-macchiato enormoso, and decide to make your article not only objective and interesting but also fun to read.

Some Suggestions:

- Discuss the Levinthal paradox.
- What is the relation among the number of conformers, timescales, and folding pathways?
- Analyze lattice or other protein simulations reported in the literature to estimate the number of possible conformers and the timescale of protein folding.

References

Please Note: Alphabetized reference items are posted on the book's website.

[1] E. O. Wilson. *Consilience. The Unity of Knowledge*. Alfred A. Knopf, New York, NY, 1998.

[2] J. N. Onuchic. Contacting the protein folding funnel with NMR. *Proc. Natl. Acad. Sci. USA*, 94:7129–7131, 1997.

[3] C. L. Brooks, III. Viewing protein folding from many perspectives. *Proc. Natl. Acad. Sci. USA*, 99:1099–1100, 2002.

[4] A. R. Fersht and V. Daggett. Protein folding and unfolding at atomic resolution. *Cell*, 108:573–582, 2002.

[5] K. Luger, A. W. Mäder, R. K. Richmond, D. F. Sargent, and T. J. Richmond. Crystal structure of the nucleosome core particle at 2.8 Å resolution. *Nature*, 389:251–260, 1997.

[6] D. A. Doyle, J.Morais Cabral, R. A. Pfuetzner, A. Kuo, J. M. Gulbis, S. L. Cohen, B. T. Chait, and R. MacKinnon. The structure of the potassium channel: Molecular basis of K+ conduction and selectivity. *Science*, 280:69–77, 1998.

[7] R. MacKinnon, S. L. Cohen, A. Kuo, A. Lee, and B. T. Chait. Structural conservation in prokaryotic and eukaryotic potassium channel. *Science*, 280:106–109, 1998.

[8] J. H. Morais-Cabral, Y. Zhou, and R. MacKinnon. Energetic optimization of ion conduction rate by the K^+ selectivity filter. *Nature*, 414:37–42, 2001.

[9] Y. Zhou, J. H. Morais-Cabral, A. Kaufman, and R. MacKinnon. Chemistry of ion coordination and hydration revealed by a K^+ channel-Fab complex at 2.0 Å resolution. *Nature*, 414:43–48, 2001.

[10] R. Dutzler, E. B. Campbell, M. Cadene, B. T. Chait, and R. MacKinnon. X-ray structure of a CIC chloride channel at 3.0 Å reveals the molecular basis of anion selectivity. *Nature*, 415:287–294, 2002.

[11] J. H. Cate, M. M. Yusupov, C. Zh. Yusupova, T. N. Earnest, and H. F. Noller. X-ray crystal structure of 70S ribosome functional complexes. *Science*, 285:2095–2104, 1999.

[12] M. M. Yusupov, G. Zh. Yusupova, A. Baucom, K. Lieberman, T. N. Earnest, J. H. D. Cate, and H. F. Noller. Crystal structure of the ribosome at 5.5 Å resolution. *Science*, 292:883–896, 2001.

[13] T. R. Cech. The ribosome is a ribozyme. *Science*, 289:878–879, 2000.

[14] N. Ban, P. Nissen, J. Hansen, P. B. Moore, and T. A. Steitz. The complete atomic structure of the large ribosomal subunit at 2.4 Å resolution. *Science*, 289:905–920, 2000.

[15] P. Nissen, J. Hansen, N. Ban, P. B. Moore, and T. A. Steitz. The structural basis of ribosome activity in peptide bond synthesis. *Science*, 289:920–930, 2000.

[16] J. R. Williamson. Small subunit, big science. *Nature*, 407:306–307, 2000.

[17] B. T. Wimberly, D. E. Brodersen, W. M. Clemons Jr., R. J. Morgan-Warren, A. P. Carter, C. Vonrhein, T. Hartsch, and V. Ramakrishnan. Structure of the 30S ribosomal subunit. *Nature*, 407:327–339, 2000.

[18] A. P. Carter, W. M. Clemons, D. E. Brodersen, R. J. Morgan-Warren, B. T. Wimberly, and V. Ramakrishnan. Functional insights from the structure of the 30S ribosomal subunit and its interactions with antibiotics. *Nature*, 407:340–348, 2000.

[19] F. Schluenzen, A. Tocilj, R. Zarivach, J. Harms, M. Gluehmann, D. Janell, A. Bashan, H. Bartels, I. Agmon, F.Franceschi, and A. Yonath. Structure of functionally activated small ribosomal subunit at 3.3 Å resolution. *Cell*, 102:615–623, 2000.

[20] R. Brimacombe. The bacterial ribosome at atomic resolution. *Structure*, 8:R195–R200, 2000.

[21] T. Strick, J.-F. Allemand, V. Croquette, and D. Bensimon. The manipulation of single biomolecules. *Physics Today*, 54:46–51, October 2001.

[22] F. N. Zaidi, U. Nath, and J. B. Udgaonkar. Multiple intermediates and transition states during protein unfolding. *Nature Struc. Biol.*, 4:1016–1024, 1997.

[23] T. Lazaridis and M. Karplus. "New view" of protein folding reconciled with the old through multiple unfolding simulations. *Science*, 278:1928–1931, 1997.

[24] C. R. Cantor and P. R. Schimmel. *Biophysical Chemistry*, volume 1–3. W. H. Freeman and Company, San Francisco, 1980.

[25] D. Crothers and D. Eisenberg. *Physical Chemistry with Applications to the Life Sciences*. Benjamin/Cummings, Menlo Park, CA, 1979.

[26] A. Fersht. *Structure and Mechanism in Protein Science: A Guide to Enzyme Catalysis and Protein Folding*. W. H. Freeman and Company, New York, NY, 1999.

[27] L. Stryer. *Biochemistry*. W. H. Freeman, New York, NY, 5 edition, 2001.

[28] C. Branden and J. Tooze. *Introduction to Protein Structure*. Garland Publishing Inc., New York, NY, second edition, 1999. (www.proteinstructure.com/).

[29] D. A. McQuarrie. *Statistical Mechanics*. Harper & Row, New York, NY, 1976. Chapters 20–21.

[30] M. P. Allen and D. J. Tildesley. *Computer Simulation of Liquids*. Oxford University Press, New York, NY, 1990.

[31] D. Frenkel and B. Smit. *Understanding Molecular Simulations. From Algorithms to Applications*. Academic Press, San Diego, CA, 1996.

[32] S. M. Ross. *A Course in Simulation*. Macmillan Publishing Company, New York, NY, 1990.

[33] P. Bratley, B. L. Fox, and L. E. Schrage. *A Guide to Simulation*. Springer-Verlag, New York, NY, 1987.

[34] H. Gould and J. Tobochnik. *An Introduction to Computer Simulation Methods: Applications to Physical Systems. Part 1*. Addison-Wesley, Reading, Massachusetts, 1988.

[35] D.C. Rapaport. *The Art of Molecular Dynamics Simulation*. Cambridge University Press, Cambridge, England, 1995.

[36] J. M. Haile. *Molecular Dynamics Simulations: Elementary Methods*. John Wiley & Sons, New York, NY, 1992.

[37] J. A. McCammon and S. C. Harvey. *Dynamics of Proteins and Nucleic Acids*. Cambridge University Press, Cambridge, MA, 1987.

[38] C. L. Brooks, III, M. Karplus, and B. M. Pettitt. *Proteins: A Theoretical Perspective of Dynamics, Structure, and Thermodynamics*, volume LXXI of *Advances in Chemical Physics*. John Wiley & Sons, New York, NY, 1988.

[39] X. Daura, B. Jaun, D. Seebach, W. F. Van Gunsteren, and A. Mark. Reversible peptide folding in solution by molecular dynamics simulation. *J. Mol. Biol.*, 280:925–932, 1998.

[40] B. Zagrovic, E. J. Sorin, and V. Pande. β-hairpin folding simulations in atomistic detail using an implicit solvent model. *J. Mol. Biol.*, 313:151–169, 2001.

[41] M. Levitt. Computer simulation of DNA double-helix dynamics. *Cold Spring Harbor Symp. Quant. Biol.*, 47:251–275, 1983.

[42] E. Tajkhorshid, P. Nollert, M. Ø Jensen, L. J. W. Miercke, J. O'Connell, R. M. Stroud, and K. Schulten. Control of the selectivity of the aquaporin water channel family by global orientational tuning. *Science*, 296:525–530, 2002.

[43] M. Levitt. The birth of computational structural biology. *Nat. Struc. Biol.*, 8:392–393, 2001.

[44] A. Rahman and F. H. Stillinger. Molecular dynamics study of liquid water. *J. Chem. Phys.*, 55:3336–3359, 1971.

[45] A. Rahman and F. H. Stillinger. Improved simulation of liquid water by molecular dynamics. *J. Chem. Phys.*, 60:1545–1557, 1974.

[46] B. J. Alder and T. E. Wainwright. Studies in molecular dynamics. I. General method. *J. Chem. Phys.*, 31:459–466, 1959.

[47] A. Jack and M. Levitt. Refinement of large structures by simultaneous minimization of energy and R factor. *Acta Crystallogr.*, A34:931–935, 1978.

[48] J. H. Konnert and W. A. Hendrickson. A restrained-parameter thermal-factor refinement procedure. *Acta Crystallogr.*, A36:344–350, 1980.

[49] G. M. Clore and A. M. Gronenborn. New methods of structure refinement for macromolecular structure determination by NMR. *Proc. Natl. Acad. Sci. USA*, 95:5891–5898, 1998.

[50] A. Korostelev, R. Bertram, and M. S. Chapman. Simulated-annealing real-space refinement as a tool in model building. *Acta Cryst.*, D58:761–767, 2002.

[51] J. P. Ryckaert, G. Ciccotti, and H. J. C. Berendsen. Numerical integration of the Cartesian equations of motion of a system with constraints: Molecular dynamics of n-alkanes. *J. Comput. Phys.*, 23:327–341, 1977.

[52] W. L. Jorgensen, J. Chandrasekar, J. Madura, R. Impey, and M. Klein. Comparison of simple potential functions for simulating liquid water. *J. Chem. Phys.*, 79:926–935, 1983.

[53] W. Wang, O. Donini, C. M. Reyes, and P. A. Kollman. Biomolecular simulations: Recent developments in force fields, simulations of enzyme catalysis, protein-ligand, protein-protein, and protein-nucleic acid noncovalent interactions. *Ann. Rev. Biophys. Biomol. Struc.*, 30:211–243, 2001.

[54] Y. Duan, P. A. Kollman, and S. C. Harvey. Protein folding and beyond. In E. Keinam and I. Schechter, editors, *Chemistry for the 21st Century*. Wiley-VCH, Weinheim, Germany, 2000.

[55] S. D. Stellman, B. Hingerty, S. B. Broyde, E. Subramanian, T. Sato, and R. Langridge. Structure of guanosine-3', 5'-cytidine monophosphate. I. Semi-empirical potential energy calculations and model-building. *Biopolymers*, 12:1731–2750, 1973.

[56] J. A. McCammon, B. R. Gelin, and M Karplus. Dynamics of folded proteins. *Nature*, 267:585–590, 1977.

[57] G. L. Seibel, U. C. Singh, and P. A Kollman. A molecular dynamics simulation of double-helical B-DNA including counterions and water. *Proc. Natl. Acad. Sci. USA*, 82:6537–6540, 1985.

[58] M. Prabhakaran, S. C. Harvey, B. Mao, and J. A. McCammon. Molecular dynamics of phenylanlanine transfer RNA. *J. Biomol. Struct. Dynam.*, 1:357–369, 1983.

[59] S. C. Harvey, M. Prabhakaran, B. Mao, and J. A. McCammon. Phenylanine transfer RNA: Molecular dynamics simulation. *Science*, 223:1189–1191, 1984.

[60] B. Tidor, K. K. Irikura, B. R. Brooks, and M. Karplus. Dynamics of DNA oligomers. *J. Biomol. Struct. Dynam.*, 1:231–252, 1983.

[61] T. E. Cheatham, III, J. L. Miller, T. Fox, T. A. Darden, and P. A. Kollman. Molecular dynamics simulations of solvated biomolecular systems: The particle mesh Ewald method leads to stable trajectories of DNA, RNA, and proteins. *J. Amer. Chem. Soc.*, 117:4193–4194, 1995.

[62] R. S. Struthers, J. Rivier, and A. T. Hagler. Theoretical simulation of conformation, energetics, and dynamics in the design of GnRH analogs. *Trans. Amer. Cryst. Assoc.*, 20:83–96, 1984. Proceedings of the Symposium on Molecules in Motion, University of Kentucky, Lexington, Kentucky, May 20–21, 1984.

[63] R. M. Levy, R. P. Sheridan, J. W. Keepers, G. S. Dubey, S. Swaminathan, and M. Karplus. Molecular dynamics of myoglobin at 298°K. Results from a 300-ps computer simulation. *Biophys. J.*, 48:509–518, 1985.

[64] J. J. Wendoloski, S. J. Kimatian, C. E. Schutt, and F. R. Salemme. Molecular dynamics simulation of a phospholipid micelle. *Science*, 243:636–638, 1989.

[65] W. E. Harte, Jr., S. Swaminathan, and D. L. Beveridge. Molecular dynamics of HIV-1 protease. *Proteins: Struc. Func. Gen.*, 13:175–194, 1992.

[66] D. Kosztin, T. C. Bishop, and K. Schulten. Binding of the estrogen receptor to DNA: The role of waters. *Biophys. J.*, 73:557–570, 1997.

[67] T. Schlick, R. D. Skeel, A. T. Brünger, L. V. Kalé, J. A. Board, Jr., J. Hermans, and K. Schulten. Algorithmic challenges in computational molecular biophysics. *J. Comput. Phys.*, 151:9–48, May 1999. (Special Volume on Computational Biophysics).

[68] M. A. Young and D. L. Beveridge. Molecular dynamics simulations of an oligonucleotide duplex with adenine tracts phased by a full helix turn. *J. Mol. Biol.*, 281:675–687, 1998.

[69] Y. Duan and P. A. Kollman. Pathways to a protein folding intermediate observed in a 1-microsecond simulation in aqueous solution. *Science*, 282:740–744, 23 October 1998.

[70] Y. Duan, L. Wang, and P. A. Kollman. The early stage of folding of villin headpiece subdomain observed in a 200-nanosecond fully solvate molecular dynamics simulation. *Proc. Natl. Acad. Sci. USA*, 95:9897–9902, 1998.

[71] S. Izrailev, A. R. Crofts, E. A. Berry, and K. Schulten. Steered molecular dynamics simulation of the Rieske subunit motion in the cytochrome bc_1 complex. *Biophys. J.*, 77:1753–1768, 1999.

[72] M. Ø Jensen, E. Tajkhorshid, and K. Schulten. The mechanism of glycerol conduction in aquaglyceroporins. *Structure*, 9:1083–1093, 2001.

[73] M. Shirts and V. Pande. Screen savers of the world unite! *Science*, 290:1903–1904, 2000.

[74] V. Daggett. Long timescale simulations. *Curr. Opin. Struct. Biol.*, 10:160–164, 2000.

[75] T. Hansson, C. Oostenbrink, and W. F. van Gunsteren. Molecular dynamics simulations. *Curr. Opin. Struct. Biol.*, 12:190–196, 2002.

[76] L. Pauling, R. B. Corey, and H. R. Branson. The structure of proteins: Two hydrogen-bonded helical configurations of the polypeptide chain. *Proc. Natl. Acad. Sci. USA*, 37:205–211, 1951.

[77] L. Pauling and R. B. Corey. Configurations of polypeptide chains with favored orientations around single bonds: Two new pleated sheets. *Proc. Natl. Acad. Sci. USA*, 37:729–740, 1951.

[78] H. F. Judson. *The Eighth Day of Creation. Makers of the Revolution in Biology*. Cold Spring Harbor Laboratory Press, Cold Spring Harbor, NY, 1996. (Expanded edition).

[79] W. G. Scott and A. Klug. Ribozymes: Structure and mechanism in RNA catalysis. *Trends Bio. Sci.*, 21:220–224, 1996.

[80] J. A. Doudna. A molecular contortionist. *Nature*, 388:830–831, 1997.

[81] K. Okada and S. Okada. X-ray crystallographic analysis and semiempirical computations. In P. von Ragué Schleyer (Editor-in Chief), N. L. Allinger, T. Clark,

J. Gasteiger, P. A. Kollman, and H. F. Schaefer, III, editors, *Encyclopedia of Computational Chemistry*, volume 5, pages 3223–3247. John Wiley & Sons, West Sussex, England, 1998.

[82] A. McPherson. Macromolecular crystals. *Sci. Amer.*, 260:62–69, 1989.

[83] W. A. Hendrickson. Determination of macromolecular structures from anomalous diffraction of synchrotron radiation. *Science*, 254:51–58, 1991.

[84] W. A. Hendrickson and C. Ogata. Phase determination from multiwavelength anomalous diffraction measurements. *Meth. Enzymol.*, 276:494–523, 1997.

[85] G. A. Petsko and D. Ringe. Observation of unstable species in enzyme-catalyzed transformations using protein crystallography. *Curr. Opin. Chem. Biol.*, 4:89–94, 2000.

[86] E. A. Galburt and B. L. Stoddard. Time-resolved macromolecular crystallography. *Phys. Today*, 54:33–39, 1989.

[87] K. Moffat. Time-resolved biochemical crystallography: A mechanistic perspective. *Chem. Rev.*, 101:1569–1581, 2001.

[88] I. Usón and G. M. Sheldrick. Advances in direct methods for protein crystallography. *Curr. Opin. Struct. Biol.*, 9:643–648, 1999.

[89] J. C. Beauchamp and N. W. Isaacs. Methods for X-ray diffraction analysis of macromolecular structures. *Curr. Opin. Chem. Biol.*, 3:525–529, 1999.

[90] S. Neidle. New insights into sequence-dependent DNA structure. *Nature Struc. Biol.*, 5:754–756, 1998.

[91] H. A. Hauptman. The phase problem of X-ray crystallography. *Phys. Today*, 42:24–29, 1989.

[92] J. Karle. Macromolecular structure from anomalous dispersion. *Phys. Today*, 42:20–22, 1989.

[93] G. Rhodes. *Crystallography Made Crystal Clear: A Guide for Users of Macromolecular Models*. Academic Press, San Diego, CA, second edition, 2000.

[94] H. Yu. Extending the size limit of protein nuclear magnetic resonance. *Proc. Natl. Acad. Sci. USA*, 96:332–334, 1999.

[95] R. R. Ernst, G. Bodenhausen, and A. Wokaum. *Principles of Nuclear Magnetic Resonance in One and Two Dimensions*, volume 14 of *International Series of Monographs on Chemistry*. Clarendon Press, Oxford, New York, NY, 1987.

[96] K. Wüthrich. *NMR of Proteins and Nucleic Acids*. (The George Fisher Baker Non-Resident Lectureship in Chemistry at Cornell University series). Wiley Interscience, New York, NY, 1986.

[97] P. Güntert. Structure calculation of biological macromolecules from NMR data. *Quart. Rev. Biophys.*, 31:145–237, 1998.

[98] G. Siegal, J. van Duynhoven, and M. Baldus. Biomolecular NMR: Recent advances in liquids, solids and screening. *Curr. Opin. Chem. Biol.*, 3:530–536, 1999.

[99] G. M. Clore and C. D. Schwieters. Theoretical and computational advances in biomolecular NMR spectroscopy. *Curr. Opin. Struct. Biol.*, 12:146–153, 2002.

[100] A. T. Brünger, P. D. Adams, and L. M. Rice. New applications of simulated annealing in X-ray crystallography and solution NMR. *Structure*, 5:325–336, 1997.

[101] D. A. Case. NMR refinement. In P. von Ragué Schleyer (Editor-in Chief), N. L. Allinger, T. Clark, J. Gasteiger, P. A. Kollman, and H. F. Schaefer, III, editors, *Encyclopedia of Computational Chemistry*, volume 3, pages 1866–1876. John Wiley & Sons, West Sussex, England, 1998.

[102] J. R. Tolman. Dipolar couplings as a probe of molecular dynamics and structure in solution. *Curr. Opin. Struct. Biol.*, 11:532–539, 2001.

[103] B. Simon and M. Sattler. De novo structure determination from residual dipolar couplings by NMR spectroscopy. *Angew. Chem. Int. Ed.*, 41:437–440, 2002.

[104] K. B. Mullis. *Dancing Naked in the Mind Field*. Pantheon Books, New York, NY, 1998.

[105] P. R. Reilly. *Abraham Lincoln's DNA and Other Adventures in Genetics*. Cold Spring Harbor Laboratory Press, Cold Spring Harbor, NY, 2000.

[106] G. T. Montelione and S. Anderson. Structural genomics: Keystone for a human proteome project. *Nature Struc. Biol.*, 6:11–12, 1999.

[107] R. Xu, B. Ayers, D. Cowburn, and T. W. Muir. Chemical ligation of folded recombinant proteins: Segmental isotopic labeling of domains for NMR studies. *Proc. Natl. Acad. Sci. USA*, 96:388–393, 1999.

[108] J. Frank. *Three-Dimensional Electron Microscopy of Macromolecular Assemblies*. Academic Press, San Diego, CA, 1996.

[109] V. M. Unger. Electron cryomicroscopy. *Curr. Opin. Struct. Biol.*, 11:548–554, 2001.

[110] E. Rothstein. DNA teaches history a few lessons of its own. *New York Times*, 1998. Sunday, May 24 (under Ideas & Trends of the Week in Review).

[111] K. Wong. The mammals that conquered the seas. *Sci. Amer.*, 286:70–79, 2002.

[112] J. B. Ristaino, C. T. Groves, and G. R. Parra. PCR amplification of the Irish potato famine pathogen from historic specimens. *Nature*, 411:695–697, 2001.

[113] E. V. Orlova, M. A. Rahman, B. Gowen, K. E. Volynski, A. C. Ashton, C. Manser, M. van Heel, and Y. A. Ushkaryov. Structure of α-latrotoxin oligomers reveals that divalent cation-dependent tetramers form membrane pores. *Nat. Struc. Biol.*, 7:48–53, 2000.

[114] Y. Tao and W. Zhang. Recent developments in cryo-electron microscopy reconstruction of single particles. *Structure*, 10:616–622, 2000.

[115] C. M. T. Spahn, P. A. Penczek, A. Leith, and J. Frank. A method for differentiating proteins from nucleic acids in intermediate-resolution density maps: Cryo-electron microscopy defines the quaternary structure of the *Escherichia coli* 70S ribosome. *Structure*, 8:937–948, 2000.

[116] E. V. Koonin, L. Aravind, and A. S. Kondrashov. The impact of comparative genomics on our understanding of evolution. *Cell*, 101:573–576, 2000.

[117] X. Nassif. A furtive pathogen revealed. *Science*, 287:1767–1768, 2000.

[118] W. A. Hasteline. Beyond chicken soup. *Sci. Amer.*, 285:56–63, 2001.

[119] M. Pizza, V. Scarlato, V. Masignani, M. M. Giuliani, B. Aricó, M. Comanducci, G. T. Jennings, L. Baldi, E. Bartolini, B. Capecchi, C. L. Galeotti, E. Luzzi, R. Manetti, E. Marchetti, M. Mora, S. Nuti, G. Ratti, L. Santini, S. Savino, M. Scarselli, E. Storni, P. Zuo, M. Broeker, E. Hundt, B. Knapp, E. Blair, T. Mason, H. Tettelin, D. W. Hood, A. C. Jeffries, N. J. Saunders, D. M. Granoff, J. C.

Venter, E. R. Moxon, G. Grandi, and R. Rappuoli. Identification of vaccine candidates against serogroup B meningococcus by whole-genome sequencing. *Science*, 287:1816–1820, 2000.

[120] K. Davies. *Cracking the Genome: Inside the Race to Unlock Human DNA*. The Free Press (A Simon & Schuster Division), New York, NY, 2001.

[121] R. H. Waterston, E. S. Lander, and J. E. Sulston. On the sequencing of the human genome. *Proc. Natl. Acad. Sci. USA*, 99:3712–3716, 2002.

[122] P. Green. Whole-genome disassembly. *Proc. Natl. Acad. Sci. USA*, 99:4143–4144, 2002.

[123] E. W. Myers, G. G. Sutton, H. O. Smith, M. D. Adams, and J. C. Venter. On the sequencing and assembly of the human genome. *Proc. Natl. Acad. Sci. USA*, 99:4145–4146, 2002.

[124] A. A. Camargo, S. J. de Souza, R. R. Brentani, and A. J. G. Simpson. Human gene discovery through experimental definition of transcribed regions of the human genome. *Curr. Opin. Chem. Biol.*, 6:13–16, 2001.

[125] P. Kapranov, S. E. Cawley, J. Drenkow, S. Bekiranov, R. L. Strausberg, S. P. A. Fodor, and T. R. Gingeras. Large-scale transcriptional activity in chromosomes 21 and 22. *Science*, 296:916–919, 2002.

[126] S. Saha, A. B. Sparks, C. Rago, V. Akmaev, C. J. Wang, B. Vogelstein, K. W. Kinzler, and V. E. Velculescu. Using the transcriptome to annotate the genome. *Nat. Biotech.*, 19:508–512, 2002.

[127] F. S. Collins and K. G. Jegalian. Deciphering the code of life. *Sci. Amer.*, 281:86–91, 1999.

[128] D. Filmore. Taming the beast. *Mod. Drug Dis.*, 4:40–46, 2001.

[129] W. E. Evans and M. V. Relling. Pharmacogenomics: Translating functional genomics into rational therapeutics. *Science*, 286:487–491, 1999.

[130] J. J. McCarthy and R. Hilfiker. The use of single-nucleotide polymorphism maps in pharmacogenomics. *Nature Biotech.*, 18:505–508, 2000.

[131] S. L. Pomeroy, P. Tamayo, M. Gaasenbeek, L. M. Sturla, M. Angelo, M. E. McLaughlin, J. Y. H. Kim, L. C. Goumnerova, P. M. Black, C. Lau, J. C. Allen, D. Zagzag, J. M. Olson, T. Curran, C. Wetmore, J. A. Biegel, T. Poggio, S. Mukherjee, R. Rifkin, A. Califano, G. Stolovitzky, D. N. Louis, J. P. Mesirov, E. S. Lander, and T. R. Golub. Prediction of central nervous system embryonal tumour outcome based on gene expression. *Nature*, 415:436–442, 2002.

[132] M. Wilson, J. DeRisi, H. H. Kristensen, P. Imboden, S. Rane, P. O. Brown, and G. K. Schoolnik. Exploring drug-induced alterations in gene expression in *Mycobacterium tuberculosis* by microarray hybridization. *Proc. Natl. Acad. Sci. USA*, 96:12833–12838, 1999.

[133] P. R. E. Mittl and M. G. Grütter. Structural genomics: Opportunities and challenges. *Curr. Opin. Chem. Biol.*, 5:402–408, 2001.

[134] S. E. Brenner. A tour of structural genomics. *Nat. Genet.*, 2:801–809, 2001.

[135] R. Bonneau, J. Tsai, I. Ruczinski, and D. Baker. Functional inferences from blind *ab initio* protein structure predictions. *J. Struc. Biol.*, 134:186–190, 2001.

[136] R. M. Karp. Mathematical challenges from genomics and molecular biology. *Notices Amer. Math. Soc.*, 49:544–553, 2002.

[137] M. R. Chance, A. R. Bresnick, S. K. Burley, J.-S. Jiang, C. D. Lima, A. Sali, S. C. Almo, J. B. Bonanno, J. A. Buglino, S. Boulton, H. Chen, N. Eswar, G. He, R. Huang, V. Ilyin, L. McMahan, U. Pieper, S. Ray, M. Vidal, and L. K. Wang. Structural genomics: A pipeline for providing structures for the biologist. *Prot. Sci.*, 11:723–738, 2002.

[138] C. Ezzell. Proteins rule. *Sci. Amer.*, 286:40–47, 2002.

[139] S. Oliver. Proteomics: Guilt-by-association goes global. *Nature*, 403:601–602, 2000.

[140] P. Uetz, L. Giot, G. Cagney, T. A. Mansfield, R. S. Judson, J. R. Knight, D. Lockshon, V. Narayan, M. Srinivasan, P. Pochart, A. Qureshi-Emili, Y. Li, B. Godwin, D. Conover, T. Kalbfleisch, G. Vijayadamodar, M. Yang, M. Johnston, S. Fields, and J. M. Rothberg. A comprehensive analysis of protein-protein interactions in *Saccharomyces cerevisiae*. *Nature*, 403:623–627, 2000.

[141] K. Howard. The bioinformatics gold rush. *Sci. Amer.*, 283:58–63, 2000.

[142] D. W. Mount. *Bioinformatics. Sequence and Genome Analysis*. Cold Spring Harbor Laboratory Press, Cold Spring Harbor, NY, 2001.

[143] M. Hann and R. Green. Cheminformatics – A new name for an old problem? *Curr. Opin. Chem. Biol.*, 3:379–383, 1999.

[144] J. U. Bowie, R. Lüthy, and D. Eisenberg. A method to identify protein sequences that fold into a known three-dimensional structure. *Science*, 253:164–170, 1991.

[145] E. Lander. The new genomics: Global views of biology. *Science*, 274:536–539, 1996.

[146] B. Al-Lazikani, J. Jung, Z. Xiang, and B. Honig. Protein structure prediction. *Curr. Opin. Struct. Biol.*, 5:51–56, 2001.

[147] D. Baker and A. Sali. Protein structure prediction and structural genomics. *Science*, 294:93–96, 2001.

[148] R. Sánchez and A. Šali. Large-scale protein structure modeling of the *saccharomyces cerevisiae* genome. *Proc. Natl. Acad. Sci. USA*, 95:13597–13602, 1998.

[149] H. H. Gan, R. A. Perlow, S. Roy, J. Ko, M. Wu, J. Huang, S. Yan, A. Nicoletta, J. Vafai, D. Sun, L. Wang, J. E. Noah, S. Pasquali, and T. Schlick. A sequence/structure map for proteins. 2002. Submitted.

[150] S. Dalal, S. Balasubramanian, and L. Regan. Protein alchemy: Changing β-sheet into α-helix. *Nature Struc. Biol.*, 4:548–552, 1997.

[151] G. Rose. Protein folding and the Paracelsus challenge. *Nature Struc. Biol.*, 4:512–514, 1997.

[152] B. Honig. Protein folding: From the Levinthal paradox to structure prediction. *J. Mol. Biol.*, 293:283–293, 1999.

[153] C. Levinthal. Are there pathways for protein folding? *J. Chim. Physique*, 65:44–45, 1969.

[154] C. Levinthal. How to fold graciously. In P. Debrunner, J. C. M. Tsibris, and E. Münch, editors, *Mossbauer Spectroscopy in Biological Systems, Proceedings of a Meeting held at Allerton House, Monticello, Illinois*, page 22, Urbana, Illinois, 1969. University of Illinois Press.

[155] W. F. van Gunsteren, R. Bürgi, C. Peter, and X. Daura. The key to solving the protein-folding problem lies in an accurate description of the denatured state. *Angew. Chem. Int. Ed.*, 40:352–355, 2001.

[156] A. R. Dinner and M. Karplus. Comment on the communication "The key to solving the protein-folding problem lies in an accurate description of the denatured state" by van Gunsteren et al. *Angew. Chem. Int. Ed.*, 40:4615–4616, 2001.

[157] W. F. van Gunsteren, R. Bürgi, C. Peter, and X. Daura. Reply. *Angew. Chem. Int. Ed.*, 40:4616–4618, 2001.

[158] K. A. Dill, S. Bromberg, K. Yue, K. M. Fiebig, D. P. Yee, P. D. Thomas, and H. S. Chan. Principles of protein folding — A perspective from simple exact models. *Protein Science*, 4:561–602, 1995.

[159] K. A. Dill and H. S. Chan. From Levinthal to pathways to funnels. *Nature Struc. Biol.*, 4:10–19, 1997.

[160] P. G. Wolynes. Folding funnels and energy landscapes of larger proteins within the capillarity approximation. *Proc. Natl. Acad. Sci. USA*, 94:6170–6175, 1997.

[161] J. N. Onuchic, Z. Luthey-Schulten, and P. G. Wolynes. Theory of protein folding: The energy landscape perspective. *Annu. Rev. Phys. Chem.*, 48:545–600, 1997.

[162] C. L. Brooks, III, J. N. Onuchic, and D. J. Wales. Statistical thermodynamics: Taking a walk on a landscape. *Science*, 293:612–613, 2001.

[163] R. A. Friesner and J. R. Gunn. Computational studies of protein folding. *Annu. Rev. Biophys. Biomol. Struc.*, 25:315–342, 1996.

[164] U. H. E. Hansmann and Y. Okamoto. New Monte Carlo algorithms for protein folding. *Curr. Opin. Struct. Biol.*, 9:177–183, 1999.

[165] D. K. Klimov and D. Thirumalai. Stretching single-domain proteins: Phase diagram and kinetics of force-induced unfolding. *Proc. Natl. Acad. Sci. USA*, 96:6166–6170, 1999.

[166] L. Mirny and E. Shakhnovich. Protein folding theory: From lattice to all-atom models. *Ann. Rev. Biophys. Biomol. Struc.*, 30:361–396, 2001.

[167] J.-E. Shea and C. L. Brooks, III. From folding theories to folding proteins: A review and assessment of simulation studies of protein folding and unfolding. *Annu. Rev. Phys. Chem.*, 52:499–535, 2001.

[168] H. J. Dyson and P. E. Wright. Insights into protein folding from NMR. *Annu. Rev. Phys. Chem.*, 47:369–395, 1996.

[169] R. Sánchez and A. Šali. Advances in comparative protein-structure modelling. *Curr. Opin. Struct. Biol.*, 7:206–214, 1997.

[170] R. L. Dunbrack, Jr., D. L. Gerloff, M. Bower, X. Chen, O. Lichtarge, and F. E. Cohen. Meeting review: the second meeting on the critical assessment of techniques for protein structure prediction (CASP2), Asilomar, California, December 13–16, 1996. *Fold. Design*, 2:R27–R42, 1997.

[171] P. Koehl and M. Levitt. A brighter future for protein structure prediction. *Nature Struc. Biol.*, 6:108–111, 1999.

[172] J. Moult, K. Fidelis, A. Zemla, and T. Hubbard. Critical assessment of methods of protein structure prediction (CASP): Round IV. *Proteins: Struc. Func. Gen.*, Suppl. 5:2–7, 2001.

[173] R. Bonneau and D. Baker. Ab initio protein structure prediction: Progress and prospects. *Ann. Rev. Biophys. Biomol. Struc.*, 30:173–189, 2001.

[174] M. A. Marti-Renom, M. S. Madhusudhan, A. Fiser, B. Rost, and A. Sali. Reliability of assessment of protein structure prediction methods. *Structure*, 10:435–440, 2002.

[175] J. Moult, K. Fidelis, A. Zemla, T. Hubbard, and A. Tramontano. The significance of performance ranking in CASP—Response to Marti-Renom et al. *Structure*, 10:291–292, 2002.

[176] A. Sali, M. A. Marti-Renom, M. S. Madhusudhan, A. Fiser, and B. Rost. Reply to Moult et al. *Structure*, 10:292–293, 2002.

[177] R. V. Pappu. Review of the fourth Johns Hopkins protein folding meeting. *Proteins: Struc. Func. Gen.*, 36:263–269, 1999.

[178] J. Wehrle and D. Barrick. Review of the fifth annual Johns Hopkins protein folding meeting. *Proteins: Struc. Func. Gen.*, 42:141–147, 2001.

[179] H. J. C. Berendsen. A glimpse of the holy grail. *Science*, 282:642–643, 1998.

[180] F. U. Hartl and M. H.-Hartl. Molecular chaperones in the cytosol: Nascent chain to folded protein. *Science*, 295:1852–1858, 2002.

[181] S. Walter and J. Buchner. Molecular chaparones—Cellular machines for protein folding. *Angew. Chem. Int. Ed.*, 41:1098–1113, 2002.

[182] W. A. Fenton and A. L. Horwich. GroEL-mediated protein folding. *Protein Sci.*, 6:743–760, 1997.

[183] D. Thirumalai and G. H. Lorimer. Chaperonin-mediated protein folding. *Ann. Rev. Biophys. Biomol. Struc.*, 30:245–269, 2001.

[184] S. Normark. Anfinsen comes out of the cage during assembly of the bacterial pilus. *Proc. Natl. Acad. Sci. USA*, 97:7670–7672, 2000.

[185] M. M. Barnhart, J. S. Pinkner, G. E. Soto, F. G. Sauer, S. Langermann, G. Waksman, C. Frieden, and S. J. Hultgren. PapD-like chaperones provide the missing information for folding of pilin proteins. *Proc. Natl. Acad. Sci. USA*, 97:7709–7714, 2000.

[186] R. J. Ellis. Molecular chaperones: Pathways and networks. *Curr. Biol.*, 9:R137–R139, 1999.

[187] R. J. Ellis. Molecular chaperones: Avoiding the crowd. *Curr. Biol.*, 7:R531–R533, 1999.

[188] H. J. Dyson and P. E. Wright. Coupling of folding and binding for unstructured proteins. *Curr. Opin. Struct. Biol.*, 12:54–60, 2002.

[189] P. E. Wright and H. J. Dyson. Intrinsically unstructured proteins: Reassessing the protein structure-function paradigm. *J. Mol. Biol.*, 293:321–331, 1999.

[190] Z. Xu, A. L. Horwich, and P. B. Sigler. The crystal structure of the asymmetric GroEL-GroES-$(ADP)_7$ chaperonin complex. *Nature*, 388:741–750, 1997.

[191] M. Shtilerman, G. H. Lorimer, and S. W. Englander. Chaperonin function: Folding by forced unfolding. *Science*, 284:822–825, 1999.

[192] W. A. Houry, D. Frishman, C. Eckerskorn, F. Lottspeich, and F. U. Hartl. Identification of *in vivo* substrates of the chaperonin GroEL. *Nature*, 402:147–154, 1999.

[193] O. Llorca, E. A. McCormack, G. Hynes, J. Grantham, J. Cordell, J. L. Carrascosa, K. R. Willison, J. J. Fernandez, and J. M. Valpuesta. Eukaryotic type II chaperonin CCT interacts with actin through specific subunits. *Nature*, 402:693–696, 1999.

[194] F. E. Cohen. Protein misfolding and prion diseases. *J. Mol. Biol.*, 293:313–320, 1999.

[195] A. Aguzzi. Prions and antiprions. *Biol. Chem.*, 378:1393–1395, 1997.

[196] M. R. Scott, R. Will, J. Ironside, H.-Oanh B. Nguyen, P. Tremblay, S. J. DeArmond, and S. B. Prusiner. Compelling transgenetic evidence for transmission of bovine spongiform encephalopathy prions to humans. *Proc. Natl. Acad. Sci. USA*, 96:15137–15142, 1999.

[197] H. Wille, M. D. Michelitsch, V. Guénebaut, S. Supattapone, A. Serban, F. E. Cohen, D. A. Agard, and S. B. Prusiner. Structural studies of the scrapie prion protein by electron crystallography. *Proc. Natl. Acad. Sci. USA*, 99:3563–3568, 2002.

[198] A. Aguzzi, F. Montrasio, and P. S. Kaeser. Prions: Health scare and biological challenge. *Nature Rev. Mol. Cell Biol.*, 2:118–126, 2001.

[199] T. L. James, H. Liu, N. B. Ulyanov, S. Farr-Jones, H. Zhang, D. G. Donne, K. Kaneko, D. Groth, I. Mehlhorn, S. B. Prusiner, and F. E. Cohen. Solution structure of a 142-residue recombinant prion protein corresponding to the infectious fragment of the scrapie isoform. *Proc. Natl. Acad. Sci. USA*, 94:10086–10091, 1997.

[200] H. Liu, S. Farr-Jones, N. B. Ulyanov, M. Llinas, S. Marqusee, D. Groth, F. E. Cohen, S. B. Prusiner, and T. L. James. Solution structure of a syrian hamster prion protein rPrP(90–231). *Biochemistry*, 38:5362–5377, 1999.

[201] R. Zahn, A. Liu, T. Lührs, R. Riek, C. von Schroetter, F. López Garcia, M. Billeter, L. Calzolai, G. Wider, and Kurt Wüthrich. NMR solution structure of the human prion protein. *Proc. Natl. Acad. Sci. USA*, 97:145–150, 2000.

[202] P. Hammarström, F. Schneider, and J. W. Kelly. *Trans*-suppression of misfolding in an amyloid disease. *Science*, 293:2459–2462, 2001.

[203] M. B. Pepys, J. Herbert, W. L. Hutchinson, G. A. Tennent, H. J. Lachmann, J. R. Gallimore, L. B. Lovat, T. Bartfai, A. Alanine, C. Hertel, T. Hoffmann, R. Jakob-Roetne, R. D. Norcross, J. A. Kemp, K. Yamamura, M. Suzuki, G. W. Taylor, S. Murray, D. Thompson, A. Purvis, S. Kolstoe, S. P. Wood, and P. N. Hawkins. Targeted pharmacological depletion of serum amyloid P component for treatment of human amyloidosis. *Nature*, 417:254–259, 2002.

[204] M. Bucciantini, E. Giannoni, F. Chiti, F. Baroni, L. Formigli, J. Zurdo, N. Taddei, G. Ramponi, C. M. Dobson, and M. Stefani. Inherent toxicity of aggregates implies a common mechanism for protein misfolding diseases. *Nature*, 416:507–511, 2002.

[205] D. M. Walsh, I. Klyubin, J. V. Fadeeva, W. K. Cullen, R. Anwyl, M. S. Wolfe, M. J. Rowan, and D. J. Selkoe. Naturally secreted oligomers of amyloid β protein potently inhibit hippocampal long-term potentiation *in vivo*. *Nature*, 416:535–539, 2002.

[206] L. Castagnoli, M. Scarpa, M. Kokkinidis, D.W. Banner, D. Tsernoglou, and G. Cesareni. Genetic and structural analysis of the CoIE1 Rop (Rom) protein. *Embo. J.*, 8:621–629, 1989.

[207] G. A. Patikoglou, J. L. Kim, L. Sun, S.-H. Yang, T. Kodadek, and S. K. Burley. TATA element recognition by the TATA box-binding protein has been conserved throughout evolution. *Genes & Devt.*, 13:3217–3230, 1999.

[208] B. Werth. *The Billion-Dollar Molecule: One Company's Quest for the Perfect Drug*. Simon & Schuster, New York, NY, 1994.

[209] D. B. Boyd. Rational drug design: Controlling the size of the haystack. *Mod. Drug Dis.*, 1:41–47, 1998.

[210] Z. Chen, Y. Li, E. Chen, D. L. Hall, P. L. Darke, C. Culberson, J. A. Shafer, and L. C. Kuo. Crystal structure at 1.9-Å resolution of human immunodeficiency (HIV) II protease complexed with L-735,524, an orally bioavailable inhibitor of the HIV proteases. *J. Biol. Chem.*, 269:26344–26348, 1994.

[211] J. Ren, R. M. Esnouf, A. L. Hopkins, E. Y. Jones, I. Kirby, J. Keeling, C. K. Ross, B. A. Larder, D. I. Stuart, and D. K. Stammers. 3′-Azido-3′-deoxythymidine drug resistance mutations in HIV-1 reverse transcriptase can induce long range conformational changes resistance. *Proc. Natl. Acad. Sci. USA*, 95:9518–9523, 1998.

[212] D. K. Stammers, D. O'N. Somers, C. K. Ross, I. Kirby, P. H. Ray, J. E. Wilson, M. Norman, J. S. Ren, R. M. Esnouf, E. F. Garman, E. Y. Jones, and D. I. Stuart. Crystals of HIV-1 reverse transcriptase diffracting to 2.2 Å resolution. *J. Mol. Biol.*, 242:586–588, 1994.

[213] D. J. Hazuda, P. Felock, M. Witmer, A. Wolfe, K. Stillmock, J. A. Grobler, A. Espeseth, L. Gabryelski, W. Schleif, C. Blau, and Michael D. Miller. Inhibitors of strand transfer that prevent integration and inhibit HIV-1 replication in cells. *Science*, 287:646–650, 2000.

[214] N. A. Roberts, J. A. Martin, D. Kinchington, A. V. Broadhurst, J. C. Craig, I. B. Duncan, S. A. Galpin, B. K. Handa, J. Kay, A. Kröhn, R. W. Lambert, J. H. Merrett, J. S. Mills, K. E. B. Parkes, S. Redshaw, A. J. Ritchie, D. L. Taylor, G. J. Thomas, and P. J. Machin. Rational design of peptide-based HIV proteinase inhibitors. *Science*, 248:358–361, 1990.

[215] J. Kling. Out of Malaysia: Finding natural products to fight AIDS. *Mod. Drug Dis.*, 2:31–36, 1999.

[216] B. A. Larder and D. K. Stammers. Closing in on HIV drug resistance. *Nature Struc. Biol.*, 6:103–106, 1999.

[217] H. Huang, R. Chopra, G. L. Verdine, and S. C. Harrison. Structure of a covalently trapped catalytic complex of HIV-1 reverse transcriptase: Implications for drug resistance. *Science*, 282:1669–1675, 1998.

[218] P. D. Kwong, R. Wyatt, J. Robinson, R. W. Sweet, J. Sodroski, and W. A. Hendrickson. Structure of an HIV gp 120 envelope glycoprotein in complex with the CD4 receptor and a neutralizing human antibody. *Nature*, 393:648–659, 1998.

[219] M. Ferrer, T. A. Kapoor, T. Strassmaier, W. Weissenhorn, J. J. Skehel, D. Oprian, S. L. Schreiber, D. C. Wiley, and S. C. Harrison. Selection of gp41-mediated HIV-1 cell entry inhibitors from biased combinatorial libraries of non-natural binding elements. *Nature Struc. Biol.*, 6:953–960, 1999.

[220] F. Gao, E. Bailes, D. L. Robertson, Y. Chen, C. M. Rodenburg, S. F. Michael, L. B. Cummins, L. O. Arthur, M. Peeters, G. M. Shaw, P. M. Sharp, and B. H.

Hahn. Origin of HIV-1 in the chimpanzee *pan troglodytes troglodytes*. *Nature*, 397:436–441, 1999.

[221] S. F. Brady, K. J. Stauffer, W. C. Lumma, G. M. Smith, H. G. Ramjit, S. D. Lewis, B. J. Lucas, S. J. Gardell, E. A. Lyle, S. D. Appleby, J. J. Cook, M. A. Holahan, M. T. Stranieri, J. J. Lynch, Jr., J. H. Lin, I.-W. Chen, K. Vastag, A. M. Naylor-Olsen, and J. P. Vacca. Discovery and development of the novel potent orally active thrombin inhibitor N-(9-Hydroxy-9-fluorenecarboxy)prolyl *trans*-4-Aminocyclohexylmethyl amide (L-372,460): Coapplication of structure-based design and rapid multiple analogue synthesis on solid support. *J. Med. Chem.*, 41(3):401–406, 1998.

[222] M. Cavazzana-Calvo, S. Hacein-Bey, G. de Saint Basile, F. Gross, E. Yvon, P. Nusbaum, F. Selz, C. Hue, S. Certain, J.-L Casanova, P. Bousso, F. Le Deist, and A. Fischer. Gene therapy of human severe combined immunodeficiency (SCID)-X1 disease. *Science*, 288:669–672, 2000.

[223] W. F. Anderson. The best of times, the worst of times. *Science*, 288:627–629, 2000.

[224] J. Vrebalov, D. Ruezinsky, V. Padmanabhan, R. White, D. Medrano, R. Drake, W. Schuch, and J. Giovannoni. A MADS-box gene necessary for fruit ripening at the tomato *ripening-inhibitor* (*Rin*) locus. *Science*, 296:343–346, 2002.

[225] B. I. Yakobson and R. E. Smalley. Fullerene nanotubes: $C_{1,000,000}$ and beyond. *American Scientist*, 85(4):324–337, 1997.

[226] R. Z. Kramer, J. Bella, B. Brodsky, and H. M. Berman. The crystal and molecular structure of a collagen-like peptide with a biologically relevant sequence. *J. Mol. Biol.*, 311:131–147, 2001.

[227] L. J. Prins, D. N. Reinhoudt, and P. Timmerman. Nonvalent synthesis using hydrogen bonding. *Angew. Chem. Int. Ed.*, 40:2382–2426, 2001.

[228] T. Schlick, S. Figueroa, and M. Mezei. A molecular dynamics simulation of a water droplet by the implicit-Euler/Langevin scheme. *J. Chem. Phys.*, 94:2118–2129, 1991.

[229] Y. Mandel-Gutfreund, H. Margalit, R. L. Jernigan, and V. B. Zhurkin. A role for CH· · ·O interactions in protein-DNA recognition. *J. Mol. Biol.*, 277:1129–1140, 1998.

[230] A. R. Dinner, R. Lazaridis, and M. Karplus. Understanding β-hairpin formation. *Proc. Natl. Acad. Sci. USA*, 96:9068–9073, 1999.

[231] J. E. Brody. What to serve for dinner, when dinner is on Mars. *The New York Times*, pages F1–2, 19 May 1998.

[232] J. Park, S. A. Teichmann, T. Hubbard, and C. Chothia. Intermediate sequences increase the detection of homology between sequences. *J. Mol. Biol.*, 273:349–354, 1997.

[233] N. M. Glykos, G. Cesareni, and M. Kokkinidis. Protein plasticity to the extreme: Changing the topology of a 4–helical bundle with a single amino acid substitution. *Struc. Fold. Design*, 7:597–603, 1999.

[234] K. Pawlowski, A. Bierzyński, and A. Godzik. Structural diversity in a family of homologous proteins. *J. Mol. Biol.*, 258:349–366, 1996.

[235] H. R. Faber and B. W. Matthews. A mutant lysozyme displays five different crystal conformations. *Nature*, 348:263–265, 1990.

[236] A. S. Edison. Linus Pauling and the planar peptide bond. *Nat. Struc. Biol.*, 8:201–202, 2001.

[237] M. W. MacArthur and J. M. Thornton. Deviations from planarity of the peptide bond in peptides and proteins. *J. Mol. Biol.*, 264:1180–1195, 1996.

[238] R. Sarma. *Ramachandran: A Biography of Gopalasamudram Narayana Ramachandran, the Famous Indian Biophysicist*. Adenine Press, Schenectady, NY, 1998.

[239] J. A. Schellman and C. Schellman. The conformation of polypeptide chains in proteins. In H. Neurath, editor, *The Proteins*, volume 2, pages 1–137. Academic Press, New York, NY, second edition, 1964.

[240] C. B. Anfinsen, E. Haber, M. Sela, and F. H. White, Jr. The kinetics of formation of native ribonuclease during oxidation of the reduced polypeptide chain. *Proc. Natl. Acad. Sci. USA*, 47:1309–1314, 1961.

[241] J. Bella, B. Brodsky, and H. M. Berman. Hydration structure of a collagen peptide. *Structure*, 3:893–906, 1995.

[242] S. K. Holmgren, K. M. Taylor, L. E. Bretscher, and R. T. Raines. Code for collagen's stability deciphered. *Nature*, 392, 1998.

[243] C. Chothia, T. Hubbard, S. Brenner, H. Barns, and A. Murzin. Protein folds in the all-β and all-α classes. *Ann. Rev. Biophys. Biomol. Struc.*, 26:597–627, 1997.

[244] C. Chothia. One thousand families for the molecular biologist. *Nature*, 357:543–544, 1992.

[245] L. Holm and C. Sander. Mapping the protein universe. *Science*, 273:595–602, 1996.

[246] Z.-X. Wang. How many fold types of protein are there in Nature? *Proteins: Struc. Func. Gen.*, 26:186–191, 1996.

[247] C. Zhang and C. DeLisi. Estimating the number of protein folds. *J. Mol. Biol.*, 284:1301–1305, 1998.

[248] J. M. Thornton, C. A. Orengo, A. E. Todd, and F. M. G. Pearl. Protein folds, functions and evolution. *J. Mol. Biol.*, 293:333–342, 1999.

[249] S. E. Brenner, C. Chothia, and T. J. P. Hubbard. Population statistics of protein structures: Lessons from structural classifications. *Curr. Opin. Struct. Biol.*, 7:369–376, 1997.

[250] J. Liu and B. Rost. Comparing function and structure between entire proteomes. *Protein Sci.*, 10:1970–1979, 2001.

[251] S. A. Teichmann, C. Chothia, and M. Gerstein. Advances in structural genomics. *Curr. Opin. Struct. Biol.*, 9:390–399, 1999.

[252] H. Erlandsen, E. E. Abola, and R. C. Stevens. Combining structural genomics and enzymology: Completing the picture in metabolic pathways and enzyme active sites. *Curr. Opin. Struct. Biol.*, 10:719–730, 2000.

[253] D. Vitkup, E. Melamud, J. Moult, and C. Sander. Completeness in structural genomics. *Nat. Struc. Biol.*, 8:559–565, 2001.

[254] A. Sali. Target practice. *Nat. Struc. Biol.*, 8:482–484, 2001.

[255] A. Sharman. The many uses of a genome sequence. *Genome Biol.*, 2:4013.1–4013.4, 2001. (URL: genomebiology.com).

[256] J. M. Grimes, J. N. Burroughs, P. Gouet, J. M. Diprose, R. Malby, S. Ziéntara, P. P. C. Mertens, and D. I. Stuart. The atomic structure of the bluetongue virus core. *Nature*, 395:470–478, 1998.

[257] F. A. Samatey, K. Imada, S. Nagashima, F. Vonderviszt, T. Kumasaka, M. Yamamoto, and K. Namba. Structure of the bacterial flagellar protofilament and implications for a switch for supercoiling. *Nature*, 410:331–337, 2001.

[258] P. Jordan, P. Fromme, H. T. Witt, O. Klukas, W. Saenger, and N. Krauβ. Three-dimensional structure of cyanobacterial photosystem I at 2.5 Å resolution. *Nature*, 411:909–917, 2001.

[259] C. A. Orengo, A. E. Todd, and J. M. Thornton. From protein structure to function. *Curr. Opin. Struct. Biol.*, 9:374–382, 1999.

[260] A. G. Murzin, S. E. Brenner, T. Hubbard, and C. Chothia. Scop: A structural classification of proteins database for the investigation of sequences and structures. *J. Mol. Biol.*, 247:536–540, 1995.

[261] L. Lo Conte, S. E. Brenner, T. J. P. Hubbard, C. Chothia, and A. G. Murzin. SCOP database in 2002: Refinements accommodate structural genomics. *Nucl. Acids Res.*, 30:264–267, 2002.

[262] J. D. Watson and F. H. C. Crick. A structure for deoxyribose nucleic acid. *Nature*, 171:737–738, 1953.

[263] J. D. Watson and F. H. C. Crick. Genetical implications of the structure of deoxyribonucleic acid. *Nature*, 171:964–967, 1953.

[264] J. D. Watson and F. H. C. Crick. The structure of DNA. *Cold Spr. Harb. Symp. Quant. Biol.*, XVIII:123–131, 1953.

[265] R. D. Knight and L. F. Landweber. The early evolution of the genetic code. *Cell*, 101:569–572, 2000.

[266] M. Ptashne. How gene activators work. *Sci. Amer.*, 260:41–47, 1989.

[267] A. Wolffe. *Chromatin Structure and Function*. Academic Press Inc., San Diego, CA, 1995.

[268] C. R. Calladine and H. R. Drew. *Understanding DNA. The Molecule and How It Works*. Academic Press, San Diego, CA, second edition, 1997.

[269] T. Schlick, D. Beard, J. Huang, D. Strahs, and X. Qian. Computational challenges in simulating large DNA over long times. *IEEE Comput. Sci. Eng.*, 2(6):38–51, November/December 2000. (Special Issue on Computational Chemistry).

[270] H. M. Berman. Crystal studies of B-DNA: The answers and the questions. *Biopolymers*, 44:23–44, 1997.

[271] H. M. Berman, W. K. Olson, D. L. Beveridge, J. Westbrook, A. Gelbin, T. Demeny, S.-H. Hsieh, A. R. Srinivasan, and B. Schneider. The nucleic acid database: A comprehensive relational database of three-dimensional structures of nucleic acids. *Biophys. J.*, 63:751–759, 1992.

[272] M. A. Young, B. Jayaram, and D. L. Beveridge. Intrusion of counterions into the spine of hydration in the minor groove of B-DNA: Fractional occupancy of electronegative pockets. *J. Amer. Chem. Soc.*, 119:59–69, 1997.

[273] M. A. Young, G. Ravishanker, and D. L. Beveridge. A 5-nanosecond molecular dynamics trajectory for B-DNA: Analysis of structure, motions, and solvation. *Biophys. J.*, 73:2313–2336, 1997.

[274] X. Qian, D. Strahs, and T. Schlick. Dynamic simulations of 13 TATA variants refine kinetic hypotheses of sequence/activity relationships. *J. Mol. Biol.*, 308:681–703, 2001.

[275] D. M. York, W. Yang, H. Lee, T. Darden, and L. G. Pederson. Toward the accurate modeling of DNA: The importance of long-range electrostatics. *J. Amer. Chem. Soc.*, 117:5001–5002, 1995.

[276] T. E. Cheatham, III and P. A. Kollman. Observation of the A-DNA to B-DNA transition during unrestrained molecular dynamics in aqueous solution. *J. Mol. Biol.*, 259:434–444, 1996.

[277] T. Darden, L. Perera, L. Li, and L. Pedersen. New tricks for modelers from the crystallography toolkit: The particle mesh Ewald algorithm and its use in nucleic acid simulations. *Structure*, 7:R55–R60, 1999.

[278] X. Shui, L. McFail-Isom, G. G. Hu, and L. D. Williams. The B-DNA dodecamer at high resolution reveals a spine of water on sodium. *Biochemistry*, 37:8341–8355, 1998.

[279] C. L. Kielkopf, S. Ding, P. Kuhn, and D. C. Rees. Conformational flexibility of B-DNA at 0.74 Å resolution: d(CCAGTACTGG)$_2$. *J. Mol. Biol.*, 296:787–801, 2000.

[280] T. Kien Chiu and R. E. Dickerson. 1 Å crystal structures of *B-DNA* reveal sequence-specific binding and groove-specific bending of DNA by magnesium and calcium. *J. Mol. Biol.*, 301:915–945, 2000.

[281] N. C. Seeman. DNA nanotechnology: Novel DNA constructions. *Annu. Rev. Biophys. Biomol. Struc.*, 27:225–248, 1998.

[282] C. Mao, T. LaBean, J. H. Reif, and N. C. Seeman. Logical computation using algorithmic self-assembly of DNA triple crossover molecules. *Nature*, 407:493–496, 2000.

[283] R. S. Braich, N. Chelyapov, C. Johnson, P. W. K. Rothemund, and L. Adleman. Solution of a 20-variable 3-SAT problem on a DNA computer. *Science*, 296:499–502, 2002.

[284] W. Saenger. *Principles of Nucleic Acid Structure*. Springer Advanced Texts in Chemistry. Springer-Verlag, New York, NY, 1984.

[285] R. R. Sinden. *DNA Structure and Function*. Academic Press, San Diego, CA, 1994.

[286] N. R. Cozzarelli and J. C. Wang, editors. *DNA Topology and Its Biological Effects*. Cold Spring Harbor Laboratory Press, Cold Spring Harbor, NY, 1990.

[287] A. V. Vologodskii. *Topology and physics of circular DNA*. CRC Press, Boca Raton, Florida, 1992.

[288] A. D. Bates and A. Maxwell. *DNA Topology*. In Focus. Oxford University Press, New York, NY, 1993.

[289] S. Neidle. *DNA Structure and Recognition*. Oxford University Press, Oxford, England, 1994.

[290] S. Neidle, editor. *Oxford Handbook of Nucleic Acid Structure*. Oxford University Press, Oxford, England, 1999.

[291] V. A. Bloomfield, D. M. Crothers, and I. Tinoco, Jr. *Nucleic Acids: Structures, Properties, and Functions*. University Science Press, New York, NY, 2000.

[292] J. D. Watson. *The Double Helix. A Personal Account of the Discovery of the Structure of DNA*. Norton Critical Edition G.S. Stent (Editor), Norton & Company, New York, NY, 1980.

[293] F. H. C. Crick. *What Mad Pursuit: A Personal View of Scientific Discovery*. Alfred P. Sloan Foundation Series. Basic Books, New York, NY, 1988.

[294] M. D. Frank-Kamenetskii. *Unravelling DNA*. VCH Publishers, New York, NY, 1993. (Translated from Russian by L. Liapin).

[295] A. R. Srinivasan and W. K. Olson. Molecular models of nucleic acid triple helixes. II. PNA and $2'$–$5'$ backbone complexes. *J. Amer. Chem. Soc.*, 120:492–499, 1998.

[296] N. B. Leontis and E. Westhof. Conserved geometrical base-pairing patterns in RNA. *Quart. Rev. Biophys.*, 31:399–455, 1998.

[297] P. E. Nielsen, M. Egholm, R. H.. Berg, and O. Buchardt. Sequence-selective recognition of DNA by strand displacement with a thymine-substituted polyamide. *Science*, 254:1497–1500, 1991.

[298] C. Altona and M. Sundaralingam. Conformational analysis of the sugar ring in nucleosides and nucleotides. a new description using the concept of pseudorotation. *J. Amer. Chem. Soc.*, 94:8205–8212, 1972.

[299] D. Cremer and J. A. Pople. A general definition of ring puckering coordinates. *J. Amer. Chem. Soc.*, 97:1354–1358, 1975.

[300] S. C. Harvey and M. Prabhakaran. Ribose puckering: Structure, dynamics, energetics and the pseudorotation cycle. *J. Amer. Chem. Soc.*, 108:6128–6136, 1986.

[301] W. K. Olson and J. L. Sussman. How flexible is the furanose ring? 1. A comparison of experimental and theoretical studies. *J. Amer. Chem. Soc.*, 104:270–278, 1982.

[302] M. A. Viswamitra, B. S. Reddy, G.H.-Y. Lin, and M. Sundaralingam. Stereochemistry of nucleic acids and their constituents. XVII. Crystal and molecular structure of deoxycytidine 5'-phosphate monohydrate. A possible puckering for the furanoside ring in B-deoxyribonucleic acid. *J. Amer. Chem. Soc.*, 93:4565–4573, 1971.

[303] A. D. MacKerell, Jr. and N. Foloppe. All-atom empirical force field for nucleic acids: I. Parameter optimization based on small molecule and condensed phased macromolecular target data. *J. Comput. Chem.*, 21:86–104, 2000.

[304] A. D. MacKerell, Jr. and N. Foloppe. All-atom empirical force field for nucleic acids: II. Application to molecular dynamics simulations of DNA and RNA in solution. *J. Comput. Chem.*, 21:105–120, 2000.

[305] A. D. MacKerell, Jr., N. Banavali, and N. Foloppe. Development and current status of the CHARMM force field for nucleic acids. *Biopolymers*, 56:257–265, 2001.

[306] N. Foloppe, B. Hartmann, L. Nilsson, and A. D. MacKerell, Jr. Intrinsic conformational energetics associated with the glycosyl torsion in DNA: A quantum mechanical study. *Biophys. J.*, 82:1554–1569, 2002.

[307] A. D. MacKerell, Jr., J. Wiorkiewicz-Kuczera, and M. Karplus. An all-atom empirical energy function for the simulation of nucleic acids. *J. Amer. Chem. Soc.*, 117:11946–11975, 1995.

[308] R. E. Dickerson, M. Bansal, C. R. Calladine, S. Diekmann, W. N. Hunter, O. Kennard, E. von Kitzing, R. Lavery, H. C. M. Nelson, W. K. Olson, W. Saenger,

Z. Shakked, H. Sklenar, D. M. Soumpasis, C.-S. Tung, A. H.-J. Wang, and V. B. Zhurkin. Definitions and nomenclature of nucleic acid structure parameters. *J. Mol. Biol.*, 208:787–791, 1989.

[309] R. E. Dickerson. Definitions and nomenclature of nucleic acid structure parameters. *EMBO J.*, 8:1–4, 1989.

[310] T. Schlick. A modular strategy for generating starting conformations and data structures of polynucleotide helices for potential energy calculations. *J. Comput. Chem.*, 9(8):861–889, 1988.

[311] K. M. Kosikov, A. A. Gorin, V. B. Zhurkin, and W. K. Olson. DNA stretching and compression: Large-scale simulations of double helical structures. *J. Mol. Biol.*, 289:1301–1326, 1999.

[312] W. K. Olson, S. K. Burley, R. E. Dickerson, M. Gerstein, S. C. Harvey, U. Heinemann, X.-J. Lu, S. Neidle, Z. Shakked, M. Suzuki, X.-S. Tung, H. Sklenar, J. Westbrook, E. Westhof, C. Wolberger, and H. Berman. A standard reference frame for the description of nucleic acid base-pair geometry. *J. Mol. Biol.*, 313:229–237, 2001. Available also through posting on NDB Archives, ndb-server.rutgers.edu/NDB/archives/index.html and www.idealibrary.com/links/doi/10.1006/jmbi.2001.4987/.

[313] R. Chandrasekaran and S. Arnott. The structure of *B*-DNA in oriented fibers. *J. Biomol. Struct. Dynam.*, 13:1015–1027, 1996.

[314] M. Levitt. How many base-pairs per turn does DNA have in solution and in chromatin? Some theoretical calculations. *Proc. Natl. Acad. Sci. USA*, 75:640–644, 1978.

[315] A. H. Wang, G. J. Quigley, F. J. Kolpak, J. L. Crawford, J. H. van Boom, G. van der Marel, and A. Rich. Molecular structure of a left-handed double helical DNA fragment at atomic resolution. *Nature*, 282:680–686, 1979.

[316] H. M. Berman. Hydration of DNA: Take 2. *Curr. Opin. Struct. Biol.*, 4:345–350, 1994.

[317] T. Schwartz, M. A. Rould, K. Lowenhaupt, A. Herbert, and A. Rich. Crystal structure of the Z domain of the human editing enzyme ADAR1 bound to left-handed Z-DNA. *Science*, 284:1841–1845, 1999.

[318] G. M. Blackburn and M. J. Gait, editors. *Nucleic Acids in Chemistry and Biology*. Oxford University Press, New York, NY, 1990.

[319] W. K. Olson, A. A. Gorin, X J. Lu, L. M. Hock, and V. B. Zhurkin. DNA sequence-dependent deformability deduced from protein-DNA crystal complexes. *Proc. Natl. Acad. Sci. USA*, 95:11163–11168, 1998.

[320] R. E. Dickerson and T. K. Chiu. Helix bending as a factor in protein/DNA recognition. *Biopolymers*, 44:361–403, 1997.

[321] J. Westbrook, Z. Feng, S. Jain, T. N. Bhat, N. Thanki, V. Ravichandran, G. L. Gilliland, W. F. Bluhm, H. Weissig, D. S. Greer, P. E. Bourne, and H. M. Berman. The Protein Data Bank: Unifying the archive. *Nucl. Acids Res.*, 30:245–248, 2002.

[322] R. E. Dickerson and H. R. Drew. Structure of a B-DNA dodecamer. II. influence of base sequence on helix structure. *J. Mol. Biol.*, 149:761–786, 1981.

[323] M. A. El Hassan and C. R. Calladine. Conformational characteristics of DNA: Empirical classifications and a hypothesis for the conformational behaviour of dinucleotide steps. *Phil. Trans. Math. Phys. Engin. Sci.*, 355:43–100, 1997.

[324] V. B. Zhurkin, Y. P. Lysov, and V. Ivanov. Anisotropic flexibility of DNA and the nucleosomal structure. *Nucleic Acids Res.*, 6:1081–1096, 1979.

[325] J. A. Schellman. The flexibility of DNA. I. Thermal fluctuations. *Biophys. Chem.*, 11:321–328, 1980.

[326] A. A. Gorin, V. B. Zhurkin, and W. K. Olson. B-DNA twisting correlates with base pair morphology. *J. Mol. Biol.*, 247:34–48, 1995.

[327] H. Rozenberg, D. Rabinovich, F. Frolow, R. S. Hegde, and Z. Shakked. Structural code for DNA recognition revealed in crystal structures of papillomavirus E2-DNA targets. *Proc. Natl. Acad. Sci. USA*, 95:15194–15199, 1998.

[328] S. C. Schultz, G. C. Shields, and T. A. Steitz. Crystal structure of a CAP-DNA complex: The DNA is bent by 90°. *Science*, 253:1001–1007, 1991.

[329] D. Crothers. DNA curvature and deformation in protein-DNA complexes: A step in the right direction. *Proc. Natl. Acad. Sci. USA*, 95:15163–15165, 1998.

[330] J. C. Marini, S. D. Levine, D. M. Crothers, and P. T. Englund. Bent helical structure in kinetoplast DNA. *Proc. Natl. Acad. Sci. USA*, 79:7664–7668, 1982.

[331] S. D. Levene, H.-M. Wu, and D. M. Crothers. Bending and flexibility of kinetoplast DNA. *Biochemistry*, 25:3988–3995, 1986.

[332] D. Crothers, T. E. Haran, and J. G. Nadeau. Intrinsically bent DNA. *J. Biol. Chem.*, 265:7093–7096, 1990.

[333] T. E. Haran, J. D. Kahn, and D. M. Crothers. Sequence elements responsible for DNA curvature. *J. Mol. Biol.*, 244:135–143, 1994.

[334] D. Strahs and T. Schlick. A-tract bending: Insights into experimental structures by computational models. *J. Mol. Biol.*, 301:643–666, 2000.

[335] A. D. DiGabriele, T. A., and Steitz. A DNA dodecamer containing an adenine tract crystallizes in a unique lattice and exhibits a new bend. *J. Mol. Biol.*, 231:1024–1039, 1993.

[336] R. Wing, H. Drew, T. Takano, C. Broka, S. Tanaka, K. Itakura, and R. E. Dickerson. Crystal structure analysis of a complete turn of B-DNA. *Nature*, 287:755–758, 1980.

[337] H. R. Drew, R. M. Wing, T. Takano, C. Broka, S. Tanaka, K. Itakura, and R. E. Dickerson. Structure of a B-DNA dodecamer: Conformation and dynamics. *Proc. Natl. Acad. Sci. USA*, 78:2179–2183, 1981.

[338] P. J. Hagerman. Straightening out the bends in curved DNA. *Biochim. Biophys. Acta*, 1131:125–132, 1992.

[339] R. E. Dickerson, D. S. Goodsell, and M. L. Kopka. MPD and DNA bending in crystals and in solution. *J. Mol. Biol.*, 256:108–125, 1996.

[340] W. K. Olson and V. B. Zhurkin. Twenty years of DNA bending. In R. H. Sarma and M. H. Sarma, editors, *Biological Structure and Dynamics: Proceedings of the Ninth Conversation in the Discipline Biomolecular Stereodynamics*, pages 341–370, Schenectady, NY, 1996. Adenine Press.

[341] D. MacDonald, K. Herbert, X. Zhang, T. Polgruto, and P. Lu. Solution structure of an A-tract DNA bend. *J. Mol. Biol.*, 306:1081–1098, 2001.

[342] J. Hizver, H. Rozenberg, F. Frolow, D. Rabinovich, and Z. Shakked. DNA bending by an adenine-thymine tract and its role in gene regulation. *Proc. Natl. Acad. Sci. USA*, 98:8490–8495, 2001.

[343] E. Selsing, R. D. Wells, C. J. Alden, and S. Arnott. Bent DNA: Visualization of a base-paired and stacked A-B conformational junction. *J. Biol. Chem.*, 254:5417–5422, 1979.

[344] M. A. Young, J. Srinivasan, I. Goljer, S. Kumar, D. L. Beveridge, and P. H. Bolton. Structure determination and analysis of local bending in an A-tract DNA duplex: Comparison of results from crystallography, nuclear magnetic resonance, and molecular dynamics simulation on d(CGCAAAAATGCG). *Methods in Enzymology*, 261:121–144, 1995.

[345] D. Sprous, M. A. Young, and D. L. Beveridge. Molecular Dynamics Studies of Axis Bending in d(G_5-(GA_4T_4C)$_2$-C_5) and d(G_5-(GT_4A_4C)$_2$-C_5): Effects of Sequence Polarity on DNA Curvature. *J. Mol. Biol.*, 285:1623–1632, 1999.

[346] E. C. Sherer, S. A. Harris, R. Soliva, M. Orozco, and C. A. Laughton. Molecular dynamics studies of DNA A-tract structure and flexibility. *J. Amer. Chem. Soc.*, 121:5981–5991, 1999.

[347] B. Schneider, K. Patel, and H. M. Berman. Hydration of the phosphate group in double-helical DNA. *Biophys. J.*, 75:2422–2434, 1998.

[348] J. Woda, B. Schneider, K. Patel, K. Mistry, and H. M. Berman. An analysis of the relationship between hydration and protein-DNA interactions. *Biophys. J.*, 75:2170–2177, 1998.

[349] A. V. Vologodskii and N. R. Cozzarelli. Modeling of long-range electrostatic interactions in DNA. *Biopolymers*, 35:289–296, 1995.

[350] T. Schlick, B. Li, and W. K. Olson. The influence of salt on DNA energetics and dynamics. *Biophys. J.*, 67:2146–2166, 1994.

[351] M. S. Babcock and W. K. Olson. The effect of mathematics and coordinate system on comparability and "dependencies" of nucleic acid structure parameters. *J. Mol. Biol.*, 237:98–124, 1994.

[352] D. Flatters, M. Young, D. L. Beveridge, and R. Lavery. Conformational properties of the TATA-box binding sequence of DNA. *J. Biomol. Struct. Dynam.*, 14:757–765, 1997.

[353] D. Flatters and R. Lavery. Sequence-dependent dynamics of TATA-box binding sites. *Biophys. J.*, 75:372–381, 1998.

[354] O. Norberto de Souza and R. L. Ornstein. Inherent DNA curvature and flexibility correlate with TATA box functionality. *Biopolymers*, 46:403–415, 1998.

[355] N. B. Ulyanov and V. B. Zhurkin. Sequence-dependent anisotropic flexibility of B-DNA: A conformational study. *J. Biomol. Struct. Dynam.*, 2:361–385, 1984.

[356] V. B. Zhurkin, N. B. Ulyanov, A. A. Gorin, and R. L. Jernigan. Static and statistical bending of DNA evaluated by Monte Carlo simulations. *Proc. Natl. Acad. Sci. USA*, 88:7046–7050, 1991.

[357] L. Ulanovsky and E. N. Trifonov. Estimation of wedge components in curved DNA. *Nature*, 326:720–722, 1987.

[358] H. C. Nelson, J. T. Finch, B. F. Luisi, and A. Klug. The structure of an oligo(dA) · oligo(dT) tract and its biological implications. *Nature*, 330:221–226, 1987.

[359] A. M. Burkhoff and T. D. Tullius. The unusual conformation adopted by the adenine tracts in kinetoplast DNA. *Cell*, 48:935–943, 1987.

[360] M. A. Price and T. D. Tullius. How the structure of an adenine tract depends on sequence context: A new model for the structure of T_nA_n DNA sequences. *Biochemistry*, 32:127–136, 1993.

[361] R. E. Dickerson, D. S. Goodsell, and S. Neidle. "... the tyranny of the lattice ...". *Proc. Natl. Acad. Sci. USA*, 91:3579–3583, 1994.

[362] D. Sprous, W. Zacharias, Z. A. Wood, and S. C. Harvey. Dehydrating agents sharply reduce curvature in DNAs containing A-tracts. *Nucl. Acids Res.*, 23:1816–1821, 1995.

[363] S. C. Harvey, M. Dlakic, J. Griffith, R. Harrington, K. Park, D. Sprous, and W. Zacharias. What is the basis of sequence-directed curvature in DNAs containing A-tracts? *J. Biomol. Struct. Dynam.*, 13:301–307, 1995.

[364] R. M. Ganunis, H. Guo, and T. D. Tullius. Effect of the crystallizing agent 2-methyl-2,4-pentanediol on the structure of adenine tract DNA in solution. *Biochemistry*, 35:13729–13732, 1996.

[365] M. Dlakic and R. E. Harrington. The effects of sequence context on DNA curvature. *Proc. Natl. Acad. Sci. USA*, 93:3847–3852, 1996. [Erratum appeared in *Proc. Natl. Acad. Sci. USA*, 93:8796, 1996.].

[366] M. Egli. DNA-cation interactions: Quo vadis? *Chem. Biol.*, 9:277–286, 2002.

[367] M. Feig and B. M. Pettitt. Crystallographic water sites from a theoretical perspective. *Curr. Biol.*, 6:1351–1354, 1998.

[368] P. Auffinger, S. Louise-May, and E. Westhof. Multiple molecular dynamics simulations of the anticodon loop of $rRNA^{Asp}$ in aqueous solution with counterions. *J. Amer. Chem. Soc.*, 117:6720–6726, 1995.

[369] P. Auffinger, S. Louise-May, and E. Westhof. Molecular dynamics simulations of the anticodon hairpin of $tRNA^{Asp}$. Structuring effects of $C\text{–}H \cdots O$ hydrogen bonds and long-range hydration forces. *J. Amer. Chem. Soc.*, 118:1181–1189, 1996.

[370] P. Brion and E. Westhof. Hierarchy and dynamics of RNA folding. *Ann. Rev. Biophys. Biomol. Struc.*, 26:113–137, 1997.

[371] P. Auffinger and E. Westhof. Simulations of the molecular dynamics of nucleic acids. *Curr. Opin. Struct. Biol.*, 8:227–236, 1998.

[372] P. Auffinger and E. Westhof. RNA solvation: A molecular dynamics simulation perspective. *Biopolymers*, 56:266–274, 2001.

[373] V. Tsui and D. A. Case. Theory and applications of the Generalized Born solvation model in macromolecular simulations. *Biopolymers*, 56:275–291, 2001.

[374] B. Schneider, D. M. Cohen, L. Schleifer, A. R. Srinivasan, W. K. Olson, and H. M. Berman. A systematic method for studying the spatial distribution of water molecules around nucleic acid bases. *Biophys. J.*, 65:2291–2303, 1993.

[375] B. Schneider and H. M. Berman. Hydration of the DNA bases is local. *Biophys. J.*, 69:2661–2669, 1995.

[376] B. Lee and F. M. Richards. The interpretation of protein structures: Estimation of static accessibility. *J. Mol. Biol.*, 55:379–400, 1971.

[377] B. J. Klein and G. R. Pack. Calculations of the spatial distribution of charge density in the environment of DNA. *Biopolymers*, 22:2331–2352, 1983.

[378] G. S. Manning. The molecular theory of polyelectrolytc solutions with applications to the electrostatic properties of polynucleotides. *Quart. Rev. Biophys.*, 179:181–246, 1978.

[379] M. O. Fenley, W. K. Olson, I. Tobias, and G. S. Manning. Electrostatic effects in short superhelical DNA. *Biophys. Chem.*, 50:255–271, 1994.

[380] G. S. Manning and J. Ray. Counterion condensation revisited. *J. Biomol. Struct. Dynam.*, 16:461–476, 1998.

[381] J. Ray and G. S. Manning. Counterion and coion distribution functions in the counterion condensation theory of polyelectrolytes. *Macromolecules*, 32:4588–4595, 1999.

[382] N. V. Hud and J. Feigon. Localization of divalent metal ions in the minor groove of DNA A-tracts. *J. Amer. Chem. Soc.*, 119:5756–5757, 1997.

[383] V. Tereshko, G. Minasov, and M. Egli. A "Hydra-Ion" spine in B-DNA minor groove. *J. Amer. Chem. Soc.*, 121:3590–3595, 1999.

[384] L. McFail-Isom, C. C. Sines, and L. D. Williams. DNA structure: Cations in charge? *Curr. Opin. Struct. Biol.*, 9:298–304, 1999.

[385] A. D. Mirzabekov and A. Rich. Asymmetric lateral distribution of unshielded phosphate groups in nucleosomal DNA and its role in DNA bending. *Proc. Natl. Acad. Sci. USA*, 76:1118–1121, 1979.

[386] G. S. Manning, K. K. Ebralidse, A. D. Mirzabekov, and A. Rich. An estimate of the extent of folding of nucleosomal DNA by laterally asymmetric neutralization of phosphate groups. *J. Biomol. Struct. Dynam.*, 6:877–889, 1989.

[387] L. J. Maher, III. Mechanisms of DNA bending. *Curr. Opin. Struct. Biol.*, 2:688–694, 1998.

[388] B. Schneider, D. Cohen, and H. M. Berman. Hydration of DNA bases: Analysis of crystallographic data. *Biopolymers*, 32:725–750, 1992.

[389] E. Liepinsh, G. Otting, and K. Wuthrich. NMR observation of individual molecules of hydration water bound to DNA duplexes: Direct evidence for a spine of hydration water present in aqueous solution. *Nucl. Acids Res.*, 20:6549–6553, 1992.

[390] A. T. Phan, J.-L. Leroy, and M. Guéron. Determination of the residence time of water molecules hydrating B'-DNA and B-DNA, by one-dimensional zero-enhancement nuclear Overhauser effect spectroscopy. *J. Mol. Biol.*, 286:505–519, 1999.

[391] K. M. Mazur. Accurate DNA dynamics without accurate long-range electrostatics. *J. Amer. Chem. Soc.*, 120:10928–10937, 1998.

[392] D. N. Chin, F. Sussman, H. M. Chun, and R. Czerminski. A simple solvation model along with a multibody dynamics strategy MBO(N)D produces stable DNA simulations that are faster than traditional atomistic methods. *Mol. Sim.*, 24:449–463, 2000.

[393] L. D. Williams and L. J. Maher, III. Electrostatic mechanisms of DNA deformation. *Annu. Rev. Biophys. Biomol. Struct.*, 29:497–521, 2000.

[394] J. F. Schildbach, A. W. Karzai, B. E. Raumann, and R. T. Sauer. Origins of DNA-binding specificity: Role of protein contacts with the DNA backbone. *Proc. Natl. Acad. Sci. USA*, 96:811–817, 1999.

[395] D. A. Leonard, N. Rajaram, and T. K. Kerppola. Structural basis of DNA bending and oriented heterodimer binding by the basic leucine zipper domains of Fos and Jun. *Proc. Natl. Acad. Sci. USA*, 94:4913–4918, 1997.

[396] E. Stofer, C. Chipot, and R. Lavery. Free energy calculations of Watson-Crick base pairing in aqueous solution. *J. Amer. Chem. Soc.*, 121:9503–9508, 1999.

[397] K. J. Breslauer, R. Frank, H. Blöcker, and L. A. Marky. Predicting DNA duplex stability from the base sequence. *Proc. Natl. Acad. Sci. USA*, 83:3746–3750, 1986.

[398] G. E. Plum and K. J. Breslauer. Calorimetry of proteins and nucleic acids. *Curr. Opin. Struct. Biol.*, 5:682–690, 1995.

[399] C. A. Gelfand, G. E. Plum, S. Mielewczyk, D. P. Remeta, and K. J. Breslauer. A quantitative method for evaluating the stabilities of nucleic acid complexes. *Proc. Natl. Acad. Sci. USA*, 96:6113–6118, 1999.

[400] M. J. van Dongen, J. F. Doreleijers, G. A. van der Marel, J. H. van Boom, C. W. Hilbers, and S. S. Wijmenga. Structure and mechanism of formation of the H-y5 isomer of an intramolecular DNA triple helix. *Nat. Struc. Biol.*, 6:854–859, 1999.

[401] M. P. Horvath and S. C. Schultz. DNA G-quartets in a 1.86 Å resolution structure of an *Oxytricha Nova* telomeric protein-DNA complex. *J. Mol. Biol.*, 310:367–377, 2001.

[402] J. Aishima, R. K. Gitti, J. E. Noah, H. H. Gan, T. Schlick, and C. Wolberger. Crystal structure of MATα2 homeodomain-DNA complex contains a Hoogsteen base pair. 2002. Submitted.

[403] J. Aishima and C. Wolberger. Crystal structure of the MATα2 homeodomain-DNA complex with nonspecifically bound homeodomains. 2002. Submitted.

[404] N. B. Leontis and E. Westhof. Geometric nomenclature and classification of RNA base pairs. *RNA*, 7:499–512, 2001.

[405] L. Pauling and R. B. Corey. A proposed structure for nucleic acids. *Proc. Natl. Acad. Sci. USA*, 39:84–97, 1953.

[406] M. D. Frank-Kamenetskii. Triplex DNA structures. *Ann. Rev. Biochem.*, 64:65–95, 1995.

[407] G. E. Plum, D. S. Pilch, S. F. Singleton, and K. J. Breslauer. Nucleic acid hybridization: Triplex stability and energetics. *Ann. Rev. Biophys. Biomol. Struc.*, 24:319–350, 1995.

[408] D. Sen and W. Gilbert. Cationic switches in the formation of DNA structures containing guanine-quartets. In R. H. Sarma and M. H. Sarma, editors, *Structure and Function: Proceedings of the Seventh Conversation in Biomolecular Stereodynamics*. Adenine Press, Schenectady, New York, 1992.

[409] D. S. Pilch, G. E. Plum, and K. J. Breslauer. The thermodynamics of DNA structures which contain lesions or guanine tetrads. *Curr. Opin. Struct. Biol.*, 5:334–342, 1995.

[410] A. Lebrun and R. Lavery. Unusual DNA conformations. *Curr. Opin. Struct. Biol.*, 7:348–356, 1998.

[411] P. E. Nielsen. DNA analogues with nonphosphodiester backbones. *Ann. Rev. Biophys. Biomol. Struc.*, 24:167–183, 1995.

[412] S. Louise-May, P. Auffinger, and E. Westhof. Calculations of nucleic acid conformations. *Curr. Opin. Struct. Biol.*, 6:289–298, 1996.

[413] D. Y. Cherny, B. P. Belotserkovskii, M. D. Frank-Kamenetskii, M. Egholm, O. Buchardt, R. H. Berg, and P. E. Nielsen. DNA unwinding upon strand-displacement binding of a thymine-substituted polyamide to double-stranded DNA. *Proc. Natl. Acad. Sci. USA*, 90:1667–1670, 1993.

[414] M. Egholm, P. E. Nielsen, O. Buchardt, and R. H. Berg. Recognition of guanine and adenine in DNA by cytosine and thymine containing peptide nucleic-acids (PNA). *J. Amer. Chem. Soc.*, 114:9677–9678, 1992.

[415] M. Egholm, O. Buchardt, P. E. Nielsen, and R. H. Berg. Peptide nucleic-acids (PNA) — Oligonucletide analogs with an achiral peptide backbone. *J. Amer. Chem. Soc.*, 114:1895–1897, 1992.

[416] S. Sen and L. Nilsson. Molecular dynamics of duplex systems involving PNA: Structural and dynamical consequences of nucleic acid backbone. *J. Amer. Chem. Soc.*, 120:619–631, 1998.

[417] Frontiers in chemistry: Single molecules. *Science* **283**: 1667–1695 (1999), special compendium of articles.

[418] A. Rich. The rise of single-molecule DNA chemistry. *Proc. Natl. Acad. Sci. USA*, 95:13999–14000, 1998.

[419] D. Bensimon, A. Simon and. A. Chiffaudel, V. Croquette, and A. Bensimon. Alignment and sensitive detection of DNA by a moving interface. *Science*, 265:2096–2098, 1994.

[420] T. T. Perkins, S. R. Quake, D. E. Smith, and S. Chu. Relaxation of a single DNA molecule observed by optical microscopy. *Science*, 264:822–825, 1994.

[421] S. B. Smith, Y. Cui, and C. Bustamante. Overstretching B-DNA: The elastic response of individual double-stranded and single-stranded DNA molecules. *Science*, 271:795–798, 1996.

[422] P. Cluzel, A. Lebrun, C. Heller, R. Lavery, J.-L. Viovy, D. Chatenay, and F. Caron. DNA: An extensible molecule. *Science*, 271:792–794, 1996.

[423] M. W. Konrad and J. I Bolonick. Molecular dynamics simulation of DNA stretching is consistent with the tension observed for extension and strand separation and predicts a novel ladder structure. *J. Amer. Chem. Soc.*, 118:10989–10994, 1996.

[424] A. Lebrun and R. Lavery. Modelling extreme stretching of DNA. *Nucl. Acids Res.*, 24:2260–2267, 1996.

[425] J. Liphardt, B. Onoa, S. B. Smith, I. Tinoco, Jr., and C. Bustamante. Reversible unfolding of single RNA molecules by mechanical force. *Science*, 292:733–737, 2001.

[426] E. H. Egelman. Does a stretched DNA structure dictate the helical geometry of Rec-A-like filaments? *J. Mol. Biol.*, 309:539–542, 2001.

[427] J. F. Allemand, D. Bensimon, R. Lavery, and V. Croquette. Stretched and overwound DNA forms a Pauling-like structure with exposed bases. *Proc. Natl. Acad. Sci. USA*, 95:14152–14157, 1998.

[428] L. A. Day, R. L. Wiseman, and C. J. Marzec. Structure models for DNA in filamentous viruses with phosphates near the center. *Nucl. Acids Res.*, 7:1393–1403, 1979.

[429] D. J. Liu and L. A. Day. Pf1 virus structure: Helical coat protein and DNA with paraxial phosphates. *Science*, 265:671–674, 1994.

[430] C. Bustamante, J. F. Marko, E. D. Siggia, and S. Smith. Entropic elasticity of λ-phage DNA. *Science*, 265:1599–1600, 1994.

[431] T. Hermann and D. J. Patel. Stitching together RNA tertiary architectures. *J. Mol. Biol.*, 294:828–849, 1999.

[432] P. B. Moore. Structural motifs in RNA. *Ann. Rev. Biochem.*, 68:287–300, 1999.

[433] D. J. Klein, T. M. Schmeing, P. B. Moore, and T. A. Steitz. The kink-turn: A new RNA secondary structure motif. *EMBO J.*, 20:4214–4221, 2001.

[434] E. Westhof and L. Jaeger. RNA pseudoknots. *Curr. Opin. Struct. Biol.*, 2:327–333, 1992.

[435] R. T. Batey, R. P. Rambo, and J. A. Doudna. Tertiary motifs in RNA structure and folding. *Angew. Chem. Int. Ed.*, 38:2326–2343, 1999.

[436] W. G. Scott, J. T. Finch, and A. Klug. The crystal structure of an all-RNA hammerhead ribozyme: A proposed mechanism for RNA catalytic cleavage. *Cell*, 81:991–1002, 1995.

[437] J. A. Doudna and J. H. Cate. RNA structure: Crystal clear? *Curr. Opin. Struct. Biol.*, 7:310–316, 1997.

[438] E. A. Schultes and D. B. Bartel. One sequence, two ribozymes: Implications for the emergence of new ribozyme folds. *Science*, 289:448–452, 2000.

[439] G. A. Soukup and R. R. Breaker. Allosteric nucleic acid catalysts. *Curr. Opin. Struct. Biol.*, 10:318–325, 2000.

[440] J. Tang and R. R. Breaker. Structural diversity of self-cleaving ribozymes. *Proc. Natl. Acad. Sci. USA*, 97:5784–5789, 2001.

[441] P. J. Hagerman. Flexibility of RNA. *Ann. Rev. Biophys. Biomol. Struc.*, 26:139–156, 1997.

[442] T. Hermann and D. J. Patel. Adaptive recognition by nucleic acid aptamers. *Science*, 287:820–825, 2000.

[443] N. D. Pearson and C. D. Prescott. RNA as a drug target. *Chem. & Biol.*, 4:409–414, 1997.

[444] G. A. Soukup and R. R. Breaker. Engineering precision RNA molecular switches. *Proc. Natl. Acad. Sci. USA*, 96:3584–3589, 1999.

[445] A. R. Ferré-D'Amaré, K. Zhou, and J. A. Doudna. Crystal structure of a hepatitis delta virus ribozyme. *Nature*, 395:567–574, 1998.

[446] H. Shi and P. Moore. The crystal structure of yeast phenylalanine tRNA at 1.93 Å resolution: A classic structure revisited. *RNA*, 6:1091–1105, 2000.

[447] A. R. Ferré-D'Amaré and J. A. Doudna. RNA folds: Insights from recent crystal structures. *Ann. Rev. Biophys. Biomol. Struc.*, 28:57–73, 1999.

[448] I. Tinoco, Jr. and C. Bustamante. How RNA folds. *J. Mol. Biol.*, 293:271–281, 1999.

[449] A. M. Pyle and J. B. Green. RNA folding. *Curr. Opin. Struct. Biol.*, 5:303–310, 1995.

[450] P. Schuster, P. F. Stadler, and A. Renner. RNA structures and folding: From conventional to new issues in structure predictions. *Curr. Opin. Struct. Biol.*, 7:229–235, 1997.

[451] M. Zuker and P. Stiegler. Optimal computer folding of large RNA sequences using thermodynamics and auxiliary information. *Nucl. Acids Res.*, 9:133–148, 1981.

[452] D. H. Mathews, J. Sabina, M. Zuker, and D. H. Turner. Expanded sequence dependence of thermodynamic parameters improves prediction of RNA secondary structure. *J. Mol. Biol.*, 288:911–940, 1999.

[453] B. Sclavi, M. Sullivan, M. R. Chance, M. Brenowitz, and S. A. Woodson. RNA folding at millisecond intervals by synchrotron hydroxyl radical footprinting. *Science*, 279:1940–1943, 1998.

[454] R. T. Batey and J. A. Doudna. The parallel universe of RNA folding. *Nature Struc. Biol.*, 5:337–340, 1998.

[455] S. R. Holbrook and S.-H. Kim. RNA crystallography. *Biopolymers*, 44:3–21, 1997.

[456] J. A. Doudna. Structural genomics of RNA. *Nature Struc. Biol.*, 7:954–956, 2000. (Structural Genomics Supplement).

[457] A. D. Ellington and J. W. Szostak. *In Vitro* selection of RNA molecules that bind specific ligands. *Nature*, 346:818–822, 1990.

[458] C. Tuerk and L. Gold. Systematic evolution of ligands by exponential enrichment: RNA ligands to bacteriophage T4 DNA polymerase. *Science*, 249:505–570, 1990.

[459] H. W. Pley, K. M. Flaherty, and D. B. McKay. Three-dimensional structure of a hammerhead ribozyme. *Nature*, 372:68–74, 1994.

[460] T. Tuschl, C. Gohlke, T. M. Jovin, E. Westhof, and F. Eckstein. A three-dimensional model for the hammerhead ribozyme based on fluorescence measurements. *Science*, 266:785–789, 1994.

[461] J. A. Doudna. Ribozymes: The hammerhead swings into action. *Curr. Biol.*, 8:R495–R497, 1998.

[462] J. B. Murray, D. P. Terwey, L. Maloney, A. Karpeisky, N. Usman, L. Beigelman, and W. G. Scott. The structural basis for hammerhead ribozyme self-cleavage. *Cell*, 92:665–673, 1998.

[463] J. H. Cate, A. R. Gooding, E. Podell, K. Zhou, B. L. Golden, C. E. Kundrot, T. R. Cech, and J. A. Doudna. Crystal structure of a group I ribozyme domain: Principles of RNA packing. *Science*, 273:1678–1685, 1996.

[464] J. H. Cate, A. R. Gooding, E. Podell, K. Zhou, B. L. Golden, A. A. Szewczak, C. E. Kundrot, T. R. Cech, and J. A. Doudna. RNA tertiary structure mediation by adenosine platforms. *Science*, 273:1696–1699, 1996.

[465] B. L. Golden, A. R. Gooding, E. R. Podell, and T. R. Cech. A preorganized active site in the crystal structure of the *tetrahymena* ribozyme. *Science*, 282:259–264, 1998.

[466] E. A. Doherty, R. T. Batey, B. Masquida, and J. A. Doudna. A universal mode of helix packing in RNA. *Nat. Struc. Biol.*, 8:339–343, 2001.

[467] P. Nissen, J. A. Ippolito, N. Ban, P. B. Moore, and T. A. Steitz. RNA tertiary interactions in the large ribosomal subunit: The A-minor motif. *Proc. Natl. Acad. Sci. USA*, 98:4899–4903, 2001.

[468] C. R. Woese, S. Winker, and R. R. Gutell. Architecture of ribosomal RNA: Constraints on the sequence of tetra-loops. *Proc. Natl. Acad. Sci. USA*, 87:8467–8471, 1990.

[469] M. Levitt and A. Warshel. Extreme conformational flexibility of the furanose ring in DNA and RNA. *J. Amer. Chem. Soc.*, 100:2607–2613, 1978.

[470] V. A. Bloomfield. DNA condensation by multivalent cations. *Biopolymers*, 44:269–282, 1997.

[471] R. Kornberg and J. O. Thomas. Chromatin structure: Oligomers of histones. *Science*, 184:865–868, 1974.

[472] G. Arents, R. W. Burlingame, B. C. Wang, W. E. Love, and E. N. Moudrianakis. The nucleosomal core histone octamer at 3.1 Å resolution: A tripartite protein assembly and a left-handed superhelix. *Proc. Natl. Acad. Sci. USA*, 88:10148–10152, 1991.

[473] D. Pruss, B. Bartholomew, J. Persinger, J. Hayes, G. Arents E. N. Moudrianakis, and A. P. Wolffe. An asymmetric model for the nucleosome: A binding site for linker histones inside the DNA gyres. *Science*, 274:614–617, 1996.

[474] A. Nicholls and B. Honig. A rapid finite difference algorithm, utilizing successive over-relaxation to solve the Poisson-Boltzmann equation. *J. Comput. Chem.*, 12:435–445, 1991.

[475] D. Beard and T. Schlick. Modeling salt-mediated electrostatics of macromolecules: The algorithm DiSCO (Discrete Charge Surface Charge Optimization) and its application to the nucleosome. *Biopolymers*, 58:106–115, 2001.

[476] J. T. Finch, L. C. Lutter, D. Rhodes, A. S. Brown, B. Rushton, M. Levitt, and A. Klug. Structure of nucleosome core particles of chromatin. *Nature*, 269:29–36, 1977.

[477] D. W. Sumners. Lifting the curtain: Using topology to probe the hidden action of enzymes. *Notices Amer. Math. Soc.*, 42:528–537, 1995.

[478] G. Călugăreanu. Sur les classes d'isotopie des noeuds tridimensionnels et leurs invariants. *Czechoslovak Math. J.*, 11:588–624, 1961.

[479] J. H. White. Self-linking and the Gauss integral in higher dimensions. *Amer. J. Math.*, 91:693–728, 1969.

[480] F. B. Fuller. The writhing number of a space curve. *Proc. Natl. Acad. Sci. USA*, 68:815–819, 1971.

[481] J. H. White. An introduction to the geometry and topology of DNA structures. In M. S. Waterman, editor, *Mathematical Methods for DNA Sequences*, chapter 9. CRC Press, Boca Raton, Florida, 1989.

[482] T. Schlick and W. K. Olson. Supercoiled DNA energetics and dynamics by computer simulation. *J. Mol. Biol.*, 223:1089–1119, 1992.

[483] R. Kanaar and N. R. Cozzarelli. Roles of supercoiled DNA structure in DNA transactions. *Curr. Opin. Struc. Bio.*, 2:369–379, 1992.

[484] A. V. Vologodskii and N. R. Cozzarelli. Conformational and thermodynamic properties of supercoiled DNA. *Ann. Rev. Biophys. Biomol. Struc.*, 23:609–643, 1994.

[485] A. V. Vologodskii and N. R. Cozzarelli. Supercoiling, knotting, looping, and other large-scale conformational properties of DNA. *Curr. Opin. Struct. Biol.*, 4:372–375, 1994.

[486] T. Schlick. Modeling superhelical DNA: Recent analytical and dynamic approaches. *Curr. Opin. Struc. Biol.*, 5:245–262, 1995.

[487] P. J. Flory. *Statistical Mechanics of Chain Molecules*. Oxford University Press, New York, NY, 1988. (Reprinted version of the 1969 text with an added excerpt from Flory's Nobel address).

[488] P. J. Hagerman. Flexibility of DNA. *Ann. Rev. Biophys. Biophys. Chem.*, 17:265–286, 1988.

[489] C. G. Baumann, S. B. Smith, V. A. Bloomfield, and C. Bustamante. Ionic effects on the elasticity of single DNA molecules. *Proc. Natl. Acad. Sci. USA*, 94:6185–6190, 1997.

[490] E. N. Trifonov, R. K.-Z. Tan, and S. C. Harvey. Static persistence length of DNA. In W. K. Olson, M. H. Sarma, R. H. Sarma, and M. Sundaralingam, editors, *Structure and Expression: DNA Bending and Curvature*, volume 3. Adenine Press, Schenectady, New York, 1987.

[491] J. A. Schellman and S. C. Harvey. Static contributions to the persistence length of DNA and dynamic contributions to DNA curvature. *Biophys. Chem.*, 55:95–114, 1995.

[492] F. B. Fuller. Decomposition of the linking number of a closed ribbon: A problem from molecular biology. *Proc. Natl. Acad. Sci. USA*, 75:3557–3561, 1978.

[493] L. D. Landau and E. M. Lifshitz. *Course of Theoretical Physics*, volume 5. Pergamon Press, Oxford, third edition, 1980.

[494] J. A. Schellman. Flexibility of DNA. *Biopolymers*, 13:217–226, 1974.

[495] J. F. Marko and E. D. Siggia. Fluctuations and supercoiling of DNA. *Science*, 265:506–508, 1994.

[496] W. K. Olson. Simulating DNA at low resolution. *Curr. Opin. Struct. Biol.*, 6:242–256, 1996.

[497] W. K. Olson and V. B. Zhurkin. Modeling DNA deformations. *Curr. Opin. Struct. Biol.*, 10:286–297, 2000.

[498] D. Beard and T. Schlick. Computational modeling predicts the structure and dynamics of the chromatin fiber. *Structure*, 9:105–114, 2001.

[499] J. Huang, T. Schlick, and A. Vologodskii. Dynamics of site juxtaposition in supercoiled DNA. *Proc. Natl. Acad. Sci. USA*, 98:968–973, 2001.

[500] G. Ramachandran and T. Schlick. Solvent effects on supercoiled DNA dynamics explored by Langevin dynamics simulations. *Phys. Rev. E*, 51:6188–6203, 1995.

[501] T. Schlick, W. K. Olson, T. Westcott, and J. P. Greenberg. On higher buckling transitions in supercoiled DNA. *Biopolymers*, 34:565–598, 1994.

[502] D. Sprous, II, R. K.-Z. Tan, and S. C. Harvey. Molecular modeling of closed circular DNA thermodynamic ensembles. *Biopolymers*, 39:248–258, 1996.

[503] R. K.-Z. Tan, D. Sprous, and S. C. Harvey. Molecular dynamics simulations of small DNA plasmids: Effects of sequence and supercoiling on intramolecular motions. *Biopolymers*, 39:259, 1996.

[504] I. Klapper. Biological applications of the dynamics of twisted elastic rods. *J. Comput. Phys.*, 125:325–337, 1996.

[505] D. J. Dichmann, Y. Li, and J. H. Maddocks. Hamiltonian formulations and symmetries in rod mechanics. In J. P. Mesirov, K. Schulten, and D. W. Sumners, editors, *Mathematical Applications to Biomolecular Structure and Dynamics*, volume 82

of *IMA Volumes in Mathematics and Its Applications*, pages 71–113, New York, NY, 1996. Springer-Verlag.

[506] Y. Shi, A. E. Borovik, and J. E. Hearst. Elastic rod model incorporating shear and extension, generalized nonlinear schrödinger equations, and novel closed-form solutions for supercoiled DNA. *J. Chem. Phys.*, 103:3166–3183, 1995.

[507] W. K. Olson, T. P. Westcott, J. A. Martino, and G.-H. Liu. Computational studies of spatially constrained DNA chains. In J. P. Mesirov, K. Schulten, and D. W. Sumners, editors, *Mathematical Approaches to Biomolecular Structure and Dynamics*, volume 82 of *IMA Volumes in Mathematics and Its applications*, New York, NY, 1996. Springer-Verlag.

[508] B. D. Coleman and D. Swigon. Theory of supercoiled elastic rings with self-contact and its application to DNA plasmids. *J. Elasticity*, 60:173–221, 2000.

[509] B. D. Coleman, D. Swigon, and I. Tobias. Elastic stability of DNA configurations: II. Supercoiled plasmids with self-contact. *Phys. Rev. E*, 61:759–770, 2000.

[510] D. Sprous and S. C. Harvey. Action at a distance in supercoiled DNA: Effects of sequences on slither, branching and intermolecular concentration. *Biophys. J.*, 70:1893–1908, 1996.

[511] A. V. Vologodskii and N. R. Cozzarelli. Effect of supercoiling on the juxtaposition and relative orientation of DNA sites. *Biophys. J.*, 70:2548–2556, 1996.

[512] H. Jian. *A Combined Wormlike-Chain and Bead Model for Dynamic Simulations of Long DNA*. PhD thesis, New York University, Department of Physics, New York, NY, October 1997.

[513] L. Ehrlich, C. Münkel, G. Chirico, and J. Langowski. A Brownian dynamics model for the chromatin fiber. *CABIOS*, 13(3):271–279, 1997.

[514] W. K. Olson, N. L. Markey, R. L. Jernigan, and V. B. Zhurkin. Influence of fluctuations on DNA curvature. A comparison of flexible and static wedge models of intrinsically bent DNA. *J. Mol. Biol.*, 232:530–554, 1993.

[515] W. K. Olson, M. S. Babcock, A. Gorin, G.-H. Liu, N. L. Markey, J. A. Martino, S. C. Pedersen, A. R. Srinivasan, I. Tobias, T. P. Westcott, and P. Zhang. Flexing and folding of double helical DNA. *Biol. Chem.*, 55:7–29, 1995.

[516] H. Jian, T. Schlick, and A. Vologodskii. Internal motion of supercoiled DNA: Brownian dynamics simulations of site juxtaposition. *J. Mol. Biol.*, 284:287–296, 1998.

[517] K. R. Benjamin, A. P. Abola, R. Kanaar, and N. R. Cozzarelli. Contributions of supercoiling to Tn3 resolvase and phage Mu Gin site-specific recombination. *J. Mol. Biol.*, 256:50–65, 1996.

[518] H. A. Benjamin and N. R. Cozzarelli. DNA-directed synapsis in recombination: Slithering and random collision of sites. *Proc. R. A. Welch Found. Conf. Chem. Res.*, 29:107–126, 1986.

[519] J. A. Martino, V. Katritch, and W. K. Olson. Influence of nucleosome structure on the three-dimensional folding of idealized minichromosomes. *Struc. Fold. Design*, 7:1009–1022, 1999.

[520] H. F. Schaefer. Methylene: A paradigm for computational quantum chemistry. *Science*, 231:1100–1107, 1986.

[521] R. J. Read and D. E. Wemmer. Biophysical methods. Bigger, better, faster and automatically too? (Editorial overview). *Curr. Opin. Struct. Biol.*, 9:591–593, 1999.

[522] P. G. Arscott, G. Lee, V. A. Bloomfield, and D. F. Evans. Scanning tunneling microscopy of Z-DNA. *Nature*, 339:484–486, 1989.

[523] R. J. Driscoll, M. G. Younquist, and J. D. Baldeschwieler. Atomic-scale imaging of DNA using scanning tunneling microscopy. *Nature*, 346:294–296, 1990.

[524] D. Rugar and P. Hansma. Atomic force microscope. *Physics Today*, 43:23–30, 1990.

[525] C. Bustamante and D. Keller. Scanning force microscopy in biology. *Physics Today*, 48:32–38, 1995.

[526] W. Kühlbrandt and K. A. Williams. Analysis of macromolecular structure and dynamics by electron cryo-microscopy. *Curr. Opin. Chem. Biol.*, 3:537–543, 1999.

[527] T. S. Anantharaman, B. Mishra, and D. C. Schwartz. Genomics via optical mapping II: Ordered restriction maps. *J. Comput. Biol.*, 4:91–118, 1997.

[528] W. Cai, J. Jing, B. Irvin, L. Ohler, E. Rose, H. Shizuya, U.-J. Kim, M. Simon, T. Anantharaman, B. Mishra, and D. C. Schwartz. High-resolution restriction maps of bacterial artificial chromosomes constructed by optical mapping. *Proc. Natl. Acad. Sci. USA*, 95:3390–3395, 1998.

[529] V. F. R. Jones. Knot theory and statistical mechanics. *Sci. Amer.*, 263:98–103, 1990.

[530] B. A. Cipra. Molecular biologists team up with mathematicians to investigate DNA. *SIAM News*, 23:16, 1990.

[531] B. Cipra. *What's Happening in the Mathematical Sciences*. American Mathematical Society, Cranston, RI, 1996. (Series Editor: P. Zorn).

[532] M. Ogihara and A. Ray. DNA computing on a chip. *Science*, 403:143–144, 2000.

[533] Q. Liu, L. Wang, A. G. Frutos, A. E. Condon, R. M. Corn, and L. M. Smith. DNA computing on surfaces. *Science*, 403:175–179, 2000.

[534] K. Sakamoto, H. Gouzu, K. Komiya, D. Kiga, S. Yokoyama, T. Yokomori, and M. Hagiya. Molecular computation by DNA hairpin formation. *Science*, 288:1223–1226, 2000.

[535] R. H. Smith. Nanotechnology gains momentum. *Mod. Drug Dis.*, 4:33–38, 2001.

[536] M. C. Roco, R. S. Williams, and P. Alivisatos, editors. *Nanotechnology Research Directions: IWGN (Interagency Working Group on Nanoscience, Engineering and Technology) Workshop Report. Vision for Nanotechnology Research and Development in the Next Decade*, Loyola College, Maryland, 1999. International Technology Research Institute, World Technology (WTEC) Division, Loyola College. URL: itri.loyola.edu/nano/IWGN.Research.Directions/; Also published in hard copy by Kluwer Academic Press, February 2000.

[537] J. Maddox. Towards the calculation of DNA. *Nature*, 339:557, 1989.

[538] L. Pauling and E. B. Wilson, Jr. *Introduction to Quantum Mechanics with Applications to Chemistry*. Dover, New York, NY, 1985.

[539] I. N. Levine. *Quantum Chemistry*. Prentice-Hall, Inc., Englewood Cliffs, New Jersey, fourth edition, 1991.

[540] S. Lifson. Theoretical foundation for the empirical force field method. *Gazzetta Chimica Italiana*, 116:687–692, 1986.

[541] V. Gogonea, D. Suárex, A. van der Vaart, and K. M. Merz, Jr. New developments in applying quantum mechanics to proteins. *Curr. Opin. Struct. Biol.*, 11:217–223, 2001.

[542] J. A. Pople. Quantum chemical models (Nobel lecture). *Angew. Chem. Int. Ed.*, 38:1894–1902, 1999.

[543] R. G. Parr and W. Yang. Density-functional theory of the electronic structure of molecules. *Ann. Rev. Phys. Chem.*, 46:701–728, 1995.

[544] W. Yang and T. Lee. A density-matrix divide-and-conquer approach for electronic structure calculations of large molecules. *J. Chem. Phys.*, 163:5674–5678, 1995.

[545] E. Schwegler and M. Challacombe. Linear scaling computation of the Hartree-Fock exchange matrix. *J. Chem. Phys.*, 105:2726–2734, 1996.

[546] A. D. Daniels and G. E. Scuseria. What is the best alternative to diagonalization of the Hamiltonian in large scale semiempirical calculations? *J. Chem. Phys.*, 110:1321–1328, 1999.

[547] P. Ordejón, D. A. Drabold, R. M. Martin, and M. P. Grumbach. Linear system-size scaling methods for electronic structure calculations. *Phys. Rev. B*, 51:1456–1476, 1995.

[548] Q. Zhao and W. Yang. Analytical energy gradients and geometry optimization in the divide-and-conquer method for large molecules. *J. Chem. Phys.*, 102:9598–9603, 1995.

[549] T.-S. Lee, D. M. York, and W. Yang. Linear-scaling semiempirical quantum calculations for macromolecules. *J. Chem. Phys.*, 105:2744–2750, 1996.

[550] D. M. York, T.-S. Lee, and W. Yang. Parameterization and efficient implementation of a solvent model for linear-scaling semiempirical quantum-mechanical calculations of biological macromolecules. *Chem. Phys. Lett.*, 263:297–304, 1996.

[551] S. Goedecker. Linear scaling electronic structure methods. *Rev. Mod. Phys.*, 71:1085–1123, 1999.

[552] J. P. Lewis, P. Ordejón, and O. F. Sankey. An electronic structure based molecular dynamics for large biomolecular systems: Applications to the 10 basepair Poly(dG)*Poly(dC) DNA double helix. *Phys. Rev. B*, 55:6880–6887, 1997.

[553] D. M. York, T.-S. Lee, and W. Yang. Quantum-mechanical study of aqueous polarization effects on biological macromolecules. *J. Amer. Chem. Soc.*, 118:10940–10941, 1996.

[554] J. Khandogin, A. Hu, and D. M. York. Electronic structure properties of solvated biomolecules: A quantum approach for macromolecular characterization. *J. Comput. Chem.*, 21:1562–1571, 2000.

[555] J. P. Lewis, C. W. Carter, Jr., J. Hermans, W. Pan, T.-S. Lee, and W. Yang. Active species for the ground-state complex of cytidine deaminase: A linear-scaling quantum mechanical investigation. *J. Amer. Chem. Soc.*, 120:5407–5410, 1998.

[556] P. Hobza and J. Šponer. Structure, energetics, and dynamics of the nucleic acid base pairs: Nonempirical Ab Initio calculations. *Chem. Rev.*, 99:3247–3276, 1999.

[557] J. P. Lewis, N. H. Pawley, and O. F. Sankey. Theoretical investigation of the cyclic peptide system cyclo[(D-Ala-Glu-D-Ala-Gln)$_{m=1-4}$]. *J. Phys. Chem. B*, 101:10576–10583, 1997.

[558] J. Åqvist and A. Warshel. Simulation of enzyme reactions using valence bond force fields and other hybrid quantum/classical approaches. *Chem. Rev.*, 93:2523–2544, 1993.

[559] H. Liu, M. Elstner, E. Kaxiras, T. Frauenheim, J. Hermans, and W. Yang. Quantum mechanics simulation of protein dynamics on long timescale. *Proteins: Struc. Func. Gen.*, 44:484–489, 2001.

[560] J. Šponer, J. Leszczyński, and P. Hobza. Nature of nucleic acid-base stacking: Nonempirical ab Initio and empirical potential characterization of 10 stacked base dimers. Comparison of stacked and H-bonded base pairs. *J. Phys. Chem.*, 100:5590–5596, 1996.

[561] C. Alhambra, F. J. Luque, F. Gago, and M. Orozco. *Ab Initio* study of stacking interactions in A- and B-DNA. *J. Phys. Chem. B*, 101:3846–3853, 1997.

[562] J. Šponer, J. Leszczyński, and P. Hobza. Structures and energies of hydrogen-bonded DNA base pairs: A nonempirical study with inclusion of electron correlation. *J. Phys. Chem.*, 100:1965–1974, 1996.

[563] Y. Zhang, T. Lee, and W. Yang. A pseudo-bond approach to combining quantum mechanical and molecular mechanical methods. *J. Chem. Phys.*, 110:46–54, 1999.

[564] Y. Zhang, H. Liu, and W. Yang. Free energy calculation on enzyme reactions with an efficient iterative procedure to determine minimum energy paths on a combined Ab Initio QM/MM potential energy surface. *J. Chem. Phys.*, 112:3483–3492, 2000.

[565] H. Liu, Y. Zhang, and W. Yang. How is the active-site of enolase organized to achieve overall efficiency in catalyzing a two step reaction. *J. Amer. Chem. Soc.*, 122:6560–6570, 2000.

[566] S. J. Weiner, P. A. Kollman, D. T. Nguyen, and D.A. Case. An all atom force field for simulations of proteins and nucleic acids. *J. Comput. Chem.*, 7:230–252, 1986.

[567] G. Némethy, M. S. Pottle, and H. A. Scheraga. Energy parameters in polypeptides. 9. Updating of geometrical parameters, nonbonded interactions, and hydrogen bond interactions for the naturally occurring amino acids. *J. Phys. Chem.*, 87:1883–1887, 1983.

[568] B. R. Brooks, R. E. Bruccoleri, B. D. Olafson, D. J. States, S. Swaminathan, and M. Karplus. CHARMM: A program for macromolecular energy, minimization, and dynamics calculations. *J. Comput. Chem.*, 4:187–217, 1983.

[569] U. Burkert and N. L. Allinger. *Molecular Mechanics*, volume 177 of *American Chemical Society Monograph*. ACS, Washington D. C., 1982.

[570] W. K. Olson. Theoretical studies of nucleic acid conformation: Potential energies, chain statistics, and model building. In S. Neidle, editor, *Topics in Nucleic Acid Structures: Part 2*, pages 1–79. Macmillan Press, London, England, 1982.

[571] S. Lifson. Potential energy functions for structural molecular biology. In D. B. Davies, W. Saenger, and S. S. Danyluk, editors, *Methods in Structural Molecular Biology*, pages 359–385. Plenum Press, London, England, 1981.

[572] N. L. Allinger. Calculation of molecular structure and energy by force-field methods. *Adv. Phys. Org. Chem.*, 13:1–85, 1976.

[573] F. A. Momany, R. F. McGuire, A. W. Burgess, and H. A. Scheraga. Energy parameters in polyeptides. VII. Geometric parameters partial atomic charges, nonbonded interactions, hydrogen bond interactions, and intrinsic torsional potentials for the naturally occurring amino acids. *J. Phys. Chem.*, 79:2361–2381, 1975.

[574] M. Bixon, H. Dekker, J. D. Dunitz, E. Eser, S. Lifson, C. Mosselman, J. Sicher, and M. Svoboda. Structural and strain-energy consequences of "intra-annular" substitution in the cyclodecane ring. *Chem. Commun.*, pages 360–362, 1967.

[575] L. S. D. Caves, J. D. Evanseck, and M. Karplus. Locally accessible conformations of proteins: Multiple molecular dynamics simulations of crambin. *Prot. Sci.*, 7:649–666, 1998.

[576] C. B. Anfinsen and H. A. Scheraga. Experimental and theoretical aspects of protein folding. *Adv. Protein Chem.*, 29:205–301, 1975.

[577] E.I. Shakhnovich and A.M. Gutin. Implications of thermodynamics on protein folding for evolution of primary sequences. *Nature*, 346:773–775, 1990.

[578] A. Holmgren and C.-I. Bränden. Crystal structure of chaperone protein PapD reveals an immunoglobulin fold. *Nature*, 342:248–251, 1989.

[579] J. Martin, T. Langer, R. Boteva, A. Schramel, A. L. Horwich, and F.-U. Hartl. Chaperonin-mediated protein folding at the surface of groEL through a 'molten globule'-like intermediate. *Nature*, 352:36–42, 1991.

[580] N. L. Allinger, Y. H. Yuh, and J.-H. Lii. Molecular mechanics. The MM3 force field for hydrocarbons. 1. *J. Amer. Chem. Soc.*, 111:8551–8566, 1989.

[581] R. S. Berry, S. A. Rice, and J. Ross. *Physical Chemistry*. Wiley, New York, NY, 1980.

[582] P. Cieplak, P. Kollman, and T. Lybrand. A new water potential including polarization: Application to gas-phase, liquid, and crystal properties of water. *J. Chem. Phys.*, 92:6755–6760, 1990.

[583] N. L. Allinger and J. T. Sprague. Calculation of the structures of hydrocarbons containing delocalized electronic systems by the molecular mechanics method. *J. Amer. Chem. Soc.*, 95:3893–3907, 1973.

[584] J. T. Sprague, J. C. Tai, Y. Yuh, and N. L. Allinger. The MMP2 calculational method. *J. Comput. Chem.*, 8:581–603, 1987.

[585] N. Nevins, J.-H. Lii, and N. L. Allinger. Molecular mechanics (MM4) calculations on conjugated hydrocarbons. *J. Comput. Chem.*, 17:695–729, 1996.

[586] J. C. Tai and N. L. Allinger. Effect of inclusion of electron correlation in MM3 studies of cyclic conjugated compounds. *J. Comput. Chem.*, 19:475–487, 1998.

[587] N. L. Allinger, M. R. Imam, M. R. Frierson, Y. H. Yuh, and L. Schäfer. The effect of electronegativity on bond lengths in molecular mechanics calculations. In N. Trinajstic, editor, *Mathematics and Computational Concepts in Chemistry*, pages 8–19. Ellis Horwood, Ltd., London, England, 1986.

[588] N. L. Allinger and K. Chen. Hyperconjugative effects on carbon-carbon bond lengths in molecular mechanics (MM4). *J. Comput. Chem.*, 17:747–755, 1996.

[589] A. D. MacKerell, Jr., D. Bashford, M. Bellott, R. L. Dunbrack, Jr., J. Evanseck, M. J. Field, S. Fischer, J. Gao, H. Guo, S. Ha, D. Joseph, L. Kuchnir, K. Kuczera, F. T. K. Lau, C. Mattos, S. Michnick, T. Ngo, D. T. Nguyen, B. Prodhom, W. E. Reiher, III., B. Roux, M. Schlenkrich, J. Smith, R. Stote, J. Straub, M. Watanabe,

J. Wiorkiewicz-Kuczera, D. Yin, and M. Karplus. An all-atom empirical potential for molecular modeling and dynamics of proteins. *J. Phys. Chem. B*, 102:3586–3616, 1998.

[590] I. K. Roterman, M. H. Lambert, K. D. Gibson, and H. A. Scheraga. A comparison of the CHARMM, AMBER and ECEPP potentials for peptides. I. Conformational predictions for the tandemly repeated peptide (Asn-Ala-Asn-Pro)$_9$. *J. Biomol. Struct. Dyn.*, 7:391–419, 1989.

[591] I. K. Roterman, M. H. Lambert, K. D. Gibson, and H. A. Scheraga. A comparison of the CHARMM, AMBER and ECEPP potentials for peptides. II. ϕ–ψ maps for N-methyl amide: Comparisons, contrasts and simple experimental tests. *J. Biomol. Struct. Dyn.*, 7:421–453, 1989.

[592] K. B. Gibson and H. A. Scheraga. Decisions in force field development: Reply to Kollman and Dill. *J. Biomol. Struct. Dyn.*, 8:1109–1111, 1991.

[593] P. A. Kollman and K. A. Dill. Decisions in force field development: An alternative to those described by Roterman *et al. J. Biomol. Struct. Dyn.*, 8:1103–1107, 1991.

[594] W. K. Olson. How flexible is the furanose ring? 2. an updated potential energy estimate. *J. Amer. Chem. Soc.*, 104:278–286, 1982.

[595] T. Schlick. *Modeling and Minimization Techniques for Predicting Three-Dimensional Structures of Large Biological Molecules*. PhD thesis, New York University, Courant Institute of Mathematical Sciences, New York, NY, October 1987.

[596] S. Lifson and A. Warshel. Consistent force field for calculations of conformations, vibrational spectra, and enthalpies of cycloalkane and *n*-alkane molecules. *J. Chem. Phys.*, 49:5116–5129, 1968.

[597] T. Schlick. A recipe for evaluating and differentiating $\cos\phi$ expressions. *J. Comput. Chem.*, 10:951–956, 1989.

[598] W. F. van Gunsteren and A. E. Mark. Validation of molecular dynamics simulation. *J. Chem. Phys.*, 108:6109–6116, 1998.

[599] T. A. Halgren. Merck molecular force field: I. Basis, form, scope, parameterization and performance of MMFF94. *J. Comput. Chem.*, 17:490–519, 1996.

[600] H. H. Gan and T. Schlick. Methods for macromolecular modeling (M^3): Assessment of progress and future perspectives. In T. Schlick and H. H. Gan, editors, *Computational Methods for Macromolecules: Challenges and Applications — Proceedings of the 3rd International Workshop on Algorithms for Macromolecular Modelling, New York, October 12–14, 2000*, volume 24 of *Lecture Notes in Computational Science and Engineering (Series Eds. M. Griebel, D.E. Keyes, R. M. Nieminen, D. Roose, and T. Schlick)*, pages 1–25, Berlin, 2002. Springer-Verlag.

[601] J.-H. Lii and N. L. Allinger. Directional hydrogen bonding in the MM3 force field: II. *J. Comput. Chem.*, 19:1001–1016, 1998.

[602] B. Ma, J.-H. Lii, and N. L. Allinger. Molecular polarizabilities and induced dipole moments in molecular mechanics. *J. Comput. Chem.*, 21:813–825, 2000.

[603] T. A. Halgren and W. Damm. Polarizable force fields. *Curr. Opin. Struct. Biol.*, 11:236–242, 2001.

[604] D. S. Maxwell, J. Tirado-Rives, and W. L. Jorgensen. A comprehensive study of the rotational energy profiles of organic systems by *ab initio* MO theory, forming a basis for peptide torsional potentials. *J. Comput. Chem.*, 16:984–1010, 1995.

[605] W. D. Cornell, P. Cieplak, C. I. Bayly, I. R. Gould, K. M. Merz, Jr., D. M. Ferguson, D. C. Spellmeyer, T. Fox, J. W. Caldwell, and P. A Kollman. A second generation force field for the simulation of proteins, nucleic acids, and organic molecules. *J. Amer. Chem. Soc.*, 117:5179–5197, 1995.

[606] T. E. Cheatham, III, P. Cieplak, and P. A. Kollman. A modified version of the cornel *et al.* force field with improved sugar pucker phases and helical repeat. *J. Biomol. Struct. Dynam.*, 16:845–862, 1999.

[607] R. A. Scott and H. A. Scheraga. Method for calculating internal rotation barriers. *J. Chem. Phys.*, 42:2209–2215, 1965.

[608] J.-H. Lii and N. L. Allinger. Intensities of infrared bands in molecular mechanics (MM3). *J. Comput. Chem.*, 13:1138–1141, 1992.

[609] A. Sandu and T. Schlick. Masking resonance artifacts in force splitting methods for biomolecular simulations by extrapolative Langevin dynamics. *J. Comput. Phys.*, 151:74–113, May 1999. (Special Volume on Computational Biophysics).

[610] R. T. Morrison and R. N. Boyd. *Organic Chemistry*. Allyn and Bacon, Inc., Newton, MA, fourth edition, 1983.

[611] B. Ma, J.-H. Lii, H. F. Schaefer, III, and N. L. Allinger. Systematic comparison of experimental, quantum mechanical, and molecular mechanical bond lengths for organic molecules. *J. Phys. Chem.*, 100:8763–8769, 1996.

[612] B. Ma, J.-H. Lii, K. Chen, and N. L. Allinger. A molecular mechanics study of the cholesteryl acetate crystal: Evaluation of interconversion among r_g, r_z, and r_α bond lengths. *J. Amer. Chem. Soc.*, 119:2570–2573, 1997.

[613] B. Ma and N. L. Allinger. Calculation of r_z structures from r_s structures. *J. Mol. Struct.*, 413–414:395–404, 1997.

[614] P. M. Morse. Diatomic molecules according to the wave mechanics. II. Vibrational levels. *Phys. Rev.*, 34:57–64, 1929.

[615] N. B. Slater. Classical motion under a morse potential. *Nature*, 180:1352–1353, 1957.

[616] P. Derreumaux and G. Vergoten. A new spectroscopic molecular mechanics force field. Parameters for proteins. *J. Chem. Phys.*, 102:8586–8605, 1995.

[617] L. Goodman, V. Pophristic, and F. Weinhold. Origin of methyl internal rotation barriers. *Acc. Chem. Res.*, 32:983–993, 1999.

[618] F. Weinhold. A new twist on molecular shape. *Nature*, 411:539–541, 2001.

[619] V. Pophristic and L. Goodman. Hyperconjugation not steric repulsion leads to the staggered structure of ethane. *Nature*, 411:565–568, 2001.

[620] R. M. Pitzer. The barrier to internal rotation in ethane. *Acc. Chem. Res.*, 16:207–210, 1983.

[621] N. L. Allinger, J. T. Fermann, W. D. Allen, and H. F. Schaefer, III. The torsional conformations of butane: Definitive energetics from *ab initio* methods. *J. Chem. Phys.*, 106:5143–5150, 1997.

[622] L. Pauling. The nature of bond orbitals and the origin of potential barriers to internal rotation in molecules. *Proc. Natl. Acad. Sci.*, 44:211–216, 1958.

[623] R. A. Scott and H. A. Scheraga. Conformational analysis of macromolecules. II. the rotational isomeric states of the normal hydrocarbons. *J. Chem. Phys.*, 44:3054–3069, 1966.

[624] D. M. Hayes, P.A. Kollman, and S. Rothenberg. A conformational analysis of H_3PO_4, $H_3PO_4^-$, HPO_4^{2-} and related model compounds. *J. Amer. Chem. Soc.*, 99:2150–2154, 1977.

[625] N. L. Allinger, M. A. Miller, F. A. VanCatledge, and J. A. Hirsch. Conformational analysis. LVII. The calculation of the conformational structures of hydrocarbons by the Westheimer-Hendrickson-Wiberg method. *J. Amer. Chem. Soc.*, 89:4345–4357, 1967.

[626] A. Larshminarayanan and V. Sasisekharan. Stereochemistry of nucleic acids and polynucleotides. IV. Conformational energy of base-sugar units. *Biopolymers*, 8:475–488, 1969.

[627] L. Pauling. *The Nature of the Chemical Bond.* third edition, Cornell University Press, New York, NY, 1960.

[628] T. A. Halgren. Merck molecular force field: II. MMFF94 van der Waals and electrostatic parameters for intermolecular interactions. *J. Comput. Chem.*, 17:520–552, 1996.

[629] R. M. Levy and E. Gallicchio. Computer simulations with explicit solvent: Recent progress in the thermodynamic decomposition of free energies, and in modeling electrostatic effects. *Annu. Rev. Phys. Chem.*, 49:531–567, 1998.

[630] B. E. Hingerty, R. H. Ritchie, T. L. Ferrell, and J. E. Turner. Dielectric effects in biopolymers: The theory of ionic saturation revisited. *Biopolymers*, 24:427–439, 1985.

[631] E. L. Mehler. The Lorentz-Debye-Sack theory and dielectric screening of electrostatic effects in proteins and nucleic acids. In J. S. Murray and K. Sen, editors, *Molecular Electrostatic Potential: Concepts and Applications*, volume 3 of *Theoretical and Computational Chemistry*, chapter 9, pages 371–405. Elsevier Science, Amsterdam, 1996.

[632] M. A. Young, B. Jayaram, and D. L. Beveridge. Local dielectric environment of B-DNA in solution: Results from a 14 ns molecular dynamics trajectory. *J. Phys. Chem. B*, 102:7666–7669, 1998.

[633] S. A. Hassan, F. Guarnieri, and E. L. Mehler. A general treatment of solvent effects based on screened Coulomb potentials. *J. Phys. Chem. B*, 104:6478–6489, 2000.

[634] A. Srinivasan and W. K. Olson. Polynucleotide conformation in real solution — a preliminary theoretical estimate. *Fed. Amer. Soc. Exp. Bio.*, 39:2199, 1980.

[635] E. L. Mehler and F. Guarnieri. A self-consistent, microenvironment modulated screened Coulomb potential approximation to calculate pH-dependent electrostatic effects in proteins. *Biophys. J.*, 77:3–22, 1999.

[636] U.C. Singh and P.A. Kollman. An approach to computing electrostatic charges for molecules. *J. Comput. Chem.*, 5:129–145, 1984.

[637] D. A. Pearlman and S. H. Kim. Determiniations of atomic partial charges for nucleic acid constituents from x-ray diffraction data. I. 2'-deoxycytidine-5'-monophosphate. *Biopolymers*, 24:327–357, 1985.

[638] J. Wang, P. Cieplak, and P. A. Kollman. How well does a restrained electrostatic potential (RESP) model perform in calculating conformational energies of organic and biological molecules? *J. Comput. Chem.*, 21:1049–1074, 2000.

[639] L. Nilsson and M. Karplus. Empirical energy functions for energy minimization and dynamics of nucleic acids. *J. Comput. Chem.*, 7:591–616, 1986.

[640] M. Feig and B. M. Pettitt. Experiment vs. force fields: DNA conformation from molecular dynamics simulations. *Phys. Chem. B*, 101(38):7361–7363, 1997.

[641] M. Feig and B. M. Pettitt. Structural equilibrium of DNA represented with different force fields. *Biophys. J.*, 75:134–149, 1998.

[642] T. E. Cheatham, III and M. A. Young. Molecular dynamics simulation of nucleic acids: Successes, limitations, and promise. *Biopolymers*, 56:232–256, 2001.

[643] M. D. Beachy, D. Chasman, R. B. Murphy, T. A. Halgren, and R. A. Friesner. Accurate Ab initio quantum chemical determination of the relative energetics of peptide conformations and assessment of empirical force fields. *J. Amer. Chem. Soc.*, 119:5908–5920, 1997.

[644] T. A. Halgren. MMFF VII. Characterization of MMFF94, MMFF94s, and other widely available force fields for conformational energies and for interaction energies and geometries. *J. Comput. Chem.*, 20:730–748, 1999.

[645] P. H. Hünenberger and J. A. McCammon. Ewald artifacts in computer simulations of ionic solvation and ion-ion interaction: A continuum electrostatics study. *J. Chem. Phys.*, 110:1856–1872, 1999.

[646] P. H. Hünenberger and J. A. McCammon. Effect of artificial periodicity in simulations of biomolecules under Ewald boundary conditions: A continuum electrostatics study. *Biophys. Chem.*, 78:69–88, 1999.

[647] C. Sagui and T. A. Darden. Molecular dynamics simulations of biomolecules: Long-range electrostatic effects. *Ann. Rev. Biophys. Biomol. Struc.*, 28:155–179, 1999.

[648] P. J. Steinbach and B. R. Brooks. New spherical-cutoff methods for long-range forces in macromolecular simulation. *J. Comput. Chem.*, 15:667–683, 1994.

[649] J. Norberg and L. Nilsson. On the truncation of long-range electrostatic interactions in DNA. *Biophys. J.*, 79:1537–1553, 2000.

[650] X. Qian and T. Schlick. Efficient multiple-timestep integrators with distance-based force splitting for particle-mesh-Ewald molecular dynamics simulations. *J. Chem. Phys.*, 116:5971–5983, 2002.

[651] T. Darden, D. York, and L. Pedersen. Particle mesh Ewald: An $N \cdot \log(N)$ method for Ewald sums in large systems. *J. Chem. Phys.*, 98:10089–10092, 1993.

[652] D. York and W. Yang. The Fast Fourier Poisson method for calculating Ewald sums. *J. Chem. Phys.*, 101:3298–3300, 1994.

[653] U. Essmann, L. Perera, M. L. Berkowitz, T. Darden, H. Lee, and L. G. Pedersen. A smooth particle mesh Ewald method. *J. Chem. Phys.*, 103:8577–8593, 1995.

[654] P. Auffinger and E. Westhof. Molecular dynamics simulations of nucleic acids. In P. von Ragué Schleyer (Editor-in Chief), N. L. Allinger, T. Clark, J. Gasteiger, P. A. Kollman, and H. F. Schaefer, III, editors, *Encyclopedia of Computational Chemistry*, volume 3, pages 1628–1639. John Wiley & Sons, West Sussex, England, 1998.

[655] O. Norberto de Souza and R. L. Ornstein. Effect of periodic box size on aqueous molecular dynamics simulation of a DNA dodecamer with particle-mesh Ewald method. *Biophys. J.*, 72:2395–2397, 1997.

[656] L. Yang, S. Wilson, W. Beard, S. Broyde, and T. Schlick. Polymerase β simulations reveal that Arg258 rotation is a slow step rather than large subdomain motions *per se*. *J. Mol. Biol.*, 317:651–671, 2002.

[657] M. Mezei. Optimal position of the solute for simulations. *J. Comput. Chem.*, 18:812–815, 1997.

[658] X. Qian, D. Strahs, and T. Schlick. A new program for optimizing periodic boundary models of solvated biomolecules (PBCAID). *J. Comput. Chem.*, 22:1843–1850, 2001.

[659] S. W. DeLeeuw, J. W. Perram, and E. R. Smith. Simulation of electrostatic systems in periodic boundary conditions. I. Lattice sums and dielectric constant. *Proc. Roy. Soc. Lond. A*, 373:27–56, 1980.

[660] W. H. Press, S. A. Teukolosky, W. T. Vetterling, and B. P. Flannery. *Numerical Recipes in Fortran 77: The Art of Scientific Computing*, volume 1 of *Fortran Numerical Recipes*. Cambridge University Press, New York, NY, second edition, 1992.

[661] D. Fincham. Optimisation of the Ewald sum for large systems. *Mol. Sim.*, 13:1–9, 1994.

[662] J. W. Perram, H. G. Petersen, and S. W. De Leeuw. An algorithm for the simulation of condensed matter which grows as the $\frac{3}{2}$ power of the number of particles. *Mol. Phys.*, 65:875–893, 1988.

[663] R. W. Hockney and J. W Eastwood. *Computer Simulation Using Particles*. Institute of Physics, London, England, 1988.

[664] B. A. Luty, M. E. David, I. G. Tironi, and W. F. Van Gunsteren. A comparison of particle-particle particle-mesh and Ewald methods for calculating electrostatic interactions in periodic molecular systems. *Mol. Sim.*, 14:11–20, 1994.

[665] B. A. Luty, I. G. Tironi, and W. F. Van Gunsteren. Lattice-sum methods for calculating electrostatic interactions in molecular simulations. *J. Chem. Phys.*, 103:3014–3021, 1995.

[666] A. Y. Toukmaji and J. A. Board, Jr. Ewald summation techniques in perspective: A survey. *Comput. Phys. Commun.*, 95:73–92, 1996.

[667] P. Batcho, D. A. Case, and T. Schlick. Optimized particle-mesh Ewald / multiple-timestep integration for molecular dynamics simulations. *J. Chem. Phys.*, 115:4003–4018, 2001.

[668] P. Procacci, M. Marchi, and G. J. Martyna. Electrostatic calculations and multiple time scales in molecular dynamics simulation of flexible molecular systems. *J. Chem. Phys.*, 108:8799–8803, 1998.

[669] S. J. Stuart, R. Zhou, and B. J. Berne. Molecular dynamics with multiple time scales: The selection of efficient reference system propagators. *J. Chem. Phys.*, 105:1426–1436, 1996.

[670] P. Batcho and T. Schlick. New splitting formulations for lattice summations. *J. Chem. Phys.*, 115:8312–8326, 2001.

[671] D. Barash, X. Qian, L. Yang, and T. Schlick. Inherent speedup limitations in multiple-timestep/particle-mesh-Ewald algorithms. Submitted, 2002.

[672] L. Greengard and V. Rokhlin. A fast algorithm for particle simulation. *J. Comput. Phys.*, 73:325–348, 1987.

[673] L. Greengard and V. Rokhlin. A new version of the fast multipole method for the Laplace equation in three dimensions. *Acta Numerica*, 6:229–269, 1997.

[674] L. Greengard and V. Rokhlin. On the evaluation of electrostatic interactions in molecular modeling. *Chemica Scripta*, 29A:139–144, 1989.

[675] L. Greengard. Fast algorithms for classical physics. *Science*, 265:909–914, 1994.

[676] A. J. Stone. *The Theory of Intermolecular Forces*. Oxford University Press, Oxford, England, 1996.

[677] H. Q. Ding, N. Karasawa, and W. A. Goddard, III. Atomic level simulations on a million particles: The cell multipole method for Coulomb and London nonbond interactions. *J. Chem. Phys.*, 97:4309–4315, 1992.

[678] C. R. Anderson. An implementation of the fast multipole algorithm without multipoles. *SIAM J. Sci. Stat. Comput.*, 13:923–947, 1992.

[679] H. Q. Ding, N. Karasawa, and W. A. Goddard, III. The reduced cell multipole method for Coulomb interactions in periodic systems with million-atom unit cells. *Chem. Phys. Lett.*, 196:6–10, 1992.

[680] P. M. W. Gill. A new expansion of the Coulomb interaction. *Chem. Phys. Lett.*, 270:193–195, 1997.

[681] H. Y. Wang and R. LeSar. An efficient fast-multipole algorithm based on an expansion in the solid harmonics. *J. Chem. Phys.*, 104:4173–4179, 1996.

[682] A. H. Boschitsch, M. O. Fenley, and W. K. Olson. A fast adaptive multipole algorithm for calculating screened Coulomb (Yukawa) interactions. *J. Comput. Phys.*, 151:212–241, 1999.

[683] M. O. Fenley, K. Chua, A. H. Boschitsch, and W. K. Olson. A fast adaptive multipole method for computation of electrostatic energy in simulations of polyelectrolyte DNA. *J. Comput. Chem.*, 17:976–991, 1996.

[684] A. W. Appel. An efficient program for many-body simulation. *SIAM J. Sci. Stat. Comput.*, 6:85–103, 1985.

[685] J. Barnes and P. Hut. A hierarchical $O(N \log N)$ force-calculation algorithm. *Nature*, 324:446–449, 1986.

[686] V. Rokhlin. Rapid solution of integral equations of classical potential theory. *J. Comput. Phys.*, 60:187–207, 1985.

[687] L. Greengard. *The Rapid Evaluation of Potential Fields in Particle Systems*. MIT Press, Cambridge, Massachusetts, 1988.

[688] J. A. Board, Jr., J. W. Causey, T. F. Leathrum, Jr., A. Windemuth, and K. Schulten. Accelerated molecular dynamics simulations with the parallel fast multiple algorithm. *Chem. Phys. Lett.*, 198:89–94, 1992.

[689] J. Shimada, H. Kaneko, and T. Takada. Performance of fast multipole methods for calculating electrostatic interactions in biomolecular simulations. *J. Comput. Chem.*, 15:28–43, 1994.

[690] J. Board, A. John, Z. S. Hakura, W. D. Elliott, and W. T. Rankin. Scalable variants of multipole-accelerated algorithms for molecular dynamics applications. In *Proceedings, Seventh SIAM Conference on Parallel Processing for Scientific Computing*, pages 295–300, Philadelphia, PA, 1995. SIAM.

[691] J. Board, A. John, C. W. Humphres, C. G. Lambert, W. T. Rankin, and A. Y. Toukmaji. Ewald and multipole methods for periodic N-body problems. In *Proceedings, Eighth SIAM Conference on Parallel Processing for Scientific Computing*, Philadelphia, PA, 1997. SIAM. CD-ROM.

[692] K. E. Schmidt and M. A. Lee. Implementing the fast multipole method in three dimensions. *J. Stat. Phys.*, 63:1223–1235, 1991.

[693] C. G. Lambert, T. A. Darden, and J. A. Board, Jr. A multipole-based algorithm for efficient calculation of forces and potentials in macroscopic periodic assemblies of particles. *J. Comput. Phys.*, 126:274–285, 1996.

[694] F. Figueirido, R. M. Levy, R. Zhou, and B. J. Berne. Large scale simulation of macromolecules in solution: Combining the periodic fast multipole method with multiple time step integrators. *J. Chem. Phys.*, 106:9835–9849, 1997. (Erratum published in *J. Chem. Phys.* 107:7002, 1997).

[695] A. Brandt and A. A. Lubrecht. Multilevel matrix multiplication and fast solution of integral equations. *J. Comput. Phys.*, 90:348–370, 1990.

[696] B. Sandak. Multiscale fast summation of long-range charge and dipolar interactions. *J. Comput. Chem.*, 22:717–731, 2001.

[697] R. D. Skeel, I. Tezcan, and D. J. Hardy. Multiple grid methods for classical molecular dynamics. *J. Comput. Chem.*, 23:673–684, 2002.

[698] Z.-H. Duan and R. Krasny. An adaptive tree code for computing nonbonded potential energy in classical molecular systems. *J. Comput. Chem.*, 22:184–195, 2001.

[699] Z.-H. Duan and R. Krasny. An Ewald summation based multipole method. *J. Chem. Phys.*, 113:3492–3495, 2000.

[700] J. Li, C. J. Cramer, and D. G. Truhlar. Application of a universal solvation model to nucleic acid bases: Comparison of semiempirical molecular orbital theory, ab initio Hartree–Fock theory, and density functional theory. *Biophys. Chem.*, 78:147–155, 1999.

[701] T. Simonson. Macromolecular electrostatics: Continuum models and their growing pains. *Curr. Opin. Struct. Biol.*, 11:243–252, 2001.

[702] D. J. Tobias. Electrostatic calculations: Recent methodological advances and applications to membranes. *Curr. Opin. Struct. Biol.*, 11:253–261, 2001.

[703] J. D. Madura, M. E. Davis, M. K. Gilson, R. C. Wade, B. A. Luty, and J. A. McCammon. Biological applications of electrostatic calculations and Brownian dynamics simulations. In K. B. Lipkowitz and D. B. Boyd, editors, *Reviews in Computational Chemistry*, volume V, pages 229–267. VCH Publishers, New York, NY, 1994.

[704] B. Roux and T. Simonson. Implicit solvent models. *Biophys. Chem.*, 78:1–20, 1999.

[705] J. G. Kirkwood. Statistical mechanics of fluid mixtures. *J. Chem. Phys.*, 3:300–313, 1935.

[706] K. T. No, S. G. Kim, K.-H. Cho, and H. A. Scheraga. Description of hydration free energy density as a function of molecular physical properties. *Biophys. Chem.*, 78:127–145, 1999.

[707] N. V. Prabhu, J. S. Perkyns, H. D. Blatt, P. E. Smith, and B. M. Pettitt. Comparison of the potentials of mean force for alanine tetrapeptide between integral equation theory and simulation. *Biophys. Chem.*, 78:113–126, 1999.

[708] V. Lounnas, S. K. Lüdemann, and R. C. Wade. Towards molecular dynamics simulation of large proteins with a hydration shell at constant pressure. *Biophys. Chem.*, 78:157–182, 1999.

[709] S. Boresch, S. Ringhofer, P. Höchtl, and O. Steinhauser. Towards a better description and understanding of biomolecular solvation. *Biophys. Chem.*, 78:43–68, 1999.

[710] Y. N. Vorobjev and J. Hermans. ES/IS: Estimation of conformational free energy by combining dynamics simulations with explicit solvent with an implicit solvent continuum model. *Biophys. Chem.*, 78:195–205, 1999.

[711] D. Bashford and D. A. Case. Generalized Born models of macromolecular solvation effects. *Ann. Rev. Phys. Chem.*, 51:129–152, 2000.

[712] B. Honig and A. Nicholls. Classical electrostatics in biology and chemistry. *Science*, 268:1144–1149, 1995.

[713] C. N. Schutz and A. Warshel. What are the dielectric "constants" of proteins and how to validate electrostatic models? *Proteins: Struc. Func. Gen.*, 44:400–417, 2001.

[714] L. Y. Zhang, E. Gallicchino, R. A. Friesner, and R. M. Levy. Solvent models for protein-ligand binding: Comparison of implicit solvent Poisson and surface generalized Born models with explicit solvent simulations. *J. Comput. Chem.*, 22:591–607, 2001.

[715] R. W. Pastor. Techniques and applications of Langevin dynamics simulations. In G. R. Luckhurst and C. A. Veracini, editors, *The Molecular Dynamics of Liquid Crystals*, pages 85–138. Kluwer Academic, Dordrecht, The Netherlands, 1994.

[716] P. E. Rouse, Jr. Dilute solutions of coiling polymers. *J. Chem. Phys.*, 21:1272–1280, 1953.

[717] F. Bueche. The viscoelastic properties of plastics. *J. Chem. Phys.*, 22:603–609, 1954.

[718] B. H. Zimm. Dynamics of polymer molecules in dilute solution: Viscoelasticity, flow birefringence and dielectric loss. *J. Chem. Phys.*, 24:269–278, 1956.

[719] H. Yamakawa. *Modern Theory of Polymer Solutions*. Harper and Row Publishers, New York, NY, 1971.

[720] T. Schlick. Modeling superhelical DNA: Recent analytical and dynamic approaches. *Curr. Opin. Struct. Biol.*, 5:245–262, 1995.

[721] G. Chirico and J. Langowski. Brownian dynamics simulations of supercoiled DNA with bent sequences. *Biophys. J.*, 71:955–971, 1996.

[722] D. Beard and T. Schlick. Inertial stochastic dynamics: II. Influence of inertia on slow kinetic properties of supercoiled DNA. *J. Chem. Phys.*, 112:7323–7338, 2000.

[723] R. C. Wade, M. E. Davis, B. A. Luty, J. D. Madura, and J. A. McCammon. Gating of the active site of triose phosphate isomerase: Brownian dynamics simulations of flexible peptide loops in the enzyme. *Biophys. J.*, 64:9–15, 1993.

[724] R. C. Wade, B.A. Luty, E. Demchuk, J. D. Madura, M. E. Davis, J. M. Briggs, and J. A. McCammon. Simulation of enzyme-substrate encounter with gated active sites. *Struct. Biol.*, 1:65–69, 1994.

[725] P. Derreumaux and T. Schlick. Long-time integration for peptides by the dynamics driver approach. *Proteins: Struc. Func. Gen.*, 21:282–302, 1995.

[726] P. Derreumaux and T. Schlick. The loop opening/closing motion of the enzyme triosephosphate isomerase. *Biophys. J.*, 74:72–81, January 1998.

[727] R. Kubo. The fluctuation-dissipation theorem. *Rep. Prog. Phys.*, 29:255–284, 1966.

[728] A. Brünger, C. L. Brooks, III, and M. Karplus. Stochastic boundary conditions for molecular dynamics simulations of ST2 water. *Chem. Phys. Lett.*, 105:495–500, 1982.

[729] W. F. van Gunsteren and H. J. C. Berendsen. Algorithms for Brownian dynamics. *Mol. Phys.*, 45:637–647, 1982.

[730] Robert D. Skeel. Integration schemes for molecular dynamics and related applications. In M. Ainsworth, J. Levesley, and M. Marletta, editors, *The Graduate Student's Guide to Numerical Analysis*, volume 26 of *Springer Series in Computational Mathematics*, pages 119–176. Springer-Verlag, New York, NY, 1999.

[731] R. J. Loncharich, B. R. Brooks, and R. W. Pastor. Langevin dynamics of peptides: The frictional dependence of isomerization rates of N-acetylalanyl-N'-methylamide. *Biopolymers*, 32:523–535, 1992.

[732] G. Ramachandran and T. Schlick. Beyond optimization: Simulating the dynamics of supercoiled DNA by a macroscopic model. In P. M. Pardalos, D. Shalloway, and G. Xue, editors, *Global Minimization of Nonconvex Energy Functions: Molecular Conformation and Protein Folding*, volume 23 of *DIMACS Series in Discrete Mathematics and Theoretical Computer Science*, pages 215–231, Providence, Rhode Island, 1996. American Mathematical Society.

[733] D. L. Ermak and J. A. McCammon. Brownian dynamics with hydrodynamic interactions. *J. Chem. Phys.*, 69:1352–1360, 1978.

[734] D. Beard and T. Schlick. Inertial stochastic dynamics: I. Long-timestep methods for Langevin dynamics. *J. Chem. Phys.*, 112:7313–7322, 2000.

[735] M. Fixman. Construction of Langevin forces in the simulation of hydrodynamic interaction. *Macromolecules*, 19:1204–1207, 1986.

[736] C. E. Hecht. *Statistical Thermodynamics and Kinetic Theory*. W. H. Freeman, New York, NY, 1990.

[737] T. L. Hill. *An Introduction to Statistical Thermodynamics*. Dover, New York, NY, 1986.

[738] S. A. Rice, M. Nagasawa, and H. Morawetz. *Polyelectrolyte Solutions: A Theoretical Introduction*, volume 2 of *Molecular Biology: An International Series of Monographs and Textbooks*. Academic Press, New York, NY, 1961.

[739] H. S. Harned and B. B. Owen. *The Physical Chemistry of Electrolytic Solutions*. American Chemical Society Monograph Series. Reinhold Publishing Corporation, New York, NY, second edition, 1950.

[740] R. A. Robinson and R. H. Stokes. *Electrolyte Solutions: The Measurement and Interpretation of Conductance, Chemical Potential and Diffusion in Solutions of Simple Electrolytes*. Butterworth & Co., London, England, second edition, 1965.

[741] D. Stigter. Interactions of highly charged colloidal cylinders with applications to double-stranded DNA. *Biopolymers*, 16:1435–1448, 1977.

[742] M. Holst, N. Baker, and E. Wang. Adaptive multilevel finite element solution of the Poisson-Boltzmann equation I. Algorithms and examples. *J. Comput. Chem.*, 21:1319–1342, 2000.

[743] N. Baker, M. Holst, and E. Wang. Adaptive multilevel finite element solution of the Poisson-Boltzmann equation II. Refinement at solvent-accessible surfaces in biomolecular systems. *J. Comput. Chem.*, 21:1343–1352, 2000.

[744] I. A. Shkel, O. V. Tsodikov, and M. T. Record, Jr. Complete asymptotic solution of cylindrical and spherical Poisson-Boltzmann equations at experimental salt concentrations. *J. Phys. Chem. B*, 104:5161–5170, 2000.

[745] W. Rocchia, S. Sridharan, A. Nicholls, E. Alexov, A. Chiabrera, and B. Honig. Rapid grid-based construction of the molecular surface and the use of induced surface charge to calculate reaction field energies: Applications to the molecular systems and geometric objects. *J. Comput. Chem.*, 23:128–137, 2002.

[746] N. Baker, D. Sept, S. Joseph, M. J. Holst, and J. A. McCammon. Electrostatics of nanosystems: Application to microtubules and the ribosome. *Proc. Natl. Acad. Sci. USA*, 98:10037–10041, 2001.

[747] M. Deserno, C. Holm, and S. May. Fraction of condensed counterions around a charged rod: Comparison of Poisson-Boltzmann theory and computer simulations. *Macromolecules*, 33:199–206, 2000.

[748] H. Qian and J. A. Schellman. Transformed Poisson-Boltzmann relations and ionic distributions. *J. Phys. Chem. B*, 104:11528–11540, 2000.

[749] M. Friedrichs, R. Zhou, S. R. Edinger, and R. A. Friesner. Poisson-Boltzmann analytical gradients for molecular modeling calculations. *J. Phys. Chem. B*, 103:3057–3061, 1999.

[750] S. Tara, A. H. Elcock, P. D. Kirchhoff, J. M. Briggs, Z. Radic, P. Taylor, and J. A. McCammon. Rapid binding of a cationic active site inhibitor to wild type and mutant mouse acetylcholinesterase: Brownian dynamics simulation including diffusion in the active site gorge. *Biopolymers*, 46:465–474, 1998.

[751] W. Rocchia, E. Alexov, and B. Honig. Extending the applicability of the nonlinear Poisson-Boltzmann equation: Multiple dielectric constants and multivalent ions. *J. Phys. Chem. B*, 105:6507–6514, 2001.

[752] P. E. Gill, W. Murray, and M. H. Wright. *Practical Optimization*. Academic Press, London, England, 1983.

[753] J. E. Dennis, Jr. and R. B. Schnabel. *Numerical Methods for Unconstrained Optimization and Nonlinear Equations*. Prentice-Hall, Inc., Englewood Cliffs, New Jersey, 1983. (Reprinted by SIAM, 1996).

[754] R. Fletcher. *Practical Methods of Optimization*. John Wiley & Sons, Tiptree, Essex, Great Britain, second edition, 1987.

[755] D.G. Luenberger. *Linear and Nolinear Programming*. Addison Wesley, Reading, Massachusetts, 1984.

[756] J. Nocedal and S. Wright. *Numerical Optimization*. Springer Verlag, New York, NY, 1999.

[757] J. Nocedal. Theory of algorithms for unconstrained optimization. *Acta Numerica*, 1:199–242, 1992.

[758] T. Schlick. Optimization methods in computational chemistry. In K. B. Lipkowitz and D. B. Boyd, editors, *Reviews in Computational Chemistry*, volume III, pages 1–71. VCH Publishers, New York, NY, 1992.

[759] T. Schlick and G. Parks. DOE computational sciences education project, 1994. Chapter on Mathematical Optimization. URL: csep1.phy.ornl.gov/.

[760] T. Schlick. Geometry optimization. In P. von Ragué Schleyer (Editor-in Chief), N. L. Allinger, T. Clark, J. Gasteiger, P. A. Kollman, and H. F. Schaefer, III, editors, *Encyclopedia of Computational Chemistry*, volume 2, pages 1136–1157. John Wiley & Sons, West Sussex, England, 1998.

[761] M. H. Wright. What, if anything, is new in optimization? In J. M. Ball and J. C. R. Hunt, editors, *ICIAM'99: Proceedings of the 4th International Congress on Industrial and Applied Mathematics*, pages 259–270, Oxford, England, 2000. Oxford University Press. Also available in modified form as Technical Report 00-4-08, Bell Laboratories, Computing Sciences Research Center, Murray Hill, New Jersey 07074; cm.bell-labs.com/cm/cs/doc/00/4-08.ps.gz.

[762] J. M. Borwein and A. S. Lewis. *Convex Analysis and Nonlinear Optimization. Theory and Examples*, volume 3 of *Canadian Mathematical Society (CMS) Books in Mathematics*. Springer-Verlag, New York, NY, 2000.

[763] J.-B. Hiriart-Urruty and C. Lemaréchal. *Convex Analysis and Minimization. Algorithms I*, volume 305 of *Grundlehren der mathematischen Wissenschaften. A Series of Comprehensive Studies in Mathematics*. Springer-Verlag, Berlin and Heidelberg, 1993.

[764] J.-B. Hiriart-Urruty and C. Lemaréchal. *Convex Analysis and Minimization. Algorithms II*, volume 306 of *Grundlehren der mathematischen Wissenschaften. A Series of Comprehensive Studies in Mathematics*. Springer-Verlag, Berlin and Heidelberg, 1993.

[765] C. A. Floudas and P. Pardalos, editors. *Optimization in Computational Chemistry and Molecular Biology: Local and Global Approaches*. Kluwer Academic Publishers, Dordrecht, The Netherlands, 2000.

[766] L. Piela, J. Kostrowicki, and H. A. Scheraga. The multiple-minima problem in conformational analysis of molecules. deformation of the potential energy hypersurface by the diffusion equation method. *J. Phys. Chem.*, 93:3339–3346, 1989.

[767] J. Nocedal, A. Sartenaer, and C. Zhu. On the behavior of the gradient norm in the steepest descent method. Technical report, CERFACS, Toulouse, France (May 2000), 2000.

[768] D. Xie and T. Schlick. A more lenient stopping rule for line search algorithms. *Opt. Math. Sci.*, 2002. In Press.

[769] A. R. Conn, N. I. M. Gould, and Ph. L. Toint. LANCELOT*: A FORTRAN Package for Large-Scale Nonlinear Optimization (Release A)*, volume 17 of *Springer Series in Computational Mathematics*. Springer-Verlag, New York, NY, 1992.

[770] M. L. Overton. *Numerical Computing with IEEE Floating Point Arithmetic*. SIAM, Philadelphia, PA, 2001.

[771] T. Schlick and M. L. Overton. A powerful truncated Newton method for potential energy functions. *J. Comput. Chem.*, 8:1025–1039, 1987.

[772] T. Schlick and A. Fogelson. TNPACK — A truncated Newton minimization package for large-scale problems: I. Algorithm and usage. *ACM Trans. Math. Softw.*, 14:46–70, 1992.

[773] T. Schlick and A. Fogelson. TNPACK — A truncated Newton minimization package for large-scale problems: II. Implementation examples. *ACM Trans. Math. Softw.*, 14:71–111, 1992.

[774] R. B. Schnabel and T. Chow. Tensor methods for unconstrained optimization. *SIAM J. Opt.*, 1:293–315, 1991.

[775] R. H. Byrd, J. Nocedal, and C. Zhu. Towards a discrete Newton method with memory for large-scale optimization. In G. Di Pillo and F. Giannessi, editors, *Nonlinear Optimization and Applications*. Plenum, 1996.

[776] P. E. Gill and M. W. Leonard. Reduced-Hessian quasi-Newton methods for unconstrained optimization. *SIAM J. Optim.*, 12:209–237, 2001.

[777] J. Nocedal. Updating quasi-Newton matrices with limited storage. *Mathematics of Computation*, 35:773–782, 1980.

[778] D. C. Liu and J. Nocedal. On the limited memory BFGS method for large-scale optimization. *Math. Prog. B*, 45:503–528, 1989.

[779] J. C. Gilbert and C. Lemarechal. Some numerical experiments with variable-storage quasi-Newton algorithms. *Math. Prog. B*, 45:407–435, 1989.

[780] J. Nocedal. Large-scale unconstrained optimization. In A. Watson and I. Duff, editors, *The State of the Art in Numerical Analysis*, pages 311–338. Oxford University Press, 1997.

[781] S. G. Nash and J. Nocedal. A numerical study of the limited memory BFGS method and the truncated-Newton method for large-scale optimization. *SIAM J. Opt.*, 1:358–372, 1991.

[782] R. H. Byrd, J. Nocedal, and R. B. Schnabel. Representations of quasi-Newton matrices and their use in limited memory methods. *Math. Prog.*, 63:129–156, 1994.

[783] R. H. Byrd, P. Lu, and J. Nocedal. A limited memory algorithm for bound constrained optimization. *SIAM J. Sci. Stat. Comput.*, 16:1190–1208, 1995.

[784] C. Zhu, R. H. Byrd, P. Lu, and J. Nocedal. Algorithm 778: L-BFGS-B, FORTRAN subroutines for large scale bound constrained optimization. *ACM Trans. Math. Softw.*, 23:550–560, 1997.

[785] J. L. Morales and J. Nocedal. Automatic preconditioning by limited memory quasi-Newton updating. *SIAM J. Opt.*, 10:1079–1096, 2000. (also Technical Report 97/07, Optimization Technology Center, Northwestern University, 1997).

[786] G. H. Golub and C. F. van Loan. *Matrix Computations*. John Hopkins University Press, Baltimore, MD, second edition, 1986.

[787] M. J. D. Powell. Restart procedures for the conjugate gradient method. *Math. Prog.*, 12:241–254, 1977.

[788] D. F. Shanno and K. H. Phua. Remark on Algorithm 500: Minimization of unconstrained multivariate functions. *ACM Trans. Math. Softw.*, 6:618–622, 1980.

[789] J. C. Gilbert and J. Nocedal. Global convergence properties of conjugate gradient methods for optimization. Technical Report 1268, Institut National de Recherche en Informatique et en Automatique, January 1991.

[790] L. Adams and J. L. Nazareth, editors. *Linear and Nonlinear Conjugate Gradient-Related Methods*. SIAM, Philadelphia, PA, 1996.

[791] R. S. Dembo and T. Steihaug. Truncated-Newton algorithms for large-scale unconstrained optimization. *Math. Prog.*, 26:190–212, 1983.

[792] P. Derreumaux, G. Zhang, B. Brooks, and T. Schlick. A truncated-Newton method adapted for CHARMM and biomolecular applications. *J. Comput. Chem.*, 15:532–552, 1994.

[793] D. Xie and T. Schlick. Efficient implementation of the truncated Newton method for large-scale chemistry applications. *SIAM J. Opt.*, 10(1):132–154, 1999.

[794] J. J. Moré and S. J. Wright. *Optimization Software Guide*, volume 14 of *Frontiers in Applied Mathematics*. SIAM, Philadelphia, PA, 1993. See www.mcs.anl.gov/otc/Guide/ and www.mcs.anl.gov/otc/Guide/SoftwareGuide/ for updated information on the software guide.

[795] B. E. Hingerty, S. Figueroa, T. L. Hayden, and S. Broyde. Prediction of DNA structure from sequence: A buildup technique. *Biopolymers*, 8:1195–1222, 1989.

[796] S. Broyde and B. E. Hingerty. Effective computational strategies for determining structures of carcinogen damaged DNA. *J. Comput. Phys.*, 151:313–332, 1999.

[797] D. Xie and T. Schlick. Remark on the updated truncated Newton minimization package, *Algorithm 702. ACM Trans. Math. Softw.*, 25(1):108–122, 1999.

[798] K. B. Lipkowitz. Abuses of molecular mechanics. Pitfalls to avoid. *J. Chem. Educ.*, 72:1070–1075, 1995.

[799] A. Griewank and G. F. Corliss, editors. *Automatic Differentiation of Algorithms: Theory, Implementation, and Applications*. SIAM, Philadelphia, PA, 1991.

[800] P. M. Pardalos, D. Shalloway, and G. Xue, editors. *Global Minimization of Nonconvex Energy Functions: Molecular Conformation and Protein Folding*, volume 23 of *DIMACS Series in Discrete Mathematics and Theoretical Computer Science*. American Mathematical Society, Providence, Rhode Island, 1996.

[801] J. Pillardy, C. Czaplewiski, A. Liwo, J. Lee, D. R. Ripoll, R. Kaźmierkiewicz, S. Oldziej, W. J. Wedemeyer, K. D. Gibson, Y. A. Arnautova, J. Saunders, Y.-J. Ye, and H. A. Scheraga. Recent improvements in prediction of protein structure by global optimization of a potential energy function. *Proc. Natl. Acad. Sci. USA*, 98:2329–2333, 2001.

[802] N. Metropolis, A. W. Rosenbluth, M. N. Rosenbluth, A. H. Teller, and E. Teller. Equation of state calculations by fast computing machines. *J. Chem. Phys.*, 21:1087–1092, 1953.

[803] M. H. Kalos and P. A. Whitlock. *Monte Carlo Methods. Volume I: Basics*. John Wiley & Sons, New York, NY, 1986.

[804] J. A. González and R. Pino. A random number generator based on unpredictable chaotic functions. *Comput. Phys. Comm.*, 120:109–114, 1999.

[805] D. E. Knuth. *The Art of Computer Programming. Volume 2: Seminumerical Methods*. Addison-Wesley, Reading, Massachusetts, second edition, 1981.

[806] A. M. Law and W. D. Kelton. *Simulation Modeling and Analysis*. McGraw-Hill Series in Industrial Engineering and Management Science. McGraw-Hill, Boston, MA, third edition, 2000.

[807] P. L'Ecuyer. Uniform random number generation. *Ann. Oper. Res.*, 53:77–120, 1994.

[808] P. L'Ecuyer. Random number generation. In J. Banks, editor, *Handbook on Simulation*, chapter 4, pages 93–137. John Wiley & Sons, New York, NY, 1998.

[809] M. Mascagni. Some methods of parallel pseudorandom number generation. In M. T. Heath, A. Ranade, and R S. Schreiber, editors, *Algorithms for Parallel Processing*, volume 105 of *IMA Volumes in Mathematics and Its Applications*, pages 277–288. Springer-Verlag, New York, NY, 1999.

[810] P. L'Ecuyer. Good parameter sets for combined multiple recursive random number generators. *Oper. Res.*, 47:159–164, 1999.

[811] S. L. Anderson. Random number generators on vector supercomputers and other advanced architectures. *SIAM Rev.*, 32:221–251, 1990.

[812] S. C. Park and K. W. Miller. Random number generators: Good ones are hard to find. *Comm. ACM*, 31:1192–1201, 1988.

[813] O. E. Percus and J. K. Percus. Intrinsic relations in the structure of linear congruential generators modulo 2^{β}. *Stat. Prob. Let.*, 15:381–383, 1992.

[814] B. L. Holian, O. E. Percus, T. T. Warnock, and P. A. Whitlock. Pseudorandom number generator for massively parallel molecular-dynamics applications. *Phys. Rev. E*, 50:1607–1615, 1994.

[815] G. Marsaglia and L.-H Tsay. Matrices and the structure of random number sequences. *Lin. Alg. Appl.*, 67:147–156, 1985.

[816] P. L'Ecuyer. Bad lattice structures for vectors of non-successive values produced by some linear recurrences. *Informs J. Comput.*, 9:57–60, 1997.

[817] P. L'Ecuyer. Maximally-equidistributed combined Tausworthe generators. *Math. Comput.*, 65:203–213, 1996.

[818] P. L'Ecuyer. Tables of maximally-equidistributed combined LFSR generators. *Math. Comput.*, 68:261–269, 1999.

[819] P. L'Ecuyer and T. H. Andres. A random number generator based on the combination of four LCGs. *Mathematics and Computers in Simulation*, 44:99–107, 1997.

[820] A. M. Ferrenberg, D. P. Landau, and Y. J. Wong. Monte Carlo simulations: Hidden errors from "good" random number generators. *Phys. Rev. Lett.*, 69:3382–3384, 1992.

[821] F. J. Resende and B. V. Costa. Using random number generators in Monte Carlo simulations. *Phys. Rev. E*, 58:5183–5184, 1998.

[822] O. E. Percus and M. H. Kalos. Random number generators for MIMD parallel processors. *J. Paral. Dist. Comput.*, 6:477–497, 1989.

[823] R. E. Odeh and J. O. Evans. The percentage points of the normal distribution. *App. Stat.*, 23:96–97, 1974.

[824] Z. Li and H. A. Scheraga. Monte Carlo-minimization approach to the multiple-minima problem in protein folding. *Proc. Natl. Acad. Sci. USA*, 84:6611–6615, 1987.

[825] V. A. Eyrich, D. M. Standley, and R. A. Friesner. Prediction of protein tertiary structure to low resolution: Performance for a large and structurally diverse test set. *J. Mol. Biol.*, 14:725–742, 1999.

[826] S. Duane, A. D. Kennedy, B. J. Pendleton, and D. Roweth. Hybrid Monte Carlo. *Phys. Lett. B*, 195:216–222, 1987.

[827] P. G. Bolhuis, D. Chandler, C. Dellago, and P. L. Geissler. Transition path sampling: Throwing ropes over rough mountain passes, in the dark. *Annu. Rev. Phys. Chem*, 53:291–318, 2002.

[828] B. J. Berne and J. E. Straub. Novel methods of sampling phase space in the simulation of biological systems. *Curr. Opin. Struct. Biol.*, 7:181–189, 1997.

[829] A. Fischer, F. Cordes, and C. Schütte. Hybrid Monte Carlo with adaptive temperature in mixed-canonical ensemble: Efficient conformational analysis of RNA. *J. Comput. Chem.*, 19:1689–1697, 1998.

[830] H. Senderowitz and W. C. Still. MC(JBW): Simple but smart Monte Carlo algorithm for free energy simulations of multiconformational molecules. *J. Comput. Chem.*, 19:1736–1745, 1998.

[831] L. J. LaBerge and J. C. Tully. A rigorous procedure for combining molecular dynamics and Monte Carlo simulation algorithms. *Chem. Phys.*, 260:183–191, 2000.

[832] J. B. Clarage, T. Romo, B. K. Andrews, B. M. Pettitt, and G. N. Philipps, Jr. A sampling problem in molecular dynamics simulations of macromolecules. *Proc. Natl. Acad. Sci. USA*, 92:3288–3292, 1995.

[833] W. L. Jorgensen and J. Tirado-Rives. Monte Carlo vs. molecular dynamics for conformational sampling. *J. Phys. Chem.*, 100:14508–14513, 1996.

[834] M. Gerstein and M. Levitt. Simulating water and the molecules of life. *Sci. Amer.*, 279(5):101–105, November 1998.

[835] P. Ferrara, J. Apostolakis, and A. Caflisch. Targeted molecular dynamics simulations of protein unfolding. *J. Phys. Chem. B.*, 104:4511–4518, 2000.

[836] R. Elber, J. Meller, and R. Olender. Stochastic path approach to compute atomically detailed trajectories: Application to the folding of C peptide. *J. Phys. Chem. B*, 103:899–911, 1999.

[837] J. M. Sanz-Serna and M. P. Calvo. *Numerical Hamiltonian Problems*. Chapman & Hall, London, England, 1994.

[838] D. Strahs, X. Qian, D. Barash, and T. Schlick. Sequence-dependent solution structure of 13 TATA/TBP complexes. 2002. Submitted.

[839] H. H. Gan, A. Tropsha, and T. Schlick. Lattice protein folding with two and four-body statistical potentials. *Proteins: Struc. Func. Gen.*, 43:161–174, 2001.

[840] A. H. Elcock, R. R. Gabdoulline, R. C. Wade, and J. A. McCammon. Computer simulation of protein-protein association kinetics: Acetylcholinesterase-Fasciculin. *J. Mol. Biol.*, 291:149–162, 1999.

[841] D. J. Patel, B. Mao, Z. Gu, B. E. Hingerty, A. Gorin, A. K. Basu, and S. Broyde. Nuclear magnetic resonance solution structures of covalent aromatic amine-DNA adducts and their magnetic relevance. *Chem. Res. Toxic.*, 11:391–407, 1998.

[842] P. S. de Laplace. *Oeuvres Complètes de Laplace. Théorie Analytique des Probabilités*, volume VII. Gauthier-Villars, Paris, France, third edition, 1820.

[843] J. Gao. Methods and applications of combined quantum mechanical and molecular mechanical potentials. In K. B. Lipkowitz and D. B. Boyd, editors, *Reviews in Computational Chemistry*, volume 7, pages 119–185. VCH Publishers, New York, NY, 1996.

[844] D. Backowies and E. Thiel. Hybrid models for combined quantum mechanical and molecular mechanical approaches. *J. Phys. Chem.*, 100:10580–10594, 1996.

[845] C. A. Schiffer, J. W. Caldwell, P. A. Kollman, and R. M. Stroud. Protein structure prediction with a combined solvation free energy-molecular mechanics force field. *Mol. Sim.*, 10:121–149, 1993.

[846] J. R. Maple, M.-J. Hwang, T. P. Stockfisch, U. Dinur, M. Waldman, C. S. Ewing, and A. T. Hagler. Derivation of class II force fields. I. Methodology and quantum force field for the alkyl functional group and alkane molecules. *J. Comput. Chem.*, 15:162–182, 1994.

[847] P. Derreumaux and G. Vergoten. Influence of the spectroscopic potential energy function SPASIBA on molecular dynamics of proteins: Comparison with the AMBER potential. *J. Mol. Struct.*, 286:55–64, 1993.

[848] J. L. McCauley. *Chaos, Dynamics, and Fractals: an Algorithmic Approach to Deterministic Chaos*. Cambridge University Press, Cambridge, 1994.

[849] X. Daura, B. Oliva, E. Querol, F. X. Avilés, and O. Tapia. On the sensitivity of MD trajectories to changes in water-protein interaction parameters: The potato carboxypeptidase inhibitor in water as a test case for the GROMOS force field. *Proteins: Struc. Func. Gen.*, 25:89–103, 1996.

[850] A. Elofsson and L. Nilsson. How consistent are molecular dynamics simulations? Comparing structure and dynamics in reduced and oxidized $Escherichia\ \ coli$ Thioredoxin. *J. Mol. Biol.*, 233:766–780, 1991.

[851] P. Deuflhard, M. Dellnitz, O. Junge, and Ch. Schütte. Computation of essential molecular dynamics by subdivision techniques: I. Basic concepts. Technical Report SC 96–45, Konrad-Zuse-Zentrum für Informationstechnik Berlin, Takustraβe 7, D-14195, Berlin-Dahlem, December 1996.

[852] H. Grubmüller, H. Heller, A. Windemuth, and K. Schulten. Generalized Verlet algorithm for efficient molecular dynamics simulations with long-range interactions. *Mol. Sim.*, 6:121–142, 1991.

[853] J. A. Board, Jr., L. V. Kalé, K. Schulten, R. D. Skeel, and T. Schlick. Modeling biomolecules: Larger scales, longer durations. *IEEE Comput. Sci. Eng.*, 1:19–30, Winter 1994.

[854] H. Frauenfelder and P. G. Wolynes. Biomolecules: Where the physics of complexity and simplicity meet. *Phys. Today*, 47:58–64, 1994.

[855] H. Frauenfelder, S. G. Sligar, and P. G. Wolynes. The energy landscapes and motions of proteins. *Science*, 254:1598–1603, 1991.

[856] P. G. Wolynes, J. N. Onuchic, and D. Thirumalai. Navigating the folding routes. *Science*, 267:1619–1620, 1995.

[857] T. Schlick. Time-trimming tricks for dynamic simulations: Splitting force updates to reduce computational work. *Structure*, 9:R45–R53, 2001.

[858] R. W. Pastor, B. R. Brooks, and A. Szabo. An analysis of the accuracy of Langevin and molecular dynamics algorithms. *Mol. Phys.*, 65:1409–1419, 1988.

[859] G. Zhang and T. Schlick. Implicit discretization schemes for Langevin dynamics. *Mol. Phys.*, 84:1077–1098, 1995.

[860] B. Mishra and T. Schlick. The notion of error in Langevin dynamics: 1. Linear analysis. *J. Chem. Phys.*, 105:299–318, 1996.

[861] T. Schlick, E. Barth, and M. Mandziuk. Biomolecular dynamics at long timesteps: Bridging the timescale gap between simulation and experimentation. *Ann. Rev. Biophys. Biomol. Struc.*, 26:179–220, 1997.

[862] W. F. van Gunsteren. Constrained dynamics of flexible molecules. *Mol. Phys.*, 40:1015–1019, 1980.

[863] W. F. van Gunsteren and H. J. C. Berendsen. Algorithms for macromolecular dynamics and constraint dynamics. *Mol. Phys.*, 34:1311–1327, 1977.

[864] S. Miyamoto and P. A. Kollman. SETTLE: An analytical version of the SHAKE and RATTLE algorithm for rigid water models. *J. Comput. Chem.*, 13:952–962, 1992.

[865] W. F. van Gunsteren and M. Karplus. Effect of constraints on the dynamics of macromolecules. *Macromolecules*, 15:1528–1543, 1982.

[866] M. Watanabe and M. Karplus. Simulations of macromolecules by multiple time-step methods. *J. Phys. Chem.*, 99:5680–5697, 1995.

[867] J. J. Biesiadecki and R. D. Skeel. Dangers of multiple-time-step methods. *J. Comput. Phys.*, 109:318–328, 1993.

[868] E. Barth and T. Schlick. Extrapolation versus impulse in multiple-timestepping schemes: II. Linear analysis and applications to Newtonian and Langevin dynamics. *J. Chem. Phys.*, 109:1632–1642, 1998.

[869] D. E. Humphreys, R. A. Friesner, and B. J. Berne. A multiple-time-step molecular dynamics algorithm for macromolecules. *J. Phys. Chem.*, 98(27):6885–6892, 1994.

[870] R. Zhou and B. J. Berne. A new molecular dynamics method combining the reference system propagator algorithm with a fast multipole method for simulating proteins and other complex systems. *J. Chem. Phys.*, 103:9444–9459, 1995.

[871] C. Störmer. Sur les trajectoires des corpuscules électrisés dans l'espace. *Archives des Sciences Physiques et Naturelles*, 24:5–18, 113–158, 221–247, 1907. (This reference is the first in a three-part essay. The second part appeared in the same journal in 1911 [pages 190, 277, 415, and 501], and the third part appeared in the 1912 volume of the journal, pages 51–69).

[872] L. Verlet. Computer 'experiments' on classical fluids: I. Thermodynamical properties of Lennard-Jones molecules. *Phys. Rev.*, 159(1):98–103, July 1967.

[873] R. W. Hockney and J. W Eastwood. *Computer Simulation Using Particles*. McGraw-Hill, New York, NY, 1981.

[874] W. C. Swope, H. C. Andersen, P. H. Berens, and K. R. Wilson. A computer simulation method for the calculation of equilibrium constants for the formation of physical clusters of molecules: Applications to small water clusters. *J. Chem. Phys.*, 76:637–649, 1982.

[875] M. E. Tuckerman, B. J. Berne, and G. J. Martyna. Reversible multiple time scale molecular dynamics. *J. Chem. Phys.*, 97:1990–2001, 1992.

[876] E. Barth and T. Schlick. Overcoming stability limitations in biomolecular dynamics: I. Combining force splitting via extrapolation with Langevin dynamics in LN. *J. Chem. Phys.*, 109:1617–1632, 1998.

[877] P. Batcho and T. Schlick. Special stability advantages of position Verlet over velocity Verlet in multiple-timestep integration. *J. Chem. Phys.*, 115:4019–4029, 2001.

[878] R. D. Skeel, G. Zhang, and T. Schlick. A family of symplectic integrators: Stability, accuracy, and molecular dynamics applications. *SIAM J. Sci. Comput.*, 18(1):202–222, January 1997.

[879] H. C. Andersen. Rattle: a 'velocity' version of the SHAKE algorithm for molecular dynamics calculations. *J. Comput. Phys.*, 52:24–34, 1983.

[880] D. J. Tobias and C. L. Brooks, III. Conformational equilibrium in the alanine dipeptide in the gas phase and aqueous solution: A comparison of theoretical results. *J. Chem. Phys.*, 89:5115–5126, 1988.

[881] B. Leimkuhler and R. D. Skeel. Symplectic numerical integrators in constrained Hamiltonian systems. *J. Comput. Phys.*, 112:117–125, 1994.

[882] E. Barth, K. Kuczera, B. Leimkuhler, and R. D. Skeel. Algorithms for constrained molecular dynamics. *J. Comput. Chem.*, 16:1192–1209, 1995.

[883] D. Xie, L. R. Scott, and T. Schlick. Analysis of the SHAKE-SOR algorithm for constrained molecular dynamics simulations. *Methods and Applications of Analysis*, 7(3):577–590, 2000. (Special Issue dedicated to Cathleen Morawetz).

[884] S. Toxvaerd. Comment on constrained molecular dynamics of macromolecules. *J. Chem. Phys.*, 87:6140–6143, 1987.

[885] E. Barth, M. Mandziuk, and T. Schlick. A separating framework for increasing the timestep in molecular dynamics. In W. F. van Gunsteren, P. K. Weiner, and A. J. Wilkinson, editors, *Computer Simulation of Biomolecular Systems: Theoretical and Experimental Applications*, volume III, chapter 4, pages 97–121. ESCOM, Leiden, The Netherlands, 1997.

[886] G. J. Martyna, M. E. Tuckerman, D. J. Tobias, and M. L. Klein. Explicit reversible integrators for extended systems dynamics. *Mol. Phys.*, 87:1117–1157, 1996.

[887] H. C. Andersen. Molecular dynamics simulations at constant pressure and/or temperature. *J. Chem. Phys.*, 72:2384–2393, 1980.

[888] M. Parrinello and A. Rahman. Crystal structure and pair potentials: A molecular-dynamics study. *Phys. Rev. Lett.*, 45:1196–1199, 1980.

[889] S. Nosé. Constant temperature molecular dynamics methods. *Prog. Theor. Phys. Suppl.*, 103:1–46, 1991.

[890] W. Hoover. Classical dynamics: Equilibrium phase-space distributions. *Phys. Rev. A*, 31:1695–1697, 1985.

[891] G. J. Martyna, A. Hughes, and M. E. Tuckerman. Molecular dynamics algorithms for path integrals at constant pressure. *J. Chem. Phys.*, 110:3275–3290, 1999.

[892] S. C. Harvey, R. K.-Z. Tan, and III T. E. Cheatham. The flying ice cube: Velocity rescaling in molecular dynamics leads to violation of energy equipartition. *J. Comput. Chem.*, 19:726–740, 1998.

[893] H. J. C. Berendsen, J. P. M. Postma, W. F. van Gunsteren, A. DiNola, and J. R. Haak. Molecular dynamics with coupling to an external bath. *J. Chem. Phys.*, 81:3684–3690, 1984.

[894] S. E. Feller, Y. Zhang, R. W. Pastor, and B. R. Brooks. Constant pressure molecular dynamics simulation: The Langevin piston method. *J. Chem. Phys.*, 103:4613–4621, 1995.

[895] S. Nosé. A molecular dynamics method for simulations in the canonical ensemble. *Mol. Phys.*, 52:255–268, 1984.

[896] G. Dahlquist and Å. Björck. *Numerical Methods*. Prentice Hall, Englewood Cliffs, New Jersey, 1974.

[897] L. Eldén and L. Wittmeyer-Koch. *Numerical Analysis*. Academic Press, Inc., San Diego, CA, 1990.

[898] R. D. Ruth. A canonical integration technique. *IEEE Trans. Nucl. Sci.*, 30:2669–2671, 1983.

[899] M. Mandziuk and T. Schlick. Resonance in the dynamics of chemical systems simulated by the implicit-midpoint scheme. *Chem. Phys. Lett.*, 237:525–535, 1995.

[900] T. Schlick, M. Mandziuk, R.D. Skeel, and K. Srinivas. Nonlinear resonance artifacts in molecular dynamics simulations. *J. Comput. Phys.*, 139:1–29, 1998.

[901] W. B. Streett, D. J. Tildesley, and G. Saville. Multiple time step methods in molecular dynamics. *Mol. Phys.*, 35:639–648, 1978.

[902] R. D. Swindoll and J. M. Haile. A multiple time-step method for molecular dynamics simulations of fluids of chain molecules. *J. Chem. Phys.*, 53:289–298, 1984.

[903] W. B. Streett, D. J. Tildesley, and G. Saville. Multiple time step methods and an improved potential function for molecular dynamics simulations of molecular liquids. In Peter Lykos, editor, *Computer Modeling of Matter*, volume 86 of *ACS Symposium Series*, pages 144–158. ACS, Washington, D. C., 1978.

[904] M. E. Tuckerman, B.J. Berne, and G. J. Martyna. Molecular dynamics algorithm for multiple time scales: Systems with long range forces. *J. Chem. Phys.*, 94:6811–6815, 1991.

[905] M. E. Tuckerman, B. J. Berne, and A. Rossi. Molecular dynamics algorithm for multiple time scales: Systems with disparate masses. *J. Chem. Phys.*, 94:1465–1469, 1991.

[906] M. E. Tuckerman and B. J. Berne. Molecular dynamics in systems with multiple time scales: Systems with stiff and soft degrees of freedom and with short and long range forces. *J. Comput. Chem.*, 95:8362–8364, 1992.

[907] B. García-Archilla, J.M. Sanz-Serna, and R.D. Skeel. Long-time-step methods for oscillatory differential equations. *SIAM J. Sci. Comput.*, 20:930–963, 1998.

[908] X. Qian. *Biomolecular Structure and Dynamics: Algorithm Development and Applications to DNA-Transcription and Promoter Elements*. PhD thesis, New York University, Department of Chemistry, New York, NY, May 2002.

[909] J. A. Izaguirre. *Longer Time Steps for Molecular Dynamics*. PhD thesis, University of Illinois at Urbana-Champaign, 1999. Also UIUC Technical Report UIUCDCS-R-99-2107. Available via www.cs.uiuc.edu/research/tech-reports.html.

[910] J. A. Izaguirre, D. P. Catarello, J. M. Wozniak, and R. D. Skeel. Langevin stabilization of molecular dynamics. *J. Chem. Phys.*, 114:2090–2098, 2001.

[911] P. Koehl and M. Levitt. Theory and simulation: Can theory challenge experiment? *Curr. Opin. Struct. Biol.*, 9:155–156, 1999.

[912] S. Doniach and P. Eastman. Protein dynamics simulations from nanoseconds to microseconds. *Curr. Opin. Struct. Biol.*, 9:157–163, 1999.

[913] T. Schlick. Computational molecular biophysics today: A confluence of methodological advances and complex biomolecular applications. *J. Comput. Phys.*, 151:1–8, May 1999. (Special Volume on Computational Biophysics).

[914] T. Simonson. Accurate calculation of the dielectric constant of water from simulations of a microscopic droplet in vacuum. *Chem. Phys. Lett.*, 250:450–454, 1996.

[915] D. Beglov and B. Roux. Finite representation of an infinite bulk system: Solvent boundary potential for computer simulations. *J. Chem. Phys.*, 100:9050–9063, 1994.

[916] D. Beglov and B. Roux. Dominant solvations effects from the primary shell of hydration: Approximation for molecular dynamics simulations. *Biopolymers*, 35:171–178, 1994.

[917] D. Beglov and B. Roux. Numerical solutions of the hypernetted chain equation for a solute of arbitrary geometry in three dimensions. *J. Chem. Phys.*, 103:360–364, 1995.

[918] M. H. Hao, M. R. Pincus, S. Rackovsky, and H. A. Scheraga. Unfolding and refolding of the native structure of bovine pancreatic trypsin inhibitor studied by computer simulations. *Biochemistry*, 32:9614–9631, 1993.

[919] D. K. Klimov and D. Thirumalai. Viscosity dependence of the folding rates of proteins. *Phys. Rev. Lett.*, 79:317–320, 1997.

[920] C. S. Peskin and T. Schlick. Molecular dynamics by the backward Euler's method. *Comm. Pure App. Math.*, 42:1001–1031, 1989.

[921] T. Schlick and C. S. Peskin. Can classical equations simulate quantum-mechanical behavior? A molecular dynamics investigation of a diatomic molecule with a Morse potential. *Comm. Pure App. Math.*, 42:1141–1163, 1989.

[922] G. Zhang and T. Schlick. LIN: A new algorithm combining implicit integration and normal mode techniques for molecular dynamics. *J. Comput. Chem.*, 14:1212–1233, 1993.

[923] G. Zhang and T. Schlick. The Langevin/implicit-Euler/Normal-Mode scheme (LIN) for molecular dynamics at large time steps. *J. Chem. Phys.*, 101:4995–5012, 1994.

[924] D. A. Case. Normal mode analysis of protein dynamics. *Curr. Opin. Struc. Biol.*, 4:385–290, 1994.

[925] T. Schlick and L. Yang. Long-timestep biomolecular dynamics simulations: LN performance on a polymerase β / DNA system. In A. Brandt, J. Bernholc, and K. Binder, editors, *Multiscale Computational Methods in Chemistry and Physics*, volume 177 of *NATO Science Series. Series III: Computer and Systems Sciences*, pages 293–305, Amsterdam, The Netherlands, 2001. IOS Press.

[926] M. E. Tuckerman, G. J. Martyna, and B. J. Berne. Molecular dynamics algorithm for condensed systems with multiple time scales. *J. Chem. Phys.*, 93:1287–1291, 1990.

[927] B. Space, H. Rabitz, and A. Askar. Long time scale molecular dynamics subspace integration method applied to anharmonic crystals and glasses. *J. Chem. Phys.*, 99:9070–9079, 1993.

[928] A. Askar, B. Space, and H. Rabitz. Subspace method for long time scale molecular dynamics. *J. Phys. Chem.*, 99:7330–7338, 1995.

[929] D. Janežič and F. Merzel. An efficient symplectic integration algorithm for molecular dynamics simulations. *J. Chem. Info. Comput. Sci.*, 35:321–326, 1995.

[930] R. Zhou, E. Harder, H. Xu, and B. J. Berne. Efficient multiple time step method for use with Ewald and particle mesh Ewald for large biomolecular systems. *J. Chem. Phys.*, 115:2348–2358, 2001.

[931] J. Rotne and S. Prager. Variational treatment of hydrodynamic interaction in polymers. *J. Chem. Phys.*, 50:4831–4837, 1969.

[932] M. Kröger, A. Alba-Perez, M. Laso, and H. C. Öttinger. Variance reduced Brownian simulation of a bead-spring chain under steady shear flow considering hydrodynamic interaction effects. *J. Chem. Phys.*, 113:4767–4773, 2000.

[933] R. M. Jendrejack, M. D. Graham, and J. J. de Pablo. Hydrodynamic interactions in long chain polymers: Application of the Chebyshev polynomial approximation in stochastic simulations. *J. Chem. Phys.*, 113:2894–2900, 2000.

[934] C. W. Gear. *Numerical Initial Value Problems in Ordinary Differential Equations*. Prentice Hall, Englewood Cliffs, New Jersey, 1971.

[935] E. Hairer and G. Wanner. *Solving Ordinary Differential Equations II. Stiff and Differential-Algebraic Problems*, volume 14 of *Springer Series in Computational Mathematics*. Springer-Verlag, New York, NY, second edition, 1996.

[936] D. Janežič and B. Orel. Implicit Runge-Kutta method for molecular dynamics integration. *J. Chem. Info. Comput. Sci.*, 33:252–257, 1993.

[937] T. Schlick. Pursuing Laplace's vision on modern computers. In J. P. Mesirov, K. Schulten, and D. W. Sumners, editors, *Mathematical Applications to Biomolecular Structure and Dynamics*, volume 82 of *IMA Volumes in Mathematics and Its Applications*, pages 219–247, New York, NY, 1996. Springer-Verlag.

[938] A. Nyberg and T. Schlick. Increasing the time step in molecular dynamics. *Chem. Phys. Lett.*, 198:538–546, 1992.

[939] T. Schlick and W. K. Olson. Trefoil knotting revealed by molecular dynamics simulations of supercoiled DNA. *Science*, 257:1110–1115, 1992.

[940] F. Kang. The Hamiltonian way for computing Hamiltonian dynamics. In R. Spigler, editor, *Applied and Industrial Mathematics*, pages 17–35. Kluwer Academic, Dordrecht, The Netherlands, 1990.

[941] J. C. Simo, N. Tarnow, and K. K. Wang. Exact energy-momentum conserving algorithms and symplectic schemes for nonlinear dynamics. *Comput. Meth. App. Mech. Engin.*, 100:63–116, 1994.

[942] J. C. Simo and N. Tarnow. The discrete energy-momentum method. Conserving algorithms for nonlinear elastodynamics. *Z. Angew. Math. Phys.*, 43:757–793, 1992.

[943] O. Gonzales and J. C. Simo. On the stability of symplectic and energy-momentum conserving algorithms for nonlinear Hamiltonian systems with symmetry. *Comput. Meth. App. Mech. Engin.*, 134:197, 1994.

[944] J. A. McCammon, B. M. Pettitt, and L. R. Scott. Ordinary differential equations of molecular dynamics. *Computers Math. Applic.*, 28:319–326, 1994.

[945] M. Zhang and R. D. Skeel. Cheap implicit symplectic integrators. *Appl. Num. Math.*, 25:297–302, 1997.

[946] A. Amadei, A. B. M. Linssen, and H. J. C. Berendsen. Essential dynamics of proteins. *Proteins: Struc. Func. Gen.*, 17:412–425, 1993.

[947] A. Amadei, A. B. M. Linssen, B. L. deGroot, D. M. F. van Aalten, and H. J. C. Berendsen. An efficient method for sampling the essential subspace of proteins. *J. Biomol. Struct. Dynam.*, 13:615–625, 1996.

[948] A. Kitao and N. Go. Investigating protein dynamics in collective coordinate space. *Curr. Opin. Struct. Biol.*, 9:164–169, 1999.

[949] G. Zou, R. D. Skeel, and S. Subramanian. Biased Brownian dynamics for rate constant calculation. *Biophys. J.*, 79:638–645, 2000.

[950] B. Isralewitz, M. Gao, and K. Schulten. Steered molecular dynamics and mechanical functions of proteins. *Curr. Opin. Struct. Biol.*, 11:224–230, 2001.

[951] B. Isralewitz, J. Baudry, J. Gullingsrud, D. Kosztin, and K. Schulten. Steered molecular dynamics investigations of protein function. *J. Mol. Graph. Model.*, 19:13–25, 2001.

[952] X. Wu and S. Wang. Self-guided molecular dynamics simulation for efficient conformational search. *J. Phys. Chem. B*, 102:7238–7250, 1998.

[953] X. Wu and S. Wang. Enhancing systematic motion in molecular dynamics simulation. *J. Chem. Phys.*, 110:9401–9410, 1999.

[954] X. Wu, S. Wang, and B. R. Brooks. Direct observation of the folding and unfolding of β-hairpin in explicit water through computer simulation. *J. Amer. Chem. Soc.*, 124:5282–5283, 2002.

[955] M. R. Sawaya, R. Prasad, S. H. Wilson, J. Kraut, and H. Pelletier. Crystal structures of human DNA polymerase β complexed with gapped and nicked DNA: Evidence for induced fit mechanism. *Biochemistry*, 36:11205–11215, 1997.

[956] M. A. Young, S. Gonfloni, G. Superti-Furga, B. Roux, and J. Kuriyan. Dynamic coupling between SH2 and SH3 domains of c-Src and Hck underlies their inactivation by C-terminal tyrosine phosphorylation. *Cell*, 105:115–126, 2001.

[957] S. C. Harvey and H. A. Gabb. Conformational transitions using molecular dynamics with minimum biasing. *Biopolymers*, 33:1167–1172, 1993.

[958] E. Paci and M. Karplus. Unfolding proteins by external forces and temperature: The importance of topology and energetics. *Proc. Natl. Acad. Sci. USA*, 97:6521–6526, 2000.

[959] S. Bernèche and B. Roux. Energetics of ion conduction through the K^+ channel. *Nature*, 414:73–77, 2001.

[960] R. Olender and R. Elber. Calculation of classical trajectories with a very large time step: Formalism and numerical examples. *J. Chem. Phys.*, 105:9299–9315, 1996.

[961] V. Zaloj and Ron Elber. Parallel computations of molecular dynamics trajectories using the stochastic path approach. *Comput. Phys. Comm.*, 128:118–127, 2000.

[962] J. Drews. Drug discovery: A historical perspective. *Science*, 287:1960–1964, 2000.

[963] D. B. Boyd. Computer-aided molecular design. In A. Kent (Executive) and C. M. Hall (Administrative), editors, *Encyclopedia of Library and Information Science*, volume 59, pages 54–84. Marcel Dekker, New York, NY, 1997. Supplement 22.

[964] A. M. Leach. *Molecular Modelling. Principles and Applications*. Pearson Education Limited, Harlow, England, second edition, 2001.

[965] N. C. Cohen, editor. *Guidebook on Molecular Modeling in Drug Design*. Academic Press, San Diego, CA, 1996.

[966] E. K. Kick, D. C. Roe, A. G. Skillman, G. Liu, T. J. A. Ewing, Y. Sun, I. D. Kuntz, and J. A. Ellman. Structure-based design of combinatorial chemistry yield low nanomolar inhibitors of cathepsin D. *Chem. Biol.*, 4:297–307, 1997.

[967] Y. Sun, T. J. A. Ewing, A. G. Skillman, and I. D. Kuntz. CombiDOCK: Structure-based combinatorial docking and library design. *J. Comput.-Aided Mol. Design*, 12:597–604, 1998.

[968] M. L. Lamb, K. W. Burdick, S. Toba, M. M. Young, A. G. Skillman, X. Zou, J. R. Arnold, and I. D. Kuntz. Design, docking, and evaluation of multiple libraries against multiple targets. *Proteins: Struc. Func. Gen.*, 42:296–318, 2001.

[969] T. Caulfield and K. Burgess. Combinatorial chemistry. Focused diversity and diversity of focus. *Curr. Opin. Chem. Biol.*, 5:241–242, 2001.

[970] S. Borman. Reducing time to drug discovery. *Chem. Eng. News*, 77:33–48, 1998.

[971] R. B. Dover, L. F. Schneemeyer, and R. M. Fleming. Discovery of a useful thin-film dielectric using a composition-spread approach. *Nature*, 392:162–164, 1998.

[972] L. M. Amzel. Structure-based drug design. *Curr. Opin. Biotech.*, 9:366–369, 1998.

[973] C. Djerassi. *The Pill, Pygmy Chimps, and Degas' Horse. The Remarkable Autobiography of the Award-Winning Scientist Who Synthesized the Birth Control Pill*. Basic Books, New York, NY, 1992.

[974] M. J. Plotkin. *Medicine Quest: In Search of Nature's Healing Secrets*. Viking Penguin, New York, NY, 2000.

[975] D. F. Veber, F. H. Drake, and M. Gowen. The new partnership of genomics and chemistry for accelerated drug development. *Curr. Opin. Chem. Biol.*, 1:151–156, 1997.

[976] E. K. Wilson. Computers customize combinatorial libraries. *Chem. Eng. News*, 76:31–37, 1998.

[977] P. R. Caron, M. D. Mullican, R. D. Mashal, K. P. Wilson, M. S. Su, and M. A. Murcko. Chemogenomic approaches to drug discovery. *Curr. Opin. Chem. Biol.*, 5:464–470, 2001.

[978] G. Wess, M. Urmann, and B. Sickenberger. Medicinal chemistry: Challenges and opportunities. *Angew. Chem. Int. Ed.*, 40:3341–3350, 2001.

[979] O. Pillai, A. B. Dhanikula, and R. Panchagnula. Drug delivery: An odyssey of 100 years. *Curr. Opin. Chem. Biol.*, 5:439–446, 2001.

[980] A. D. Rodrigues and J. H. Lin. Screening of drug candidates for their drug–drug interaction potential. *Curr. Opin. Chem. Biol.*, 5:396–401, 2001.

[981] G. M. Downs and P. Willett. Similarity searching in databases of chemical structures. In K. B. Lipkowitz and D. B. Boyd, editors, *Reviews in Computational Chemistry*, volume VII, pages 1–66. VCH Publishers, New York, NY, 1996.

[982] A. C. Good and J. S. Mason. Three-dimensional structure database searches. In K. B. Lipkowitz and D. B. Boyd, editors, *Reviews in Computational Chemistry*, volume VII, pages 66–117. VCH Publishers, New York, NY, 1996.

[983] M. G. Bures and Y. C. Martin. Computational methods in molecular diversity and combinatorial chemistry. *Curr. Opin. Chem. Biol.*, 2:376–380, 1998.

[984] P. Willett. Structural similarity measures for database searching. In P. von Ragué Schleyer (Editor-in Chief), N. L. Allinger, T. Clark, J. Gasteiger, P. A. Kollman, and H. F. Schaefer, III, editors, *Encyclopedia of Computational Chemistry*, volume 4, pages 2748–2756. John Wiley & Sons, West Sussex, England, 1998.

[985] D. K. Agrafiotis, J. C. Myslik, and F. R. Salamme. Advances in diversity profiling and combinatorial series design. *Mol. Div.*, 4:1–22, 1999.

[986] J. M. Schafer, E.-S. Lee, R. C. Dardes, D. Bentrem, R. M. O'Regan, A. De Los Reyes, and V. C. Jordan. Analysis of cross-resistance of the selective estrogen receptor modulators arzoxifene (LY353381) and LY117018 in tamoxifen-stimulated breast cancer xenografts. *Clin. Cancer Res.*, 7:2505–2512, 2001.

[987] C. Korth, B. C. H. May, F. E. Cohen, and S. B. Prusiner. Acridine and phenothiazine derivatives as pharmacotherapeutics for prion disease. *Proc. Natl. Acad. Sci. USA*, 98:9836–9841, 2001.

[988] S. Makino, T. J. A. Ewing, and I. D. Kuntz. DREAM++: Flexible docking program for virtual combinatorial libraries. *J. Comput.-Aided Mol. Design*, 13:513–532, 1999.

[989] C. T. Vogelson. Advances in drug delivery systems. *Mod. Drug Dis.*, 4:49–52, 2001.

[990] L. Pauling, H. A. Itano, S. J. Singer, and I. C. Wells. Sickle cell anemia, a molecular disease. *Science*, 110:543–548, 1949.

[991] V. M. Ingram. Hemoglobin: The chemical difference between normal and sickle cell hemoglobin. *Nature*, 180:326–328, 1957.

[992] C. M. Henry. Pharmacogenomics. *Chem. Engin. News*, 79:37–42, 2001.

[993] M. M. Shi, D. Mehrens, and K. Dacus. Pharmacogenomics: Changing the health care paradigm. *Mod. Drug Disc.*, 4:27–32, 2001.

[994] Molconn-X version 2.0, 1995. Hall Associates Consulting, Quincy, MD.

[995] Molconn-Z version 3.1, 1998. Hall Associates Consulting, Quincy, MD.

[996] D. K. Agrafiotis. Diversity of chemical libraries. In P. von Ragué Schleyer (Editor-in Chief), N. L. Allinger, T. Clark, J. Gasteiger, P. A. Kollman, and H. F. Schaefer, III, editors, *Encyclopedia of Computational Chemistry*, volume 1, pages 742–761. John Wiley & Sons, West Sussex, England, 1998.

[997] S. L. Dixon and H. O. Villar. Investigation of classification methods for the prediction of activity in diverse chemical libraries. *J. Comput.-Aided Mol. Design*, 13:533–545, 1999.

[998] P. A. Hunt. QSAR using 2D descriptors and TRIPOS' SIMCA. *J. Comput.-Aided Mol. Design*, 13:453–467, 1999.

[999] A. K. Ghose, V. N. Viswanadhan, and J. J. Wendoloski. A knowledge-based approach in designing combinatorial or medicinal chemistry libraries for drug discovery. 1. A qualitative and quantitative characterization of known drug databases. *J. Comb. Chem.*, 1:55–68, 1999.

[1000] P. Gedeck and P. Willet. Visual and computational analysis of structure-activity relationships in high-throughput screening data. *Curr. Opin. Chem. Biol.*, 5:389–395, 2001.

[1001] D. K. Agrafiotis. Stochastic algorithms for maximizing molecular diversity. *J. Chem. Inf. Comput. Sci.*, 37:841–851, 1997.

[1002] W. J. Cook, W. H. Cunningham, W. R. Pulleyblank, and A. Schrijver. *Combinatorial Optimization*. John Wiley & Sons, New York, NY, 1998.

[1003] J. W. Demmel. *Applied Numerical Linear Algebra*. SIAM, Philadelphia, PA, 1997.

[1004] R. B. Lehoucq, D. C. Sorensen, and C. Yang. *ARPACK Users' Guide: Solution of Large-Scale Eigenvalue Problems with Implicitly Restarted Arnoldi Methods*. SIAM, Philadelphia, PA, 1998. www.caam.rice.edu/software/ARPACK/indexold.html.

[1005] T. Pinou, T. Schlick, B. Li, and H. G. Dowling. Addition of Darwin's third dimension to phylectic trees. *J. Theor. Biol.*, 182:505–512, 1996.

[1006] D. Xie, A. Tropsha, and T. Schlick. A data projection approach using the singular value decomposition and energy refinement. *J. Chem. Inf. Comput. Sci.*, 40(1):167–177, 2000.

[1007] D. Xie and T. Schlick. Visualization of chemical databases using the singular value decomposition and truncated-Newton minimization. In C. A. Floudas and P. Pardalos, editors, *Optimization in Computational Chemistry and Molecular Biology: Local and Global Approaches*, pages 267–286. Kluwer Academic Publishers, Dordrecht, The Netherlands, 2000.

[1008] E. Hodgkin and K. Andrew-Cramer. Compound collections get focused. *Modern Drug Discovery*, 3:55–60, 2000.

[1009] G. F. Joyce, W. C. Still, and K. T. Chapman. Combinatorial chemistry. Searching for a winning combination (Editorial overview). *Curr. Opin. Chem. Biol.*, 1:3–4, 1997.

Index

Index Key
(suffixes or font types may point to specific location of item)

Suffix	Font	Meaning
b		box
f		figure
h		homework
n		footnote
q		quote
t		table
	bold	chapter

Interdisciplinary Applied Mathematics

1. *Gutzwiller:* Chaos in Classical and Quantum Mechanics
2. *Wiggins:* Chaotic Transport in Dynamical Systems
3. *Joseph/Renardy:* Fundamentals of Two-Fluid Dynamics:
   Part I: Mathematical Theory and Applications
4. *Joseph/Renardy:* Fundamentals of Two-Fluid Dynamics:
   Part II: Lubricated Transport, Drops and Miscible Liquids
5. *Seydel:* Practical Bifurcation and Stability Analysis:
   From Equilibrium to Chaos
6. *Hornung:* Homogenization and Porous Media
7. *Simo/Hughes:* Computational Inelasticity
8. *Keener/Sneyd:* Mathematical Physiology
9. *Han/Reddy:* Plasticity: Mathematical Theory and Numerical Analysis
10. *Sastry:* Nonlinear Systems: Analysis, Stability, and Control
11. *McCarthy:* Geometric Design of Linkages
12. *Winfree:* The Geometry of Biological Time (Second Edition)
13. *Bleistein/Cohen/Stockwell:* Mathematics of Multidimensional
    Seismic Imaging, Migration, and Inversion
14. *Okubo/Levin:* Diffusion and Ecological Problems: Modern Perspectives
    (Second Edition)
15. *Logan:* Transport Modeling in Hydrogeochemical Systems
16. *Torquato:* Random Heterogeneous Materials: Microstructure and
    Macroscopic Properties
17. *Murray:* Mathematical Biology I: An Introduction (Third Edition)
18. *Murray:* Mathematical Biology II: Spatial Models and Biomedical
    Applications (Third Edition)
19. *Kimmel/Axelrod:* Branching Processes in Biology
20. *Fall/Marland/Wagner/Tyson:* Computational Cell Biology
21. *Schlick:* Molecular Modeling and Simulation: An Interdisciplinary Guide